Geological Survey of Canada

Geology of Canada, no. 3

GEOLOGY OF THE INNUITIAN OROGEN AND ARCTIC PLATFORM OF CANADA AND GREENLAND

edited by

H.P. Trettin

1991

This is volume E of the Geological Society of America's Geology of North America series produced as part of the Decade of North American Geology project.

© Minister of Supply and Services Canada 1991

Available in Canada through

authorized bookstore agents and other bookstores

or by mail from

Canada Communication Group — Publishing
Ottawa, Canada K1A 0S9

and from

Geological Survey of Canada offices:

601 Booth Street
Ottawa, Canada K1A 0E8

3303-33rd Street N.W.,
Calgary, Alberta T2L 2A7

100 West Pender Street,
Vancouver, B.C. V6B 1R8

A deposit copy of this publication is also available for reference in public libraries across Canada

Cat. No. M40-49/1E
ISBN 0-660-13131-5

Price subject to change without notice

Technical Editor
E.G. Snow

Design and layout
M.J. Kiel

Cartography
Geological Survey of Canada, Calgary and Ottawa
Geological Survey of Greenland, Copenhagen, Denmark

Cover
Photo by D.W. Esson see Frontispiece description

Printed in Canada

PREFACE

The Geology of North America series has been prepared to mark the Centennial of The Geological Society of America. It represents the cooperative efforts of more than 1000 individuals from academia, state and federal agencies of many countries, and industry to prepare syntheses that are as current and authoritative as possible about the geology of the North American continent and adjacent oceanic regions.

This series is part of the Decade of North American Geology (DNAG) Project which also includes eight wall maps at a scale of 1:5 000 000 that summarize the geology, tectonics, magnetic and gravity anomaly patterns, regional stress fields, thermal aspects, seismicity, and neotectonics of North America and its surroundings. Together, the synthesis volumes and maps are the first coordinated effort to integrate all available knowledge about the geology and geophysics of a crustal plate on a regional scale.

The products of the DNAG Project present the state of knowledge of the geology and geophysics of North America in the 1980s, and they point the way toward work to be done in the decades ahead.

From time to time since its foundation in 1842 the Geological Survey of Canada has prepared and published overviews of the geology of Canada. This volume represents a part of the seventh such synthesis and besides forming part of the DNAG Project series is one of the nine volumes that make up the latest *Geology of Canada*.

J.O. Wheeler
General Editor for the volumes
published by the
Geological Survey of Canada

A.R. Palmer
General Editor for the volumes
published by the
Geological Society of America

CONTENTS

Chapter

1. Introduction 1
2. Geographic and geological exploration 5
3. Geomorphic regions 27
4. Tectonic framework 57
5. Geophysical characteristics 67
6. Precambrian successions in the northernmost part of the Canadian Shield 101
7. Cambrian to Silurian basin development and sedimentation, North Greenland 109
8. Cambrian to early Devonian basin development, sedimentation, and volcanism, Arctic Islands 163
9. The Proterozoic to Late Silurian record of Pearya 239
10. Middle-Upper Devonian clastic wedge of the Arctic Islands 261
11. Devonian - Early Carboniferous deformation and metamorphism, North Greenland 281
12. Silurian - Early Carboniferous deformational phases and associated metamorphism and plutonism, Arctic Islands 293
13. Carboniferous and Permian history of the Sverdrup Basin, Arctic Islands 343
14. Mesozoic history of the Arctic Islands 369
15. Late Cretaceous and Tertiary basin development and sedimentation, Arctic Islands 435
16. Late Cretaceous - Early Tertiary deformation, North Greenland 459
17. Late Cretaceous - Early Tertiary deformation, Arctic Islands 467
18. Middle and Late Tertiary tectonic and physiographic developments 491
19. The Quaternary record 497
20. Resources 515
21. Summary and remaining problems 545

    Index 555
    Photo index 569

Frontispiece. Cliffs northeast of head of Yelverton Inlet, northwestern Ellesmere Island, view from the northwest. An angular unconformity separates early Paleozoic volcanics, carbonates, and slate (Yelverton assemblage, shades of grey) from Upper Carboniferous red beds (Borup Fiord Formation). The latter were deposited during the rifting event that created the Sverdrup Basin, and are overlain by Upper Carboniferous – Lower Permian shelf carbonates (Nansen Formation). The unconformity represents the terminal deformation of the Franklinian mobile belt, which extended from latest Silurian to Early Carboniferous time (Ellesmerian Orogeny and preceding events). (Photograph: D.W. Esson)

Chapter 1
INTRODUCTION

Chapter 1

INTRODUCTION

H.P. Trettin

This volume describes and interprets the geology of the northern margin of the North American continent in the Canadian Arctic Archipelago and North Greenland[1]. The northwestern part of this region is underlain by the Innuitian Tectonic Province, which comprises strata of the Late Proterozoic to Late Devonian Franklinian mobile belt, the Carboniferous to Paleogene Sverdrup Basin, and the Cretaceous-Cenozoic Arctic Continental Terrace Wedge (Fig. 4.1). The Franklinian mobile belt, deformed in Late Silurian to Early Carboniferous time, and the Sverdrup Basin, deformed in Paleogene time, together constitute the Innuitian Orogen. The junction of the Franklinian mobile belt with the Caledonides is concealed by the Wandel Sea off northeastern Greenland, and also by the upper Paleozoic and younger strata of the Wandel Sea Basin, the junction with the Cordillera and the Arctic Alaska Plate is covered by the Beaufort Sea.

The southeastern part of the region discussed in this volume is underlain by the Arctic Platform, characterized by Cambrian to Paleogene strata that are contiguous with those in the Innuitian Orogen but relatively undisturbed. The Arctic Platform is a part of the sedimentary cover of the North American craton that is separated from other parts, such as the Interior and Hudson platforms, by Recent seaways or erosional gaps.

Until World War II, explorers moved freely between the archipelago and Greenland, but in the post-war period the international boundary has become increasingly effective, and geological investigations have been carried out separately in the two areas. As a result, somewhat different stratigraphic and structural frameworks have developed — in spite of exchange of personnel and information — that have typically been presented in the context of national geological compilations (e.g. Thorsteinsson and Tozer, 1970; Dawes, 1976). Efforts at correlation have focussed mainly on Nares Strait, the tectonically problematic seaway that constitutes the political and geographic boundary (Dawes and Kerr, 1982). A new synthesis that includes all of North Greenland and the Arctic Archipelago is attempted in this volume. It was prompted by the stimulating project of the Geological Society of America, to prepare an integrated overview of North America, Greenland, and the adjacent oceans and was feasible only because of the co-operation of the Geological Survey of Greenland who provided all the necessary information on North Greenland[2].

The organization of the book follows a format designed for all volumes on the regional geology of Canada, with some variation. The introductory part includes an account of the fascinating, often tragic history of exploration of the Arctic (Chapter 2); a well illustrated survey of the geomorphic subdivisions (Chapter 3); a summary of the tectonic framework (Chapter 4); and succinct accounts of the geophysical characteristics of the region (Chapter 5). The main part (Chapters 6 to 19) gives an overview of the geological history as recorded in the rocks. The final part discusses resources (Chapter 20) and presents a brief, interpretative summary of geological history and broader regional relationships (Chapter 21).

The subject matter treated here borders or overlaps that of several other volumes in the series, especially those dealing with the Precambrian of Canada and Greenland (Hoffman et al., in prep.), the cratonic cover in Canada (Stott and Aitken, in press), the Quaternary of Canada and Greenland (Fulton, 1989), the Labrador Sea and Baffin Bay (Keen and Williams, 1990), the Arctic Ocean region (Grantz et al., in press), and Canadian mineral deposits (Thorpe and Eckstrand, in prep.). Also relevant are volumes on the Appalachians (Williams and Neale, in prep.), the Canadian Cordillera (Gabrielse and Yorath, in press), and northern Alaska (Plafker and Jones, in prep.).

The contributions in this volume were prepared by more than forty experts in Canada and Europe, employed by government agencies, universities, and industry. It has been attempted to weld these articles into a coherent whole, but differences in approach, organization, and nomenclature were unavoidable, and perhaps even desirable. The original manuscripts, received from 1984 onward, have generally been updated to reflect the state of knowledge in 1986 or 1987.

The volume editor is indebted to the authors for their co-operation and patience. Special thanks are due to N. Henriksen, who co-ordinated the contributions from the Geological Survey of Greenland and to H.R. Balkwill, S.E. Jenness, and J.O. Wheeler who reviewed the volume and made numerous suggestions for its improvement. The technical and production editing was in the capable hands of Elspeth Snow.

Trettin, H.P.
1991: Introduction; Chapter 1 in Geology of the Innuitian Orogen and Arctic Platform of Canada and Greenland, H.P. Trettin (ed.); Geological Survey of Canada, Geology of Canada, no. 3; (also Geological Society of America, The Geology of North America, v.E).

[1] Greenland achieved a form of self-government and independence from Denmark in 1979 and changed its legal name to Kalaallit Nunaat. The name Greenland, long established in the literature, is used throughout this volume.

[2] Chapters 2, 3, 7, 11, 16, 20B, and 20E are published by permission of the Director of the Geological Survey of Greenland.

REFERENCES

Dawes, P.R.
1976: Precambrian to Tertiary of northern Greenland; in Geology of Greenland, A. Escher and W.S. Watt, (ed.); The Geological Survey of Greenland, Copenhagen, p. 248-303.

Dawes, P.R. and Kerr, J.W. (ed.)
1982: Nares Strait and the Drift of Greenland: a Conflict in Plate Tectonics; Meddelelser om Grønland, Geoscience 8, 392 p.

Fulton, R.J. (ed.)
1989: Quaternary Geology of Canada and Greenland; Geological Survey of Canada, Geology of Canada, no. 1 (also Geological Society of America, The Geology of North America, v. K-1), 848 p.

Gabrielse, H. and Yorath, C.J. (ed.)
in press: Geology of the Cordilleran Orogen in Canada; Geological Survey of Canada, Geology of Canada, no. 4 (also Geological Society of America, The Geology of North America, v. G-2).

Grantz, A., Johnson, L., and Sweeney, J.F. (ed.)
in press: The Arctic Ocean Region; Geological Society of America, The Geology of North America, v. L.

Hoffman, P.F., Card, K.D., and Davidson, A. (ed.)
in prep.: Precambrian Geology of the Craton in Canada and Greenland; Geological Survey of Canada, Geology of Canada, no. 7 (also Geological Society of America, The Geology of North America, v. C-1).

Keen, M.J. and Williams, G.L. (ed.)
1990: Geology of the Continental Margin of Eastern Canada; Geological Survey of Canada, Geology of Canada, no. 2 (also Geological Society of America, The Geology of North America, v. I-1).

Plafker, G. and Jones, D.L. (ed.)
in prep.: The Cordilleran Orogen: Alaska; Geological Society of America, The Geology of North America, v. G-1.

Stott, D.F. and Aitken, J.D. (ed.)
in press: Sedimentary Cover of the Craton in Canada; Geological Survey of Canada, Geology of Canada, no. 5 (also Geological Society of America, The Geology of North America, v. D-1).

Thorpe, R.I., Eckstrand, O.R. (ed.)
in prep.: Mineral Deposits of Canada; Geological Survey of Canada, Geology of Canada, no. 8 (also Geological Society of America, The Geology of North America, v. P-1).

Thorsteinsson, R. and Tozer, E.T.
1970: Geology of the Arctic Archipelago; in Geology and Economic Minerals of Canada, R.J.W. Douglas (ed.); Geological Survey of Canada, Economic Geology Report no.1, p. 547-590.

Williams, H. and Neale, E.R.W.
in prep.: Geology of the Appalachian Orogen in Canada and Greenland; Geological Survey of Canada, Geology of Canada, no. 6 (also Geological Society of America, The Geology of North America, v. F-1).

Author's address:

H.P. Trettin
Geological Survey of Canada
3303-33rd St. N.W.
Calgary, Alberta
T2L 2A7

Chapter 2
GEOGRAPHIC AND GEOLOGICAL EXPLORATION

Introduction

I. Paleo-eskimo habitation and the Norsemen, 2000 B.C. — A.D. 1500

II. Early mariners and the Franklin Search, 1576-1859

III. Exploration along the Polar Route, 1852-1909

IV. Geological exploration by ship and dog sledge, 1903-1947

V. The aircraft age of geological study

Acknowledgments

References

Chapter 2

GEOGRAPHIC AND GEOLOGICAL EXPLORATION

R.L. Christie and P.R. Dawes

INTRODUCTION

The region discussed in this chapter encompasses the Canadian Arctic and northern Greenland. A dividing seaway — Davis Strait, Baffin Bay, Nares Strait — forms a prominent route by which generations of explorers, from the Norsemen to the classical late 19th century voyages of geographic discovery, pushed north to explore the mysteries of "Ultima Thule". Ardent pursuit of two geographic objectives, the Northwest Passage and the North Pole, resulted in the discovery and outline mapping of much of the region. Ship and sledge expeditions directed at specific scientific programs followed in the 20th century; these activities were superseded by aircraft-supported operations in the years following World War II.

Geographic and geological research continues, now supported by Canadian and Danish government agencies and private operators. Paradoxically, it was as recently as 1978 that the region's (and the Earth's) most northerly land was discovered: a Danish helicopter party that year landed on Oodaaq Ø, a small island at about 83°40'N off the north Greenland coast.

This account of exploration in the Innuitian Region is divided into five roughly chronological sections: I. Paleo-eskimo habitation and Norsemen, 2000 B.C. – A.D. 1500; II. Early mariners and the Franklin Search, 1576-1859; III. Exploration along the Polar Route, 1852-1909; IV. Geological exploration by ship and dog sledge, 1903-1947; V. The aircraft age of geological study, 1947 to the present. Brief notes on the development of geological understanding of the region follow sections II to IV[1].

I. PALEO-ESKIMO HABITATION AND THE NORSEMEN, 2000 B.C. – A.D. 1500

Archeological finds indicate that the Innuitian Region was inhabited by paleo-eskimos as far back as 4300 years ago. Several Eskimo cultures have been recognized: the oldest, the Independence I culture, extended to the northernmost parts of the region onto the shores of the Arctic Ocean (Knuth, 1967, 1981; Maxwell, 1960). Successive cultures — Independence II, 700 BC; Dorset, 100 B.C. to A.D. 700, and Thule, A.D. 900 to A.D. 1200 — indicate a long history of Inuit[2] habitation.

European exploration and habitation started, just 1000 years ago, with the voyages by Erik Thorvaldssøn (Erik the Red) from Iceland in 982-986. The finding of the Kingigtorssuaq runic stone at about 73°N indicates that by the 1300s the Norsemen reached as far north as the Upernavik district. There is also evidence of Norse influence in the archeological sites of the Thule district (77°N).

The earliest landings by Europeans in mainland North America probably were made in the Canadian Arctic and sub-Arctic. The visitors were "Northmen", from settlements in western Greenland, who were exploring westward, possibly searching for timber and good fishing grounds. Evidence of direct or indirect Norse-Inuit contact has been found in the region between east-central Ellesmere Island and southern Baffin Island, as early as A.D. 1250-1350 (Schledermann, 1981). The Greenland Norse colonies declined and disappeared in the late 1400s. About a century passed before the arrival of the English navigator, Martin Frobisher, in Arctic waters (Knuth, 1968).

II. EARLY MARINERS AND THE FRANKLIN SEARCH, 1576-1859

Mariners in the 14th and 15th centuries sailed westward from Europe in the hope of reaching the near-legendary riches of the Orient. For these voyages, Greenland and the Americas were an obstacle. Exploration inevitably turned northward as each gulf or channel was navigated, in vain, in the search for a passage to the west. The tides and currents in the northern waterways gave rise to the conviction that there must be a "Northwest Passage" to the Pacific Ocean. The network of immense channels of the Canadian Arctic excited the navigators and their sponsors, who pushed ever north and westward: Martin Frobisher to Baffin Island and Greenland in 1576, 1577, and 1578; John Davis to Davis Strait and Baffin Bay (up to 73°N) in 1585, 1586 and 1587; George Weymouth into Davis and Hudson straits in 1602; Henry Hudson into Hudson Bay in 1610; and William Baffin to Smith, Jones and Lancaster sounds (up to 77°30'N) in 1616.

The idea of a northeasterly route to China, north of Asia, also began to attract attention. On one such voyage, in 1607, Henry Hudson observed the coast of East Greenland at about 73°N.

Christie, R.L. and Dawes, P.R.
1991: Geographic and geological exploration; Chapter 2 in Geology of the Innuitian Orogen and Arctic Platform of Canada and Greenland, H.P. Trettin (ed.); Geological Survey of Canada, Geology of Canada, no. 3; (also Geological Society of America, The Geology of North America, v. E).

---

[1] Sources include historical accounts by Koch (1940), Taylor (1955), Wright (1959), Neatby (1958-1970), Zaslow (1975), Dawes and Haller (1979), Christie and Kerr (1981), Jones (1981), Dawes and Christie (1982), and Greene (1982). For geographic names see Figure 1 (in pocket).

[2] "Inuit" (singular, Inuk) is the term modern Eskimo people apply to themselves. "Innuitian" retains an earlier spelling.

The search for a navigable route from Hudson Bay to the Pacific drew Danish and British sailors during the 17th and 18th centuries; these explorers were encouraged by the Spanish claim to have discovered a strait on the Pacific Ocean side of the continent. Navigators of the Hudson's Bay Company (a commercial venture established in 1668 as the "Governor and Company of Adventurers of England Tradeing into Hudson's Bay") were making routine voyages through Hudson Strait into Hudson Bay by the early 1700s to service their fortified trading posts. The Company was charged with searching for a passage to the Pacific but the bleak and ice-bound west coast of Hudson Bay did not improve hopes that such a gateway could be found. The Company, furthermore, was occupied with efforts to turn a profit from furs collected by Indians of the region draining into the Bay.

Criticism of the Company's failure to pursue the search for "the Passage" arose in England; this helped begin, by 1736, a renewal of exploration, which now was combined with a search for minerals. Chesterfield Inlet and Wager Bay (on the northwest side of Hudson Bay) were discovered and explored, but no Northwest Passage could be found. Finally, between 1770 and 1772, Samuel Hearne settled the question of westward navigation from Hudson Bay with his arduous overland journey from Churchill, on Hudson Bay, to the mouth of the Coppermine River, on the Arctic coast.

The arrival of Hans Egede in Greenland in 1721 started a new phase of exploration. Karl Ludwig Giesecke, a mineralogist assisted by the Royal Greenland Trading Company, completed a traverse of the colonized parts of the west coast of Greenland by 1813. This completed mapping of the coast as far north as Melville Bugt.

The Arctic regions were visited for many years only by whalers, mainly British, Dutch and American, most of whose exploits went unrecorded. However, we know that these whalers negotiated Melville Bugt and reached at least as far north as Kap York (76°N) in Greenland. On one trip in 1817, Bernard O'Reilly, serving as surgeon on the whaler *Thomas* of Hull, England, produced a description of Melville Bugt that included a map (O'Reilly, 1818). In 1817 the respected British whaling captain, William Scoresby, Jr., suggested that, because of improving climatic conditions, the time was propitious for further exploration, and attention was turned again to the far north. He succeeded in 1822 in mapping the east coast of Greenland from 69°N to 75°N. The British Government in 1818 sent Captain John Ross into Baffin Bay to search for passages that might lead to the Pacific. Ross, in H.M.S. *Isabella*, accompanied by H.M.S. *Alexander*, with W.E. Parry commanding, retraced William Baffin's track of 1616 and, as had Baffin, found Smith and Jones sounds choked by ice. The ships' names were given to opposing capes of Smith Sound and these became important navigational landmarks for generations of travellers. The highest latitude attained was 76°54'N. Lancaster Sound was clear of ice but, perhaps due to thick weather or perhaps in haste, preconceiving a closed bay, Ross retreated from the mouth of the sound and returned to England. Controversy ensued over whether the sound had been sufficiently explored and, indeed, a new expedition organized the following year "ran quickly up the Sound" all the way to Prince Leopold Island before being stopped by ice. Ross (1819) had much cause, over the years, to regret his error of judgment.

In fact, several notable advances were achieved by Ross's navigation of Baffin Bay: Baffin's and Bylot's discoveries of some 200 years earlier, later distorted or even disbelieved, were confirmed; the large "James's Island" in Baffin Bay (a feature claimed in the early 17th century by Captain James of Bristol) was expunged from the charts; the Inuit (named the "Arctic Highlanders" by Ross) of the Kap York region were encountered, and geological samples were obtained in northern Baffin Bay. In any case, Ross's voyage heralded a vigorous renewal of exploration in the North American Arctic. By this time the efficiency and safety of sailing ships had improved, the Napoleonic Wars had ended and England's proud and dominant navy was available, the science of geology was developing, and in Europe an interest in world geography was awakening.

The new British expedition to Lancaster Sound departed in 1819 under W.E. Parry (who had been second-in-command with Ross), and was sent at a time of rising expectations and continuing faith in the existence of a Northwest Passage. As it happened, 1819 was a good season for sailing vessels in the Archipelago and his two ships, H.M.S. *Hecla* and *Griper*, were able to sail almost to the 113th meridian along what is now named Parry Channel. Parry took refuge in Winter Harbour on Melville Island. By their return in 1820, Parry and his crews had seen and named the major islands of Somerset, Devon, Cornwallis, Bathurst, Melville and Banks. The large bay indenting the north side of Melville Island, reached by an overland journey, was named Hecla and Griper Bay after the two ships. Parry's voyage, the first to enter the Canadian Archipelago, ranks among the most successful of Arctic journeys under sail, and the descriptions of the islands and great channels fired the enthusiasm of the British so that for some years expeditions sailed from England in quick succession. Parry returned to the Arctic in 1821 to search for a passage in the lower latitudes of Foxe Basin; he searched the east coast of Melville Peninsula from Repulse Bay to Fury and Hecla Strait (again, named after his ships). After wintering twice in Foxe Basin and failing in a second attempt to navigate Fury and Hecla Strait, Parry retreated to his home port. In 1824 he sailed into Prince Regent Inlet (again in H.M.S. *Fury* and *Hecla*) but was frustrated by exceptionally wintry conditions and heavy ice. The *Fury* was wrecked in 1825 on Somerset Island and, after landing the ship's stores, both crews sailed home in the *Hecla*.

The adventures experienced later by the *Griper*, the ship commanded by M. Liddon on Parry's first voyage, are of note: in 1824 she was sent, under G.F. Lyon, to winter at the base of Melville Peninsula in order to begin a shore-survey (overland across the peninsula) of the northern coasts toward Coppermine River. The heavily laden ship, however, was severely battered by northerly gales in Roes Welcome Sound and, with the *Griper* crippled and partly flooded, Lyon abandoned the survey and returned home.

The difficulties of a passage across the top of North America were now becoming evident and the British Government concluded that a Northwest Passage was perhaps not of sufficient interest to risk more ships and men. Popular interest remained, however, and in 1829 John Ross, privately financed, sailed the *Victory* (equipped with both sail and paddles, the first ship to use steam power in the Arctic Archipelago), into Prince Regent Inlet. The *Victory* had in tow a small sloop, the *Krusenstern*, and the two ships were navigated without difficulty to a small inlet on southeastern Boothia Peninsula. This expedition proved to be one of both successes and failures: Ross failed to discover a Northwest Passage (he twice contrived to miss the narrow passage, since named Bellot Strait, that separates Somerset Island and Boothia Peninsula); on the other hand, thanks to the energies of his nephew, James Ross, considerable lengths of coastline were mapped and the then position of the north magnetic pole was located (Chapter 5F). Ross's party spent four years in the Arctic and retreated on foot and by small boat after abandoning the *Victory* and *Krusenstern*. In spite of privations and hardships, only two men died. After wintering at Fury Beach, consuming stores and fuel providentially landed there in 1825 by Parry, the sailors embarked in three of *Fury*'s boats and were picked up in Lancaster Sound by whalers in Captain Ross's old ship, the *Isabella*.

The achievements of the voyages by sailing ship in the Canadian Arctic Archipelago tended, then as now, to attract much notice. Simultaneously, however, (during the interval 1819 to 1839), the northern coast of the Canadian mainland was being explored by canoe and boat by parties descending the northern rivers. These parties were led by John Franklin, George Back (both of the British Admiralty), and by Peter Dease and Thomas Simpson (both of the Hudson's Bay Company). Franklin and Back, with Dr. John Richardson, descended first the Coppermine River, then the Mackenzie, in the years 1819-22 and mapped the coastline between Return Reef (then named Cape Return) in Alaska and Cape Turnagain, east of the mouth of the Coppermine. Back descended the "Great Fish River" (since named Back River) between 1833 and 1835 and explored the vicinity of its mouth. Dease and Simpson, 1837 to 1839, descended the Mackenzie and Coppermine rivers and extended Franklin's mapping both westward and eastward.

In 1839, the Company explorers, Dease and Simpson, carried out a remarkable boat voyage of some 2200 km (1400 miles): they descended the Coppermine and travelled the coasts of the mainland, Victoria Island and King William Island. In August of 1839, they viewed the Boothia Peninsula region from a point east of Cape Britannia. All these voyages involved cold, wet conditions of sailing in small boats, and, in the travel of Back River, the descent and ascent of large waterfalls and rapids. Franklin's overland return from his first expedition nearly ended in death by starvation of the entire party, due to complete dependence for provisions and support upon the then competing local trading companies and native guides and hunters.

Her Britannic Majesty's Ships *Erebus* and *Terror* (Fig. 2.1) had been occupied in a highly successful expedition to the Antarctic under the command of James Ross. It was proposed that these now famous and proven ships should be fitted with steam propulsion (the screw propeller had come into use) and sent to complete the Northwest Passage, of which only about 1500 km (900 miles) remained to be delineated. So confident were some supporters that it was suggested that the ships might effect the return trip from Bering Strait in the single season! Age and experience were given high premiums, and John Franklin, by now knighted, was granted a request to lead an expedition of two ships and 128 men, equipped for three years. The command of the second ship fell to Captain F.R.M. Crozier, who had served on Parry's second and third expeditions, and had commanded the *Terror* during the recent Antarctic expedition. The "Franklin Expedition", as it has become named, thus comprised the best of the Royal

Figure 2.1. H.M.S. *Erebus* and *Terror*, of the Franklin Expedition; Royal Navy barque-rigged "bomb-vessels" of 370 and 340 tons; their disappearance sparked the prolonged "Franklin Search". It was later determined that the ships were abandoned in the ice in 1848 (from Wright, 1959, p. 99).

Navy's men and ships. The expedition ships were provisioned to capacity with stores, and the supplies were "topped up" from a supply ship that accompanied them to Davis Strait. The ships were sighted by a whaler in Melville Bugt in late July, 1845, then they vanished. Whalers sailing into Lancaster Sound in following years found no messages or signs of the expedition, and in 1847 the Admiralty became uneasy at the continuing absence of news. The *Plover* was sent that fall, with stores, to Bering Strait, and the *Investigator* and *Enterprise*, under James Ross, were sent in 1848 to search for the missing expedition from bases in Barrow Strait. Similar naval searches were carried out, some on a dramatic scale, during the following twelve years. A great deal of the Archipelago was explored in the course of the "Franklin Search". An unfortunate series of poor decisions, guesses, and interpretations resulted in almost all of the southern parts of the Archipelago, including the southern Queen Elizabeth Islands, being searched *except* the very area, ironically, to which Sir John's principal instructions had directed him. It is particularly touching that of all the suggestions as to the whereabouts of the missing ships (by people ranging from "experts", through eminent scientists, to well meaning cranks), the person who was nearest to the mark was Lady Franklin. She reasoned simply that her husband, following his Admiralty orders, would have sailed southward from Barrow Strait, that his ships had become beset between King William Island and Boothia Peninsula and had been abandoned, and that the crews would have had to retreat up the Great Fish River. When the few remains of the expedition were found (by Leopold M'Clintock, in 1859) near and on King William Island, it was clear that her surmise had essentially been correct. Among the information gleaned from two written messages — the only ones that so far have been found — was the statement that Sir John Franklin died on June 11, 1847. None of the diaries or records have been discovered, so we know little of the accomplishments of Franklin's last voyage; we do know, however, that Franklin did not witness the terrible disintegration, probably by scurvy, of his entire expedition.

Some geological discoveries were achieved in the Archipelago and northwestern Greenland, along with major advances in geographical knowledge, during the early 1800s, or to about the end of the Franklin Search in 1859. A period of rapid evolution of geological understanding had taken place; "neptunist" concepts of the Wernerian school, for example, are evident in some of the terms used to describe the rocks — "primary" for basement gneisses and granitoid rocks of the Precambrian Shield, and "secondary" for overlying beds of Ordovician to Jurassic ages. The early expeditions were charged with collecting rocks and fossils and this was done, often, by the ship's surgeon, who was described as the "naturalist". Descriptions of fossils and interpretations were written, following the return of the expedition, by authorities in England. Thus Dr. J. M'Culloch (1819) described the rocks collected by Ross in Baffin Bay, Charles Konig (1823, 1824) described the rocks of the region explored by Parry, J.W. Salter (1852, 1853, 1855) wrote a geological account and described fossils collected by the squadron led by Horatio Austin and William Penny (in 1850-51), and the Reverend Samuel Haughton (1858, 1859, 1860) wrote geological accounts from M'Clintock's journeys. Haughton's work (1858) included a map on which he compiled the then known geology of the Arctic Islands — the first regional geological synthesis of the Arctic.

The measurement and study of the behaviour of the earth's magnetic field were of great interest to the early explorers, for whom the ship's compass was a principal instrument of navigation. Thus James Ross's locating of the north magnetic pole (then near Cape Adelaide, western Boothia Peninsula) was an important advance for the time.

III. EXPLORATION ALONG THE POLAR ROUTE, 1852-1909

This period concerns exploration along the channels between Greenland and Ellesmere Island (Nares Strait), which became known as the Polar Route during the period of journeys to reach the North Pole.

During the Franklin Search, ships had sailed northward from Baffin Bay into Smith Sound. One of these, the *Isabel*, equipped by Lady Franklin and led by Comdr. Edward A. Inglefield, had achieved a new northing in 1852 at 78°28'N, just entering Kane Basin before being driven south by storm and ice. The surgeon-naturalist Peter C. Sutherland made notes and sketches on the geology of the coasts traversed.

Interest in poleward exploration was renewed, and other expeditions, chiefly American, pushed into the long, irregular, but conspicuously north-trending channel that separates Greenland and Ellesmere Island. Open water to the north of Kane Basin was described as an "Open Polar Sea" by Elisha K. Kane's expedition, 1853 to 1855, and the ancient, recurrent and fascinating concept was revived that an open sea, or even a maelstrom, lay in the polar region. Kane took the brigantine *Advance* into winter quarters on the Greenland coast at Rensselaer Bugt and succeeded in the following sledging seasons in charting the coasts of Kane Basin and southern Kennedy Channel. One sledge party reached a new farthest north at 80°34'N on the Washington Land coast, from where observations on the "Open Polar Sea" were made. After abortive attempts to sail south in 1854, the ship was finally abandoned in 1855 and the expedition reached safety in western Greenland in small boats.

Isaac I. Hayes, who had been surgeon on Kane's expedition, took up the challenge of the Polar Route and sailed the schooner *United States* into the same region in 1860. Forced to winter on the Greenland shore some way south of Kane's winter quarters, he succeeded in sledging up the east coast of Ellesmere Island, from where he claimed also to have seen an Open Polar Sea to the north. The actual northing attained by Hayes is still a matter of discussion, but it is clear that he could not have seen Lincoln Sea, at the northern opening of Nares Strait. In any case, the report of open water undoubtedly brought attention to this northward route to the Pole, and other shipborne expeditions soon followed: C.F. Hall in 1871; G.S. Nares in 1875; A.W. Greely in 1881; and R.E. Peary in 1898 and later.

Charles F. Hall had travelled in southern Baffin Island, northwestern Hudson Bay, and the Igloolik region between 1860 and 1869. In the course of his travels he learned Inuktitut (Eskimo) and sledged to King William Island to confirm some intriguing tales that had been related by his Inuit friends. Hall obtained from the Inuit articles that had been carried by Franklin's crew and confirmed the tragedy reported by Dr. John Rae and by M'Clintock. In 1871 Hall led an expedition "to the North Pole" in the U.S.S. *Polaris*, backed by the U.S. Government. The steam-powered vessel reached 82°11'N, the opening of Nares Strait into the Arctic Ocean, but retreated to winter quarters at Thank God Harbour on the Greenland coast (Hall Land). However, the expedition ended in disorder after Hall's death in November 1871 — a death probably due to poisoning (Loomis, 1971). The surrounding country in Greenland (parts of Washington, Hall and Nyeboe lands) was explored, with Emil Bessels, chief of the scientific staff, being responsible for collection of geological material. During the retreat in *Polaris* in summer 1872, the expedition was split into two during a "nip". One group of 19 people spent over six months on an ice floe, drifting over 3000 km before being rescued off the coast of Labrador; the group remaining on the ship was forced to winter on the Greenland shore north of Kap Alexander, where the *Polaris* sank. This group finally escaped south in small boats.

Captain George S. Nares sailed H.M.S. *Alert* and *Discovery* northward in 1875 to attempt an assault on the North Pole. This was the last British Admiralty expedition for geographic exploration in the High Arctic. The *Discovery* wintered at Discovery Harbour in northern Ellesmere Island, while the *Alert* pushed on to 82°27'N (not far from present-day Alert weather station), where it wintered. The north coasts of Ellesmere Island (as far west as Alert Point) and Greenland (as far east as Wulff Land) were explored by man-hauled sledge parties in 1876. However, these discoveries were not made without loss of life; both the Ellesmere and Greenland sledge parties, considerably weakened by scurvy, lost men. Important geographic and geological results were achieved and new maps were compiled. Nares withdrew his expedition before the crew could be severely depleted by scurvy, and the ships reached England in 1876. Nares became the first to make a successful return navigation of the Polar Route; the strait now bears his name.

Lieut. Adolphus W. Greely's expedition to Discovery Harbour in 1881 was part of the "International Polar Year", the earliest co-ordinated international scientific study of the physical sciences in the polar regions. Greely, of the U.S. Signal Corps, was left at the *Discovery*'s wintering site with his party of 25 by the ship *Proteus*; a substantial building was erected and named Fort Conger. During two years, the party explored and collected scientific, including geological, data both on Ellesmere Island and Greenland. In 1882 Greely discovered and named Lake Hazen, and in the spring of the same year, Lieut. J.B. Lockwood, accompanied by Sgt. D.L. Brainard and Inuk Frederick Christiansen, mapped the Greenland coast as far east as Kap Washington to attain a new world northing on Lockwood Ø at 83°23'N. Relief ships had not arrived by 1883, however, so Greely and his party, following their orders, retreated southward in small boats. The party reached Cape Sabine, where a few supplies from the unsuccessful relief ship *Proteus* had been landed. There, under an overturned boat, starvation and death took all but seven of the party of 26. Only six survived to return to the United States.

It should be noted that the Geological Survey of Canada's first venture into the Arctic Islands was that of Robert Bell, who in 1884 and later years accompanied Canadian government expeditions to determine conditions of navigation in Hudson Strait and Hudson Bay. Bell reported on well banded hornblende- and mica-gneisses, and noted numerous and extensive bands of crystalline limestone (marble) associated with the gneisses.

Lieut. Robert E. Peary (Fig. 2.2) undertook seven polar expeditions between 1891 and 1909, spending four winters in Greenland and five in Ellesmere Island. Peary's appetite for polar expeditions had been whetted on his reconnaissance of the Greenland Inland Ice in 1886. His early exploration from winter bases in the Thule district was mainly concerned with the mapping of that region, recovery of the Kap York meteorites, and in 1891 and 1895 two traverses of the Inland Ice (Peary, 1898). From his "landfall" positions at Navy Cliff on the northeast margin of the Inland Ice, Peary looked northward to Independence Fjord and to the vast land beyond that now bears his name. He also was rewarded with the first glimpses of the east side of North Greenland — the eastern part of the Innuitian Region. Peary observed a deep channel to the north, which he surmised to be an arm of the sea. This "Peary Channel", with the presumed island to the north, later became a focus of Danish expeditions.

Peary, during his early work, fostered a close association with the Thule Inuit, who became an integral part of his expeditions. In 1898 he turned his attention to the Polar Route, and this occupied him and his companions for over a decade. In 1899, from his ship *Windward*, at Cape d'Urville on the east coast of Ellesmere Island, Peary visited Greely's Fort Conger (Fig. 2.3). In 1900, from a base at Etah, Greenland, Peary made a long traverse of the Greenland coast, during which he discovered Kap Morris Jesup and continued some way to the south and east, thereby fixing the outlines of the northern coast and establishing the insularity of Greenland. With the 1898-1902 expedition, Peary concluded his main geographic work in Greenland and his primary goal became the North Pole — to be attained by the now well travelled Polar Route along Nares Strait. In 1905 and again in 1908, he pushed the *Roosevelt*, with Captain Bob Bartlett (Fig. 2.4) as skipper, to about where the *Alert* had wintered, on the edge of the Polar Sea. Peary made his last attempt to achieve the Pole in 1909, aided by well organized support parties and accompanied

Figure 2.2. Commander Robert E. Peary (1856-1920) of the U.S. Navy. His extensive geographical exploration in the High Arctic of Greenland and Canada culminated in his "push for the pole" in 1909 (from Hayes, 1929, facing p. 262; also Library of Congress 12244 US262 8234).

by Mathew Henson, Oodaaq and three other Inuit companions.

In the course of these adventures, Peary and his supporting sledge parties added considerably to geographic knowledge of the coasts of the Polar Sea. Peary traversed the northern coast of Ellesmere Island in 1906 to reach the northern tip of Axel Heiberg Island, passing Aldrich's farthest west of 1876, thus linking with the discoveries of Sverdrup's 1898-1902 expedition (see below). In 1909, G. Borup and D.B. MacMillan of Peary's party modified the map of the Greenland coast en route to Kap Morris Jesup. Although earth scientists participated in some of Peary's early expeditions, in general little geological work was done during his far-reaching adventures.

An unfortunate period for polar history ensued: a former companion of Peary's, Dr. Frederick A. Cook, also claimed to have reached the North Pole — but in 1908! The controversy over which, or if either, explorer reached the vicinity of the Pole continues to this day, but the accounts of both men are so clouded by contradictions and mystery

Figure 2.3. Peary's houses at Fort Conger. These houses were built in Inuit style using material from Greely's large building. (Photograph: R.L. Christie, Geological Survey of Canada GSC-115426)

that the rival claims may never be resolved. Cook, during his journey in 1908-09, travelled from Inglefield Land, Greenland, across Ellesmere Island, to the north end of Axel Heiberg Island.

Otto Sverdrup in 1898 took *Fram* (of Fridtjof Nansen fame), into Smith Sound to investigate the then unexplored northern coastline of Greenland; circumnavigation of Greenland was considered. When northward travel by ship through Kane Basin became impossible due to ice conditions, Sverdrup directed his attention westward. The hardy Norwegians spent four winters in the harbours of southern Ellesmere Island and made extensive journeys by dog sledge and ski. Axel Heiberg, Ellef Ringnes, and Amund Ringnes islands were discovered, and large areas of southern and western Ellesmere Island were mapped. The geologist, Per Schei (Fig. 2.5), studied the Precambrian metamorphic complex and the overlying Paleozoic and Mesozoic sedimentary rocks. Schei died soon after his return to Europe, but his rock collections, preliminary reports and sketch maps, with fuller reports written by others, formed a solid framework for the geology of the region that is incorporated into the geological understanding of today (Holtedahl, 1917).

Exploration of the Polar Route between Canada and Greenland was, as noted above, frequent and vigorous, and geological data were soon published. The early exploits of Inglefield, Kane, Hayes and Hall provided basic information on the Precambrian Shield and overlying cover sequences; of particular note are Sutherland's (1853a, b) vivid descriptions, which include the first account of the geology of Smith Sound. Kane (1856) and Hayes (1867) discovered the flat-lying Paleozoic fossiliferous strata (Arctic Platform), fossils from which were described by Prof. J.B. Meek (1865). Hall's men discovered the deformed clastic strata north of the carbonate platform (Bessels, 1879).

An extensive area of Ellesmere Island and adjacent Greenland was mapped geologically by Capt. Henry W. Feilden (Fig. 2.6), naturalist to the Nares expedition.

Figure 2.4. Captain R.A. (Bob) Bartlett (1875-1946), famed Newfoundland sealing skipper, a most skilled ice-pilot who guided several Arctic expeditions (from Peary, 1910, p. 25).

Feilden's map and report, co-authored with C.E. de Rance (1878), were the first such on Arctic geology by a geologist from his own fieldwork. The fairly extensive fossil collections (Paleozoic, Mesozoic and Cenozoic) were described by Etheridge (1878) and Jeffries (1877). Astutely, Feilden mapped the tectonic margin of what is now known as the Ellesmere-Greenland fold belt, and he traced it on both sides of Nares Strait. A modern flavour in Feilden's work is the introduction of the first stratigraphic names: "Cape Rawson beds" and "Dana Bay beds".

Greely's party was obliged to abandon its rock collections and instruments in preparation for the retreat southward in small boats. However, some geological information was obtained: in Ellesmere Island coaly debris was noted at several localities, and petrified trees were discovered by D.L. Brainard (his "petrified forest") on Judge Daly Promontory; in Greenland, a notable contribution was the discovery of volcanic rocks at Lockwood Ø — the first record of the Kap Washington Group. By this time the fundamental division of the northern terranes into southern, homoclinal carbonate strata and northern, folded clastic rocks (i.e. Feilden's Cape Rawson beds) had been realized; moreover, Peary's observations from Navy Cliff in 1892 and 1895 indicated that the vast, unfolded sedimentary terrane continued across northern Greenland to the east coast. Specimens of metamorphic rock from northernmost Ellesmere Island, brought back to civilization by the Nares expedition and by Robert Peary, undoubtedly contributed to Charles Schuchert's later (1923) proposals for North American geosynclines. The northern sedimentary belt was named the Franklinian Geosyncline and a hypothetical borderland of crystalline rocks was named Pearya (Chapter 4).

Per Schei's reconnaissance of the sedimentary rocks resulted in a greatly improved understanding of what are now named the Arctic Platform, Franklinian Geosyncline, and Sverdrup Basin; Schei, for example, was the first to recognize that the Mesozoic section was strongly folded. Schei's divisions of the stratigraphic column have been retained, with formational names subsequently applied, and an early account of the Precambrian metamorphic complex in both Ellesmere Island and Greenland (Bugge, 1910) is derived directly from Schei's fieldwork. Schei's extensive fossil collections were systematically described and published by chosen specialists. The four-year Norwegian expedition in the *Fram*, returning in 1902, advanced the geographic and geological understanding of

Figure 2.6. Capt. Henry Wemyss Feilden (1838-1921), naturalist to the Nares expedition; Feilden was the first person to compile a geological map of Arctic geology from own observations. (Photograph: courtesy of The Museum of King's Own Royal Lancaster Regiment, U.K.)

the eastern Arctic Islands more than any expedition up to that time (Dawes and Christie, 1986).

IV. GEOLOGICAL EXPLORATION BY SHIP AND DOG SLEDGE, 1903-1947

The first few years of this century saw a change in the motivation of polar exploration. While territorial claims and the unfurling of national flags over newly-discovered land have always been the unchallenged right of explorers, many of the exploits described in the preceding sections, carried out by British, Americans, and Norwegians, were inspired by prestige in the quest for objectives like the North Pole, circumnavigation of Greenland, and navigation of the Northwest Passage. Now, however, sovereignty became a prime concern, and a number of Canadian and Danish expeditions were dispatched in which consolidation of territorial rights became the overriding motive. These government-backed expeditions operated within defined areas: Danish expeditions between 1906-18 were concerned only with northern Greenland, and Canadian expeditions

Figure 2.5. Per Schei (1875-1905), of the Norwegian expedition in the *Fram*, contributed greatly to geological knowledge of the northeastern Arctic Islands. (Photograph: Jacob Schei and family, Trondheim, Norway)

in the same period rarely crossed Nares Strait. It should be noted that Britain's claims to sovereignty in the North American Arctic were transferred to Canada in 1880, but Peary's geographical discoveries in the far north, particularly relevant to Greenland, naturally held a strong political implication. Canada, therefore, took definite steps to express new claims; Denmark did likewise, and in 1916 proclaimed sovereignty over all of Greenland. As it happened, the United States relinquished all territorial claims over northern Greenland in 1917 as part of negotiations over the purchase of the Danish West Indies (now the Virgin Islands).

The main expeditions of this period (both government and private) were characterized by small parties often ranging widely by ship and by dog team to explore remote parts of the region, where coasts were surmised or only known in outline. Geologists travelled with these parties and even led them, and by the end of the period a much improved understanding of the areal geology of the islands had been obtained.

The "sovereignty period" was heralded by the Canadian expedition under the command of A.P. Low, Geological Survey of Canada, that sailed into Hudson Bay, Smith Sound, and Lancaster Sound in 1903 and 1904. Establishment of sovereignty over the Arctic Islands was achieved by acts of declaration at several points, by the establishment of permanent stations for customs purposes, by the collection of whaling fees, and by the "patrol" of the vessel, the steam-powered sealer, *Neptune*. Low (1906) reported on the geology at scattered localities in the Nares Strait, Lancaster Sound, and Baffin Island regions, but his main contribution was to summarize the then known geology of the region traversed by his ship. His geological map included most of the Arctic Islands and northwestern coastal Greenland.

At the time of Low's voyage in the *Neptune*, Roald Amundsen, a Norwegian, completed the first navigation of the Northwest Passage in the little 47 ton herring boat, the *Gjøa*. Sailing into the Archipelago from Greenland in 1903, the ship wintered twice on King William Island at a harbour now named Gjoa Haven. Amundsen recorded the new position of the north magnetic pole, which had drifted northwestward from its earlier (Ross) position on Boothia Peninsula.

J.G. McMillan, geologist on Captain J.E. Bernier's voyages in the D.G.S. *Arctic* (Fig. 2.7) between 1906 and 1911, wrote an account of his geological observations and reviewed the earlier syntheses of Haughton, Low and Edward Suess on Arctic Islands geology (McMillan, 1910). The *Arctic* was navigated through several of the sounds and channels, including Parry Channel, and wintered three times in the islands: at Pond Inlet, Baffin Island; at Winter Harbour (Parry's winter quarters), Melville Island; and at Arctic Bay, Baffin Island. Bernier's party collected many historical documents and relics, and in addition "annexed" the "whole of the Arctic Archipelago" for Canada, unveiling a tablet on the occasion (in 1909) at Winter Harbour, Melville Island.

Danish authorities, at the conclusion of Robert Peary's geographic work in Greenland in 1902, began to show real concern for their northernmost inhabitants — the Thule Inuit (as noted, Peary's later ventures were directed to the Pole). The close association between Peary and his many auxiliary expeditions and the Inuit of the Thule district had continued for a decade. Peary's departure ended the supply

Figure 2.7. The C.G.S. *Arctic*, 650 tons, (formerly the *Gauss*, built in Germany and used in the Antarctic) sailed in Canada's northern, ice-filled waters between 1904 and 1925. The ship is here anchored in Dundas Harbour, 1925. (Photograph: G.H. Valiquette; Public Archives Canada, PA-102494)

of such modern material as arms, ammunition and utensils, material on which the Inuit had become dependent. The Danish Literary Expedition (1902-04) led by Ludwig Mylius-Erichsen, and the Greenland Administration's expedition of 1905 and 1906-08 led to the re-establishment of vital trading facilities. In 1910, following Peary's retirement from Greenland, Peter Freuchen and Knud Rasmussen (Fig. 2.8) established the now famous trading station at Thule. This station became the operational base for a series of "Thule" expeditions to the far north in the period 1912 to 1924.

The Thule district was the northerly limit of Danish activity in Greenland for some years. The coasts of the Polar Route had been mapped by the British and Americans, and only parts of the northeast corner of Greenland were still uncharted. The Danish government therefore began supporting expeditions to northern Greenland, both by the Polar Route and by way of the Greenland Sea. Surveying and cartography were of highest priority on these expeditions, but specific geological aims were also outlined.

This assault on the far north began with the prestigious Danmark Expedition, which left Copenhagen in the steam bark *Danmark* in 1906 to survey the unknown part of northeast Greenland, the eastern part of the Innuitian Region. Winter quarters for 28 men were established at 77°N at a site that is now the weather station Danmarkshavn. Sledge journeys in 1907 reached as far north as 83°30'N and the deep fiord system of Independence, Hagen and Danmark fjords was explored, linking up with the land discovered by Peary. The expedition discovered that the Peary Channel was not an arm of the sea (as Peary had surmised), and that the northern part of Greenland was not an island (Amdrup, 1913a, b). However, in achieving these results Mylius-Erichsen and two companions, Lieut. N.P. Høeg Hagen and the Greenlander Jørgen Brønlund (Fig. 2.9) delayed their retreat southward and this delay cost them their lives. Their fate became known in 1908 when a relief sledge party found the body of Brønlund, his diary, and some of Hagen's sketch-maps. Brønlund's words told of the earlier deaths of his two companions, but their bodies and diaries have never been found.

Figure 2.8. Knud Rasmussen (1879-1933), Danish-Greenlandic ethnographer who in the course of leading several "Thule" expeditions between 1912 and 1924 traversed the entire North American Arctic from Alaska to Peary Land, Greenland. (Photograph taken 1910, Thule, Greenland — from "Bogen om Knud", "written by friends", Westermanns Forlag, Copenhagen, 1943)

exploration in the far north. Plans had been formulated for surveying and scientific investigation of northern Greenland to be achieved by small, Inuit-style sledge parties that were to "live off the land". The first expedition, in 1912, was to include the exploration of the Peary Channel by way of the Polar Route. However, nothing had been heard of Mikkelsen for two years and plans were hurriedly changed so that the First Thule Expedition, led by Knud Rasmussen and with Peter Freuchen, went directly across the Inland Ice toward Danmark Fjord. No traces were found of Mikkelsen, but the expedition rediscovered that the Peary Channel does not exist (unaware of Mylius-Erichsen's discoveries five years previously). The tantalizing questions of the whereabouts of the missing members of Mylius-Erichsen's party, and their routes, remain unanswered today and expeditions are still launched to the region in search of clues.

On the return to Denmark in 1908, the expedition — successful in its geographic achievements but tragic by the loss of life — prompted Ejnar Mikkelsen to seek government funds for an expedition with the aim of recovering the bodies and records of Mylius-Erichsen and Hagen and investigating the "Peary Channel". The Alabama Expedition, led by Mikkelsen, left Copenhagen in 1909 in the 50 ton yacht *Alabama*. Progress through the ice of the Greenland Sea was slow and winter quarters had to be established as far south as 75°N. Mikkelsen, accompanied by Iver Iversen, reached the Danmark Fjord area on a long sledge trip in 1910. Two cairns with reports were discovered, but no other vital traces. The expedition carried out geological and surveying work, but nearly ended in disaster: the *Alabama* was "nipped" and sank while Mikkelsen and Iversen were up north and the other members of the expedition were rescued by a Norwegian sealer in August, 1910. Mikkelsen and Iversen were forced to stay in a lonely outpost until picked up in 1912 (Mikkelsen, 1922).

The loss of the *Alabama* is significant to our story because the incident changed the course of Danish

Figure 2.9. Jørgen Brønlund (1877-1907), Greenlander, companion of Mylius-Erichsen and Høeg Hagen. All perished in the course of exploration in North Greenland, 1907-1908. (Sketch by Achton Friis, 1906; from Astrup, 1913a, facing page 192)

Another Peary claim, that he had sighted Crocker Land in the Arctic Ocean, prompted renewed American enterprise: Donald B. MacMillan's Crocker Land Expedition of 1913 to 1917. MacMillan had earlier provided sledging support for Peary's massive assault on the Pole, and now he proposed a search of the section of the Arctic Ocean in which the explorer had thought he had seen the loom of land and "snowy peaks". From a base at Etah in Greenland, wide-ranging travels in the eastern Arctic Islands were effected by MacMillan and by an exploring sub-party led by the geologist, W.E. Ekblaw (MacMillan, 1918).

One of the largest and most colourful of expeditions to the Canadian Arctic was Vilhjalmur Stefansson's Canadian Arctic Expedition of 1913-18, the first major scientific project in the Archipelago to be supported by the Canadian government. Stefansson had been carrying out private projects on the mainland and on Victoria Island between 1908 and 1912 and he proposed a multi-faceted program that incorporated many other studies, including geological, with the geographic exploration. Support for the expedition soon was organized: undiscovered islands, it was thought, might lie in the Beaufort Sea, northwest of the Archipelago, and sovereignty over the islands was of increasing concern. The expedition was divided into two parties; a scientific group under R.M. Anderson (second-in-command) operated in the region of the northern mainland, Coronation Gulf, and southern Victoria Island, and a northern party, led by Stefansson himself, carried out wide-ranging geographic exploration in the Beaufort Sea and the western islands. The experienced ice-skipper Bob Bartlett (Fig. 2.4) commanded the principal expedition ship, the *Karluk*. The considerable scientific results of the expedition were published as a series of volumes over many years, and Stefansson and his northern party discovered the last major islands of the Archipelago: Brock, Borden, Mackenzie King, Meighen, and Lougheed. The expedition, however, was marred by mishap and discord. One of northern Canada's more tragic stories is that of the drift, and sinking in Siberian waters, of the *Karluk*, and the loss of more than half of the ship's company of 28.

The geographic outline of both the Canadian Archipelago and Greenland were known by the time of World War I. Some years passed with little activity in the Canadian Arctic, but Danish effort was directed to systematic cartographic and geological mapping of northern Greenland, from Melville Bugt to Danmark Fjord. This major task was essentially carried out between 1916 and 1923 on two expeditions: viz. the Second Thule Expedition, 1916-18, and the Bicentenary Jubilee Expedition, 1920-23. Exploration in this period is associated with the name of Lauge Koch (Fig. 2.10) who was cartographer and geologist on both expeditions and who for the next forty years participated in scientific work in Greenland, organizing and leading in the field dog sledge, boat, and aircraft geological operations.

The Second Thule and the Bicentenary Jubilee expeditions engaged the services of the Thule Inuit. The former expedition, led by Knud Rasmussen, surveyed the coast between Melville Bugt and Peary Land (Rasmussen, 1927); the Jubilee Expedition, led by Koch, continued the survey around the northern coast of Peary Land to the Wandel Sea — "the only portion of the coast of Greenland that has yet never been traversed by a Dane" (Koch, 1926). Both expeditions came under the threat of starvation, and dogs had to be killed for food; the expeditions, in depleted state, returned to Thule district by way of the Inland Ice. The scientific results (cartographic, physiographic, geological, glaciological) of the expeditions were immense: Koch compiled his basic data into two map series: 19 map sheets of topographic coverage at a scale of 1:300 000 (Koch, 1932), and 5 map sheets of geological coverage at varying

Figure 2.10. Lauge Koch (1892-1964), Danish geologist and cartographer who between 1916 and 1923 carried out by dog sledge the first regional survey of northern Greenland. In 1933 he made the first air sortie over northern Greenland reaching about 82°30'N from his base in East Greenland. (Photograph taken in the early 1930s, Arktisk Institut, Copenhagen)

scales (Koch, 1929b, 1933; Dawes and Haller, 1979). Much basic mapping remained to be accomplished, particularly in the inner fiord regions, and Koch had plans of returning north. However, an adverse political climate developed, and Norway challenged Denmark's sovereignty over all of Greenland (in 1924 a formal territorial treaty between Norway and Denmark had been signed, but in 1931 Norway formally annexed part at East Greenland). Danish presence on the east coast was necessary, and in 1926 Koch's operations were redirected to East Greenland, where he continued until the late 1950s. Other Danish activities were also concentrated in northeast Greenland, and one of these, Eigil Knuth's and Ebbe Munck's expedition of 1938-39 (Nielsen, 1941), reached the region of 81°N while surveying Kronprins Christian Land. However, apart from Lauge Koch's pioneer air forays (see below), northern Greenland remained unvisited between 1921 (Koch's sledging) and the period following World War II when aircraft were used.

In the west, the Thule expeditions continued to be organized, and two of these reached the Innuitian Region. The Third Thule Expedition, led by Comdr. Godfred Hansen, placed caches of supplies along the Polar Route, in northernmost Ellesmere Island and on Victoria Island, in support of the proposed trans-Arctic polar exploration in the *Maud* by Amundsen (Hansen, 1921). Hansen collected some fossils from Greenland (Teichert, 1937a). The Fifth Thule Expedition, 1921-24, led by Rasmussen, undertook the great traverse of the entire North American Arctic — from Greenland to Alaska — carrying out ethnological and other scientific work, including geology.

Several other wide-ranging parties travelled in the Arctic islands during the '20s and '30s, and geological knowledge was advanced by many of them. For example the Canadian government's "Eastern Arctic Patrol" — annual voyages into the eastern Arctic Islands by Captain Bernier in his now aging ship, the *Arctic* (Fig. 2.7) — commenced in 1922; in the course of these patrols the R.C.M.P. and various newly constituted branches of the government were enabled to administer the vast and sparsely peopled region that was at last the subject of some official concern. Bache Peninsula (Fig. 2.11), Craig Harbour, Dundas Harbour, and other police posts were established by ship, and the R.C.M.P. carried out far-ranging patrols by dog sledge. In the Queen Elizabeth Islands, these patrols were aided by Greenland Inuit. One of the more famous of such sledge journeys, some 2200 km long, was carried out by Corporal H.W. Stallworthy in 1932 in search of the lost German geologist, H.K.E. Krueger and his Danish and Greenlandic companions. Stallworthy was again in action with the Oxford University Ellesmere Land Expedition 1934-35, which wintered at Etah, Greenland; several sledge parties with Inuit made journeys in Ellesmere Island. Of note is a northern traverse that led to the first ascent of the icefields and of mountains of the United States Range (Moore, 1936); the expedition's geologist, Robert Bentham, carried out studies in both Ellesmere Island and Greenland, and in 1936-38 continued his work by accompanying R.C.M.P. patrols.

The '20s and '30s also saw the opening of many commercial trading posts in both the eastern (Baffin, Devon, Somerset islands) and the western Arctic (Victoria and King William islands, and along the mainland coast). The Hudson's Bay Company ships *Nascopie* and *Aklavik* met in 1937 at Bellot Strait, one ship from the east and the other from the west, thus marking the first commercial use of the Northwest Passage. (Bellot Strait was navigated from west to east in 1940-42 by Staff-sergeant Henry Larsen in the wooden police schooner *St. Roch*. In 1944 Larsen took the same vessel from east to west in a single open water season).

The last pre-World War II visitors to the islands were members of the Danish Thule and Ellesmere Land Expedition, who set up winter quarters on the Greenland side of Smith Sound. Helped by Inuit dog teams and drivers, the expedition travelled widely in Greenland and Ellesmere Island and reached Axel Heiberg Island, with Johannes C. Troelsen (Fig. 2.12) making important geological observations. Sledge parties returning to the winter base in 1940 were met by the news of war and the impossibility of returning to Denmark!

The science of geology matured rapidly during the period 1903-47. Edward Suess of Austria had written, and published in the course of several years, a comprehensive synthesis of the geology of the world. Suess's (1885-1924) massive work, in which he assimilated virtually all geological knowledge in previous literature, became a benchmark of geological thought; the science, worldwide, assumed essentially its modern form. The geologists, listed below, who travelled the Arctic Islands in the period from 1903 to 1947, viewed the beds and structures with the "new tectonics" in mind, and the observations and interpretations accelerated the understanding of arctic North America. W.E. Ekblaw, geologist of the Crocker Land Expedition (1913-17), traversed central and northern Ellesmere Island and recovered some fossils (recorded by Troelsen, 1950). J.J. O'Neill (1924) of the Geological Survey of Canada accompanied the southern party of the Canadian Arctic Expedition (1913-16) and reported on the mainland coastal regions of the western Arctic. Geological collections from Baffin Island, Melville Peninsula, and King William Island, made during the Fifth Thule Expedition, were reported on by a member of the expedition, Therkel Mathiassen (1933), and by Curt Teichert (1937b). L.J. Weeks (1927) of the Geological Survey of Canada visited shore points on Baffin, Devon and Ellesmere islands while on Captain Bernier's patrol ship, *Arctic* (and later from the *Beothic*), and Weeks and M.H. Haycock, wintering at Pangnirtung, studied the region of the head of Cumberland Sound and Nettilling Lake. In 1927 an American group, the Putnam Baffin Island Expedition, explored in the Foxe Peninsula – Amadjuak Lake region and made geological collections in Paleozoic terrane. R. Bentham (1936, 1941) in the period 1934 to 1938 studied the Precambrian terrane and the overlying cover in southern Ellesmere Island and adjacent Greenland; H. Drever and J.M. Wordie (Wordie, 1938) also studied the Bache Peninsula section seen by Bentham; D.A. Nichols (1936) collected marine shells from raised beaches at Dundas Harbour (at that time a Hudson's Bay post). In the southern parts of the Archipelago, A.L. Washburn (1947) provided an extensive account of the geology of the coastal regions between southern Banks Island and King William Island, based on fieldwork between 1938 and 1941.

All the above mentioned geological observations in the Arctic Islands were made by a series of unassociated expeditions in widely separated areas, expeditions which, moreover, had varying aims. Not surprisingly, the geological work lacked continuity, and few systematic geological syntheses were written. On the other hand, in northern Greenland, the travels of Lauge Koch took one geologist across the entire region from Melville Bugt to the Wandel Sea to link with the observations made by expeditions from East Greenland. Regional maps and syntheses resulted (e.g. Koch, 1920, 1923, 1925, 1929a, 1929b, 1932, 1933, 1935a). Koch (1929a) was able to divide the Proterozoic-Paleozoic succession into some 20 formations, which he portrayed on his maps, and he traced the main structural units throughout the region. The Precambrian Shield, the Arctic Platform (the "Great sediment plain" of Koch) and the Franklinian Geosyncline (the "Smith Sound geosyncline" of Koch) with its belt of "Caledonian folding" in the north, were also joined with units of the Arctic Islands that were beginning to emerge from the work of Per Schei. In addition, Koch's travels resulted in large fossil collections, which formed the basis of treatises published in Copenhagen in the series "Meddelelser om Grønland" by Christian Poulsen, Curt Teichert, Gustav Troedsson, and others.

Syntheses of the Innuitian Region at this time were compiled by Frebold (1934, 1942), Koch (1935a), Teichert (1939) and Kindle (1939). J.C. Troelsen's work (1939-41) in Greenland, Ellesmere Island and Axel Heiberg Island was based both on his own field studies and on the syntheses just

Figure 2.11. Bache Peninsula Police Post at the time of construction, 1926. The R.C.M.P. and ship's party pose with Greenland Inuit, who aided the Canadian police in travel and living in the High Arctic. M.H. Haycock and L.J. Weeks stand 7th and 13th from the left. Cambrian strata in the background. (Photograph: M.H. Haycock)

noted. Troelsen recognized the sub-Pennsylvanian unconformity at the base of the upper Paleozoic-Mesozoic-Tertiary succession (the Sverdrup Basin). His comprehensive synthesis of the geology of the region, illustrated by a map (Troelsen, 1950), marked the end of an era in which geological knowledge was obtained entirely without air support.

V. THE AIRCRAFT AGE OF GEOLOGICAL STUDY[3]

The post World War II appearance of aircraft in the Arctic Islands marked a turning point in Arctic geological exploration. Small "bush" aircraft had pushed north of the

---

[3] This period extends from 1947 to the present day. This historical account, however, deals primarily with early investigations and is taken in detail only to the early '70s. Modern work is summarized in following chapters of this volume.

Figure 2.12. Johannes C. Troelsen, Danish geologist, who spent several years in northern Greenland and travelled in Ellesmere Island and Axel Heiberg Island. Troelsen was a member of the last ship and dog-sledge expedition (pre-World War II) and also a member of the first air-borne expedition to the Innuitian Region, in 1947. (Photograph: Eigil Knuth, 1948)

Canadian mainland on occasion during the '30s, and several flights reached the Innuitian Region. For example, the MacGregor Arctic Expedition 1937-38, with headquarters at Etah, Greenland, made reconnaissance flights in a Waco hydroplane, including a trip to the Polar Basin northwest of Ellesmere Island, in search of Crocker Land (MacGregor, 1939). However, prior to this, in 1933 during the Danish Geodetic Institute's pioneering photogrammetric survey of East Greenland, Lauge Koch was able to make flights in a Heinkel seaplane over eastern North Greenland and made important revisions to the map (Koch, 1935b). Prompted by this success and by his air-supported geological program on the east coast, in 1938 Koch operated a Dornier-Wal aircraft "Perssuaq" from Svalbard, and succeeded in mapping the hitherto unknown inner part of Peary Land. During these flights he discovered at last the real nature of the "Peary Channel" that had figured so dramatically in the aims of early Danish expeditions (Koch, 1940): that it is a long, narrow valley, with lakes, that joins the northern and eastern coasts.

The pre-war Danish work with aeroplanes and the experiments with photogrammetry in the High Arctic were a noteworthy success and a large step forward in mapping techniques. Aerial photography had been used for mapping purposes since the middle of the '20s, and reliable topographic maps, with aerial photographs, were fast becoming the basic requirements for modern geological field recording. The wartime development of reliable, long-range freighter and passenger aircraft enabled, firstly, the establishment of Arctic weather stations in remote parts of the region and secondly, the transport of geologists and field supplies in and out of the Arctic during a single season, obviating the necessity of wintering one or more times for each full summer season of fieldwork.

The period of air-supported geological work started in 1947, when the first major airborne expedition to the Innuitian Region (the Danish Peary Land Expedition) descended on Jørgen Brønlund Fjord (82°N), southern Peary Land, in a Catalina amphibian (PBY) of the Royal Danish Navy (Fig. 2.13) (poor weather had prevented a landing at a site north of 83°N). This multidisciplinary venture (1947-50), led by archeologist Eigil Knuth, operated in northern Greenland for three years, with personnel being exchanged each year by Catalina aircraft: J.C. Troelsen worked there in 1947 and 1948 and K. Ellitsgaard-Rasmussen took over in 1949. Geological work away from the station, "Brønlundhus" (Fig. 2.14), was achieved by dog sledge. Troelsen (1949) worked mainly in southern Peary Land on the Arctic Platform and upper Paleozoic-Mesozoic-Tertiary section (which he compared to the Sverdrup Basin deposits with which he was familiar), Ellitsgaard-Rasmussen (1955) travelled north to investigate the North Greenland fold belt, and B. Fristrup (1952) undertook glaciological work.

Figure 2.13. Catalina (PBY) of the Royal Danish Navy on Jørgen Brønlund Fjord in 1949. The use of PBY's by Eigil Knuth's Peary Land Expedition (1947-50) heralded a new exploration epoch in the far north — air-borne geological parties. (Photograph: Børge Fristrup, 1949)

In the Canadian Arctic Islands in 1947, Y.O. Fortier of the Geological Survey of Canada carried out an aerial geological reconnaissance while accompanying a Dominion Observatory magnetic party in a Royal Canadian Air Force Canso flying boat (Fortier, 1948). The year 1947 also saw the beginnings of the system of six Arctic weather stations (Joint Arctic Weather Stations, or JAWS, operated by Canadian and United States authorities), each with a graded airstrip and occupied the full year. Two joint U.S. – Danish weather stations were established in northern Greenland: one at Thule (76°30'N) and a second at Station Nord, to the north and east (82°N). A radio/weather station had functioned at Thule during the Second World War, and in 1951-52 the site was developed into a U.S. military airbase. Conventional means of travel — ship, boat or canoe, and dog sledge — continued to be used by geologists for many years, as noted below, but the availability of air

Figure 2.14. "Brønlundhus", Jørgen Brønlund Fjord, Peary Land (83°10'N), the wintering station of the Danish Peary Land expedition 1947-50; the first air-borne expedition to the Innuitian Region. Cambrian strata in background. (Photograph: Børge Fristrup, 1948)

inch (1:506 880) scale geological maps and accompanying stratigraphic records (Fortier et al., 1963).

Air travel became increasingly useful in geological fieldwork, yet a small group, mainly of the Geological Survey of Canada, continued to work during the '50s and '60s using traditional Inuit sledging methods (Fig. 2.15), motor-driven canoes, and long traverses on foot. Vigorous reconnaissance exploration was carried out at remote and widely scattered localities by Y.O. Fortier, R. Thorsteinsson, W.L. Davison, G.C. Riley, W.W. Heywood, R.G. Blackadar, G. Hattersley-Smith and R.L. Christie. Geological contributions were also made in this period by T.H. Manning.

Figure 2.15. Travel by dog sledge; an Inuk dog driver here ices the sled runner using a specially made polar bear mitt. Dolphin and Union Strait, March, 1931. (Photograph: R.S. Finnie; Public Archives of Canada, PA-100611)

support soon resulted in major changes in style and pace of Arctic fieldwork. Helicopter-supported parties, working from Thule Air Base or from ice-breakers that reached as far north as the Lincoln Sea, entered the scene in the late '40s and '50s. The U.S.S. *Edisto* (see Kurtz and Wales, 1951; Kurtz et al., 1952), the U.S.S. *Eastwind* (Prest, 1952), and the U.S.S. *Atka* (Davies et al., 1959) supported fieldwork in this period. Also, a number of military "operations" included studies in geology, geomorphology and engineering geology in the '50s and early '60s; noteworthy here are William E. Davies and Daniel B. Krinsley of the U.S. Geological Survey, who carried out bedrock and surficial geology studies across northern Greenland (Davies, 1972; Davies et al., 1959, 1963). With the military interest in the far north came the completion of coverage by aerial photography; the first aerial surveys for use in map making were carried out by the U.S. Air Force in 1947 and the resultant trimetrogon photographs were used in the preparation of contoured topographic maps.

Danish geologists participated in the military "operations" just noted and in addition, in 1952 and 1953, field parties including P.J. Adams, J.W. Cowie and E. Fränkl were put out by float plane in sorties organized by Lauge Koch from his base in East Greenland. Fränkl (1955), accompanied by Fritz Müller, reached the northern coast of Peary Land (83°30'N).

In Arctic Canada the aircraft age of geological exploration began in earnest in 1955 with the Geological Survey's Operation Franklin, a combined geological and aerial geophysical reconnaissance. Two-man geological teams were put out at widely scattered, key localities using Sikorsky S-55 helicopters. The small areas of study on the ground, combined with extensive study of aerial photographs, enabled the group of 11 geologists to compile 8-miles-to-the-

The helicopter, with its speed and capability to land almost anywhere, is clearly the nearest to ideal transport for a geologist in a remote and large region. In 1955, however, helicopters were expensive and of limited range, and for some years light fixed-wing aircraft were adapted to fieldwork, achieving advantages in cost. The well known Canadian Arctic pilot, W.W. Phipps devised light, oversized and low-pressure, or "balloon" tires and wheels for the Piper Super-Cub (PA-18A) aeroplane (Fig. 2.16). This aircraft was designed for "short take-off and landing" (STOL) and the "big wheels" enabled the Cub, and later other aircraft, especially the de Havilland Beaver, Otter, and Twin Otter, to land at level but unprepared sites throughout the islands. For some 10 years these adapted "bush" aircraft and a small number of skilled Canadian pilots and engineers aided in exploration of even the most remote and mountainous islands, and on occasion in neighbouring Greenland. Improved models (i.e., faster, greater range and reliability) of helicopters appeared in the mid-1960s and the Super-Cub was quickly displaced. The larger fixed-wing aircraft, particularly the Twin Otter, continue to serve in transport of fuel, passengers, and matériel from main supply points to field base camps. In northern Greenland, in recent years, STOL flying has been undertaken by Icelandic aircraft and crews.

Figure 2.16. Piper Super-Cub, with pilot W.W. (Weldy) Phipps, on southern Ellesmere Island, 1960. The Super-Cub on oversize wheels could be landed at many localities and for some years served as the geologist's "work horse" in the Islands. (Photograph: R.L. Christie, Geological Survey of Canada, GSC-115233)

Combinations of helicopter and fixed-wing aircraft support (on occasion using float aircraft during the open water season) became the norm for the Arctic Islands after 1957. Some of the projects, and their leaders, were: R. Thorsteinsson and E.T. Tozer in the western Queen Elizabeth Islands, 1958; "Operation Eureka", Thorsteinsson, on central Ellesmere Island in 1961-62; "Operation Prince of Wales", R.G. Blackadar, in the Boothia Peninsula region in 1962; "Operation Admiralty", Blackadar, on northern Baffin Island in 1962-63; "Operation Bathurst", J.W. Kerr, on Bathurst and adjacent islands in 1963-64; "Operation Amadjuak", Blackadar, on southern Baffin Island in 1965; "Operation Grant Land", R.L. Christie, on northern Ellesmere Island and adjacent Greenland in 1965-66; an unnamed operation on Ellef Ringnes Island, D.F. Stott, in 1967; "Operation Bylot", G.D. Jackson, on northeastern Baffin and Bylot islands in 1968; "Operation Peel Sound", Thorsteinsson, on Prince of Wales Island in 1970; and "Operation Grinnell", Kerr, on northwestern Devon Island in 1971-72. The list of "operations" just noted refers to the larger teams of geologists that were supported by one or more aircraft, usually for seasons of five weeks to two or more months. Other, smaller parties, often from universities, were given air support for short periods, or intermittently. Scientific studies, including geological work, were also supported by the Department of National Defence, especially in northern Ellesmere Island under the leadership of G. Hattersley-Smith (1974).

Geological studies in the islands became more specialized in the late '60s and early '70s, in contrast to earlier work, much of which was devoted to first-stage, reconnaissance mapping. One result of this was the eventual disappearance of the large, named "operation": the air support that was provided was dispersed among a plethora of specialists, some of whom might spend only a part of the field season in a given area.

An important element of Arctic scientific research (including a wide range of sciences) since 1958 has been the Polar Continental Shelf Project (PCSP), a combined logistical support and scientific branch of the Department of Energy, Mines and Resources of Canada. This sister organization of the Geological Survey of Canada, begun under the leadership of E.F. Roots and later directed by G.D. Hobson, was organized primarily to support field research on and near the Continental Shelf off the Arctic Islands, but in time was able to provide increasingly capable and efficient air support and other logistical services for geological fieldwork throughout the islands and on the northernmost mainland. In addition the PCSP has supported co-operative Canadian-Danish ventures across Nares Strait.

Government-financed geological exploration in northern Greenland in this period showed a rather different development. The Geological Survey of Greenland (GGU; established in 1946) directed its main effort to the Precambrian crystalline terranes of the west coast, while the Lauge Koch expeditions operated in East Greenland. In the Innuitian Region, J.H. Allaart and P.R. Dawes joined the Canadian Operation Grant Land (1965-66) to work in the Hall Land – Wulff Land area, H.F. Jepsen worked in southern Peary Land during Eigil Knuth's expeditions of 1966, 1968, and 1970, and Dawes and N.J. Soper studied the North Greenland fold belt in northern Peary Land in 1969 as members of the British Joint Services Expedition led by J.D. Peacock. In the 1970s GGU began mapping of selected areas of northern Greenland using limited air support; these investigations stretched from the Thule – Inglefield Land region to eastern Peary Land. Systematic regional mapping of northernmost Greenland at a scale of 1:500 000 was begun in 1978; this five-season, air-supported project, led by N. Henriksen in conjunction with the Danish Geodetic Institute, was carried out each season by a geological crew of about 20 persons.

Geophysical work was carried out in the Canadian islands in 1955, when an airborne magnetic and radiometric reconnaissance coincided with Operation Franklin. Aeromagnetic surveys of Foxe Basin, Baffin Bay, Lancaster Sound, and the Arctic Ocean have been carried out in co-operation with Canada's National Aeronautical Establishment since 1964; the geophysical data were compiled, interpreted, and published by P.J. Hood, M.E. Bowers, A.F. Gregory, and L.W. Morley. The initial aeromagnetic survey of northern Greenland was undertaken in the early 1970s by Greenarctic Consortium — a Canadian-financed commercial organization. Deep seismic refraction surveys were conducted in 1960 and later by G.D. Hobson and A. Overton of the Geological Survey of Canada and by G.W. Sander, Dominion Observatory (later Earth Physics Branch, now melded with the Geological Survey of Canada).

The search for resources has played an important role in the exploration and "opening up" of the Arctic and also has contributed significantly to the basic geological knowledge of the region (Nassichuk, 1987).

The Inuit recovered native copper in the Coppermine River region (on the mainland south of Victoria Island) and meteoric iron in the Kap York district of Greenland, and fashioned implements from these materials. Martin

Frobisher's several voyages to southern Baffin Island were motivated by the desire for gold, a deposit of which he thought he had found. However, it was not until the 1960s that commercial deposits of metallic minerals were found and developed. A zinc deposit in Proterozoic sediments of northwestern Baffin Island (Nanisivik Mine) was brought into production in 1976, and a giant lead-zinc deposit on Little Cornwallis Island (Polaris Mine) in 1981. Noteworthy, also, is the discovery of large iron deposits on northern Baffin Island and on the eastern and western coasts of Melville Peninsula, all in Precambrian crystalline rocks (Christie and Kerr, 1981). In addition, various noncommercial mineral showings or geochemical anomalies have been discovered in the Arctic Islands and Greenland.

Cretaceous coal near Pond Inlet on northeastern Baffin Island was mined by whalers, and Tertiary coal in northern Ellesmere Island by the Nares and Greely expeditions. A systematic inventory of the coal resources of the Arctic Islands by Petro-Canada in the late 1970s and early 1980s revealed that the Arctic Islands contain about 20% of the total inferred coal resources in Canada.

The presence, in the Arctic Islands, of large basins with thick successions of sedimentary rocks that might favour hydrocarbon accumulations was demonstrated in the 1950s, mainly by the Geological Survey of Canada (e.g. Fortier et al., 1954; Thorsteinsson, 1958). Exploration by petroleum companies began in the early 1960s, and noteworthy is the pioneer work of J.C. Sproule, whose confidence in the economic potential of the High Arctic resulted in geological and geophysical studies across the entire Innuitian Region, from the Beaufort Sea to northeastern Greenland. The development of specific prospects usually comprised three phases: geological surface work, seismic reflection surveys, and drilling.

Seismic surveys for petroleum exploration began on Melville Island in 1968, and were extended to other islands in later years. Offshore techniques were developed in the late 1960s, especially by the Suncor and Panarctic companies, and, by the mid-1970s, surveys had covered much of the oil-prospective areas of the Arctic Islands. To date some 65 000 km of seismic reflection lines have been shot.

The first wildcat well (Winter Harbour No. 1) was spudded in 1962 on southwestern Melville Island. It encountered gas, but not in commercial quantity. Between 1961 and 1986, 176 wells were drilled in the Arctic Islands, and 18 hydrocarbon fields have been discovered, mainly by Panarctic Oils Ltd.

In North Greenland, surface reconnaissance by the Greenarctic Consortium in 1968-73 was followed by systematic investigations by the Geological Survey of Greenland from 1978 onward. They included organic geochemistry and shallow drilling in addition to stratigraphic and structural work.

The study of the respective parts of the region — the Arctic Islands and northern Greenland — has been dealt with mainly by Canadian and Danish government agencies. Thus, the Geological Survey of Canada, with its continuing field and support research staff, has published an extended series of reports, maps and syntheses, with the geology of the nation reviewed in "Geology and Economic Minerals of Canada", the fifth edition of which appeared in 1970. The Geological Survey of Greenland in 1976 published "Geology of Greenland", and the geology across Nares Strait was reviewed in a volume of Meddelelser om Grønland (Dawes and Kerr, 1982). Some earlier Canadian syntheses of the Innuitian Region include those of Fortier et al. (1954, 1963); Douglas et al. (1963), Thorsteinsson and Tozer (1970); and Trettin et al. (1972). Early Danish syntheses are those of Dawes (1971, 1976). Intensive study of the petroleum and mineral potential by exploration companies during the 1960s (see below) also led to several summary accounts such as those by P.A. Ziegler (1969), B.P. Plauchut (1971, 1972), K.J. Drummond (1973), H.D. Daae and A.T. Rutgers (1975), and R.G. McCrossan and J.W. Porter (1973). Similarly, field research by university groups, accelerated by easier access and shared logistical support, has led to publication of syntheses and theses such as those by D.L. Dineley (1971), and A.F. Embry and J.E. Klovan (1976).

The cumulative geological understanding of the Innuitian Region now available is the subject of this volume.

ACKNOWLEDGMENTS

The authors thank R. Thorsteinsson and Q.H. Goodbody (Geological Survey of Canada), T.H. Levere (University of Toronto), and A.K. Higgins and W.S. Watt (Geological Survey of Greenland) for critical review of the manuscript.

REFERENCES

Amdrup, G.
1913a: Report on the Danmark Expedition to the north-east coast of Greenland 1906-1908; Meddelelser om Grønland, v. 41, 270 p.
1913b: Mylius-Erichsen's report on the non-existence of the Peary Channel; information brought home by Ejnar Mikkelsen; Meddelelser om Grønland, v. 41, p. 469-474.

Bentham, R.
1936: Appendix I, Geology; in Oxford University Ellesmere Land Expedition; Geographical Journal, v. 87, no. 5, p. 427-431.
1941: Structure and glaciers of southern Ellesmere Island; Geographical Journal, v. 97, p. 36-45.

Bessels, E.
1879: Die amerikanische Nordpol-Expedition; Wilhelm Engelmann, Leipzig, 647 p.

Bugge, C.
1910: Petrographische Resultate der 2ten Fram-Expedition; in Report of the Second Norwegian Arctic Expedition in the *Fram* 1898-1902; Videnskabs-Selskabet, Kristiania (Oslo), Report no. 22.

Christie, R.L. and Kerr, J.W.
1981: Geological exploration of the Canadian Arctic Islands; in A Century of Canada's Arctic Islands, 1880-1980, M. Zaslow (ed.); Royal Society of Canada, p. 187-202.

Daae, H.D. and Rutgers, A.T.
1975: Geological history of the Northwest Passage; in Canada's continental margins and offshore petroleum exploration, C.J. Yorath, E.R. Parker, and D.J. Glass (ed.); Canadian Society of Petroleum Geologists, Memoir 4, p. 477-500.

Davies, W.E.
1972: Landscape of northern Greenland; U.S. Army Cold Regions Research Engineering Laboratory, Hanover, New Hampshire, Special Report 164, 54 p.

Davies, W.E., Needleman, S.M., and Klick, D.W.
1959: Report on Operation Groundhog (1958) North Greenland. Investigation of ice-free sites for aircraft landing, Polaris Promontory, North Greenland; U.S. Air Force Cambridge Research Center, Bedford, 45 p.

Davies, W.E., Krinsley, D.B., and Nicol, A.H.
1963: Geology of the North Star Bugt area, Northwest Greenland; Meddelelser om Grønland, v. 162, no. 12, 68 p.

Dawes, P.R.
1971: The North Greenland fold belt and environs; Geological Society of Denmark, Bulletin, v. 20, p. 197-239.
1976: Precambrian to Tertiary of northern Greenland; in Geology of Greenland, A. Escher and W.S. Watt (ed.); Geological Survey of Greenland, Copenhagen, p. 248-303.

Dawes, P.R. and Christie, R.L.
1982: History of exploration and geology in the Nares Strait region; in Nares Strait and the Drift of Greenland: a Conflict in Plate Tectonics, P.R. Dawes and J.W. Kerr (ed.); Meddelelser om Grønland, Geoscience, no. 8, p. 19-36.
1986: Per Schei (1875-1905); Arctic, v. 39, no. 1, p. 106-107.

Dawes, P.R. and Haller, J.
1979: Historical aspects in the geological investigation of northern Greenland. Part 1: New maps and photographs from the 2nd Thule Expedition 1916-1918 and the Bicentenary Jubilee Expedition 1920-1923; Meddelelser om Grønland, v. 200, no. 4, 38 p.

Dawes, P.R. and Kerr, J.W. (ed.)
1982: Nares Strait and the Drift of Greenland: a Conflict in Plate Tectonics; Meddelelser om Grønland, Geoscience, no. 8, 392 p.

Dineley, D.L.
1971: Arches and basins of the southern Arctic Islands of Canada; Proceedings of Geologists' Association, London, v. 82, pt. 4, p. 411-444.

Douglas, R.J.W., Norris, D.K., Thorsteinsson, R., and Tozer, E.T.
1963: Geology and petroleum potentialities of northern Canada; Geological Survey of Canada, Paper 63-31, 28 p.

Drummond, K.J.
1973: Canadian Arctic Islands; in The Future Petroleum Provinces of Canada, their Geology and Potential, R.G. McCrossan (ed.); Canadian Society of Petroleum Geologists, Memoir 1, p. 443-472.

Ellitsgaard-Rasmussen, K.
1955: Features of the geology of the folding range of Peary Land, North Greenland; Meddelelser om Grønland, v. 127, no. 7, 56 p.

Embry, A.F. and Klovan, J.E.
1976: The Middle-Upper Devonian clastic wedge of the Franklinian Geosyncline; Bulletin of Canadian Petroleum Geology, v. 24, p. 485-639.

Etheridge, R.
1878: Palaeontology of the coasts of the Arctic lands visited by the late British Expedition under Captain Sir George Nares, R.N., K.C.B., F.R.S.; Quarterly Journal of the Geological Society of London, v. 34, p. 568-636.

Feilden, H.W. and de Rance, C.E.
1878: Geology of the coasts of the Arctic lands visited by the late British expedition under Captain Sir George Nares, R.N., K.C.B., F.R.S.; Quarterly Journal of the Geological Society of London, v. 34, p. 556-567.

Fortier, Y.O.
1948: Flights in 1947 over the region of the North Magnetic Pole and the mainland between the Arctic coast, Great Slave Lake and Hudson Bay, Northwest Territories; Geological Survey of Canada, Paper 48-23, 11 p.

Fortier, Y.O., Blackadar, R.G., Glenister, B.F., Greiner, H.R., McLaren, D.J., McMillan, N.J., Norris, A.W., Roots, E.F., Souther, J.G., Thorsteinsson, R., and Tozer, E.T.
1963: Geology of the north-central part of the Arctic Archipelago, Northwest Territories (Operation Franklin); Geological Survey of Canada, Memoir 320, 671 p.

Fortier, Y.O., McNair, A.H., and Thorsteinsson, R.
1954: Geology and petroleum possibilities in Canadian Arctic Islands; Bulletin of the American Association of Petroleum Geologists, v. 38, p. 2075-2109.

Fränkl, E.
1955: Rapport über die Durchquerung von Nord Peary Land (Nordgrönland) im Sommer 1953; Meddelelser om Grønland, v. 103, no. 8, 61 p.

Frebold, H.
1934: Tatsachen und Deutungen zur Geologie der Arktis; Meddelelser fra Dansk geologisk forening, v. 8, p. 301-326.
1942: Die Arktis; Geologische Jahresberichte, v. 48, p. 1-34.

Fristrup, B.
1952: Climate and glaciology of Peary Land, North Greenland; Association Internationale d'Hydrologie Scientifique, Assemblé Générale de Bruxelles, 1951, tome 1, p. 185-19.

Greene, M.T.
1982: Geology in the nineteenth century: changing views in a changing world; Cornell University Press, New York, 324 p.

Hansen, G.
1921: Den tredje Thuleekspedition; Norges depotekspedition til Roald Amundsen; in Nordostpassagen, R.E.G. Amundsen; Kristiania (Oslo), p. 437-462.

Hattersley-Smith, G.
1974: North of latitude eighty; Defence Research Board, Canada, 121 p.

Haughton, S.
1858: Geological notes and illustrations; in Reminiscences of Arctic Ice-Travel in Search of Sir John Franklin and his Companions, F.L. M'Clintock; Journal of the Royal Dublin Society, v. 1, p. 183-250.
1859: Geological account of the Arctic Archipelago: in The Voyage of the Fox in the Arctic Seas; a Narrative of the Discovery of the Fate of Sir John Franklin and his Companions, F.L. M'Clintock; John Murray, London, Appendix IV, p. 372-399.
1860: Geological results of the voyage of the Fox, notes and appendices in Reminiscences of Arctic Ice-Travel in Search of Sir John Franklin and his Companions, F.L. M'Clintock; Journal of the Royal Dublin Society, v. 3, p. 53-58.

Hayes, I.I.
1867: The Open Polar Sea: a Narrative of a Voyage of Discovery towards the North Pole in the Schooner United States; Hurd and Houghton, New York, 454 p.

Hayes, J.G.
1929: Robert Edwin Peary. A Record of his Explorations 1886-1909; Grant Richards and Humphry Toulmin, London, 299 p.

Holtedahl, O.
1917: Summary of the geological results; in Report of the Second Norwegian Arctic Expedition in the Fram 1898-1902; Videnskabs-Selskabet, Kristiania (Oslo), Report no. 36, 27 p.

Jeffries, J.G.
1877: The post-tertiary fossils procured in the late Arctic Expedition; with notes on some of the recent or living Mollusca from the same expedition; Annual Magazine of Natural History, Series 4, v. 20, p. 229-242.

Jones, G.H.
1981: Economic development — oil and gas; in A Century of Canada's Arctic Islands, 1880-1980, M. Zaslow (ed.); Royal Society of Canada, p. 221-230.

Kane, E.K.
1856: Arctic Explorations: The Second Grinnell Expedition in Search of Sir John Franklin, 1853, '54, '55; Childs and Peterson, Philadelphia, v. 1, 464 p., v. 2, 467 p.

Kindle, E.M.
1939: Geology of the Arctic Archipelago and the Interior Plains of Canada; in Geology of North America, R. Ruedemann and R. Balk (ed.); Geologie der Erde, v.1, Gebrüder Borntraeger, Berlin, p. 176-231.

Knuth, E.
1967: Archaeology of the Musk-ox Way, École Pratique des Hautes Études, sixième section; Contributions du Centre d'Études Arctiques et Finno-Scandinaves 5, Sorbonne, Paris, 78 p.
1968: Âlut Kangermio — Aron of Kangek; K'avdlunâtsianik — the Norsemen and the Skraelings; Det Grønlandiske Forlag, 111 p.
1981: Greenland news from between 81° and 83° North; Folk, v. 23, p. 106-107.

Koch, L.
1920: Stratigraphy of Northwest Greenland; Meddelelser fra Dansk geologisk forening, v. 5, no. 17, 78 p.
1923: Preliminary report upon the geology of Peary Land, Arctic Greenland; American Journal of Science, 5th Series, v. 5, p. 189-199.
1925: The geology of North Greenland; American Journal of Science, v. 9, p. 271-285.
1926: Report on the Danish Bicentenary Jubilee Expedition north of Greenland 1920-1923; Meddelelser om Grønland, v. 70.1, no.1, p. 1-232.
1929a: Stratigraphy of Greenland; Meddelelser om Grønland, v. 73.2, no. 2, p. 205-320.
1929b: The geology of the south coast of Washington Land; Meddelelser om Grønland, v. 73.1, no. 1, 39 p.
1932: Map of North Greenland, scale 1:300,000, 19 sheets; Geodetic Institute, Copenhagen.
1933: The geology of Inglefield Land; Meddelelser om Grønland, v. 73.1, no. 2, 38 p.
1935a: Geologie von Grönland; in Geologie der Erde, E. Krenkel (ed.); Gebrüder Borntraeger, Berlin, 159 p.
1935b: A day in North Greenland; Geografiska Annaler 1935; Sven Hedin volume, p. 609-620.
1940: Survey of North Greenland; Meddelelser om Grønland, v. 130, no. 1, 364 p.

Konig, C.
1823: An account of the rock specimens collected by Captain Parry, during the Northern Voyage of Discovery, performed in the years 1819 and 1820; Quarterly Journal of Science, v. XV, p. 11-22.

1824: Rock specimens; in A supplement to the Appendix of Captain Parry's voyage for the discovery of a Northwest Passage, in the years 1819-1820, W.E. Parry; John Murray, London, p. ccxlvii-cclvii.

Kurtz, V.E. and Wales, D.B.
1951: Geology of the Thule area, Greenland; Proceedings of the Oklahoma Academy of Science for 1950, v. 31, p. 83-89.

Kurtz, V.E., McNair, A.H., and Wales, D.B.
1952: Stratigraphy of the Dundas Harbour area, Devon Island, Arctic Archipelago; American Journal of Science, v. 250, p. 636-655.

Loomis, C.C.
1971: Weird and Tragic Shores — The Story of Charles Francis Hall, Explorer; Alfred A. Knopf, New York, 367 p.

Low, A.P.
1906: Report on the Dominion Government Expedition to Hudson Bay and the Arctic Islands on Board the D.G.S. *Neptune*, 1903-1904; Government Printing Bureau, Ottawa, 355 p.

MacGregor, C.J.
1939: The MacGregor Arctic expedition to Etah, Greenland, July 1, 1937 to October 4, 1938; U.S. Weather Bureau Monthly Weather Review, v. 67, p. 366-382.

MacMillan, D.B.
1918: Four Years in the White North; Harper & Brothers, New York, 426 p. (Also: revised edition, 1925, The Medici Society, London and Boston; pagination unchanged).

Mathiassen, T.
1933: Contributions to the geography of Baffin Land and Melville Peninsula; Report of the Fifth Thule Expedition, 1921-24, v. 1, no. 3, Gyldendal Nordisk Forlag, Copenhagen.

Maxwell, M.S.
1960: An archeological analysis of eastern Grant Land, Ellesmere Island, Northwest Territories; National Museum of Canada, Bulletin 170, 109 p.

McCrossan, R.G. and Porter, J.W.
1973: The geology and petroleum potential of the Canadian sedimentary basins — a synthesis; in The Future Petroleum Provinces of Canada, their Geology and Potential, R.G. McCrossan (ed.); Canadian Society of Petroleum Geologists, Memoir 1, p. 589-720.

McMillan, J.G.
1910: Geological Report; in Report on the Dominion of Canada Government Expedition to the Arctic Islands and Hudson Strait on Board the D.G.S. *Arctic* (1908-09), J.E. Bernier; Queen's Printer, Ottawa, p. 382-479.

M'Culloch, J.
1819: Appendix III, Geological memoranda; in A Voyage of Discovery, Made under the Orders of the Admiralty in His Majesty's Ships *Isabella* and *Alexander*, for the Purpose of Exploring Baffin's Bay, and Inquiring into the Possibility of a North-West Passage, J. Ross; John Murray, London, v. 2, p. 121-141.

Meek, F.B.
1865: Preliminary notice of a small collection of fossils found by Dr. Hays on the west shore of Kennedy Channel at the highest northern localities ever explored; American Journal of Science, v. 40, p. 31-34.

Mikkelsen, E.
1922: Alabama-Expeditionen til Grønlands Nordøstkyst 1909-1912; I. Report on the Expedition; Meddelelser om Grønland, v. 52, no. 1, 142 p.

Moore, A.W.
1936: The sledge journey to Grant Land; in Oxford University Ellesmere Land Expedition; Geographical Journal, v. 87, p. 419-427.

Nassichuk, W.W.
1987: Forty years of northern non-renewable natural resource development; Arctic, v. 40, p. 274-284.

Neatby, L.H.
1958: In Quest of the North West Passage; Constable and Company, London, 194 p.
1970: The Search for Franklin; Arthur Barker Ltd., London, 281 p.

Nielsen, E.
1941: Remarks on the map and the geology of Kronprins Christians Land; Meddelelser om Grønland, v. 126, no. 2, 34 p.

Nichols, D.A.
1936: Post-Pleistocene fossils of the uplifted beaches of the eastern Arctic regions of Canada; Canadian Field Naturalist, v. L., p. 127-129.

O'Neill, J.J.
1924: The geology of the Arctic Coast of Canada, west of the Kent Peninsula; Part A, in vol. XI; Geology and Geography, Report of the Canadian Arctic Expedition 1913-18, King's Printers, Ottawa, 107 p.

O'Reilly, B.
1818: Greenland, the Adjacent Seas and the North-West Passage to the Pacific Ocean, illustrated in a Voyage to Davis's Strait, during the Summer of 1817; Baldwin, Craddock and Joy, London.

Peary, R.E.
1898: Northward over the "Great Ice" — A Narrative of Life and Work along the Shores and upon the Interior Ice-Cap of Northern Greenland in the Years 1886 and 1891-1897; Frederick A. Stokes, New York, v. 1, 521 p., v. 2, 618 p.
1910: The North Pole; Hodder and Stoughton, London, 326 p.

Plauchut, B.P.
1971: Geology of the Sverdrup Basin; Bulletin of Canadian Petroleum Geology, v. 19, p. 659-679.
1972: Géologie de l'Archipel Arctique Canadien; Revue de l'association Française des techniciens de pétrole, n. 216, p. 23-64.

Prest, V.K.
1952: Notes on the geology of parts of Ellesmere and Devon Islands, Northwest Territories; Geological Survey of Canada, Paper 52-32, 15 p.

Rasmussen, K.
1927: Report on the II. Thule-Expedition for the exploration of Greenland from Melville Bay to de Long Fjord, 1916-1918; Meddelelser om Grønland, v. 65, no. 1, p. 1-180.

Ross, J.
1819: A voyage of discovery made under the orders of the Admiralty in His Majesty's Ships *Isabella* and *Alexander*, for the purpose of exploring Baffin's Bay and inquiring into the probability of a North-West Passage; John Murray, London, 252 p.

Salter, J.W.
1852: Geology; in Journal of a Voyage in Baffin's Bay and Barrow Straits, in the Year 1850-51, performed by H.M. Ships *Lady Franklin* and *Sophia* under the Command of William Penny, P.C. Sutherland; Longman, Brown, Green, and Longman, London, v. 2, appendix, p. 217-233.
1853: On Arctic Silurian fossils; Quarterly Journal of the Geological Society of London, v. 9, p. 313-317.
1855: Account of the Arctic Carboniferous fossils; in The Last of the Arctic Voyages, E. Belcher; Lovell Reeve, London, v. II, p. 377-399.

Schledermann, P.
1981: Inuit prehistory and archeology; in A Century of Canada's Arctic Islands, M. Zaslow (ed.); Royal Society of Canada, p. 245-256.

Schuchert, C.
1923: Sites and nature of the North American geosynclines; Bulletin of the Geological Society of America, v. 34, p. 151-230.

Suess, E.
1885-
1924: The Face of the Earth (Das Antlitz der Erde); Translation by Hertha B.C. Sollas; Clarenden Press, Oxford, 1904-24; v. I to V.

Sutherland, P.C.
1853a: On the geological and glacial phaenomena of the coasts of Davis' Strait and Baffin's Bay; Quarterly Journal of the Geological Society of London, v. 9, p. 296-312.
1853b: Appendix II. A few remarks on the physical geography, & c., of Davis Strait and its east and west shores; in A Summer Search for Sir John Franklin, Commander E.A. Inglefield; Thomas Harrison, London, p. 147-192.

Taylor, A.
1955: Geographical discovery and exploration in the Queen Elizabeth Islands; Geographical Branch, Department of Mines and Technical Surveys, Canada, Memoir 3, 172 p.

Teichert, C.
1937a: A new Ordovician fauna from Washington Land, North Greenland; Meddelelser om Grønland, v. 119, no. 1, 48 p.
1937b: Ordovician and Silurian faunas from Arctic Canada; Report of the Fifth Thule Expedition, 1921-24, v. I, no. 5, Gyldendal Nordisk Forlag, Copenhagen, 169 p.
1939: Geology of Greenland; in Geology of North America, R. Ruedemann and R. Balk (ed.); Geologie der Erde, v. 1, Gebrüder Borntraeger, Berlin, p. 100-175.

Thorsteinsson, R.
1958: Cornwallis and Little Cornwallis Islands, District of Franklin, Northwest Territories; Geological Survey of Canada, Memoir 294, 134 p.

Thorsteinsson, R. and Tozer, E.T.
1970: Geology of the Arctic Archipelago; in Geology and Economic Minerals of Canada, R.J.W. Douglas (ed.); Geological Survey of Canada, Economic Geology Report no. 1, p. 547-590.

Trettin, H.P., Frisch, T.O., Sobczak, L.W., Weber, J.R., Niblett, E.R., Law, L.K., DeLaurier, I., and Witham, K.
1972: The Innuitian Province; in Variations in Tectonic Styles in Canada, R.A. Price and R.J.W. Douglas (ed.); Geological Association of Canada, Special Paper 11, p. 83-179.

Troelsen, J.C.
1949: Contributions to the geology of the area round Jørgen Brønlunds Fjord, Peary Land, North Greenland; Meddelelser om Grønland, v. 149, no. 2, 28 p.
1950: Contributions to the geology of Northwest Greenland, Ellesmere Island, and Axel Heiberg Island; Meddelelser om Grønland, v. 149, no. 7, 86 p.

Washburn, A.L.
1947: Reconnaissance geology of portions of Victoria Island and adjacent regions, Arctic Canada; Geological Society of America, Memoir 22, 142 p.

Weeks, L.J.
1927: The geology of parts of eastern Arctic Canada; Geological Survey of Canada, Summary Report, 1925, Part C, p. 136-141.

Wright, N.
1959: Quest for Franklin; William Heinemann, London, 258 p.

Wordie, J.M.
1938: An expedition to Northwest Greenland and the Canadian Arctic in 1937; Geographical Journal, v. 92, p. 385-421.

Zaslow, M.
1975: Reading the Rocks; the Story of the Geological Survey of Canada, 1842-1972; MacMillan Co., Toronto, 599 p.

Ziegler, P.A.
1969: The Development of Sedimentary Basins in Western and Arctic Canada; Alberta Society of Petroleum Geologists, 89 p. (figures and legends; guidebook).

Authors' addresses

R.L. Christie
Geological Survey of Canada
3303-33rd St. N.W.
Calgary, Alberta
T2L 2A7

P.R. Dawes
Geological Survey of Greenland
Øster Voldgade 10
DK-1350, Copenhagen K
Denmark

CHAPTER 2

Printed in Canada

Chapter 3
GEOMORPHIC REGIONS

Introduction

Canadian and Greenland shields

Arctic Platform

Innuitian Orogen

Coastal plains and continental shelves

Acknowledgments

References

Chapter 3

GEOMORPHIC REGIONS

P.R. Dawes and R.L. Christie

INTRODUCTION

The geomorphic regions of the Canadian Arctic Archipelago[1], one of the great archipelagos of the world, and of northern Greenland are described and illustrated in this chapter. Also included in the description are two large peninsulas, Melville and Boothia, that extend from mainland North America northward into the Arctic Islands. The land forms of this region are controlled, to some extent, by the lithology and structure of the bedrock, and therefore the geomorphic subdivisions distinguished here (Fig. 3.1) correspond fairly closely to the stratigraphic-structural provinces delineated in Chapter 4 (Fig. 4.3). The influence of Tertiary to recent events (uplift, erosion, glaciation, and sea level changes) on the genesis of these land forms is discussed in Chapters 18 and 19.

The region measures about 3000 km from east to west and 2000 km from the southern islands to the northern tip of Ellesmere Island. Several large ice caps occur on the eastern Canadian Arctic Islands, and the region is bordered to the east by the great Greenland ice sheet — The Inland Ice. The land area described here exceeds 1 450 000 km^2. A network of channels divides the Canadian islands; many of the channels are broad seaways whereas others are narrow and resemble drowned river systems. Baffin Bay and Nares Strait separate the islands of Canada from Greenland.

Northern Greenland and the Arctic Islands are characterized by glacier-covered mountains, towering sea cliffs, dark canyons, rolling lowlands, and immense, ice-filled channels. Bedrock features, easily discerned in the arctic desert, make evident the links between landscape and underlying geology.

Geological and geomorphic unity of the Arctic regions can be seen in geological and topographic maps (Fig. 1, 2 [in pocket] and Fig. 3.1). The Innuitian Orogen forms a continental rim in northernmost Greenland and the Queen Elizabeth Islands[2], and is bounded successively to the south, toward the continental interior, by the Arctic Platform and the Canadian-Greenland Shield. The Shield and the Arctic Platform are parts of the Central Stable Region, or fundamental craton, of North America – Greenland, the outstanding characteristic of which is its relative stability since the end of Precambrian time. The Innuitian Orogen, in contrast, has been a region of tectonic mobility. The different tectonic histories of the two regions are expressed in their geomorphic characteristics: uplands and mountainous highlands in the ancient shield terrane, plains and plateaus in the Arctic Platform, and lowlands, uplands and mountains in the young, Innuitian Orogen. The continental masses are fringed by coastal plains and continental shelves, and brief accounts of these are also given.

Landscape studies in northern Greenland have been reported on by Paterson (1951), Colton and Holmes (1954), Victor (1955), Ahnert (1959, 1962), Malaurie (1968) and Nichols (1969), while Koch (1928), Davies (1972), and Sugden (1974) have attempted regional classification and synthesis. Topographic and geomorphological descriptions of part or all of the Arctic Islands have been provided by Dunbar and Greenaway (1956), Taylor (1956), Fortier (1957), Roots (1963), Bird (1967), and Bostock (1970). In the following text, a simplified nomenclatural scheme is followed, adapted mainly from Fortier (1957) and Davies (1972).

Many of the terms used to describe landscapes have dual meanings: a common, or popular one, usually with a purely "topographic" connotation, and a geomorphological one, which relates to the underlying geology. An example is "plain": in a topographic sense a plain is any surface with small differences in elevation (i.e. level or nearly so), whereas in a geotectonic context it is a flatland that may be of various origins, all relating to "platform structures" of, or surrounding, the ancient, crystalline shields that form the foundations of continents. Distinctions in meaning must be made by context or by using modifying words, e.g. flood plain, or peneplain. One should also note that, in the text that follows, the geomorphic regions may not match exactly the tectonic regions or the geographic names now in use. Thus, the Eureka Upland (a geomorphic region) includes the Hazen Plateau (a geographic name), which is an elevated area of low relief underlain by folded strata (that is, an upland).

The distinguishing characteristics, as noted in the definitions below, are the nature of the bedrock — undeformed in plains and plateaus — and the topographic relief — flat to mountainous.

Dawes, P.R. and Christie, R.L.
1991: Geomorphic regions; Chapter 3 in Geology of the Innuitian Orogen and Arctic Platform of Canada and Greenland, H.P. Trettin (ed.); Geological Survey of Canada, Geology of Canada, no. 3; (also Geological Society of America, The Geology of North America, v. E).

---

[1] The term, Canadian Arctic Archipelago (synonyms: Canadian Arctic Islands, Arctic Islands), is here used for all islands north of the Arctic Circle that are east of longitude 126°W, i.e. east of the western extremity of Banks Island.

[2] The Queen Elizabeth Islands comprise that part of the Arctic Archipelago lying north of Parry Channel, the major seaway extending from northern Baffin Bay westward to the Beaufort Sea (Fig. 1, in pocket).

CHAPTER 3

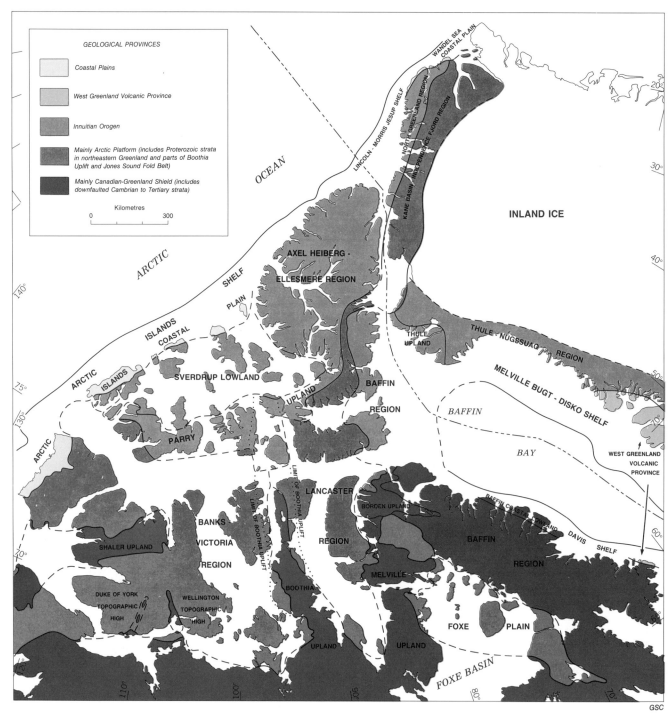

Figure 3.1. Geomorphic regions of the Canadian Arctic Archipelago and northern Greenland.

Definitions

Plains, as already noted, are flat lands, usually areas of unmetamorphosed sedimentary rocks that are essentially or nearly flat lying (e.g. Foxe Plain). A plain is not simply a peneplain (defined below), but may comprise remnants of such plains with various other "abrasion surfaces" such as stripped surfaces of sedimentary beds (see Mescherikov, 1968, p. 850). Other plains are underlain by extensive surficial deposits, e.g. Polaris Plain of central Hall Land, Greenland.

A **plateau** is an elevated tract of comparatively flat ground — a high, level region. A plateau in the broad sense may be an uplifted peneplain underlain by, say, an ancient tectonic complex, but "plateau" is here reserved for high plains of flat-lying rocks (see Fairbridge, 1968, p. 856).

Lowlands are areas of low relief (e.g. rolling, scarped) with underlying deformed rocks (e.g. Sverdrup Lowlands). Fairbridge (1968, p. 625) described lowlands as "disturbed structures, now degraded . . . mountain roots, degraded metamorphic belts". This terrain lies at low elevations — below 200 m according to one authority (see Gentilli, 1968, p. 933).

Upland applies to dissected terrain of rather irregular, nonlinear character (Fairbridge, 1968, p. 745). Uplands can be described as intermediate between "lowland" and "highland", where all three terms describe topography of disturbed structures that is reduced to an irregular, subdued relief, i.e. hills, less than about 600 m high (Fairbridge, 1968, p. 625).

A **highland** is an elevated, dissected terrain, higher than an upland, but with similar relief. The absolute relief may be that of a mountain region — e.g. 1000 m mean height for the Scottish Highlands — but the slopes are more mature.

Mountains is a term generally used where relative relief is over about 700 m (Fairbridge, 1968, p. 737) and structures are disturbed or complicated. Great height and relief of mountains indicate a relatively recent uplift — whatever the ages of the rocks and tectonic disturbance.

The nature of mountain relief derives from erosive processes and the geological history of the bedrock. The erosive processes are dependent upon a variety of factors including topographic gradient, climate, and latitude. Mountains usually cause increased precipitation, and frost-riving may produce saw-toothed crests and large talus slopes at higher elevations. Mountain character is affected by bedrock lithology: thus, in the Arctic, thick carbonate units give rise to massive fault and fold structures and are likely to produce cliffs and scarps, whereas thinly stratified clastic sediments produce rolling uplands and ridge-and-valley terrain. Volcanic-plutonic complexes generally have irregular topography. The scale and pattern of the joints, fractures, and faults have considerable effect on relief and drainage pattern.

The following terms also are useful in the context of this paper.

A **peneplain**, "almost a geographic plain", is, in its most mature development, an almost featureless plain that truncates bedding and both soft and resistant rocks. A peneplain is controlled only by its close approach to its erosional base level. Many peneplains, however, are less than perfect and contain residual hills that owe their presence to hardness of rock or to remoteness from the controlling base level. Many peneplains have been uplifted, and some warped or faulted. Subsequent dissection results in scattered upland plains (or, broadly, plateaus). At later stages, the unity of accordant, flat-topped summits is all that may remain to indicate a former peneplain. Accordant mountain summit levels (without remnants of the plain) are often, but with less certainty, taken to represent a former peneplain (Brown, 1968).

Ancient peneplains may be exhumed by the stripping of younger, soft rocks, as, for example, the exhumed Precambrian peneplain of Inglefield Land, Greenland (Fig. 3.3).

The **continental shelves** are described by Heezen and Wilson (1968) as shallow and gently sloping, typically with a gradient of 1:1000 and depths up to 200 m. At the edge of the shelf the slope increases to 1:40 (the **continental slope**). Shelves usually are constructed on unconsolidated sediments or sedimentary rocks. Shelf morphology reflects various submarine and coastal erosional and depositional processes (e.g. erosion by ice or current scour; deposition by wave, river, or biological activity). In some cases, subaerial erosion may have taken place during low stands of sea level.

CANADIAN AND GREENLAND SHIELDS

The shields of Canada and Greenland form the exhumed Precambrian crystalline core of North America. They comprise complexes of rocks of a wide range of ages — Archean to Late Proterozoic — and of variable metamorphic grade — granulites to unmetamorphosed rocks. In the combined Canadian-Greenland Shield four main periods of orogeny have been established; these have resulted in repeated cycles of deposition, orogeny, epeirogeny, and deep erosion. Some of the youngest sediments of the Shield — Early, Middle, or Late Proterozoic in age — occur as unmetamorphosed cratonic cover in scattered basins on deeply eroded crystalline basement. These rocks (e.g. Thule Basin, Minto inlier) are separated in age from the widespread Paleozoic platformal rocks by the pre-Cambrian hiatus, and are here included with the Shield.

The Canadian-Greenland Shield is flanked on the west, north and east by a Phanerozoic cover. Shield rocks are inferred to underlie most of the Inland Ice of Greenland, and are bordered on the east by the East Greenland Caledonides. The platformal cover may have, at one time, extended over the whole of the Canadian-Greenland Shield, but the cover is now stripped away except for some erosional remnants and "grabens" such as the Foxe-Baffin structural depression (Fig. 17.1), which underlies the Foxe Plain. The present surface of the Shield, at least where it dips northward beneath the platform cover, appears to be an exhumed pre-Paleozoic erosion surface (Ambrose, 1964). However, little-disturbed Proterozoic sedimentary rocks rest on a similar widespread erosion surface, so it is evident that the surface of the Precambrian Shield has had a long history of repeated burials, exhumation, and erosion. Finally, during the Pleistocene, the Shield was scoured and modified by ice sheets.

The main areas of Precambrian basement rocks are the Thule – Nûgssuaq Region, forming the east coast of Baffin Bay, and the Canadian counterpart, the Baffin Region of Baffin Island, eastern Devon Island, and southeastern Ellesmere Island. Other areas of basement rocks (including Proterozoic cover) occur as mainland extensions into the Arctic Islands: Melville and Boothia uplands; or as inliers in the Arctic Platform: Shaler Upland, and the Wellington and Duke of York topographic highs.

Thule – Nûgssuaq Region

The Greenland Precambrian terrane facing Baffin Bay is referred to here as the Thule – Nûgssuaq Region (Fig. 3.1). Like its counterpart, the Baffin Region, the crystalline rocks are overlain by unmetamorphosed strata of late Precambrian and Phanerozoic ages. These cover rocks are concentrated in two areas: in the north around Thule (Middle and Upper Proterozoic) and in the Disko – Nûgssuaq area of the south (Cretaceous-Tertiary).

The Thule – Nûgssuaq Region forms a narrow coastal strip behind which lies the Inland Ice. In the south in the Disko area the width of the region reaches about 150 km; in Melville Bugt the ice-free area is extremely narrow or absent. The region, sculptured by glacial erosion, is dominated by uplands and mountainous highlands. The notable lowland areas are some narrow coastlands and wide Quaternary-filled valleys in the two areas of the younger cover strata (Thule Air Base lies on one such lowland). The land is generally higher in the east, and nunataks appear near the fringe of the Inland Ice. The coast has the character of a faulted or rifted margin: younger rocks overlying the basement occupy the coastal areas, and downfaulted blocks with the younger strata are present on the continental shelf (cf. Chapter 17).

Mountainous highlands lie along the edge of the Inland Ice in the southern part of the region. The general elevation of the highlands is 1000 m, with individual peaks reaching 2000 m or more, the highest elevations of the Thule – Nûgssuaq Region. Flat-lying plateau basalts (the West Greenland Volcanic Province of Clarke and Pedersen, 1976) overlie sediments of the Nûgssuaq embayment (Henderson et al., 1976) to form a high, dissected plateau landscape. Downcutting by cirques has produced a rugged terrain of sharp ridge and valley topography. The Inland Ice reaches the sea on a broad front in Melville Bugt, in the central part of the region, where the coast consists for the most part of islands, small peninsulas and nunataks (Fig. 3.2). Nunataks lie at about 1000 m, with the highest peak near 1500 m.

In the Thule Upland of the Thule district the inner, mountainous highlands have a dissected "plateau" character, with dome-shaped mountains rising above a general elevation of about 800 m. The outer area is formed of tilted fault blocks of unmetamorphosed Proterozoic sediments and igneous rocks of the Thule Basin that in many places form spectacular, steep, cliffed coasts. Koch (1926) drew attention to fault control in the Thule district, where faults and the appreciable lithological variation of the strata produce a landscape that varies from rolling uplands with some flatter "plateau" areas to ice-covered mountains reaching over 1000 m. Major valleys and fiords are fault-controlled, and resistant basalt sills in the sedimentary sequence produce miniature "mesa" or "cuesta" landscapes at low elevations.

Several paleo-erosion surfaces have been recognized in the Thule – Nûgssuaq Region, although reliable regional correlation is still lacking. Synthesis of essential data in Koch (1923a, 1928, 1929, 1935), Birket-Smith (1928), Paterson (1951), Malaurie (1955), Cowie (1961), Davies et al. (1963), Bendix-Almgreen et al. (1967), Nichols (1969), and Weidick (1975) indicates that at least two main planation surfaces are preserved: a Precambrian level and a Tertiary or early Pleistocene level. Other uplifted surfaces that have been described may be local features. The Precambrian surface underlies Middle Proterozoic (Thule) strata. This surface, much faulted, slopes gently westward; it is below sea level at the outer coast and at about 1000 m elevation in the east, near the Inland Ice. To the north in Inglefield Land, an extensive inland, level upland roughly coincides with this paleoplain (Fig. 3.3), a surface that is also prominent in adjacent Ellesmere Island. This ancient surface may correlate with a prominent summit level between 700 m and 1000 m described by Paterson (1951) in southern Melville Bugt, and with the sub-Mesozoic peneplain defined in the Disko region (Koch, 1935). The Tertiary-Pleistocene level has been identified in Inglefield Land (a plateau surface on sedimentary rocks, Fig. 3.3), at Carey Øer (at 200 m elevation) and in the Thule district (300-500 m). The young surface may also equate with a surface developed on the West Greenland Volcanic Province (Weidick, 1975).

Baffin Region

The predominantly Precambrian terrane lying along the eastern edge of the Canadian Archipelago, from southeastern Ellesmere Island to Baffin Island, is here described as the Baffin Region. The region also includes downfaulted lower Paleozoic strata on Ellesmere Island and Cretaceous-Tertiary strata on southern Bylot Island and adjacent northeastern Baffin Island. The Baffin Region corresponds to the northern part of the Davis Region of Dunbar and Greenaway (1956), or the "highland rim" of Bird (1972, Fig. 13-2). The Baffin Region is mainly uplands and mountainous highlands, but includes lowland areas such as the Baffin Coastal Lowland and lowlands on the west coast of Baffin Island.

The Baffin Region is an uplifted Precambrian belt of highlands and uplands that appears to be the rifted eastern margin of the Arctic Islands (the Eastern Coastal Uplift, Christie, 1972). The land is high to the northeast, where the terrain is deeply dissected to form mountainous highlands, and the general surface slopes gently southwestward through less dissected uplands and lowlands to disappear beneath the Paleozoic platformal cover.

Bird (1967) identified several planation surfaces on Baffin Island, the principal ones named the Penny and the Baffin surfaces. The Penny Surface is represented by eastern mountain summit levels at about 1500 m elevation and, in places, higher; the Baffin Surface is a lower (typically 600-800 m) surface of subdued topography that covers most of Baffin Island. The relationships between the several surfaces are considered to be complex, but two conclusions that seem to be of general importance are: (a) the lower, younger Baffin Surface penetrates along valleys into the eastern highlands, and (b) the Baffin and other interior and western surfaces appear to conform, in large part, to an exhumed erosion surface that underlies the Paleozoic cover.

The ancient erosion surfaces of southern Baffin Island appear to be broken by faults into three regions (each of which has been given "upland" names; [see Bird, 1967; Bostock, 1970]). In each case the upland surface slopes westward, from heights of 900-1200 m along the northeastern sides of the great promontories, toward sea level along the southwestern sides. The repeated westward tilting of the (actual or near) pre-Paleozoic erosion surface suggests block faulting, and this is confirmed by rifts in the Paleozoic strata underlying southeastern Foxe Plain.

Small areas of lower Tertiary, flat-lying volcanic rocks and related sediments, resting on eroded Precambrian basement, lie between Cape Dyer and Cape Searle, a distance of about 90 km, on eastern Baffin Island. These areas are remnants of a youthful basalt plateau province that is related to the more extensive volcanic terrane of West Greenland (Clarke and Upton, 1971; West Greenland Volcanic Province of Fig. 3.1). A general post-volcanic uplift of about 400-500 m is suggested by the relationships of certain flows and (submarine) breccias. This uplift along a narrow coastal strip tends to confirm the youthfulness of

Figure 3.2. Nunataks, peninsulas, and islands of the Precambrian Shield terrane of Melville Bugt, northern part of Thule – Nûgssuaq Region, Greenland. Inland Ice reaches the sea along a broad front; main mountain summits are between 800-1000 m elevation. View is to the northeast. (Air photograph 542 N-Ø, no. 1890, July 1949, copyright Geodetic Institute, Copenhagen.)

Figure 3.3. Two peneplains are evident in coastal Inglefield Land, Greenland, where flat-lying Cambrian sediments overlie crystalline rocks of the Precambrian Shield. The upper plateau surface at about 200 m elevation (cut into sedimentary rocks) is a Tertiary(?) erosion surface; a parallel, lower peneplain on Precambrian Shield rocks is elsewhere overlain by Proterozoic strata. The view is to the west, and Marshall Bugt crosses the centre of the photograph, with Kane Basin and Ellesmere Island visible in the distance. (Air photograph 544 C-N, no. 11829, copyright Geodetic Institute, Copenhagen.)

substantial tectonic uplift along the eastern edge of the Baffin Region (Chapter 19).

An area of plateaus, mesas, and cuestas in a Proterozoic sedimentary and volcanic terrane on Borden Peninsula, northern Baffin Island, is here named the Borden Upland.

Spectacular, multicoloured cliffs are a feature of the area. The thick, unmetamorphosed Proterozoic sequence (Lemon and Blackadar, 1963), gently deformed, faulted, and intruded by a swarm of diabase dykes, is exposed in a broad structural uplift (see Christie, 1972). The gneissic basement and a stripped pre-Proterozoic peneplain are exposed in places and the sedimentary rocks form conspicuous mesas. The plateaus of the Borden Upland are approximately continuous (at about 550 m elevation) with those of the Lancaster Region and the upland was included in that region by Bostock (1970, Map 1254A). Bird (1967), however, noted contrasting elevations of eastern extensions of the Lancaster Region ("Barrow Surface", about 300 m) and the main plateau surface of Borden Peninsula ("Baffin Surface", about 600 m).

Melville Upland

The Melville Upland is an irregular, linear belt of Precambrian crystalline rocks that forms Melville Peninsula and parts of northern Baffin Island. The upland is the geomorphic expression of the Melville Arch (or Melville Horst of Chapter 17), a moderate tectonic uplift that lacks major border faults (Christie, 1972). However, southern Melville Arch has a faulted contact with the lower Paleozoic strata of the Foxe Plain. Downthrow is to the east some 250 m (Trettin, 1975), so that this part of the uplift structure appears to be horst-like (Fig. 3.4).

The upland of Melville Peninsula is a rolling surface with broad, widely spaced valleys and low, dome-shaped hills; the surface stands at 410-460 m elevation (the Melville Surface; Bird, 1967). Eastern parts of the surface slope gently toward Foxe Plain, but the surface drops more abruptly, to the west, into Committee Bay. The upland becomes discontinuous northward and is transected by the geomorphic feature, Fury and Hecla Channel (Sanford and Grant, 1976). A broad lowland depression (an extension of Foxe Plain topography) separates an upland surface north of Fury and Hecla Channel from uplands and plateaus of Borden Peninsula (see Blackadar, 1970, Fig. 2).

Boothia Upland

The Boothia Upland coincides with an elongate, northerly extension of the Canadian Shield on Boothia Peninsula and on Somerset and Prince of Wales islands. The Precambrian terrane is part of a larger tectonic unit, the Boothia Uplift, a structural feature that includes folded and faulted younger rocks and extends northward to Cornwallis and neighbouring islands (Chapters 4, 12C).

A deep channel, Peel Sound, forms a longitudinal trench in the northern part of Boothia Upland.

The Boothia Upland is an irregular surface at about 200 to 600 m elevation. Much of the western boundary (e.g. on Prince of Wales Island) is marked by steep slopes descending to surrounding lowland hills or plateaus, the resistant Precambrian rocks standing high topographically due to faulting and differential erosion. The eastern boundary is less distinct: on Boothia Peninsula the Precambrian terrane stands about 200-500 m above a Paleozoic lowland or plain; but on Somerset Island, the Precambrian surface is continuous with the adjacent plateau surface of the Lancaster Region. The upland surfaces of Boothia Peninsula, Somerset Island, and eastern Prince of Wales Island were included in the Barrow Surface by Bird (1967).

The surface features of the upland vary from place to place: glacial drift deposits are widespread in the central part (Boothia Peninsula), but felsenmeer, little displaced from its bedrock source, is characteristic of northern regions (Somerset Island) (Blackadar, 1967). Deeply incised, lake- and inlet-filled valleys form prominent lineaments in the southern and northern extremities of the peninsula (Bellot Strait, the "Northwest Passage", is such a valley). These are evidently fault zones that have been excavated by the action of streams and glacial ice. Peel Sound, a channel probably deepened by north-flowing ice, must follow a zone of structural weakness — perhaps a fault zone relating to the western margin of the Boothia Uplift.

Inliers in the Arctic Platform

The Arctic Platform of Phanerozoic beds is divided into broad "basinal" areas by several linear belts of "basement" Precambrian rocks. Some of the older rocks are sedimentary; others are crystalline. The more prominent belts are tectonic uplifts, or arches: the Boothia Uplift, the Melville Arch (already noted), and the Minto Arch. The Minto Arch, a Proterozoic inlier (Chapter 6) surrounded by Paleozoic platformal rocks, is dominated by the Shaler Mountains and is here described as the Shaler Upland. Two smaller "tongues" of Proterozoic rocks on Victoria Island (Chapter 6) are merely partially exhumed low hills named the Wellington and the Duke of York topographic highs.

Shaler Upland

Shaler Upland is a broad inlier of Middle(?) and Upper Proterozoic sedimentary and volcanic rocks on Victoria Island. The upland is presumed to be continuous (but deeply dissected by large marine channels) with the southern tip of Banks Island and with the mainland east of Darnley Bay, both areas of Precambrian inlier rocks. The upland is an arcuate belt of cuestas and mesas; the higher parts, with greatest elevations over 600 m, are named the Shaler Mountains.

The dissected cuesta- and mesa-terrain of the upland is characterized by ridges, escarpments, and plateaus (Fig. 3.5). Minor topography is controlled by extensive, resistant diabase-gabbro sills, some of which are up to 100 m thick. The highest part of the upland is occupied by Upper Proterozoic basaltic flows; these essentially flat-lying strata form an elongate plateau surface at 400 to 600 m elevation. The Kuujjua River, trending subparallel to the Minto Arch, has its headwaters in the upland region southeast of a major synclinal axis, crosses the axis, then empties into Minto Inlet northwest of the syncline. The river evidently is an antecedent stream in its general trend, although its course is disorganized due to glacial scour of the contrasting soft sedimentary rocks and the resistant, intercalated sills.

Precambrian topographic highs

Two narrow belts of Proterozoic rocks form inliers in the Arctic Platform of southern Victoria Island; these are the Wellington Topographic High, trending north some 100 km from Wellington Bay, and the Duke of York Topographic High, a line of partially exhumed hills that trends north-northeasterly about 150 km from the Richardson Islands (Christie, 1972, p. 77-80).

Bedrock of the Wellington Topographic High is mainly reddish weathering quartzite and conglomerate with diabase, and these rocks form low hills in the nearly flat Paleozoic terrane. The surrounding beds (of dolomite) abut

Figure 3.4. Flat-lying lower Paleozoic rocks of the Foxe Plain (Foxe-Baffin structural depression) are downfaulted against Precambrian rocks (Canadian Shield) of the Melville Upland; view is to the west of the western edge of the Foxe Plain. Note the "flights" of raised beaches on the nearer, Paleozoic rocks. (Air photograph T247R-111, National Air Photo Library, Department of Energy, Mines and Resources, Ottawa.)

Figure 3.5. A view northwestward across the Shaler Upland, Victoria Island. Note cuestas formed by resistant diabase sills, and central plateau underlain by the Natkusiak volcanic formation. (Air photograph T486R-66, National Air Photo Library, Department of Energy, Mines and Resources, Ottawa.)

the Precambrian hills. The Wellington High may be a remnant of an ancient arch or uplift that has been reduced to a low ridge, now partly exhumed through erosion of the younger sedimentary cover.

The hills of the Duke of York Topographic High, mainly diabasic rock, stand up to about 60 m above the surrounding flat-lying beds of Ordovician-Silurian dolomite. The hills probably represent a pre-Paleozoic topographic ridge, perhaps a cuesta. A small amount of arching of the Paleozoic beds is evident along the south coast of Victoria Island, so that a tectonic component is not entirely absent.

Ridges of Precambrian rock surrounded by flat-lying Paleozoic beds can be seen on Air Force Island in Foxe Basin (Burns, 1952, p. 8); the "basement hills" stand about 40 m above the present sedimentary plain.

ARCTIC PLATFORM

The Arctic Platform tectonic province coincides with a broad belt of plains and plateaus that includes much of the southern and central Arctic Islands and a broad tract across northern Greenland (Fig. 3.1). The northern boundary, against the Innuitian Orogen, conforms with structural grain to form a gentle curve that trends east and northeast; the southern boundary, however, is extremely irregular due to the effects of topography and the gentle dips of platformal rocks. In Greenland the southern boundary is mainly covered by the Inland Ice. The Paleozoic sedimentary rocks of the Arctic Platform rest on a peneplain surface of the Precambrian, mainly crystalline basement, the Canadian-Greenland Shield. This surface is a gently undulating, nearly horizontal one in the continental interior but dips gently to great depths "outward" toward the continental margins (to the northwest in the Arctic).

The main part of the Arctic Platform is here divided, for purposes of geomorphic description, into three large regions: the Kane Basin – Independence Fjord Region, stretching across northern Greenland and including a small "sliver" of plateau on east-central Ellesmere Island; the Lancaster Region, east of Boothia Upland; and the Banks – Victoria Region, west of the upland. As mentioned below, these geomorphic regions also include some strata from adjacent geological provinces. For example, Upper Proterozoic sedimentary and volcanic units occur in the Kane Basin – Independence Fjord region, parts of the Jones Sound Fold Belt and Boothia Uplift in the Lancaster Region, and in the Banks – Victoria Region. The southernmost part of the Arctic Platform, the Foxe Plain, forms an outlier in the Canadian Shield.

Kane Basin – Independence Fjord Region

The plateau terrain that stretches across Greenland from Kane Basin to Peary Land and the area southeast of Independence Fjord (Fig. 3.1) is here named the Kane Basin – Independence Fjord Region. This geomorphic unit, somewhat wedge-shaped in plan, is 1000 km long, increasing in width to the east: 25 km wide at Bache Peninsula, 100 km in Wulff Land, and more than 200 km near Independence Fjord. The region is bounded on the south by crystalline Precambrian rocks in the Kane Basin area and in southern Wulff Land, and by the northern part of the Caledonian mountains in the extreme east.

The Kane Basin – Independence Fjord Region roughly coincides with the North Greenland Foreland plateaus and plains province and the Northeast Foreland plateaus province of Davies (1972). Davies drew a distinction between plateau terrain "associated" with the Innuitian deformation to the north, and that of the East Greenland Caledonian deformation to the east. The two plateau provinces are continuous, however, and they are here so grouped. Some mountainous areas of folded clastic sediments, such as the Polaris Highlands included by Davies (1972) in his North Greenland plateaus and plains province, are included here with the geomorphic regions of the Innuitian Orogen.

Much of the region is a homocline of flat-lying to shallow-dipping strata with a regional dip to the north; in the extreme east, in western Kronprins Christian Land, the strata dip gently east toward the East Greenland Caledonian fold belt. The dominant topography is that of a plateau or concordant upland, with the topographic surface parallel to regional bedding. Elevations of the plateau surface vary from near sea level to about 1500 m; regionally the surface rises southward, and some flat-topped nunataks occur within the margin of the Inland Ice.

The strata composing the main part of the region are mainly lower Paleozoic carbonates, with Proterozoic clastic and igneous rocks present in the area between Independence Fjord and Danmark Fjord.

Striking variations in the landscape are evident from south to north, the different features due to differences in physical competence of the rocks: resistant carbonates in the south, and less resistant clastic sequences along the northern, transitional boundary with the Innuitian Orogen. A southern sub-region consists of high tablelands (plateaus), whereas the northern area is one of lower, more subdued relief. A third sub-region, of coastal lowlands, is recognized in eastern North Greenland.

The **southern sub-region**, which forms the main part of the Kane Basin – Independence Fjord Region, includes Bache Peninsula, Washington Land, the southern parts of Hall Land, Nyeboe Land, Warming Land, Wulff Land, Nares Land, Freuchen Land, southernmost Peary Land and terrain to the southeast. The sub-region is mainly a dissected plateau, now appearing as mesa-like tablelands separated by large, flat-floored glaciated valleys and steep-sided fiords (Fig. 3.6). Relief is commonly in the order of 700 to 800 m. The mesa-like blocks vary considerably in size and reach tens of kilometres in length; they are generally bounded by high scarps, often with spectacular cliffs at the feet of which lie prominent talus slopes. Plateau surfaces are often covered by till and frost-shattered bedrock rubble; many of the higher parts support ice caps. Some valleys transecting the plateau are narrow and canyon-like, with long, narrow lakes (Apollo Sø in Wulff Land is 50 km long and 2.5 km wide). Sydpasset, a wide east-west valley in southern Peary Land, also contains a long lake. This conspicuous low-lying depression, which joins inner Independence Fjord and inner J.P. Koch Fjord, was earlier named the Peary Channel — surmised by the explorer Robert Peary to be a channel that made Peary Land an island and so indicated on atlases for several decades (Chapter 2).

Figure 3.6. Dissected tableland developed in lower Paleozoic strata; plateau surface at about 700-800 m in southern part of Kane Basin – Independence Fjord Region, Arctic Platform. View is east across southern Warming Land, Greenland, with Inland Ice in background. (Air photograph 546L-Ø, no. 4245, copyright Geodetic Institute, Copenhagen.)

The southern plateau region descends into rolling or hilly plains in southernmost Washington Land, at the head of Danmark Fjord, and in southwestern Kronprins Christian Land. In some areas toward the north, for example in Wulff Land, scattered remnants of the plateau are separated by wide valleys and plains.

A distinctive feature of the plateau region in the west, well developed between Washington Land and Wulff Land, is a belt of mountains with smooth rounded topography and with a relief that reaches 1000 m. Across Hall Land, such mountains form a discrete linear chain about 5 km wide — the Hauge Bjerge. The conspicous, dome-shaped hills are exhumed Silurian reefs that reach a thickness of several hundred metres; the slopes typically are subparallel to the inclined reefal beds and the intervening lowlands or valleys are excavated from less resistant, off-reefal mudstones (Fig. 3.7; Chapter 7). Some north-south valleys across the Hauge Bjerge, in which steep reefal sections are exposed, represent former deep water passes through the reef zone (Dawes, 1987).

The **northern sub-region**, in which Silurian clastic rocks are prominent, is an east-west tract of rolling, dissected plateaus and plains, with some wide lowlands of Quaternary deposits. Two areas of somewhat contrasting landscape occur, in the west and east, separated in Freuchen Land and western Peary Land by a northern extension of the high plateau southern sub-region.

In the west, between Hall Land and Nares Land (north of the belt of exhumed reefs), fine grained turbidites and off-reef argillites give rise to a tract of low plains and rolling, dissected plateaus (up to 700 m elevation) that grades to the north into the uplands of the Innuitian Orogen. Bedrock exposures are generally poor and relief is generally less than 200 m, although higher cliff sections occur along some fiords.

The Polaris Plain in central Hall Land (Davies, 1972) is a distinctive feature, much of it below 100 m elevation. Large areas of this plain are level and smooth surfaces of lacustrine and marine clays, and have proved suitable as natural air strips for heavy aircraft (Davies et al., 1959).

The northern plateau surface rises eastward toward central Peary Land to levels above 1000 m. Much of the area is characterized by a peculiar stepped topography due to differential erosion of alternating sandstone and shale intervals of the Silurian turbidite bedrock (Fig. 3.8). Low, flat-topped hills characterize the plateau surface, with resistant sandstone beds forming conspicuous, tiered ledges on the slopes.

The **eastern sub-region** comprises coastal lowlands and low plateaus around Independence Fjord (ca. 3000 km^2) in southern Peary Land, northern J.C. Christensen Land and Valdemar Glückstadt Land (includes the Independence lowlands and Valdemar Glückstadt plain of Davies [1972]). This landscape is one of heavily glaciated, dissected, low plateaus and rolling plains that have a widespread cover of glacial and glaciofluvial deposits.

A curious hilly area on the north side of Independence Fjord is marked by prominent ridges 15 km long and 150 m high. This landscape, earlier regarded as a glacial terminal moraine (Koch, 1923b; Davies, 1963), comprises marine sediments and represents eroded remnants of glacially induced thrust sheets (Funder et al., 1984).

Structural control of fiords and valleys

Two dominant directions of fiords and major valleys occur in the Kane Basin – Independence Fjord Region: a north-northwesterly trend characterizes the region between Washington Land and Peary Land, and an easterly one is evident mainly in Peary Land. Both trends appear to be fault-controlled. A northeasterly linear scarp pattern, followed mainly by the outer coast and parallel to Nares Strait, was emphasized by Monahan and Johnson (1982), and a northeasterly fracture pattern was recognized throughout the Arctic Islands by Sobczak (1982). The faulting discussed here probably was mainly Paleogene in age (Chapters 17, 18) and followed by erosion that probably caused some scarp retreat.

Nares Strait, like some other linear seaways of the Arctic Platform (e.g. Lancaster Sound) is clearly fault-controlled (Chapter 18); however, the roles played by normal or strike-slip faulting have yet to be elucidated (for discussion of the origin of Nares Strait, see Dawes and Kerr, 1982).

One hypothesis for the geomorphology of Nares Strait is considered here. There is clear evidence for appreciable submergence along Nares Strait. Drowned coastlines have been described, for example along Kane Basin (Nichols, 1969), and the seaway may owe its existence to the submergence of a deeply dissected plateau or peneplain. Koch (1925, 1928, 1929) emphasized the importance of drainage patterns in the landscape of northern Greenland (for example, fiord directions along the north coast) and he considered river erosion to have played an important part in the development of Nares Strait. Kane Basin appears to be the shallowest part of the seaway; thus prior to submergence the area may have acted as a divide between two (or more) drainage systems, one flowing south into Baffin Bay and the other north into the Lincoln Sea (Pelletier, 1966; Nichols, 1969). Such drainage systems may have produced the various channels and basins that now constitute Nares Strait. Faults would naturally contribute greatly in guiding the erosion of the various depressions. Any earlier drainage system has, of course, been extensively modified by glacial ice streams, the youngest of which were active just prior to deglaciation (in the Kane Basin region, about 9000 BP — Blake, 1981).

Lancaster Region

The Lancaster Region is an area of plateaus and plains encompassing parts of northwestern Baffin Island, Somerset Island, central Devon Island, and southern Ellesmere Island. Lancaster Sound is a prominent marine feature in this region. The topographic surfaces lie at altitudes ranging from near sea level (southern Gulf of Boothia) to about 800 m (southern Ellesmere Island).

Bedding is flat lying or nearly so over much of the region, but homoclines with low dips to both east and west produce a regional, broad synclinal or "basin" structure. On Ellesmere Island, the strata dip north-northwesterly toward the axis of the Schei Syncline (Chapter 12E). Farther southeast, near the coast of Jones Sound, they are faulted and folded (Jones Sound Fold Belt).

The Lancaster Region is transected by numerous faults and lineaments. Coastal blocks are downfaulted along the shores of Jones Sound so that the sound resembles a graben. Geophysical and geological evidence suggests that Lancaster

Figure 3.7. Dome-shaped uplands on Arctic Platform; Silurian reefs, up to 500 m thick, form rounded summits separated by low areas of off-reef shales. View southeast with Kap Constitution and Kap Independence (left and right in foreground), Washington Land, and Kennedy Channel. (Air photograph 545 K1-SØ, no. 2258, copyright Geodetic Institute, Copenhagen.)

Sound is broadly a half-graben with the south coast of Devon Island marking a major faultline. Faults of lesser vertical displacement parallel to and near the south side of Lancaster Sound are also indicated (Kerr, 1980). The faults in Lancaster Sound probably were active in the Miocene but may date back to the middle Cretaceous (Chapters 18, 19).

The plateau surface of the Lancaster Region slopes gently southward and westward from the adjacent Precambrian terranes of Ellesmere, Devon, and Baffin islands to become a near-plain in the Gulf of Boothia. This surface forms the Lancaster Plateau and the Boothia Plain of Bostock (1970). The region is deeply dissected by canyon-like valleys and by wide marine inlets and channels. Coastlines of the region are notable for their sweeping, uniform curves (e.g. Brodeur Peninsula, and the south coast of Devon Island) or their straight lines (e.g. the east coast of Somerset Island), so that the locations and trends of these valleys and channels appear to be fault-controlled. The channels are undoubtedly parts of a drainage system that became established during Tertiary time and was modified by glacial ice during the Pleistocene (Fortier and Morley, 1956; Bird, 1967).

The Lancaster Plateau is most conspicuously developed on northern Somerset Island, Brodeur Peninsula of Baffin Island, and central and western Devon Island[3], where elevations of 350 to 400 m are typical (Fig. 3.9). The plateau surface is a mature one that truncates the gently west-dipping beds on Devon Island, and extends westward across

---

[3] Two divisions of this plateau surface were named by Bird (1967, Fig. 16): the Barrow Surface, a major one at 350-400 m and the Baffin Surface, a higher, northward extension of the Precambrian uplands of Baffin Island. The plateau surface of Devon Island rises continuously eastward, however, and the "abrupt" division between the two surfaces of Bird (1967, p. 69) seems not discernible (see Roots, 1963, p. 167).

Figure 3.8. Stepped concordant uplands developed in Silurian turbidites. Southern Peary Land, Greenland, northern sub-region of Arctic Platform. View is south to Independence Fjord. (Air photograph 548 C-S, no. 1954, copyright Geodetic Institute, Copenhagen.)

Figure 3.9. Plateau surface of southwest Devon Island; note youthful, dendritic drainage pattern and deeply incised valleys; lower Paleozoic strata of the Arctic Platform. View is to the northwest. (Air photograph T441R-106, National Air Photo Library, Department of Energy, Mines and Resources, Ottawa.)

the resistant, Precambrian rocks of the Boothia Uplift (Boothia Upland) on Somerset Island and Boothia Peninsula.

The Lancaster Plateau passes westward, in the Cornwallis Island region, to an area of mixed plateaus and ridged uplands that reflect the variably disturbed sedimentary rocks of the northern Boothia Uplift.

Banks – Victoria Region

The Banks – Victoria Region is an area of plateaus and low-lying land that includes parts or all of several islands: Victoria, Banks, Melville, Prince of Wales, and King William. The land surface lies at various elevations and wide areas are characterized by conspicuous glacial landforms and by marine strand lines.

The Paleozoic beds of the Banks – Victoria Region are horizontal, or nearly so, over very wide areas. Gentle homoclines, however, are apparent in the vicinity of the structural and topographic highs: northwesterly and southeasterly dips flank the Minto Arch, and westerly and easterly dips the Duke of York Topographic High.

Bedding and other structures are obscured over large parts of the region by thick glacial and fluvial deposits of the last (Wisconsinan) and earlier glacial advances. Belts of hilly moraine are prominent on western Victoria and eastern Banks islands, and drumlin terrain covers much of eastern and southern Victoria Island (Fig. 3.10) and nearly all of Stefansson Island (Fyles, 1962, 1963).

The general topographic surface in the Banks – Victoria Region is predominantly low — within 150 m of sea level — but rises gradually to the east or northeast and northwest toward the Precambrian uplifts and arches (the eastern Victoria lowland of Fortier, 1957). Central and eastern Banks Island and northwestern Victoria Island, lying at 300 to 700 m elevation and underlain by slightly deformed sedimentary rocks, is a plateau region. The plateau surface of northwestern Victoria Island dips gently northwestward toward Prince of Wales Strait. On Banks Island, the surface descends to the west to merge with the Arctic Islands Coastal Plain. Many of the coasts in the plateau region just described are characterized by cliffs and bluffs, the most spectacular of which are the coastal cliffs of Cape Lambton and Nelson Head on Banks Island (these cliff exposures are flat-lying Proterozoic beds; in fact the cliffs form an exposed margin of the Minto Arch).

The sounds, channels, and straits are mainly irregular and broad; these features undoubtedly reflect the shallow surface relief and the "drowned" state of the region. The overall pattern of seaways probably is inherited from a Tertiary drainage system (Chapter 19). The former valleys have undoubtedly been modified, both through excavation and infilling, by lobes of continental glacial ice that crossed the region in a generally westerly and northwesterly direction. The lineation of drumlinoid and fluted terrain (see Fyles, 1963, Fig. 2) trends toward, or parallels, many of the major channels (e.g. Dolphin and Union Strait, Prince Albert Sound, and M'Clintock Channel) and also smaller depressions (Richard Collinson Inlet, Hadley Bay, and others). These channels and depressions appear, therefore, to have undergone extensive modification, including widening and deepening, by glacial ice. The narrow, straight-sided channel, Prince of Wales Strait, may be primarily fault-controlled and, lying in what may have been a region of thinned, marginal ice (Fyles, 1963, p. 34), perhaps was relatively little modified by glacial processes.

Foxe Plain (outlier)

The geomorphological expression of the Paleozoic outlier that underlies Foxe Basin and environs is the Foxe Plain (Bostock, 1970). This nearly inundated plain includes low-lying islands in Foxe Basin, the Great Plain of the Koukdjuak on Baffin Island, and some coastal and inland sedimentary terrane on Baffin Island and Melville Peninsula, mainly lying below 100 m elevation.

The strata underlying the Foxe Plain include limestone, dolomite, and sandstone. The relatively thin lower Paleozoic sedimentary succession rests, with nearly horizontal attitude, on the almost planar surface of the eroded crystalline Shield. The beds lie in a structural depression (the Foxe-Baffin structural depression; Fig. 17.1) that, although broad and shallow, is a complex of grabens and half-grabens. The depression is bounded by a series of normal faults both on the southwest and the northeast (Fig. 3.4) (Trettin, 1975).

The Foxe Basin terrane, with its horizontal strata and low topographic relief, is a geomorphic plain. The topographic relief — 60 to 90 m on the smaller islands (plus about 50 m of water in the basin), 120 to 150 m on eastern Melville Peninsula, and about 200 m on Baffin Island — is due to Cenozoic erosion. The fractures and faults noted earlier are expressed locally as rectilinear coastlines and drainage patterns, and as small escarpments (Trettin, 1975).

INNUITIAN OROGEN

The Innuitian Orogen, in the northern part of the region here described, comprises sinuous belts of folded sedimentary rocks and some older, metamorphic terrane. Minor basaltic lavas are present, and in northern Ellesmere Island a wide variety of metamorphic, plutonic and volcanic rocks occur.

Fold structures form compound, sweeping curves through the archipelago: easterly in the Parry Islands, northerly to northwesterly on Axel Heiberg and western Ellesmere islands, northeasterly on northern Ellesmere Island, and east-northeasterly in northern Greenland. The style of folding varies from place to place: broad, open folds of the Parry Islands continue into Ellesmere Island, but folding becomes near-isoclinal in a flysch terrane in north-central Ellesmere Island (Hazen Fold Belt of Fig. 4.3); in northern Greenland, early isoclinal folds are refolded by open structures.

The orogenic belt includes mountains, uplands and lowlands, and is here divided into four regions: the North Greenland Region and the Axel Heiberg – Ellesmere Region of mountains and uplands, the Parry Upland, and the Sverdrup Lowland (see Fig. 3.1). The belt of uplands and mountains is more or less continuous from northern Greenland to Melville Island; the belt is broken only by deep channels, of which the most prominent is Nares Strait. The Sverdrup Lowland, a topographically depressed area occupied by the western Sverdrup Islands, is flanked on two sides by uplands and mountains and opens northwestward to the Arctic Ocean.

The complex tectonic history of the orogen is summarized in Chapters 4 and 21 and described in more detail in Chapters 11, 12, 17, 18, and 19.

Figure 3.10. Drumlinized plain of southwestern Victoria Island, Banks – Victoria Region; view eastward. Drumlins average 2.5 km in length; glacial flow was toward the observer. Rough-textured areas are felsenmeer and rubbly outcrop of Paleozoic dolomite. (Air photograph T321L-180, National Air Photo Library, Department of Energy, Mines and Resources, Ottawa.)

North Greenland Region

This region includes the coastal area of northern Greenland, from the Wandel Sea to the Lincoln Sea; it is widest in the east in Peary Land where it approaches 100 km. Included in the region are the North Greenland mountains province of Davies (1972) and lower plateau areas in both the west and east that were included by Davies in his North Greenland Foreland plateaus and plains province. The region is both geologically and geomorphically continuous with the eastern part of the Axel Heiberg – Ellesmere Region but lacks the metamorphic and granitoid rocks exposed in its western conterpart. The bedrock comprises mainly sandstone, siltstone, and mudstone with subsidiary, thick carbonate units. The terrane is dominated by easterly-trending sedimentary and structural grains and this consistent trend dominates the landscape, both in gross and detail character. The intensity of deformation and metamorphism generally increase northward: thus in the south (north of the Arctic Platform), unmetamorphosed strata lie in open folds; in northern Peary Land, more complex fold patterns are present in low grade metamorphic rocks (see Chapters 11, 17).

Two main east-west mountainous sub-regions can be recognized; in addition, a smaller sub-region of block-faulted uplands occurs in eastern Peary Land.

A mountainous belt that crosses the northern parts of the peninsulas between Hall Land and Freuchen Land and across Peary Land, north of the Arctic Platform, forms a southern sub-region. The main landscape is that of rounded, rolling uplands with mainly dome-shaped hills (an eastern continuation of the Hazen Plateau of northeastern Ellesmere Island). Major fold structures, such as the Wulff Land anticline (traceable for 200 km) produce areas of steeply inclined beds that form ridged uplands (Fig. 3.11). The upland is relatively high, with summits of ice-covered highlands in Peary Land attaining altitudes over 1600 m. The surface descends westward to elevations of 1000 m or higher.

The **northern sub-region**, composed of the Roosevelt Fjelde and mountain ranges to the east and west, occupies northern Peary Land and Nansen Land. The landscape is dominated by alpine-type mountains with large ice fields (Fig. 3.12) — the Nansen-Jensen alps of Davies (1972). Deep erosion has produced angular topography with sharp ridges and steep cliffs; relief is well over 1000 m. The highest peak of northern Greenland, Helvetia Tinde, at about 1920 m, lies in central Roosevelt Fjelde. Several prominent piedmont areas are characterized by rolling, ridged topography — for example around Frederick E. Hyde Fjord — and several areas of subdued topography and lowlands coincide with outcrops of important mudstone units.

The **eastern upland sub-region** occurs in eastern Peary Land, mainly on the little-deformed upper Paleozoic, Mesozoic and Cenozoic strata of the Wandel Sea Basin (Chapter 16), a sedimentary basin equivalent to the Sverdrup Basin (Chapters 14, 15) in Canada. Block faulting in this region has produced landscapes that vary from rounded plateaus to linear hills and ridges — a sort of miniature basin and range topography (the Herluf Trolle upland of Davies, 1972). Upper surfaces stand at about 700 m, with the highest ice-covered plateau at 1000 m elevation.

Axel Heiberg – Ellesmere Region

The Axel Heiberg – Ellesmere Region includes northern and west-central Ellesmere Island and most of Axel Heiberg Island. Three areas of ice-covered mountains are separated by deeply dissected, irregular uplands. The mountain systems are named the Grantland, Axel Heiberg, and Victoria and Albert Mountains, and the upland areas between the mountains, the Eureka Upland (Bostock, 1970).

As observed by Taylor (1956, v. 3) for Grant Land (an old name for northern Ellesmere Island), the region is one of superlatives: the largest lake of the Queen Elizabeth Islands, the highest peaks of eastern North America, and one of the great fiord systems of the Canadian Arctic, and indeed, of the world, all lie in this region.

The Grantland Mountains occupy a deeply dissected, ice-covered mountainous highland area dominated by the United States Range. A strong, northeasterly sedimentary and structural grain characterizes the main ranges, but trends are more varied in the north-coastal mountains and there, the terrain deeply indented by fiords, the mountain topography is more irregular. Nearly continuous icefields feed numerous glaciers, which flow radially into fiords and onto the adjacent uplands. The mountain peaks commonly reach 1700 m, and the highest summit, Barbeau Peak, lies at 2604 m.

The Victoria and Albert Mountains form a region of uniform northeasterly fold and fault trends and a well developed ridge and valley topography due to contrasting sedimentary rocks (Fig. 3.13). The ridges typically reach 1500 to 1800 m, and the relief is about 750 m. The Agassiz Ice Cap drains northwestward onto the adjacent upland and southeastward into a coastal belt of crosscutting valleys and fiords.

The Axel Heiberg Mountains are a dissected, mountainous highland of younger sedimentary rocks with diabase sills (the Sverdrup Basin) that have been deformed into north-northwesterly trending structures. Fiords and deep, glacially modified valleys radiate from the ice-covered central highlands (the peaks of which reach 2000 m) so that some valleys parallel the regional structure and some cross it. The highland surface descends eastward to hilly uplands and westward, more abruptly, to the Sverdrup Lowlands.

The irregular depression that lies between the mountain systems, here described as an upland complex (Eureka Upland), is a dissected, high terrane of folded and faulted sedimentary rocks. The surface is rolling and ridged in the west (eastern Axel Heiberg and western Ellesmere islands), where it rises gradually into the surrounding mountainous highlands. In the northeast, a relatively uniform, rolling surface (the Hazen Lake upland of Fortier, 1957, and the Hazen Plateau of Christie, 1964), rises gradually from its lowest point, at the foot of the Grantland Mountains, southeastward to become the Victoria and Albert Mountains. The relatively abrupt boundary with the Grantland Mountains coincides with the Lake Hazen Fault Zone, along which the mountain terrain to the north was uplifted to stand above the upland southeast of Lake Hazen. The upland surface is marked by a uniform, northeasterly trend of tightly folded, flyschoid sedimentary rocks; this trend is reflected by minor topography, and smaller streams follow it in spite of the disorganizing effect of glacier ice that flowed

Figure 3.11. Ridged mountainous uplands of the Innuitian Orogen in northern Hendrik Ø (foreground) and Nyeboe Land, Greenland. View is west along the regional strike of folded, mainly Silurian clastic strata, with highest peak in the linear chain at about 1000 m. Lincoln Sea and Ellesmere Island are visible in right background. (Air photograph 546 L-V, no. 5523, copyright Geodetic Institute, Copenhagen.)

Figure 3.12. Alpine mountains of the Roosevelt Fjelde, northern Peary Land, Greenland, developed in lower Paleozoic, mainly low-grade marbles and psammites. View is to the west along the regional strike with highest summits at about 1500 m elevation. The Arctic Ocean is visible in the right distance; the prominent outer cape is Kap Washington. (Air photograph 548 F-V, no. 10145, copyright Geodetic Institute, Copenhagen.)

Figure 3.13. Victoria and Albert Mountains, Ellesmere Island, view west in the Axel Heiberg – Ellesmere Region. The Agassiz Ice Cap and the upper parts of glaciers draining the ice to east-coastal fiords are evident. Note the linear ridges (nunataks). (Air photograph T398R-222, National Air Photo Library, Department of Energy, Mines and Resources, Ottawa.)

southeastward across the upland. The upland is a pre-Pennsylvanian peneplain that was stripped in early Tertiary time and again in late Tertiary or Pleistocene time. Today's drainage is southeastward through deep valleys and fiords; these antecedent valleys have been excavated by both fluvial and glacial processes (Fig. 3.14).

Parry Upland

The Parry Upland is a rolling surface of low to moderate elevations in which the strong fold trends of the deformed lower Paleozoic and Devonian strata are evident. It includes the Parry Islands and Canrobert Hills fold belts, the northern Boothia Uplift (Cornwallis Fold Belt), and western Jones Sound Fold Belt, along with a salient of the Arctic Platform north of the western Jones Sound Fold Belt (Fig. 12A.1). The upland passes southward into plateaus of unfolded lower Paleozoic sedimentary rocks, and is bordered northward by the Sverdrup Lowlands of relatively little-deformed, less resistant sediments of the Sverdrup Basin. The upland is underlain mainly by clastic rocks on Melville and western Bathurst islands, and by both clastic and carbonate rocks on eastern Bathurst, Cornwallis, northwestern Devon, and southwestern Ellesmere islands.

A graben structure containing younger beds crosses the Parry Upland at Eglinton Island to form a southwest extension of the Sverdrup Lowland.

The upland surface lies at about 150 to 250 m elevation on southeastern Melville Island and rises gradually to the west and east. The surface reaches about 700 m on western Melville Island, where it supports a few small ice caps. The upland joins the distinctly ridged uplands of the Ellesmere – Axel Heiberg Region at about 600 to 900 m on southwestern Ellesmere Island. The Parry Upland is characterized by subdued ridges with the locations of ridges and valleys influenced by the fold trends and the contrasting competence of the sedimentary strata. On western Melville Island, the upland is deeply dissected by valleys and fiords that radiate from a central highland. The fiord (i.e. glacial) origin of these features (e.g. Murray Inlet, on the south coast) is clear from the steepness of the valley walls, the U-shape of the valleys, and the "hanging" relationship of certain tributary valleys. A small island at the head of Murray Inlet is ice-grooved parallel to the valley.

Sverdrup Lowland

The Sverdrup Lowland, a shallow topographic basin in the western Queen Elizabeth Islands north of the Parry Islands, coincides with the less deformed parts of the Sverdrup Basin. The topographic basin is bounded on the south and east by the Parry Uplands and Axel Heiberg Mountains, but opens northwestward to the Arctic Islands Coastal Plain and the Arctic Ocean.

Fold trends of the western Sverdrup Basin are variable, and gentle dips prevail. This structural style, with the weakness of the rocks, has produced a low-lying region (generally less than 150 m elevation) of irregular hills and scarps. Dissected gypsum domes and related structures form distinctive landmarks ("badlands") in the lowlands (Fig. 3.15). A former drainage system, undoubtedly modified by glacial ice (Pelletier, 1966) is now deeply drowned so that about half the lowland is water covered. Shallow shelves with water depths less than 200 m surround individual islands or groups of islands. Between the shelves are broad channels with depths between 200 and 500 m that locally contain narrow troughs with depths between 500 and some 700 m (e.g. Peary Channel and Massey Sound west of Axel Heiberg Island, Hassel Sound between Amund Ringnes and Ellef Ringnes islands, and Maclean Strait west of Ellef Ringnes Island).

COASTAL PLAINS AND CONTINENTAL SHELVES

The coastal plains and continental shelves are young features that border the Innuitian Orogen and Arctic Platform on the north or northwest (Arctic Ocean) and the southeast (Baffin Bay). The northern features are here named the Wandel Sea Coastal Plain and the Lincoln – Morris Jesup Shelf; and the Arctic Islands Coastal Plain and the Arctic Islands Shelf. The continental shelves of Baffin Bay include the Melville Bugt – Disko Shelf (off Greenland) and the Davis Shelf (off Baffin Island).

Wandel Sea Coastal Plain

The mountains along the northern coast of Greenland (as in northern Ellesmere Island) drop abruptly into the sea, a feature that suggests fault control. However, in eastern Peary Land, a narrow lowland plain faces the Arctic Ocean and Wandel Sea. This is referred to here as the Wandel Sea Coastal Plain; it includes the Arctic strand and Herlufsholm strand of Davies (1972).

The coastal plain is 15 km across its widest part, and is made up of unconsolidated Quaternary deposits: glacial, marine and glaciofluvial sediments that form a flat to rolling, in places ridged, topography. The surface of the plain rises inland gradually from sea level to, in places, 200 m. To the east of Kap Morris Jesup, the coastal plain is limited landward by an abrupt highland scarp along the Kap Bridgman fault; in easternmost Peary Land the plain merges without apparent tectonic control into the rolling topography of the eastern sub-regions of the Innuitian Orogen and Arctic Platform.

Marine deposits and features, dominated in places by spectacular strand lines, indicate that glacial rebound in pre-Holocene and Holocene time has been in the order of 100 m or more (Funder and Hjort, 1980). The coastal plain is essentially an emergent feature that is probably continuous offshore with the continental shelf.

Arctic Islands Coastal Plain

The Arctic Islands Coastal Plain (or Islands Coastal Plain of Bostock, 1970) is a strip of land of very low relief, up to 50 km wide, that borders the coast between Axel Heiberg Island and the Mackenzie Delta. The plain is underlain by unconsolidated upper Tertiary and Quaternary deposits; like its Greenland counterpart, it is presumably continuous offshore with the inner part of the continental shelf.

The gently-sloping surface of the coastal plain lies nearly, but, not quite, parallel to the bedding of the upper Tertiary sediments, which in places dip seaward more steeply than the plain. Landward dips also occur. The sediments may be slightly deformed, though the dips may be partly or wholly depositional features. In any case, the emergent plain is evidently an erosional surface that

Figure 3.14. Hazen Plateau, a dissected upland area in the Axel Heiberg – Ellesmere Region that represents a stripped peneplain; view west. Lake Hazen and the United States Range are visible in the right distance, and Lady Franklin Bay and Conybeare Fiord in the left foreground and distance, respectively. The irregular bay in the middle foreground is Discovery Harbour. (Air photograph T397L-180, National Air Photo Library, Department of Energy, Mines and Resources, Ottawa.)

Figure 3.15. Sverdrup Lowland of gently folded sedimentary rocks. The anticlinal structure here is pierced by an elongate gypsum mass. Eastern Ellef Ringnes Island; view east across Hassel Sound to Amund Ringnes Island. (Air photograph T428R-121, National Air Photo Library, Department of Energy, Mines and Resources, Ottawa.)

truncates the youngest beds, which are also dissected by the existing system of valleys, channels, and straits. Lineaments and associated scarps in young beds indicate late Tertiary or Recent movement, perhaps due to reactivation of older faults in the underlying, deformed bedrock. (Earthquakes occur in this region but are small in magnitude [Fig. 5C.1, 5C.4, 5C.5]). The plain is a very gently-sloping, uniform surface, characteristically with slightly incised consequent streams draining into the Arctic Ocean (Fig. 3.16).

Lincoln – Morris Jesup Shelf

The continental shelf off northern Greenland, here named the Lincoln – Morris Jesup Shelf, extends from northeastern Ellesmere Island (Lincoln Sea) to Kap Morris Jesup, the northernmost cape of Greenland. This shelf is perhaps the least known of the continental shelves of North America (see Dawes, in press, for review). Sparse bathymetric data suggest that the shelf "break", or change in slope, occurs at a depth of about 500 m (Johnson et al., 1979). A prominent indentation in the submarine contours off Ellesmere Island reduces the width of the shelf to less than 20 km, and this is taken as the boundary with the Arctic Islands Shelf, described below (Fig. 3.1). The shelf is about 50 km wide off eastern Peary Land, and about 250 km wide beneath the Lincoln Sea, where it is characterized by an undulating surface without prominent rises. A distinct, narrow linear trench, close to land and more than 500 m deep, parallels the Greenland coast for more than 200 km east of Robeson Channel. This is on line with the submarine depression of Lady Franklin Bay on the Ellesmere Island coast; its linearity suggests fault control. Off both Ellesmere Island and Peary Land, the shelf falls away into submarine plateau areas (rather than ocean deeps) formed by the Lomonosov Ridge and the Morris Jesup Plateau.

Geophysical data suggest that the shelf in the Lincoln Sea represents a large depositional province with a sedimentary sequence several kilometres thick (perhaps Sverdrup Basin deposits); in addition, the shelf must be underlain by substantial fluvial detritus derived from the mountainous areas of the surrounding Innuitian Orogen. The shelf in the Lincoln Sea probably represents a downfaulted region, possibly associated with the intersection of east-west faults (parallel to many that occur on land) and faults parallel to Nares Strait.

Arctic Islands Shelf

The continental shelf offshore from the Arctic Islands and Yukon Territory is a gently-sloping surface that lies between the shore and the continental slope. The "break" in slope, where the shelf (with a slope of about 1:450) falls away more steeply into the Arctic Ocean, lies at depths of about 200 m off the Yukon Territory and about 500 m off the islands. The shelf is about 120 km wide (but only about 50 km off Yukon), and narrows gradually northeastward to about 80 km off Axel Heiberg Island. Off northern Ellesmere Island the shelf and slope demarcation is complicated by the presence of plateau areas associated with the Alpha Ridge.

The Arctic Islands Shelf appears to be at greater depth (500 m compared to the typical 200 m) and to be of greater slope (1:450 compared to 1:1000) than shelves considered by Heezen and Wilson (1968) to be typical of the world's oceans. It appears, from this, that the Arctic Islands Shelf is in a "drowned" state at present.

Melville Bugt – Disko Shelf

An extensive shelf area off western Greenland attains a maximum width of 300 km (Fig. 3.1), with the shelf break defined by the 500 m contour. The shelf is characterized by widespread Quaternary deposits, in places up to 300 m thick.

The shelf is dissected by transverse, glacially eroded channels at the mouths of major river systems. The most prominent of these are in the Disko – Nûgssuaq area and in Melville Bugt, where channels reach 500 m in depth. Large sediment fans mark the shelf edge at the distal ends of the channels. These semicircular fans can reach 200 km in diameter, as does the fan represented by a sinuous bulge in the shelf edge off Disko Bugt. Although clearly controlled by major fluvial systems, the channels have been considerably sculptured and deepened by glaciers. The sediment fans may well date from before the onset of the main glaciation, when sea level was several hundred metres lower than at present.

The Melville Bugt – Disko Shelf is underlain by an extensive sedimentary basin of Mesozoic and Cenozoic sediments (the West Greenland Basin, Henderson et al., 1981). The basin is bounded by faulted Precambrian "highs", and an easterly embayment of the basin (the Nûgssuaq embayment) is exposed in the Disko – Nûgssuaq Peninsula region. To the north, in Melville Bugt, a coast-parallel graben, 400 km long and 50-75 km wide, contains a 10 km thick sedimentary sequence, the age and structure of which are as yet little known.

Davis Shelf

The narrow continental shelf on the Baffin Bay shores of Baffin, Devon, and Ellesmere islands was named the Davis Shelf by Bostock (1970). The eastern, Precambrian coasts of the archipelago evidently lack a coastal plain, possibly because they are young and underlain by resistant rocks. Seismic evidence indicates that the adjacent shelf is underlain by a drowned pediment and that there are no significant faults at the boundary between inner shelf and coast (H.R. Balkwill, pers. comm., 1987).

The Davis Shelf, considerably narrower than its counterpart in western Greenland, is variable in both depth and width, and becomes indefinite in the vicinity of Davis Strait. Little evidence of a shelf is apparent from the bathymetry of the broad, southeastern end of Baffin Island; rather, a long gentle slope with some low rises descends into Labrador Sea, possibly with a narrow, steeper slope at about 1500 to 2000 m. Northward along Baffin Island, a sinuous shelf edge lies at about 200 m depth and the shelf is only about 40 km wide. The slopes become indefinite off Bylot Island, presumably due to the presence of a huge depositional fan that lies off the mouth of Lancaster Sound. The northwest end of Baffin Bay appears to be a long, gentle slope with irregularities, not unlike that at the head of Labrador Sea, and no distinctive shelf morphology is recognizable off Devon and Ellesmere islands, where the Melville Bugt – Disko Shelf and the Davis Shelf fuse. Three major seaways, Lancaster, Jones and Smith sounds, all with depths below 500 m, enter this area. These three channels, bordered by steep, cliffed coasts, are fault-controlled structures that have been deepened by fluvial and glacial processes.

Figure 3.16. Arctic Islands Coastal Plain, Banks Island, view west to the Beaufort Sea. Note the fault line in the middle distance, parallel to the shore. (Air photograph T429R-169, National Air Photo Library, Department of Energy, Mines and Resources, Ottawa.)

ACKNOWLEDGMENTS

We are grateful to the following for critical appraisal of the manuscript and for suggestions for its improvements: A.K. Higgins and W.S. Watt, Geological Survey of Greenland, and D.A. Hodgson, Geological Survey of Canada.

Aerial photographs are reproduced as plates with the permission of the Geodetic Institute, Copenhagen, Denmark (A-495/79) and the National Air Photo Library, Ottawa, Canada.

REFERENCES

Ahnert, F.
1959: Notes on the physical geography of Northwestern Greenland; in U.S. Army Transportation Corps Arctic Projects: Greenland 1959, Final Report; U.S. Army Transportation Environmental Operations Group, Fort Eustis, Virginia, p. 68-92.
1962: The physical environment of Nyeboe Land, North Greenland; U.S. Army Transportation Board, Fort Eustis, Virginia, 58 p. (Published as a supplement to U.S. Army Transpagation Corps Arctic Projects: Greenland 1959, Final Report.)

Ambrose, J.W.
1964: Exhumed paleoplains of the Precambrian Shield of North America; American Journal of Science, v. 262, p. 817-857.

Bendix-Almgreen, S.E., Fristrup, B., and Nichols, R.L.
1967: Notes on the geology and geomorphology of the Carey Øer, North-West Greenland; Meddelelser om Grønland, v. 164, no. 8, 19 p.

Bird, J.B.
1967: The Physiography of Arctic Canada with special reference to the area south of Parry Channel; The John Hopkins Press, Baltimore, Maryland.
1972: The Natural Landscapes of Canada, a Study in Regional Earth Science; Wiley, Toronto, 191 p.

Birket-Smith, K.
1928: Physiography of West Greenland; in Greenland, M. Vahl, G.C. Amdrup, L. Bobé, and A.S. Jensen (ed.); C.A. Reitzel, Copenhagen, v. 1, p. 423-490.

Blackadar, R.G.
1967: Geological reconnaissance, southern Baffin Island, District of Franklin; Geological Survey of Canada, Paper 66-47, 32 p.
1970: Precambrian geology, northwestern Baffin Island, District of Franklin; Geological Survey of Canada, Bulletin 191, 89 p.

Blake, W., Jr.
1981: Lake sediment coring along Smith Sound, Ellesmere Island and Greenland; in Current Research, Part A, Geological Survey of Canada, Paper 81-1A, p. 191-200.

Bostock, H.S.
1970: Physiographic subdivisions of Canada; in Geology and Economic Minerals of Canada, R.J.W. Douglas (ed.); Geological Survey of Canada, Economic Geology Report no. 1, p. 9-30. (see also, Physiographic regions of Canada; Geological Survey of Canada, Map 1254A, scale 1:5 000 000).

Brown, E.H.
1968: Peneplain; in The Encyclopedia of Geomorphology, R.W. Fairbridge (ed.); Reinhold Book Corporation, New York, p. 821-822.

Burns, C.A.
1952: Geological notes on localities in James Bay, Hudson Bay, and Foxe Basin visited during an exploration cruise in 1949; Geological Survey of Canada, Paper 52-25, 16 p.

Christie, R.L.
1964: Geological reconnaissance of northeastern Ellesmere Island, District of Franklin (120, 340, parts of); Geological Survey of Canada, Memoir 331, 79 p.
1972: Central Stable Region; in Guidebook for Field Excursion A-66, 24th International Geological Congress, Montreal, 1972, p. 40-87.

Clarke, D.B. and Pedersen, A.K.
1976: Tertiary volcanic province of West Greenland; in Geology of Greenland, A. Escher and W.S. Watt (ed.); Geological Survey of Greenland, Copenhagen, p. 364-385.

Clarke, D.B. and Upton, B.G.J.
1971: Tertiary basalts of Baffin Island: field relations and tectonic setting; Canadian Journal of Earth Sciences, v. 8, p. 248-258.

Colton, R.B. and Holmes, C.D.
1954: Geomorphology of the Nunatarssuak area; in Final Report, Program B, Operation Ice Cap 1953; U.S. Army, Stanford Research Institute, California, p. 27-52.

Cowie, J.W.
1961: Contributions to the geology of North Greenland; Meddelelser om Grønland, v. 164, no. 3, 47 p.

Davies, W.E.
1963: Glacial geology of northern Greenland; Polarforschung, v. 5, p. 94-103.
1972: Landscape of northern Greenland; U.S. Army, Cold Regions Research and Engineering Laboratory, Hanover, New Hampshire, Special Report 164, 67 p.

Davies, W.E., Krinsely, D.B., and Nicol, A.H.
1963: Geology of the North Star Bugt area, Northwest Greenland; Meddelelser om Grønland, v. 162, no. 12, 68 p.

Davies, W.E., Needleman, S.M. and Klick, D.W.
1959: Report on Operation Groundhog (1958) North Greenland. Investigation of ice-free sites for aircraft landing, Polaris Promontory, North Greenland; U.S. Air Force, Cambridge Research Center, Bedford, 45 p.

Dawes, P.R.
1987: Topographical and geological maps of Hall Land, North Greenland. Description of a computer-supported photogrammetrical research programme for production of new maps, and the Lower Palaeozoic and surficial geology; Grønlands Geologiske Undersøgelse, Bulletin 155.
in press: The North Greenland Continental Margin; in The Arctic Ocean region; Geological Society of America, The Geology of North America, v. L.

Dawes, P.R. and Kerr, J.W. (ed.)
1982: Nares Strait and the Drift of Greenland; a Conflict in Plate Tectonics; Meddelelser om Grønland, Geoscience no. 8, 392 p.

Dunbar, M. and Greenaway, K.R.
1956: Arctic Canada from the Air; Canada, Defence Research Board, Queen's Printer, Ottawa, 541 p.

Fairbridge, R.W. (ed.)
1968: The Encyclopedia of Geomorphology; Reinhold Book Corporation, New York, 1295 p.

Fortier, Y.O.
1957: The Arctic Archipelago; in Geology and Economic Minerals of Canada, C.H. Stockwell (ed.); Geological Survey of Canada, Economic Geology Series No. 1 (fourth edition), p. 393-442.

Fortier, Y.O. and Morley, L.W.
1956: Geological unity of the Arctic Islands; Royal Society of Canada, Transactions, v. 50, p. 3-12.

Funder, S., Bennike, O., Mogensen, G.S., Noe-Nygaard, B., Pedersen, S.A.S., and Petersen, K.S.
1984: The Kap København Formation, a late Cainozoic sedimentary sequence in North Greenland; Grønlands Geologiske Undersøgelse, Report no. 120, p. 9-18.

Funder, S. and Hjort, C.
1980: A reconnaissance of the Quaternary geology of eastern North Greenland; Grønlands Geologiske Undersøgelse, Report no. 99, p. 99-105.

Fyles, J.G.
1962: Physiography; in Banks, Victoria, and Stefansson Islands, Arctic Archipelago, R. Thorsteinsson and E.T. Tozer; Geological Survey of Canada, Memoir 330, p. 8-17.
1963: Surficial geology of Victoria and Stefansson Islands, District of Franklin; Geological Survey of Canada, Bulletin 101, 38 p.

Gentilli, J.
1968: Regions, natural and geographical; in The Encyclopedia of Geomorphology, R.W. Fairbridge (ed.); Reinhold Book Corporation, New York, p. 932-933.

Henderson, G., Rosenkrantz, A., and Schiener, E.J.
1976: Cretaceous-Tertiary sedimentary rocks of West Greenland; in Geology of Greenland, A. Escher and W.S. Watt (ed.); Geological Survey of Greenland, Copenhagen, p. 340-362.

Henderson, G., Schiener, E.J., Risum, J.B., Croxton, C.A., and Andersen, B.B.
1981: The West Greenland Basin; in Geology of the North Atlantic Borderlands, J.W. Kerr and A.J. Fergusson (ed.); Canadian Society of Petroleum Geologists, Memoir 7, p. 399-428.

Heezen, B. and Wilson, L.
1968: Submarine geomorphology; in The Encyclopedia of Geomorphology, R.W. Fairbridge (ed.); Reinhold Book Corporation, New York, p. 1079-1097.

Johnson, G.L., Monahan, D., Grønlie, G., and Sobczak, L.
1979: General bathymetric chart of the oceans (GEBCO), 1:6 000 000, sheet 5.17, The Arctic Ocean; Canadian Hydrographic Service, Ottawa.

Kerr, J.W.
1980: Structural framework of Lancaster Aulacogen, Arctic Canada; Geological Survey of Canada, Bulletin 319, 24 p.

Koch, L.
1923a: Some new features in the physiography and geology of Greenland; Journal of Geology, v. 31, p. 42-65.
1923b: Preliminary report upon the geology of Peary Land, Arctic Greenland; American Journal of Science, v. 5, p. 189-199.
1925: The geology of North Greenland; American Journal of Science, v. 9, p. 271-285.
1926: A new fault zone in Northwest Greenland; American Journal of Science, v. 12, p. 301-310.
1928: The physiography of North Greenland; in Greenland, M. Vahl, G.C. Amdrup, L. Bobé, and A.S. Jensen (ed.); C.A. Reitzel, Copenhagen, v. 1, p. 491-518.
1929: Stratigraphy of Greenland; Meddelelser om Grønland, v. 73, no. 2, p. 205-320.
1935: Geologie von Grönland; in Geologie der Erde, E. Krenkel (ed.); Gebrüder Borntraeger, Berlin, 159 p.

Lemon, R.R.H. and Blackadar, R.G.
1963: Admiralty Inlet area, Baffin Island, District of Franklin; Geological Survey of Canada, Memoir 328, 84 p.

Malaurie, J.
1955: The French Geographical Expedition to Thule, 1950-51, a preliminary report; Arctic, v. 8, no. 4, p. 203-214.
1968: Thèmes de recherche géomorphologique dans le nord-ouest du Groenland; Editions du Centre national de la Recherche scientifique, Paris, 495 p.

Mescherikov, Y.A.
1968: Plains; in The Encyclopedia of Geomorphology, R.W. Fairbridge (ed.); Reinhold Book Corporation, New York, p. 850-856.

Monahan, D. and Johnson, G.L.
1982: Physiography of Nares Strait: importance to the origin of the Wegener Fault; in Nares Strait and the Drift of Greenland: a Conflict in Plate Tectonics, P.R. Dawes and J.W. Kerr (ed.); Meddelelser om Grønland, Geoscience 8, p. 53-64.

Nichols, R.L.
1969: Geomorphology of Inglefield Land, North Greenland; Meddelelser om Grønland, v. 188, no. 1, 109 p.

Paterson, T.T.
1951: Physiographic studies in North West Greenland; Meddelelser om Grønland, v. 151, no. 4, 60 p.

Pelletier, B.R.
1966: Development of submarine physiography in the Canadian Arctic and its relation to crustal movement; in Continental Drift, G.D. Garland (ed.); Royal Society of Canada, Special Publication No. 9, University of Toronto Press, Toronto, p. 77-101.

Roots, E.F.
1963: Physiography; in Geology of the North-Central Part of the Arctic Archipelago, Northwest Territories (Operation Franklin), Y.O. Fortier et al.; Geological Survey of Canada, Memoir 320, p. 107-117, p. 164-179, p. 266-275, p. 418-426, p. 580-585.

Sanford, B.V. and Grant, G.M.
1976: Physiography, eastern Canada and adjacent areas; Geological Survey of Canada, Map 1399A (4 sheets, scale 1:2 000 000).

Sobczak, L.W.
1982: Fragmentation of the Canadian Arctic Archipelago, Greenland, and surrounding oceans; in Nares Strait and the Drift of Greenland: a Conflict in Plate Tectonics, P.R. Dawes and J.W. Kerr (ed.); Meddelelser om Grønland, Geoscience no. 8, p. 221-236.

Sugden, D.
1974: Landscapes of glacial erosion in Greenland and their relationship to ice, topographic and bedrock conditions; Institute of British Geographers, Special Publication no. 7, p. 177-195.

Taylor, A.
1956: Physical Geography of the Queen Elizabeth Islands, Canada; 12 volumes; American Geographical Society, New York.

Trettin, H.P.
1975: Investigations of lower Paleozoic geology, Foxe Basin, northeastern Melville Peninsula, and parts of northwestern and central Baffin Island; Geological Survey of Canada, Bulletin 251, 177 p.

Victor, P.E.
1955: Geography of Northeast Greenland; U.S. Army, Corps of Engineers, Snow, Ice and Permafrost Research Establishment, Wilmette, Illinois, Special Report 15, 51 p.

Weidick, A.
1975: A review of Quaternary investigations in Greenland; Institute of Polar Studies, Ohio State University Research Foundation, Report No. 55, 161 p.

Authors' addresses

R.L. Christie
Geological Survey of Canada
3303-33rd St. N.W.
Calgary, Alberta
T2L 2A7

P.R. Dawes
Geological Survey of Greenland
Øster Voldgade 10
DK-1350, Copenhagen K
Denmark

Printed in Canada

Chapter 4
TECTONIC FRAMEWORK

Introduction

Canadian Shield

Arctic Platform

Innuitian Tectonic Province

References

Chapter 4

TECTONIC FRAMEWORK

H.P. Trettin

INTRODUCTION

This chapter summarizes the major geological components and inferred tectonic events of the Arctic Islands and North Greenland and explains the currently used terms. References are mostly restricted to historical remarks and original definitions; additional references and details are given in later chapters.

Basic elements of the present framework were introduced by Schuchert (1923), who made the first synthesis of North American geology that included the Arctic (Fig. 4.1). The results of the vigorous exploration in the Canadian Arctic Islands in the 1950s and early 1960s were compiled by Fortier et al. (1954), Fortier (1957), Douglas et al. (1963), and by Thorsteinsson and Tozer (1960, 1970), whose "geological provinces" (Fig. 4.2) have served as the basis for all subsequent syntheses. Comprehensive later accounts of this region include Trettin et al. (1972), Drummond (1973, 1974), Stuart Smith and Wennekers (1977), Trettin and Balkwill (1979), and Kerr (1981). Syntheses of the geology of North Greenland were made by Koch (1925, 1935, 1961), Dawes (1971, 1976), Dawes and Soper (1973), and Dawes and Peel (1981).

The first-order elements of the present tectonic framework, all introduced previously, are Canadian Shield, Arctic Platform, and Innuitian Tectonic Province.

CANADIAN SHIELD

The Canadian Shield borders the Arctic Platform on the southeast and extends into it as salients and inliers, such as the core of the Boothia Uplift[1], Minto Arch, Melville Horst, and the Wellington and Duke of York highs, exhumed hills on the Late Proterozoic–Cambrian erosion surface (Chapter 3). It consists of a metamorphic-plutonic basement and unconformably overlying sedimentary and volcanic successions (Chapter 6). The metamorphic basement rocks in the northernmost part of the Shield are generally in the amphibolite or granulite facies and represent a group of Archean terranes welded together in a series of Early Proterozoic ("Hudsonian") orogenies (Hoffman et al., in prep.).

The oldest sedimentary units resting on the crystalline basement are sporadic in distribution. They include thin clastic and carbonate units of presumed Early and Middle Proterozoic ages in Victoria Island and a Middle Proterozoic clastic unit with a 1.36 Ga Rb-Sr age of clay mineral diagenesis (Independence Fiord Group) in northeasternmost Greenland.

Figure 4.1. Paleozoic tectonic elements of North America according to Schuchert (1923, Fig. 3).

---

[1] The term, Boothia Uplift, is here used for both the crystalline basement and the deformed sedimentary cover of this major tectonic element (Thorsteinsson and Tozer, 1970), rather than the basement alone (Kerr and Christie, 1965). The term, Cornwallis Fold Belt (Fortier, 1957), is applied to the deformed sedimentary cover along the entire length of the uplift (Kerr and Christie, 1965).

Trettin, H.P.
1991: Tectonic Framework; Chapter 4 in Geology of the Innuitian Orogen and Arctic Platform of Canada and Greenland, H.P. Trettin (ed.); Geological Survey of Canada, Geology of Canada, no. 3; (also Geological Society of America, The Geology of North America, v. E).

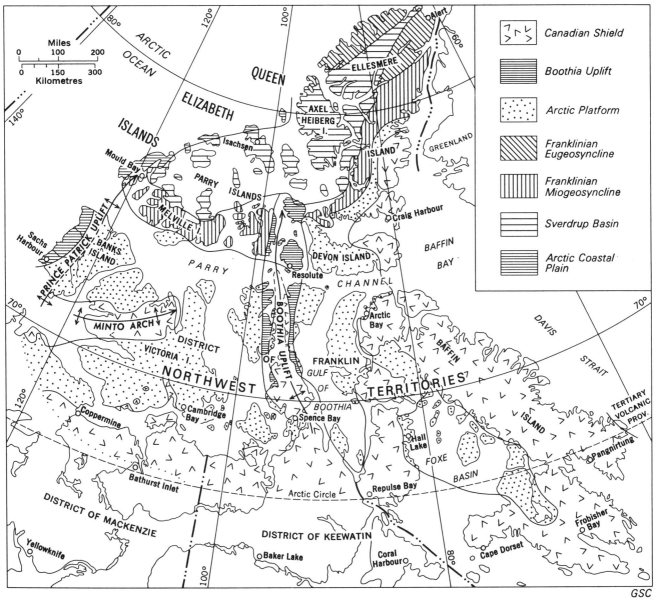

Figure 4.2. Geological provinces of the Arctic Archipelago according to Thorsteinsson and Tozer (1970, Fig. X-1).

A 4 to 6 km thick, slightly deformed succession of Middle and Late Proterozoic age is exposed in a discontinuous belt of Late Proterozoic structural "basins" that is subparallel with the southeastern margin of the Franklinian mobile belt. The succession comprises a lower coarse clastic and basaltic unit that indicates an extensive rifting event about 1.2 to 1.3 Ga ago, and an upper, predominantly shallow marine sedimentary unit of probable late Middle and Late Proterozoic age — the precise range is uncertain because of conflicts between paleomagnetic and micropaleontological age determinations.

The Proterozoic sedimentary strata of Victoria Island are disconformably overlain by a basaltic unit (Natkusiak Formation), that has been correlated with the extensive Franklin diabase dyke swarms dated at about 750 Ma (Fahrig and West, 1986), although it has yielded a somewhat younger (635 Ma) K-Ar age.

ARCTIC PLATFORM

The term, Arctic Platform, introduced by Douglas (1970) and Thorsteinsson and Tozer (1970) in the structural sense of King (1969) as opposed to more recent sedimentological uses (e.g. Read, 1982), designates Phanerozoic successions in the northern part of the continent that have not been folded or thrust faulted on a regional scale. The boundary with the Interior Platform in the region south and southwest of Victoria Island is arbitrary.

The sedimentary rocks of the Arctic Platform are stratigraphically divisible into a lower-middle Paleozoic succession and a Mesozoic-Paleogene succession. The lower-middle Paleozoic exposures (Chapters 7, 8B) represent a once continuous cover that ranges in age from Early Cambrian to Late Devonian in the north and from Middle Ordovician to Early Silurian in the south; a decrease in thickness accompanies the decrease in age range. The succession is mainly of shallow marine origin but includes some nonmarine units as well as Silurian deep water units that transgressed the shelf in northern Greenland. This cover is preserved in post-depositional structural basins (Christie, 1972) that are related in age and origin to the bounding Precambrian highs. The stratigraphic record indicates that the Boothia Uplift was positive in Late Silurian – Early Devonian time. The Minto Arch is flanked on the northwest by a northwest-dipping homocline of Lower Cambrian to Upper Devonian (Famennian) strata that is truncated on the northwest by Middle Jurassic and younger units; the arch therefore developed during an unspecified time within the latest Devonian – Middle Jurassic span (Thorsteinsson and Tozer, 1962; Miall, 1979). The Melville Horst is flanked by lower Paleozoic strata both on the east and west. The eastern boundary consists of normal faults of presumed Cretaceous-Paleogene age; the western contact is concealed. The Wellington and Duke of York highs of Victoria Island appear to be paleotopographic highs of Early Cambrian or older age. On the northwest and north, the lower Paleozoic deposits of the Arctic Platform are contiguous with the folded strata of the Franklinian Shelf. Strata adjacent to uplifted Precambrian crystalline rocks in southeastern Ellesmere Island, eastern Devon Island, and Bylot Island, form an arcuate, northwesterly- to southwesterly-dipping homocline.

The most extensive outcrop area of Mesozoic and Tertiary strata (Chapters 14, 15) is on Banks Island (Banks Basin), where Jurassic to Paleogene strata are preserved. Cretaceous and Paleogene sediments are exposed in the Eclipse Trough of Bylot Island and northeastern Baffin Island. Two basins in Lancaster Sound, known only from seismic surveys, contain sediments of uncertain Mesozoic-Cenozoic age.

INNUITIAN TECTONIC PROVINCE

The term, Innuitian Orogenic System, was coined by Fortier et al. (1954) for the "complex assemblage of mountain structures extending from the Beaufort Sea across the Princess Elizabeth Islands and which continues through north Greenland". The adjective, Innuitian, was derived from the noun "innuit" (present spelling "Inuit"), meaning people — the name that the "Eskimo" (a now largely abandoned, originally pejorative, Indian word) gave themselves. The name was simplified to Innuitian Orogen by Douglas (1970).

The related term, Innuitian Tectonic Province (synonym: Innuitian Province), places emphasis on the entire geological history, including patterns of subsidence and deposition (cf. Price and Douglas, 1972; Trettin et al., 1972). The Innuitian Tectonic Province is characterized by its position on the northern margin of the North American craton and by southwesterly to westerly depositional and structural trends. Its southeastern and southern limit coincides with the limit of Phanerozoic folding or thrust faulting and its northwestern limit with the boundary between continental and Mesozoic-Tertiary oceanic crust.

The province comprises four major stratigraphic successions, named for the tectonic setting in which they were deposited: (1) a concealed upper Middle and Upper Proterozoic succession, known only from cratonward equivalents in the northernmost part of the Canadian Shield (see above and Chapter 6); (2) the uppermost Proterozoic to Upper Devonian succession of the Franklinian mobile belt; (3) the Carboniferous to Neogene successions of the Sverdrup Basin and of part of the Wandel Sea Basin; and (4) the Lower Cretaceous to Recent Arctic Continental Terrace Wedge. Deposition in the Franklinian mobile belt was terminated by the Ellesmerian Orogeny and deposition in the Sverdrup Basin by the Eurekan Orogeny.

The two terms, Innuitian Orogen and Innuitian Tectonic Province, have different age connotations. The Innuitian Orogen. does not include the undeformed Cretaceous to Holocene Arctic Continental Terrace Wedge, but must encompass, in the subsurface, the entire Precambrian succession, including the crystalline basement. The Innuitian Tectonic Province, on the other hand, ranges in age from upper Middle Proterozoic to Recent. The first term is more useful for regional thematic maps; the second for discussion of geological history.

Franklinian mobile belt
Earlier geosynclinal concepts

The geosynclinal concept, developed in the 19th century by J. Hall and J.D. Dana for the sedimentary succession of the Appalachian region, was first applied to the entire continent by Schuchert (1923), who recognized four major types of Paleozoic tectonic elements (Fig. 4.1). The Canadian Shield in the centre of the continent was bordered on the south by a "neutral" region, underlain by thin and undeformed sedimentary sequences. Shield and "neutral" area were surrounded by geosynclines and embayments characterized by relatively thick, deformed sedimentary successions. These included the Franklinian Geosyncline in the north, which merged with the Cordilleran Geosyncline east of the Mackenzie Delta. On the oceanic sides of the geosynclines were borderlands of crystalline rocks, assumed to be Precambrian (Archean) in age and to be contiguous with the Canadian Shield beneath the geosynclines. The existence of the borderland Pearya on the northwest side of the Franklinian Geosyncline was inferred from the occurrence of metamorphic rocks of assumed Precambrian age in northernmost Ellesmere Island, first mapped by Feilden and de Rance (1878). Schuchert concluded that the clastic sediments of the Franklinian Geosyncline were derived from Pearya because they seemed to become coarser and thicker toward the north. To account for the large sediment volume in the geosyncline, he extended Pearya northward into the present offshore area and southwestward to Prince Patrick Island, the geology of which was unknown at the time.

Thorsteinsson and Tozer (1957, 1960) established that deposition in the Franklinian Geosyncline was terminated by a deformation of latest Devonian – Early Carboniferous age, for which they introduced the name, Ellesmerian Orogeny (1970). Following the practice of Kay (1951) for the Appalachian and Cordilleran geosynclines, they subdivided the Franklinian into a miogeosyncline and a eugeosyncline, which included Pearya, no longer recognized as a separate entity. Fossil evidence from the Appalachians had shown that at least some crystalline terranes previously interpreted as Precambrian borderlands, in fact are metamorphic belts of early Paleozoic age.

Present modifications

Fieldwork in the northern regions, new methods of age determination, and the general progress from geosynclinal to plate tectonic concepts have affected primarily the concept of the eugeosyncline. Schuchert's borderland Pearya now is considered to be an exotic continental fragment with an internal suture of mid-Ordovician age (Trettin, 1987). The region between Pearya and the original miogeosyncline is now interpreted as vestige of a deep water basin, which may include exotic units in its northern part. The term, geosyncline, has been replaced by the term, mobile belt, to allow for the presence of "suspect" and exotic elements (Trettin, 1987). The concept of the miogeosyncline (or miogeocline of some writers) has essentially remained intact, but the name has been changed to "shelf province" or "Franklinian Shelf", in keeping with other changes in nomenclature.

The term, Franklinian Basin, introduced by Stuart Smith and Wennekers (1979) as a substitute for Franklinian Geosyncline, has been accepted widely in the literature (including Chapter 7). It is applicable to most of the mobile belt but should not be used for the Pearya Terrane.

The base of the Franklinian mobile belt is not exposed, and the event that initiated it has not been defined previously. It is here proposed to equate it with a Late Proterozoic (0.75 Ga?) rifting episode represented by the Natkusiak volcanics of Victoria Island and the extensive Franklin diabase dyke swarms.

Paleozoic sequences and depositional provinces

The coherent record of the Franklinian mobile belt begins in the mid-Early Cambrian. The Lower Cambrian to Upper Silurian or Lower Devonian successions were deposited in two major provinces, a southern or southeastern shelf and a northern deep water basin (Chapters 7, 8).

The **shelf province** (Franklinian Shelf) includes all those strata of Early Cambrian (or older) to Early Devonian age lying between the slope of the deep water basin and the limit of mid-Paleozoic folding or thrust faulting. The northwestern or northern boundary of the shelf retreated cratonward in several major steps, but was relatively stable from Middle Cambrian to late Middle Ordovician time. This position of the shelf edge, which also is significant from a structural point of view, is used to separate shelf and deep water basin in generalized maps such as Figure 4.3. The southeastern or southern limit of the shelf province, as mentioned, is not a stratigraphic but a structural boundary, and Franklinian Shelf and Arctic Platform therefore are discussed together in the chapters dealing with depositional history.

The record of the shelf province is dominated by carbonate sediments but also includes craton-derived clastic sediments and evaporites. The deposits thicken toward the shelf edge where they attain a thickness of more than 9 km. Positive movements of the Boothia Uplift in Late Silurian – Early Devonian time, and of the Inglefield Uplift in Early Devonian time, produced angular unconformities and clastic sediments.

The **deep water basin**, characterized mainly by sediment gravity flows, graptolitic mudrock, and radiolarian chert, is divisible into two parts. The southeastern part of the basin, which is exposed from northeastern Greenland to northwestern Melville Island, is filled with sediments of Early Cambrian to Early Devonian age (sedimentary subprovince). These strata are contiguous, through facies changes, with the shelf sediments and hence do not represent an exotic terrane.

The northwestern part, exposed only in northern Ellesmere and Axel Heiberg islands, contains arc-type volcanics and associated shallow marine carbonates in addition to the predominant deep water sediments of Early Cambrian to early Late Silurian age (sedimentary-volcanic subprovince). Some units in this belt are similar to units of the sedimentary subprovince and probably linked with the shelf province; others may be exotic.

The strata of the Franklinian Shelf are conformably overlain by a "clastic wedge" deposited in a syntectonic foreland basin (Chapter 10). The preserved deposits range in age from early Middle to latest Devonian (Eifelian to Famennian) and are up to 5 km thick; another 3-6 km of clastic sediments of latest Devonian and Early Carboniferous (Famennian-Tournaisian) age are inferred to have been removed by erosion (Chapter 10). The sediments are mainly of shallow marine and nonmarine origin but include turbidites and submarine slope deposits of early Middle Devonian (Eifelian) age in the western part. The basin received sediments from northern parts of the Franklinian mobile belt, the Caledonides, and the Canadian Shield.

The sedimentary-volcanic subprovince of the deep water basin is unconformably overlain by Upper Silurian and Devonian carbonate and clastic successions (Chapter 8C) and poorly exposed Middle(?) and Upper Devonian clastic sediments (Chapter 10).

Pearya Terrane

Pearya, located north of the deep water basin in Ellesmere Island, is divisible into four major successions (Chapter 9). Succession I comprises sedimentary and(?) volcanic rocks, deformed, metamorphosed and intruded by granitic plutons at 1.0 to 1.1 Ga. Succession II, of Late Proterozoic(?) to earliest Ordovician age, consists mainly of platformal (or miogeoclinal) sediments such as carbonate, quartzite, and mudrock, with smaller proportions of volcanics, diamictite, and chert. Succession III (Early to Middle Ordovician?) includes arc-type and(?) ocean-floor volcanics, chert, mudrock and carbonates, and is associated with fault slices of ultramafic-mafic complexes, possibly dismembered ophiolites. The faulted contact of succession III and the ultramafic-mafic complexes with succession II is unconformably overlapped by succession IV, 7 to 8 km of volcanic and sedimentary rocks, ranging in age from late Middle Ordovician (early Caradoc) to Late Silurian. The angular unconformity at the base of this succession represents the early Middle Ordovician M'Clintock Orogeny, a collision that was accompanied by metamorphism and granitic plutonism.

Pearya is related to the Appalachian-Caledonian mobile belt by the Grenville age of its crystalline basement, the Early Ordovician age of its ultramafic-mafic complexes, and evidence for an early Middle Ordovician orogeny, comparable to the Taconian in age and character. By contrast,

TECTONIC FRAMEWORK

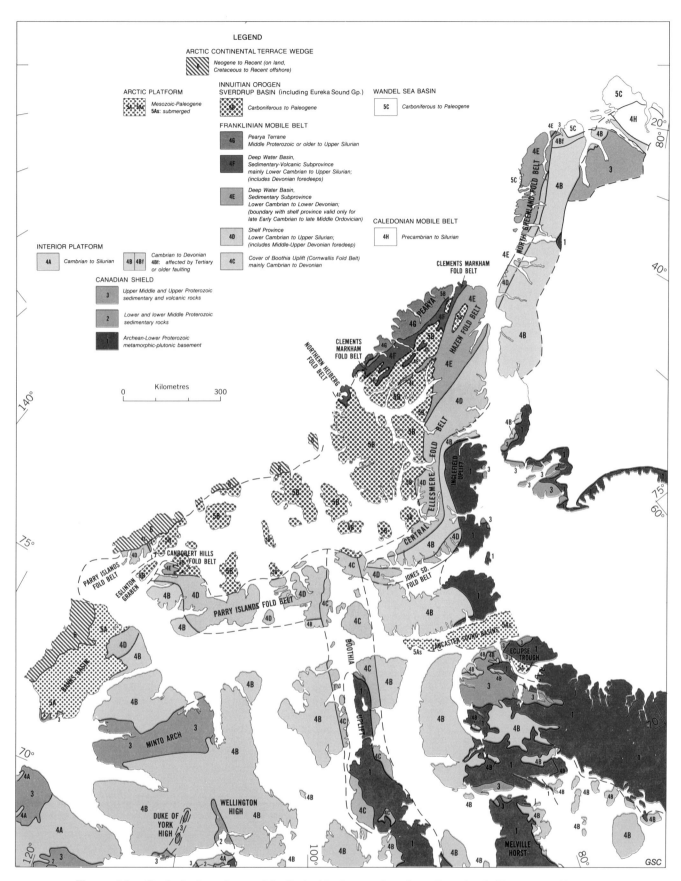

Figure 4.3. Geological provinces of Arctic Archipelago and northern Greenland. (For geographic names, see Fig. 1, in pocket.)

the Franklinian Shelf has a Lower Proterozoic – Archean basement, and the shelf and deep water basin were not deformed in the Ordovician. An Early Ordovician disconformity in northeasternmost Greenland may represent the development of a "peripheral bulge" caused by a collision in the Caledonian mobile belt (Chapter 7).

Late Silurian – Early Carboniferous orogenic pulses

The Franklinian mobile belt was affected by intermittent deformation of limited extent from Late Silurian to Middle Devonian time and was terminated by an extensive compressional event of latest Devonian – Early Carboniferous age, the Ellesmerian Orogeny in the most restricted sense; in a wider sense the term has also been used for the movements that produced the Middle-Upper Devonian clastic wedge (Thorsteinsson and Tozer, 1970; Chapters 11, 12).

Stratigraphic and structural evidence suggests that Pearya was transported by sinistral strike slip as three or more slices and accreted to the deep water basin in Late Silurian – Early Devonian time. Intense deformation, including horizontal bending, appears to have occurred when the motion was arrested and the terrane became locked in its present position. The sinistral movement probably also affected the sedimentary-volcanic subprovince of the deep water basin in Ellesmere Island (Clements Markham Fold Belt) and(?) adjacent parts of the sedimentary subprovince (Hazen Fold Belt). A Late Silurian – Early Devonian compressional deformation of the sedimentary-volcanic subprovince in Axel Heiberg Island (Northern Heiberg Fold Belt), characterized by northwesterly structural trends, may have been caused by collision with northwestern Ellesmere Island. These movements were accompanied or followed by emplacement of basement-derived granitic plutons, one of which has an Emsian (390 ± 10 Ma) U-Pb age.

Positive movements of the Boothia Uplift in Late Silurian – Early Devonian time and of the Inglefield Uplift in Early Devonian time probably are due to reverse faulting rather than mantle-controlled vertical uplift, as previously thought. These motions were restricted to areas with northerly basement trends and evidently were caused by east-west compression acting over large distances.

A Givetian(?)-Frasnian uplift within northern parts of the Hazen Fold Belt is apparent from stratigraphic and sedimentological relationships. If caused by renewed sinistral motion, it may also account for a second compressional deformation in the Northern Heiberg Fold Belt that occurred some time between the mid-Early Devonian and the late Early Carboniferous (Viséan).

The compressional event of latest Devonian – Early Carboniferous age (Ellesmerian Orogeny sensu stricto) had its most severe effects in those regions not affected by the earlier movements, i.e. the southwestern part of the deep water basin (northwestern part of North Greenland Fold Belt, southwestern part of Hazen Fold Belt, and Canrobert Fold Belt) and those parts of the Franklinian Shelf lying northeast and west of the Boothia Uplift (southwestern part of North Greenland Fold Belt, Central Ellesmere Fold Belt, Jones Sound Fold Belt, Parry Islands Fold Belt).

Regional metamorphism was restricted to parts of northern Ellesmere Island and North Greenland, where it ranges up to amphibolite grade. The metamorphism is of uncertain Late Silurian to latest Devonian age.

The cause of the Late Devonian – Early Carboniferous movements is problematic. Pure strike slip can probably be ruled out because of the great width and sinuous trend of the deformed belt. If the orogeny was caused by a collision, the corresponding suture must be hidden in the present offshore area.

Sverdrup Basin

The concept of the Sverdrup Basin was introduced by Thorsteinsson and Tozer (1960) and has remained unchanged. The Sverdrup Basin, which overlies the central and northwestern parts of the Franklinian mobile belt with angular unconformity, is about 1300 km long in a northeast-southwest direction, and up to 400 km wide. Depth to mantle is 37 to 41 km in the axial region, as compared with 48 km beneath the Canadian Shield and the Sverdrup Rim, a horst-like feature that bounds the basin proper on the northwest (Chapter 5B). The axial region contains up to 3 km of upper Paleozoic strata, up to 9 km of Mesozoic strata, and locally more than 3 km of syntectonic clastic sediments of latest Cretaceous and Tertiary age.

Late Paleozoic deposition (Chapter 13) was characterized by carbonate, evaporite and clastic sediments; Mesozoic and Tertiary deposition (Chapters 14, 15) by clastic sediments except for some impure carbonates of Middle Triassic age. The absence of carbonate sediments and evaporites from the Mesozoic is due to drift into more northerly latitudes, apparent from paleomagnetic studies. The clastic sediments were derived mainly from the south and east, and to a lesser extent from the northwestern side of the basin. The stratigraphic succession is disrupted by numerous disconformities. Some, for example a pronounced disconformity at the base of Triassic shallow marine deposits, are due to a eustatic fall in sea level, others to local or regional uplift.

Volcanism and hypabyssal intrusion were limited to northern parts of the basin in the Carboniferous and Permian but were more extensive in the Cretaceous. The rocks are mostly mafic in composition and commonly tholeiitic, but alkali basalt and bimodal suites of basement-derived felsic components are also present.

The tectonic history of the basin (Chapters 13, 14) is divisible into the following, partly overlapping, phases: (1) initial rifting (late Early Carboniferous to Early Permian), locally accompanied by deformation (Melvillian Disturbance); (2) thermal subsidence (Late Permian to earliest Cretaceous; (3) renewed rifting at the present continental margin (Middle Jurassic to Early Cretaceous or later); (4) thermal subsidence (Late Cretaceous); and (5) deformation and syntectonic deposition (latest Cretaceous to early Oligocene — see below). The Jurassic – Early Cretaceous phase of rifting was followed by seafloor spreading in the Amerasian Basin (cf. Grantz et al., in press). Flowage of Carboniferous evaporites, caused by differential loading, extension, or compression, has occurred through much of Triassic to Neogene time.

Wandel Sea Basin

This basin, named by Dawes and Soper (1973), overlies parts of Franklinian mobile belt, Arctic Platform, and Caledonides with angular unconformity. It contains more than 7 km of Lower Carboniferous to Paleogeone strata deposited in a variety of rift- and strike-slip related subbasins (Håkansson and Stemmerik, 1983, and pers. comm., January, 1988). Coarse clastic sediments of early Late Carboniferous age are overlain by Carboniferous and Permian carbonate and clastic sediments with minor amounts of evaporites. The Mesozoic to Paleogene succession consists mostly of clastic sediments, but a thick peralkaline volcanic unit (Kap Washington Group) of latest Cretaceous and(?) Paleocene age occurs in northernmost Greenland (Chapter 16). The basin has some affinities with the Sverdrup Basin but is more closely related to Svalbard and the North Atlantic region and will not be discussed in this volume.

Latest Cretaceous – Paleogene deposition and Tertiary deformation

During the late Campanian to Late Eocene or Early Oligocene, a complex succession of sandstone, mudrock, conglomerate, and coal of shallow marine and nonmarine origin, locally more than 3 km thick, was deposited in parts of the Innuitian Orogen and Arctic Platform (Eureka Sound Group; Chapter 15). A compressive deformation, known as the Eurekan Orogeny (Thorsteinsson and Tozer, 1970) affected northern parts of the islands from middle Eocene or earlier, perhaps latest Cretaceous or Paleocene time, to early Oligocene(?) time (Chapter 17). It began with the rise of fault-controlled arches and uplifts within the Sverdrup Basin, that shed sediments into adjacent basins, and culminated in extensive thrust faulting and folding.

In northeasternmost Greenland (Chapter 16), intrusion of alkaline dyke swarms, and extrusion of more than 3 km of peralkaline volcanics during a Late Cretaceous – Early Paleocene extensional regime were followed by thrust faulting prior to about 35 Ma (Early Oligocene).

The Eurekan Orogeny coincided with late phases of the development of the Labrador Basin (Labrador Sea – Baffin Bay region), where Early Cretaceous rifting was succeeded by middle Cretaceous to Early Oligocene seafloor spreading. This extension resulted in counterclockwise rotation and northeastward translation of Greenland, which, in turn, caused intra-plate compressive deformation in those parts of the Innuitian Orogen lying outboard of the Precambrian crystalline basement. The development of Baffin Bay probably was accompanied by graben faulting in surrounding areas.

The early Tertiary horizontal movements were succeeded by middle to late Tertiary vertical movements, which, in combination with erosion, gave rise to the present physiography (Chapter 18). Thermal(?) uplift affected Precambrian crystalline terranes around the Labrador Basin, and isostatic uplift the main sites of Eurekan compression and crustal thickening in Ellesmere and Axel Heiberg islands.

Arctic Continental Terrace Wedge

The term, Arctic Coastal Plain, introduced by Fortier (1957), was used by Thorsteinsson and Tozer (1960, 1970) for a narrow strip on the northwestern margin of the Arctic Archipelago where nearly undisturbed sediments of Neogene to Recent age, dipping gently toward the Arctic Ocean, lie unconformably on Mesozoic and Paleogene strata (Chapter 2, "Arctic Islands Coastal Plain", and Chapter 15). The Neogene Beaufort Formation consists mainly of fluvial sediments but includes marine strata of Pliocene age. The deposits on land represent the thin edge of the Arctic Continental Terrace Wedge (Balkwill in Trettin and Balkwill, 1979), which, in the offshore area, dates back to the Cretaceous and includes distal equivalents of the Eureka Sound Group.

REFERENCES

Christie, R.L.
1972: Central Stable Region; in The Canadian Arctic Islands and the Mackenzie Region, XXIV International Geological Congress, Guidebook Excursion A-66, D.J. Glass (ed.), p. 40-87.

Dawes, P.R.
1971: The North Greenland fold belt and environs; Bulletin of the Geological Society of Denmark, v. 20, p. 197-239.
1976: Precambrian to Tertiary of northern Greenland; in Geology of Greenland, A. Escher and W.S. Watt, (ed.); The Geological Survey of Greenland, Copenhagen, p. 248-303.

Dawes, P.R. and Peel, J.S.
1981: The northern margin of Greenland from Baffin Bay to the Greenland Sea; in The Ocean Basins and Margins, v. 5, The Arctic Ocean, A.E.M. Nairn, M. Churkin, Jr., and F.G. Stehli (ed.); Plenum Press, New York and London, p. 201-264.

Dawes, P.R. and Soper, N.J.
1973: Pre-Quaternary history of Greenland; in Arctic Geology, M.G. Pitcher (ed.); American Association of Petroleum Geologists, Memoir 19, p. 117-134.

Douglas, R.J.W.
1970: Introduction; in Geology and Economic Minerals of Canada, R.J.W. Douglas (ed.); Geological Survey of Canada, Economic Geology Report no. 1, p. 2-8.

Douglas, R.J.W., Norris, D.K., Thorsteinsson, R., and Tozer, E.T.
1963: Geology and petroleum potentialities of northern Canada; Geological Survey of Canada, Paper 63-31.

Drummond, K.J.
1973: Canadian Arctic Islands; in Future Petroleum Provinces of Canada — their Geology and Potential, R.G. McCrossan (ed.); Canadian Society of Petroleum Geologists, Memoir 1, p. 442-472.
1974: Paleozoic margin of North America; in The Geology of Continental Margins, C.A. Burk and C.L. Drake (ed.); Springer-Verlag, Berlin, Heidelberg, New York, p.797-810.

Fahrig, W.F. and West, T.D.
1986: Diabase dyke swarms of the Canadian Shield; Geological Survey of Canada, Map 1627A.

Feilden, H.E. and de Rance, C.E.
1878: Geology of the coasts of the Arctic lands visited by the late British expedition under Captain Sir George Nares, R.N., K.C.B., F.R.S.; The Quarterly Journal of the Geological Society of London, v. 34, p. 556-567.

Fortier, Y.O.
1957: The Arctic Archipelago; in Geology and Economic Minerals of Canada, C.H. Stockwell (ed.); Geological Survey of Canada, Economic Geology Series no. 1 (fourth edition), p. 393-442.

Fortier, Y.O., McNair, A.H., and Thorsteinsson, R.
1954: Geology and petroleum possibilities in Canadian Arctic Islands; Bulletin of the American Association of Petroleum Geologists, v. 38, p. 2075-2109.

Grantz, A., Johnson, L., and Sweeney, J.F. (ed.)
in press: The Arctic Ocean Region; Geological Society of America, The Geology of North America, v. L.

Håkansson, E. and Stemmerik, L.
1983: Wandel Sea Basin — the North Greenland equivalent to Svalbard and the Barents Shelf; in Petroleum Geology of the North European Margin, A.M. Spencer et al. (ed.); Norwegian Petroleum Society (Graham and Trotman), Chapter 7, p. 97-107.

Hoffman, P.F., Card, K.D., and Davidson, A. (ed.)
in prep.: Precambrian Geology of the Craton in Canada and Greenland; Geological Survey of Canada, The Geology of Canada, no. 7 (also Geological Society of America, The Geology of North America, v. C-1).

Kay, M.
1951: North American Geosynclines; Geological Society of America, Memoir 48, 143 p.

Kerr, J. W.
1981: Evolution of the Canadian Arctic Islands: a transition between the Atlantic and Arctic Oceans; in The Ocean Basins and Margins, v. 5, The Arctic Ocean, A.E.M. Nairn, M. Churkin, Jr., and F.G. Stehli (ed.), Plenum Press, New York and London, p. 105-199.

Kerr, J.W. and Christie, R.L.
1965: Tectonic history of Boothia Uplift and Cornwallis Fold Belt, Arctic Canada; Bulletin of the American Association of Petroleum Geologists, v. 49, p. 905-926.

King, P.B.
1969: The tectonics of North America — a discussion to accompany the Tectonic Map of North America, scale 1:5 000 000; United States Geological Survey, Professional Paper 628, 95 p.

Koch, L.
1925: The geology of north Greenland; American Journal of Science, v. 9, p. 271-286.
1935: Geologie von Grönland; in Geologie der Erde, E. Krenkel (ed.); Gebrüder Borntraeger, Berlin, 158 p.
1961: A summary of the geology of North and East Greenland; in Geology of the Arctic, G.O. Raasch (ed.); University of Toronto Press, v. 1, p. 148-154.

Miall, A.D.
1979: Mesozoic and Tertiary geology of Banks Island, Arctic Canada: the history of an unstable cratonic margin; Geological Survey of Canada, Memoir 387, 235 p.

Price, R.A. and Douglas, R.J.W. (ed.)
1972: Variations in Tectonic Styles in Canada; Geological Association of Canada, Special Paper 11, 688 p.

Read, J.F.
1982: Carbonate platforms of passive (extensional) continental margins: types, characteristics and evolution; Tectonophysics, v. 81, p. 195-212.

Schuchert, C.
1923: Sites and nature of the North American geosynclines; Bulletin of the Geological Society of America, v. 34, p. 151-229.

Stuart Smith, J.H. and Wennekers, J.H.N.
1979: Geology and hydrocarbon discoveries of Canadian Arctic Islands; The American Association of Petroleum Geologists Bulletin, v. 61, p.1-27.

Thorsteinsson, R. and Tozer, E.T.
1957: Geological investigations in Ellesmere and Axel Heiberg Islands, 1956; Arctic, v. 10, p. 2-31.
1960: Summary account of structural history of the Canadian Arctic Archipelago since Precambrian time; Geological Survey of Canada, Paper 60-7, 23 p.
1962: Banks, Victoria, and Stefansson Islands, Arctic Archipelago; Geological Survey of Canada, Memoir 330, 85 p.
1970: Geology of the Arctic Archipelago; in Geology and Economic Minerals of Canada, R.J.W. Douglas (ed.); Geological Survey of Canada, Economic Geology Report no.1, p. 547-590.

Trettin, H.P.
1987: Pearya: a composite terrane with Caledonian affinities in northern Ellesmere Island; Canadian Journal of Earth Sciences, v. 24, p. 224-245.

Trettin, H.P. and Balkwill, H.R.
1979: Contributions to the tectonic history of the Innuitian Province, Arctic Canada; Canadian Journal of Earth Sciences, v. 16, p. 748-769.

Trettin, H.P., De Laurier, I., Frisch, T.O., Law, L.K., Niblett, E.R., Sobczak, L.W., Weber, J.R., and Witham, K.
1972: The Innuitian Province; in Variations in Tectonic Styles in Canada, R.A. Price and R.J.W. Douglas (ed.); Geological Association of Canada, Special Paper 11, p. 83-179.

Author's address

H.P. Trettin
Geological Survey of Canada
3303-33rd St. N.W.
Calgary, Alberta
T2L 2A7

Chapter 5
GEOPHYSICAL CHARACTERISTICS

A. Gravity field — *L.W. Sobczak*

B. Crustal structure from seismic and gravity studies — *L.W. Sobczak, D.A. Forsyth, A. Overton, and I. Asudeh*

C. Seismicity — *D.A. Forsyth, H.S. Hasegawa, and R.J. Wetmiller*

D. Aeromagnetic field — *R.L. Coles*

E. Conductivity anomalies — *E.R. Niblett and R.D. Kurtz*

F. Motions of the north magnetic pole — *L.R. Newitt and E.R. Niblett*

G. Heat flow — *A.M. Jessop*

References

Chapter 5

GEOPHYSICAL CHARACTERISTICS

A. GRAVITY FIELD

L.W. Sobczak

INTRODUCTION

The first measurements of gravity in Arctic Canada were made on Melville Island in 1819-1820 (Sabine, 1821). Gravity measurements in the Arctic did not resume until 1957 when five gravity control stations were established by the Canadian Government (Bancroft, 1958), and in 1958 about 200 gravity observations were made over Gilman Glacier in northeastern Ellesmere Island (Weber, 1961). The same year, as a contribution to the International Geophysical Year (IGY), a subtense bar and gravity traverse was carried out from Clements Markham Inlet on the Arctic coast across the United States Range to Archer Fiord on Kennedy Channel (Weber, 1961). In 1958 the Polar Continental Shelf Project (PCSP) was established, in the then Department of Mines and Technical Surveys, to coordinate most of the scientific studies carried out in the Arctic by the Canadian Government.

From 1960, under the aegis of the PCSP, systematic gravity surveys have been carried out in the Arctic by the Dominion Observatory (later renamed the Earth Physics Branch [EPB], of the Department of Energy, Mines and Resources, and now part of the Geological Survey of Canada) primarily over the relatively flat areas of the Innuitian Region and adjacent continental margin (Sobczak, 1963; Sobczak et al., 1963; Picklyk, 1969; Sobczak and Weber, 1970; Sobczak and Stephens, 1974; Sobczak and Sweeney, 1978). Except for small areas completed in the southern, central, and northern parts of Ellesmere Island, gravity surveys of the mountainous areas of Axel Heiberg and Ellesmere islands have been deferred until reliable elevations can be obtained routinely during the survey. The surveyed and unsurveyed areas of Innuitian Orogen and Arctic Platform are indicated in the gravity map, Figure 5A.1 (in pocket).

During 1960 to 1968, about 8800 gravity measurements on land and sea ice were made at a grid interval of about 10 km over the western half of the region, an area of approximately one million square kilometres (Sobczak and Weber, 1970). During 1967 about 450 gravity measurements were taken on land at the same grid interval over the southern end of Ellesmere Island and the eastern two-thirds of Devon Island (Picklyk, 1969). During the period 1957-1967 more than 700 gravity observations were taken at variable intervals on land, glaciers, and sea ice in the northeastern part of the region (northeastern Ellesmere Island and northern Greenland; Sobczak and Stephens, 1974). The Geodetic Institute of Denmark carried out regional land gravity surveys in North Greenland during 1978-1980 (Forsberg, 1979, 1981; Weng, 1980) and in eastern Greenland in 1982, making in all about 600 gravity observations.

Gravity surveys by EPB in the Arctic Platform, south of M'Clure Strait, Viscount Melville Sound, and Lancaster Sound are described in the following reports: (1) Somerset and Prince of Wales islands: Berkhout and Sobczak, 1967; (2) Southampton Island, Melville Peninsula, King William Island and Boothia Peninsula: Gibb and Halliday, 1975; (3) southern Victoria Island and the mainland to the south: Hornal and Boyd, 1972; (4) Banks Island: Stephens et al., 1972; (5) Beaufort Sea, Banks Island and Mackenzie Delta: Sobczak et al., 1973; (6) northern Baffin Island, Somerset Island, and Prince of Wales Island: Berkhout, 1970. Additional surveys were carried out over Baffin Island in 1962, over eastern and western sides of Victoria Island in 1972 and 1975, over Amundsen Gulf in 1973, 1974, and 1976, and over western and eastern parts of Viscount Melville Sound in 1977 and 1978. The methods of survey are similar to the procedures described in the above reports and the results are shown on the Gravity Map of Canada (Earth Physics Branch, 1980). Since that time D. Halliday of the EPB has completed the gravity surveys of northern and southern M'Clintock Channel in 1980 and 1981, Prince of Wales Strait between Banks and Victoria islands in 1982, and northern Prince Regent Inlet between Somerset and Baffin islands in 1984.

GRAVITY ANOMALIES

Gravity anomalies, Bouguer on land and free-air offshore, are shown in Figure 5A.1 (in pocket). These anomalies were reduced to sea level using a standard density of 2.67 Mg/m^3, the Geodetic Reference System 1967 (1971), and the International Gravity Standardization Net (Morelli et al., 1974).

Gravity anomalies have a range of over 200 mGal from "highs" of about +100 mGal along the shelf break northwest of the Queen Elizabeth Islands and around the northern end of Baffin Bay, to "lows" of below -100 mGal over the mountainous areas of northwestern Greenland, central

---

Sobczak, L.W.
1991: Gravity field; *in* Chapter 5 of Geology of the Innuitian Orogen and Arctic Platform of Canada and Greenland, H.P. Trettin (ed.); Geological Survey of Canada, Geology of Canada, no. 3; (*also* Geological Society of America, The Geology of North America, v. E).

Ellesmere Island, and Baffin Island, and water-covered basin areas such as Kane Basin and Lancaster Sound. Accuracy of these anomalies varies from about 0.2 mGal for detailed surveys such as along a line between Melville Island and Axel Heiberg Island to about 2 mGal for most regional surveys over relatively flat areas (Sobczak and Overton, 1984). However, in the mountainous areas such as those of Ellesmere Island and Baffin Island where terrain corrections have yet to be applied, uncertainties of up to 30 mGal can be expected.

GENERAL CHARACTER OF THE GRAVITY FIELD

Gravity anomalies indicate areas of mass excesses ("highs") and mass deficiencies ("lows") with respect to a standard crust.

The most important gravity "highs" can be summarized as follows. (1) A series of elongated "highs" overlie the shelf break to the northwest of the Queen Elizabeth Islands and around the northern end of Baffin Bay. (2) Pronounced circular to oval singular "highs" are located near the coastal regions such as at the southeastern end of Devon Island, eastern side of Bylot Island, south of Darnley Bay (on the mainland south of Amundsen Gulf), southwestern end of Victoria Island, and at the head of Prince Albert Sound on the west side of Victoria Island. They are usually associated with mass excesses (dense intrusive bodies) within the crust. (3) East- to northeast-trending belts of gravity "highs" of lesser amplitude than the ones described above overlie the Sverdrup Rim (cf. Chapters 5B, 13, 14) and Parry Islands Fold Belt (Melville Island and Bathurst Island; cf. Fig. 4.3; Chapter 12G). (4) North-trending belts of "highs" occur over uplifted and folded areas such as the Boothia Uplift (Fig. 4.3; Chapter 12C), the Cornwall Arch (Chapter 17, Fig. 17.1) and a structure in the Prince Gustaf Adolf Sea – Lougheed Island region referred to as the "Gustaf-Lougheed Arch" by Forsyth et al. (1969) and Hea et al. (1980, Fig. 7; name not adopted in this volume).

Gravity "lows" generally coincide with basin, channel, and mountainous areas. They are present, for example, over sub-basins of the Sverdrup Basin north of Melville Island (the "Sabine Basin" of Hea et al., 1980, Fig. 3; name not adopted in this volume), between Bathurst and Ellef Ringnes islands ("Ellef-Edinburgh Basin") and south and southwest of Eureka Weather Station, Ellesmere Island ("Eureka Basin"); and over Cretaceous-Tertiary basins in Lancaster Sound (cf. Chapter 17).

The negative anomalies in the mountainous areas are caused in part by terrain effects, which have yet to be calculated, and by the removal of the effect of land mass above sea level in the Bouguer anomaly reduction. As a result they are not necessarily indicative of anomalous zones in the gravity field and are not discussed any further here. Anomalous zones in these areas would be better defined by isostatic residual anomalies such as are presently being calculated for the United States of America (R.W. Simpson, pers. comm., 1983).

Negative free-air anomalies over the channel and basin areas in part reflect the mass deficiency of the water, which for small areas is not isostatically compensated. These channels are up to 500 m deep in places, producing a gravity effect of up to 34 mGal.

ISOSTASY

Gravity anomalies are affected by mass distributions within the crust and by isostasy. Regionally, isostatic conditions prevail for the inter-island and coastal regions but locally, areas of disequilibrium exist as is indicated by local gravity "highs" and contemporary seismicity along the shelf break and over folded and uplifted regions such as the Sverdrup Rim, "Gustaf-Lougheed Arch" (see above), and Boothia Uplift (Sobczak et al., 1986). For the western Queen Elizabeth Islands, Sobczak and Weber (1973) determined an average free-air anomaly of 7.3 mGal, equivalent to -2.7 mGal if corrected to the current gravity datum (International Gravity Standardization Net; Morelli et al., 1971; International Association of Geodesy, 1974). This value indicates regional isostatic equilibrium and shows that the vast thickness of sedimentary rocks (up to 18 km south of Ellef Ringnes Island), which would cause gravity to be lowered by about 120 mGal, is regionally compensated by crustal thinning by as much as 7 km (Sobczak and Overton, 1984). The regional gravity anomaly level for the Arctic Platform immediately south of the Innuitian Orogen appears to be similar to that of the western Queen Elizabeth Islands and also indicates isostatic equilibrium. Farther south, over the Canadian mainland, the gravity anomalies become progressively more negative with elevation, which indicates some form (Airy or Pratt) of isostatic compensation. To the northwest, over the Canadian continental margin, near-equilibrium conditions are indicated; a mean free-air anomaly of +27 mGal (Trettin et al., 1972, p. 153) recalculated to the new datum would be +17 mGal. This value can be reduced to 10 mGal if the continental margin edge effect (about 7 mGal; Sobczak, 1975) is taken into consideration. The remaining +10 mGal may indicate an additional load which is not fully compensated. Along the seaward side of the continental margin, the mass deficiency for the water (over 3 km deep) and sediments (up to 7 km thick) has a total gravity effect of up to -360 mGal and has been compensated by crustal thinning (Sobczak et al., 1986).

Bathymetry also indicates areas of disequilibrium. The shelf break from the Beaufort Sea northeastward to Lincoln Sea varies from a depth of 200 m off the Mackenzie Delta, to 500 m northwest of Banks Island, M'Clure Strait and Prince Patrick Island, to 650 m northwest of Ellef Ringnes Island and back to about 500 m northwest of Ellesmere Island. In addition, there are deep (in excess of 500 m) troughs along and seaward of the fiords on the northeastern coast of Ellesmere Island (CESAR Project, J.R. Weber, pers. comm., 1984), as well as deep troughs within the inter-island areas (Johnson et al., 1979). According to Pelletier (1966) these waterways represent a drowned early Tertiary fluvial system that has been smoothed and fluted by later glaciation (cf. Chapter 18). If the equilibrium level along the shelf break is taken as 200 m (a world average value — Worzel, 1968) then the shelf has been drowned by 300 to 450 m. Sedimentation has not kept pace with subsidence, indicating a region starved for sediments.

ACKNOWLEDGMENTS

Thanks are extended to D. Cleary for contouring the gravity data, to J. Halpenny for providing colour Applicon gravity maps, and to J.R. Weber, R.K. McConnell, D.B. Hearty, A.M. Jessop, and M.R. Dence for reading and commenting on the manuscript.

B. CRUSTAL STRUCTURE FROM SEISMIC AND GRAVITY STUDIES

L.W. Sobczak, D.A. Forsyth, A. Overton, and I. Asudeh

INTRODUCTION

Studies of the structure of the crust down to the Mohorovičic discontinuity have to date been confined to the southwestern part of the Innuitian Orogen and parts of the Arctic Platform, where five deep seismic refraction surveys (Fig. 5B.1, lines 1-5) and extensive gravity surveys have been carried out. The present discussion is concerned mainly with the most recent interpretations of the refraction lines 2 and 3, and associated gravity analyses, and a 1300 km long geological and geophysical transect, partly based on lines 1, 2, and 3. Also mentioned are preliminary results from a refraction seismic survey conducted on the continental shelf northwest of Axel Heiberg Island in 1985.

Earlier interpretations of deep crustal structure, not reviewed here, include two extensive cross-sections, roughly parallel with the transect, that are based on gravity surveys in conjunction with lines 1 and 4 (Sobczak and Weber, 1973), and a profile across Banks Island and the adjacent continental margin, based on gravity data alone (Sobczak, 1975). Structures in the upper crust and supracrustal sedimentary successions are discussed in later chapters.

DEEP SEISMIC REFRACTION SURVEYS

During 1962 and 1963 Sander and Overton (1965) prepared three refraction seismic profiles, which formed a continuous section from Cornwallis Island to northwesternmost Ellef Ringnes Island (Fig. 5B.1, line 1 and Fig. 5B.2). The base of the crust was determined to be at a depth of about 38 km. The crustal succession for the Sverdrup Basin was divided into a lower "basaltic" crust (v = 7.3 km/s) and an undifferentiated layer with a seismic velocity of 6.0 km/s that includes Precambrian granitic basement, Proterozoic basinal sediments, and rocks belonging to the Franklinian mobile belt. The Sverdrup Basin was found to be about 10 km thick in the axial region.

During 1964 and 1965, Overton (1970) carried out a series of refraction surveys extending from Melville Island to Prince Patrick, Mackenzie King and Brock islands, and over the continental margin (Fig. 5B.1, line 4). He found high-velocity layers within the sedimentary section, which made estimates of the velocity and depth to the underlying basement complex uncertain. The inferred crustal thickness varied from 29 to 42 km. The mantle was found to be overlain by an "intermediate" crustal layer with velocities of 6.1-6.25 km/s, the top of which generally occurs 10-20 km below the surface.

In 1967 a seismic refraction experiment was conducted by Berry and Barr (1971) (Fig. 5B.1, line 5 and Fig. 5B.3) for 220 km northwest across the continental shelf and slope from the southwest tip of Prince Patrick Island. They estimated the Mohorovičic discontinuity at a depth of 28 ± 4 km at the coast, rising to 15 ± 9 km at the northeast end of the profile. Midway down the continental slope, about 5 km of sediments lie on a basement complex.

Two seismic profiles, one extending from northeastern Melville Island to Amund Ringnes Island (Fig. 5B.1, line 3) and the other along the east coast of Amund Ringnes Island (line 2), were established by Overton in 1972-1974. These results were published by Overton (1982) who used the shallow, reversed refraction data in a model with velocities gradually increasing with depth, and determined delay time-velocity profiles showing five velocity layers at intervals of 0.75 km/s for the sediment cover. Analysis of the deep seismic portion of the study by Forsyth et al. (1979) revealed crustal thickness variations from 40 km beneath eastern Sabine Peninsula to 32 km west of King Christian Island, but did not support a distinct mid-crustal layer or the presence of a Conrad discontinuity.

INTERPRETATION OF SEISMIC AND GRAVITY INFORMATION IN RECENT STUDIES

Lines 2 and 3

Gravity data along lines 2 and 3 were analyzed using the seismic refraction information from Overton (1982) and Forsyth et al. (1979) and density data from logs of 17 drill holes near the profile lines. The data were reduced by successively removing the gravity effects due to the water, the five shallow sedimentary layers, and the thickness variations of the crust (Sobczak and Overton, 1984). The analysis showed that negative gravity effects of about 60-120 mGal, due to the mass deficiency of water and sediments (upper three layers approximate the Sverdrup Basin and lower two layers the underlying Franklinian mobile belt), are offset by positive gravity effects of similar magnitude due to crustal thinning (Fig. 5B.4, 5B.5). Calculation of the

---

Sobczak, L.W., Forsyth, D.A., Overton, A., and Asudeh, I.
1991: Crustal structure from seismic and gravity studies; in Chapter 5 of Geology of the Innuitian Orogen and Arctic Platform of Canada and Greenland, H.P. Trettin (ed.); Geological Survey of Canada, Geology of Canada, no. 3; (also Geological Society of America, The Geology of North America, v. E).

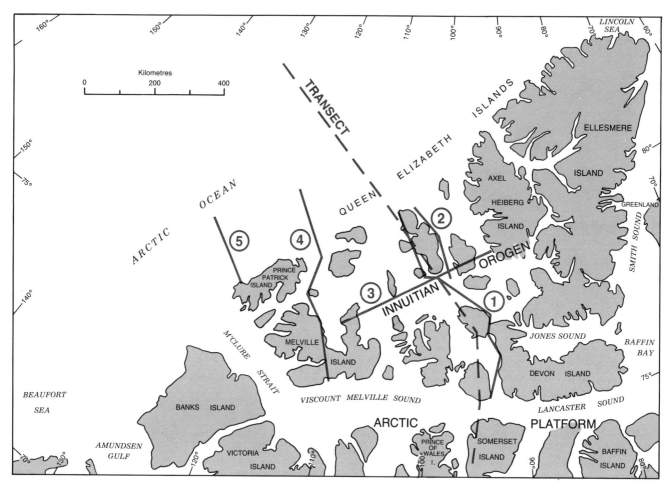

Figure 5B.1. Location of the deep seismic refraction surveys and the transect discussed in the text.

gravity effects of the seismic model left residual gravity anomalies that could be explained by low density evaporites, high density mafic intrusions and density variations due to lithological changes, features that characterize the known geology of the sedimentary section. No anomalous masses within the crystalline crust were required to account for the residual gravity anomalies. These studies suggested that the sediment thickness varies inversely with crustal thickness.

Transect and data from northwest of Axel Heiberg Island

A crustal-scale section, extending from Somerset Island on the Arctic Platform to the Arctic Ocean (Fig. 5B.1), has been compiled in Figures 5B.6 and 5B.7 (from Sobczak et al., 1986, and Sweeney et al., 1986). It is based on deep seismic refraction surveys (Sander and Overton, 1965; Forsyth et al., 1979; Overton, 1982), shallow refraction and reflection surveys (Hobson and Overton, 1967; Meneley et al., 1975), gravity studies (Sobczak, 1963; Sobczak and Weber, 1973; Sobczak and Overton, 1984), well data, surface geology, and proprietary information from Panarctic Oils Ltd. (H.R. Balkwill, pers. comm., 1983). The section is accompanied by plots of: (1) the observed gravity anomaly; (2) the total gravity effect of the sedimentary layers (from shallow seismic data); (3) the gravity effect of the mantle (from deep seismic data); (4) the calculated anomaly (sum of [2] and [3]); and (5) the residual anomaly ([1] – [4]), which can be explained by structures within the sediments. Homogeneous continental and oceanic crystalline crusts with constant densities are assumed. Generally the mass deficiency of the surface layers is assumed to be isostatically compensated by mantle excess due to crustal thinning. The transect is best controlled where it intersects the refraction survey of Forsyth et al. (1979); the uncertainty about the crustal thickness here is in the order of 10%.

The depths to the top of the mantle, calibrated from seismic refraction data where line 3 intersects the transect at a control depth of 41 km, were calculated from gravity data assuming Airy isostatic compensation, and found to be 48 km for the areas of the Canadian Shield and the Sverdrup Rim, 48-43 km for the "Franklinian Basin" (Franklinian mobile belt of Chapter 4), 41-37 km for the Sverdrup Basin, and 16.5 km for the present continental shelf margin, where the transition from continental to oceanic crust appears to begin according to the gravity-density model.

The following major crustal successions are recognized (Fig. 5B.6, 5B.7): Archean continental crust; a Proterozoic basin (layer 15) that predates the "Franklinian Basin"; the Cambrian and possibly Upper Proterozoic to Devonian

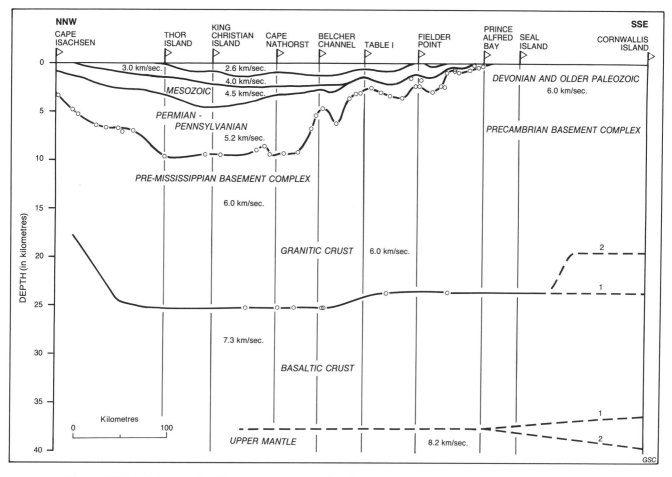

Figure 5B.2. Seismic refraction cross-section of Sander and Overton (1965) (Fig. 5B.1, line 1).

"Franklinian Basin" (layers 14-13); the Sverdrup Basin (layers 12-3); and the Tertiary to Recent "Canadian Arctic Margin Basin" (Arctic Continental Terrace Wedge of Chapter 4). It must be emphasized that Precambrian basement, Proterozoic basin and Franklinian mobile belt have not been distinguished in the seismic refraction surveys; their distinction in the transect is based on gravity models developed from geological interpretations. These interpretations are necessarily speculative for the northern parts of the transect, which are far removed from relevant outcrops; it is conceivable, for example, that the layers labelled "Franklinian Basin" and "Archean Basement" include exotic terranes below the present "Queen Elizabeth Shelf" (Arctic Islands Shelf of Chapter 3) and Sverdrup Rim.

The present model assumes that the Archean crystalline basement of the Precambrian Shield extends northward to the limit of the continental crust; its thickness is inferred to decrease from 48 km for the exposed Canadian Shield to 8 km at its northern limit below the shelf in the transitional zone. This is geometrically equivalent to a "stretch factor" of 1.5, but several phases of stretching and orogenic compression have occurred. Superimposed upon this general trend is thinning beneath the southernmost part of the "Franklinian Basin" (Barrow Strait) and below the axial region of the Sverdrup Basin (Ellef Ringnes Island). The Sverdrup Basin is about 10 km deep in this region. Its arched northern rim is greatly intruded by mafic rocks, from the base of the basin to the surface. The entire succession appears to be truncated laterally below the continental shelf, where a prominent belt of gravity highs (up to 100 mGal) marks the transition zone between continental and oceanic crust.

Preliminary results from the 1985 Canadian Ice Island refraction survey over the Queen Elizabeth Shelf, from about 100 km to 145 km northwest of Axel Heiberg Island, indicate from 0.7 to 4.5 km of 2.1 km/s unconsolidated sediments, overlying 8 km or more of 4.4 km/s consolidated sediments. These sediments overlie lower crustal material with a velocity of about 6.2 km/s down to a Mohorovičic discontinuity near a depth of 25 km at about 145 km offshore. The upper crustal section is similar in dip and thickness to the section offshore of Ellef Ringnes Island in Figure 5B.7. The total crustal thickness at 145 km off Axel Heiberg Island is comparable to the thickness at 125 km off Ellef Ringnes Island. This difference is an indication of lateral variation in crustal thicknesses along the shelf, which is reflected by variations in the gravity field (Sobczak and Weber, 1987).

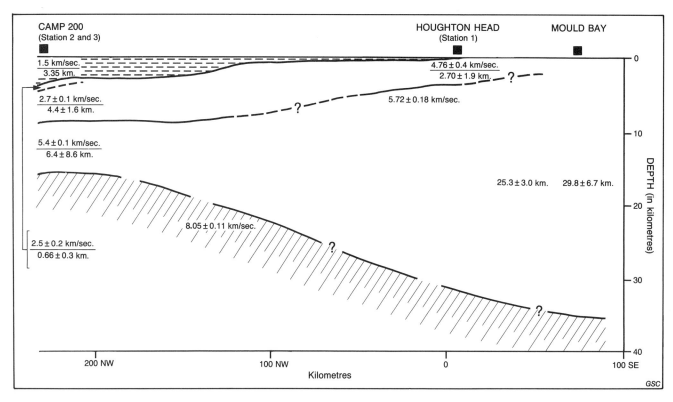

Figure 5B.3. Seismic refraction cross-section of Berry and Barr (1971) (Fig. 5B.1, line 5).

SEISMIC REFLECTION SURVEYS

Apart from the transect (Fig. 5B.6, 5B.7), seismic reflection surveys, carried out by oil companies, have not been included in this discussion because the subject is more closely related to other chapters, dealing with shallow structure and stratigraphy. However, the available data are briefly mentioned here for general information.

To date over 65 000 km of reflection seismic lines have been shot and seismic records, geophysical reports, and location maps are available from the Canadian Oil and Gas Administration (COGLA) of the Federal Government. Although data have been obtained from most islands and intervening ice-covered channels, the majority are from the Sverdrup Basin in the central and western part of the region (about 40 000 km). Line spacing varies greatly and is closest (3-5 km) in the area between Sabine Peninsula, Melville Island, and western Ellef Ringnes Island where most hydrocarbon fields occur. Data quality is usually good in the upper 3-4 seconds, and, in general, better on land than over channels.

This vast data base is presently being interpreted by the Geological Survey of Canada, but only a small proportion of the information emerging now (mainly that on the Parry Islands Fold Belt [Chapter 12G]) was available when this volume was written. A few seismic reflection lines have been published by Meneley et al. (1975), Meneley (1977), Waylett (1979), Hea et al. (1980), Rayer (1981), Embry (1982), Texaco Canada Resources Ltd. (1983), and Fox (1983, 1985).

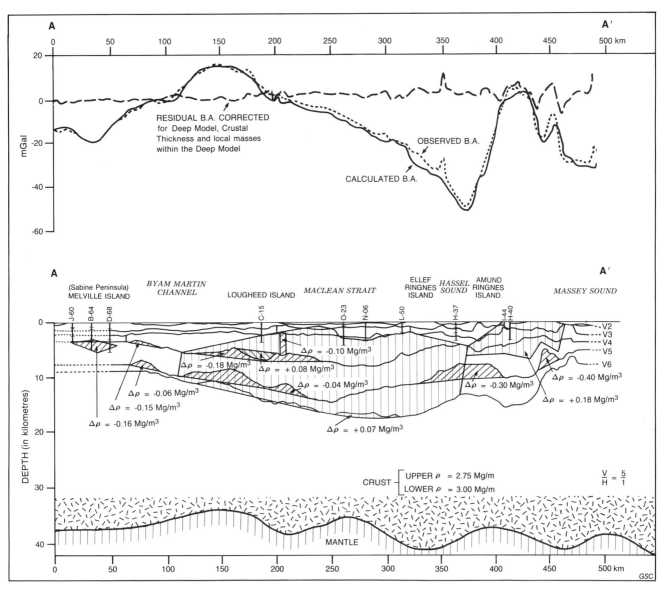

Figure 5B.4. Gravity-density shallow and deep cross-section by Sobczak and Overton (1984, Fig. 12) based partly on Forsyth et al. (1979) (Fig. 5B.1, line 3).

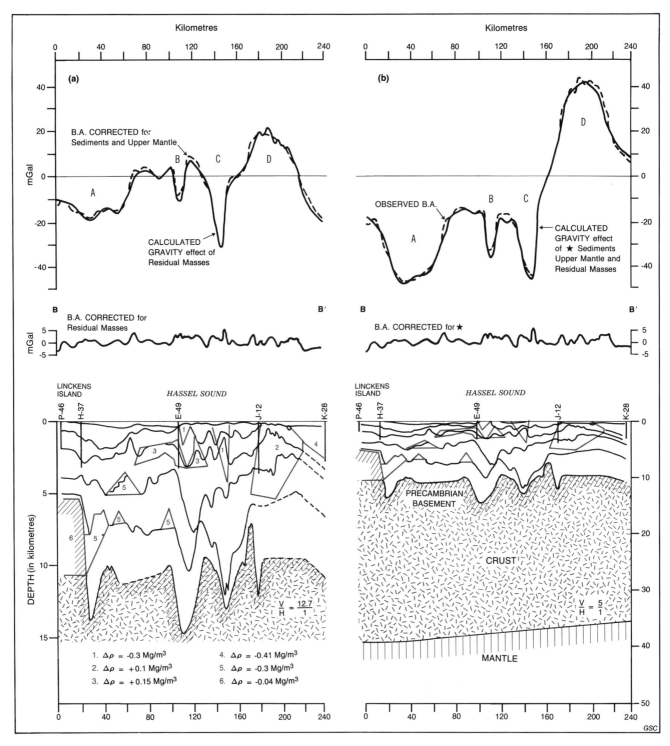

Figure 5B.5. Gravity-density and shallow and deep cross-section by Sobczak and Overton (1984, Fig. 15) (Fig. 5B.1, line 2).

Figure 5B.6. Southern part of the transect, showing gravity, topography and bathymetry and density by Sobczak et al. (1986, Fig. 3). Layers 13 and 14 represent the "Franklinian Basin", layer 15 represents the Proterozoic basin. The Archean basement between A and B is lower in density, possibly the result of fracturing.

GEOPHYSICAL CHARACTERISTICS

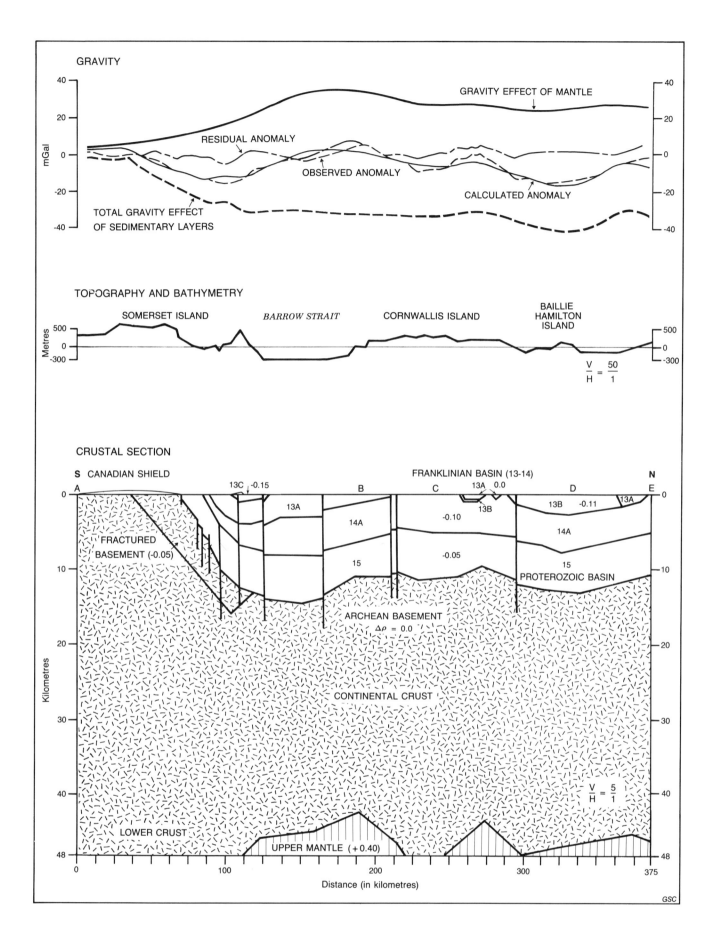

CHAPTER 5

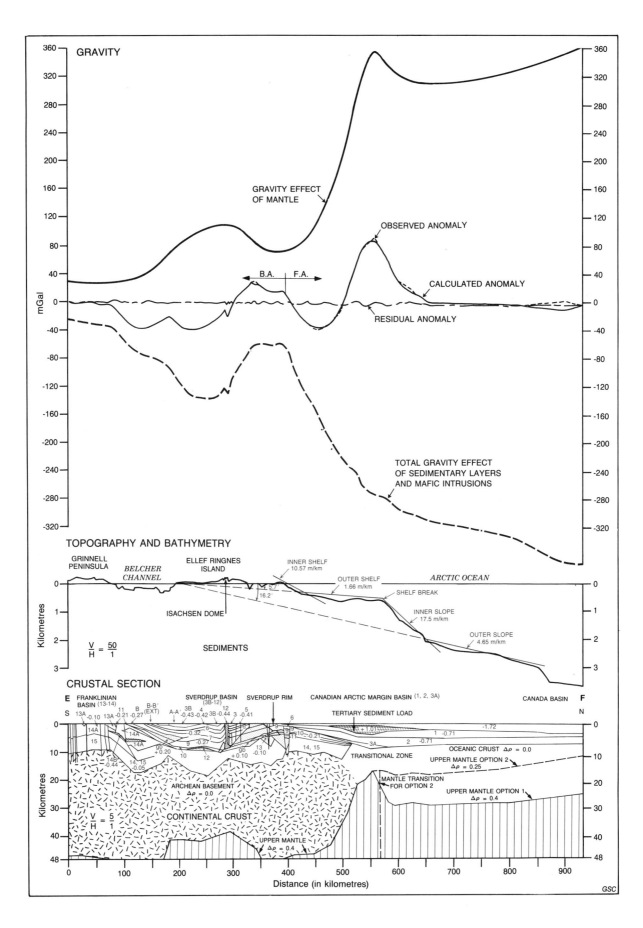

C. SEISMICITY

D.A. Forsyth, H.S. Hasegawa, and R.J. Wetmiller

INTRODUCTION

The Canadian Arctic Archipelago and adjacent offshore area continue to manifest the "surprising" number of local events first noted by Milne and Smith (1961) from seismograms recorded at the Resolute Bay station (RES in Fig. 5C.1). Available seismic information has been listed by Meidler (1961) for the years 1899-1955, Smith (1961) for the years 1956-1959, and in annual catalogues of the Earth Physics Branch (Department of Energy, Mines and Resources) for the last 23 years. The distribution of epicentres of seismic events with a magnitude (M) greater than 3 is shown in Figure 5C.2, superimposed on the geology of the Canadian arctic and subarctic region in general. Two points should be noted with respect to the main epicentral trends evident on this map. Firstly, the trends are controlled by a lower detection-location threshold of about M3.5-4 in general, although some more densely instrumented areas such as the southern Beaufort Sea have a lower detection level near M3. And secondly, the data represent only a 13 year snapshot, whereas controlling tectonic forces operate on much longer time scales. The following discussion is restricted to two areas of fairly active seismicity, the Boothia Uplift and the Sverdrup Basin, and also to Nares Strait, which is of special tectonic interest.

BOOTHIA UPLIFT

The Boothia Uplift, which trends for about 1000 km northward from the Canadian Shield through the Arctic Platform into the Franklinian mobile belt, was tectonically active mainly in Late Silurian – Early Devonian time (Chapter 12C). At present the uplift is characterized by a much higher level of seismic activity than adjacent areas, the concentration of earthquakes being highest in Parry Channel where crossfaults may be present (Fig. 5C.3). The seismicity continues from the Boothia Uplift along the Bell Arch (BA in Fig. 5C.2) to northeastern Hudson Bay. It is attributed to reactivation of ancient zones of weakness by differential postglacial uplift and neotectonic stresses (Okulitch et al., 1986).

Figure 5B.7. Northern part of transect, showing gravity — Bouger anomalies (B.A.) landward from the Arctic Ocean and free-air anomalies (F.A.) seaward; total gravity effect of water, sediments and intrusions; gravity effect of compensation of the mantle above 48 km; topography, bathymetry, and crustal section to a depth of 48 km. Density contrasts are in Mg/m^3. The Tertiary sediment load and layers 1, 2, and 3A form the "Canadian Arctic Margin Basin"; layers 3-12 the Sverdrup Basin; and layers 13 and 14 the "Franklinian Basin". Combined layers 14-15 may be a mix of Paleozoic and Proterozoic strata with possible inclusions of fractured crystalline basement rocks. Seaward crustal thinning is based on two alternative options: option 1 assumes a constant density contrast (0.40 Mg/m^3) between mantle and lower crustal rocks; option 2 assumes a thin oceanic crust, after Mair and Lyons (1980) and Sobczak et al. (1986).

SVERDRUP BASIN

Seismicity

Seismicity within the Sverdrup Basin is characterized by low to moderate magnitude, with occasional swarm activity (Fig. 5C.2, 5C.4; for geographic names see Fig. 1 and 5A.1, in pocket). Currently, most seismic activity is in the west-central part of the basin. Intense low-magnitude swarms of earthquakes occurred on Prince Patrick Island in 1965 (Smith et al., 1968). There are earthquake clusters on northwestern Ellesmere Island and on western Axel Heiberg Island where an M5.2 event took place in 1975.

From 1962 to 1972 the region northeast of Melville Island had an earthquake occurrence rate of about one to nine per year. However, from November 1972 to January 1973, 74 events were located, nine of which had magnitudes greater than 4.5. More events were detected at Resolute Bay but were not locatable. Since 1973 the rate of activity has declined, but is continuing to the present with the recent addition of more events both in the region of the original swarm and also to the north along an en echelon continuation of a structural high in the Prince Gustaf Adolf Sea – Lougheed Island region, referred to as the "Gustaf-Lougheed Arch" by Forsyth et al. (1979) and Hea at al. (1980; cf. p. 452-463; see Fig. 5A.1).

Seismotectonics

The southwestern part of the Sverdrup Basin (including the "Gustaf-Lougheed Arch") is characterized by northeast-striking normal faults, evaporite diapirs, and magnetic anomalies, the last probably caused by dykes of Late Cretaceous and older ages (cf. Chapter 14). Combined, these elements represent a "long-lasting domain of crustal dilation", especially during the Carboniferous to Cretaceous (Balkwill and Fox, 1982). A seismic refraction survey across the "Gustaf-Lougheed Arch" (Forsyth et al., 1979), together with earthquake swarm activity showed a zone, the incipient rift of Balkwill and Fox (1982), extending to lower crustal depths, where seismic energy is greatly attenuated.

Fault-plane solutions of four of the larger events in the Byam Martin Channel earthquake swarm in November and December of 1972 all indicate dextral strike-slip motion in a northeasterly direction on steeply-dipping faults (Hasegawa, 1977). Focal depths of the four larger events range from 10 to 30 km and magnitudes from 5.1 to 5.7. The directions of deviatoric tensional stress from nodal plane solutions of four of the largest earthquakes in the swarm (Hasegawa, 1977) are oriented about normal to magnetic anomalies, i.e. parallel to the direction of dilation in Cretaceous and earlier time. Relatively high heat flow over the region of the "Gustaf-Lougheed Arch" (A. Judge, pers.

Forsyth, D.A., Hasegawa, H.S., and Wetmiller, R.J.
1990: Seismicity; in Chapter 5 of Geology of the Innuitian Orogen and Arctic Platform of Canada and Greenland, H.P. Trettin (ed.); Geological Survey of Canada, Geology of Canada, no. 3; (also Geological Society of America, The Geology of North America, v. E).

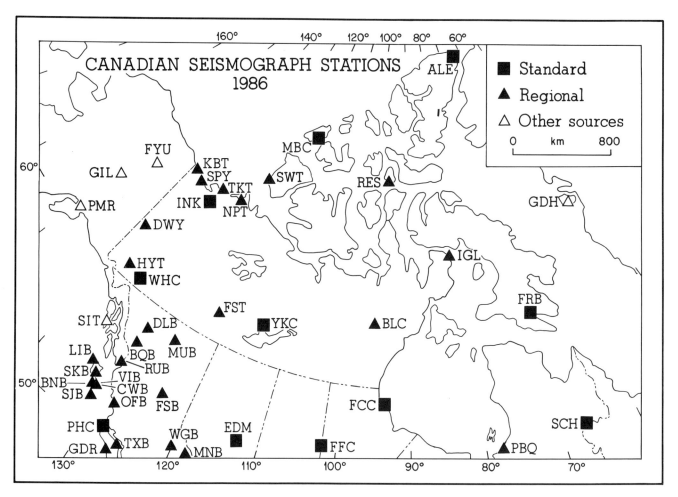

Figure 5C.1. Canadian seismograph stations, 1986.

comm., 1984) is in agreement with relatively shallow sources of stress adjustment and earthquake focal depths.

The occurrence of earthquake swarm activity in regions which appeared essentially aseismic prior to about 1970 shows that our understanding of the seismic activity may be rather incomplete in the western Sverdrup Basin.

NARES STRAIT

Nares Strait, the linear channel separating Greenland from eastern Ellesmere Island, has been a topic of tectonic debate in recent years (Chapter 17). On the one hand, it had generally been assumed that middle Cretaceous to early Oligocene seafloor spreading in the Labrador Sea was accommodated by perhaps 200 km or more of sinistral strike slip along the strait. On the other hand, correlation of stratigraphic-structural markers on either side suggested that the postulated offset, if present at all, does not exceed a few tens of kilometres (Dawes and Kerr, 1982). Likewise, neither geological evidence nor the isobase curves in Figure 5C.5 indicate Nares Strait has been the locus of significant vertical movement.

Available epicentre information indicates that the channel has been aseismic at least for the last two decades or so, as shown in Figure 5C.5 (Wetmiller and Forsyth, 1982; Gregersen, 1982). Prior to about 1960, sparse information is available for magnitudes less than 5, and epicentral uncertainties are of the order of a few hundred kilometres. However, by the late 1960s, events with magnitudes as low as 3 3/4 are almost completely detected and epicentre uncertainties are less than 50 km (Basham et al., 1977; Gregersen, 1982).

Fault-plane solutions of two earthquakes to the south (in northern Baffin Bay) and of one earthquake to the north, in the Lincoln Sea, indicate approximately N-S deviatoric compression. On the basis of this observation, Wetmiller and Forsyth (1982) suggest that the intervening aseismic Nares Strait could be experiencing N-S compression, that is, in a direction subparallel to the strait, but point out that the hypocentres may not lie within the Nares Strait stress regime.

Thus, if a fault zone of Late Cretaceous – early Tertiary age indeed underlies the strait, it is not being reactivated at present by local glacial adjustment or regional tectonic activity. It is possible that the present aseismic regime may be a quiescent period in a non-stationary seismic history.

Alternatively, the difference in the nature of Lg propagation between northern Baffin Bay to the north across Nares Strait and west to Resolute Bay, suggests the crust beneath Nares Strait is not characterized by a major

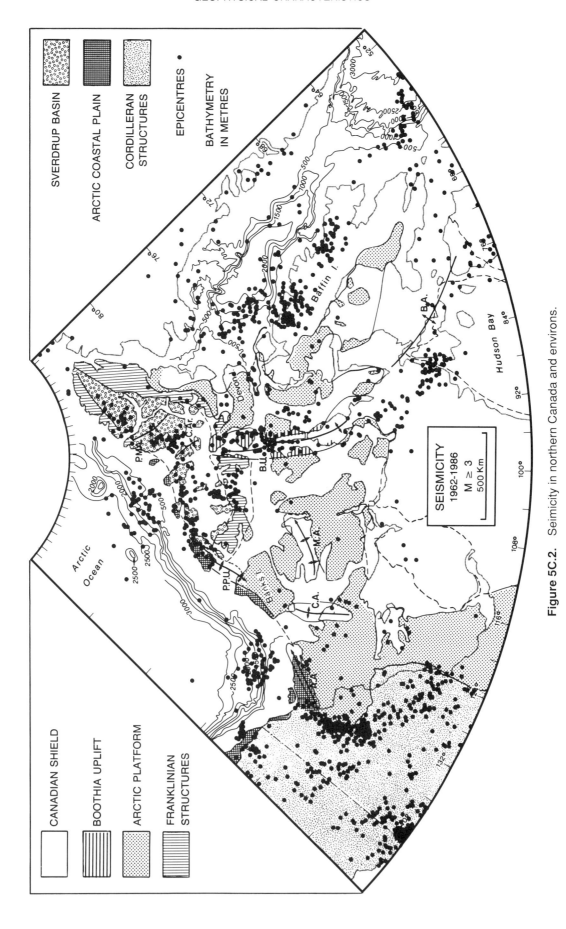

Figure 5C.2. Seismicity in northern Canada and environs.

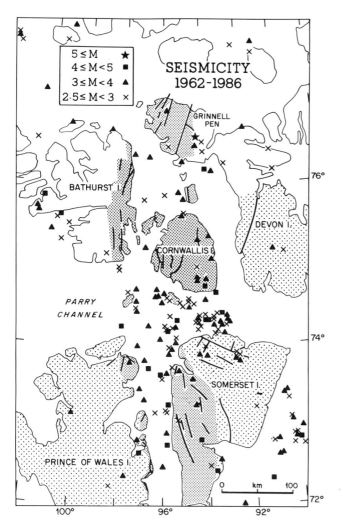

Figure 5C.3. Seismicity of the Boothia Uplift. (Boothia Uplift: fine stippling; Arctic Platform: coarse stippling)

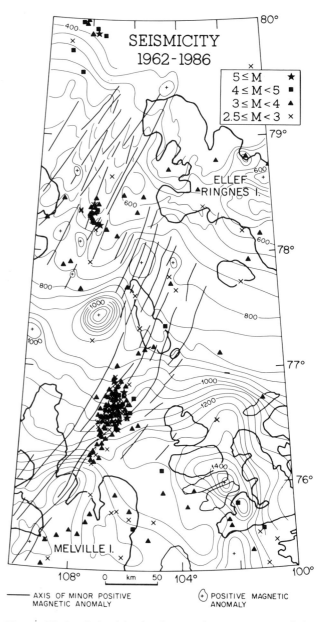

Figure 5C.4. Seismicity in the southwestern part of the Sverdrup Basin.

discontinuity. Rather, such a feature is more likely beneath Jones or Lancaster sounds to the west. Accordingly, the lithospheric adjustment required by North Atlantic – North America plate reconstructions (Srivastava, 1985) must then be accommodated along Jones Sound, Lancaster Sound, and structures in the eastern Sverdrup Basin (Wetmiller and Forsyth, 1982).

ACKNOWLEDGMENTS

We wish to thank reviewers from the Geophysics Division, the Lithosphere and Canadian Shield Division, and the Institute of Sedimentary and Petroleum Geology for their most constructive comments, and J. Drysdale for very expediently producing updated seismicity maps.

GEOPHYSICAL CHARACTERISTICS

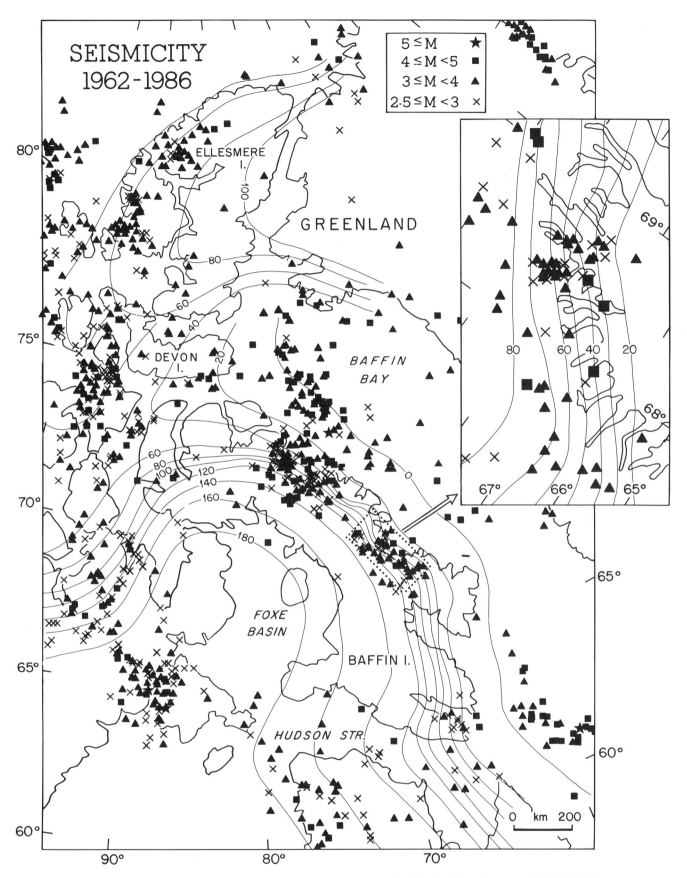

Figure 5C.5. Seismicity and glacial isobases in the eastern part of the Innuitian Orogen and Arctic Platform and adjacent parts of Canadian Shield and Baffin Bay.

CHAPTER 5

D. AEROMAGNETIC FIELD

R.L. Coles

AN INTRODUCTORY PERSPECTIVE

The magnetic anomaly field over the Innuitian Orogen and Arctic Platform will be considered on a broad scale to relate the region as a whole to its surroundings, and then on a more localized scale to study in more detail features specific to the region itself.

On the broadest scale, maps exist that have been derived from POGO and MAGSAT satellite data (Langel et al., 1980; Langel and Thorning, 1982a; Coles et al., 1982; Coles, 1985; Haines, 1985). Figure 5D.1 shows an extract from a MAGSAT map from Haines (1985). These maps show that, regionally, the Innuitian Tectonic Province is magnetically low, i.e. its net crustal magnetization is below the global average, as is also the southern Canada Basin (probably oceanic crust). In contrast, broad regions of high magnetic field exist over northern Greenland, the Alpha Ridge of the Arctic Ocean, and parts of the Canadian Shield to the south and west. These differences can result from changes in the depth of the Curie isotherm (normally taken as the temperature above which magnetite, the commonest magnetic mineral, loses its magnetization) and thus in the thickness of the "magnetic crust". Alternatively, they can result from significant compositional differences (i.e. the amount of magnetite present) from region to region in the crust shallower than the Curie isotherm. Known great depths of sedimentary rocks (which tend usually to be low in magnetite) in the Innuitian region, particularly in the Sverdrup Basin, contrast with the crystalline rocks (often high in magnetite) of the adjacent shield regions (Sander and Overton, 1965).

On a somewhat more detailed scale, the high-altitude surveys by the Earth Physics Branch, Ottawa, (Riddihough et al., 1973; Haines and Hannaford, 1974; Coles et al., 1976) and by the U.S. Coast and Geodetic Survey (King et al., 1966) again demonstrate the relatively low relief of Innuitian magnetic anomalies compared to that over the Precambrian Shield to the south and over the Alpha Ridge and northern Canada Basin to the north.

These broad scale surveys, by satellite and high-altitude aircraft, provide the regional setting for more detailed surveys at lower flight altitudes and closer line spacings. The Geological Survey of Canada began these surveys in 1955 with a 14 000 line-kilometre reconnaissance program, covering the Arctic Islands with a network of lines centred at Resolute Bay (Gregory et al., 1961a, 1961b). Some of the survey tracks flown by the University of Wisconsin (Ostenso, 1963; Ostenso and Wold, 1971) also cross parts of the Innuitian Province. In 1961, the Canadian government initiated a program of aeromagnetic surveys over the Arctic Islands and continental shelf, using Decca navigational chains, and flight lines ranging from one-half to over two mile spacings. By 1961, the western half of the region was covered in this way. These detailed, low-level surveys provide us with the most complete, readily available magnetic anomaly data set for the Innuitian region (GSC Aeromagnetic Maps in NTS areas 69, 78, 79, 89, 98, 99, 108, 109, 780, 1000). Other low-level magnetic surveys have been carried out by industry. The most extensive of these was initiated by Pallister and Associates, and flown by Geoterrex Ltd., beginning in 1969. Much of the area was flown at a reconnaissance line spacing of about 15 kilometres, although a large portion of Axel Heiberg Island and part of western Ellesmere Island were surveyed on a one by three mile grid pattern (for geographic names, see Fig. 1, in pocket).

A shipborne survey of Lancaster Sound and adjacent Baffin Bay was conducted in 1964 (Barrett, 1966), with tracks about 10 km apart. Shipborne data have also been obtained in the Amundsen Gulf region (Currie and Tiffin, 1974).

THE MAGNETIC ANOMALIES

The most complete, readily available presentation of magnetic anomalies over the Innuitian Orogen and Arctic Platform is in the Magnetic Anomaly Map of Arctic Canada (McGrath and Fraser, 1981), obtainable from the Geological Survey of Canada. This colour map displays in a vivid form the main features of the anomaly field. The Magnetic Anomaly Map of Canada, 4th edition (Dods et al., 1984) and the Magnetic Anomaly Map of North America (to be published as part of the DNAG contributions) display these anomalies in the context of the rest of the continent, again in colour. Figure 5D.2 is a simplified version (derived from the above maps with the assistance of D.J. Teskey and J. Halpenny) prepared for this volume. The anomalies are referenced to the International Geomagnetic Reference Field as defined in Peddie (1982).

At first impression, the region is characterized by several large domains of low magnetic relief, of 200 nT or less, with generally broad anomalies. Other domains, for example over northern Axel Heiberg Island, north of Ellef Ringnes and Amund Ringnes islands, contain many short-wavelength, relatively high relief anomalies (see, for example, Bhattacharyya, 1968).

To the south and east of the Innuitian Orogen proper, the magnetic signatures of the Precambrian Shield show clearly through the platform cover on southern Ellesmere Island, Jones Sound and Devon Island. The weak anomalies over much of the Arctic Platform, however, suggest the presence of deep sedimentary sections. An arcuate sequence of low anomalies occurs over the southern part of Melville Island and south of Bathurst Island. Gregory et al. (1961b) reported that the Precambrian Minto Arch has little effect on the magnetic field, being largely covered by a few kilometres of Proterozoic sedimentary rocks. The Natkusiak basalt at the top of the arch, with surrounding diabasic sills, might give a magnetic signature — unfortunately, there are no published magnetic anomaly data for that region. An anomaly of about 800 nT was found by Currie and Tiffin

---

Coles, R.L.
1991: Aeromagnetic field; in Chapter 5 of Geology of the Innuitian Orogen and Arctic Platform of Canada and Greenland, H.P. Trettin (ed.); Geological Survey of Canada, Geology of Canada, no. 3; (also Geological Society of America, The Geology of North America, v. E).

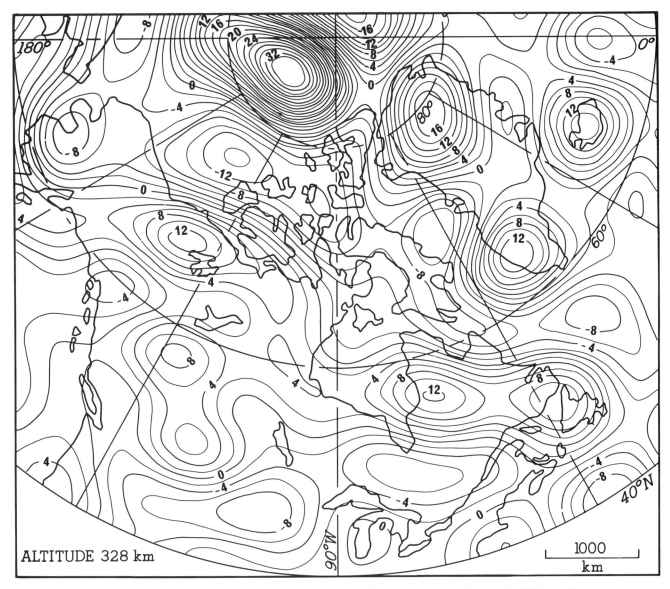

Figure 5D.1. Magnetic anomaly map of Arctic Canada and Greenland derived from MAGSAT satellite data, after Haines (1985). Contour interval is 2 nanotesla.

(1974) near the intersection of the Coppermine and Minto Arches (Fig. 5D.2, near bottom left corner). They suggested that this feature, the Pearce Point anomaly of southern Amundsen Gulf, may represent a major diabasic intrusion, similar to one suggested by Riddihough and Haines (1972) about 100 km to the southwest. Currie and Tiffin also reported a system of dyke-like anomalies to the southeast of the Pearce Point anomaly, with trends at variance with trend of dykes outcropping on land to the south. The prominent north-trending anomalies associated with the Boothia Uplift are interrupted, unlike the structural trends, at southern Bathurst Island and to the north of Cornwallis Island, by a deep northeast-trending low anomaly. This trend can be extended to the northeast to Ellesmere Island and thence, with a change to a more northerly direction, along a narrow linear low anomaly, marking the eastern limit of the Sverdrup Basin, from Bay Fiord to Tanquary Fiord. The higher field north and west of Bathurst Island and extending to Melville Island marks the southern limit of the Sverdrup Basin.

Over the central Sverdrup Basin, the magnetic relief is weak and of long wavelength, except for anomalies associated with features introduced into the sedimentary strata (Forsyth et al., 1979). In fact, within the central part of the basin, the Precambrian basement is so deep, and so uniform magnetically, that basement structure cannot be interpreted from the magnetic data. The basement surface may lie below the Curie point depth, so that the rocks are nonmagnetic (Reford et al., 1972). Electrical conductivity studies show high conductivities at upper crustal depths below Prince Patrick and Melville islands (Niblett and Kurtz, Chapter 5E), suggesting the existence there of thick marine sedimentary deposits. A change in electrical conductivity

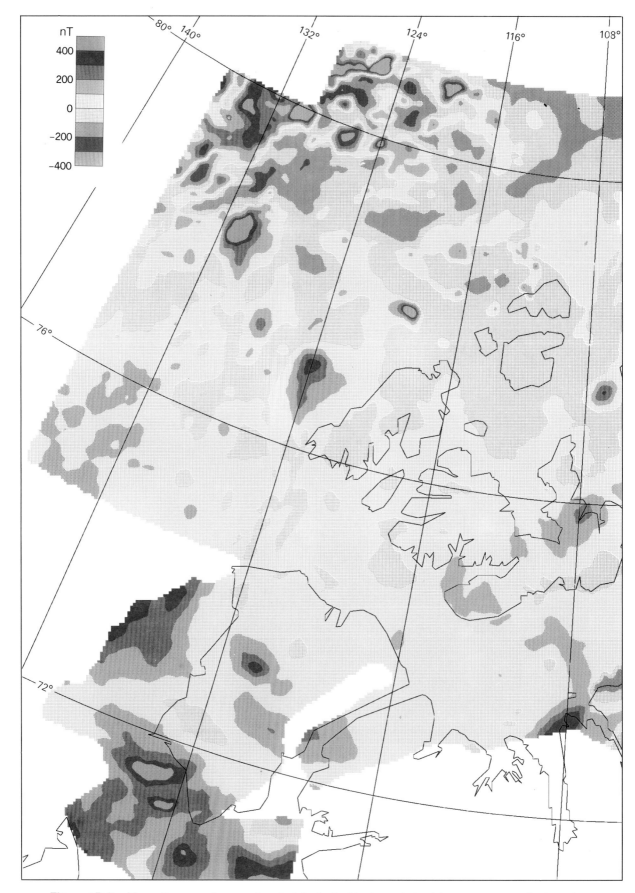

Figure 5D.2. Magnetic anomaly map of part of Arctic Archipelago derived from aeromagnetic data after McGrath and Fraser (1981). Contour interval is 100 nanotesla.

GEOPHYSICAL CHARACTERISTICS

structure appears to correlate with the southward increase in magnetic field, perhaps related to shallower crystalline basement, at the southern edge of the Sverdrup Basin. High temperatures at mid- to upper-crustal depths in parts of the region (A. Judge, pers. comm., 1985) may also be a factor resulting in low magnetic field.

Reford et al. (1972) noted that because of the regionally weak magnetic field originating from the deep basement under the Sverdrup Basin, it is relatively easy to identify anomalies with magnetic material within the sedimentary section. This magnetic material consists mostly of intrusive rocks of several types. Basic plugs and larger bodies are present and some are associated with sheets of magnetic material rising to near the present surface. Reford et al. (1972) distinguished at least three types of dykes, some exposed or shallow, some subsurface, and some associated with particular faults. In the eastern Sverdrup Basin, magnetic features are related mainly to folded sills. In a belt sweeping across the north-central part of the basin, anomalies are commonly associated with diapirs. Many of these latter anomalies have classic simple forms.

The Sverdrup Rim (Chapters 4, 5B, 13, 14), also known as the Sverdrup Ridge (Hea et al., 1980), appears to be the northwestern limit of a region of high relief anomalies on and to the north of Ellef Ringnes Island (Forsyth et al., 1979). This zone of magnetic highs crossing the shelf and extending seaward northwest of Ellef Ringnes Island correlates with a distinct physiographic section of the continental slope, as well as with anomalous free-air gravity and seismicity. The high anomalies on central-northeastern Ellef Ringnes Island have been related to known domes (Pallister and Bourne, 1972; Bourne and Pallister, 1973). Gregory et al. (1961b) discussed other examples.

A weak north-northwest trending ridge of high magnetic field extends along the eastern side of Axel Heiberg Island; it is most clearly seen on the map of McGrath and Fraser (1981). It separates a region of weak north-northwest trending anomalies on the west from weak north-northeast to northeast-trending anomalies on the east. This ridge of higher magnetic field may be associated with the eastern flank of the Princess Margaret Arch, a structural high (Chapter 17, Fig. 17.1). These different anomaly trends appear to relate to structural trends as described in Trettin et al. (1972, p. 134).

Anomalies of greater than 200 nT in the Strand Bay region of west-central Axel Heiberg Island may be related to basalt of the Upper Cretaceous Strand Fiord Formation (Ricketts et al., 1985).

Closer examination of the anomaly field over the western Sverdrup Basin shows numerous linear features. Bhattacharyya (1968) interpreted a series of linear anomalies north of Ellef Ringnes Island to be a system of dykes. Reford (1967) and Reford et al. (1972) delineated a series of weak linear anomalies extending from Melville Island to Ellef Ringnes Island. Forsyth et al. (1979) discussed these anomalies in relation to gravity and seismic information. They concluded that the "Gustaf-Lougheed Arch" (cf. Chapters 5A, 5C) is cut by extensive dyke-like features. Balkwill and Fox (1982) noted the coincidence between anomalies and normal faults and grabens. Many of the features have positive magnetic anomalies of about 10 nT with widths of more than 3 km and extend for up to 160 km in length. Forsyth et al. (1979) noted that whereas many of these features have a consistent north-northeasterly trend, a group of anomalies between Borden Island and Ellef Ringnes Island has a more northeasterly trend. One of these latter anomalies is coincident with a Tertiary graben on Borden Island whereas on the Sabine Peninsula to the south an anomaly is coincident with a gabbroic dyke of Early Cretaceous age. Forsyth et al. (1979) concluded that the northernmost structures may therefore be younger features. Balkwill and Fox (1982) have suggested that the generally northeast-trending features are evidence for a long-lasting incipient rift zone in the Carboniferous – Upper Cretaceous rocks of the western Sverdrup Basin.

A particularly prominent feature west of Lougheed Island is a 300 nT anomaly of classic form. McGrath and Hood (1973) interpreted this as a basic intrusion with a high magnetization of 4 A/m and a depth to the top of the body of about 15 km. The lateral extent of the intrusion is approximately 15 by 30 km. It is a large intrusive plug, possibly related to diapiric activity, and could, as McGrath and Hood point out, be expected to have had tectonic and possibly thermal effects on the sedimentary rocks that it intruded.

The Lancaster Sound survey data (Barrett, 1966) indicate that the sound represents a major structural feature, in the form of a half-graben, with vertical movement of as much as 8 km on the north side and implying some north-south movement between Baffin and Devon islands.

SUMMARY

Magnetically, the Innuitian Orogen and Arctic Platform are one of the quieter parts of the earth's crust. This is largely a consequence of extremely thick deposits of marine sediments known to exist in much of the Sverdrup Basin, but also thick sedimentary sequences in the Arctic Platform contribute to the regionally low magnetic field. Broad regions of higher magnetic field south of the Sverdrup Basin suggest the presence of crystalline basement ridges.

E. CONDUCTIVITY ANOMALIES

E.R. Niblett and R.D. Kurtz

INTRODUCTION

With the exception of semi-metallic minerals such as sulphides and graphite, most rock-forming materials are extremely good electrical insulators. However, the electrical conductivity of rocks in the upper crust is controlled not so much by their chemical composition as by the amount of moisture they contain. Thus, sedimentary rocks and unconsolidated overburden tend to be two or more orders of magnitude more conductive than crystalline metamorphic and igneous rocks, on account of their higher porosity and water content. Clay minerals, when moist, and hydrous substances such as serpentine are also reasonably good conductors. The critical factor in determining the electrical properties of earth materials is the conductivity of the fluid (electrolyte) which permeates the rock or soil.

Spaces within a rock body created by cracks and fractures are rendered conductive when filled with ground water. Such structural conductors depend very little on rock type or mineral content but they depend a great deal on the manner in which component fractures are interconnected to form a continuous path for the passage of electric current. Linear features such as faults, shear zones, and contact fractures, some of which extend over distances of hundreds of kilometres, may, for this reason, possess a much higher effective conductivity than the surrounding host rock. Conductors of this type represent zones of weakness in the crust, sometimes extending to substantial depths, though not necessarily evident in the surface geology. However, their widespread occurrence has been demonstrated by airborne electrical methods in near-surface formations, and at greater depths by natural electromagnetic induction methods such as magnetic variation analysis and magnetotellurics.

Deep sounding electromagnetic methods have frequently located high-conductivity zones in the lower crust and upper mantle (Edwards et al., 1980; Garland, 1981; Kurtz, 1982). Fluids probably play a dominant role in the deep crust, though, until recently, it was usually assumed that interconnected fractures and pore conduction would be inhibited by high pressures. Deep drilling on the Kola Peninsula in the Soviet Union has revealed substantial amounts of water and fractured rock at depths to 9 km. Partial melting can increase bulk conductivity of rock by easily two orders of magnitude. This phenomenon can explain the high conductivities in the crust often found in regions of high geothermal gradients. Partial melting also is thought to be responsible for the rise in conductivity which many workers have found in the upper mantle and associated with the asthenosphere.

Natural or transient time-variations in the earth's magnetic field arise as a consequence of electric currents generated by solar radiation in the ionosphere/magnetosphere region of the upper atmosphere. They represent the low-frequency end of the electromagnetic wave spectrum with periodicities ranging from one second to hours, days, and months. Horace Lamb (1883) formulated the problem of electromagnetic induction in a spherical earth by external field variations more than 100 years ago. Later, Schuster (1908) applied this theory to daily variation data (Sq) obtained from magnetic observatories. He was able to show that magnetic variations observed at the surface could be separated into parts of external and internal origin, and that the internal part must be attributed to electrial currents induced in the earth by the external portion of the field. These induced (telluric) currents and their associated magnetic fields can penetrate into the earth to a depth which depends on their frequency and on the electrical conductivity of the rock. Calculations of the kind which Schuster first attempted, and which subsequent workers have greatly refined, provide a means of estimating the electrical conductivity of the earth's interior to depths of several hundred kilometres if long-period data (more than 24 hours) are included in the analysis.

In the early 1950s, local geomagnetic variation anomalies were discovered in Japan by Rikitake and his colleagues, and in northern Germany by Bartels and co-workers (Rikitake, 1966, Chapter 19). Such anomalies are characterized by large differences in recorded geomagnetic variations at certain periods, especially the vertical component, over distances as small as a few tens of kilometres. The observations indicate local disturbances in the pattern of telluric currents flowing in the crust or upper mantle, which are caused by prominent inhomogeneities in electrical conductivity. If data are available from an array of magnetometers, each recording three components of magnetic field variations, regions where telluric currents are concentrated can be identified and mapped within the array. Experiments of this kind in the Innuitian Province have shown that this region contains broad zones of high electrical conductivity, which must be associated with major structural features in the crust.

CHARACTER AND DISTRIBUTION OF ARCTIC ANOMALIES

Prominent geomagnetic induction anomalies were discovered in the Innuitian Province when magnetic observatories were first established at Alert, on northeastern Ellesmere Island (Whitham and Andersen, 1962) and at Mould Bay (MLB in Fig. 5E.2) on Prince Patrick Island (Whitham, 1963).

At Mould Bay, the most obvious induction effect is a strong and persistent attenuation of vertical magnetic force (Z) with periodicities of about 5-10 min (DeLaurier et al., 1974, 1980). At longer periods the Z energy rises fairly rapidly, levelling off at a period close to 100 min. The analogue Z magnetograms from this observatory therefore

---

Niblett, E.R. and Kurtz, R.D.
1991: Conductivity anomalies; in Chapter 5 of Geology of the Innuitian Orogen and Arctic Platform of Canada and Greenland, H.P. Trettin (ed.); Geological Survey of Canada, Geology of Canada, no. 3; (also Geological Society of America, The Geology of North America, v. E).

have a smoothed or filtered appearance when compared to the horizontal force (H) traces. The effect is illustrated in Figure 5E.1 in which $|\frac{Z}{H}|$ energy densities are compared for a number of stations in the Queen Elizabeth Islands. Note that Isachsen (ISN) shows a similar type of response to Mould Bay. Such a strong suppression of the shorter-period Z variations is typical of magnetograms recorded on sea ice over ocean depths of 1000 m or more. When observed on land, the effect implies the existence of substantial thicknesses of exceptionally highly conductive strata in the upper or middle crust.

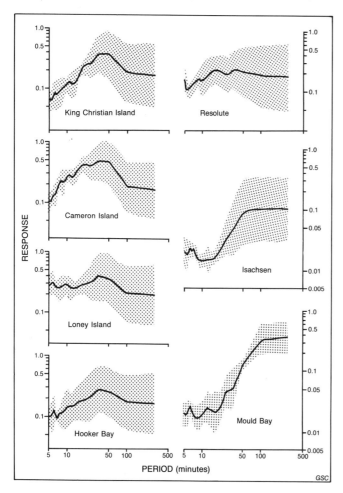

Figure 5E.1. Observed $|\frac{Z}{H}|$ energy density ratios at seven stations in the Arctic Islands. The shaded areas represent ±1 standard deviation (after DeLaurier et al., 1980, p. 1645).

Strongly suppressed Z variations at periods near 10 min have been found at several locations in the western Arctic Islands and are marked by closed circles in Figure 5E.2. At other locations where magnetic field variations have been recorded, this "Mould Bay effect" is not seen, implying that the underlying crust at these stations is much more resistive. These data therefore indicate that major differences in crustal structure or composition exist within the Innuitian Province.

The Alert induction anomaly is characterized by a persistent tendency for the horizontal variation field to be both strongly enhanced and strongly polarized in a northwesterly direction. As a consequence, magnetograms for the north (X) and east (Y) components tend to mirror each other at any locality in its vicinity. The anomaly has been attributed to an elongated, roughly two-dimensional conducting body located in the crust and striking northeast across the northern portion of Ellesmere Island and beyond the coast into the Lincoln Sea (Niblett et al., 1974). Vertical force variations (Z) are suppressed near the axis of the anomaly but are very large at its flanks and reversed on opposite sides. Near Alert the conductor is about 100 km wide. The anomaly has been mapped over a distance exceeding 500 km from Greeley Fiord to the Lincoln Sea continental shelf. Its location is shown in Figure 5E.2.

GEOMAGNETIC TRANSFER FUNCTIONS AND CRUSTAL STRUCTURES

In studies of natural electromagnetic induction in the earth, the concept of a transfer function is associated with the observation that at many locations the three components of the geomagnetic variation field show a persistent tendency of linear interdependence. Such a tendency was first demonstrated around the Australian coast by Parkinson (1959) and since then empirical relationships such as

$$Z = AX + BY$$

have been used by many workers to study local induction anomalies and their structural implications. In this relation, X, Y, Z are frequency domain representations of the true north, true east, and vertically downward field components. The number pair (A, B) is called a geomagnetic transfer function. Generally speaking, these numbers are complex and frequency dependent. They may be readily converted to quasi-vectors or "induction arrows" with magnitudes which express the degree of correlation between vertical and horizontal variations and directions which are controlled by the local distribution of electrical conductivity in the crust (Everett and Hyndman, 1967). The in-phase induction arrow tends to point toward regions of high current density in the vicinity of the recording station, i.e. toward regions of high electrical conductivity. Induction arrows or transfer functions have found wide application in the study of local and regional crustal features such as faults, dykes, grabens, coastlines, geothermal reservoirs, and other prominent structural inhomogeneities. Such features are usually associated with strong contrasts in electrical conductivity, which influence the distribution of electrical currents in the crust or upper mantle and therefore affect the magnetic field variations observed on the surface. When magnetic variations are recorded on structures which are more or less uniform over horizontal distances that are large compared to the penetration depth, there is normally very little correlation between Z and H variations and the induction arrows tend to vanish.

TRANSFER FUNCTION RESULTS IN THE ARCTIC ISLANDS

In-phase induction arrows are shown in Figure 5E.2 for four regions:

(1) At a period of 64 min for a profile of 15 stations extending across Victoria Island, Banks Island, and the continental shelf to the deep ocean (Beaufort Sea).

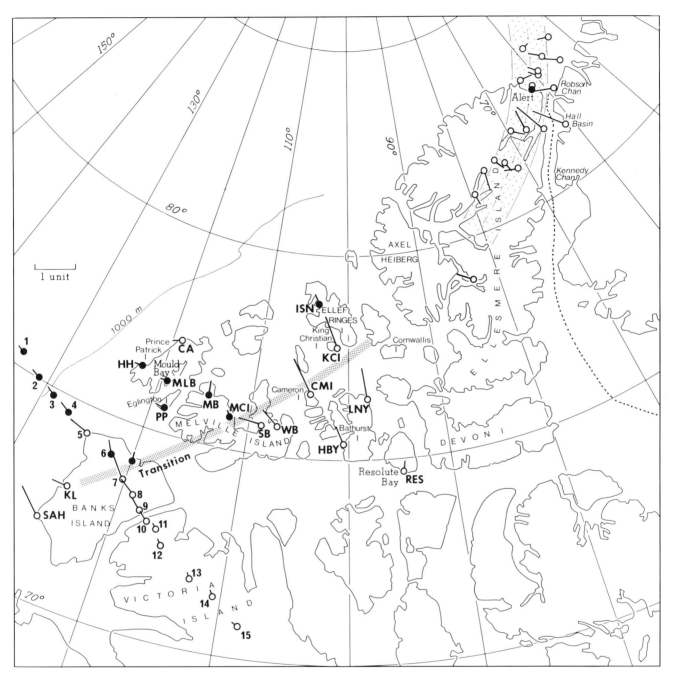

Figure 5E.2. The circles indicate stations at which geomagnetic variations in X, Y, and Z have been recorded for brief periods (a few days to several weeks). Permanent magnetic observatories are located at Alert, Resolute Bay, and Mould Bay. Stations where short-period Z variations are strongly suppressed are represented by closed circles; all others by open circles; shaded line marks transition. Lengths and azimuths of the induction arrows corresponding to longer periods (see text) are represented by radial lines at each station.

(2) At a period of 37 min for a set of 9 stations extending across Melville, Eglinton, and Prince Patrick islands, and including Resolute Bay Observatory (RES) on Cornwallis Island (DeLaurier et al., 1974).

(3) At a period of 90 min for a set of 6 stations extending from Resolute Bay to Isachsen. The 90 min arrow at Mould Bay Observatory is also shown for comparison with these stations (DeLaurier et al., 1980).

(4) At an average period of about 70 min for a set of 18 field stations on Ellesmere Island and the Lincoln Sea near the Alert anomaly (Praus et al., 1971; Niblett et al., 1974).

For both the Beaufort – Banks – Victoria profile and the Isachsen – Resolute profile all the induction arrows tend to point north or northwest toward the deep ocean. For the Prince Patrick – Melville – Resolute profile the majority of the arrows are small and point in a westerly direction. The largest arrows at Sabine Bay (SB) and Weatherall Bay (WB) are directed west and northwest respectively. Figure 5E.3 shows for these three profiles the amplitudes of the in-phase arrows resolved along a direction perpendicular to the 1000 m isobath (i.e. the continental slope). Maximum amplitudes lie at station 7 in the middle of Banks Island for the first profile, at Sabine Bay for the second, and at Cameron Island for the third. These data indicate that an abrupt change in the electrical conductivity of the crust must occur near these stations, with conductive material lying to the northwest and a more resistive crust to the southeast. The Beaufort – Banks – Victoria profile indicates very little change in the in-phase transfer functions across the continental slope (stations 1 to 4), but a secondary maximum occurs at station 5 at Cape Prince Alfred. The enhanced response here is almost certainly caused by concentration of induced currents in the sea water near this prominent coastal feature (Chan et al., 1981).

By drawing a smooth line through the positions on each profile where the induction arrows achieve their maximum amplitudes we identify a narrow transition zone extending northeastward from central Banks Island through the Sverdrup Basin and roughly 400 km inland from the continental slope. It is also evident that those stations which display several attenuated short-period Z amplitudes (those represented by black dots in Fig. 5E.2) all occur on the oceanic side of this transition. This confirms the presence of substantial thicknesses of highly conductive material in the crust of this region. The fact that the Beaufort Sea continental slope is not associated with a significant increase in the in-phase transfer function suggests that both the deep ocean and the Banks Island shelf are underlain by a conducting crust, which may extend inland as far as the transition zone.

The induction arrows shown in Figure 5E.2 at recording stations near the Alert anomaly on Ellesmere Island and over the Lincoln Sea were vectorially averaged from results obtained at periods of 22, 36, 65, 90, and 144 min. Observed effects were found to vary little with frequency over this range. Z variations are subject to phase reversal on opposite sides of the anomaly as the arrows indicate. These data reveal the presence of persistent highly concentrated electric currents in the crust between Greeley Fiord and Lincoln Sea, the width of the anomaly being approximately 100 km. The Z anomaly can be fitted by a uniform current flowing in a conductor of rectangular cross-section with a resistivity of

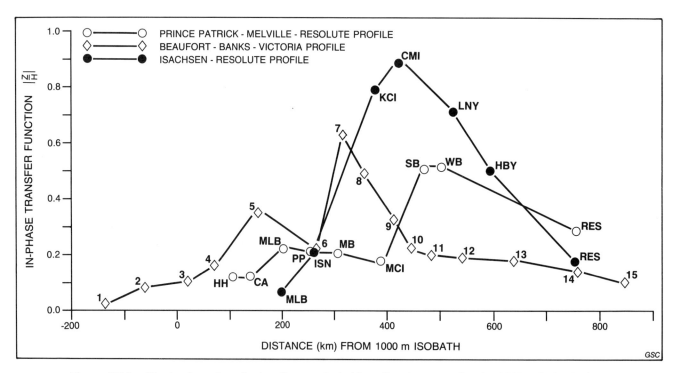

Figure 5E.3. The in-phase transfer functions projected in a direction normal to the 1000 m isobath along three profiles in the western Arctic Islands. Transfer functions correspond to a period of 64 min for the Beaufort – Banks – Victoria profile, 37 min for the Prince Patrick – Melville – Resolute profile, and 90 min for the Isachsen – Resolute profile.

4 Ωm or less (Niblett et al., 1974). Such a conductor would be about 10 km thick and 100 km in width with its upper surface at a depth near 10 km. Depths are very poorly resolved for models of this kind, however, and a conductor lying much closer to the surface would also provide an adequate fit to the data. In reviewing electrical conductivity anomalies on a global scale, Porath and Dziewonski (1971) and Hutton (1976) have suggested that the Alert anomaly may be caused by currents channelled through the narrow sea passage separating Ellesmere Island and Greenland. The field observations do not support this conclusion. The anomaly axis lies well to the west of Robeson and Kennedy channels as is shown in Figure 5E.2, and it may be traced for at least 100 km over the Lincoln Sea shelf, well beyond the channel entrance. A model conductor representing sea water in these channels with resistive crustal blocks on either side cannot reproduce the observed distribution of geomagnetic transfer functions and induction arrows (Niblett et al., 1974).

It is not known how far the anomaly extends to the southwest across Ellesmere and Axel Heiberg islands or to the northeast across the continental shelf. However, that part which has been mapped, closely follows the trend of a Bouguer gravity high striking northeast across Ellesmere Island and into the Lincoln Sea (Sobczak and Stephens, 1974; L.W. Sobczak, pers. comm., 1985). The axis of the induction anomaly also coincides exactly with the elongated magnetic low over northern Ellesmere Island derived from POGO satellite measurements, and which separates the North Greenland magnetic high to the east from the Alpha Ridge high to the northwest (Langel and Thorning, 1982b). Together these geophysical data provide evidence for a prominent northeasterly-trending structure in the crust lying somewhat to the west of Nares Strait and extending from Greeley Fiord through Alert to the Lincoln Sea.

DISCUSSION

It is clear that the magnetic variation data obtained so far in the Canadian Arctic Islands have successfully delineated some major crustal anomalies. A narrow transition zone separating a continental regime (at least in the context of induction effects) to the south from a more highly conducting regime to the north has been traced across northern Banks Island into the central part of the Sverdrup Basin. A prominent, narrow, and elongated crustal conductor has been mapped across northern Ellesmere Island and part of the Lincoln Sea shelf. Available data are insufficient to show whether or not the Ellesmere Island-Lincoln Sea anomaly links up with the transition zone mapped in the western part of the Arctic Islands. Unusually high electrical conductivity in the crust and possibly also in the upper mantle is associated with both the Ellesmere anomaly and with the Sverdrup Basin lying to the north of the transition zone. The thicknesses of these conducting bodies must be substantial, perhaps 5-10 km, but their depths are not well resolved. It is possible that a highly conductive crust extends from the Lincoln Sea in the northeast right across the Arctic Archipelago to the northwest corner of Banks Island.

Relationships of these anomalies with the regional geology, as discussed in other chapters, will be considered next.

The Alert anomaly coincides closely with the central part of the Hazen Fold Belt and parallels its structural trend (Chapters 8C, 12E). Exposures in this area consist of tightly folded flysch of the Silurian Danish River Formation and of starved-basin sediments of the upper Lower Cambrian to Lower Silurian Hazen Formation. The anomaly may coincide with the concealed facies change between Lower Cambrian shelf sediments of the Ellesmere Group, exposed in the southeastern part of the fold belt, and deep water sediments of the Lower Cambrian Grant Land Formation, exposed in the northwestern part of the fold belt. (Both units underlie the Hazen Formation). The entire known succession is probably more than 5 km in stratigraphic thickness and could be more than 15 km in structural thickness. The succession has lost its formation waters and porosity during intense mid-Paleozoic deformations and is not known to contain significant amounts of hydrated minerals. It does contain, however, dark grey slates, the organic matter in which may have become "graphitic" and hence somewhat conductive at depth. The basement of these rocks, stratigraphic or structural, is unknown, but the possibility that conductive oceanic basalt, marking an ancient plate boundary, is present at depth (Niblett et al., 1974) cannot be excluded.

From a structural standpoint, it is unlikely to be mere coincidence that a Bouguer gravity high, a major aeromagnetic low, and a well defined induction anomaly all lie along the Hazen Fold Belt. Miall (1983) has suggested that much of the strike-slip movement required to satisfy geophysical models of seafloor spreading in the Labrador Sea – Baffin Bay region could have occurred along faults parallel to the regional structural grain in northern and central Ellesmere Island (Chapter 17). If the Hazen Plateau were the scene of crustal deformation of this kind during the Paleogene, it would explain in a general way why the region is so anomalous in a geophysical sense, and why Paleozoic geological markers were not substantially displaced on opposite sides of Nares Strait. It would then be reasonable to speculate that the source of high electrical conductivity is associated with a major network of fluid-filled fractures extending to depths of 10 km or more in the crust. However, whereas location and trend of the induction anomaly are compatible with the hypothesis of Paleogene transcurrent movement, geological field evidence for such movements still is lacking (Chapter 17).

In the southwestern part of the region, the relationship between the mapped conductivity boundary and the surface geology is ambiguous. On Banks Island, the boundary between conductive and nonconductive crust coincides to some extent with the geological boundary between Cretaceous and Tertiary strata in the northeastern part. In the subsurface, it may coincide roughly with the southeastern boundary of a Lower – Middle Devonian chert-shale unit of deep water origin (Nanuk Formation of Miall, 1976) that lies on shelf carbonates. On the other hand, on Melville Island and in the central part of the Sverdrup Basin, it cuts across stratigraphic and structural trends. Sobczak and Overton (1984) have found that sedimentary formations in the central Sverdrup Basin can extend to depths exceeding 15 km. Well logs indicate that these sediments tend to have higher porosities, higher fluid content, and substantially higher conductivities than the Devonian and older formations to the south (DeLaurier et al., 1980).

Farther west, on Melville, Eglinton, Prince Patrick, and Banks islands, the Phanerozoic sedimentary sequences are much thinner, but the magnetic variation data still require high conductivity in parts of the crust (including Proterozoic sediments) and possibly the upper mantle. Though depth estimates are unreliable with these data, the forward models of DeLaurier et al. (1974) indicate that at the anomalous stations the short-period Z suppression is compatible with a thin (1-2 km) conductive layer at a depth of about 5 km, and with a much thicker conductor in the upper mantle. The transition zone as mapped in Figure 5E.2 is based on induction arrows derived from fairly long-period magnetic variations (37-90 min), and these could be influenced by either shallow or deeply-seated conductors. The absence of any strong correlation with surface geology in the western Arctic leaves the impression of a deep, perhaps very old Precambrian feature that is only locally expressed in the Phanerozoic geology.

Thick successions of permeable marine sediments are probably responsible for at least part of the high conductivity observed in the southwestern Arctic Islands. Additional magnetotelluric data, permitting more reliable depth estimates and improved modelling, will be required to determine whether or not the lower crust or upper mantle contribute to the observed anomalies.

F. MOTIONS OF THE NORTH MAGNETIC POLE

L.R. Newitt and E.R. Niblett

The north and south magnetic dip poles are points on the earth's surface at which the magnetic inclination is 90°. Their positions are constantly varying, being subject to diurnal (daily), seasonal, and secular (long period) variations. The first direct determination of the north magnetic pole was made in 1831 by James Ross (1834), who located it on the west coast of Boothia Peninsula, north of King William Island. During the 20th century, five additional determinations have been made (Barraclough and Malin, 1981; Newitt and Niblett, 1986). These observations show that from 1904 to 1984 the pole has moved about 750 km in a northerly or northwesterly direction, an average of 9.4 km per year (Fig. 5F.1). This motion is a result of the secular variation of the earth's magnetic field, defined as a slow change in strength and direction that varies in time and space. The cause of the secular variation is still a subject of active research, but is believed to be linked to the same fluid motions in the outer core of the earth that cause the magnetic field itself.

The average diurnal motion about the mean position for 1984 is shown in Figure 5F.2 (Newitt and Niblett, 1986). On an average "quiet" day the pole is displaced 12 km from the mean position, but on an average "disturbed" day the displacement may be up to 50 km. These displacements are attributed to the effect of ionospheric current systems (Dawson and Newitt, 1978, 1982).

Because of the mobility of the north magnetic pole and the weakness of the magnetic field in the Arctic, the magnetic compass is an unreliable tool for navigation or detailed studies in structural geology or sedimentology. The observed movements of the pole also have important implications for the evaluation of paleomagnetic data.

Newitt, L.R. and Niblett, E.R.
1991: Motions of the north magnetic pole; in Chapter 5 of Geology of the Innuitian Orogen and Arctic Platform of Canada and Greenland, H.P. Trettin (ed.); Geological Survey of Canada, Geology of Canada, no. 3; (also Geological Society of America, The Geology of North America, v. E).

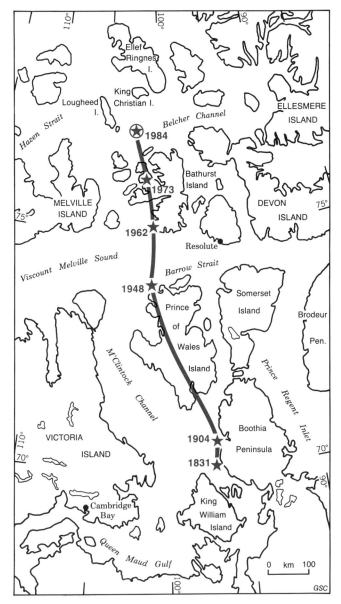

Figure 5F.1. Secular motion of the north magnetic dip pole. Dates indicate determinations of the pole position from local surveys (from Newitt and Nibblett, 1986).

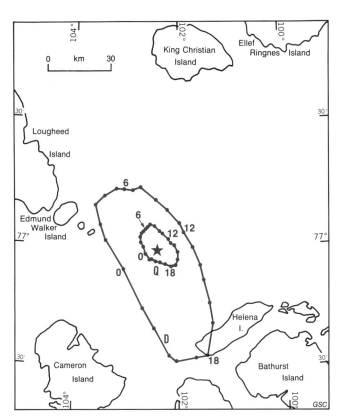

Figure 5F.2. Diurnal motion of the north magnetic dip pole. Q denotes quiet days, D disturbed days (from Newitt and Niblett, 1986).

CHAPTER 5

G. HEAT FLOW

A.M. Jessop

During exploratory drilling for oil and gas in the Arctic Islands, 37 wells have been preserved for geothermal research (Judge et al., 1981; Taylor et al., 1982). Repeated temperature logs over periods of up to ten years have permitted the elimination of drilling disturbance from the observed temperatures, and the calculated equilibrium temperatures have revealed the thickness of permafrost (Fig. 19.3) and the temperature gradients. For technical and logistic reasons it has been possible to preserve only the upper part of each hole, generally to a depth of about 600 to 1000 m. Distortions of the thermal field by recent climatic changes, thermodynamic effects of the phase boundary of ice and water, and proximity to coastlines create considerable problems in the calculations of terrestrial heat flow from these data.

The first attempt to measure heat flow in the Canadian Arctic, at Resolute Bay, Cornwallis Island, produced a high value of 121 mW/m^2 (Misener, 1955), which was later reduced to 52 mW/m^2, after correction for the distortions produced by the present shoreline and the recent geological history of shoreline change (Lachenbruch, 1957). The magnitude of the correction illustrates the importance of corrections to the calculated heat flow and the need for careful analysis of each site.

Thermal conductivity of the drill cuttings has been measured by the "cell" method in a divided bar (Sass et al., 1971), a method that includes some problems in the conversion from the laboratory data to properties of the undisturbed rock. Combination of temperature and conductivity data has given an uncorrected heat flow for each well, which falls in the range 60 to 160 mW/m^2. In general, the highest values are found in the Sverdrup Basin, and the lowest in the Arctic Platform. Because of rebound from the weight of glaciation, much of the coastal area has emerged from the sea in the last few thousand years. The corrections for this can be greater than the heat flow originating from great depth, and variations of apparent heat flow can be used to derive indications of the shoreline history (Taylor et al., 1983). With the restriction of shallow data, it is not possible to examine the possible variations of heat flow with depth or the implications for water movement throughout the basin, as has been possible for some basins with more well information and less distortion of heat flow by surface effects (Majorowicz et al., 1987).

An offshore well near Ellef Ringnes Island (Cape Allison C-47; Fig. 19.3) has been equipped with a multi-thermistor cable and a data-recovery system (Taylor et al., in press). The best estimate of heat flow from this well is 39 mW/m^2. Variations of heat flow between different formations are observed, but it is possible that these are the result of the difficulties of determination of the true conductivity of the rock by laboratory measurement on small samples of cuttings. In contrast with nearby onshore wells, this site appears to be in reasonable equilibrium with its environment. The difference between this low value of heat flow and the higher values yielded by land-based wells remains to be resolved, but it is believed that much of this discrepancy will be accounted for by the shortcomings of present measurement and correction techniques, rather than by a significant difference in the heat flow from depth.

REFERENCES

Balkwill, H.R. and Fox, F.G.
1982: Incipient rift zone, western Sverdrup Basin, Arctic Canada; in Arctic Geology and Geophysics, A.F. Embry and H.R. Balkwill (ed.); Canadian Society of Petroleum Geologists, Memoir 8, p. 171-187.

Bancroft, A.M.
1958: Gravity measurements in the Queen Elizabeth Islands of Arctic Canada; Transactions of the American Geophysical Union, v. 39, p. 615-619.

Barraclough, D.R. and Malin, S.R.C.
1981: 150 years of the north magnetic pole; Nature (London), v. 291, p. 377.

Barrett, D.L.
1966: Lancaster Sound shipborne magnetometer survey; Canadian Journal of Earth Sciences, v. 3, p. 223-235.

Basham, P.W., Forsyth, D.A., and Wetmiller, R.J.
1977: The seismicity of northern Canada; Canadian Journal of Earth Sciences, v. 14, p. 1646-1667.

Berkhout, A.W.J.
1970: The gravity anomaly field of Prince of Wales, Somerset and northern Baffin Islands, District of Franklin, Northwest Territories; Department of Energy, Mines and Resources, Canada, Publications of the Dominion Observatory, v. 39, no. 7, p. 179-209.

Berkhout, A.W.J. and Sobczak, L.W.
1967: A preliminary investigation of gravity observations in the Somerset and Prince of Wales islands, Arctic Canada, with map; Department of Energy, Mines and Resources, Canada, Dominion Observatory, Gravity Map Series, no. 81, 10 p.

Berry, M.J. and Barr, K.G.
1971: A seismic refraction profile across the polar continental shelf of the Queen Elizabeth Islands; Canadian Journal of Earth Sciences, v. 8, p. 347-360.

Bhattacharyya, B.K.
1968: Analysis of aeromagnetic data over the Arctic Islands and continental shelf of Canada; Geological Survey of Canada, Paper 68-44, 14 p.

Bourne, S.A. and Pallister, A.E.
1973: Offshore areas of Canadian Arctic Islands — geology based on geophysical data; in Arctic Geology, M.G. Pitcher (ed.); American Association of Petroleum Geologists, Memoir 19, p. 48-56.

Chan, G.H., Dosso, H.W., and Law, L.K.
1981: An analogue model study of electromagnetic induction for cape and bay coast lines; Physics of the Earth and Planetary Interiors, v. 25, p. 167-176.

Coles, R.L.
1985: Magsat scalar magnetic anomalies at northern high latitudes; Journal of Geophysical Research, v. 90, p. 2576-2582.

Coles, R.L., Haines, G.V., and Hannaford, W.
1976: Large scale magnetic anomalies over western Canada and the Arctic: a discussion; Canadian Journal of Earth Sciences, v. 13, p. 790-802.

Coles, R.L., Haines, G.V., Jansen van Beek, G., Nandi, A., and Walker, J.K.
1982: Magnetic anomaly maps from 40°N to 83°N derived from Magsat satellite data; Geophysical Research Letters, v. 9, p. 281-284.

Currie, R.G. and Tiffin, D.L.
1974: Preliminary results of a shipborne magnetic survey in Amundsen Gulf, District of Franklin; in Report of Activities, Part B, Geological Survey of Canada, Paper 74-1B, p. 65-67.

Jessop, A.M.
1991: Heat flow; in Chapter 5 of Geology of the Innuitian Orogen and Arctic Platform of Canada and Greenland, H.P. Trettin (ed.); Geological Survey of Canada, Geology of Canada, no. 3; (also Geological Society of America, The Geology of North America, v. E).

Dawes, P.R. and Kerr, J.W. (ed.)
1982: Nares Strait and the Drift of Greenland: a Conflict in Plate Tectonics; Meddelelser om Grønland, Geoscience 8, 392 p.

Dawson, E. and Newitt, L.R.
1978: An analytical representation of the geomagnetic field in Canada for 1975. Part III: the north magnetic pole; Canadian Journal of Earth Sciences, v. 15, p. 994-1001.
1982: The magnetic poles of the Earth; Journal of Geomagnetism and Geoelectricity, v. 34, p. 225-240.

DeLaurier, J.M., Law, L.K., Niblett, E.R., and Plet, F.C.
1974: Geomagnetic variation anomalies in the Canadian Arctic. II. Mould Bay anomaly; Journal of Geomagnetism and Geoelectricity, v. 26, p. 223-245.

DeLaurier, J.M., Niblett, E.R., Plet, F.C., and Camfield, P.A.
1980: Geomagnetic depth sounding over the central Arctic Islands, Canada; Canadian Journal of Earth Sciences, v. 17, p. 1642-1652.

Dods, S.D., Hood, P.J., Teskey, D.J., and McGrath, P.H.
1984: Magnetic anomaly map of Canada, 14th edition; Geological Survey of Canada, Map 1255A.

Earth Physics Branch
1980: The gravity map of Canada; Department of Energy, Mines and Resources, Canada, Gravity Map Series 80-1.

Edwards, R.N., Bailey, R.C., and Garland, G.D.
1980: Crustal and upper mantle electrical conductivity studies with natural and artificial sources; in The Continental Crust and its Mineral Deposits, D.W. Strangway (ed.); Geological Association of Canada, Special Paper 20, p. 255-271.

Embry, A.F.
1982: The Upper Triassic – Lower Jurassic Heiberg Deltaic Complex of the Sverdrup Basin; in Arctic Geology and Geophysics, A.F. Embry and H.R. Balkwill (ed.); Canadian Society of Petroleum Geologists, Memoir 8, p. 189-217.

Everett, J.E. and Hyndman, R.D.
1967: Geomagnetic variations and electrical conductivity structure in south-western Australia; Physics of the Earth and Planetary Interiors, v. 1, p. 24-34.

Forsyth, D.A., Mair, J.A., and Fraser, I.
1979: Crustal structure of the central Sverdrup Basin; Canadian Journal of Earth Sciences, v. 16, p. 1581-1598.

Forsberg, R.
1979: A gravity map of Peary Land, North Greenland; Grønlands Geologiske Undersøgelse, Report no. 88, p. 93-94.
1981: Preliminary Bouguer anomalies of north-east Greenland; Grønlands Geologiske Undersøgelse, Report no. 106, p. 105-107.

Fox, F.G.
1983: Structure sections across Parry Islands Fold Belt and Vesey Hamilton salt wall, Arctic Archipelago, Canada; in Seismic Expression of Structural Styles, a Picture and Work Atlas, A.W. Bally (ed.); American Association of Petroleum Geologists, Studies in Geology, Series 15, v. 3, p. 3.4.1-54 to 3.4.1-72.
1985: Structural geology of the Parry Islands Fold Belt; Bulletin of Canadian Petroleum Geology, v. 33, p. 306-340.

Garland, G.D.
1981: The significance of terrestrial electrical conductivity variations; Annual Review of Earth and Planetary Science, v. 9, p. 147-174.

Gibb, R.A. and Halliday, D.W.
1975: Gravity measurements in northern District of Keewatin and parts of District of Mackenzie and District of Franklin, N.W.T., with maps; Department of Energy, Mines and Resources, Canada, Earth Physics Branch, Gravity Map Series, nos. 139-148, 8 p.

Gregersen, S.
1982: Earthquakes in Greenland; Bulletin of the Geological Society of Denmark, v. 31, p. 11-27.

Gregory, A.F., Bower, M.E., and Morley, L.W.
1961a: Geological interpretation of aerial magnetic and radiometric profiles, Arctic Archipelago, Northwest Territories; Geological Survey of Canada, Bulletin 73, 148 p.

Gregory, A.F., Morley, L.W., and Bower, M.E.
1961b: Airborne geophysical reconnaissance in the Canadian Arctic Archipelago; Geophysics, v. 26, p. 727-737.

Haines, G.V.
1985: Magsat vertical-field anomalies above 40°N from spherical cap harmonic analysis; Journal of Geophysical Research, v. 90, p. 2593-2598.

Haines, G.V. and Hannaford, W.
1974: A three-component aeromagnetic survey of the Canadian Arctic; Department of Energy, Mines and Resources, Canada, Publications of the Earth Physics Branch, v. 44, no. 8, p. 213-234.

Hasegawa, H.S.
1977: Focal parameters of four Sverdrup Basin, Arctic Canada, earthquakes in November and December of 1972; Canadian Journal of Earth Sciences, v. 4, p. 2481-2494.

Hea, J.P., Arcuri, J., Campbell, G.R., Fraser, I., Fuglem, M.O., O'Bertos, J.J., Smith, D.R., and Zayat, M.
1980: Post-Ellesmerian basins of Arctic Canada: their depocentres, rates of sedimentation and petroleum potential; in Facts and Principles of World Petroleum Occurrence, A.D. Miall (ed.); Canadian Society of Petroleum Geologists, Memoir 6, p. 447-488.

Hobson, G.D. and Overton, A.
1967: A seismic section of the Sverdrup Basin, Canadian Arctic Islands; in Seismic Refraction Prospecting, A.W. Musgrave (ed.); Society of Exploration Geophysicists, Tulsa, Oklahoma, p. 550-562.

Hornal, R.W. and Boyd, J.B.
1972: Gravity measurements in the Slave and Bear structural provinces, Northwest Territories, with maps; Department of Energy, Mines and Resources, Canada, Earth Physics Branch, Gravity Map Series, nos. 89-95, 12 p.

Hutton, V.R.S.
1976: The electrical conductivity of the earth and planets; Reports on Progress in Physics, v. 39, p. 487-572.

International Association of Geodesy
1971: Geodetic reference system 1967; Special Publication no. 3, 116 p.

Johnson, G.L., Monahan, D., Gronlie, G., and Sobczak, L.W.
1979: General bathymetric chart of the oceans, Arctic Ocean; Canadian Hydrographic Service, Chart 5.17.

Judge, A.S., Taylor, A.E., Burgess, M., and Allen, V.S.
1981: Canadian geothermal data collection — northern wells 1978-80; Energy, Mines and Resources, Canada, Earth Physics Branch, Geothermal Series, no. 12, 190 p.

King, E.R., Zietz, I., and Alldredge, L.R.
1966: Magnetic data on the structure of the central Arctic region; Geological Society of America Bulletin, v. 77, p. 619-646.

Kurtz, R.D.
1982: Magnetotelluric interpretation of crustal and mantle structure in the Grenville Province; Geophysical Journal of the Royal Astronomical Society, v. 70, p. 373-397.

Lachenbruch, A.H.
1957: Thermal effects of the ocean on permafrost; Bulletin of the Geological Society of America, v. 68, p. 1515-1529.

Lamb, H.
1883: On electrical motions in a spherical conductor; Philosophical Transactions of the Royal Society of London, Series A, v. 174, p. 519-549.

Langel, R.A., Coles, R.L., and Mayhew, M.A.
1980: Comparisons of magnetic anomalies of lithospheric origin measured by satellite and airborne magnetometers over western Canada; Canadian Journal of Earth Sciences, v. 17, p. 876-887.

Langel, R.A. and Thorning, L.
1982a: A satellite magnetic anomaly map of Greenland; Geophysical Journal of the Royal Astronomical Society, v. 71, p. 599-602.
1982b: Satellite magnetic field over the Nares Strait region; in Nares Strait and the Drift of Greenland, a Conflict in Plate Tectonics, P.R. Dawes and J.W. Kerr (ed.); Meddelelser om Grønland, Geoscience 8, p. 291-293.

Mair, J.A. and Lyons, J.A.
1980: Crustal structure and velocity anisotropy beneath the Beaufort Sea; Canadian Journal of Earth Sciences, v. 18, p. 724-741.

Majorowicz, J.A., Jones, F.W., and Jessop, A.M.
1987: Geothermics of the Williston Basin in Canada in relation to hydrodynamics and hydrocarbon occurrences; Geophysics, v. 51, p. 767-779.

McGrath, P.H. and Fraser, I.
1981: Magnetic anomaly map of Arctic Canada, scale 1:3.5 million; Geological Survey of Canada, Map 1512A.

McGrath, P.H. and Hood, P.J.
1973: An automatic least-squares multimodal method for magnetic interpretation; Geophysics, v. 38, p. 349-358.

Meidler, S.S.
1961: Seismic activity in the Canadian Arctic 1899-1955; Department of Mines and Technical Surveys, Canada, Dominion Observatory, Seismological Series, 1961-3, 9 p.

Meneley, R.A.
1977: Prospects in the Canadian Arctic Islands; Panarctic Oils Ltd., Calgary, Alberta, 16 p.

Meneley, R.A., Henao, D., and Merritt, R.K.
1975: The northwest margin of the Sverdrup Basin; in Canada's Continental Margins and Offshore Petroleum Exploration, C.J. Yorath, E.R. Parker, and D.J. Glass (ed.); Canadian Society of Petroleum Geologists, Memoir 4, p. 531-544.

Miall, A.D.
1976: Devonian geology of Banks Island, Arctic Canada, and its bearing on the tectonic development of the circum-Arctic region; Geological Society of America Bulletin, v. 87, p. 1599-1608.
1983: The Nares Strait problem: a re-evaluation of the geological evidence in terms of a diffuse oblique-slip plate boundary between Greenland and the Canadian Arctic Islands; Tectonophysics, v. 100, p. 227-239.

Milne, W.G. and Smith, W.E.T.
1961: Canadian earthquakes — 1960; Department of Mines and Technical Surveys, Canada, Dominion Observatory, Seismological Series, 1960-2, 23 p.

Misener, A.D.
1955: Heat flow and depth of permafrost at Resolute Bay, Cornwallis Island, N.W.T., Canada; Transactions, American Geophysical Union, v. 36, p. 1055-1060.

Morelli, C., Ganatar, C., Honkasalo, T., McConnell, R.K., Tanner, J.G., Szabo, B., Uotila, U., and Whalen, C.T.
1974: The international gravity standardization net; International Association of Geodesy, Special Publication no. 4, 194 p.

Newitt, L.R. and Niblett, E.R.
1986: Relocation of the north magnetic dip pole; Canadian Journal of Earth Sciences, v. 23, p. 1062-1067.

Niblett, E.R., DeLaurier, J.M., Law, L.K., and Plet, F.C.
1974: Geomagnetic variation anomalies in the Canadian Arctic. I. Ellesmere Island and Lincoln Sea; Journal of Geomagnetism and Geoelectricity, v. 26, p. 203-221.

Okulitch, A.V., Packard, J.J., and Zolnai, A.I.
1986: Evolution of the Boothia Uplift, arctic Canada; Canadian Journal of Earth Sciences, v. 23, p. 350-358.

Ostenso, N.A.
1963: Aeromagnetic survey of the Arctic Ocean Basin; Proceedings of Thirteenth Alaskan Science Conference, Juneau, Alaska, p. 115-148.

Ostenso, N.A. and Wold, R.J.
1971: Aeromagnetic survey of the Arctic Ocean: techniques and interpretations; Marine Geophysical Researches, v. 1, p. 178-219.

Overton, A.
1970: Seismic refraction surveys, western Queen Elizabeth Islands and polar continental margin; Canadian Journal of Earth Sciences, v. 7, p. 346-365.
1982: Seismic reconnaissance profiles across the Sverdrup Basin, Canadian Arctic Islands; in Current Research, Part B, Geological Survey of Canada, Paper 82-1B, p. 139-145.

Pallister, A.E. and Bourne, S.A.
1972: Geology of offshore areas in the Arctic Islands based on geophysical data; Journal of Canadian Petroleum Technology, v. 11, p. 75-79.

Parkinson, W.D.
1959: Directions of rapid geomagnetic fluctuations; Geophysical Journal of the Royal Astronomical Society, v. 2, p. 1-14.

Peddie, N.W.
1982: International Geomagnetic Reference Field: the third generation; Journal of Geomagnetism and Geoelectricity, v. 34, p. 309-326.

Pelletier, B.R.
1966: Development of submarine physiography in the Canadian Arctic and its relation to crustal movements; in Continental Drift, G.D. Garland (ed.); The Royal Society of Canada, Special Publication no. 9, p. 77-101.

Picklyk, D.D.
1969: A regional gravity survey of Devon and southern Ellesmere islands, Canadian Arctic Archipelago, with map; Department of Energy, Mines and Resources, Canada, Dominion Observatory, Gravity Map Series, no. 87, 10 p.

Porath, H. and Dziewonski, A.
1971: Crustal resistivity anomalies from geomagnetic deep sounding studies; Reviews of Geophysics and Space Physics, v. 9, p. 891-915.

Praus, O., DeLaurier, J.M., and Law, L.K.
1971: The extension of the Alert geomagnetic anomaly through northern Ellesmere Island, Canada; Canadian Journal of Earth Sciences, v. 8, p. 50-64.

Rayer, F.G.
1981: Exploration prospects and future petroleum potential of the Canadian Arctic Islands; Journal of Petroleum Geology, v. 3, p. 367-412.

Reford, M.S.
1967: Aeromagnetic interpretation – Sverdrup Basin; Department of Indian Affairs and Northern Development, Oil and Gas Technical Reports, no. 678-7-10-1 to 678-7-10-5.

Reford, M.S., Leridon, J., and Fraser, I.
1972: Geological interpretations from aeromagnetic surveys over the Arctic Islands of Canada (abstract); International Geological Congress, 24th Session, Montreal, Section 9, Exploration Geophysics, p. 34.

Ricketts, B., Osadetz, K.G., and Embry, A.F.
1985: Volcanic style in the Strand Fiord Formation (Upper Cretaceous), Axel Heiberg Island, Canadian Arctic Archipelago; Polar Research 3 n.s, p. 107-122.

Riddihough, R.P. and Haines, G.V.
1972: Magnetic measurements over Darnley Bay, N.W.T.; Canadian Journal of Earth Sciences, v. 9, p. 972-978.

Riddihough, R.P., Haines, G.V., and Hannaford, W.
1973: Regional magnetic anomalies of the Canadian Arctic; Canadian Journal of Earth Sciences, v. 10, p. 157-163.

Rikitake, T.
1966: Electromagnetism and the Earth's Interior; Elsevier Publishing Company, New York, 308 p.

Ross, J.C.
1834: On the position of the north magnetic pole; Philosophical Transactions of the Royal Society of London, v. 124, p. 47-51.

Sabine, E.
1821: An account of experiments to determine the acceleration of the pendulum in various latitudes; Philosophical Transactions of the Royal Society of London, Part 2, p. 163-190.

Sander, G.W. and Overton, A.
1965: Deep seismic refraction investigation in the Canadian Arctic Archipelago; Geophysics, v. 30, p. 87-96.

Sass, J.H., Lachenbruch, A.H., and Munro, R.J.
1971: Thermal conductivity of rocks from measurements on fragments and its application to heat-flow determinations; Journal of Geophysical Research, v. 76, p. 3391-3401.

Schuster, A.
1908: The diurnal variation of terrestrial magnetism; Philosophical Transactions of the Royal Society of London, Series A, v. 208, p. 163-204.

Smith, W.E.T.
1961: Earthquakes of the Canadian Arctic 1956-1959; Department of Mines and Technical Surveys, Canada, Dominion Observatory, Seismological Series, 1961-2, 8 p.

Smith, W.E.T., Whitham, K., and Piché, W.T.
1968: A microearthquake swarm in 1965 near Mould Bay, N.W.T., Canada; Bulletin of the Seismological Society of America, v. 58, p. 1991-2011.

Sobczak, L.W.
1963: Regional gravity survey of the Sverdrup Islands and vicinity, with map; Department of Energy, Mines and Resources, Canada, Dominion Observatory, Gravity Map Series, no. 11, 19 p.
1975: Gravity and deep structure of the continental margin of Banks Island and Mackenzie Delta; Canadian Journal of Earth Sciences, v. 12, p. 378-394.

Sobczak, L.W., Hearty, D.B., Forsberg, R., Kristoffersen, Y., Edholm, O., and May, S.D.
in press: Gravity from 64°N to the North Pole; in The Arctic Ocean Region, A. Grantz, L. Johnson, and J.F. Sweeney, (ed.); Geological Society of America, The Geology of North America, v. L, Plate 3.

Sobczak, L.W., Mayr, U., and Sweeney, J.F.
1986: Crustal section across the polar continent-ocean transition in Canada; Canadian Journal of Earth Sciences, v. 23, p. 608-621.

Sobczak, L.W. and Overton, T.
1984: Shallow and deep crustal structure of the western Sverdrup Basin, Arctic Canada; Canadian Journal of Earth Sciences, v. 21, p. 902-919.

Sobczak, L.W. and Stephens, L.E.
1974: The gravity field of northeastern Ellesmere Island, part of northern Greenland and Lincoln Sea, with map; Department of Energy, Mines and Resources, Canada, Earth Physics Branch, Gravity Map Series, no. 114, 9 p.

Sobczak, L.W., Stephens, L.E., Winter, P.J., and Hearty, D.B.
1973: Gravity measurements over the Beaufort Sea, Banks Island and Mackenzie Delta; Department of Energy, Mines and Resources, Canada, Earth Physics Branch, Gravity Map Series, no. 151, 16 p.

Sobczak, L.W. and Sweeney, J.F.
1978: Gravity of the Arctic Ocean; in Arctic Geophysical Review, J.F. Sweeney (ed.); Publications of the Earth Physics Branch, v. 45., no. 4, p. 7-14.

Sobczak, L.W. and Weber, J.R.
1970: Gravity measurements in the Queen Elizabeth Islands, with maps; Department of Energy, Mines and Resources, Canada, Earth Physics Branch, Gravity Map Series, nos. 115-116, 14 p.
1973: Crustal structure of the Queen Elizabeth Islands and polar continental margin; in Arctic Geology, M.G. Pitcher (ed.); American Association of Petroleum Geologists, Memoir 19, p. 517-525.
1987: Gravity and bathymetry taken along seismic refraction lines from the Canadian Ice Island during 1985 and 1986; in Current Research, Part A, Geological Survey of Canada, Paper 87-1A, p. 299-304.

Sobczak, L.W., Weber, J.R., Goodacre, A.K., and Bisson, J.L.
1963: Preliminary results of gravity surveys in the Queen Elizabeth Islands, with maps; Department of Energy, Mines and Resources, Canada, Dominion Observatory, Gravity Map Series, nos. 12-15, 2 p.

Srivastava, S.P.
1985: Evolution of the Eurasian Basin and its implications to the motion of Greenland along Nares Strait; Tectonophysics, v. 114, p. 29-53.

Stephens, L.E., Sobczak, L.W., and Wainwright, E.S.
1972: Gravity measurements on Banks Island, N.W.T., with map; Department of Energy, Mines and Resources, Canada, Earth Physics Branch, Gravity Map Series, no. 150, 4 p.

Sweeney, J.F., Balkwill, H.R., Franklin, R., Mayr, U., McGrath, P., Snow, E., Sobczak, L.W, Wetmiller, R.J., and Panarctic Oils Ltd.
1986: Transect G. Somerset Island to Canada Basin; Geological Society of America, North America Continent-Ocean Transects Program, Centennial Continent/Ocean Transect #11, 5 p.

Taylor, A.E., Burgess, M., Judge, A.S., and Allen, V.S.
1982: Canadian geothermal data collection — northern wells 1981; Energy, Mines and Resources, Canada, Earth Physics Branch, Geothermal Series, no. 13, 153 p.

Taylor, A.E., Judge, A.S., and Desrochers, D.
1983: Shoreline regression: its effect on permafrost and the geothermal regime, Canadian Arctic Archipelago; in Permafrost: Fourth International Conference, Proceedings, National Academy Press, Washington, D.C., p. 1239-1244.

Taylor, A.E., Judge, A.S. and Allen, V.S.
in press: The automatic well temperature measuring system installed at Cape Allison C-47, offshore well, Arctic Islands of Canada. 2. Data retrieval and analysis of the thermal regime; Journal of Canadian Petroleum Technology.

Texaco Canada Resources Ltd.
1983: Melville Island, Northwest Territories, Canada Line no. 7; in Seismic Expression of Structural Styles, a Picture and Work Atlas, A.W. Bally (ed.); American Association of Petroleum Geologists, Studies in Geology, Series 15, v.3, p. 3.4.1-73 to 3.4.1-78.

Trettin, H.P., De Laurier, I., Frisch, T.O., Law, L.K., Niblett, E.R., Sobczak, L.W., Weber, J.R., and Witham, K.
1972: The Innuitian Province; in Tectonic Styles in Canada, R.A. Price and R.J.W. Douglas (ed.); Geological Association of Canada, Special Paper 11, p. 83-179.

Waylett, D.C.
1979: Gas in the Arctic Islands: discovered reserves and future potential; Journal of Petroleum Geology, v. 1, no. 3, p. 21-34.

Weber, J.R.
1961: Comparison of gravitational and seismic depth determinations on the Gilman Glacier and adjoining ice cap in northern Ellesmere Island; in Geology of the Arctic, G.O. Raasch, (ed.); University of Toronto Press, v. II, p. 781-790.

Weng, W.L.
1980: Preliminary Bouguer anomalies of western North Greenland; Grønlands Geologiske Undersøgelse, Report no. 99, p. 153-154.

Wetmiller, R.J. and Forsyth, D.A.
1982: Review of seismicity and other geophysical data near Nares Strait; in Nares Strait and the Drift of Greenland: a Conflict in Plate Tectonics, P.R. Dawes and J.W. Kerr (ed.); Meddelelser om Grønland, Geoscience 8, p. 261-274.

Whitham, K.
1963: An anomaly in geomagnetic variations at Mould Bay in the Arctic Archipelago of Canada; Geophysical Journal of the Royal Astronomical Society, v. 8, p. 26-43.

Whitham, K. and Andersen, F.
1962: The anomaly in geomagnetic variation at Alert in the Arctic Archipelago of Canada; Geophysical Journal of the Royal Astronomical Society, v. 7, p. 220-243.

Worzel, J.L.
1968: Advances in marine geophysical research of continental margins; Canadian Journal of Earth Sciences, v. 5, p. 458-469.

Authors' addresses

I. Asudeh
Geological Survey of Canada
1 Observatory Crescent
Ottawa, Ontario
K1A 0Y3

R.L. Coles
Geological Survey of Canada
1 Observatory Crescent
Ottawa, Ontario
K1A 0Y3

D.A. Forsyth
Geological Survey of Canada
1 Observatory Crescent
Ottawa, Ontario
K1A 0Y3

H.S. Hasegawa
Geological Survey of Canada
1 Observatory Crescent
Ottawa, Ontario
K1A 0Y3

A.M. Jessop
Geological Survey of Canada
3303-33rd St. N.W.
Calgary, Alberta
T2L 2A7

R.D. Kurtz
Geological Survey of Canada
1 Observatory Crescent
Ottawa, Ontario
K1A 0Y3

L.R. Newitt
Geological Survey of Canada
1 Observatory Crescent
Ottawa, Ontario
K1A 0Y3

E.R. Niblett
Geological Survey of Canada
1 Observatory Crescent
Ottawa, Ontario
K1A 0Y3

A. Overton
Geological Survey of Canada
601 Booth Street
Ottawa, Ontario
K1A 0E8

CHAPTER 5

L.W. Sobczak
Geological Survey of Canada
1 Observatory Crescent
Ottawa, Ontario
K1A 0Y3

R.J. Wetmiller
Geological Survey of Canada
1 Observatory Crescent
Ottawa, Ontario
K1A 0Y3

Printed in Canada

Chapter 6
PRECAMBRIAN SUCCESSIONS IN THE NORTHERNMOST PART OF THE THE CANADIAN SHIELD

Introduction

Metamorphic-plutonic basement

Proterozoic sedimentary and volcanic successions

References

Chapter 6

PRECAMBRIAN SUCCESSIONS IN THE NORTHERNMOST PART OF THE CANADIAN SHIELD

T. Frisch and H.P. Trettin

INTRODUCTION

This chapter summarizes information on the northernmost parts of the Shield relevant to the understanding of the early history of the Innuitian region; a systematic description of the Canadian Shield and Greenland can be found in another volume of this series (Hoffman et al., in prep.). From a stratigraphic point of view, the Shield is divisible into crystalline basement, metamorphosed in Archean and/or Early Proterozoic time, and unconformably overlying sedimentary and volcanic successions of Proterozoic age that are only slightly deformed.

METAMORPHIC-PLUTONIC BASEMENT

The following section describes briefly the crystalline basement rocks at the margin of the Arctic Platform, proceeding from southwest to northeast. Regional metamorphic facies of the Canadian part have been compiled by Fraser et al. (1978).

Victoria Island. Granodiorite on the northeast side of the Minto Arch yielded a' K-Ar age of 2391 ± 125 Ma (recalculated from Lowdon et al., 1963), suggesting that it may be part of the original Slave Province. This interpretation is supported by the inferred Early Proterozoic age of unconformably overlying sediments (Campbell, 1981; see below). A later thermal event, however, is apparent from the 1645 ± 20 Ma K-Ar age on a pegmatitic granite in the Wellington High (W.A. Gibbins, quoted by Campbell and Cecile, 1979).

Boothia Peninsula, Somerset Island and Prince of Wales Island. The Precambrian core of the Boothia Uplift (Fig. 4.2, 4.3) consists of quartzofeldspathic, pelitic, calcareous, and mafic rocks metamorphosed largely in the transitional granulite facies (Blackadar, 1967; Brown et al., 1969). Structural trends are northerly on Somerset Island and northern Boothia Peninsula; farther south, they appear to be truncated by northeasterly trends that continue to Melville Peninsula (Heywood, 1961, 1967). K-Ar ages of 1617-1715 Ma probably are cooling ages related to Early Proterozoic events; analogy with adjacent regions suggests that the rocks are Archean in age.

Melville Peninsula and Baffin and Bylot islands. The crystalline terrane in this region, which includes Archean greenstone belts, has a complex history of deformation, granitic intrusion, and metamorphism, mainly between 2.9 and 1.9 Ga but extending to 1.6 Ga (Jackson and Morgan, 1978; Frisch, 1982, Henderson, 1983). The regional metamorphism is mostly of amphibolite or granulite facies. The northeasterly-trending Foxe and Committee fold belts strike across Melville Peninsula into central Baffin Island; elsewhere in Baffin Island, trends are mainly NW-SE, subparallel with Baffin Bay (Jackson and Taylor, 1972).

Southeastern Ellesmere Island to eastern Devon Island. This area is of particular interest for this volume: it is close to the Innuitian Orogen, includes the Inglefield Uplift, which was positive in the Devonian (Fig. 4.3), and is significant for correlations across Nares Strait.

Precambrian crystalline basement is exposed on southern Ellesmere Island and eastern Devon Island eastward of longitude 86° to an increasing extent, concomitant with a general rise in the land surface. Easternmost Ellesmere and Devon islands and Coburg Island form a rugged terrain, whose relief exceeds 1500 m commonly and 2300 m locally, and which is covered extensively by permanent ice and snow. On Ellesmere Island, the western margin of the icefields essentially coincides with the contact between Precambrian Shield and Arctic Platform. The northernmost exposures of crystalline rocks are found south and east of Bache Peninsula.

The entire crystalline terrane is in the granulite facies, with only local retrogression to amphibolite facies. Supracrustal rocks abound and tend to form distinct belts, such as on southernmost Ellesmere Island and on Devon Island. These belts exemplify contrasting structural trends: mainly northerly on Ellesmere and Coburg islands, easterly on Devon Island. Quartzofeldspathic gneisses of granitic (*sensu lato*) composition occur abundantly in close association with, and independently of, the supracrustal rocks attached. Plutonic intrusive rocks, equivalent in composition and metamorphic facies to the quartzofeldspathic gneisses, occupy large tracts on Jones Sound and at the northern edge of the Shield. In addition, synmetamorphic anatectic granite is widely distributed on Ellesmere and Coburg islands.

Frisch (1983) has described the main lithological features and structural relations of the basement rocks; what follows here is chiefly a summary of more recent work. Quartzofeldspathic gneisses contain the assemblage: orthopyroxene + biotite + feldspar + quartz ± hornblende ± garnet, with tonalitic compositions predominating. Gneiss from southern Devon Island has a U-Pb zircon age of 2518

Frisch, T. and Trettin, H.P.
1991: Precambrian successions in the northernmost part of the Canadian Shield; Chapter 6 in Geology of the Innuitian Orogen and Arctic Platform of Canada and Greenland, H.P. Trettin (ed.); Geological Survey of Canada, Geology of Canada, no. 3; (also Geological Society of America, The Geology of North America, v. E).

+56/-33 Ma. The gneisses are interpreted as representing a continental margin calc-alkaline suite. By far the most abundant metasediment is migmatitic garnet-cordierite-sillimanite gneiss, in which cordierite is a secondary mineral formed by reaction between garnet and sillimanite, caused by falling pressure during uplift. Widespread partial melting of the aluminous metasediments produced intrusive bodies, ranging in size from veins to stocks, of peraluminous granite containing garnet, cordierite and/or sillimanite. Associated with the cordierite-bearing gneisses are marble, quartzite, and metabasalt. Marble, rich in diopside or forsterite and locally with wollastonite, is relatively common on Ellesmere and Coburg islands but scarce on Devon. Metabasalt, garnet-free on Ellesmere but garnetiferous locally on Devon, is chemically of oceanic (island arc?) affinity. The marble and quartzite are thought to have been shelf sediments, the cordierite gneisses deeper water mudstones and greywackes, all formed at a continental margin.

Crudely gneissic to locally massive orthopyroxene-bearing plutonic bodies of tonalitic to granitic composition have intruded the gneisses syntectonically. U-Pb zircon ages of two of these bodies are 1951 +43/-33 Ma and 1927 +18/-17 Ma. A U-Pb age of 1913 +20/-10 Ma determined on zircon from a garnet- and cordierite-rich anatexite is essentially in agreement, and provides further evidence of a major thermal event in mid-Early Proterozoic time.

The moderate pressure-temperature regime of granulite metamorphism indicated by the widespread occurrence of cordierite, the common absence of garnet from metamafic rocks, and the presence of wollastonite in marble is confirmed by mineral geothermobarometry. Inferred pressures range approximately from 4 to 6 kb on Ellesmere Island and up to 7 kb on Devon Island, at temperatures between 600° and 750°C. It is emphasized that the retrograde conditions attending uplift have strongly influenced these estimates, and peak conditions of metamorphism may have been significantly higher.

Deformation was intense throughout the terrane and remains poorly known. Lithological layering and foliation generally dip steeply. Two periods of isoclinal folding were followed by broad open folding. Major shear zones and mylonite belts are particularly common on Devon Island, where deformation appears to have been greater than on Ellesmere Island.

Northwestern Greenland. Continuity of the crystalline terrane of northwestern Greenland with that of southeastern Ellesmere Island is indicated by the overall similarity in lithology and metamorphic grade of the rocks, the presence of similar intrusive bodies cutting gneisses and supracrustal strata, and the distribution of on-strike marble units, on both sides of Nares Strait (Dawes, 1976; Frisch and Dawes, 1982). Rb-Sr and K-Ar analyses on two meta-igneous complexes suggest intrusion at circa 2.6 Ga, and metamorphism and deformation in Early to early Middle Proterozoic ("Hudsonian") time (Larsen and Dawes, 1974; Kalsbeek and Dawes, 1980; P.R. Dawes, pers. comm., 1986). Similar results were obtained from Rb-Sr studies of ice-transported boulders and a giant inclusion in a dolerite dyke in northeastern Greenland (Kalsbeek and Jepsen, 1980).

PROTEROZOIC SEDIMENTARY AND VOLCANIC SUCCESSIONS

Slightly deformed and almost unmetamorphosed sedimentary and volcanic successions of different Proterozoic ages overlie the basement rocks with high angular unconformity or nonconformity. The contrast in the metamorphic grade above and below this unconformity indicates that profound erosion has occurred in the intervening interval; it has been estimated, for example, that about 30 km of crust have been removed from the Baffin Island region (Jackson and Morgan, 1978). For simplicity in this brief account, these successions can be assigned to two major age groups.

(1) **Lower and lower Middle Proterozoic** deposits are confined to Victoria Island, where they occur in the northeastern part of the Minto Uplift and in the Wellington High (Fig. 4.3).

(a) The Hadley Bay Formation (Campbell, 1981), which appears to lie unconformably on Archean(?) basement, consists of quartzite, dolostone, and siltstone, deposited on a southwest-dipping shelf. It has been correlated with the Western River Formation of the Early Proterozoic Kilohigok Basin (on the mainland south of Victoria Island).

(b) An unconformably overlying unit of conglomerate and quartzite, in turn unconformably overlain by the Shaler Group (see below), was derived from southeasterly sources. It is comparable to the lower Middle Proterozoic Ellice Formation of the Coppermine Homocline (on the mainland southwest of Victoria Island).

(2) Deposits of **Middle and Late Proterozoic** age occur in structural depressions known as the Amundsen, Fury and Hecla, Borden, and Thule basins and in unnamed depressions on Somerset Island and northeastern Greenland (Fig. 6.1, 6.2, 6.3; cf. Christie, 1972). Preserved thicknesses in these areas are as follows: Amundsen Basin, about 4 km (Young, 1981; Campbell, 1985; Jefferson, 1985); Somerset Island, 2.4 km (Stewart, 1987); Fury and Hecla Basin, about 6 km (Chandler et al., 1980; Chandler and Stevens, 1981); Borden Basin, 6.1 km (Jackson and Ianelli, 1981); Thule Basin, 4.5 km (Dawes, 1976; Dawes et al., 1982; Frisch and Christie, 1982; Jackson, 1986); northeastern Greenland, 4+ km (Clemmensen, 1979; Jepsen and Kalsbeek, 1979; Collinson, 1980). These deposits are divisible into four major units or successions (herein referred to as 2a to 2d).

(2a) The oldest unit is the Independence Fjord Group of northeastern Greenland (Collinson, 1980), which consists of from 1.7 km to probably more than 2.5 km of sandstone and siltstone. Clay fractions have a Rb-Sr age of 1380 Ma, interpreted as the age of diagenesis (Larsen and Graff-Petersen, 1980).

(2b) A second succession consists of: (1) basalt and interstratified coarse clastic sediments, included in the Nauyat Formation of the Borden Basin, the lower redbed formation of the Fury and Hecla Strait Basin, and the lower two members of the Wolstenholme Formation of the Thule Basin (assigned to the Nauyat Formation by Jackson, 1986); and (2) of basalt assigned to the Zig-Zag Dal Basalt Formation of northeastern Greenland, which is up to 1350 m thick (Jepsen et al., 1980). Granophyre genetically related to the latter has an eleven-point Rb-Sr isochron age of 1230 ± 25 Ma; less precise K-Ar whole-rock ages from the Thule

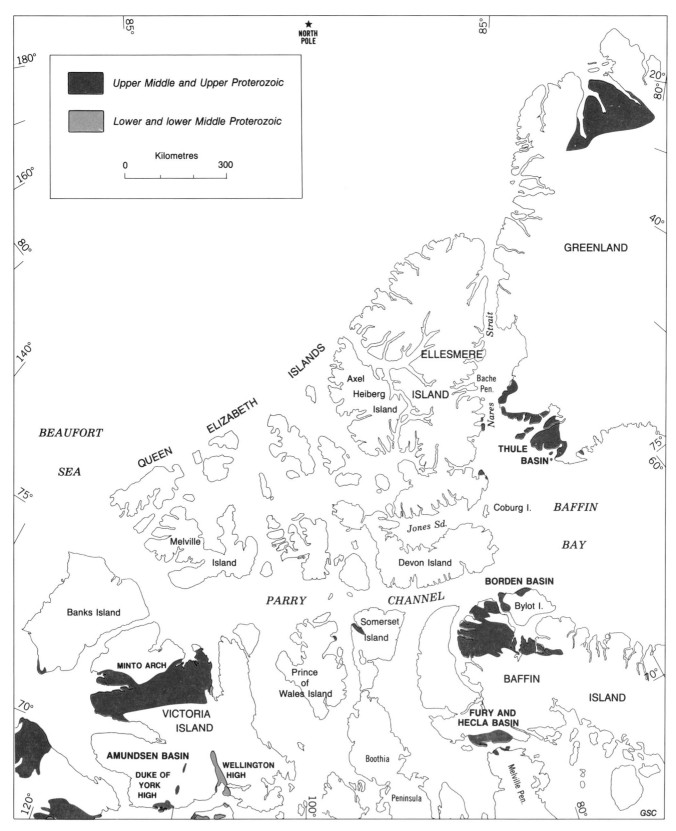

Figure 6.1. Generalized outcrop areas of Proterozoic volcanic and sedimentary successions in the northernmost part of the Shield (units in the Caledonian and Franklinian mobile belts not included).

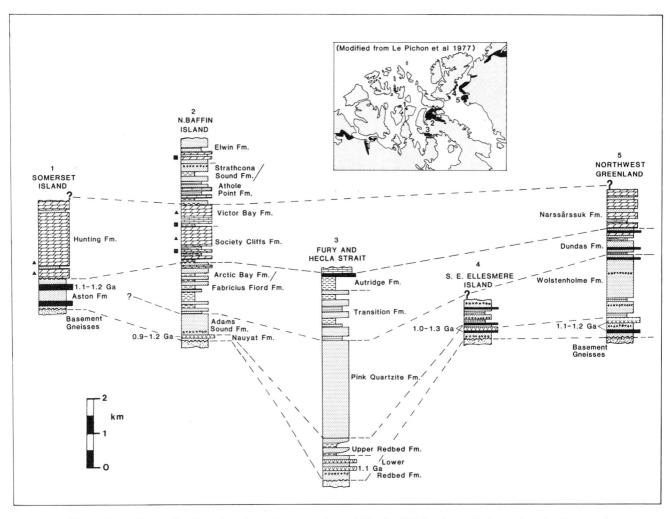

Figure 6.2. Generalized stratigraphy and proposed correlation for Middle-Upper Proterozoic units, Somerset Island to northwestern Greenland (Fig. 33 in Jackson and Ianelli, 1981).

and Borden basins are broadly comparable with this determination (Jackson and Ianelli, 1981; Dawes and Rex, 1985). The age of the Zig-Zag Dal Formation is identical — within the confidence limits stated — with a 6-point Rb-Sr isochron age of 1257 ± 47 Ma for Coppermine River basalt (Victoria Island; recalculated from Wanless and Loveridge, 1972). Paleomagnetic studies (Fahrig et al., 1981) confirm correlation of the Nauyat Formation with the Mackenzie intrusions, which are related to the Coppermine volcanics. Combined, these data indicate a major rifting event at circa 1.25 Ga (Jackson and Ianelli, 1981; Fahrig et al., 1981).

(2c) The volcanic and clastic units are followed by thick successions of clastic, carbonate, and locally evaporitic sediments, which include all strata above unit 2b in the Borden and Thule basins; the Campanuladal and Fins Sø formations of northeastern Greenland; the Hunting Formation of Somerset Island; and the Shaler Group and equivalent Rae Group of the Amundsen Basin. The latter overlies the basaltic Coppermine River Group on the mainland and has been correlated, formation by formation, with the Mackenzie Mountains Supergroup of the northeastern Cordillera.

The position of succession 2c within the 1.2-0.7 Ga span, defined by the isotopic ages of successions 2b and 2d, is uncertain, as conflicting results have been obtained. Paleomagnetic studies have dated the entire Borden Basin succession (Bylot Supergroup) at 1220-1202 Ma (Fahrig et al., 1981) and the Mackenzie Mountains Supergroup (equivalent to the Shaler Group) at circa 880-770 Ma (Park and Aitken, 1986). However, acritarchs from the Thule Group, which has been correlated with the Bylot Supergroup (Jackson, 1986), are late Riphean and early Vendian in age (Dawes and Vidal, 1985), suggesting that these two units fill much of the 1.2-0.7 Ga span.

The stratigraphy and sedimentology of this succession are too complex to be discussed here. Suffice it to state that it seems to have been deposited in a southwesterly-trending belt of nonmarine to subtidal environments, that may represent the passive margin of an ocean ("Poseidon") on the northwest (Jackson and Ianelli, 1981; Jackson, 1986).

(2d) In the Arctic Islands, the upper succession (2d) is represented only by the Natkusiak Formation of Victoria Island, which overlies the Shaler Group with disconformity (Thorsteinsson and Tozer, 1962). It is about 740 m thick

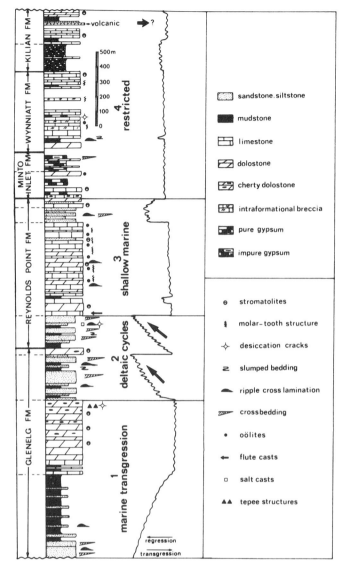

Figure 6.3. Stratigraphy of the Shaler Group on Victoria Island. The formational names of Thorsteinsson and Tozer (1962) have been retained but other mappable units are indicated by dashed lines on left side of diagram. Wavy line on the right is an interpretation in terms of interplay between clastic influx and marine transgression. The depositional history has been divided into four phases. Arrows at phase 2 indicate northwesterly transport; small arrow at top of phase 4 indicates possible westerly provenance (Fig. 1 in Young, 1981).

and composed mainly of basaltic flows and pyroclastic deposits. Flow rocks have given a K-Ar isochron age of 635 Ma (recalculated from Palmer and Hayatsu, 1975). Lithologically related diabase intrusions yielded K-Ar ages of 635 and 640 Ma (recalculated from Christie, 1964). Paleomagnetic investigations (Palmer et al., 1983) demonstrate correlation with the Franklin diabases, which intrude strata of succession 2c in most areas. The K-Ar ages on these rocks are scattered, but a fair number of determinations from the Thule, Borden, and Fury and Hecla basins, lying between 700 and 800 Ma (Dawes and Rex, 1985; Jackson and Ianelli, 1981; Chandler and Stevens, 1981;), are comparable to Rb-Sr and U-Pb ages of about 770-780 Ma on diabase sheets and related quartz diorite from the Mackenzie Mountains (Armstrong et al., 1982; C.W. Jefferson and R. Parrish, pers. comm., 1988). The dykes occur near the base of the uppermost Proterozoic Windermere Supergroup, which includes abundant rift-related clastic sediments, along with glaciogenic clastic sediments and shelf carbonates. Glaciogenic sediments in northeastern Greenland (Morænsø Formation; Clemmensen, 1979) may be correlative with those in the Cordillera.

The Natkusiak Formation and underlying units of the Minto Arch were deformed into broad open folds with dips of 5° or less on the flanks, prior to the deposition of unconformably overlying clastic sediments of late Early Cambrian age (Thorsteinsson and Tozer, 1962; 1970). In other parts of the Arctic Platform, for example on northern Baffin Island, slight tilting is apparent beneath the sub-Cambrian unconformity, which truncates the entire Bylot Supergroup and cuts into the crystalline basement. These structures are attributed to the last spasms of rifting.

REFERENCES

Armstrong, R.L., Eisbacher, G.H., and Evans, P.D.
1982: Age and stratigraphic-tectonic significance of Proterozoic diabase sheets, Mackenzie Mountains, northwestern Canada; Canadian Journal of Earth Sciences, v. 19, p. 316-323.

Blackadar, R.G.
1967: Precambrian geology of Boothia Peninsula, Somerset Island, and Prince of Wales Island, District of Franklin; Geological Survey of Canada, Bulletin 151, 62 p.

Brown, R.L., Dalziel, I.W.D., and Rust, B.R.
1969: The structure, metamorphism, and development of the Boothia Arch, Arctic Canada; Canadian Journal of Earth Sciences, v. 6, p. 525-543.

Campbell, F.H.A.
1981: Stratigraphy and tectono-depositional relationships of the Proterozoic rocks of the Hadley Bay area, northern Victoria Island, District of Franklin; in Current Research, Part A, Geological Survey of Canada, Paper 81-1A, p. 15-22.
1985: Stratigraphy of the upper part of the Rae Group, Johansen Bay area, northern Coronation Gulf area, District of Franklin; in Current Research, Part A, Geological Survey of Canada, Paper 85-1A, p. 693-696.

Campbell, F.H.A. and Cecile, M.P.
1979: The northeastern margin of the Aphebian Kilohigok Basin, Melville Sound, Victoria Island, District of Franklin; in Current Research, Part A, Geological Survey of Canada, Paper 79-1A, p. 91-94.

Chandler, F.W., Charbonneau, B.W., Ciesielski, A., Maurice, Y.T., and White, S.
1980: Geological studies of the Late Precambrian supracrustal rocks and underlying granitic basement, Fury and Hecla Strait area, Baffin Island, District of Franklin; in Current Research, Part A, Geological Survey of Canada, Paper 80-1A, p. 125-132.

Chandler, F.W. and Stevens, R.D.
1981: Potassium-argon age of the Late Proterozoic Fury and Hecla Formation, northwest Baffin Island, District of Franklin; in Current Research, Part A, Geological Survey of Canada, Paper 81-1A, p. 37-40.

Christie, R.L.
1964: Diabase-gabbro sills and related rocks of Banks and Victoria islands, Arctic Archipelago; Geological Survey of Canada, Bulletin 105, 11 p.

Clemmensen, L.B.
1979: Notes on the paleogeographical setting of the Eocambrian tillite-bearing sequence of southern Peary Land, North Greenland; Grønlands Geologiske Undersøgelse, Report no. 88, p. 15-22.

Collinson, J.D.
1980: Stratigraphy of the Independence Fjord Group (Proterozoic) of eastern North Greenland; Grønlands Geologiske Undersøgelse, Report no. 99, p. 7-23.

Dawes, P.R.
1976 Precambrian to Tertiary of northern Greenland; in Geology of Greenland, A. Escher and W.S. Watt, (ed.), The Geological Survey of Greenland, Copenhagen, p. 248-303.

Dawes, P.R., Frisch, T., and Christie, R.L.
1982 The Proterozoic Thule Basin of Greenland and Ellesmere Island: importance to the Nares Strait debate; in Nares Strait and the Drift of Greenland; a Conflict in Plate Tectonics, P.R. Dawes and J.W. Kerr (ed.); Meddelelser om Grønland, Geoscience 8, p. 89-104.

Dawes, P.R. and Rex, D.C.
1985: Proterozoic basaltic magmatic periods in North-West Greenland: evidence from K/Ar ages; Grønlands Geologiske Undersøgelse, Report no. 130, p. 24-31.

Dawes, P.R. and Vidal, G.
1985: Proterozoic age of the Thule Group: new evidence from microfossils; in Grønlands Geologiske Undersøgelse, Report no. 125, p. 22-28.

Fahrig, W.F., Christie, K.W., and Jones, D.L.
1981 Paleomagnetism of the Bylot basins: evidence for Mackenzie continental tensional tectonics; in Proterozoic Basins of Canada, F.H.A. Campbell (ed.); Geological Survey of Canada, Paper 81-10, p. 303-312.

Fraser, J.A., Heywood, W.W., and Mazurski, M.A. (compilers)
1978: Metamorphic map of the Canadian Shield; Geological Survey of Canada, Map 1475A.

Frisch, T.
1982: Precambrian geology of the Prince Albert Hills, western Melville Peninsula, Northwest Territories; Geological Survey of Canada, Bulletin 346, 70 p.
1983: Reconnaissance geology of the precambrian Shield of Ellesmere, Devon and Coburg islands, Arctic Archipelago: a preliminary account; Geological Survey of Canada, Paper 82-10, 11 p.

Frisch, T. and Christie, R.L.
1982: Stratigraphy of the Proterozoic Thule Group, southeastern Ellesmere Island, Arctic Archipelago; Geological Survey of Canada, Paper 81-19, 13 p.

Frisch, T. and Dawes, P.R.
1982: The Precambrian Shield of northernmost Baffin Bay: correlation across Nares Strait; in Nares Strait and the Drift of Greeland: a Conflict in Plate Tectonics, P.R. Dawes and J.W. Kerr (ed.); Meddelelser om Grønland, Geoscience 8, p. 79-88.

Henderson, J.R.
1983: Structure and metamorphism of the Aphebian Penrhyn Group and its Archean basement complex in the Lyon Inlet area, Melville Peninsula, District of Franklin; Geological Survey of Canada, Bulletin 324, 50 p.

Heywood, W.W.
1961: Geological notes, northern District of Keewatin; Geological Survey of Canada, Paper 61-18, 9 p.
1967: Geological notes, northeastern District of Keewatin and southern Melville Peninsula, District of Franklin, Northwest Territories; Geological Survey of Canada, Paper 66-40, 20 p.

Hoffman, P.F., Card, K.D., and Davidson, A.
in prep.: Precambrian Geology of the Craton in Canada and Greenland; Geological Survey of Canada, Geology of Canada, no. 7 (also Geological Society of America, The Geology of North America, v. C-1).

Jackson, G.D.
1986: Notes on the Proterozoic Thule Group, northern Baffin Bay; in Current Research, Part A, Geological Survey of Canada, Paper 86-1A, p. 541-552.

Jackson, G.D. and Ianelli, T.R.
1981: Rift-related cyclic sedimentation in the Neohelikian Borden Basin, northern Baffin Island; in Proterozoic Basins of Canada, F.H.A. Campbell (ed.); Geological Survey of Canada, Paper 81-10, p. 269-302.

Jackson, G.D. and Morgan, W.C.
1978: Precambrian metamorphism on Baffin and Bylot islands; in Metamorphism in the Canadian Shield, J.A. Fraser and W.W. Heywood (ed.), Geological Survey of Canada, Paper 78-10, p. 249-267.

Jackson, G.D. and Taylor, F.C.
1972: Correlation of major Aphebian rock units in the northeastern Canadian Shield; Canadian Journal of Earth Sciences, v. 9, p. 1650-1669.

Jefferson, C.W.
1985: Uppermost Shaler Group and its contact with the Natkusiak basalts, Victoria Island, District of Franklin; in Current Research, Part A, Geological Survey of Canada, Paper 85-1A, p. 103-110.

Jepsen, H.F. and Kalsbeek, F.
1979: Igneous rocks in the Proterozoic platform of eastern North Greenland; Grønlands Geologiske Undersøgelse, Report no. 88, p. 11-14.

Jepsen, H.F., Kalsbeek, F., and Suthren, R.J.
1980: The Zig-Zag Dal Basalt Formation, North Greenland; Grønlands Geologiske Undersøgelse, Report no. 99, p. 25-32.

Kalsbeek, F. and Dawes, P.R.
1980: Rb-Sr whole-rock measurements of the Kap York meta-igneous complex, Thule district, North-West Greenland; Grønlands Geologiske Undersøgelse, Report no. 100, p. 30-33.

Kalsbeek, F. and Jepsen, H.F.
1980: Preliminary Rb-Sr isotope evidence on the age and metamorphic history of the North Greenland crystalline basement; Grønlands Geologiske Undersøgelse, Report no. 99, p. 107-110.

Larsen, O. and Dawes, P.R.
1974: K/Ar and Rb/Sr age determinations on Precambrian crystalline rocks in the Inglefield Land – Inglefield Bredning region, Thule district, western North Greenland; Grønlands Geologiske Undersøgelse, Report no. 66, p. 5-8.

Larsen, O. and Graff-Petersen, P.
1980: Sr-isotopic studies and mineral composition of the Hagen Brae Member in the Proterozoic clastic sediments at Hagen Brae, eastern North Greenland; Grønlands Geologiske Undersøgelse, Report no. 99, p. 111-118.

Lowden, J.A., Stockwell, C.H., Tipper, H.W., and Wanless, R.K.
1963: Age determinations and geological studies (including isotopic ages – Report 3); Geological Survey of Canada, Paper 62-17, 140 p.

Palmer, H.C., Baragar, W.R.A., Fortier, M., and Foster, J.H.
1983: Paleomagnetism of Late Proterozoic rocks, Victoria Island, Northwest Territories, Canada; Canadian Journal of Earth Sciences, v. 20, p. 1456-1469.

Palmer, H.C. and Hayatsu, A.
1975: Paleomagnetism and K-Ar dating of some Franklin lavas and diabases, Victoria Island; Canadian Journal of Earth Sciences, v. 12, p. 1439-1447.

Park, J.K. and Aitken, J.D.
1986: Paleomagnetism of the Katherine Group in the Mackenzie Mountains: implications for post-Grenville (Hadrynian) apparent polar wander; Canadian Journal of Earth Sciences, v. 23, p. 308-323.

Stewart, W.D.
1987: Late Proterozoic to early Tertiary stratigraphy of Somerset Island and northern Boothia Peninsula, District of Franklin, N.W.T.; Geological Survey of Canada, Paper 83-26, 78 p.

Thorsteinsson, R. and Tozer, E.T.
1962: Banks, Victoria, and Stefansson islands, Arctic Archipelago; Geological Survey of Canada, Memoir 330, 85 p.
1970: Geology of the Arctic Archipelago; in Geology and Economic Minerals of Canada, R.J.W. Douglas (ed.); Geological Survey of Canada, Economic Geology Report no.1, p. 547-590.

Wanless, R.K. and Loveridge, W.D.
1972: Rubidium-strontium isochron age studies, Report 1; Geological Survey of Canada, Paper 72-23, 77 p.

Young, G.M.
1981: The Amundsen Embayment, Northwest Territories; relevance to the Upper Proterozoic evolution of North America; in Proterozoic Basins of Canada, F.H.A. Campbell (ed.); Geological Survey of Canada, Paper 81-10, p. 203-218.

Authors' addresses

T. Frisch
Geological Survey of Canada
601 Booth Street
Ottawa, Ontario
K1A 0E4

H.P. Trettin
Geological Survey of Canada
3303-33rd St. N.W.
Calgary, Alberta
T2L 2A7

Printed in Canada

Chapter 7
CAMBRIAN TO SILURIAN BASIN DEVELOPMENT AND SEDIMENTATION, NORTH GREENLAND

Geological setting

Tectonic lineaments

Basin evolution

Caledonian events in North Greenland

Acknowledgments

References

Chapter 7

CAMBRIAN TO SILURIAN BASIN DEVELOPMENT AND SEDIMENTATION, NORTH GREENLAND

A.K. Higgins, J.R. Ineson, J.S. Peel, F. Surlyk, and M. Sønderholm

GEOLOGICAL SETTING

The Franklinian Basin extended from northern Ellesmere Island across North Greenland, where its sedimentary infill is exposed from Inglefield Land and Washington Land in the west to Kronprins Christian Land in the east (Fig. 7.1-7.3). The segment of the basin exposed in North Greenland is approximately 800 km long and has a maximum preserved north-south width of 200 km. The thickness of the sedimentary column reaches about 8 km; the main part of the succession is of Cambrian-Silurian age, but it may extend down into the latest Precambrian and up into the earliest Devonian. A distinction into a shelf sequence and deep water trough sequence can be recognized in northern Ellesmere Island and in North Greenland, and the variations in facies with time in the two regions show close parallels. Correlation on group or formation level is often possible across Nares Strait (Peel and Christie, 1982; Peel et al., 1982). Detailed knowledge of the North Greenland sequences, however, permits an integrated account of shelf and trough development in the North Greenland segment of the Franklinian Basin.

A craton composed of Archean and Upper Proterozoic crystalline basement rocks, overlain by Middle and Upper Proterozoic sedimentary and volcanic rocks, lies to the south of the Franklinian Basin (Chapter 6). This is now exposed intermittently along the margin of the Inland Ice, and more extensively in eastern North Greenland (Fig. 7.1). In the early Paleozoic, this craton was fringed to the north by an east-west trending shallow marine shelf. Two main facies belts characterize the shelf, a southern shallow water carbonate-dominated platform (*sensu* Schlager, 1981), and a northern shale-dominated outer shelf. The boundary between these two regimes fluctuated considerably; in some periods the platform was almost drowned, while in others the platform prograded and the platform margin coincided with the shelf-slope break. A deep water basin, or trough, characterized by deposition of fine grained sediments, sand turbidites, and carbonate conglomerates was situated north of this zone. Shelf-parallel turbidite transport directions suggest that the deep water basin was two-sided for most of its existence (Surlyk et al., 1980; Surlyk and Hurst, 1984); for this reason the term trough is used in the following description. Furthermore, the Franklinian Basin includes the shelf as well as the deep water sequence. It should be noted, however, that there is no direct evidence in North Greenland of a northern margin to the Franklinian Basin.

The Franklinian Basin in North Greenland is thus interpreted as recording the evolution of a carbonate-dominated, east-west trending, passive continental margin. Progressive closure of the Iapetus Ocean and Caledonian mountain building to the east had a profound influence on the stratigraphic development of the trough sequence.

Deposition in the trough was brought to a close in North Greenland, as in northern Ellesmere Island, by the Ellesmerian Orogeny (Chapters 11, 12). In Greenland this produced the North Greenland fold belt; deformation is largely confined to the trough sequence, and was intense in the extreme north where it was accompanied by low amphibolite facies metamorphism. Deformation decreases southward, and dies out in a belt of thrusts and buckle folds located slightly north of the Silurian shelf-trough transition (see Chapter 11).

Because of the remoteness of North Greenland, most geological work until recent years has been on a reconnaissance level. General reviews of the geology and of earlier work are given by Dawes (1971, 1976), Dawes and Soper (1973), Dawes and Peel (1981), Dawes and Christie (1982), Peel (1982), and Christie and Dawes (Chapter 2). In 1978-80 and 1984-85, regional systematic mapping and general geological investigations were carried out by the Geological Survey of Greenland, resulting in a considerable increase in knowledge of virtually every aspect of North Greenland geology. The lower Paleozoic shelf stratigraphy (Fig. 7.2) has been described by Christie and Peel (1977), Hurst (1980a, 1984), Ineson and Peel (1987, in press), Peel (1982), Peel et al. (1981), Sønderholm and Due (1985), and Sønderholm et al. (1987). The stratigraphy of the lower Paleozoic deep water sequence has been outlined within a framework of six groups by Friderichsen et al. (1982) (Fig. 7.2). Preliminary descriptions of the Cambrian and Ordovician sequences were given by Soper et al. (1980), Higgins et al. (1981), Bengaard et al. (1987), Surlyk and Ineson (1987a), and Davis and Higgins (1987). The stratigraphy of the Cambrian-Ordovician trough sequence in Peary Land was established by Surlyk et al. (in press), and the stratigraphy of the Silurian deep water rocks was described by Hurst (1980a), Hurst and Surlyk (1982), Larsen and Escher (1985, 1987), and Surlyk and Ineson (1987b).

The main aspects of the tectonic-sedimentological evolution of the early Paleozoic deep water basin in North Greenland have been described by Surlyk and Hurst (1983, 1984). A series of stages in the evolution of the basin are

---

Higgins, A.K., Ineson, J.R., Peel, J.S., Surlyk, F., and Sønderholm, M.
1991: Cambrian to Silurian basin development and sedimentation, North Greenland; Chapter 7 in Geology of the Innuitian Orogen and Arctic Platform of Canada and Greenland, H.P. Trettin (ed.); Geological Survey of Canada, Geology of Canada, no. 3; (also Geological Society of America, The Geology of North America, v. E).

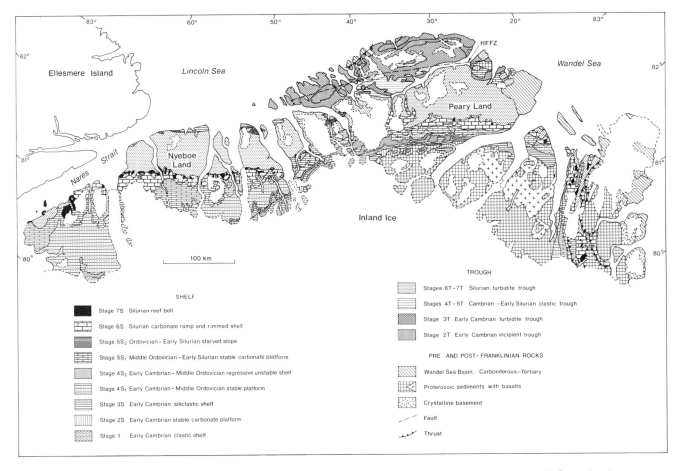

Figure 7.1. Geological map showing the subdivision of the Franklinian Basin sequence in North Greenland. See Figure 7.2 for localities named in the text.

distinguished, and related to intra-basinal tectonic lineaments which govern, or influence, the boundaries of the shelf, slope, and trough of the basin at different times (Fig. 7.3, 7.4).

TECTONIC LINEAMENTS

The evolution of the North Greenland early Paleozoic basin and its differentiation into a southern shelf and a northern trough was interpreted in terms of control by tectonic lineaments by Surlyk et al. (1980), a model elaborated and presented in more detail by Surlyk and Hurst (1983, 1984). It is envisaged that the basin expanded in several episodes by southward shift of the southern margin to new east-west trending lineaments. This concept is adopted here, and forms the basis for the integrated description of shelf and trough sequences presented in this paper. Some of the lineaments responsible for control of sedimentation during the early Paleozoic are easily defined. The identity of others, however, is not clear, since some of the structures have been shown to be main elements of Ellesmerian thin skinned thrust zones or Eurekan fault systems. These later thrusts and faults may have had precursors which were active in the early Paleozoic and influenced sedimentation.

The earliest stage of the Franklinian Basin recognized in North Greenland (Skagen Group) is loosely dated as Late Precambrian – Early Cambrian and its base is not known. For the next stage, of Early Cambrian age, a facies boundary can be defined that separates shelf carbonates of the Portfjeld Formation from equivalent trough carbonates of the Paradisfjeld Group (Fig. 7.2). This boundary runs from Depotbugt in the east, westward to outer J.P. Koch Fjord and along the northern coast of Greenland as far as northern Nyeboe Land (Fig. 7.4, stage 2). However, the original position of this lineament is obscured by the north-south shortening due to Ellesmerian thrusting. It may have corresponded to a precursor of the present Harder Fjord fault zone (Surlyk et al., 1980; Surlyk and Hurst, 1984; Soper and Higgins, 1987). The present distribution of succeeding deposits of the Buen Formation (shelf sandstones and shales) and Polkorridoren Group (trough turbidites) suggests that the shelf margin was located slightly farther south (Fig. 7.4, stage 3); in logged sections, deeper, more offshore facies overlie shallow shelf facies. The original position of the shelf margin lineament, however, is again obscured by Ellesmerian thrusting.

Carbonate accumulation resumed on the platform in the late Early Cambrian and continued through the Ordovician, while shale deposition dominated in the slope area which can be characterized as "starved". In some areas the total sequence deposited on the outer shelf and slope during this period is less than 100 m thick; 300-500 m is

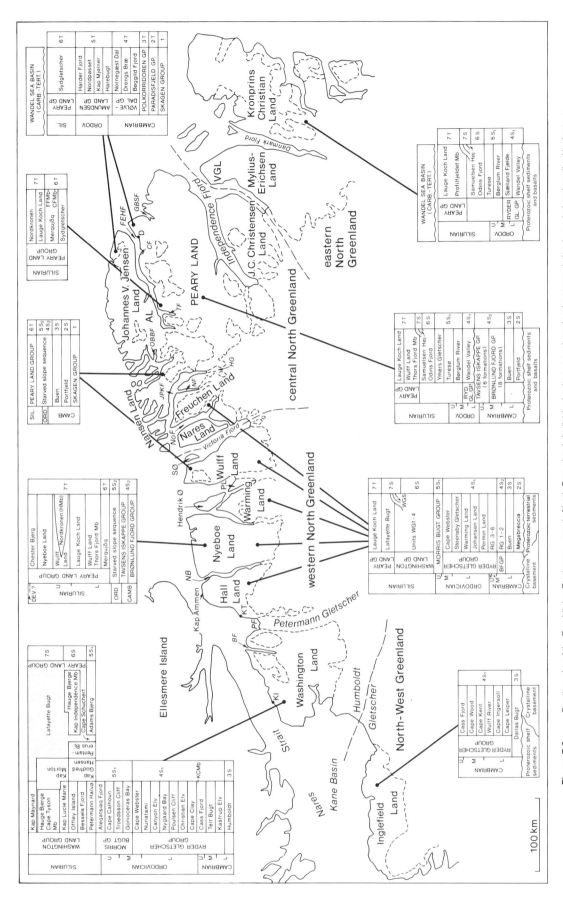

Figure 7.2. Stratigraphy of the Franklinian Basin sequence in North Greenland. Individual units are assigned to the stages in basin evolution described in the text (see Fig. 7.1). AL = Amundsen Land; BF = Bessels Fjord; BFGP = Brønlund Fjord Group; CF = Citronens Fjord; CFMb = Citronens Fjord Member; D = Depotbugt; FEHF = Frederick E. Hyde Fjord; FFMb = Freja Fjord Member; GBSF = G.B. Schley Fjord; HG = Henson Gletscher; HMb = Hendrik Ø Member; JPKF = J.P. Koch Fjord; KCMb = Kap Coppinger Member; KI = Kap Independence; KT = Kap Tyson; NB = Newman Bugt; NF = Navarana Fjord; NoF = Nordenskiöld Fjord; OBBF = O.B. Bøggild Fjord; PF = Petermann Fjord; PL = Permin Land; Ryd.Gl.Gp. = Ryder Gletscher Group; SØ = Stephenson Ø; TF = Thor Fjord; VGL = Valdemar Glückstadt Land.

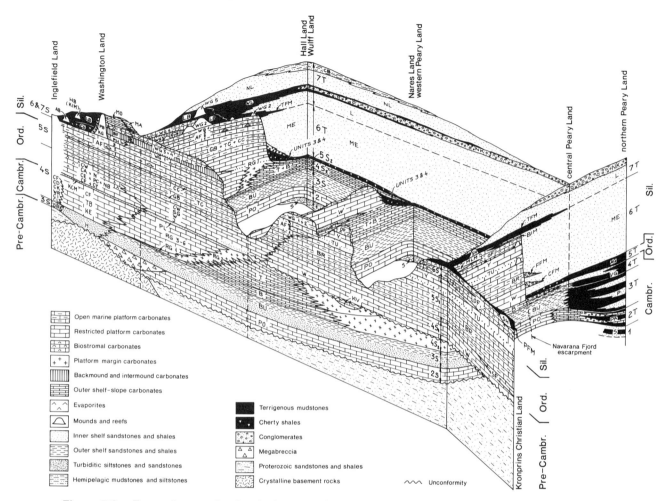

Figure 7.3. Fence diagram showing the interpreted relationships of shelf, slope and trough deposits of the various stages in the evolution of the Franklinian Basin sequence of North Greenland. Abbreviations denote formations unless otherwise specified. AB = Adams Bjerg; AF = Aleqatsiaq Fjord; AG = Amundsen Land Group; B = Brønlund Fjord Group; BF = Bessels Fjord; BIM = Bure Iskappe Member; BR = Børglum River; BU = Buen; CA = Canyon Elv; CB = Chester Bjerg; CC = Cape Calhoun; CD = Cape Wood; CE = Christian Elv; CF = Cass Fjord; CFM = Citronens Fjord Member; CI = Cape Ingersoll; CK = Cape Kent; CL = Cape Leiper; CS = Cape Schuchert; CW = Cape Webster; CY = Cape Clay; DB = Dallas Bugt; FFM = Freja Fjord Member; GB = Gonioceras Bay; H = Humboldt; HB = Hauge Bjerge; J = Johansen Land; KCM = Kap Coppinger Member; KE = Kastrup Elv; KG = Kap Godfred Hansen; KIM = Kap Independence Member; KV = Koch Vaeg; L = Lauge Koch Land; LB = Lafayette Bugt; MA = Kap Maynard; ME = Merqujôq; MLM = Melville Land Member; MO = Kap Morton; N = Nunatami; NB = Nygaard Bay; NK = Nordkronen; NL = Nyeboe Land; OF = Odins Fjord; OI = Offley Island; P = Polkorridoren Group; PA = Paradisfjeld Group; PB = Pentamerus Bjerge; PC = Poulsen Cliff; PFM = Profilfjeldet Member; PH = Petermann Halvø; PL = Permin Land; PO = Portfjeld; RG = Ryder Gletscher Group; RLM = Røhling Land Member; S = Skagen Group; SF = Sjaelland Fjelde; SG = Steensby Gletscher; SH = Samuelsen Høj; T = Tavsens Iskappe Group; TB = Telt Bugt; TC = Troedsson Cliff; TFM = Thors Fjord Member; TU = Turesø; VG = Vølvedal Group; W = Wandel Valley; WG = Washington Land Group; WL = Wulff Land; WR = Wulff River; Y = Ymers Gletscher.

more normal. The trough sequence, up to 1200 m thick, is only known from northern Peary Land (Johannes V. Jensen Land) where it includes the thickly developed turbiditic units of the Vølvedal Group. Two lineaments seem to have determined the boundaries between shelf and trough deposition during this period. The northern of these, extending from Depotbugt to Nyeboe Land, may have marked the north margin of the shelf during the first part of this period (Fig. 7.4, stage 4). The turbidites of the Vølvedal Group occur only north of this line. The southern lineament is the Navarana Fjord lineament, which is exposed as a platform margin scarp in the Navarana Fjord – J.P. Koch Fjord area (Surlyk and Hurst, 1984; Escher and Larsen, 1987; Surlyk and Ineson, 1987a, b). It extends eastward to the vicinity of Depotbugt and westward beneath younger rocks to northern Nyeboe Land and just offshore Hall Land. Carbonate-dominated outer shelf sediments accumulated in the area between the two lineaments during the early part of the period (Fig. 7.4, stage 4), but these are overlain by deeper water slope or trough sediments (Amundsen Land Group and equivalent chert-shale sequences), suggesting that the basin margin had retreated southward in the Ordovician to the Navarana Fjord lineament (Fig. 7.4, stage 5).

In the Late Ordovician and Early Silurian, the Navarana Fjord lineament was the facies boundary separating shelf carbonate deposition to the south from turbidite deposition of the Peary Land Group to the north (Fig. 7.4, stage 6). The lineament took the form of a pronounced escarpment, and the lowest formations of the turbiditic Peary Land Group (Sydgletscher and Merqujôq formations) occur only to the north. In Late Llandovery time, the outer platform foundered, probably in response to loading of the trough by thick turbidite sequences and to orogenic activity in the Caledonides in eastern North Greenland. The trough filled to the brim with turbidites and expanded southward to cover most of what is now North Greenland (Fig. 7.4, stage 7) (Hurst and Surlyk, 1982; Hurst et al., 1983). A new shelf-trough boundary was established farther to the south, corresponding to the line of impressive carbonate mounds running from central Hall Land to Kronprins Christian Land. The shelf was almost totally drowned at about the Llandovery-Wenlock boundary, and carbonate deposition only persisted in a few isolated mounds. The succession of formations in the Peary Land Group — Wulff Land Formation (mudstone), Lafayette Bugt Formation (mudstone and carbonate conglomerates), Lauge Koch Land Formation (sandstone turbidites), Nordkronen Formation (mainly chert pebble conglomerates) and Chester Bjerg Formation (turbiditic mudstone and siltstone) — reflects facies variations governed by the supply of detritus to the deep water trough from the source areas in the east, principally the rising mountains of the East Greenland Caledonides (Hurst et al., 1983; Surlyk and Hurst, 1984).

BASIN EVOLUTION

The development of the Franklinian Basin in North Greenland is described here in terms of a sequence of stages, defined on the basis of sedimentary infill and position and structural style of the shelf-trough boundary (cf. Surlyk and Hurst, 1983, 1984). The stages are broadly analogous to seismic stratigraphic sequences (Mitchum et al., 1977; Vail et al., 1977) or to groups of temporally related depositional systems (e.g. Galloway and Hobday, 1983). The stages are described below in ascending order; in each stage description of the shelf development being followed by description of the equivalent slope and deep water trough development.

Stage 1: basin initiation (Late Proterozoic? – Early Cambrian)

The oldest representative of the early Paleozoic sedimentary cycle is the Skagen Group (Fig. 7.1-7.3, 7.5). It is recognized in northeast Peary Land, Johannes V. Jensen Land, and in northern parts of central and western North Greenland, notably northern Wulff Land (Fig. 7.5). The base of the group is not seen, although stratigraphic relationships in northeast Peary Land suggest that it rests unconformably on Proterozoic quartzites and volcanics (Christie and Ineson, 1979). In its northern outcrop in Johannes V. Jensen Land, the Skagen Group is overlain by the basinal Paradisfjeld Group (see stage 2, below), whereas south of Frederick E. Hyde Fjord and in central and western North Greenland, it is overlain by platform carbonates of the Portfjeld Formation. The age of the Skagen Group is uncertain, but a late Early Cambrian fauna has been collected from the Kennedy Channel Formation of Ellesmere Island (Chapter 8) from strata considered by Dawes and Peel (1984) to be equivalent to the Skagen Group of northern Wulff Land.

In its type area in northeast Peary Land, the Skagen Group consists of tightly folded quartzitic sandstones and mudstones divided into three units by Friderichsen et al. (1982); the depositional environment is not known, but upper beds of the sequence in the G.B. Schley Fjord area include stromatolitic, mudcracked dolomites of peritidal origin. The lowest unit comprises structureless quartzitic sandstones, the middle unit dark phyllitic mudstones, and the upper unit consists of thick bedded quartzitic sandstones with phyllitic mudstone interbeds and rare pebble conglomerates. The upper unit is also exposed on the north side of the mouth of Frederick E. Hyde Fjord, where it has a conformable, but highly deformed, contact with the overlying Paradisfjeld Group.

Higgins and Soper (1985) suggested similarities between the Skagen Group and the clastic sequence which underlies the Portfjeld Formation between northern Wulff Land and J.P. Koch Fjord. This sequence is here included in the Skagen Group (see also Surlyk and Ineson, 1987a). In northern Wulff Land, where deformation is less intense, the Skagen Group consists of over 500 m of mudstones, sandstones and dolomites. The lower half of the sequence is dominated by mudstones with thin, crosslaminated, fine- to medium-grained sandstone beds and laminae. Sandstone bed thickness increases upward and hummocky cross-stratification becomes more common toward the middle of the group. This lower interval records sedimentation on an offshore, storm-influenced shelf. It is succeeded by a more varied sequence of coarse pebbly sandstones, hummocky cross-stratified sandstones, mudstones, and intraclastic oolitic dolomites, reflecting shallow water, inshore environments. The coarse sandstones show large-scale trough crossbedding and parallel, even lamination of storm-dominated shoreface and possibly beach origin. Grey, hummocky cross-stratified oolitic-intraclastic dolomites dominate in the upper 100 m of the group and grade up into massive pale dolomites of the Portfjeld Formation (see stage 2 below).

The Skagen Group represents the earliest phase in the development of the Franklinian Basin in North Greenland. It records the initial subsidence and transgression of Proterozoic basement, to produce a storm-dominated shelf that was restricted to the northernmost parts of North Greenland. The succession reflects an overall shallowing and the first differentiations of the shelf into a carbonate platform and a mainly siliciclastic, storm-influenced shelf, leading to the development of a stable carbonate platform represented by the Portfjeld Formation (stage 2S) (Surlyk and Ineson, 1987a). The Kennedy Channel Formation of Ellesmere Island shows a similar depositional history (Chapter 8).

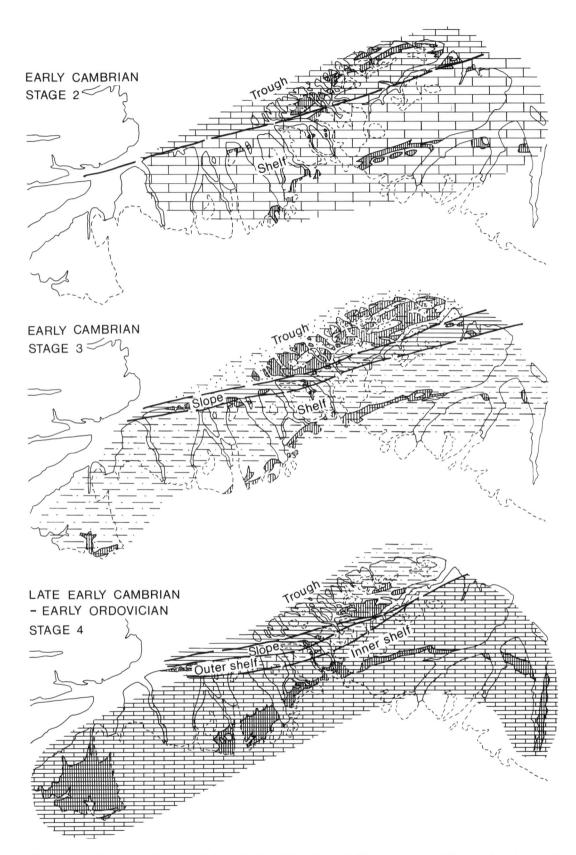

Figure 7.4. Stages 2-7 in the evolution of the Franklinian Basin in North Greenland showing the relationships of major sedimentary regimes. The different signatures indicate schematically the distribution of lithologies dominated by carbonate (bricks), sand (dots) and mud (horizontal lines). Vertical ruling indicates present day exposures. The position of the Navarana Fjord escarpment is shown in stage 6.

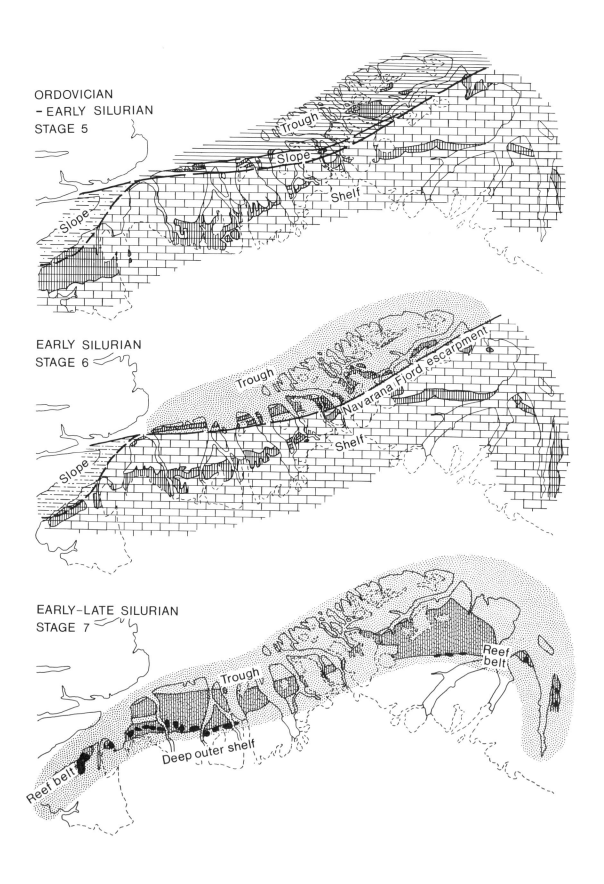

Figure 7.5. Mudstones, sandstones and dolomites of the Skagen Group (S, stage 1) overlain by a 340 m thick sequence of light coloured Portfjeld Formation dolomites (PO, stage 2S), northern Wulff Land.

Figure 7.6. Stable platform dolomites, about 200 m thick, of the Portfjeld Formation (PO, stage 2S) unconformably overlying Proterozoic red sandstones with dark weathering dolerite intrusions (d), southern Peary Land. Succeeding siliciclastic sediments of the Buen Formation (BU, stage 3S) can be divided into a lower, inshore, sandstone unit and an upper unit of outer shelf mudstones and siltstones. The cliff is capped by carbonates of the Brønlund Fjord Group (B, stage $4S_2$).

Stage 2: Early Cambrian stable platform — incipient trough

The transition from the first to the second stage in basin evolution is marked by the change from mainly siliciclastic to carbonate sediments, and by a clear differentiation into shelf and deeper water trough environments (Fig. 7.4, stage 2).

This stage in the evolution of the region is represented by the Portfjeld Formation on the shelf and the Paradisfjeld Group in the trough (Fig. 7.1-7.3).

2S: *Stable Platform*

The Portfjeld Formation outcrops across eastern and central North Greenland but is not known to the west of Wulff Land and to the east of Danmark Fjord. In its northern exposure it conformably overlies the shelf sediments of the Skagen Group (Fig. 7.5); farther south it rests unconformably on Proterozoic strata (Fig. 7.2, 7.6). The formation is poorly fossiliferous but has yielded blue-green algae of probable Early Cambrian age.

The Portfjeld Formation is 200-280 m thick in southern Peary Land (O'Connor, 1979) and typically comprises crossbedded oolitic and intraclastic dolomites, flat-pebble conglomerates, wave-rippled silty dolomites, and algal-laminated dolomites displaying planar, crinkly, domal, digitate and columnar stromatolites. Dark grey bituminous, cherty dolomites form a distinctive, laterally persistent unit (10-15 m thick) near the base of the formation in this area (O'Connor, 1979). Oncolitic and pisolitic dolomites are present at some levels, and desiccation cracks and irregular, brecciated surfaces (paleokarstic?) occur locally. The latter are sometimes associated with breccias containing well rounded quartz grains in the dolomitic matrix. Together with intervals of well rounded, medium- to coarse-grained sandstone (O'Connor, 1979), they probably represent periods of regression and local emergence (Hurst and Surlyk, 1983b). The association of facies suggests deposition in shallow subtidal and intertidal environments on a marine carbonate platform. Ooid and intraclast grainstones were deposited on subaqueous carbonate sand banks in shallow turbulent water, while fetid and algal-laminated carbonate muds and

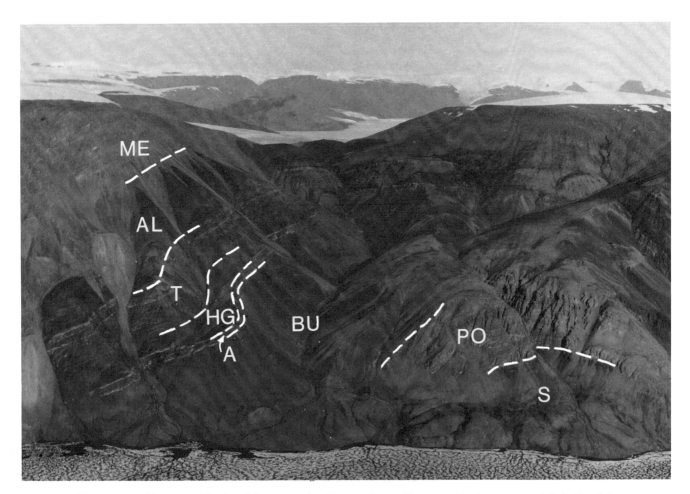

Figure 7.7. Navarana Fjord anticline, exposing Skagen Group (S, stage 1) in the core, overlain by Portfjeld Formation dolomites (PO, stage 2S), dark coloured sandstones and shales of the Buen Formation (BU, stage 3S), and outer shelf representatives of the Aftenstjernesø Formation (A), Henson Gletscher Formation (HG), and Tavsens Iskappe Group (T), (all stage $4S_2$). The uppermost dark unit comprises dark cherts and cherty shales of a new formation of the Amundsen Land Group (AL, stage 5T) and is overlain by turbiditic sandstones of the Merqujôq Formation (ME, stage 6T). Plateau icecap is 1100 m high.

silts accumulated in protected shallow subtidal and low- to moderate-energy intertidal environments.

The Portfjeld Formation thickens appreciably toward the north. In northeast Peary Land it is 400-700 m thick (Christie and Ineson, 1979); in the northwest Peary Land region the formation thickens dramatically northward over less than 20 km from 260 m at Navarana Fjord (Fig. 7.7) to 500-700 m east of J.P. Koch Fjord. In these northern exposures the formation is composed almost entirely of dolomitized ooid-intraclast grainstones, locally showing hummocky cross-stratification, with rare intervals of columnar stromatolites and oncolitic grainstones. This thickened succession of high-energy carbonates in the north clearly represents the outer rim of the platform which accreted rapidly in response to greater subsidence adjacent to the platform-trough boundary. The thickness difference and the occurrence of mixed carbonate-siliciclastic deposits in the upper part of the underlying Skagen Group suggest that deposition of Portfjeld Formation carbonates was initiated in the outer shelf, while the inner shelf still received terrigenous sand and mud (Surlyk and Ineson, 1987a). Farther north, the Portfjeld Formation is replaced by the carbonate-dominated, basinal Paradisfjeld Group (see below).

The Portfjeld Formation is overlain by siliciclastic sediments of the Buen Formation without angular discordance. The contact is sharp and upper levels of the Portfjeld Formation in several areas may be brecciated and stained red (Davis and Higgins, 1987); clasts of Portfjeld Formation are incorporated into the base of the Buen Formation (O'Connor, 1979). It is likely that the boundary with the overlying Buen Formation indicates exposure and resulting demise of much of the carbonate platform; siliciclastic sedimentation was established following rapid transgression of the platform (see stage 3, below).

The Portfjeld Formation can be readily correlated with the Ella Bay Formation of Ellesmere Island (Peel and Christie, 1982; Chapter 8) to which it shows striking similarities in both facies and evolution.

An important tectonic event along the southern outcrop margin of the Franklinian Basin in western North Greenland, following deposition of the Portfjeld Formation, is reflected by the occurrence of a remarkable megabreccia around the head of Victoria Fjord. This unit varies from 85-270 m in thickness and was essentially formed by the collapse, breakup and mass transport of the entire Portfjeld Formation (Fig. 7.8). It rests upon an irregular topography of crystalline basement or thin remnants of the Upper Proterozoic Moraenso Formation, and is itself overlain by undisturbed sandstone and mudstone of the Buen Formation. The breccia consists mainly of blocks and large slabs of various types of dolomite and cherty dolomite of the Portfjeld Formation, together with large clasts of quartzite, gneissic basement rocks, and red siltstone, set in a carbonate matrix containing abundant, very well rounded quartz grains. Clasts are angular and very poorly sorted. The largest slabs are more than 100 m long and may be tens of metres thick; they are mainly subhorizontal but show all degrees of deformation, from weak bending and doming to strong folding, faulting and shearing. Slabs may have sharp boundaries or they may pass both upward and laterally into progressively more deformed strata, and eventually into disorganized matrix-supported conglomerate. Large-scale deformation, with folds of high amplitude, as seen in the land area south of Nares Land, may represent contortions at the head of the debris sheet.

In spite of the ubiquitous occurrence of folded and tilted slabs, directional phenomena are not immediately obvious in the field. However, measurements of fold axes and the general southerly dip of slab surfaces suggest a transport direction approximately toward the north.

It is envisaged that violent earthquake activity associated with faults along the southern hinge line of the Franklinian Basin shattered and mobilized Proterozoic sediments and the already lithified Portfjeld Formation, generating megadebris flows which moved downslope to the north. Neither the geographic location nor the geological origin of the presumed faults are known. The brecciated unit extends northeastward into southernmost Wulff Land and to the central part of the land area south of Nares Land, but the geometry of the megabreccia and its relationship to undisturbed Portfjeld Formation are likewise uncertain. However, undisturbed Portfjeld Formation is known as far west as Nordenskiöld Fjord, with isolated outcrops in fold cores in northern Freuchen Land and northern Wulff Land (Surlyk and Ineson, 1987a).

2T: Incipient trough

The Paradisfjeld Group (Friderichsen et al., 1982), a sequence of carbonate mudstones, dolomites and conglomerates at least 1 km thick, is the deep water equivalent of the Portfjeld Formation (Surlyk et al., 1980). This group is widely exposed in Johannes V. Jensen Land and Nansen Land (Fig. 7.1-7.4) where it is strongly deformed and outcrops generally in anticlinal fold cores. Along strike from east to west there are few facies changes, whereas more marked variations occur from south to north. The group is probably entirely of Early Cambrian age, although fossils (*Chancelloria* and inarticulate brachiopods) are only known from the upper part (Peel and Higgins, 1980). A two-fold division can be recognized in most of the region of outcrop (Fig. 7.9): a thick, lower unit of dark grey, uniform, carbonate rocks, and a thinner upper unit of light coloured, more varied carbonate lithologies (Friderichsen et al., 1982). Preliminary descriptions of the sequence are given by Dawes and Soper (1973), Friderichsen and Bengaard (1985), Higgins et al. (1981), and Soper et al. (1980).

The lower division comprises largely dark coloured, rubbly weathering carbonates and calcareous mudstones. The dark carbonates show poor bedding; textures are often deformed and recrystallized, but many seem to have been poorly sorted calcarenites. The overall uniform nature of the lower division is occasionally relieved by bands of light coloured calcareous turbidites and calcareous shales, but these are not sufficiently continuous to permit any subdivision. Limestone conglomerates with outsize blocks (up to 50 m across) occur locally.

The upper unit is conspicuous on account of its yellow and orange weathering and varied lithologies; the generally well bedded sequence includes light grey and yellow dolomites, calcareous turbidites, a sequence of up to six limestone conglomerate beds, and a transitional 20-40 m unit of orange weathering shale and siltstone, which passes up into the overlying Polkorridoren Group.

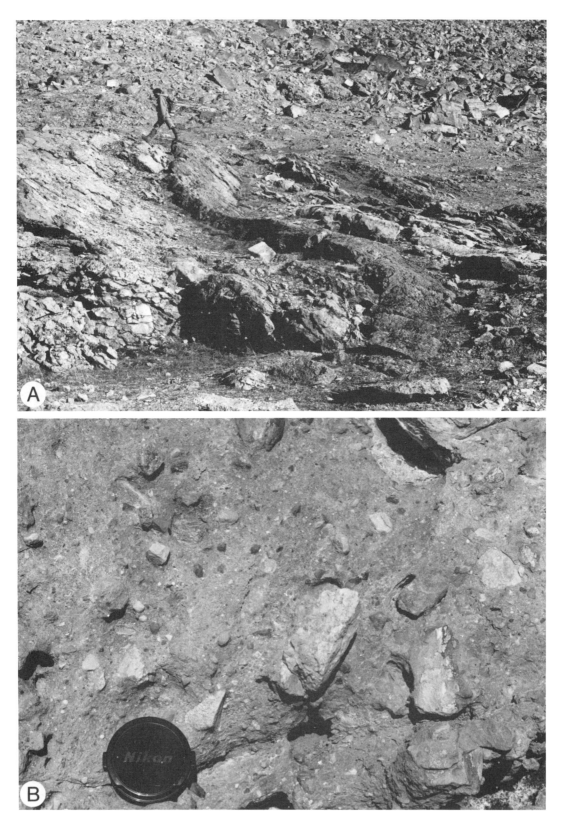

Figure 7.8. A. Large bent and folded slabs of dolomite derived from the Portfjeld Formation (stage 2S). Southern Wulff Land. B. Disorganized matrix-supported conglomerate, which itself forms the matrix between the large slabs shown in A.

Figure 7.9. Major, northerly overturned folds in eastern Nansen Land. The Paradisfjeld Group (stage 2T) is divided into a lower sequence of dark rubbly carbonates (PAd) and an upper sequence of light coloured dolomites (PAl). Overlying dark weathering rocks are siltstones and sandstones of the Polkorridoren Group (P, stage 3T). The near summits are 650 m above the fjord.

Calcareous shales appear and become increasingly important in the Paradisfjeld Group sequence toward the north, while the conglomerates thin out and disappear. However, details of the lithological variations are difficult to untangle in the extreme north due to the intense deformation.

Limestone conglomerates occur in the top parts of the upper division in a broad belt over a length of 250 km along the southern outcrop margin. Both clast supported and matrix supported types are present, the latter dominating and showing a greater variety of clast size. Clasts are mainly of light grey or pale grey carbonate, usually from a few centimetres to tens of centimetres in length, set in a fine grained carbonate matrix; deformed rip-up clasts occur in some areas. The highest conglomerate bed characteristically has a quartz sand matrix. Grading is common. The beds are planar, with sharp, often erosive, boundaries. Bed thickness normally varies between 15 cm and 3 m, but composite units as much as 15 m thick have been observed. The sum of sedimentary features suggests transport by high-density turbidity currents and debris flows. The original pebble fabric commonly has a strong tectonic overprint, and transport directions have not been measured. Lithology and distribution, however, clearly indicate a source area of the conglomerates in the southern carbonate platform area.

The facies association of mainly grey limestones with calcareous turbidites and related deposits points toward a relatively deep slope and trough environment.

The southward transition of the Paradisfjeld Group into the shallow marine sediments of the Portfjeld Formation of the carbonate shelf has not been directly observed. The present line of demarcation between outcrops of the two units (Fig. 7.4, stage 2) coincides with the location of a low-angle Ellesmerian thrust zone, and the transition is hidden beneath the hanging wall sequence. A strong tectonic control of the original carbonate shelf margin, which was probably an escarpment, has been suggested, and may have coincided with a precursor to the present Harder Fjord fault zone (Surlyk et al., 1980; Surlyk and Hurst, 1984). Huge olistoliths derived from the Portfjeld Formation occur in the lower part of the turbiditic Polkorridoren Group (Fig. 7.10) (Friderichsen and Bengaard, 1985). Thus the Portfjeld Formation scarp persisted or was revived by later faulting so that parts of it were still exposed in Buen Formation – Polkorridoren Group times. Subsequent to burial beneath the Ellesmerian thrust sheets, the ancient fault line may have been reactivated in post-Ellesmerian time to form the present Harder Fjord fault (Soper and Higgins, 1987).

Stage 3: Early Cambrian siliciclastic shelf — turbidite trough

The passage from the second to the third stage in basin evolution is characterized by a marked differential subsidence leading to the formation of the deep water turbidite trough. This was accompanied by the southward shift of the shelf-trough boundary to a position probably controlled by new faults (Fig. 7.4, stage 3). The resultant shift in facies belts caused slope and trough margin deposits of the Paradisfjeld Group to be overlain by true deep water trough deposits of the Polkorridoren Group, whereas the slope area moved southward. The marked increase in subsidence correlates with an interruption of carbonate shelf deposition. The water depth over the shelf, however, was not very great, and the area still formed a well defined shelf contrasting with the deeper slope and trough environments. The tectonic-sedimentological setting at this stage can thus be visualised as follows. Outside a shelf area receiving terrigenous mud and sand, a wide slope was formed, characterized by deposition in a well oxygenated environment of red and green mudstones with occasional thin turbidites. Coarse clastic material was probably funneled across the shelf into submarine canyons that still remain to be found. This material was deposited by turbidity currents onto several submarine fans, the outer parts of the fans showing deflection westward following the east-west axis of the trough.

3S: Siliciclastic shelf

The shelf record of stage 3 is represented by a sequence of siliciclastic sediments termed the Buen Formation (Jepsen, 1971) in the region between Warming Land, and Danmark Fjord and the Humboldt and Dallas Bugt formations in Washington Land and Inglefield Land, respectively (Fig. 7.1-7.3; Peel and Christie, 1982). The Kap Holbaek Formation

Figure 7.10. Yellow weathering dolomite olistoliths of the Portfjeld Formation (stage 2S) within a siltstone sequence of the Polkorridoren Group (stage 3T), Nansen Land. The largest blocks are 150 m across. Inset photograph is of a block 2 m in diameter.

of easternmost areas of North Greenland has previously been considered to be a direct correlative of the Buen Formation, but acritarchs suggest a Late Proterozoic age (Peel, 1985). Northward, these Lower Cambrian shelf clastics grade into the turbiditic Polkorridoren Group (see below). The shelf sediments overlie the platform carbonates of stage 2 in the area east of Wulff Land (Fig. 7.6, 7.7), but are seen to overstep onto Proterozoic sedimentary rocks and Archean gneisses to the southwest in Inglefield Land; their extent in the western part of the basin is not known, as only younger rocks are exposed in these areas. Shelf sediments of stage 3 are conformably overlain by carbonates of stage 4 throughout North Greenland.

The Buen Formation (middle-late Early Cambrian) ranges between 425 and 500 m in thickness in its southern outcrop, and about half this amount (250 m) in the outer shelf outcrops in Navarana Fjord and northern Wulff Land. It thickens rapidly northward at the transition into the deep water basin (see below). It consists of a sequence of sandstones and mudstones, with rare conglomerates and skeletal limestones, and typically can be subdivided into a lower, sand-dominated unit and an upper, mud-dominated unit (Fig. 7.6). In general, the proportion of sandstone in the formation decreases northward; hence, the formation is sandstone-dominated in southeast Peary Land and southern Wulff Land (Jepsen, 1971; Hurst and Peel, 1979) but consists mainly of thick mudstone-dominated sequences in the northeast Peary Land to northern Nyeboe Land area (Christie and Ineson, 1979; Davis and Higgins, 1987; Soper and Higgins, 1985; Surlyk and Ineson, 1987a). The sandstone unit contains minor conglomerate and silty mudstone interbeds. In its southern outcrop, medium- to coarse-grained sandstones show abundant cross-stratification; large-scale compound sets are common and locally show siltstone drapes along low-angle foresets (3-15°); compound sets characteristically show a unimodal (northward) paleocurrent pattern whereas thin to medium bedded, glauconitic sandstones often show herringbone crossbedding. Bioturbation is common, particularly near the top of the sandstone unit where *Skolithos* is locally abundant.

Figure 7.11. Regularly bedded sandstone turbidite sequence (Ps) overlying recessive siltstone and mudstone sequence (Pm), Polkorridoren Group (stage 3T), northeast Nansen Land. Folds are overturned northward. Highest summits reach 1100 m.

Hummocky cross-stratified sandstone occurs scattered in southern outcrops but is the dominant facies in northern Wulff Land. The upper fine grained member is composed mainly of laminated or bioturbated, grey-green silty mudstones with subordinate interbeds (2-50 cm thick) of medium- to fine-grained sandstone showing parallel and ripple cross lamination and hummocky cross-stratification.

The Buen Formation represents a siliciclastic marine shelf environment, the sandstone unit recording inshore environments influenced by tides and high-energy storms, whereas the mudstone unit reflects a low-energy, deep outer shelf environment in which episodic high-energy events were the result of storm-induced currents. The ubiquitous upward transition from sand-dominated inner shelf to fine grained outer shelf facies reflects a transgressional or deepening event during this depositional phase.

The Dallas Bugt Formation (25-150 m) of Inglefield Land clearly records the marine transgression of stage 3 during a period of abundant sediment supply. Feldspathic, conglomeratic, coarse grained red sandstones of fluviatile origin unconformably overlie Proterozoic siliciclastics and crystalline basement and are succeeded by well sorted, large-scale crossbedded white sandstones with reactivation surfaces and massive beds containing *Skolithos*, deposited under tidal influence. The sandstones become interbedded with bioturbated siltstones with *Cruziana* prior to grading up into dolomites of the overlying Cape Leiper Formation (Peel et al., 1982). The Dallas Bugt Formation is also recognized in immediately adjacent Ellesmere Island, where it contains the Bache Peninsula and Sverdrup members of Christie (1967; see Peel et al., 1982). Correlative strata in southern Washington Land are referred to the Humboldt Formation.

3T: Turbidite trough

The change in trough sedimentation, from the Paradisfjeld Group carbonates to the Polkorridoren Group sandstone turbidites and mudstones, is a major step in the evolution of the trough, and reflects a dramatic increase in subsidence, probably controlled by trough-parallel faulting (Surlyk and Hurst, 1984). The corresponding change on the shelf from the Portfjeld Formation dolomites to the Buen Formation

Figure 7.12. Thick, poorly or nongraded mid-fan sandstone turbidites of the Polkorridoren Group (stage 3T), central Johannes V. Jensen Land.

Figure 7.13. Sequence of prograding slope and platform carbonates (stage $4S_2$) on the east side of Henson Gletscher, west Peary Land. Slope sediments of the Perssuaq Gletscher Formation (PG) of the Tavsens Iskappe Goup are overlain by restricted platform interior carbonates of the Koch Vaeg Formation (KV) and Wandel Valley Formation (W) of the Ryder Gletscher Group (stage $4S_1$). Northward progradation of units within the Perssuaq Gletscher Formation produced the inclined depositional surfaces seen to the right. A dolomite mound (m) to the left is draped by well bedded dolomites and siltstones of the Koch Vaeg Formation, which are overlain by dolomites of the Wandel Valley Formation. The cliff is 600 m high, and is cut by a dyke (d).

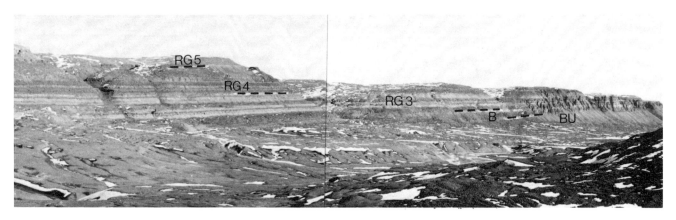

Figure 7.14. Carbonates and siltstones of formations RG3-RG5 of the Ryder Gletscher Group (stage $4S_1$) overlying dolomites of formations RG1-RG2 of the Brønlund Fjord Group (B, stage $4S_2$) and recessive siliciclastics of the Buen Formation (BU, stage 3S), southwest Wulff Land. Formation RG3 is 100 m thick.

siliciclastics was initiated by shallowing of the shelf sea and exposure of much of the carbonate platform (see above). The earliest Polkorridoren Group sandstones and mudstones may correspond to a hiatus at the base of the Buen Formation (Davis and Higgins, 1987).

The Polkorridoren Group is widely exposed in Johannes V. Jensen Land and Nansen Land, in the intensely deformed part of the North Greenland fold belt. The total thickness of the group is estimated to be 2-3 km, and it is considered to be entirely of Early Cambrian age (Surlyk et al., 1980; Friderichsen et al., 1982). Two main facies associations are present: black, green and purple mudstone, and sandy turbidites. No formal lithostratigraphic subdivision has yet been made.

The alternation of mud-dominated and sand-dominated units, generally on a scale of 100-500 m thick, has proved useful in defining local, informal stratigraphies (Fig. 7.11). A subdivision into six or seven parts has been possible in eastern Johannes V. Jensen Land and the islands east of Nansen Land (Soper et al., 1980; Higgins et al., 1985). Correlation of these units between regions is difficult, however, and in some areas, such as western Johannes V. Jensen Land, thinly bedded, finer grained units are poorly developed and the sand-dominated sequence can be indivisible. Sand-rich units typically consist of thick, often structureless beds that are commonly amalgamated; fining-upward cycles on a scale of 50-100 m are recognized locally. Intervening mudstone units include thin bedded sandstone turbidites which may show abundant trace fossils at some levels (Pickerill et al., 1982). Lateral variation in thickness can be marked, and sand-rich units locally coalesce or wedge out laterally. Carbonate conglomerate beds occur at several levels in the sequence and sometimes form persistent mapping units (see below).

The uppermost unit of the group, the Frigg Fjord mudstone (Fränkl, 1955; Dawes and Soper 1973. Friderichsen et al., 1982), has a more widespread, regional distribution and can be recognized in various guises between Johannes V. Jensen Land and southeast Nansen Land. Along the southern trough margin, where the sandstone turbidites have wedged out, the Frigg Fjord mudstone extends stratigraphically further downward. This distinctive unit is up to 400 m thick and consists mainly of purple and green mudstones. A sequence of similarly coloured mudstones and thin, flaser bedded siltstones and sandstones, which forms the uppermost unit in northwest Nansen Land, is probably the lateral equivalent of the Frigg Fjord mudstones.

Paleocurrent readings from Johannes V. Jensen Land show three distinct modes toward the west, north and, less commonly, east. The longitudinal transport toward the west is interpreted as representing a combination of westerly directed submarine fans and deflection of the outer parts of northerly directed fans along the trough axis. In Nansen Land paleocurrent readings give a dominant northward mode, with a narrow range between northeast and northwest.

A persistent and spectacular carbonate conglomerate level occurs about 600 m above the base of the Polkorridoren Group, and is traceable over a distance of 125 km throughout southern Nansen Land and the islands to the east and the west (Friderichsen and Bengaard, 1985). Individual clasts range from tens of centimetres in size to outsize blocks up to 300 m in diameter (Fig. 7.10); many blocks can be identified as Portfjeld Formation lithologies. The yellow weathering dolomite blocks, which have been described as olistoliths, are totally enveloped by dark weathering siltstones of the Polkorridoren Group. Relationships between the blocks and the surrounding sediments are masked, however, by the strong deformation. Most of the large blocks occur in two clusters in southern Nansen Land, spread over areas of 2-3 km and 8 km, respectively. Friderichsen and Bengaard (1985) suggest the blocks may have been derived from a submarine canyon excavated in Portfjeld Formation dolomites at the trough-shelf transition, whereas Soper and Higgins (1987) consider their wide east-west distribution more indicative of a line source, such as a faulted platform margin.

The strong deformation precludes detailed facies analysis and environmental reconstruction in many areas. The general picture, however, is clear. At the shelf-slope break, the black and green mudstones of the upper part of the Buen Formation gradually give way to variegated, thickly developed Frigg Fjord mudstone facies characterizing the wide slope area. Farther north, thin sandstone turbidites start to appear in the Frigg Fjord mudstones, reflecting the approach of the transition of the base-of-slope area and the trough floor.

Sand was funneled from the shelf across the slope probably through a number of canyons. These, however, have not been observed in the field, most likely due to the intense deformation of the Frigg Fjord mudstones, or to a cover of younger rocks. Many of the sandy turbidites were deposited in a system of relatively small borderland fans which were deflected toward the west or less commonly toward the east along the trough axis. Thick sequences of poorly sorted, massive sandstone turbidites occur in some areas and possibly represent rapid sediment dumping in a channel mouth – proximal lobe setting (Fig. 7.12). The dominance of sandstones in areas such as western Johannes V. Jensen Land and Nansen Land would then represent major depositional lobe systems of individual fans, while intervening mudstone-dominated areas reflect interfan and basin plain environments.

Stage 4: Late Early Cambrian – Middle Ordovician platform — starved basin

From the late Early Cambrian to the Early Ordovician, sedimentation in North Greenland was influenced by two major tectonic developments: (1) differential subsidence and southward expansion of the east-west deep water trough, and (2) uplift in eastern North Greenland, probably the effect of an early Caledonian event. The influence of the latter was most pronounced in eastern North Greenland and decreased westward (Fig. 7.3). Hence, the evolution of the shelf during this period is best envisaged in terms of two components, a western, uniformly subsiding, stable shelf (stage $4S_1$) in areas west of Nordenskiöld Fjord (represented by the Ryder Gletscher Group; Fig. 7.2), and an unstable shelf (stage $4S_2$), subject to uplift in southern Freuchen Land, Peary Land, and Kronprins Christian Land (recorded by the Brønlund Fjord and Tavsens Iskappe groups; Fig. 7.2). Uplift in the latter areas ceased in the Early Ordovician, and more uniform, stable platform conditions extended across North Greenland. Thus, the Wandel Valley Formation, which unconformably overlies Cambrian strata assigned to stage $4S_2$ in the Freuchen Land – Peary Land area (Fig. 7.1-7.3), is described here as part of the stable shelf sequence (stage $4S_1$).

$4S_1$: Stable shelf

The stable shelf sequence is represented by a thick and varied succession of carbonates, mainly of restricted platform aspect (*sensu* Wilson, 1975) with subordinate siliciclastics and evaporites, which are collectively referred to the Ryder Gletscher Group. This term was originally introduced by Peel and Wright (1985) to contain six informally designated, mainly carbonate formations (RG1-RG6) of Early Cambrian – Early Ordovician age in the Warming Land and southern Wulff Land area. Ineson and Peel (1987) subsequently transferred formations RG1-RG2 to the Brønlund Fjord Group of the regressive unstable shelf sequence (stage $4S_2$), described below, and introduced the stratigraphically widely expanded concept of the Ryder Gletscher Group employed here (Fig. 7.2, 7.3).

The Ryder Gletscher Group is sedimentologically best known in its type area, namely the Warming Land – Nares

Land region, where the succession between formation RG3 and the Cape Webster Formation (Fig. 7.2) attains a total thickness of about 1100 m (Bryant and Smith, 1985; Ineson and Peel, 1987; Peel and Wright, 1985; Sønderholm and Due, 1985). To the east, the group is represented by the Wandel Valley and Sjaelland Fjelde formations (320-520 m) of late Early – Middle Ordovician age, which unconformably overlie strata assigned to stage $4S_2$ and older deposits (Fig. 7.2; Ineson et al., 1986; Peel and Smith, in press). Summaries of the stratigraphy of the Ryder Gletscher Group in Inglefield Land and Washington Land, where it is approximately 1500 m thick, are contained in overviews by Dawes (1976), Peel (1982), and Peel and Christie (1982).

The stable shelf sequence in the Warming Land – Nares Land area overlies carbonate ramp and platform margin deposits of formations RG1-RG2 of the Brønlund Fjord Group. The Koch Vaeg Formation of southwest Peary Land forms an isolated but analogous outcrop of Ryder Gletscher Group, overlying strata assigned to the Tavsens Iskappe Group (Fig. 7.13; Ineson and Peel, in press). In Inglefield Land and Washington Land, to the west, the Ryder Gletscher Group sequence conformably overlies Lower Cambrian mudstones and sandstones of stage 3S. Throughout North Greenland, strata of stage $4S_1$ are conformably overlain by the stable platform sequence of stage $5S_1$.

Formations RG3-RG6 of the Ryder Gletscher Group (combined thickness 470 m) in the type area (Fig. 7.14) consist of mudcracked stromatolitic and cryptalgal dolomites, dark burrow-mottled dolomites, silty lime mudstones, and flat-pebble conglomerates. Thrombolitic and stromatolitic bioherms with relief of several metres are developed at several levels in this sequence. Mudcracked, wave-rippled and crossbedded fine- to medium-grained sandstones and siltstones occur locally in the upper formation (RG6, Peel and Wright, 1985). This poorly fossiliferous sequence records sedimentation on a restricted carbonate platform in shallow subtidal and intertidal environments. The increased siliciclastic content, relative to the equivalent carbonates of the Cass Fjord Formation of Washington Land (Fig. 7.2, 7.3), reflects proximity to the emergent eastern shelf in Late Cambrian times (see below and stage $4S_2$).

Contemporaneous platform strata of the Ryder Gletscher Group in Washington Land are referred to the Kastrup Elv, Telt Bugt, and Cass Fjord formations (Henriksen and Peel, 1976; see Fig. 7.2). The Kastrup Elv Formation (Early-Middle Cambrian) is approximately 140 m thick and dominated by well bedded, mottled, crystalline dolomite. Paler units of more finely crystalline, laminated dolomite, with stromatolites, occur in the upper part of the formation. The overlying Middle Cambrian Telt Bugt Formation (5-100 m) consists mainly of thin, irregularly bedded, mottled micritic limestone with a wavy silty lamination. The formation grades up into the overlying Cass Fjord Formation, which is widely distributed in Washington Land, Inglefield Land and adjacent Ellesmere Island (Peel and Christie, 1982). The latter formation attains a thickness of 470 m in southern Washington Land, but this includes some 30 m of the Kap Coppinger Member and overlying carbonates at the top of the formation, discussed below. The typical Cass Fjord lithology consists of thin, irregularly bedded, micritic limestone with a wavy, silty lamination; scours, ripple marks, mudcracks, and burrows are abundant. Laterally extensive beds of flat-pebble conglomerate, up to 50 cm in thickness, are common. Calcarenites, thinly bedded, finely recrystallized massive dolomites, thin beds of siltstone and well sorted sandstone, rare oolites, and thin beds of evaporites may be locally prominent.

The Kastrup Elv and Telt Bugt formations resemble formations RG3 and RG4 of Wulff Land and Nares Land, while the Cass Fjord Formation below the Kap Coppinger Member is very similar to formations RG5 and RG6. They represent environments ranging from restricted subtidal lagoons to intertidal and supratidal carbonate mud flats that are locally evaporitic.

In Inglefield Land, large-scale crossbedded to structureless dolomites of the Cape Leiper and Cape Ingersoll formations (about 40 m) conformably overlie siliciclastics of the Dallas Bugt Formation (stage 3). Succeeding limestones of the Wulff River and Cape Wood formations (Lower-Middle Cambrian; thickness about 150 m) are lithologically very similar to the Telt Bugt Formation of Washington Land. The Cape Kent Formation (Lower Cambrian; thickness 5-10 m) lies between the latter two formations in Inglefield Land. It is a distinctive unit of oolitic limestone, recognized on both sides of Kane Basin, and represents deposition on high-energy ooid sand banks in shallow turbulent water.

The Permin Land Formation conformably overlies formation RG6 between Warming Land and the land area south of Nares Land and is of earliest Ordovician age (Bryant and Smith, 1985). It comprises up to 53 m of well sorted, well rounded, medium- to fine-grained sandstone with subordinate grey dolomite and siltstone interbeds. Sedimentary structures include trough and tabular crossbedding, wave-ripple lamination, flat and low-angle lamination; paleocurrent directions are widely dispersed (Bryant and Smith, 1985). The formation represents a shallow marine sand sheet of considerable lateral extent, which can be traced westward from southwestern Freuchen Land into Washington Land where it forms the Kap Coppinger Member of the Cass Fjord Formation (Fig. 7.2). This widespread clastic incursion during the earliest Ordovician is correlated with the time of maximum regression and exposure in eastern North Greenland (Fig. 7.3). Broadly contemporaneous siliciclastic incursions also occurred in outer shelf environments (upper Tavsens Iskappe Group) and in trough environments (Vølvedal Group).

The eastern part of the North Greenland shelf was progressively exposed from the Early Cambrian to the Early Ordovician (see stage $4S_2$ below), hence the upper Lower Ordovician – Middle Ordovician Wandel Valley Formation unconformably oversteps Lower Ordovician and Cambrian strata from west to east, ultimately resting on Proterozoic strata in easternmost North Greenland (Fig. 7.2, 7.3, 7.13, 7.16, 7.22). West of Freuchen Land, however, shelf sedimentation persisted through the later Cambrian into the Early Ordovician and is represented by the upper half of the Ryder Gletscher Group (Fig. 7.3).

Subsequent to the deposition of the shallow marine sandstone of the Permin Land Formation, sedimentary environments during this period alternated between restricted and more open-marine carbonate platform environments and the succession can be regarded in terms of two major shallowing-upward cycles. Each cycle begins with carbonates, mainly of shallow subtidal aspect, which pass up into peritidal facies, commonly associated with evaporitic deposits. The thickness of these deposits increases from 335 m in Nares Land to approximately 850 m in Washington Land, reflecting decreasing influence of the area of uplift in the Peary Land region.

The lower, dominantly subtidal portion of the first cycle is represented by the Johansen Land Formation (15-35 m) in central and more western parts of North Greenland (Fig. 7.15, 7.16), and by the uppermost Cass Fjord (30 m), Cape Clay (50 m) and Christian Elv (140 m) formations in Washington Land (Henriksen and Peel, 1976; Sønderholm and Due, 1985). These mainly comprise burrow-mottled, nodular dolomites and lime mudstones with subordinate wave- and current-rippled lime grainstones and flat-pebble conglomerates. Occasional algal-laminated dolomites, columnar stromatolites, scalloped surfaces, and siliciclastic intervals in the Washington Land sequence reflect periodic shallowing into the intertidal zone. The upper portion of this cycle in the Warming Land – Nares Land area (Warming Land Formation; 50-150 m) consists of alternating dark and light grey fine grained dolomites (Fig. 7.15) showing current- and wave-ripple crosslamination, flaser and wavy bedding, and abundant bioturbation. Scalloped surfaces, cryptalgal dolomites with desiccation cracks and chert cones possibly pseudomorphous after gypsum occur in places. These sediments were deposited on prograding tidal flats, dominated by very shallow subtidal to supratidal, locally probably evaporitic, environments (Sønderholm and Due, 1985). Farther west, in Washington Land, however, this interval is dominated by evaporitic deposits of the Poulsen Cliff (100-125 m) and Nygaard Bay (40 m) formations (Fig. 7.2, 7.17), which are direct correlatives of the thick evaporitic sequence of Ellesmere Island (Baumann Fiord Formation; see Peel and Christie, 1982). These gypsum deposits may show fine irregular lamination, current- or wave-ripple lamination or may consist of nodular or nodular-mosaic gypsum; they commonly occur interbedded with laminated shaly dolomites on a 10 m scale. By analogy with the Ellesmere Island deposits (cf. Chapter 8), these evaporites are probably subaqueous deposits that accumulated in a broad, shallow, possibly segmented saline basin. The lower half of the Nygaard Bay Formation is

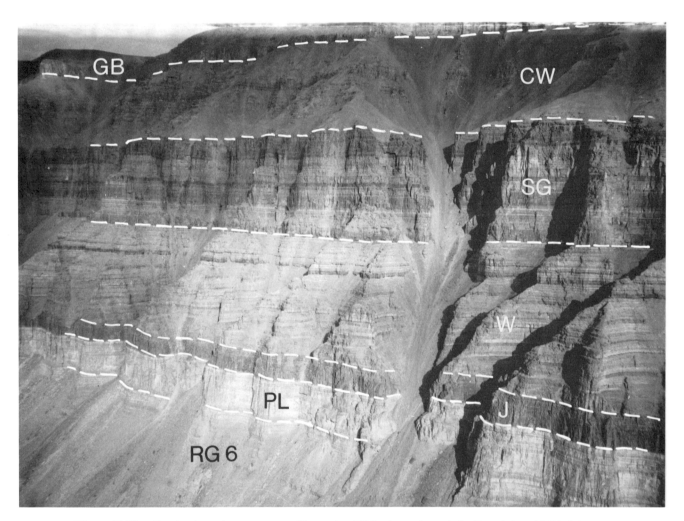

Figure 7.15. Section from the upper part of formation RG6 at base through the Permin Land (PL), Johansen Land (J), Warming Land (W), Steensby Gletscher (SG), and Cape Webster (CW) formations, all Ryder Gletscher Group (stage $4S_1$) of platform interior origin. The Gonioceras Bay Formation (GB) of the Morris Bugt Group (stage $5S_1$) forms the top of cliff, which is 600 m high. Southern Permin Land.

Figure 7.16. Restricted platform sediments of the Ryder Gletscher Group (stage $4S_1$) east of the southern tip of Nares Land, showing the western extension of the sub-Warming Land Formation (WL) unconformity (u). The underlying Johansen Land (JL) and Permin Land (PL) formations are truncated over the crest of a shallow east-west trending anticline. RG6 = formation RG6; SG = Steensby Gletscher Formation; CW = Cape Webster Formation. The Permin Land Formation is 23 m thick.

Figure 7.17. Ryder Gletscher Group (stage $4S_1$), overlain by the Gonioceras Bay (GB) and Troedsson Cliff (TC) formations of the Morris Bugt Group (stage $5S_1$) in central Washington Land. The transition from recessive dolomitic sediments (stage $4S_1$) to cliff-forming limestones (stage $5S_1$) is a conspicuous marker traceable across most of the North Greenland shelf. PC = Poulsen Cliff Formation; NB = Nygaard Bay Formation (45 m thick); CA = Canyon Elv Formation; N = Nunatami Formation; CW = Cape Webster Formation.

dominated by dark limestones with intraformational conglomerate interbeds and is probably of shallow subtidal or intertidal origin.

The basal portion of the second cycle represents a marked transgression of probable eustatic origin, which can be traced from Ellesmere Island (Eleanor River Formation; Barnes, 1984) across Washington Land (Canyon Elv and Nunatami formations; 40-60 m and 150 m, respectively) and the rest of western and central North Greenland (Steensby Gletscher Formation; 90-135 m) to eastern North Greenland (Wandel Valley Formation; 320-410 m). In the western parts, the basal part of the second shallowing-upward cycle is typified by dark, thick bedded, burrow-mottled lime mudstones containing a rich open-marine fauna of gastropods, trilobites, and cephalopods. These low-energy, subtidal deposits are dominant, but pale dolomites, locally algally laminated and mudcracked, occur in some southern localities and toward the top of the Nunatami and Steensby Gletscher formations (Peel and Cowie, 1979; Sønderholm and Due, 1985); they reflect periodic progradation of shallow subtidal to intertidal environments. The overlying Cape Webster Formation (175-280 m) marks the return to restricted conditions on the carbonate platform. It consists of grey to dark grey, fine grained dolomites with horizontal lamination, often with desiccation cracks and fluidization features. Evaporites occur in the lower part of the formation in southern Washington Land and near the top in Nares Land, while breccias, possibly of solution origin, are common in Warming Land and central Washington Land (Sønderholm and Due, 1985).

In eastern Freuchen Land, Peary Land and Kronprins Christian Land, the late Early Ordovician transgression resulted in submergence of the previously uplifted and exposed eastern shelf region and deposition of shallow water platform carbonates assigned to the Wandel Valley and Sjaelland Fjelde formations (Fig. 7.3, 7.13, 7.22; Troelsen, 1949; Christie and Peel, 1977, Ineson et al., 1986; Peel and Smith, in press). The Wandel Valley Formation (320-410 m) is dominated by pale dolomites. These may be burrow-mottled with an irregular, parallel, discontinuous lamination, or exhibit cryptalgal and horizontal lamination, and wave- and current-formed crosslamination, as well as desiccation and water escape features. In the lower part of the formation, dark, nodular, burrow-mottled, pelletal lime mudstones, containing *Ceratopea* and other gastropods are interbedded with the dolomites in 2-5 m, rarely up to 15 m thick, shallowing-upward cycles (Hurst and Surlyk, 1983b; Sønderholm and Due, 1985). The Wandel Valley Formation thus probably represents the progradation of autocyclic deposits of low-energy tidal flat and lagoonal environments over a slowly, but continuously subsiding platform (cf. Ginsburg, 1971).

In Kronprins Christian Land, the Wandel Valley Formation is overlain by the Sjaelland Fjelde Formation (105 m), which in the lower part is dominated by bioturbated, skeletal lime mudstones and burrow-mottled dolomites, with flat-pebble conglomerates occurring throughout. The upper part is characterized by pale weathering, cryptalgal laminated dolomites. Spar-filled vugs and fenestrae, desiccation cracks, and stromatolite domes are common in places. Bioturbation and halite pseudomorphs are only rarely observed. The sediments reflect deposition in generally low-energy, shallow subtidal to low supratidal, locally evaporitic, environments (Ineson et al., 1986).

$4S_2$: Regressive unstable shelf

The Brønlund Fjord and Tavsens Iskappe groups are a complex, diachronous array of carbonates with subordinate siliciclastics. Collectively, they attain their maximum development (900 m in thickness) in southwest Peary Land and southeast Freuchen Land, and range in age from late Early Cambrian to earliest Ordovician (Fig. 7.18; Ineson and Peel, 1980 and in press). The succession in the type area of southern Peary Land and southeastern Freuchen Land is subdivided into twelve formations, divided equally between the two groups. Two additional formations of the Brønlund Fjord Group are recognized from southwest Freuchen Land to Warming Land, and a further two formations are known from highly faulted outcrops in northeast Peary Land, around G.B. Schley Fjord. The Brønlund Fjord and Tavsens Iskappe groups also outcrop in anticlinal fold cores in the northern coastal regions, from Nyeboe Land, in the west, to the J.P. Koch Fjord area, in the east (Fig. 7.1, 7.7, 7.19). The two lowest formations of the Brønlund Fjord Group in its type area are recognized in these northern outcrops, while an overlying new formation (unit 3 of Higgins and Soper, 1985) is assigned to the Tavsens Iskappe Group (Fig. 7.19).

Formations RG1-RG2 of the Brønlund Fjord Group record the evolution from the siliciclastic outer shelf (Buen Formation, stage 3S) to restricted platform (Ryder Gletscher Group, stage $4S_1$) in outcrops between southwest Freuchen Land and Warming Land (Fig. 7.13, 7.20). Formation RG1 (maximum thickness 65 m) represents a storm-dominated carbonate ramp that developed during initial stages of carbonate sedimentation on the shelf. The succeeding formation (RG2; 80 m thick) consists essentially of a lower sequence of dolomitized mass-flow carbonate breccias and carbonate turbidites, overlain by an upper succession of trough crossbedded and hummocky cross-stratified carbonate grainstones. This formation records the progressive development and northward progradation of a flat-topped, rimmed carbonate platform with a high-energy margin and flanking carbonate slope deposits. In southwest Freuchen Land, at the transition from the stable platform into the outer shelf-slope sequence ($4S_2$), the platform margin facies thickens markedly (Fig. 7.21) and includes a 200 m sequence of dolomitized algal boundstones with archaeocyathans.

To the east of Freuchen Land, progressive uplift and erosion resulted in an unconformity at the base of the succeeding Wandel Valley Formation (Early-Middle Ordovician), which consequently oversteps on to older units of the Brønlund Fjord and Tavsens Iskappe groups (Fig. 7.3, 7.13, 7.22). Thus, the Tavsens Iskappe Group has been eroded or was never deposited in areas to the east of central southern Peary Land, while the Brønlund Fjord Group persists eastward to Independence Fjord.

The complex nature of the lithostratigraphic scheme applied to the sediments of the Brønlund Fjord and Tavsens Iskappe groups reflects the marked diachronism and rapid lateral facies variations within the succession (Ineson, 1985). Three major environments are recognized: platform margin, outer shelf-slope and starved ramp.

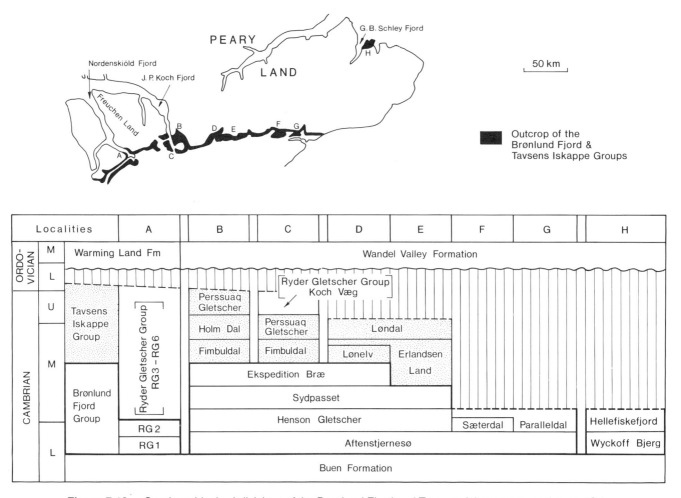

Figure 7.18. Stratigraphical subdivisions of the Brønlund Fjord and Tavsens Iskappe groups (stage $4S_2$). The northeastern outer shelf equivalents have not been formally defined and are not shown.

Starved ramp facies form a thin (2-7 m) unit at the base of the Brønlund Fjord Group, marking the transition from the siliciclastic-dominated marine shelf (Buen Formation, stage 3S) to the succeeding carbonate regime. The remainder of the Brønlund Fjord Group and the Tavsens Iskappe Group form a progradational platform margin to outer shelf system (Fig. 7.20). Platform interior rocks associated with the prograding complex are referred to the Ryder Gletscher Group (stage $4S_1$, above).

The starved ramp sequence at the base of the Aftenstjernesø Formation consists of phosphoritic, glauconitic skeletal grainstone, packstone, and wackestone (and dolomitized equivalents) with subordinate burrowed siltstone. Bioturbation is widespread and bedding is typically nodular or wavy; sedimentary structures such as shell imbrication and current ripple crosslamination are preserved locally. Black or dark brown phosphorite occurs as detrital grains in winnowed shell lags, as irregular nodules, and within impregnated, pyritic hardground surfaces (Frykman, 1980).

This highly condensed sequence clearly records a period of sediment starvation on the Cambrian shelf (Hurst and Surlyk, 1983b; Ineson, 1985). As the supply of terrigenous sediment declined due to transgression and drowning of sediment sources, carbonate production gradually took over. The starved, muddy carbonate ramp sequence represents the transitional phase, preceding the establishment of a carbonate-producing platform.

As carbonate sedimentation became established, the shelf environment was resolved into a shallow water platform that passed northward into a deeper water, outer shelf region. Rocks of the platform interior (Ryder Gletscher Group) pass northward into thick bedded or massive dolomites of the **carbonate platform margin** (Fig. 7.20, 7.21). Typically, the dolomites are trough crossbedded, well sorted, ooid, peloid and intraclast grainstones that record sedimentation in a high-energy, shallow water setting at the outer margin of the platform. Associated facies in this setting include archaeocyathid-algal biostromes in the Lower Cambrian Paralleldal Formation and formation RG2, and thrombolitic algal mounds in Upper Cambrian sandy foreslope deposits of the Perssuaq Gletscher Formation. Mature quartzitic sandstones with well rounded grains form a large proportion of the latter formation, particularly in more northerly localities, where trough crossbedded, medium grained sandstones alternate with bioturbated, parallel-laminated sandstones and sandy mass-flow deposits (Fig. 7.21).

Figure 7.19. Outer shelf shales and siltstones of the Henson Gletscher Formation (HG), overlain by lime mudstones and limestone breccias assigned to a new formation of the Tavsens Iskappe Group (T); both stage $4S_2$. The sequence has a near vertical dip and youngs southward, to the left of the photograph. Cliffs are about 200 m high, coast of northern Nyeboe Land.

The transition from platform edge to outer shelf-slope is commonly marked by spectacular foreslope deposits that show northward primary dips of up to 30°. The foreslope grainstones and carbonate mass-flow deposits wedge out northward, interdigitating with dark carbonates of the outer shelf and slope; this transition is superbly illustrated within formation RG2 and the Perssuaq Gletscher and Løndal formations (Fig. 7.21, 7.22).

Outer shelf and slope rocks make up over two-thirds of the Brønlund Fjord and Tavsens Iskappe groups (Fig. 7.19, 7.20), forming continuous sections up to 450 m thick in southwestern Peary Land. Typically, they consist of thinly bedded, dark, hemipelagic lime mudstones and nodular limestones, interbedded with graded limestones and chaotic carbonate breccia beds. Secondary dolomitization affects nearly three-quarters of the succession, and occurs preferentially within thick bedded coarse grained intervals. The hemipelagites are commonly bituminous, argillaceous, and cherty, and often contain small phosphatic brachiopods, trilobites, and siliceous sponges represented by spicules; this facies is typical of the Henson Gletscher, Ekspedition Brae, Erlandsen Land, and Holm Dal formations (Fig. 7.21, 7.22). Even, parallel, occasionally micrograded, lamination is the dominant sedimentary structure. *Chondrites* burrows occur at certain levels. Some sections show a regular, cyclic alternation of burrowed and nonburrowed, laminated carbonate mudstone, reflecting rhythmic variations in the degree of oxygenation of bottom waters. The interbedded graded limestone beds range from silty lime mudstones less than 5 cm thick, to thick bedded (<60 cm), coarse grained peloidal grainstones showing Tab and Tabc Bouma divisions; they are the deposits of low-density turbidity currents derived from the upper slope and platform margin.

Very thinly bedded, nodular carbonates are a striking feature of the outer shelf and slope rocks, and are particularly well developed in the Aftenstjernesø, Sydpasset and Fimbuldal formations (Fig. 7.20). Now composed mainly of dolomite or neomorphic spar (formed by recystallization of lime mud), these beds show a complex, platy, nodular banding that probably reflects primary interbedding of turbiditic and hemipelagic lime mud, enhanced by early diagenetic, differential cementation. Interstratal sliding within such partially lithified sediments resulted in a spectacular array of slope creep structures. These range from pull-aparts, boudins, microfaults and discontinuous, locally discordant interstratal breccias to folds and

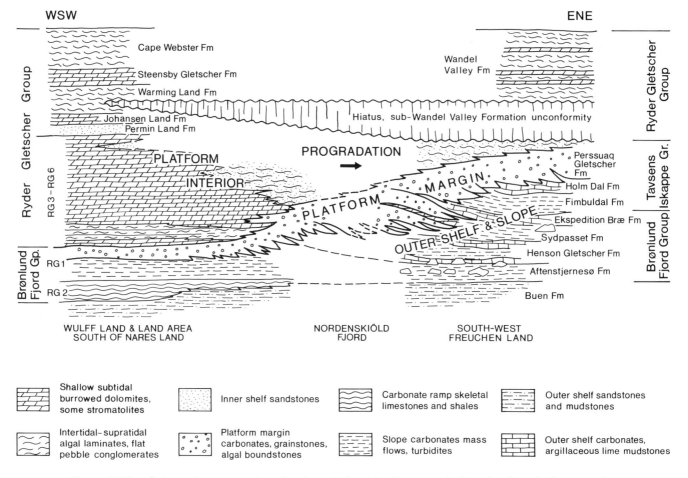

Figure 7.20. Schematic cross-section showing relationships between platform interior, platform margin and outer shelf-slope rocks of stage 4S$_2$, at the head of Nordenskiöld Fjord.

hummocks with relief of up to 10 m and wavelength up to several tens of metres.

Carbonate breccia beds, commonly dolomitized, range from 0.2 to 45 m in thickness and are mainly sheet-like, nonchanneled deposits. Typically clast-supported, these beds are internally unstratified, chaotic and poorly sorted, with an interstitial carbonate mud matrix; they are interpreted as the deposits of submarine debris flows (Ineson, 1980, 1985). Platform-derived blocks of ooid grainstone, up to several tens of metres across, are prominent in some breccia beds, particularly within the Aftenstjernesø, Sydpasset, Lønelv, and Fimbuldal formations (Fig. 7.18-7.22). More commonly, however, the breccias consist of coarse pebble to cobble-size, platy, angular fragments derived from slope environments (Fig. 7.23). All stages of slope mass wastage are preserved in the succession, from slope creep through sliding and slumping to mass flow.

A siliciclastic sand wedge, up to 124 m thick, interrupts the outer shelf-slope carbonates near the Lower to Middle Cambrian boundary, forming the Saeterdal Formation and much of the Henson Gletscher Formation (Fig. 7.18, 7.20). The sand wedge is persistent along depositional strike (roughly east-west) for over 300 km, from northern Nyeboe Land to central Peary Land and, despite a general northward thinning (see Fig. 7.20), it extends over 40 km from south to north, across depositional strike. It consists of thin to very thick bedded, fine grained sheet sandstones, interbedded with burrowed, laminated sandy siltstones (Fig. 7.21). Although commonly structureless, some sandstone beds show dewatering structures, parallel lamination and hummocky cross-stratification. The sand was clearly transported seaward during severe storms, but succeeding remobilization and mass flow transport farther down the slope can be documented in several cases.

Coeval outermost shelf and slope rocks of late Early Cambrian to Early Ordovician age outcrop between northern Nyeboe Land and J.P. Koch Fjord in a series of anticlinal fold cores. This sequence has been provisionally described in terms of three units by Higgins and Soper (1985); their fourth unit, equivalent to the Amundsen Land Group, forms part of stage 5S$_2$. The lower two units are distal, outermost shelf correlatives of the Aftenstjernesø and Henson Gletscher formations, respectively. The third unit (upper Middle Cambrian – lowermost Ordovician) is considered to be a new formation of the Tavsens Iskappe Group, and shows a somewhat different development from its more proximal shelf equivalents.

It is generally between 150 and 300 m thick, and is mainly composed of dark lime mudstones, although lime turbidites in beds from 10 cm to 1 m thick are conspicuous in eastern outcrops. Alternations of thin bedded, grey,

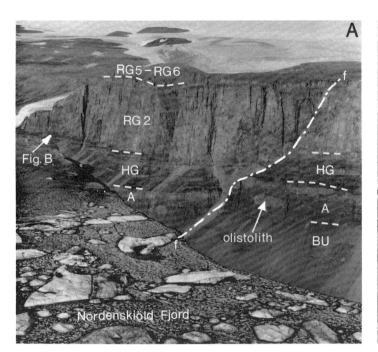

Figure 7.21. Cambrian platform margin sequence, at the head of Nordenskiöld Fjord. A. Outer shelf sandstones and mudstones of the Buen Formation (BU, stage 3S) are overlain by slope carbonates and debris flow units of the Aftenstjerneso (A) and Henson Gletscher (HG) formations of the Brønlund Fjord Group (stage $4S_2$). Succeeding platform margin carbonate grainstones of formation RG2 (Brønlund Fjord Group, stage $4S_2$) show steeply inclined depositional dips (Fig. B), and are overlain by platform carbonates and siliciclastic rocks of formations RG5-RG6 (Ryder Gletscher Group, stage $4S_1$). Cliff 700 m high. f-f = fault.

Figure 7.22. Prograding shelf carbonates of the Perssuaq Gletscher Formation (PG) overlying outer shelf carbonates and shales of the Holm Dal (HD) and Fimbuldal (F) formations (all Tavsens Iskappe Group, stage $4S_2$). Platform dolomites of the Wandel Valley Formation (W, stage $4S_1$) unconformably overlie the Perssuaq Gletscher Formation. East side of J.P. Koch Fjord, west Peary Land. Height of cliff about 1000 m.

Figure 7.23. Intraformational breccia (debris flow) formed of platy light and dark coloured lime mudstones; this rock type is abundant in the outer shelf equivalent of the Tavsens Iskappe Group (stage $4S_2$) of northern Nyeboe Land.

nodular or banded limestone and yellow-orange dolomite dominate the western sequence, while carbonate breccia beds are common in every section.

Traced northward toward the limit of exposure, in northern Nares Land, northern Freuchen Land and along the east side of central J.P. Koch Fjord, the three part division of the sequence fails and the total thickness is less than 100 m. The sequence in these areas consists almost entirely of chert and cherty shales, indicating a starved slope environment throughout the late Early Cambrian to Early Ordovician. Farther north still, in southern Johannes V. Jensen Land, the sequence thickens again, but now represents the trough deposits of the Vølvedal Group.

It is clear from Figure 7.20 that the Brønlund Fjord and Tavsens Iskappe groups together reflect a major, shallowing-upward, regressive depositional episode, probably controlled by regional uplift of central and eastern North Greenland during the Middle and Late Cambrian. The uplift is interpreted as a reflection of westward migration of a peripheral bulge during an early collisional phase along the western margin of the Iapetus Ocean (Surlyk and Hurst, 1984).

Outer shelf-slope rocks dominate the sequence but are everywhere overlain by, and grade southward into, platform carbonates (Fig. 7.18-7.21) reflecting northward progradation of shallow water environments. In southern Peary Land, paleocurrents, paleoslopes, and facies patterns indicate progradation toward the north-northwest, whereas in southwest Freuchen Land the platform prograded toward the northeast. Clearly, shelf subsidence was most pronounced in the region of southwest Peary Land and Freuchen Land during the Early and Middle Cambrian.

Progradation in more eastern areas of North Greenland was accompanied by progressive exposure of the carbonate platform and the development of a regional hiatus that decreases in stratigraphic importance toward the west (Fig. 7.3, 7.18). Karstic breccias and solution vugs are common beneath this unconformity. In southwestern Peary Land and Freuchen Land, shelf emergence was preceded by deposition of a thick sequence of northward prograding, quartz-rich siliciclastic sediments (Perssuaq Gletscher Formation) of Late Cambrian to earliest Ordovician age. This regressive clastic influx is represented farther west by the shallow-shelf clastics of the Permin Land Formation (see stage $4S_1$ and Fig. 7.20) and to the north, in the trough, by the sandstone turbidites of the Vølvedal Group (Surlyk and Hurst, 1984; Surlyk et al., in press).

4T: Trough

In southern Johannes V. Jensen Land the slope-and-rise, red and green mudstones at the top of the Polkorridoren Group are overlain by an outer slope and trough-floor sequence including dark mudstones, cherts, turbidites and base-of-slope conglomerates (Surlyk and Hurst, 1984). This late Early Cambrian to Early Ordovician sequence is placed in the Vølvedal Group (Friderichsen et al., 1982; Surlyk et al., in press). The group differs markedly from the outer shelf and slope sequence described above by its much greater thickness and by the presence of thick sandstone turbidite sequences. Higgins and Soper (1985) suggested that the turbidites of the Vølvedal Group did not extend westward much beyond the west point of Amundsen Land; it is possible that these deposits were in fact restricted to the eastern part of the trough, if their deposition was linked to uplift of eastern areas in the Cambrian as suggested by Surlyk and Hurst (1984). However, the absence of outcrops of the trough sequence in Nansen Land makes this difficult to verify, and farther west in western North Greenland this part of the basin is situated north of the present day land areas. Comparable turbidites are not known from the equivalent strata of the Hazen Formation in the outcrops of northeast Ellesmere Island.

The present outcrops of the Vølvedal Group are almost confined to southern Johannes V. Jensen Land, including Amundsen Land, and the region adjacent to Frederick E. Hyde Fjord (Fig. 7.24).

The group is generally 600-700 m thick. It begins with a sequence of dark grey and black, nonbioturbated mudstones with thin bedded sandstone turbidites comprising the Nornegaest Dal Formation. Deposition of the formation took place on the trough floor under poorly aerated conditions. Fine grained sedimentation continued in the form of greenish chert, cherty mudstones and siltstones, locally with fine sandstone turbidites. These deposits form the approximately 50 m thick Drengs Brae Formation.

The quiet, partly anoxic trough-floor sedimentation was eventually interrupted by the influx from the south of medium to thick bedded quartzitic turbidites of the Bøggild Fjord Formation (Fig. 7.25). The turbidites alternate with

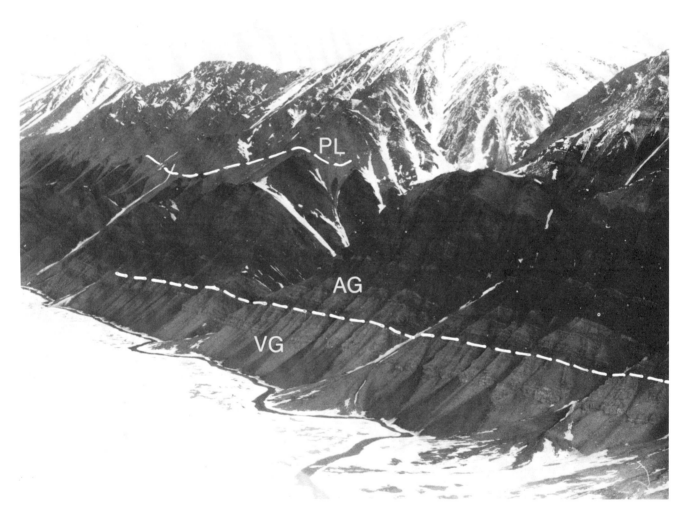

Figure 7.24. North side of O.B. Bøggild Fjord, showing turbidites of the Vølvedal Group (VG, stage 4T), overlain by black cherts and shales with carbonate and chert conglomerates of the Amundsen Land Group (AG, stage 5T), and turbidites of the Merqujôq Formation, Peary Land Group (PL, stage 6T). The contact between the two latter units is disturbed by thrusting.

thin beds of black mudstones. The formation is approximately 240 m thick and is interpreted as representing one or more relatively small borderland fans prograding northward into the trough, where black and green bedded cherts and black mudstones with thin bedded fan fringe and basin-plain turbidites were deposited. The more proximal southern occurrences display channelized mid-fan and outer fan lobe features, and the turbidite sequence decreases dramatically in bed thickness and in sand/mud ratio toward the north.

Along strike, the fan turbidites pass into interfan mudstones of the trough floor. All mudstones and fine grained, thin bedded turbidites are chertified to varying degrees. In many cases the chertification is so intense that the original turbidite features are completely masked.

Toward the termination of submarine-fan deposition, the upper slope underwent slumping on a large scale. A composite debris sheet in southern Johannes V. Jensen Land is up to 20 m thick and can be traced along strike for 75 km. The mainly clast-supported conglomerates are composed largely of tabular, angular clasts of laminated carbonate that range in size from pebbles to boulders up to 1 m long, set in a lime mudstone matrix. Clasts often show varying degrees of plastic deformation, indicating that the parent rock was only partially lithified at the time of slumping and subsequent mass-flow transport. Clast lithologies suggest derivation from the upper slope and outer shelf rather than from the carbonate platform. One conspicuous boulder type comprises alternations of grey limestone and yellow dolomite, known to form thick developments in the equivalent Cambrian sequence of the outer shelf and slope (new formation of the Tavsens Iskappe Group; unit 3 of Higgins and Soper, 1985).

There is little faunal control of the age of the Vølvedal Group, and the late Early Cambrian age of the base is fixed by comparison with the outer shelf and slope sequence. Early Ordovician graptolites are fairly common in mudstones in the middle and upper levels of the group.

A local development of the submarine fan sequence occurs in southern Johannes V. Jensen Land. It consists of thin to medium bedded calcarenitic turbidites, carbonate conglomerates and black shale. The carbonate-dominated lithology probably reflects the progressive uplift and erosion of the eastern platform region.

Figure 7.25. Stacked, thinning-upward turbidite sequences representing mid-fan channels. Vølvedal Group (stage 4T), O.B. Bøggild Fjord. Person for scale (ringed).

Stage 5: Middle Ordovician and Early Silurian stable platform, starved slope, and trough

This stage in basin evolution is marked by a back-stepping of the platform margin to approximately the line of the Navarana Fjord escarpment (Fig. 7.4, stage 5). Carbonate deposition continued to build up on the platform to the south (stable platform), with the platform margin forming a scarp-like feature (Fig. 7.3, 7.26). North of the scarp a broad slope received very restricted amounts of sediment (starved slope). Farther north, in the deep water trough, deposition was also slow, but considerably greater than in the slope areas. The relatively starved phase of slope and trough deposition lasted until the Early Silurian, when it was terminated by the abrupt commencement of turbidite deposition of the Peary Land Group (stage 6T).

$5S_1$: Stable platform

The phase in the evolution of the shelf recognized as stage 5 commenced with the late Middle Ordovician transgression (Christie and Peel, 1977; Fortey, 1984) and persisted into the Early Silurian (Early to Middle Llandovery). It is represented by the Morris Bugt Group and the basal part of the Washington Land Group in the Washington Land – Freuchen Land region (Hurst, 1980a; Peel and Hurst, 1980; Sønderholm et al., 1987) and by the Børglum River, Turesø and Ymers Gletscher formations and equivalent strata in Freuchen Land and Peary Land (Fig. 7.2, 7.3; Christie and Peel, 1977; Hurst, 1984; Sønderholm et al., 1987; Smith et al., in press).

The constant total thickness of these units throughout the region (ca. 650 m) reflects uniform subsidence over the area (Smith et al., in press).

The shelf was bounded by an escarpment hundreds of metres high, which can be seen in the inner part of Navarana Fjord and J.P. Koch Fjord (Fig. 7.26), where it forms an erosional sediment by-pass zone (Surlyk and Hurst, 1984; Surlyk and Ineson, 1987a). The extension of the escarpment farther to the west as far as northern Nyeboe Land and north of Hall Land is inferred from stratigraphic considerations and from changes in the style of tectonic deformation (Escher and Larsen, 1987; Hurst and Surlyk, 1982; Larsen and Escher, 1985). Hurst and Kerr (1982) suggested that the isolated outcrop of carbonates around Kap Ammen in northern Hall Land (Fig. 7.1, 7.2, 7.27) represented an Early Llandovery carbonate horst surrounded by deep water shelf clastic sediments. However, the presence of the same facies and lithostratigraphic units in the southern part of Nyeboe Land seems to indicate that the carbonate shelf extended unbroken from the southern outcrop belt to northern Hall Land and Nyeboe Land until late in the Middle Llandovery (Dawes and Peel, 1984; Sønderholm et al., 1987).

The shelf margin itself is only clearly exposed in the inner parts of J.P. Koch Fjord and Navarana Fjord (Fig. 7.26, 7.28) and these outcrops also closely resemble the equivalent exposures at Kap Ammen in northern Hall Land (Fig. 7.27). Hurst and Surlyk (1983b) described mud mounds up to 200 m in diameter and 50-100 m high from the shelf margin at J.P. Koch Fjord. The mounds apparently had considerable relief above the surrounding level bedded sediments.

Elsewhere in Peary Land, shelf margin facies equivalent to the Turesø Formation are only known from large blocks contained in contemporaneous base-of-slope conglomerates of the Citronens Fjord Member (Fig. 7.2), occurring along the southern coast of Frederick E. Hyde Fjord (Hurst and Surlyk, 1982; Hurst, 1984). The boulders consist of distinctive lithoclastic-bioclastic grainstones rich in fragments of virginiid brachiopods, stromatoporoids, gastropods and algae. Some small dolomite clasts are red stained. This suggests that the shelf was rimmed by high-energy carbonate sand shoals; the abundance of algae may indicate the presence of algal reefs well within the photic zone with possible episodes of subaerial exposure (Hurst, 1984).

Figure 7.26. Platform margin escarpment (Navarana Fjord escarpment; NFE) on the east side of J.P. Koch Fjord. Sandstone turbidites of the Merqujôq Formation (ME, stage 6T) abut against platform carbonates of stages $5S_1$ and 6S forming the escarpment. Note that bedding in the carbonates becomes diffuse as the escarpment is approached. W = Wandel Valley Formation; Ryder Gletscher Group (stage $4S_1$); BR = Børglum River Formation (stage $5S_1$); AF2, AF3 = units 2 and 3 of the Aleqatsiaq Fjord Formation (stage $5S_1$); WG 1/2 = units 1 and 2 of the Washington Land Group (stage $5S_1$ and 6S) with a small mound (m). Cliff 800 m.

Figure 7.27. Carbonate shelf sequence (stage $5S_1$ and 6S) at Kap Ammen, north Hall Land, close to the shelf margin (Fig. 7.4). The generally dark, open marine deposits of the Aleqatsiaq Fjord Formation (AF1-3) are overlain by pale carbonates (WGl) and shelf edge mounds (WG2) with intermound deposits (WG3) of the Washington Land Group. The shelf carbonates are overlain by deep sea fan turbidite sequences of the Peary Land Group (PL; stage 7T), which begin with the dark mudstones of the Thors Fjord Member (TFM). Small mounds within unit AF2 are arrowed. In the background the linear belt of reefs (stage 7S) forming the Hauge Bjerge. Aerial photograph 546 K-S no. 2190, copyright Geodetic Institute, Denmark. Reproduced by permission (A.200/87).

The sediments of the Gonioceras Bay and Troedsson Cliff formations (the Kap Jackson Formation of Smith et al., in press), together with the Cape Calhoun Formation and the lower part of the Aleqatsiaq Fjord Formation (Unit AFl of Sønderholm et al., 1987), with a total thickness ranging from 300 to 390 m, and the corresponding Børglum River Formation (approximately 430 m) (Fig. 7.2, 7.3, 7.13, 7.26, 7.29) are dominantly dark, nodular, burrow-mottled lime mudstones, skeletal wackestones and packstones. Epifauna and infauna are abundant and Hurst and Surlyk (1983b) report that algal remains are common. These level bedded sediments are of typical subtidal carbonate-shelf aspect. The lack of grainstones suggests low-energy environments probably below normal wave base, while the abundant fauna and algae indicate an oxygenated substrate within the euphotic zone.

This stable and uniform phase of deposition came to an end in latest Ordovician to earliest Silurian time as a result of a pronounced shallowing, followed by a later deepening of the entire shelf. The event is represented by the upper two units (AF2 and AF3; 115-160 m) of the Aleqatsiaq Fjord Formation (Sønderholm et al., 1987) and the 115-200 m thick Turesø Formation (Fig. 7.2, 7.3; Hurst, 1984). The great lateral extent of these units and the timing of the shallowing and deepening may indicate eustatic events related to the Late Ordovician glaciation (Brenchley and Newall, 1980; Fortey, 1984; McKerrow, 1979). Unit AF2 of the Aleqatsiaq Fjord Formation consists of light grey, massively bedded, variably biostromal and coralliferous skeletal limestones. *In situ* stromatoporoids and corals are locally very abundant and contribute to the formation of small mounds in some areas, e.g. Bessels Fjord, Petermann Fjord, Newman Bugt, and Kap Ammen (Fig. 7.27). The sediments are interpreted as being of agitated, shallow, open shelf origin, reflecting a relative lowering of sea level compared to the underlying unit. Toward the east, in the Nares Land to J.P. Koch Fjord region, unit AF2 of the Aleqatsiaq Fjord Formation becomes progressively more dolomitic and less fossiliferous, suggesting a more restricted environment (Sønderholm et al., 1987). Farther to the east

Figure 7.28. Platform margin relationships on the west side of Navarana Fjord. Top of the drowned Silurian platform margin scarp to the left (S, stage 6S), onlapped by turbidites of the Merqujôq Formation to the right (ME, stage 6T). Carbonate conglomerates of the Navarana Fjord Member (NF) drape the platform carbonates, interfinger with the highest turbidites of the Merqujôq Formation, and are overlain by black shales and thin bedded turbidites of the Thors Fjord Member (TFM) and turbidites of the Lauge Koch Land Formation (L, stage 7T).

in Peary Land, it grades into the light weathering dolomites of the lower part of the Turesø Formation (Hurst, 1984).

The Turesø Formation in central Peary Land displays a well developed cyclicity (Fig. 7.30), each cycle varying from 2 to 20 m in thickness and consisting of burrow-mottled, dolomitic wackestones with occasional silt ripples and rare benthic fauna, overlain by laminated and cryptalgally laminated or fenestral dolomites (Armstrong and Lane, 1981; Fürsich and Hurst, 1980; Hurst, 1984). The scarcity of intraclast conglomerates and shelly coquinas suggests that the whole sequence was formed under very low-energy conditions, and the possibility of periodically raised salinities has been discussed (Fürsich and Hurst, 1980; Hurst, 1984). In the G.B. Schley Fjord region to the northeast the environment again became more open marine. Here, the Turesø Formation is darker, less dolomitic, and consists nearly entirely of nodular lime wackestones.

The deepening which followed this phase of shallow water deposition is represented by the sediments of the upper unit of the Aleqatsiaq Fjord Formation (unit AF3) in the Washington Land to J.P. Koch Fjord region, and the upper part of the Turesø Formation in Peary Land and Kronprins Christian Land (Fig. 7.2, 7.3). Unit AF3 of the Aleqatsiaq Fjord Formation is uniformly developed over the entire area and comprises very dark, strongly mottled, bituminous, fine grained skeletal limestones with abundant large pentamerid brachiopods and frequent stromatoporoids and corals of Silurian age. These sediments originated on an open, mainly low-energy shelf. The upper part of the Turesø Formation in central Peary Land consists of dolomites much like the lower unit but dominated by the darker subtidal facies (Armstrong and Lane, 1981; Christie and Peel, 1977; Hurst, 1984). To the northeast, in the G.B. Schley Fjord region, the upper part of the Turesø Formation is more reminiscent of the upper unit of the Aleqatsiaq Fjord Formation, again suggesting a deepening of the shelf toward the east (Christie and Ineson, 1979).

Figure 7.29. Dark, cliff-forming carbonates of the Morris Bugt Group, representing the open marine shelf deposits of stage $5S_1$. Gonioceras Bay (GB), Troedsson Cliff (TC), and Cape Calhoun (CC) formations, and units AF1 and AF2 of the Aleqatsiak Fjord Formation. The Steensby Gletscher (SG) and Cape Webster (CW) formations of the Ryder Gletscher Group (stage $4S_1$) form the lower half of these 700 m high cliffs in southern Permin Land. Base camp of the Geological Survey of Greenland expedition 1984-85 in foreground.

This phase in the evolution of the stable platform came to a close during the Early Silurian, which was a period of general shallowing of the platform sea, probably due to rapid vertical accretion. This period is represented by the Ymers Gletscher Formation in Peary Land, the informal unit WGl, and the Petermann Halvø Formation between J.P. Koch Fjord and Bessels Fjord (Fig. 7.2, 7.3). In Washington Land, however, carbonate deposition took place on a homoclinal ramp, which evolved in response to initial subsidence and tilting of the Ordovician shallow water platform (Hurst and Surlyk, 1983a). The boundary between the carbonate ramp and slope developed along a sinuous flexure parallel to the present coastline (Fig. 7.4; stage 5).

The Ymers Gletscher Formation (25-45 m) comprises grey, thinly laminated and well bedded, fenestral lime mudstones. Coquinas dominated by pentamerid brachiopods are present as well as desiccation cracks (Armstrong and Lane, 1981; Hurst, 1984). The sediments reflect deposition in peritidal environments, possibly with raised salinities (Fürsich and Hurst, 1980). The sediments of unit WGl and the Petermann Halvø Formation are dominated by light grey to dark, flat bedded, skeletal limestones, usually rich in laterally extensive coral and stromatoporoid colonies with locally abundant crinoidal debris. Facies variations within unit WGl and the lower part of the Petermann Halvø Formation, such as dolomitization in southerly areas, probably reflect a slight deepening from peritidal, possibly partly emergent environments, in the south, to shallow outer platform in the north (Sønderholm et al., 1987).

Sedimentation patterns in Washington Land west of Bessels Fjord reflect the development of a homoclinal carbonate ramp (Hurst and Surlyk, 1983a). In the westernmost areas, this event is represented by the Cape Schuchert Formation (55-80 m), which is dominated by thin bedded, black, bituminous, cherty, limy mudstones, occasionally faintly laminated, and of slope and starved basin origin (Hurst, 1980a). Contemporaneously, fringing reefs developed at the carbonate ramp margin, represented by the sediments of the Adams Bjerg and Pentamerus Bjerge formations of Hurst (1980a). The initiation of the carbonate ramp fringing reefs promoted the development of a scarp between the carbonate ramp margin and the slope.

Figure 7.30. Alternations of pale and dark coloured dolomitic limestones of the Tureso Formation (T) overlying dark limestones of the Børglum River Formation (BR), central Kronprins Christian Land (both stage $5S_1$). Caledonian deformation has disturbed the sequence, and has emplaced a thrust sheet of Proterozoic sandstone turbidites (RS = Rivieradal Sandstone) over the Tureso Formation. Cliff height 800 m.

The introduction of resedimented carbonates of the Kap Godfred Hansen Formation along the slope correlates with this greater ramp-to-slope differentiation and initial reef building (Figs. 7.2, 7.3; Hurst, 1980a; Hurst and Surlyk, 1983a).

$5S_2$: Starved slope

The outer shelf and slope dolomitic mudstones of stage $4S_2$ between northern Nyeboe Land and central J.P. Koch Fjord are overlain by a sequence of chert and cherty shale, with partly chertified sequences of siltstone, black limestone, and dolomitic mudstone (Fig. 7.7; unit 4 of Higgins and Soper, 1985). This starved slope sequence is generally between 50 and 150 m in thickness, and is assigned to a new formation of the Amundsen Land Group. Rich graptolite faunas occur throughout, and show an age range from Tremadoc to Late Llandovery.

Throughout the area between northern Nyeboe Land and J.P. Koch Fjord, chert and mudstone deposition was brought to a close in the Late Llandovery with the incoming from the east of the first sand turbidites of the Peary Land Group (Higgins and Soper, 1985; Hurst and Surlyk, 1982). Deposition of fine grained turbidites was initiated slightly earlier, in latest Ordovician time, toward the east in the Peary Land area. These distal turbidites are, however, interbedded with black mudstones and cherts characteristic of stage 5T, and they are accordingly not included in the Peary Land Group (Hurst and Surlyk, 1982; Surlyk et al., 1980; Surlyk et al., in press).

The Ordovician starved slope sequence passes downslope into similar, but more basinal and more thickly developed sediments of the same age. These are, however, only preserved in the type area of the Amundsen Land Group.

5T: Trough

The relatively small-sized borderland fans of the Vølvedal Group (stage 4T) were eventually abandoned and a long period of basin starvation and, at times, also stagnation set in (Surlyk and Hurst, 1984). In southern Johannes V. Jensen Land, including Amundsen Land, and the areas along the south side of Frederick E. Hyde Fjord, the deposits of this period are included in the Lower Ordovician – Lower Silurian Amundsen Land Group (Friderichsen et al., 1982; Surlyk et al., in press). The sedimentary facies are dominated by black and green radiolarian cherts and black and green mudstones; they include thin bedded turbidites and thick sequences of carbonate conglomerates (Fig. 7.31).

Turbidite deposition of the previous stage (Bøggild Fjord Formation of the Vølvedal Group) rapidly waned, and stage 5T was heralded by a change to slow sedimentation of mud and siliceous ooze mainly under anoxic conditions (Harebugt Formation of the Amundsen Land Group). In latest Tremadoc and Arenig times the quiet deposition of fine grained sediments was interrupted by a period dominated by deposition of carbonate conglomerates and turbidites of the Kap Mjølner Formation, which reaches a thickness of up to 200 m in the southernmost localities around inner Frederick E. Hyde Fjord. The conglomerates commonly form the base of fining- and thinning-upward cycles, considered to reflect waning deposition resulting from episodes of slumping and successive development of mass flows and turbidity currents. The conglomerate sheets can be traced over a wide area of Amundsen Land, but

Figure 7.31. Sheet of base-of-slope carbonate conglomerates and calcarenites in the lower part of Amundsen Land Group, stage 5T. Individual beds are up to 10 m thick; the height of the wall is about 100 m. Southern Johannes V. Jensen Land.

Figure 7.32. Shelf-trough transition of stage 6, eastern Wulff Land. Unit WG 1/2 (stage 6) thickens toward the steep, indented shelf margin. Dark shales (S) occur close to the shelf margin between and behind mound complexes of unit WG 1/2, and carbonate mud and wackestone farther back on the shelf (unit WG3). The shelf margin of stage 6S is dominated by the characteristic, striped biostromal unit WG4 and does not here show mound development. Carbonate conglomerates (C) embedding foreslope mounds occur in a base-of-slope setting and form together with the interbedded mudstones the Lafayette Bugt Formation (LB). Trough deposits (stage 7T) are represented by dark mudstones of the Wulff Land Formation (WL) and pale turbidites of the Nyeboe Land Formation (NL). TC = Troedsson Cliff Formation; CC = Cape Calhoun Formation; AFl, AF2, AF3 = units 1, 2 and 3 of the Aleqatsiaq Fjord Formation; all stage 6S. Height of mountain in centre about 900 m.

become less important westward and northward. The northernmost localities of the Amundsen Land Group, just north of the present Harder Fjord fault in Johannes V. Jensen Land, are characterized by thick units of black chert and mudstone, reflecting increasing stagnation of the trough.

The redeposited material represented by the Lower Ordovician conglomerate sheets is of interest in the links that can be made with conditions in the source areas of the shelf (Surlyk and Hurst, 1984). The dominant clast composition of the conglomerates changes upward from flat-pebble carbonates to black chert, reflecting a change in source area and lithology. Also, while boulder and cobble grades dominate the lower conglomerates, pebble grades become dominant in the higher parts of the sequence. The eroded material is considered to have been transported across the outer shelf and shelf-slope break in a series of debris flows. On the shelf to the south, a correlation can be made with a period of uplift and erosion preceding deposition of the Wandel Valley Formation (see also above); the latter unconformity overlies and oversteps rocks from Late Proterozoic to earliest Ordovician age. Although much of the hiatus is believed to be due to nondeposition, there was evidently significant Early Ordovician erosion.

Eventually conglomerate and turbidite deposition faded out and the quiet, slow sedimentation of fine grained deposits was resumed. They include green and green-grey ribbon cherts, chertified mudstones and siltstones, and thin bedded silty turbidites, all referred to the Nordpasset Formation.

The youngest deposits of stage 5T include black cherts and mudstones, dark silty mudstones, siltstones and thin bedded silty turbidites of the Harder Fjord Formation (Late Ordovician – Early Llandovery). These deposits herald an important event in the evolution of the trough, with the starved basin deposition of the Amundsen Land Group being brought to a close in the latest Ordovician to Early Silurian (Hurst and Surlyk, 1982; Surlyk and Hurst, 1984). The abrupt onset of sandstone turbidite sequences of the Peary Land Group seems, however, to have started nearly everywhere in the Late Llandovery (Higgins and Soper, 1985; Hurst and Surlyk, 1982).

Stage 6: Early Silurian ramp and rimmed shelf, and longitudinal turbidite trough

In the early Late Llandovery, deposition of a major system of sand turbidites derived from the rising Caledonian mountain belt to the east was initiated in the trough. The earliest pulses of deposition of distal fine grained turbidites were, however, initiated in the latest Ordovician (see stage 5T). The trough filled rapidly; by about the Llandovery-Wenlock boundary, the Navarana Fjord scarp, which had formerly marked the platform margin, was drowned, and the trough expanded southward (stage 7) (Fig. 7.3, 7.4). The loading effect of the thick sequences of turbidites accumulating in the trough caused downflexing of the outer part of the platform. Carbonate deposition continued during this general deepening of the shelf, but variation in facies became more complex, especially in the western part of the region.

6S: Ramp and rimmed shelf

The early Late Llandovery downflexing of the outer shelf can be traced all across North Greenland. In Kronprins Christian Land and Peary Land it resulted in a widespread deepening of the carbonate platform and deposition of the Odins Fjord Formation (Hurst, 1984). Farther west, the outer part of the shelf was drowned slightly earlier and covered by mudstones and turbidites of the Peary Land

Group, although some outlying carbonate mounds survived the initial phase of drowning, and the carbonate shelf margin changed to a new, more southerly position. In this area, shelf sediments of stage 6S are referred to units WG2, WG3, and WG4, and the Bessels Fiord, Offley Island, Kap Godfred Hansen, Pentamerus Bjerge, and Hauge Bjerge (Kap Independence Member) formations, while slope sediments have been referred to the Lafayette Bugt and Cape Schuchert formations of the Peary Land Group (Fig. 7.2, 7.3; Hurst, 1980a; Sønderholm et al., 1987).

The Odins Fjord Formation (200-350 m) comprises in the lower part dark, massive, nodular or wavy bedded skeletal wackestone and lime mudstone with weak horizontal lamination. Burrow-mottling is common, but algal remains are rare. The sediments reflect deposition in low-energy subtidal environments (Hurst, 1984).

In southern Peary Land, peritidal environments prograded northward, represented by the Melville Land Member (15-30 m), which consists of grey fenestral and laminated lime mudstones (Hurst, 1984). A further rise in relative sea level took place in the late Middle to early Late Llandovery (middle Odins Fjord Formation), and sediments of low-energy subtidal environments were deposited both in northern and southern shelf areas. In the Late Llandovery, a continuing relative sea level rise in the northern platform areas directly south of the platform margin is indicated by the Bure Iskappe Member (30-60 m), which comprises cherty, thin bedded, nodular wackestones, horizontally laminated lime mudstones, and rare interbeds of terrigenous mudstone (Fig. 7.3). Algae have not been recorded from this member. These features suggest deposition in very low-energy, relatively deep water environments below the photic zone, possibly under partly anoxic conditions (Hurst, 1984).

In the southern platform areas, the contemporaneous sediments of the Odins Fjord Formation are much like the lower part of this formation, but contain abundant biostromal units rich in *in situ* stromatoporoids and skeletal debris. These units are lenticular on a kilometre scale, suggesting sedimentation on extensive high-energy shoals in the euphotic zone, with smaller intervening lower-energy, subtidal areas.

Figure 7.33. Dark intermound sediments (unit WG3) and patch reefs (unit WG2) belonging to stage 6S. Sediments of stage $5S_1$ are represented by units AFl to AF3, and unit WGl. Looking east, in southern Nyeboe Land. Height of mountain about 550 m.

In essence, these large-scale shoals and the earlier peritidal environments of the Melville Land Member did not prograde farther north than to an east-west line running through central Peary Land, suggesting that some pre-existing shelf topography with an increase in dip at this line, prevented further progradation of the shallow marine facies (Hurst, 1984).

Sediments of the Odins Fjord Formation in the G.B. Schley Fjord region and in Kronprins Christian Land have not been investigated in detail. However, as reported by Hurst (1984), there is no evidence to suggest major differences in sediment characteristics and hence environments were probably dominantly subtidal.

The rim of the Peary Land shelf during this late stage is poorly known, but some information on the rim facies of the Odins Fjord Formation is yielded by boulders contained in contemporaneous base-of-slope conglomerates of the Freja Fjord Member (Fig. 7.2). These show clasts up to 10 m in diameter, which are composed of grainstone and algal-bryozoan boundstone, suggesting that the platform was rimmed by a series of high-energy shoals and incipient algal-bryozoan reefs. The platform margin in Peary Land was still developed as the steep sediment bypass zone of the Navarana Fjord scarp, although stratigraphic considerations indicate that the height of the scarp was gradually reduced due to trough infill during the Early Silurian (Hurst, 1984; Surlyk and Ineson, 1987a).

The drowning of the northern part of the shelf in western North Greenland was accompanied by a retreat of the shelf margin to a more southerly position running from south of Nares Land through central Wulff Land to central Hall Land (Fig. 7.3, 7.34). As a consequence of the increased subsidence, unit WG1 and the Petermann Halvø Formation (stage $5S_1$) are locally overlain by a series of mounds (unit WG2 and basal mounds of the Bessels Fjord Formation), which increase in size and numbers toward the north (Fig. 7.26, 7.27; Sønderholm et al., 1987). However, the mound complexes of unit WG2 did not form a complete barrier across the region, and a steep indented shelf margin developed. Between and behind these major mound

Figure 7.34. Biostromal deposits of stage 6S (Offley Island Formation, unit WG4) overlain by mound-complexes (Hauge Bjerge Formation, unit WG5, stage 7S) and siliciclastic sediments of the Peary Land Group (PL, stage 7T). The mountains to the east, forming the background, are part of the impressive reef belt extending across western North Greenland. Kap Tyson, western Hall Land. Height of cliff is 740 m.

Figure 7.35. Carbonate mounds (m) of the Kap Independence Member (Hauge Bjerge Formation) embedded in slope deposits of the Cape Schuchert (CS) and Lafayette Bugt (LB) formations (stage 6S). AF = Aleqatsiaq Fjord Formation; KI = Kap Independence. Westernmost Washington Land, looking across Nares Strait toward Ellesmere Island. Aerial photograph 545 H-N, no. 11803. Copyright Geodetic Institute, Denmark. Reproduced by permission (A.200/87).

complexes an essentially flat carbonate platform was maintained, probably sloping gently toward the north; this was the site of accumulation of units WG3 and WG4 and the Bessels Fjord and Offley Island formations (Fig. 7.2, 7.3, 7.32; Hurst, 1980a, b; Sønderholm et al., 1987).

Unit WG3 and the equivalent Bessels Fjord Formation (250-600 m) consist of dark grey to black, back-mound and inter-mound sediments (Fig. 7.32), and may include isolated mounds up to 200 m thick. The sequence comprises thinly bedded, variably mottled, skeletal lime mudstones with thin interbeds of graded bioclastic limestones, pebble conglomerates, and slump sheets. Chert layers and nodules occur throughout. These sediments drape laterally onto, or terminate against, light grey to dark grey carbonate mound sediments. The mounds are mostly small patch reefs but may interconnect with large mounds of unit WG2, especially to the north (Fig. 7.33).

Unit WG3 and the Bessels Fjord Formation interfinger with, and are overlain by, unit WG4 and the equivalent Offley Island Formation, which consist of 200-450 m of very variable limestones, mainly composed of interbedded dark and light stromatoporoid and coral biostromes (Fig. 7.32,

Figure 7.36. Sandstone turbidites of the Merqujôq Formation (ME, stage 6T), overlain by a thin mudstone unit, the Thors Fjord Member of the Wulff Land Formation (TFM), and sandstone turbidites of the Lauge Koch Land Formation (L, stage 7T). Crosscutting black bands are Upper Cretaceous dykes. West side of inner J.P. Koch Fjord, cliffs 900 m high.

7.34). The dark beds contain *in situ* stromatoporoids and corals, while the pale beds contain abundant reworked material. Crinoid material occurs virtually throughout, and beds up to 10 m thick consisting almost entirely of crinoid columnals are interbedded with biostromes. In exceptional cases, major mound complexes are flanked by sequences up to 200 m thick, almost exclusively composed of crinoid columnals. These sediments originated on a rapidly subsiding shallow marine platform dominated by high-energy environments.

In western Washington Land, the continuing relative sea level rises during late Middle to latest Llandovery led to inundation of the Washington Land ramp by starved basin and slope sediments of the Cape Schuchert and Lafayette Bugt formations, whereby a greater differentiation of the carbonate ramp from the slope was achieved (Fig. 7.3, 7.4, 7.35). Concomitantly, the southwestern part of Washington Land was drowned during the earliest Late Llandovery, while new mounds were initiated contemporaneously on the slope farther to the east (Kap Independence Member of the Hauge Bjerge Formation). Meanwhile the impressive carbonate ramp-margin reef belt represented by the Pentamerus Bjerge and Kap Godfred Hansen formations continued its development (Hurst, 1980a; Hurst and Surlyk, 1983a).

6T: *Longitudinal trough*

The most dramatic depositional change in the deep water trough took place in the early Late Llandovery when a major longitudinal sand-rich, turbidite system was developed following the late Early Cambrian – earliest Silurian starved basin phase. An early phase of fine grained dilute turbidity current deposition was initiated close to the Ordovician-Silurian boundary. Significant turbidite deposition, however, started abruptly in Late Llandovery time, and a sand-rich elongate submarine fan system rapidly developed (Hurst and Surlyk, 1982; Surlyk, 1982; Surlyk Hurst, 1984).

Figure 7.37. Turbidite filled mid-fan channels in the top part of the Merqujôq Formation (stage 6T), Thor Fjord.

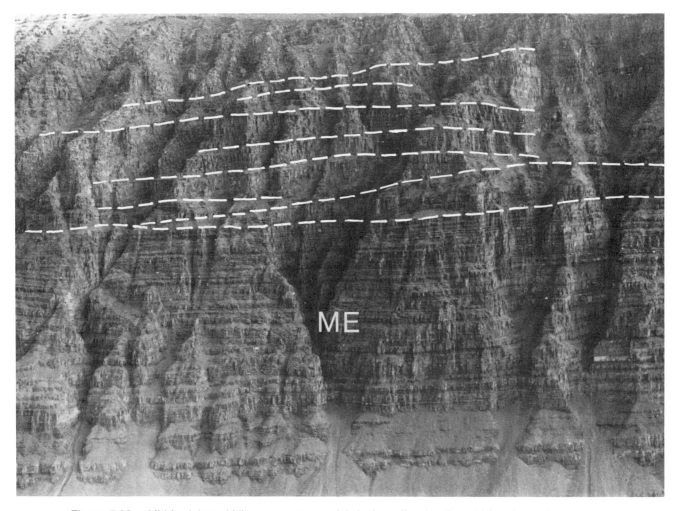

Figure 7.38. Mid-fan lobe turbidite sequences overlain by laterally migrating mid-fan channel sequences. Merqujôq Formation (ME; stage 6T). Section is about 400 m high and is oriented N-S (north to the left). Current toward the observer. Western Peary Land.

Figure 7.39. Thick sequence of muddy carbonate turbidites (A) overlying carbonate boulder conglomerates of the Citronens Fjord Member shown (B). Merqujôq Formation (stage 6T), Citronens Fjord, Peary Land.

Figure 7.40. Carbonate boulder conglomerate of the Navarana Fjord Member showing giant-scale loading into sandstone turbidites of the Merqujôq Formation (stage 6T). West coast of Navarana Fjord.

Turbidite deposition continued throughout the Silurian, and the resulting sequence is placed in the Peary Land Group. The formations of the Peary Land Group, dominated by sand turbidites, mudstones or conglomerates (Hurst and Surlyk, 1982; Larsen and Escher, 1985, 1987; Surlyk and Ineson, 1987a, b), are interpreted here as stages 6T-7T in the evolution of the basin. In northern Ellesmere Island, a parallel change is recorded, with starved basin deposition of the Hazen Formation giving way in the Early Silurian to the sandstone turbidites of the Danish River Formation (Chapter 8B).

The initiation of stage 6T is represented by the very thick sequence of sandstone turbidites of the Merqujôq Formation (>500-2800 m), which appears to be largely Late Llandovery in age (Higgins and Soper, 1985; Hurst and Surlyk, 1982; Larsen and Escher, 1985). These turbidites are widely distributed in North Greenland between Frederick E. Hyde Fjord in the east and northern Nyeboe Land in the west (Fig. 7.7, 7.36). In some areas there is a gradual increase in grain size at the top of the starved slope sequence passing into the first sandstone turbidites, while in other areas the contact with thick sandstone turbidites is abrupt. In several areas the first sandstone turbidites interdigitate with the starved slope facies (Higgins and Soper, 1985; Hurst and Surlyk, 1982; Surlyk et al., in press). Turbidite deposition commenced abruptly in central Johannes V. Jensen Land in the Llandovery with the incoming of extremely thick (up to 30 m) sandstone turbidite beds, distinguished as the Sydgletscher Formation (Hurst and Surlyk, 1982, Fig. 9).

The main part of the Merqujôq Formation represents deposition in the outer-fan and trough-floor environment in a highly elongate east-west submarine fan system. Paleocurrents are uniform toward the west-southwest, parallel to the Navarana Fjord scarp, which formed the margin of the carbonate platform. Turbidite deposition took place right up to the scarp, and the steep profile of the scarp may have been maintained by scouring. The top part of the formation is characterized by complex channeling and scouring, interpreted as representing a braided mid-fan environment (Fig. 7.37). Some channels show point bar-like features suggestive of lateral migration and meandering (Fig. 7.38). The largest observed channels have widths of several hundred metres and depths of about 50 m. Turbidite sedimentation was punctuated by several episodes of carbonate conglomerate deposition; beds range from 0.5 to

Figure 7.41. Type locality of the Samuelsen Høj Formation (SH, stage 7S) in central Peary Land, showing the isolated carbonate mound forming Samuelsen Høj, which is rooted in Silurian shelf carbonates of the Odins Fjord Formation. Siliciclastic sediments of the Peary Land Group (WL = Wulff Land Formation; L = Lauge Koch Formation, stage 7T) envelope the carbonates and form the characteristic terraced hills to the north. Samuelsen Høj is about 2 km long. Aerial photograph 548 C-N, no. 4280A, copyright Geodetic Institute, Denmark. Reproduced by permission (A.200/87).

50 m in thickness and, in type, from well sorted pebble beds to totally disorganized boulder beds with individual clasts often reaching a diameter of several metres (Fig. 7.39, 7.40). In contrast to the longitudinally derived turbidites, the conglomerates were derived from the southern platform area and from the platform margin scarp as reflected in the concave erosional surface which truncates the top of the scarp at Navarana Fjord (Surlyk and Ineson, 1987a, b).

Stage 7: Final drowning of the platform

The last phase in the evolution of the Silurian carbonate ramp and rimmed shelf was a consequence of continued foundering and associated sea level rise. The trough expanded southward as a result of the loading effect of the thick turbidite sequence to the north, although platform foundering in eastern areas is also attributed to downflexing caused by encroaching Caledonian nappe sheets. Eventually the shelf was drowned completely during the latest Llandovery (C5 to C6) and carbonate sedimentation was only locally maintained around mound complexes; the shelf was inundated by hemipelagic mudstones and siltstone turbidites of the Thors Fjord Member (Wulff Land Formation), Profilfjeldet Member (Lauge Koch Land Formation), and Lafayette Bugt and Kap Lucie Marie formations (Fig. 7.2-7.4).

7S: *Late Llandovery – Early Ludlow reef belt*

Carbonate sedimentation of this final phase was only locally maintained and led to formation of isolated mounds. These sediments are referred to the Samuelsen Høj Formation

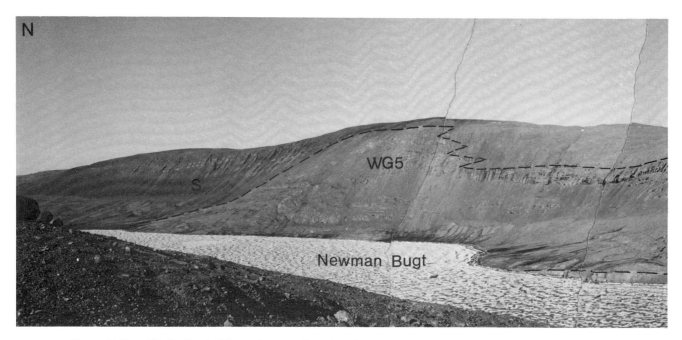

Figure 7.42. Shelf edge buildups of stage 7S (unit WG5) on top of biostromal platform carbonates (unit WG4; stage 6S), surrounded by shales (S) of the Peary Land Group (stage 7T). Arrows indicate possible slump scars. Western Nyeboe Land, cliffs 550 m high.

Figure 7.43. Unit WG5 reefs (stage 7S) on the east side of Hall Land. Steeply-dipping beds indicate a syndepositional relief of 300 m for the reef. The reefs are surrounded by carbonate conglomerates (C) and shales of the Peary Land Group (PL, stage 7T). WG4 signifies unit of Offley Island Formation.

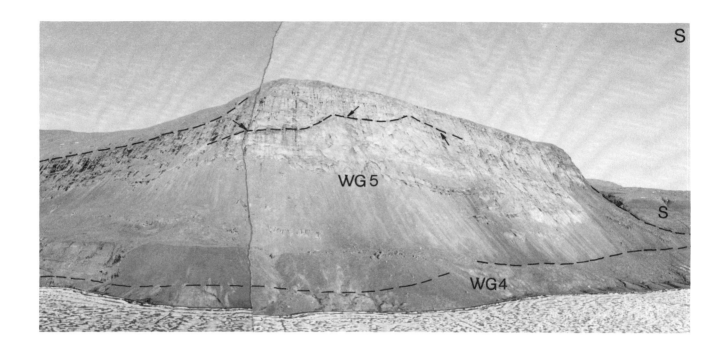

Figure 7.44. Stacked thinning- and fining-upward inner fan valley sequence of the Lauge Koch Land Formation (stage 7T). Eastern Peary Land.

between Kronprins Christian Land and Peary Land, unit WG5 and the Cape Tyson Member of the Hauge Bjerge Formation in the area between western Wulff Land and eastern Washington Land, and by the Pentamerus Bjerge, Kap Godfred Hansen, Kap Morton, and Kap Maynard formations in Washington Land (Hurst, 1980a, 1984; Sønderholm et al., 1987; Fig. 7.2, 7.3). Carbonates of this late stage are not represented between western Peary Land and western Wulff Land, probably due to later erosion.

The approximately 300 m high pinnacle reefs of the Samuelsen Høj Formation (Fig. 7.41) occur on top of the Odins Fjord Formation where they form an east-west linear belt extending for 50 km in eastern Peary Land. Mound complexes also occur in Valdemar Glückstadt Land and in Kronprins Christian Land (Fig. 7.1, 7.4) where they are widespread and larger than in Peary Land. In Kronprins Christian Land they occur stratigraphically lower (Late Llandovery, C5) in the succession than in Peary Land (Late Llandovery, C6), since there is lateral interdigitation of the lower part of the mound complexes with the level bedded sediments of the Odins Fjord Formation (stage 6S). Furthermore, mound formation probably terminated during the Late Llandovery (C6); sparse graptolite evidence suggests that it occurred slightly earlier in the east than in the west (Hurst, 1984; Hurst and Surlyk, 1982; Peel et al., 1981).

In western North Greenland, carbonate sedimentation was only maintained around the major mounds of the previous phase of deposition (Fig. 7.42, 7.43). Here, mound complex or associated flank sediments of unit WG5, the Cape Tyson Member of the Hauge Bjerge Formation, and the Pentamerus Bjerge, Kap Godfred Hansen, Kap Morton and Kap Maynard formations form sequences from approximately 200-1000 m thick (Dawes, 1987; Hurst, 1980a; Sønderholm et al., 1987). These sediments form a linear belt extending from northern Washington Land, through central Hall Land, Nyeboe Land and Warming Land to western Wulff Land, indicating the position of the former shelf edge (Hurst, 1980b; Sønderholm et al., 1987; see Fig. 7.1, 7.4). Due to present-day erosion it is not known how long carbonate deposition continued. Hurst (1980a) and Dawes and Peel (1984) considered that it extended at least into the Early Ludlow, but thin beds of resedimented carbonates occur within the Chester Bjerg Formation of Pridoli age.

Figure 7.45. Chert-pebble conglomerate showing inverse grading. Hendrik Ø Member (stage 7T), Hendrik Ø.

7T: Trough expansion

The complex foundering of the shelf described above was associated with a southward expansion of the trough during the Late Llandovery (Hurst and Surlyk, 1982; Larsen and Escher, 1985; Surlyk, 1982; Surlyk and Hurst, 1984; Sønderholm et al., 1987). The foundering of more than 30 000 km² of carbonate shelf was probably caused by the combined effects of normal passive margin tectonics, loading of the eastern shelf by Caledonian nappes, loading of the trough by several kilometres of mainly Upper Llandovery turbidites and associated downflexing of the outer shelf, and by the influx from the Caledonian mountains over the platform of terrigenous muds. The northern shelf margin was inundated by clastic sediments in the Late Llandovery, whereas the southern part first started to receive terrigenous clastics at the Llandovery-Wenlock boundary, indicating that shelf foundering was in part accomplished by flexuring. In western North Greenland, the inundation of the outer platform was probably initiated in the early Late Llandovery (C1-3), while it commenced somewhat later in the Late Llandovery (C4) in central North Greenland. The extensive submergence resulted in deposition of a thick, uniform sequence of black mudstone on top of the foundered shelf carbonates; the sequence of up to about 100 m of mudstone makes up the Thors Fjord Member of the Wulff Land Formation (Hurst and Surlyk, 1982; Larsen and Escher, 1985). The depositional regime can be characterized as a slope and rise system passing northward into the deep water trough.

Following the Late Llandovery – Wenlock phase of trough expansion and submarine fan starvation, caused by foundering of the carbonate platform, the turbidite depositional systems rapidly built up again, as an extensive, westward prograding submarine fan system.

The turbidite deposits of this phase are mainly placed in the Lauge Koch Land Formation (Fig. 7.36; Hurst and Surlyk, 1982; Larsen and Escher, 1985; Surlyk and Hurst, 1984). Easternmost localities in eastern Peary Land are characterized by thick, upward-fining cycles, about 50-100 m thick, composed of thick, locally pebbly, structureless sandstones passing upward into dark, laminated mudstones (Fig. 7.44). The cycles can be traced laterally for at least 5 km. They are interpreted as representing a system of stacked, wide inner fan valleys where the upper muddy part reflects channel abandonment and passive filling. Farther westward the inner fan valleys give way to a system of complex channels characteristic of a braided mid-fan environment. From western Peary Land and westward to

Figure 7.46. Large-scale flute casts on the sole of vertical chert-pebble conglomerate of the Hendrik Ø Member (stage 7T). Current toward the top of picture. Hammer to the right for scale. Stephenson Ø.

Hall Land, only outer fan, fan fringe and basin plain environments are represented. West of Victoria Fjord, the sandstone/mudstone ratio, sand grain size, and turbidite bed thickness decreased in a downcurrent direction.

The entire depositional setting is interpreted as an elongate submarine fan imperceptibly passing into a basin plain (Hurst and Surlyk, 1982; Surlyk and Hurst, 1984). Pulses in turbidite deposition alternated with periods of fan starvation resulting in alternations of turbidite sandstone and mudstone packets and in changes in the southern extent of the main turbidite body. These variations probably mainly reflect changes in sea level, with low levels resulting in increased influx of turbidity currents, and high levels resulting in fan starvation and mud deposition.

During the middle Wenlock, a major phase of conglomeratic deposition was initiated at the eastern, proximal end of the basin (Hurst and Surlyk, 1982; Larsen and Escher, 1985, 1987; Surlyk and Hurst, 1984). The conglomeratic depositional system prograded westward and reached northern Nyeboe Land in the early Ludlow. The conglomeratic sequences are placed in the Nordkronen Formation. At the type locality south of Frederick E. Hyde Fjord the sequence is incomplete, and only 100 m are preserved; farther west a 600-700 m sequence occurs on Stephenson Ø, 281 m on Hendrik Ø, and in northern Nyeboe Land a feather edge of the unit measures 32 m.

The conglomerates differ markedly from all of the earlier deep water conglomerates in the basin sequence described above. The latter are all derived from the upper slope and carbonate shelf to the south; they are often chaotic, disorganized types with lime mud matrix, poor clast sorting, and a variety of clast lithologies. In contrast, the conglomerates of the Nordkronen Formation are more organized and are interbedded with sandstone turbidites. They occur in relatively thin, sheet-like beds, with a sandstone matrix identical to the sandy turbidites. The clasts are well sorted and rounded, pebble-size, green, black and grey chert (Fig. 7.45). The chert conglomerates were deposited from high-density turbidity currents travelling westward along the trough axis from a source area in the rising Caledonian mountains (Fig. 7.46) (Surlyk and Hurst, 1984).

The generally well rounded and sorted appearance of the chert pebbles indicates that the parent rock was lithified at the time of erosion and that considerable transport and sorting had taken place in the coastal zone before redeposition by density currents. Some chert pebbles have yielded radiolarian remains suggestive of an Ordovician age, and the chert pebbles were probably derived from uplifted, thick cherty sequences of mainly Ordovician age (Vølvedal and Amundsen Land groups and correlatives); no other known units contain sufficient volumes of chert.

The last phase of the trough fill is only preserved in the western part of North Greenland, west of Freuchen Land. It is represented by the Nyeboe Land Formation and the Chester Bjerg Formation (Hurst and Surlyk, 1982; Larsen and Escher, 1985, 1987). This part of the sequence totals about 1000 m in thickness and is of latest Silurian age, possibly reaching into the earliest Devonian.

In the transition zone of the Nyeboe Land Formation and the Chester Bjerg Formation, packages of fine sandstone turbidites alternate with packages of silty mudstones, often with starved ripples. This facies association reflects a fan fringe depositional environment characteristic of the transition between outer fan and basin plain deposits. The bulk of the Chester Bjerg Formation comprises laminated light green weathering mudstones or siltstones, deposited from muddy contour currents, very dilute turbidity currents, or nepheloid (i.e. mud-charged) layers in a distal basin plain.

CALEDONIAN EVENTS IN NORTH GREENLAND

The early Paleozoic Franklinian Basin in North Greenland seems to have formed by rifting. It may thus be of fully ensialic nature, or an aulacogen, or it may represent continental breakup and formation of a narrow ocean (Surlyk and Hurst, 1984). The nature of the crust beneath the basin is not well known, but is probably of continental or transitional type in most of its preserved, southern part. The basin reflects then the evolution of a normal, passive continental margin. A number of events during basin evolution, however, can be related to the progressive closure of the Iapetus Ocean and the formation of the East Greenland Caledonides (Surlyk and Hurst, 1984).

The earliest and most conjectural event was the uplift of the eastern shelf areas in Cambrian and Early Ordovician time which resulted in a marked hiatus, which decreases progressively toward the west away from the Iapetus margin. This phase was followed by the incoming in the Early Silurian of enormous amounts of sandy turbidites forming the submarine fan system of the Peary Land Group. Sedimentation rate increased dramatically and the bulk of the turbidite sequence was deposited during a short time interval in the Late Llandovery. The turbidites had their source area in the rising Caledonian mountain belt to the east; their initiation gives the most precise time record of Iapetus Ocean closure and orogenic uplift in the present northern North Atlantic region. In the Late Llandovery huge areas of carbonate platform foundered, probably due to the combined effect of downflexing of the outer platform caused by loading of the trough with several kilometres of sediments derived from the rising Caledonides, and loading by westward-advancing Caledonian nappes and thrust sheets.

The next phase was characterized by the incoming in mid-Wenlock time of westward-prograding chert-pebble conglomerates, perhaps originating from upthrust Ordovician chert sequences exposed to erosion in Caledonian thrust sheets.

The Franklinian Basin in North Greenland thus records the evolution of a carbonate-dominated, passive, east-west trending continental margin and the influence of the closure of the Iapetus Ocean and Caledonian mountain building to the east. The Franklinian Basin finally closed in Devonian – Early Carboniferous times resulting in strong deformation of the northern part of the Franklinian trough sequence.

ACKNOWLEDGMENTS

This paper is an outcome of the expeditions of the Geological Survey of Greenland to North Greenland in 1978, 1979, 1980, 1984 and 1985. We are greatly indebted to Niels Henriksen for his dedicated leadership of the expeditions. We acknowledge the benefit of stimulating discussions with colleagues in GGU and elsewhere. We thank especially Merete Bjerreskov for graptolite identifications, and J.H. Aldridge, H.A. Armstrong, M.P. Smith and S.J. Tull for information concerning conodont biostratigraphy. Birgitte Larsen is thanked for typing the manuscript, Bodil Sikker Hansen for artwork, and Jakob Lautrup and Jakob Fernqvist for photographic illustrations.

REFERENCES

Armstrong, H.A. and Lane, P.D.
1981: The un-named Silurian(?) dolomite formation, Børglum Elv, central Peary Land; Grønlands Geologiske Undersøgelse, Report no. 106, p. 29-34.

Barnes, C.R.
1984: Early Ordovician eustatic events in Canada; in Aspects of the Ordovician System, D.L. Bruton (ed.); Palaeontological Contributions from the University of Oslo, no. 295, Universitetsforlaget, Oslo, Norway, p. 51-63.

Bengaard, H-J., Davis, N.C., Friderichsen, J.D., and Higgins, A.K.
1987: Lithostratigraphy and structure of the North Greenland fold belt in Nansen Land; Grønlands Geologiske Undersøgelse, Report no. 133, p. 99-106.

Brenchley, P.J. and Newall, G.
1980: A facies analysis of Upper Ordovician regressive sequences in the Oslo region, Norway — a record of glacio-eustatic changes; Palaeogeography, Palaeoclimatology, Palaeoecology, v. 31, p. 1-38.

Bryant, I.D. and Smith, M.P.
1985: Lowermost Ordovician sandstones in central and western North Greenland; Grønlands Geologiske Undersøgelse, Report no. 126, p. 25-30.

Christie, R.L.
1967: Bache Peninsula, Ellesmere Island, Arctic Archipelago; Geological Survey of Canada, Memoir 347, 63 p.

Christie, R.L. and Dawes, P.R.
1991: Geographic and geological exploration; in Geology of the Innuitian Orogen and Arctic Platform of Canada and Greenland; Geological Survey of Canada, Geology of Canada, no. 3 (also Geological Society of America, The Geology of North America, v. E), Chapter 2.

Christie, R.L. and Ineson, J.R.
1979: Precambrian-Silurian geology of the G.B. Schley Fjord region, eastern Peary Land, North Greenland; Grønlands Geologiske Undersøgelse, Report no. 88, p. 63-71.

Christie, R.L. and Peel, J.S.
1977: Cambrian-Silurian stratigraphy of Børglum Elv, Peary Land, eastern North Greenland; Grønlands Geologiske Undersøgelse, Report no. 82, 48 p.

Davis, N.C. and Higgins, A.K.
1987: Cambrian-Lower Silurian stratigraphy in the fold and thrust zone between northern Nyeboe Land and J.P. Koch Fjord, North Greenland; Grønlands Geologiske Undersøgelse, Report no. 133, p. 91-98.

Dawes, P.R.
1971: The North Greenland fold belt and environs; Bulletin of the Geological Society of Denmark, v. 20, p. 197-239.

1976: Precambrian to Tertiary of northern Greenland; in Geology of Greenland, A. Escher and W.S. Watt (ed.); Geological Survey of Greenland, Copenhagen, p. 248-303.

1987: Topographical and geological maps of Hall Land, North Greenland. Description of a computer-supported photogrammetrical research programme for production of new maps, and the Lower Palaeozoic and surficial geology; Grønlands Geologiske Undersøgelse, Bulletin no. 155, 88 p.

Dawes, P.R. and Christie, R.L.
1982: History of exploration and geology in the Nares Strait region; in Nares Strait and the Drift of Greenland: a Conflict in Plate Tectonics, P.R. Dawes and J.W. Kerr (ed.); Meddelelser om Grønland, Geoscience 8, p. 19-36.

Dawes, P.R. and Peel, J.S.
1981: The northern margin of Greenland from Baffin Bay to the Greenland Sea; in The Ocean Basins and Margins, v. 5, The Arctic Ocean, A.E.M. Nairn, M. Churkin, Jr., and F.G. Stehli (ed.), Plenum Press, New York and London, p. 201-264.

1984: Biostratigraphic reconnaissance in the Lower Palaeozoic of western North Greenland; Grønlands Geologiske Undersøgelse, Report no. 121, p. 19-51.

Dawes, P.R. and Soper, N.J.
1973: Pre-Quaternary history of North Greenland; in Arctic Geology, M.G. Pitcher (ed.); American Association of Petroleum Geologists, Memoir 19, p. 117-134.

Escher, J.C. and Larsen, P-H.
1987: The buried western extension of the Navarana Fjord escarpment in central and western North Greenland; Grønlands Geologiske Undersøgelse, Report no. 133, p. 81-89.

Fortey, R.A.
1984: Global earlier Ordovician transgressions and regressions and their biological implications; in Aspects of the Ordovician System, D.L. Bruton (ed.); Palaeontological Contributions from the University of Oslo, no. 295, Universitetsforlaget, Oslo, Norway, p. 37-50.

Friderichsen, J.D. and Bengaard, H-J.
1985: The North Greenland fold belt in eastern Nansen Land; Grønlands Geologiske Undersøgelse, Report no. 126, p. 69-78.

Friderichsen, J.D., Higgins, A.K., Hurst, J.M., Pedersen, S.A.S., Soper, N.J., and Surlyk, F.
1982: Lithostratigraphic framework of the Upper Proterozoic and Lower Palaeozoic deep water clastic deposits of North Greenland; Grønlands Geologiske Undersøgelse, Report no. 107, 20 p.

Frykman, P.
1980: A sedimentological investigation of the carbonates at the base of the Brønlund Fjord Group (Early-Middle Cambrian), Peary Land, eastern North Greenland; Grønlands Geologiske Undersøgelse, Report no. 99, p. 51-55.

Fränkl, E.
1955: Rapport über die Durchquerung von Nord Peary Land (Nordgrönland) im Sommer 1953; Meddelelser om Grønland, v. 103, no. 8, 61 p.

Fürsich, F.T. and Hurst, J.M.
1980: Euryhalinity of Palaeozoic articulate brachiopods; Lethaia, v. 13, p. 303-312.

Galloway, W.E. and Hobday, D.K.
1983: Terrigenous Clastic Depositional Systems. Applications to Petroleum, Coal, and Uranium Exploration; Springer Verlag, Berlin, 423 p.

Ginsburg, R.N.
1971: Landward movement of carbonate mud: new model for regressive cycles in carbonates; The American Association of Petroleum Geologists Bulletin, v. 155, p. 340.

Henriksen, N. and Peel, J.S.
1976: Cambrian – Early Ordovician stratigraphy in south-western Washington Land, western North Greenland; Grønlands Geologiske Undersøgelse, Report no. 80, p. 17-23.

Higgins, A.G., Friderichsen, J.D., and Soper, N.J.
1981: The North Greenland fold belt between central Johannes V. Jensen Land and eastern Nansen Land; Grønlands Geologiske Undersøgelse, Report no. 106, p. 35-45.

Higgins, A.K. and Soper, N.J.
1985: Cambrian – Lower Silurian slope and basin stratigraphy between northern Nyeboe Land and western Amundsen Land, North Greenland; Grønlands Geologiske Undersøgelse, Report no. 126, p. 79-86.

Higgins, A.K., Soper, N.J., and Friderichsen, J.D.
1985: North Greenland fold belt in eastern North Greenland; in The Caledonide Orogen — Scandinavia and Related Areas, D.G. Gee and B.A. Sturt (ed.); John Wiley and Son Ltd., London, p. 1017-1029.

Hurst, J.M.
1980a: Silurian stratigraphy and facies distribution in Washington Land and western Hall Land, North Greenland; Grønlands Geologiske Undersøgelse, Bulletin no. 138, 95 p.

1980b: Paleogeographic and stratigraphic differentiation of Silurian carbonate buildups and biostromes of North Greenland; The American Association of Petroleum Geologists Bulletin, v. 64, p. 527-548.

1984: Upper Ordovician and Silurian carbonate shelf stratigraphy, facies and evolution, eastern North Greenland; Grønlands Geologiske Undersøgelse, Bulletin no. 148, 73 p.

Hurst, J.M. and Kerr, J.W.
1982: Upper Ordovician to Silurian facies patterns in eastern Ellesmere Island and western North Greenland and their bearing on the Nares Strait lineament; in Nares Strait and the Drift of Greenland: a Conflict in Plate Tectonics, P.R. Dawes and J.W. Kerr (ed.); Meddelelser om Grønland, Geoscience 8, p.137-145.

Hurst, J.M., McKerrow, W.S., Soper, N.J., and Surlyk, F.
1983: The relationship between Caledonian nappe tectonics and Silurian turbidite deposition in North Greenland; Journal of the Geological Society of London, v. 140, p. 123-132.

Hurst, J.M. and Peel, J.S.
1979: Late Proterozoic(?) to Silurian stratigraphy of southern Wulff Land, North Greenland; Grønlands Geologiske Undersøgelse, Report no. 91, p. 37-56.

Hurst, J.M. and Surlyk, F.
1982: Stratigraphy of the Silurian turbidite sequence of North Greenland; Grønlands Geologiske Undersøgelse, Bulletin no. 145, 121 p.

1983a: Depositional environments along a carbonate ramp to slope transition in the Silurian of Washington Land, North Greenland; Canadian Journal of Earth Sciences v. 20, p. 473-499.

1983b: Initiation, evolution, and destruction of an early Paleozoic carbonate shelf, eastern North Greenland; Journal of Geology, v. 91, p. 671-691.

Ineson, J.R.
1980: Carbonate debris flows in the Cambrian of south-west Peary Land, eastern North Greenland; Grønlands Geologiske Undersøgelse, Report no. 99, p. 43-49.

1985: The stratigraphy and sedimentology of the Brønlund Fjord and Tavsens Iskappe Groups (Cambrian) of Peary Land, eastern North Greenland; unpublished Ph.D. thesis, University of Keele, England, 310 p.

Ineson, J.R. and Peel, J.S.
1980: Cambrian stratigraphy in Peary Land, eastern North Greenland; Grønlands Geologiske Undersøgelse, Report no. 99, p. 33-42.

1987: Cambrian platform – outer shelf relationships in the Nordenskiöld Fjord region, central North Greenland; Grønlands Geologiske Undersøgelse, Report no. 133, p. 13-26.

in press: Cambrian shelf stratigraphy of the Peary Land region, central North Greenland; Grønlands Geologiske Undersøgelse, Bulletin.

Ineson, J.R., Peel, J.S., and Smith, M.P.
1986: The Sjaelland Fjelde Formation: a new Ordovician formation from eastern North Greenland; Grønlands Geologiske Undersøgelse, Report no. 132, p. 27-37.

Jepsen, H.F.
1971: The Precambrian, Eocambrian and early Palaeozoic stratigraphy of the Jørgen Brønlund Fjord area, Peary Land, North Greenland; Grønlands Geologiske Undersøgelse, Bulletin no. 96, 42 p.

Larsen, P-H. and Escher, J.C.
1985: The Silurian turbidite sequence of the Peary Land Group between Newman Bugt and Victoria Fjord, western North Greenland; Grønlands Geologiske Undersøgelse, Report no. 126, p. 47-67.

1987: Additions to the lithostratigraphy of the Peary Land Group in western and central North Greenland; Grønlands Geologiske Undersøgelse, Report no. 133, p. 65-80.

McKerrow, W.S.
1979: Ordovician and Silurian changes in sea level; Journal of the Geological Society of London, v. 136, p. 137-145.

Mitchum, R.M., Vail, P.R., and Thompson, S.
1977: Seismic stratigraphy and global changes of sea level; Part 2: the depositional sequence as a basic unit for stratigraphic analysis; in Seismic Stratigraphy; Applications to Hydrocarbon Exploration, C.E. Payton (ed.); American Association of Petroleum Geologists, Memoir 26, p. 53-62.

O'Connor, B.
1979: The Portfjeld Formation (?early Cambrian) of eastern North Greenland; Grønlands Geologiske Undersøgelse, Report no. 88, p. 23-28.

Peel, J.S.
1982: The Lower Paleozoic of Greenland; in Arctic Geology and Geophysics, A.F. Embry and H.R. Balkwill (ed.); Canadian Society of Petroleum Geologists, Memoir 8, p. 309-330.
1985: Cambrian-Silurian platform stratigraphy of eastern North Greenland; in The Caledonide Orogen — Scandinavia and Related Areas, D.G. Gee and B.A. Sturt (ed.); John Wiley and Son Ltd., London, p. 1047-1063.

Peel, J.S. and Cowie, J.W.
1979: New names for Ordovician formations in Greenland; Grønlands Geologiske Undersøgelse, Report no. 91, p. 17-124.

Peel, J.S. and Christie, R.L.
1982: Cambrian-Ordovician platform stratigraphy: correlations around Kane Basin; in Nares Strait and the Drift of Greenland: a Conflict in Plate Tectonics, P.R. Dawes and J.W. Kerr (ed.); Meddelelser om Grønland, Geoscience 8, p. 117-135.

Peel, J.S., Dawes, P.R., Collinson, J.D., and Christie, R.L.
1982: Proterozoic – basal Cambrian stratigraphy across Nares Strait: correlation between Inglefield Land and Bache Peninsula; in Nares Strait and the Drift of Greenland: a Conflict in Plate Tectonics, P.R. Dawes and J.W. Kerr (ed.); Meddelelser om Grønland, Geoscience 8, p. 105-115.

Peel, J.S. and Higgins, A.K.
1980: Fossils from the Paradisfjeld Group, North Greenland fold belt; Grønlands Geologiske Undersøgelse, Report no. 101, p. 28.

Peel, J.S. and Hurst, J.M.
1980: Late Ordovician and early Silurian stratigraphy of Washington Land, western North Greenland; Grønlands Geologiske Undersøgelse, Report no. 100, p. 18-24.

Peel, J.S., Ineson, J.R., Lane, P.D., and Armstrong, H.A.
1981: Lower Palaeozoic stratigraphy around Danmark Fjord, eastern North Greenland; Grønlands Geologiske Undersøgelse, Report no. 106, p. 21-27.

Peel, J.S. and Smith, M.P.
in press: The Wandel Valley Formation (Early-Middle Ordovician) of North Greenland and its correlatives; Grønlands Geologiske Undersøgelse, Report.

Peel, J.S. and Wright, S.C.
1985: Cambrian platform stratigraphy in the Warming Land – Freuchen Land region, North Greenland; Grønlands Geologiske Undersøgelse, Report no. 126, p. 17-24.

Pickerill, R.K., Hurst, J.M., and Surlyk, F.
1982: Notes on Lower Palaeozoic flysch trace fossils from Hall Land and Peary Land, North Greenland; in Palaeontology of Greenland: Short Contributions, J.S. Peel (ed.); Grønlands Geologiske Undersøgelse, Report no. 108, p. 25-29.

Schlager, W.
1981: The paradox of drowned reefs and carbonate platforms; Geological Society of America Bulletin, v. 92, p. 197-211.

Smith, M.P., Sønderholm, M., and Tull, S.J.
in press: The Morris Bugt Group (Middle Ordovician – Early Silurian) of North Greenland and its correlatives; Grønlands Geologiske Undersøgelse, Report.

Soper, N.J., Higgins, A.K., and Friderichsen, J.D.
1980: The North Greenland fold belt in eastern Johannes V. Jensen Land; Grønlands Geologiske Undersøgelse, Report no. 99, p. 89-98.

Soper, N.J. and Higgins, A.K.
1985: Thin-skinned structures at the basin-shelf transition in North Greenland; Grønlands Geologiske Undersøgelse, Report no. 126, p. 87-94.
1987: A shallow detachment beneath the North Greenland fold belt: implications for sedimentation and tectonics; Geological Magazine, v. 124, p. 441-450.

Surlyk, F.
1982: Nares Strait and the down-current termination of the Silurian turbidite basin of North Greenland; in Nares Strait and the Drift of Greenland: a Conflict in Plate Tectonics, P.R. Dawes and J.W. Kerr (ed.); Meddelelser om Grønland, Geoscience 8, p. 147-150.

Surlyk, F., Hurst, J.M., and Bjerreskov, M.
1980: First age-diagnostic fossils from the central part of the North Greenland foldbelt; Nature, v. 286, no. 5775, p. 800-803.

Surlyk, F. and Hurst, J.M.
1983: Evolution of the early Paleozoic deep-water basin of North Greenland — aulacogen or narrow ocean?; Geology, v. 11, p. 77-81.
1984: The evolution of the early Paleozoic deep-water basin of North Greenland; Geological Society of America Bulletin, v. 95, p. 131-154.

Surlyk, F., Hurst, J.M., and Pedersen, S.A.S.
in press: Stratigraphy of the Cambro-Ordovician deep-water sediments of Peary Land, North Greenland; Grønlands Geologiske Undersøgelse, Bulletin.

Surlyk, F. and Ineson, J.R.
1987a: Aspects of Franklinian shelf, slope and trough evolution and stratigraphy in North Greenland; Grønlands Geologiske Undersøgelse, Report no. 133, p. 41-58.
1987b: The Navarana Fjord Member (new) — an Upper Llandovery platform derived carbonate conglomerate; Grønlands Geologiske Undersøgelse, Report no. 133, p. 59-63.

Sønderholm, M. and Due, P.H.
1985: Lower and Middle Ordovician platform carbonate lithostratigraphy of Warming Land, Wulff Land and Nares Land, North Greenland; Grønlands Geologiske Undersøgelse, Report no. 126, p. 31-46.

Sønderholm, M., Harland, T.L., Due, P.H., Jørgensen, L.N., and Peel, J.S.
1987: Lithostratigraphy and depositional history of Upper Ordovician – Silurian shelf carbonates in central and western North Greenland; Grønlands Geologiske Undersøgelse, Report no.133, p. 27-40.

Troelsen, J.C.
1949: Contributions to the geology of the area around Jørgen Brønlund Fjord, Peary Land, North Greenland; Meddelelser om Grønland, v. 149, no. 2, 29 p.

Vail, P.R., Mitchum, R.M., Todd, R.G., Widmier, J.M., Thompson, S., Sangree, J.B., Bubb, J.N., and Hatlelid, W.G.
1977: Seismic stratigraphy and global changes of sea level; in Seismic Stratigraphy — Applications to Hydrocarbon Exploration, C.E. Payton (ed.); American Association of Petroleum Geologists, Memoir 26, p. 49-212.

Wilson, J. L.
1975: Carbonate Facies in Geologic History; Springer-Verlag, Berlin, 471 p.

Authors' addresses

A.K. Higgins
Geological Survey of Greenland
Øster Voldgade 10
DK-1350, Copenhagen K
Denmark

J.R. Ineson
Geological Survey of Denmark
Thoravej 8
DK-2400, Copenhagen NV
Denmark

J.S. Peel
Geological Survey of Greenland
Øster Voldgade 10
DK-1350, Copenhagen K
Denmark

F. Surlyk
Geological Survey of Greenland
Øster Voldgade 10
DK-1350, Copenhagen K
Denmark

M. Sønderholm
Geological Survey of Greenland
Øster Voldgade 10
DK-1350, Copenhagen K
Denmark

CHAPTER 7

Printed in Canada

Chapter 8
CAMBRIAN TO EARLY DEVONIAN BASIN DEVELOPMENT, SEDIMENTATION, AND VOLCANISM, ARCTIC ISLANDS

Introduction

Franklinian Shelf and Arctic Platform

Franklinian Deep Water Basin

Acknowledgments

References

Chapter 8

CAMBRIAN TO EARLY DEVONIAN BASIN DEVELOPMENT, SEDIMENTATION, AND VOLCANISM, ARCTIC ISLANDS

H.P. Trettin, U. Mayr, G.D.F. Long and J.J. Packard

A. INTRODUCTION

Two first-order depositional provinces are distinguished — a southeastern shelf, which encompasses nearly all of the Arctic Platform and a large part of the Franklinian mobile belt, and a northwestern deep water basin. The latter, in turn, is divisible into a southeastern sedimentary subprovince, and a northwestern sedimentary and volcanic subprovince. The unstable boundary between deep water basin and shelf migrated cratonward from late Early Cambrian to Early Devonian time.

In the preceding chapter, the Cambrian to Silurian depositional history of North Greenland has been divided into seven evolutionary phases that are applicable to both shelf and deep water basin as they are based, to some extent, on shifts of their mutual boundary. This organization was feasible because in Greenland only northern parts of the shelf province, which have been affected markedly by the cratonward expansion of the basin, are exposed.

Separate schemes for shelf and basin, however, are required in the Arctic Islands where both provinces are more extensive and more complex stratigraphically. It is most convenient to discuss the stratigraphy of Franklinian Shelf and Arctic Platform in terms of informal time-rock slices, and that of the deep water basin in terms of unrelated, variably diachronous, informal rock units. The basic differences in the stratigraphic record of the two provinces reflects the fact that different geological processes are important in them. Eustatic fluctuations in sea level, for example, affect primarily the shelf; tectonic events in "outboard" orogenic belts affect primarily the basin.

The different classifications chosen in Chapters 7 and 8, however, should not obscure the stratigraphic continuity between Greenland and Ellesmere Island. This continuity is stressed throughout the following discussion and also is evident from the correlation chart (Fig. 4, in pocket), the regional lithofacies maps (Fig. 8B.27-8B.37), and Figure 8C.2, which summarizes the deep water units.

Trettin, H.P., Mayr, U., Long, G.D.F., and Packard, J.J.
1991: Cambrian to Early Devonian basin development, sedimentation, and volcanism, Arctic Islands; Chapter 8 in Geology of the Innuitian Orogen and Arctic Platform of Canada and Greenland, H.P. Trettin (ed.); Geological Survey of Canada, Geology of Canada, no. 3; (also Geological Society of America, The Geology of North America, v. E).

B. FRANKLINIAN SHELF AND ARCTIC PLATFORM

Introduction

This subchapter deals with the shallow marine and subordinate nonmarine facies of the Franklinian Shelf and Arctic Platform in the Arctic Archipelago; no distinction will be made between these two structural divisions, as many rock units or facies belts straddle the boundary between them.

Biostratigraphic standards. The Proterozoic-Cambrian boundary, and strata of earliest Cambrian (Tommotian) age are not known to be exposed, and therefore only trilobite zones (as listed, for example, in Fritz, 1981) have been used as standards for the Cambrian. Cambrian trilobites, however, are extremely sparse in the Arctic Islands and only a few zones are known to be present.

Conodont zones or faunas provide the chronostratigraphic framework for the Lower Ordovician to lower Middle Devonian carbonate successions in the Arctic Islands. Relevant publications are by Ethington and Clark (1971, 1982), Ethington et al. (1987), Ethington and Repetski (1984), Klapper (1977), Klapper and Murphy (1975), Landing et al. (1987), McCracken and Barnes (1981), Miller (1980), Nowlan (1981), Repetski (1982), Ross et al. (1982), Sweet (1979, 1984) and Sweet et al. (1971). Comprehensive studies of conodont biostratigraphy in the Arctic Islands have been made by Barnes (1974), Mirza (1976), Nowlan (1976, 1985), and Uyeno (1980, and in press), and Uyeno and Klapper (1980).

Until recently (e.g. Thorsteinsson and Mayr, 1987, Table 2), late Early to Late Ordovician age assignments in the Arctic Islands were keyed to the numbered Conodont Faunas (1-12) of Sweet et al. (1971). These numbered faunas have now been replaced by formal zones (based on species), and their age ranges have become somewhat uncertain. Fortunately, however, several formations in the Arctic Islands can be correlated lithologically with certain units in Greenland (cf. Thorsteinsson and Mayr, 1987, Table 2), the age of which has been established on the basis of the new conodont zones by M.P. Smith (pers. comm., 1987), and these ages have been accepted for the present compilation. The numbered conodont faunas will nevertheless be mentioned in order to relate this summary to the publications on which it is based.

The correlation of Wenlock to Emsian conodont zones with standard graptolite zones is shown in Thorsteinsson, 1980, Table 1. The extensive and valuable studies carried out during the last three decades on shelly faunas from the Arctic Islands will not be enumerated here. Suffice it to state that, in particular, the widespread Red River fauna (named for the Red River Formation of Manitoba) and the

Thaerodonta-Bighornia fauna have been very useful for the correlation of upper Middle and Upper Ordovician formations respectively (e.g. Barnes et al., 1981).

Time-rock slices and facies belts. For purposes of description, the Cambrian to Lower Devonian shelf succession of the "main" shelf region has been divided into six time-rock slices, informally labelled P1 to P6 (P for platform). Some are bounded by well established unconformities and therefore are comparable to the Sloss-Vail sequences of the continental interior, but others are defined by lithological changes that resulted from changes in shelf architecture; local or problematic disconformities are present within most. They are only partly equivalent to major units distinguished in the shelf succession of northern Greenland (Chapter 7), and to the (in part highly diachronous) major rock units B1-B5 of the deep water basin. Most time-rock slices have been subdivided into lithological units, and most time-rock slices or units have been divided laterally into facies belts, which are different in nature from the recurrent assemblages of sediment types referred to as "lithofacies". The relationships of the time-rock slices, units, or facies belts to established rock units (groups, formations or members) is summarized in Figures 8B.1-8B.4, and briefly mentioned in the text. The proper stratigraphic names also appear in the correlation chart (Fig. 4) and, where possible, on the geological map (Fig. 2, in pocket). This scheme is valid only for those parts of the shelf that have not been affected by local tectonic events; units related to movements of the Boothia and Inglefield uplifts are described separately. The generalized cross-sections (Fig. 8B.2-8B.4) give an overview of the basic stratigraphic framework of the region. Its development is portrayed in a series of lithofacies maps at the end of the subchapter (Fig. 8B.27-8B.37).

Lower Cambrian (P1)

This time-rock slice, of unspecified Early Cambrian age, contains the oldest known strata of the Franklinian Shelf and Arctic Platform (Kerr, 1967a; Long, unpub. ms.). Restricted to northeastern central Ellesmere Island (Fig. 8B.5, 8B.27), it has a maximum exposed thickness of nearly 2 km and comprises a lower unit (P1a) of clastic and carbonate sediments (Kennedy Channel Formation) and an upper unit (P1b) of predominantly carbonate sediments (Ella Bay Formation). The lower boundary of the time-rock slice is concealed; its upper boundary is a regional disconformity. Age control for both units is discussed below. The Kennedy Channel Formation is correlated with the Skagen Group of Peary Land (stage 1 of Chapter 7), and the Ella Bay Formation with the Portfjeld Formation (stage 2S).

The **Kennedy Channel Formation** has a maximum exposed thickness of 1223 m at the type section (Fig. 8B.5, section 3) and is composed of sandstone, mudrock, and limestone, reflecting repetitive progradations of shallow water facies into deeper environments. The lowest two progradational hemicycles culminate in quartz arenite units that accumulated in tide- and storm-influenced shallow marine and barrier island settings. The next major hemicycle is heralded by carbonate grain flow deposits including minor bioclastic units (marked "trilobite" in Fig. 8B.5, section 3), and is capped by massive, intensely bioturbated dark grey limestone. At the type section, the top of the formation contains abundant oolitic and oncolitic grain flow deposits, that suggest progradation of a rimmed shelf. Stromatolites are present cratonward of the type section (section 6), as are thin siliciclastic units.

The **Ella Bay Formation,** which conformably overlies the Kennedy Channel Formation, consists predominantly of dolostone, locally with abundant mudrock and minor amounts of quartz sandstone. Measured thicknesses range from 409 to 762 m. The dolostone is massive or planar stratified and locally shows ripple crosslamination. In places, stromatolites are well preserved (Fig. 8B.6), as are beds containing abundant algal lumps and ooids. Mudrock is locally abundant and forms up to fifty percent of the section at Radmore Harbour (Fig. 8B.5, section 5).

Stratiform breccias at the base of the formation (section 3) and filled caverns at its top suggest karst development. The partial solution may have occurred during emergence prior to the deposition of the Ellesmere Group.

Stromatolite bioherms in the northern outcrops (sections 2, 3) show preferred orientation in response to northwest-southeast directed tides, while those in more southerly locations (sections 5, 6) are more equant. Abundant mudrock at section 5 may have been deposited in a protected local lagoon, landward of the northern bioherms. Dolomitization possibly occurred during the same emergence that caused the partial solution. Mudrock at the base of the formation probably prevented the penetration of the dolomitizing solutions into the underlying Kennedy Channel Formation.

Vertical burrows (*Skolithos*) of unspecified Phanerozoic age occur at the base of the Kennedy Channel Formation and trilobite fragments, indicative of a post-Tommotian age, in its lower middle part (section 3). The next diagnostic fossils are trilobites of the middle(?) part of the late Early Cambrian *Bonnia-Olenellus* Zone, collected about 1.5 km above the top of the Ella Bay Formation at section 1. If the *Bonnia-Olenellus* Zone is no more than 1.5 km thick, the upper two thirds of the time-rock slice must be in the *Nevadella* and/or *Fallotaspis* Zones; the lower one third would be that age and older(?).

Lower-Middle Cambrian (P2)

Time-rock slice P2 is exposed in the eastern part of central and southern Ellesmere Island and on Devon Island. It overlies the Ella Bay Formation in the north and sedimentary or metamorphic rocks of the Canadian Shield farther south. About 2.8 km thick in northeastern central Ellesmere Island, it thins to the south and pinches out between Baffin and Devon islands. The shelf edge, concealed by younger strata, lay somewhere between Archer Fiord and Lake Hazen (Fig. 8B.28, 8B.29).

P2 consists of a Lower Cambrian clastic unit (P2a) and an upper Lower Cambrian – lower Middle Cambrian carbonate unit (P2b). A disconformity seems to exist between the Lower and Middle Cambrian parts of the time-rock slice on Devon Island where the upper Lower Cambrian part of the carbonate unit probably is absent but this hiatus has not been identified with certainty farther north. The Lower Cambrian clastic unit of this chapter corresponds to the deposits of stage 3S in northern Greenland and the carbonate unit to those of the early part of stage $4S_1$.

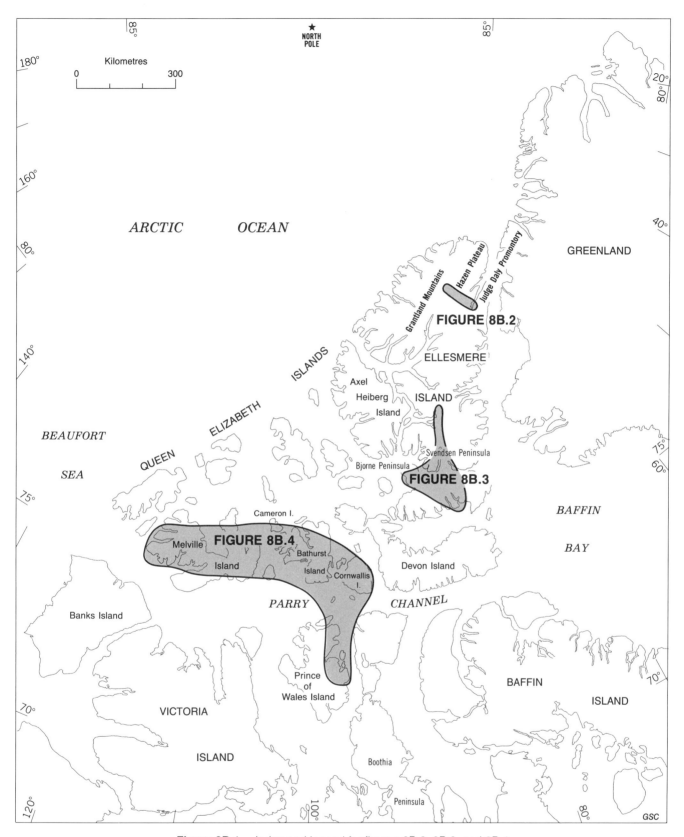

Figure 8B.1. Index and legend for figures 8B.2, 8B.3, and 8B.4.

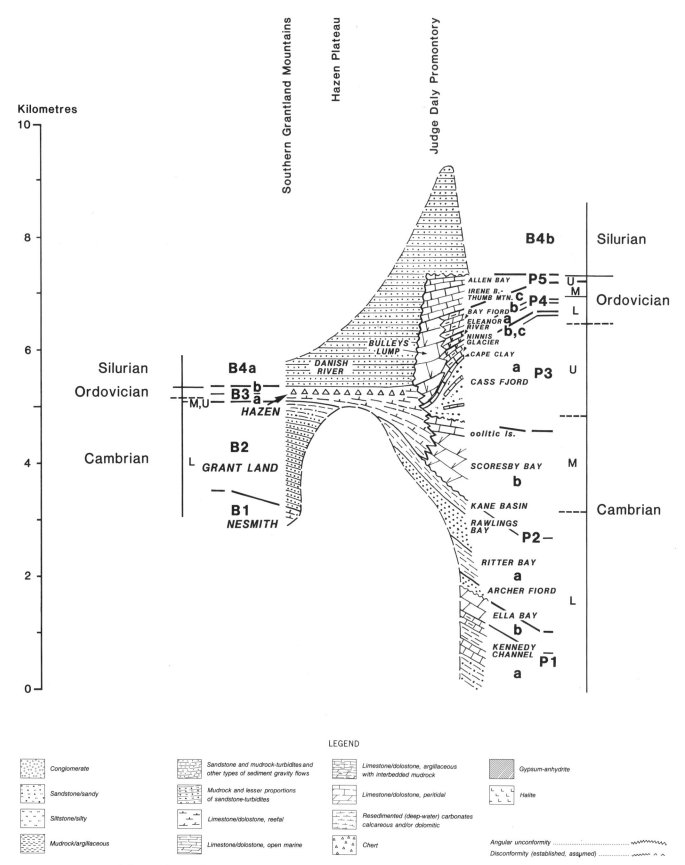

Figure 8B.2. Generalized stratigraphic cross-section, part of northern Ellesmere Island.

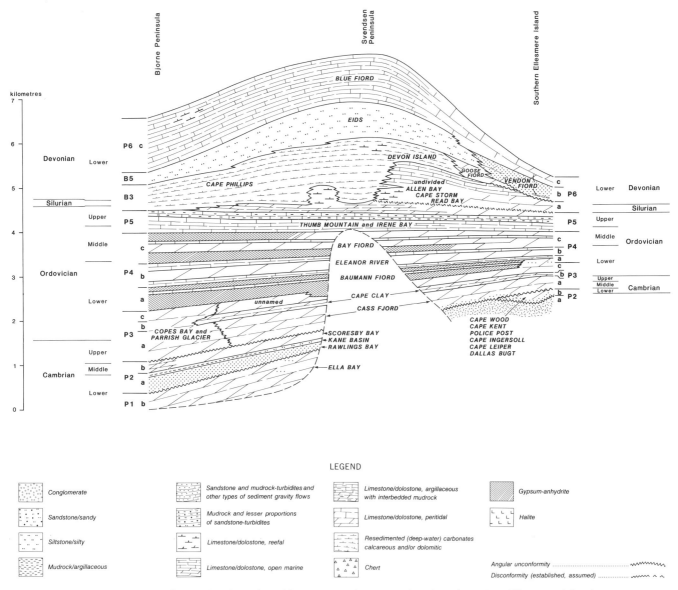

Figure 8B.3. Generalized stratigraphic cross-section, central and southwestern Ellesmere Island.

Clastic unit (P2a)

This wedge-shaped deposit, which is correlative with the Dallas Bugt, Humboldt and Buen formations of northern Greenland, thickens from 0-19 m on Devon Island to about 1.5 km on Judge Daly Promontory. In the central and northern parts of the Franklinian Shelf (Fig. 8B.5) it is identical with the Ellesmere Group, which has been divided into the predominantly sandy Archer Fiord Formation (45-184 m), the predominantly argillaceous Ritter Bay Formation (275-586 m), the sandy and argillaceous Rawlings Bay Formation (548-767 m), and the argillaceous Kane Basin Formation (0-270 m) (Kerr, 1967a; Long, unpub. ms.; Trettin, unpub. ms.). Archer Fiord and Rawlings Bay formations cannot be distinguished south of Radmore Harbour where the Ritter Bay Formation pinches out (Kerr, 1967a, Fig. 3, sections 2-3).

The Archer Fiord Formation is of shallow marine origin in the northern part of the area where it locally contains ooids and oncoids, but may include fluvial deposits farther south. The Ritter Bay and Rawlings Bay formations are of marine-deltaic aspect in the northern part of their distributional area. There the distal prodelta facies is represented by the Archer Fiord Formation, which typically consists of plane-laminated mudrock and very fine grained argillaceous sandstones with rare sediment gravity flows of coarser grade. The proximal delta front facies is well displayed by the upper part of the Rawlings Bay Formation at section 1 (Fig. 8B.5), composed of horizontally interstratified mudrock and fine grained sandstone, both with internal horizontal lamination and bedding-parallel trace fossils (Fig. 8B.7). As the delta front is approached, crosslamination, trough crossbedding and vertical burrows become more common and the proportion of sandstone increases. A 163 m thick unit of mostly massive sandstone with rare trough crossbedding (middle part of Rawlings Bay Formation, section 1) probably represents the delta front facies.

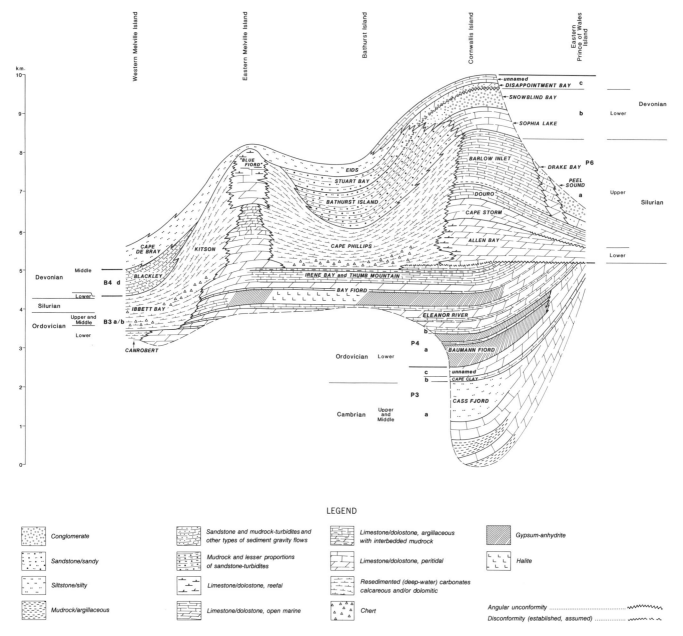

Figure 8B.4. Generalized stratigraphic cross-section, Prince of Wales Island to Melville Island.

In the northern sections, the Kane Basin Formation consists of predominantly flat-laminated, rarely crosslaminated glauconitic mudrock with minor proportions of very fine grained sandstone that locally shows hummocky cross-stratification (Fig. 8B.9). Sandstone is more common in southerly areas. Trilobites of the *Bonnia-Olenellus* Zone have been reported from the upper part of the formation at three localities. The Kane Basin Formation accumulated in shelf environments of moderate to shallow depth. The abrupt decrease in grain size at the Rawlings Bay – Kane Basin boundary indicates a relatively rapid marine transgression.

Farther south, the clastic unit has been assigned to the Dallas Bugt Formation on Bache Peninsula (Christie, 1967) and in southern Ellesmere Island (Packard, unpub. ms.) and to the Rabbit Point Formation on Devon Island (Kurtz et al., 1952; Thorsteinsson and Mayr, 1987). The thickness of these deposits decreases from 114 m on Bache Peninsula to 0-19 m on Devon Island. They commonly consist of crossbedded, pebbly sandstone in the lower part, and *Skolithos*-rich sandstone (Fig. 8B.8) in the upper part. This trace fossil, as well as *Olenellus* sp. from Devon Island, indicate that at least parts of this succession are shallow marine in origin. Crossbedded pebbly sandstones at Makinson Inlet also are interpreted as shallow marine (Packard, unpub. ms.). Olenellid trilobites have also been reported from basal Phanerozoic strata on Victoria Island (Thorsteinsson and Tozer, 1970).

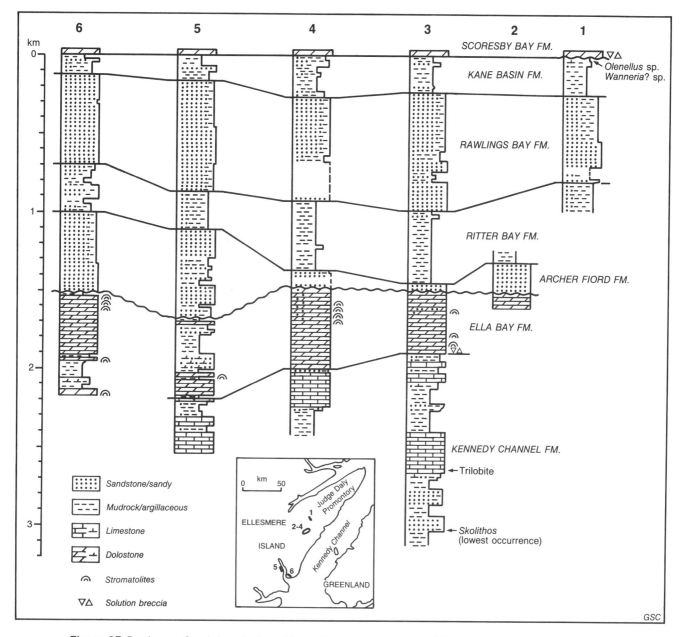

Figure 8B.5. Lower Cambrian stratigraphic sections, central eastern Ellesmere Island (P1a, P1b, P2a).

Arkosic sandstones of the clastic unit were derived directly from crystalline Precambrian basement whereas predominantly quartzose sandstones may have been derived partly from Lower Cambrian(?) and older sandstones. In sandstones from northeastern central Ellesmere Island, pervasive alteration and destruction of feldspar has produced a sparse argillaceous matrix.

Carbonate unit (P2b)

The overall stratigraphy and facies development of the carbonate unit are uncertain because of paucity of fossils in the region north of Bache Peninsula and absence of fossils on Judge Daly Promontory. Instead of a synthesis, therefore, summaries for four separate areas are presented, proceeding from southwest to northeast.

On **Devon Island** (Thorsteinsson and Mayr, 1987) the carbonate unit is represented by the Bear Point Formation, 25-86 m of burrow-mottled dolostone with minor amounts of lime mudstone, shale and sandstone, deposited in intertidal to predominantly subtidal settings. The unit has yielded trilobites of early Middle Cambrian age only (*Albertella*(?), *Glossopleura*, and(?) *Bathyuriscus* Zones) but the possibility that it contains Lower Cambrian strata cannot be excluded. It overlies the Rabbit Point Formation, presumably with disconformity.

The stratigraphic framework for **southeastern Ellesmere Island** has been adapted from Inglefield Land,

Figure 8B.6. Stromatolites in the Ella Bay Formation.

Figure 8B.8. *Skolithos*-like tubes, perpendicular to bedding, in quartz sandstone of Marshall Bugt Member, Dallas Bugt Formation, Makinson Inlet, Ellesmere Island.

Figure 8B.9. Hummocky cross-stratification in Kane Basin Formation (section 3, Fig. 8B.5).

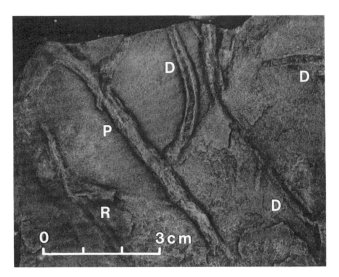

Figure 8B.7. Bedding-parallel trace fossils *Rusophycus* (R), *Planolites* (P), and *Didymaulichnus* (D) from the Rawlings Bay Formation, (section 1, Fig. 8B.5, 61-112 m below top of formation).

northeastern Greenland. On Bache Peninsula (Christie, 1967; Peel and Christie, 1982; Brunton, 1986), the carbonate unit lies stratigraphically between the Dallas Bugt and Cass Fjord formations, is about 140 m thick, and has been divided into five formations. The lower four (Cape Leiper, Cape Ingersoll, Police Post, Cape Kent) are assigned to the late Early Cambrian on the basis of olenellid trilobites from this area and northwestern Greenland; the Cape Wood Formation has yielded early Middle Cambrian fossils of the *Glossopleura* Zone. The apparent absence of the early Middle Cambrian *Plagiura-Poliella* and *Albertella* Zones may indicate a disconformity but lithological evidence for a hiatus has not been reported. The Cape Leiper and Cape Ingersoll formations consist mainly of dolostone; the Police Post Formation is composed of trilobite-bearing lime mudstone; the Cape Kent Formation of dolomitized oolite;

and the Cape Wood Formation of burrow-mottled dolomitic limestone and dolostone. The depositional environments range from supratidal (stromatolites in the Cape Leiper Formation) to subtidal (burrow-mottled units in all formations) and represent several transgressive-regressive cycles. A comparable stratigraphic succession occurs at Fram Fiord, southeastern Ellesmere Island (Fig. 8B.10; Packard, unpub. ms.).

In **east-central Ellesmere Island** (Kerr, 1967a, sections 3, 13, 18) the carbonate unit overlies the Kane Basin Formation. It comprises the Scoresby Bay Formation and unspecified lower parts of the Parrish Glacier Formation and is correlative with the combined Kastrup Elv and Telt Bugt formations of Washington Land (cf. Peel and Christie, 1982). Sections of the Scoresby Bay Formation (Brunton, 1986) are about 450 m thick and consist of dolostone, interpreted to represent mainly original lime mudstone with minor oolite, intraformational conglomerate, and bioclastic grainstone. Noteworthy are the local occurrence of algal mounds, up to 6 m across and 3 m high, and of intraformational breccias interpreted as karst. The age range of the Scoresby Bay Formation in this region is defined by the occurrence of *Olenellus praenuntius* Cowie (lower(?) part of *Bonnia-Olenellus* Zone) from a carbonate-rich unit in the upper part of the Kane Basin Formation that should perhaps be included in the Scoresby Bay Formation (Brunton, 1986), and by trilobites of the *Glossopleura* Zone, found 15-20 m above the base of the Parrish Glacier Formation (Thorsteinsson, 1963; Kerr, 1967, section 13; W.H. Fritz, pers. comm., 1986). The lower part of the Parrish Glacier Formation consists mainly of "aphanitic limestone" with lesser amounts of dark grey shale and very small amounts of quartzite. The lowest multicoloured beds, reminiscent of the Cass Fjord Formation, occur 77 m above the base of the Parrish Glacier Formation but the relationship between the Parrish Glacier and Cass Fjord formations in this region still is uncertain.

On **southwestern Judge Daly Promontory,** close to the margin of the deep water basin, the succession is about 1200 m thick and divisible into four major units (Kerr, 1967a; Trettin, unpub. ms.). The lower three have been assigned to the Scoresby Bay Formation; the fourth was originally included in the Parrish Glacier Formation. In ascending order, the lithology and origin of the units can be summarized as follows:

(1) Interbedded dolostone and dolomitic solution breccia (145 m), chaotic in the lower part, and stratiform in the upper part. It is uncertain whether the brecciation is due to removal of interstratified evaporites or to karst development in original carbonate rocks.

(2) Dolostone, with about 10% lime mudstone (458 m). Lamination, relict birdseye structures and intraclasts, and a unit of club-shaped stromatolites suggest peritidal, predominantly intertidal settings for the dolostone.

(3) Limestone, mainly peloidal packstone with rare intraclasts, trilobite fragments and birdseye structures, interbedded with about 30% massive dolostone (349 m). This unit is intertidal to perhaps subtidal in origin.

(4) Dolomitic limestone, mainly grainstone composed of ooids and intraclasts with lesser proportions of pelletal packstone and lime mudstone (279 m). Partial dolomitization is related to burrows. Deposition occurred in a shallow, agitated shelf environment.

Figure 8B.10. Lower and Middle Cambrian succession near Fram Fiord, southern Ellesmere Island (p€ = Precambrian crystalline basement; €DB = Dallas Bugt Formation; €CL = Cape Leiper Formation; €CI = Cape Ingersoll Formation; €UN = unnamed units, probably equivalent to Wulff River, Cape Kent, and Police Post formations; €CW = Cape Wood Formation; €FP = Fairman Point Formation).

No diagnostic fossils have been found in this succession. Its age is bracketed by late Early Cambrian olenellids from the underlying Kane Basin Formation and the presumed late Middle Cambrian age of the base of the overlying succession. The actual range, however, may be shorter because both upper and lower contacts are disconformable.

Upper Middle Cambrian to mid-Lower Ordovician (P3)

The base of time-rock slice P3 (Fig. 8B.30) is of unspecified late Middle Cambrian age; the top is early Arenig (within Conodont Fauna C). It lies disconformably on Middle Cambrian strata or unconformably on Precambrian sedimentary or metamorphic rocks, continuing the pattern of onlap displayed by the upper Lower and Middle Cambrian strata. The upper boundary is defined by the appearance of a carbonate buildup at the shelf margin (Bulleys Lump Formation) and of a related extensive evaporite unit in the shelf interior (Baumann Fiord Formation). The maximum recorded thickness of 2567+ m is attained in the subsurface of Cornwallis Island, some distance south of the shelf margin. The time-rock slice tapers cratonward and has a minimum recorded thickness of 64 m in the Rowley Island well.

It is divisible into three major informal units: (1) a thick lower unit of subtidal to supratidal carbonate and clastic sediments, locally with diachronous and probably discontinuous, nonmarine to shallow marine deposits at the base (P3a); (2) a relatively thin carbonate unit of subtidal origin (P3b); and (3) an upper unit of supratidal to subtidal carbonate and clastic sediments (P3c). The succession of depositional environments indicates a major transgressive-regressive-transgressive cycle. Facies relationships suggest that the shelf was a gently inclined ramp.

Time-rock slice P3 is correlative with middle parts of the stage $4S_1$ stable shelf deposits of western North Greenland (upper part of Ryder Gletscher Group and Permin Land Formation), and three important formations discussed below, the Cass Fjord, Cape Clay, and Castrup Elv, have been established in that region (Washington Land). It also is correlative with the upper part of the unstable shelf deposits of stage $4S_2$ of eastern North Greenland (Tavsen Iskappe Group).

Lower carbonate and clastic unit (P3a)
Basal siliciclastic deposits

The basal clastic sediments of time-rock slice P3 have been included in the Parrish Glacier Formation in central Ellesmere Island (Kerr, 1967a, sections 18 and 36) and in map unit 10a in Victoria Island (Thorsteinsson and Tozer, 1962). In southern Ellesmere Island, they have been assigned to the Fairman Point Formation (Packard, unpub. ms.). They consist of white, dusky red, and green, texturally mature quartz arenites and minor variegated dolomitic shales and carbonates, 46-427 m thick. The Fairman Point Formation is considered to be shallow marine in origin because of the consistent recurrence of interbedded carbonates. This unfossiliferous formation probably is

Figure 8B.11. Faulted Cambro-Ordovician succession on Precambrian (p€) crystalline basement (€O$_{CF}$ = Cass Fjord Formation; O$_{CC}$ = Cape Clay Formation; O$_{CE}$ = Christian Elv Formation).

Middle Cambrian in age because it lies above shales of the Cape Wood Formation with *Glossopleura* and below typical Cass Fjord Formation whose base is regarded as late Middle Cambrian in age (see below).

The unfossiliferous Gallery Formation of northern Baffin Island (Lemon and Blackadar, 1963), previously considered as Early Cambrian in age (Trettin, 1969a, 1975), more likely is late Middle or early Late Cambrian (Stewart, 1987). The formation varies in thickness between 0 and 340 m and consists mainly of coarse grained, red and grey quartz sandstone with common planar and trough crossbedding. Absence of carbonates from most sandstones and persistent unimodal paleocurrent directions in given areas suggest a predominantly fluvial origin, although local dolostone attests to the presence of shallow marine strata. Combined paleocurrent and isopach maps outline the western half of an east-plunging basin that received sediments from sources to the south, west and northwest. If indeed fluvial in origin, the Gallery Formation must be older than any marine sediments in those directions.

On adjacent Boothia Peninsula, the lower 24 m of the Turner Cliffs Formation (previously Boothia Felix Formation; Christie, 1973) consist of fine grained, dolomitic quartz sandstone with some ripple marks. These shallow marine strata have yielded trilobites of the Dresbachian *Crepicephalus* Zone (Palmer et al., 1981).

The southernmost occurrence of these deposits is in the Rowley M-O4 well of Foxe Basin, where 20 m of quartzose and dolomitic sandstone with a basal pebble conglomerate lie between Archean crystalline basement and the Turner Cliffs Formation. Carbonate cement and phosphatic material indicate that most of the unit is shallow marine in origin. Although assigned to the Gallery Formation, these deposits probably are younger than the Gallery Formation of northern Baffin Island.

Carbonate and clastic sediments

The carbonate and clastic sediments that constitute the bulk of unit P3a are relatively uniform throughout the eastern Arctic, which permits lithological correlation into those areas where fossils are absent. The age range of these strata is identical with that of the Cass Fjord Formation, late Middle Cambrian to early Tremadoc. Trilobites of unspecified late Middle Cambrian age have only been reported from northwestern Greenland (Palmer and Peel, 1981) but Dresbachian to Trempealeauan trilobites and Tremadoc conodont faunas (Faunas B and C) are known from Canada. The sediments are divisible into a predominantly subtidal facies, developed along the northwestern margin of the shelf, and a largely peritidal facies in the rest of the region.

The **mainly subtidal facies belt** is best known from Judge Daly Promontory (Kerr, 1967a; Trettin, unpub. ms.) and the Central Dome H-40 well of Cornwallis Island (Mayr, 1978). Previously assigned to parts of the Parrish Glacier and Copes Bay formations, it now is referred to the Cass Fjord Formation. It is more than 2176 m thick in the Central Dome well and 727 m on Judge Daly Promontory, where two lithological assemblages are distinguished. The first, predominant in the lower 577 m, comprises thinly interstratified, variegated, impure dolostone, limestone, siltstone and sandstone and minor flat-pebble conglomerate. Flat lamination is the predominant primary structure but undulating lamination, crosslamination (including herringbone and ripple drift structures), convolute lamination, desiccation cracks and minor truncation surfaces also are present. These features indicate intertidal to predominantly subtidal conditions. The second assemblage, predominant in the upper 150 m, consists of discrete units of medium grey, predominantly massive limestone, mainly lime mudstone and pelletal packstone, that range in thickness up to 93 m and probably are subtidal in origin.

The **peritidal facies belt** is best known in an area extending from Washington Land, northwestern Greenland (Henriksen and Peel, 1976; Palmer and Peel, 1981) through Bache Peninsula (Christie, 1967) and southeastern Ellesmere Island (Packard, unpub. ms.; Packard and Mayr, unpub. ms.) to Devon Island (Thorsteinsson and Mayr, 1987). It has been mapped as Cass Fjord Formation (Fig. 8B.11) and ranges in thickness from 170 to 488 m. The formation is divisible into four members that can be traced throughout this region (Fig. 8B.12). Overall, the characteristic rock type is flat-pebble conglomerate, which at Makinson Inlet constitutes 34% of the recorded units and 11% of the total thickness. Beds and mounds of stromatolitic boundstone form extensive markers (Fig. 8B.13). Other conspicuous rock types are rhythmites composed of peloidal calcisiltite and marl; gypsum; herringbone-crossbedded quartzarenite; and massive, faintly banded to variegated dolosiltite. The sediments represent a complex mosaic of shallow subtidal to predominantly peritidal deposits, such as shallow subtidal peloidal muds and storm deposits (flat-pebble conglomerates), tidal sand bars, supratidal mud cays, algal reefs, beaches and ooid shoals.

Similar rocks are referred to as the lower member of the Turner Cliffs Formation in Somerset Island and Boothia Peninsula (Miall and Kerr, 1980; Stewart, 1987); and as the Turner Cliffs Formation in northwestern Baffin Island (Lemon and Blackadar, 1963; Trettin, 1969a) and in the Rowley M-O4 well of Foxe Basin (Trettin, 1975). Recorded thicknesses south of Lancaster Sound decrease from 250 m in the north to 44 m in the Rowley Island well. Significant differences from the Cass Fjord Formation are the absence of limestone throughout this region and the absence of evaporites in the Rowley well and Baffin Island(?).

In Victoria and Stefansson islands, strata equivalent to the peritidal facies belt occur in the lowermost part of map unit 10b, which consists mainly of dolostone and ranges in age from late Middle or Late Cambrian into the Ludlow (Thorsteinsson and Tozer, 1962). The stratigraphy of this dolomitic succession is poorly known and will not be discussed in the following sections.

Carbonate unit (P3b)

The carbonate unit is mostly co-extensive with the carbonate and clastic unit but absent from the Rowley Island well (see references quoted for the carbonate and clastic unit). It consists of massive, bioturbated limestone (lime mudstone, pelletal or peloidal limestone, grapestone) that is dolomitized to varying extent or of massive dolostone, both commonly with chert nodules. Clastic impurities seem to be significant only on Judge Daly Promontory. The rocks are of subtidal shelf origin, and variations in water depth cannot be demonstrated. Reported thicknesses are generally between 70 and 125 m. The unit has yielded conodonts of Fauna C. From northwestern Greenland to the subsurface of

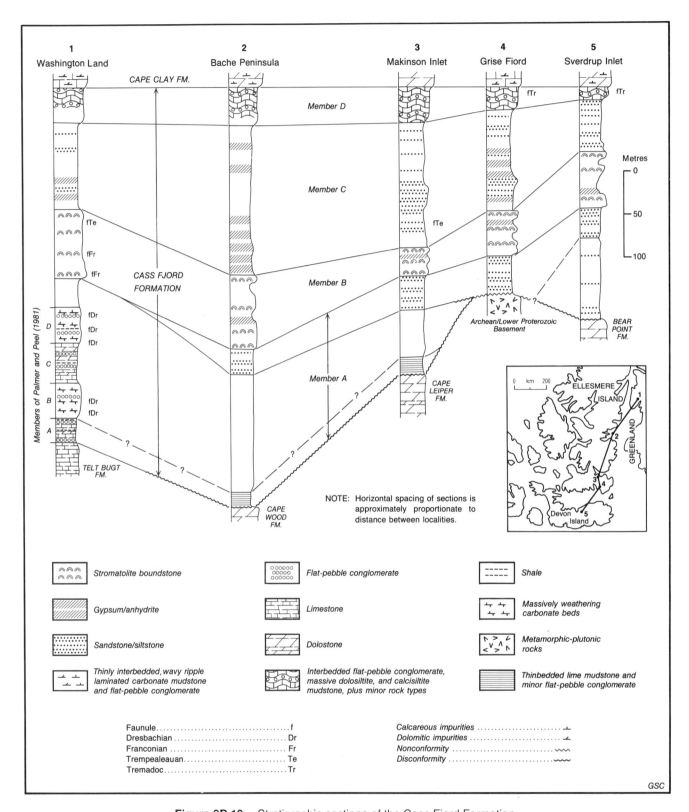

Figure 8B.12. Stratigraphic sections of the Cass Fjord Formation.

Cornwallis Island it has been mapped (or logged) as Cape Clay Formation (Fig. 8B.14). It corresponds to parts of the original Copes Bay Formation in east-central Ellesmere Island; to most of the upper member of the Turner Cliffs Formation on Somerset Island; and probably also to the lowermost part of the Ship Point Formation in northern Baffin Island. Substitution of the term Cape Clay Formation for these local names in the most recent work (Mayr et al., 1980; Thorsteinsson and Mayr, 1987) simplifies the regional stratigraphic picture.

Upper carbonate and clastic unit (P3c)

To date this unit is recognized as a distinct entity only in parts of central and southern Ellesmere Island, on Devon Island, and in the Central Dome H-40 well of Cornwallis Island. Stratigraphic position and a few conodont collections (Fauna C) indicate that it falls into the middle Tremadoc. The limited information available permits distinction of two lithofacies that are comparable to those in the lower carbonate and clastic unit.

The **subtidal lithofacies** is best known from the outermost margin of the shelf on Judge Daly Promontory, where it lies stratigraphically between Cape Clay and Bulleys Lump formations and is informally referred to as the Ninnis Glacier beds (Trettin, unpub. ms.). The unit is in the order of 100 m thick and consists mainly of interlaminated silty limestone and dolostone and calcareous and dolomitic mudrock that are similar to those in the adjacent slope facies of the Hazen Formation (Chapter 8C). Also present are packstones, grainstones and intraformational conglomerates that may represent storm deposits.

The **peritidal facies belt** has been traced from northwestern Greenland (Christian Elv Formation; Henriksen and Peel, 1976) through Bache Peninsula (map-unit 6, Christie, 1967) and southern Ellesmere Island (Christian Elv Formation; Packard, unpub. ms.; Packard and Mayr, unpub. ms.) to Devon Island (lower Blanley Bay Formation; Thorsteinsson and Mayr, 1987). At Makinson Inlet it is 158 m thick and similar to the Cass Fjord Formation, consisting of lime mudstone, flat-pebble

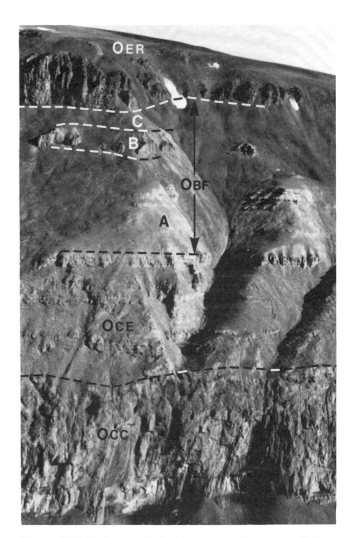

Figure 8B.14. Lower Ordovician succession near Grise Fiord, southern Ellesmere Island (O$_{CC}$ = Cape Clay Formation; O$_{CE}$ = Christian Elv Formation; O$_{BF}$ = Baumann Fiord Formation; O$_{ER}$ = Eleanor River Formation); tripartite division of Baumann Fiord Formation (A, B, C) is evident.

conglomerate, laminated dolosiltite, algal boundstone with clotted fabric (thrombolite), quartz sandstone with calcareous intraclasts, and medium crystalline dolostone. The sediments were deposited on a restricted mud flat in shallow subtidal to supratidal environments. At Grise Fiord (Fig. 8B.14) the same unit has 13% gypsum-anhydrite (Packard and Mayr, 1982). Because of these evaporites, the unit could alternatively be placed into time-rock slice P4.

In central Ellesmere Island, strata equivalent to the Christian Elv Formation have been included in the upper Copes Bay Formation (Kerr, 1968).

Upper Lower and lower Middle Ordovician (P4)

Time-rock slice P4 ranges in age from middle Tremadoc (early Canadian; undifferentiated faunas C/D) to latest Llanvirn (Chazyan; Conodont Faunas 6-7). The lower contact

Figure 8B.13. Coalesced cryptalgal mounds in the Cass Fjord Formation, south coast of Ellesmere Island.

is mostly gradational but probably disconformable in Foxe Basin. The upper contact is a disconformity toward the craton where Conodont Faunas 8-9 or 7-9 are missing; in the more basinward areas of the shelf the contact is sharp but there is no gap in conodont zones.

Three major depositional belts with somewhat unstable boundaries are apparent in the Arctic Islands: a northwestern rim, an intra-shelf basin, and a southern shelf. The intra-shelf basin continues through Washington Land, western North Greenland (upper part of stage $4S_1$ stable shelf deposits), but is terminated on the northeast by an unstable shallow shelf (stage $4S_2$) that was uplifted and eroded in mid-Early Ordovician time. The intra-shelf basin is not interpreted as a structural feature but as a large lagoon, i.e. as a transient paleobathymetric depression, caused by differential sedimentation rates. The thickness of the time-rock slice is greatest in the axial region of the intra-shelf basin, exceeding 2 km in depocentres on Bjorne Peninsula and in the Bathurst Island – Melville Island region. Overall, the succession consists mainly of carbonates and lesser amounts of fine grained clastic sediments with evaporites in the axial region of the intra-shelf basin and sandy carbonates and minor sandstones in the southern shelf area.

An outline of the basic stratigraphy will be followed by a brief discussion of the origin of the evaporites.

Shelf rim

At present, the shelf rim is best known from Judge Daly Promontory, central eastern Ellesmere Island, where it is about 10 km wide (Fig. 8C.26, sections EB2, EB3). The rim sediments are assigned to the Bulleys Lump Formation, which is about 1.6 km thick and consists mainly of thin bedded to predominantly massive carbonate rocks — peloidal packstone with lesser proportions of peloidal grainstone, intraformational conglomerate, lime mudstone and dolostone (Trettin, unpub. ms.). Less common are impure dolostone and limestone and calcareous and dolomitic siltstone and minor sandstone that show flat lamination, small-scale crosslamination and rare trough crossbedding. Birdseye structure is characteristic of the entire formation but most common in the nonlaminated rocks. The impoverished fauna consists mainly of gastropods, ostracodes and sparse conodonts.

The rim is probably also represented by parts of the thick carbonate succession of the Raglan Range in northwestern Melville Island, studies of which have not yet been completed.

Intra-shelf basin and southern and northeastern shelves

The stratigraphy of intra-shelf basin and southern shelf is described in terms of three major units with slightly diachronous mutual boundaries that correspond to the main stratigraphic subdivisions in the axial region of the intra-shelf basin, the Baumann Fiord, Eleanor River, and Bay Fiord formations.

Lower carbonate and evaporite unit (P4a)

This unit, which includes the Baumann Fiord Formation and its equivalents, is middle Tremadoc in age, lying approximately in the range of the upper part of Conodont Fauna C and the lower part of D (Fig. 8B.31).

Intra-shelf basin

Two facies belts are distinguished in the intra-shelf basin. A **non-evaporitic facies belt** in the northwestern part of the basin is exposed only on southwestern Judge Daly Promontory (Trettin, unpub. ms.). The unit equivalent to the Baumann Fiord Formation, difficult to distinguish from the underlying Christian Elv equivalent, is roughly 200 m thick (both are included in the informal Ninnis Glacier beds, shown in Fig. 8C.26). The predominant sediments are impure, mostly medium dark grey limestone (mainly lime mudstone and pelletal packstone) and microcrystalline dolostone that show plane lamination. The lamination is caused mainly by vertical variations in the concentration of carbonaceous matter and argillaceous impurities but also of the calcite/dolomite ratio. These sediments probably were deposited from mud-charged, turbid layers generated by storms in adjacent shallow water areas. Dolomitization probably occurred, lamina by lamina, at the seafloor-sediment interface, under fluctuating, commonly superhaline conditions.

In the Arctic Islands the **evaporitic facies belt** is identical with the Baumann Fiord Formation, exposed in Ellesmere Island (Kerr, 1968; Mossop, 1979; Packard, unpub. ms.; Packard and Mayr, unpub. ms.) and Cornwallis Island (Thorsteinsson and Kerr, 1968; Mayr, 1978). Thicknesses range between 184 and 500 m. In much of this region three members can be distinguished, a lower and an upper member consisting of carbonates and evaporites, and a middle member consisting of carbonates only (Fig. 8B.14). The partly correlative Poulsen Cliff Formation of Washington Land is similar to the Baumann Fiord Formation of east-central Ellesmere Island (Peel and Christie, 1982).

The carbonate rocks include: lime mudstone, laminated to massive with rare crosslamination; argillaceous and evaporitic, fissile dolosiltite; flat and domal algal stromatolites; flat-pebble conglomerate and related grainstone; and solution-collapse breccia. They are of shallow subtidal and peritidal origin.

The evaporites include: massive anhydrite; alabastrine gypsum; satin-spar gypsum veins; and flat-laminated, rarely crosslaminated gypsum (Fig. 8B.17).

Southern shelf

This depositional belt is best known from Devon Island where it corresponds approximately to the upper member of the Blanley Bay Formation (Thorsteinsson and Mayr, 1987). The 80 to 120 m thick member consists mainly of interstratified dolostone and medium- to coarse-grained, in part crossbedded sandstone with rare flat-pebble conglomerate, sandy oolite, solution breccia, gypsiferous shale, gypsiferous dolostone and thin bedded gypsum.

On the surface of Somerset Island (Miall and Kerr, 1980; Stewart, 1987) and northern Baffin Island (Trettin, 1969a), the Baumann Fiord equivalents have not been distinguished from the Eleanor River equivalents and have collectively been mapped as Ship Point Formation, but in the subsurface (Mayr, 1978; Mayr et al., 1980) a separation appears possible. The strata consist of dolostone, sandy dolostone and quartzose and dolomitic sandstone with rare anhydrite in the subsurface samples.

In the Rowley M-O4 well of Foxe Basin, the lower 18.5 m of the Ship Point Formation (member A) consist mainly of quartzose and dolomitic sandstone and sandy dolostone (Trettin, 1975). The unit has yielded conodonts (Arenig according to an earlier identification) that permit correlation with either Baumann Fiord Formation or lower Eleanor River Formation.

Victoria Island also was part of the southern shelf as evaporites are absent here but the units equivalent to the Baumann Fiord, Eleanor River and Bay Fiord formations have not yet been identified in the Middle Cambrian to Ludlow dolomitic succession of that island (map unit 10b of Thorsteinsson and Tozer, 1962).

Northeastern shelf

In Warming and Nares lands of central North Greenland, equivalents of the Baumann Fiord Formation consist of shallow marine to peritidal dolostone included in the middle part of the Warming Land Formation (Chapter 7).

Carbonate Unit (P4b)

This unit, which coincides with the Eleanor River Formation and its equivalents, ranges in age from late Tremadoc to Arenig (Conodont Faunas D to 1 or 2).

Intra-shelf basin

The Eleanor River Formation is exposed in Ellesmere Island (Fig. 8B.14, 8B.15; Kerr, 1968; Packard, unpub. ms.; Packard and Mayr, unpub. ms.; Trettin, unpub. ms.), Devon Island (Thorsteinsson and Mayr, 1987), and Cornwallis Island (Thorsteinsson and Kerr, 1968) and has been penetrated by several wells in Melville Island (Fox, 1985). Recorded thicknesses range from 150 to 1000 m and are greatest in the axial region of the intra-shelf basin. The formation consists of a considerable variety of limestone types with smaller amounts of dolostone and very small amounts of anhydrite. Typical subtidal limestones are medium grey, bioturbated and dolomitic, and contain varying amounts of skeletal material and fecal pellets. Dark grey, plane-laminated limestones and dolostones, comparable to those in the Baumann Fiord equivalent, occurring in the northern part of the intra-shelf basin (Fig. 8C.26, section EB4), also originated in subtidal settings below wave base. Shallow subtidal and intertidal conditions are indicated by less abundant intraformational conglomerates and stromatolitic limestones. Overall, the stratigraphy of the Eleanor River Formation indicates fluctuations in water depth between intertidal and predominantly subtidal levels, and mostly unrestricted conditions.

Southern shelf

In Somerset Island and northern Baffin Island, strata correlative with the Eleanor River Formation have been included in the upper part of the Ship Point Formation,

Figure 8B.15. Eleanor River (OER) and Bay Fiord (OCB) formations at Makinson Inlet, southeastern Ellesmere Island.

Figure 8B.16. Middle Ordovician succession near Grise Fiord, southern Ellesmere Island (OER = Eleanor River Formation; OCB = Bay Fiord Formation; OCT = Thumb Mountain Formation); the Bay Fiord Formation is divisible into three members (A, B, C).

composed largely of dolostone and sandy dolostone with sparse anhydrite in the subsurface (Mayr, 1978; Miall and Kerr, 1980). Stromatolites, intraformational conglomerates, and sparse oolites indicate that at least part of the unit was deposited in shallow water settings.

In Foxe Basin the unit includes either the entire Ship Point Formation, or its middle and upper parts (member B). In the Rowley well, member B is 84 m thick and consists mainly of bioturbated dolostone with numerous flat-pebble conglomerates and minor amounts of dolomitic siltstone and sandstone and of secondary chert (Trettin, 1975).

Figure 8B.17. Ripple crosslaminated detrital evaporite and dolomite in the Baumann Fiord Formation, Muskox Fiord, southern Ellesmere Island.

Figure 8B.18. Large domal stromatolite on top of Ordovician carbonate mound, northeastern Melville Peninsula.

Northeastern shelf

This region was dominated by shallow marine or peritidal dolostones, assigned to the upper Warming Land Formation and Steensby Gletscher Formation in the Warming Land – Nares Land region, and to parts of the Wandel Valley Formation in Peary Land (Chapter 7).

Upper carbonate and evaporite unit (P4c)

Intra-shelf basin

During the early Middle Ordovician the basin was wider on the south, including Somerset and northern Baffin islands (Fig. 8B.32). The deposits have mostly been assigned to the Bay Fiord Formation but also to member A of the Baillarge Formation in northern Baffin Island (Trettin, 1969a). The Bay Fiord Formation ranges in age approximately from late Arenig to latest Llanvirn (earliest Whiterockian to Chazyan; Conodont Faunas 2 to 6/7) and is exposed in central and southern Ellesmere Island (Kerr, 1968; Packard, unpub. ms.; Packard and Mayr, unpub. ms; Trettin, unpub. ms.), Devon Island (Morrow and Kerr, 1977; Thorsteinsson and Mayr, 1987), Cornwallis Island (Thorsteinsson and Kerr, 1968), Somerset Island (Miall and Kerr, 1980) and Bathurst Island (Kerr, 1974). Subsurface information on this region has been published by Mayr and others (Mayr, 1978, 1980; Mayr et al., 1978, 1980). The formation also occurs in the subsurface of Melville Island (Fox, 1985). Recorded surface thicknesses range from 43.5 m on Somerset Island to 640 m on Ellesmere Island. The unit is composed of two major carbonate lithofacies with or without evaporites.

The predominant lithofacies consists of laminated impure dolostone and subordinate dolomitic mudrock. The dolostone is medium grey to medium dark grey, cryptocrystalline to very finely crystalline, mostly microcrystalline and contains variable amounts of calcareous and siliciclastic material largely of silt grade. Plane or slightly undulating lamination is the predominant primary structure, but crosslamination of very small scale also is present. This lithofacies is attributed to settling from storm-generated turbid layers, followed by penecontemporaneous dolomitization under superhaline conditions.

A less commonly occurring carbonate lithofacies consists of thin bedded to massive dolostones and dolomitic limestones that are generally bioturbated and locally contain abundant bioclastic material. These deposits are of unrestricted subtidal aspect. Stratigraphically, they seem to form a distinct, if thin, middle member in the Bay Fiord Formation in parts of Ellesmere (Fig. 8B.16) and Devon islands.

Evaporites exposed at the surface consist entirely of calcium sulphates, commonly anhydrite with nodules or veins of secondary gypsum. They occur mainly as thin, thick or massive beds (up to 27 m thick on Devon Island) with or without internal plane lamination. Halite is known only from the subsurface of Bathurst and Melville islands. Stratigraphically, the evaporites are most abundant in the lower part of the formation and die out upsection. On Ellesmere Island these deposits seem to form a series of narrow, discontinuous lenses within the axial part of the intra-shelf basin. A longer and wider sub-basin underlies eastern Mellville Island, Bathurst Island, Cornwallis Island and northwesternmost Devon Island (Daae and Rutgers, 1975, Fig. 4).

In western and central North Greenland (Washington Land to Nares Land), equivalents of the Bay Fiord Formation have been assigned to the Cape Webster Formation, composed mostly of dolostone with local evaporite lenses and solution breccias. They are tentatively portrayed as part of the intra-shelf basin in Figure 8B.32 but may be of a shallower origin than those in the Arctic Islands.

Southern shelf
The southern shelf is represented only by the uppermost part of the Ship Point Formation, which has yielded conodonts of Fauna 4 (Barnes in Trettin, 1975). The strata consist of bioturbated dolostone, locally with relicts of benthonic fossils, and dolomitic flat-pebble conglomerate.

Northeastern shelf
In Peary Land, the Wandel Valley Formation of late Early to early Middle Ordovician age disconformably oversteps Cambrian to Lower Ordovician formations. Middle and upper parts of this unit appear to be correlative with the Cape Webster and Bay Fiord formations. The Wandel Valley Formation is composed of dolostone and minor limestone and shows abundant emergence features (Chapter 7).

Origin of evaporites
The origin of the calcium sulphate is difficult to decipher because of pervasive recrystallization to secondary gypsum. Mossop (1979) interpreted the evaporites of the Baumann Fiord Formation as (supratidal) sabkha deposits, whereas Packard (unpub. ms.) interpreted them as subaqueous deposits formed in a broad, shallow, possibly segmented, saline basin. The same interpretation is here applied to the Bay Fiord Formation.

Evidence derived from the evaporites themselves includes: (1) the extreme thickness of individual units of calcium sulphate (commonly more than 10 m); (2) the great lateral extent of the deposits; (3) common, predominantly plane lamination (see Warren and Kendall, 1985), the primary origin of which is apparent from the local preservation of associated subordinate crosslamination (Fig. 8B.17). Moreover, at least the carbonate strata associated with the evaporites of the Bay Fiord Formation clearly are of subtidal origin.

Because the appearance of extensive evaporites coincides with the development of a shelf rim on Judge Daly Promontory, it is likely that the superhaline conditions were caused by this barrier. Evaporitic phases in the basin probably represent phases during which the rim was relatively extensive and continuous, and normal carbonate sedimentation phases during which it was rather discontinuous.

Middle and Upper Ordovician (P5)
Time-rock slice P5 (Fig. 8B.33) ranges in age from earliest Llandeilo to mid-Ashgill (Chazyan to late Richmondian; Conodont Faunas 6/7 to 12), and corresponds to the lower part of the stage $5S_1$ stable shelf deposits of northern Greenland. Both lower and upper contacts are disconformable. The lower hiatus encompasses Conodont Faunas 8 and 9 on Devon Island but narrows or disappears northwestward (Thorsteinsson and Mayr, 1987) and also narrows in Foxe Basin and on Melville Peninsula. Evidence for a disconformity at the Ordovician-Silurian boundary includes the apparent absence of Conodont Fauna 13 and of conodonts and macrofossils of early Llandovery age as well as local signs of erosion (Mirza, 1976; Thorsteinsson and Mayr, 1987).

Two major processes brought about fundamental changes in sedimentary regime. The first is a marked rise in sea level that caused a major transgression during which eventually most of the continent was flooded. As a result, the region was below fair-weather wave base most of the time.

The second process was the accelerated subsidence of outer parts of the shelf, apparent during this interval mainly in a retreat of the shelf margin in eastern Melville, Bathurst, northern Cornwallis, and central western Ellesmere islands, where deep water sediments were deposited in a broad embayment (see below).

A predominantly subtidal shelf carbonate facies covered the remaining part of the Franklinian Shelf and all of the Arctic Platform. It is rather uniform in lithology, both horizontally and vertically. Three main units are recognized in Ellesmere Island (Kerr, 1974), Devon Island (Morrow and Kerr, 1977; Thorsteinsson and Mayr, 1987), Cornwallis Island (Thorsteinsson and Kerr, 1968), Bathurst Island (Kerr, 1974), Somerset Island and Boothia Peninsula (Miall and Kerr, 1980; Stewart, 1987).

The **Thumb Mountain Formation,** earliest Llandeilo to earliest Ashgill (Chazyan to mid-Edenian) in age and correlative with the Gonioceras Bay and Troedsson Cliff formations of Washington Land and the lower part of the Børglum River Formation of Peary Land (Peel and Christie, 1982), comprises intensely burrowed skeletal packstone, pelletal wackestone or packstone and lime mudstone, the amount of skeletal material increasing stratigraphically upward. Where the formation is fully developed it is 300-400 m thick, the lower part showing shallow water features such as birdseye structures and laminated dolostones. The lower part is absent in parts of the Arctic Platform, where subtidal strata lie disconformably on the Bay Fiord Formation or its equivalents.

The Thumb Mountain Formation is overlain, generally conformably, by the **Irene Bay Formation,** which in most of the region is a thin (± 50 m) unit of nodular limestone and greenish grey shale with abundant and diverse fossils of the "Arctic Ordovician Fauna". The older component of this fauna, referred to as the Red River Fauna (for Red River Formation of Manitoba) also occurs in the Thumb Mountain Formation; and the younger component, known as the *Thaerodonta-Bighornia* Fauna, in the lower Allen Bay Formation, but in lower concentrations. The Irene Bay Formation is correlative with the Cape Calhoun Formation of Washington Land and the upper middle part of the Børglum River Formation of Peary Land (Peel and Christie, 1982). On Somerset Island and Devon Island, disconformities are present at the base of the Irene Bay Formation and within it, but their paleogeographic significance is unclear (Stewart, 1987; Thorsteinsson and Mayr, 1987). At the shelf margin, on Judge Daly Promontory, the formation is about

100 m thick and consists of skeletal wackestone without shale.

The Irene Bay Formation is overlain with generally gradational contact by the lower member of the **Allen Bay Formation** or its equivalents. The unit is similar to the upper Thumb Mountain Formation but contains more stromatoporoids and is dolomitized locally. The lower Allen Bay Formation has been correlated with the Aleqatsiak Fjord Formation of Washington Land and with the uppermost part of the Børglum River Formation of Peary Land.

On northwestern Baffin Island, the entire time-rock slice is included in member B of the **Baillarge Formation,** which extends into the Llandovery. Equivalents of Thumb Mountain, Irene Bay and lower Allen Bay formations can be distinguished in the sea cliffs of Brodeur Peninsula (Trettin, 1969a; 1975).

Farther south, in Foxe Basin, the time-rock slice consists of some 200 m of bioturbated, dolomitic, skeletal lime wackestone, ranging in age approximately from early Caradoc to late Ashgill (Blackriveran to Richmondian).

On Melville Peninsula (Trettin, 1975; Bolton et al., 1977), upper Middle Ordovician strata, correlative with the Bad Cache Rapids Group of the Hudson Platform, are overlain by an unnamed reefal unit of uncertain late Middle or Late Ordovician age. About 23 identified mounds, 300-1500 m in diameter and up to about 30 m high, are composed of microcrystalline limestone and dolostone. Large domal stromatolites (Fig. 8B.18) and favositid corals have locally been identified as reef builders, but the bulk of the reefs is of uncertain algal origin. Inter-reefal areas are underlain by laminated limestone and dolostone with relicts of algal mats.

Silurian and Lower Devonian (P6)

Time-rock slice P6 includes all Silurian and Lower Devonian strata. The base is the disconformity at the Ordovician-Silurian boundary mentioned above, and the top coincides approximately with the base of the Middle and Upper Devonian clastic wedge (Chapter 10).

Shelf carbonate sedimentation, governed to some extent by eustatic changes in sea level, continued through much of the area as previously, but the paleogeography was complicated by four different events: (1) the continued downwarping of outer parts of the Franklinian Shelf (accompanied by development of carbonate buildups), which now affected parts of Ellesmere and Devon islands (see below); (2) activity of the Boothia Uplift, from late Ludlow to Pragian or early Emsian time; and (3) activity of the Inglefield Uplift in Pragian–early Emsian time. Deposition in the main shelf area will be discussed first, and then the clastic sediments produced by Boothia Uplift, Inglefield Uplift and perhaps other sources.

Deposition in the main shelf region

Carbonate sedimentation prevailed in those parts of the shelf area that were not affected by the Boothia and Inglefield uplifts, and by accelerated subsidence of the outer shelf. The Silurian - Lower Devonian carbonate succession is divisible into three major units (P6a, P6b, P6c), which are separated, to some extent, by tongues of deep water or syntectonic clastic sediments.

Lower carbonate unit (P6a)

The lower carbonate unit comprises, in ascending order, the upper two members of the Allen Bay Formation (or their equivalents), and the Cape Storm, Douro, and Barlow Inlet formations. Although these units probably are widely distributed, they are well known only in the central part of the Arctic Islands (Cornwallis, Somerset, Devon, Prince of Wales, and southern Ellesmere islands) where they have been studied mainly by Thorsteinsson (1958, 1980), Jones (1974), Morrow and Kerr (1977), Narbonne (1981), Mortensen (1985), Packard (1985), and Thorsteinsson and Mayr (1987). Equivalent units have been recognized in surface exposures or wells in the Victoria – Banks – Stefansson islands region (Thorsteinsson and Tozer, 1962; Miall, 1976; Mayr et al., 1980). In central Ellesmere Island they have commonly been included in the undivided Allen Bay and Read Bay formations (McGill, 1974; Kerr, 1976). Equivalents in northern Greenland, assigned to stage 6A, include the upper part of the Turesø Formation and the Ymers Gletscher, Odins Fiord, and Samuelson Høj formations.

The age range, inferred water depth, and shelf architecture of these units are shown in Figure 8B.19 and further discussed below. The cyclical changes of water depth are interrupted by three events. Event 1, at the Ordovician-Silurian boundary, represents a eustatic drop in sea level that probably is related to a glaciation in the Sahara (McClure, 1978), but also coincides with a change from ramp to rimmed shelf architecture. Event 2 is a local disconformity, interpreted by Kerr (1977) as the first, weak pulse of the Cornwallis Disturbance. Event 3 is marked by a widespread and abrupt change in lithofacies, which in Somerset and Cornwallis islands (i.e. in the vicinity of the Boothia Uplift) indicate uplift from subtidal to peritidal depth and in northwestern Devon Island and southern Ellesmere Island subsidence into a "starved-basin" setting. Subsequently, only 200-300 m of argillaceous and carbonate sediments were deposited in the latter area, probably in submarine slope environments, while 1300 m of shallow marine carbonates accumulated in the former area. It coincides with the rise of the Boothia Uplift in the region south of Barrow Strait and evidently is related to it (Packard, 1985).

Combined, the upper two members of the Allen Bay Formation (Fig. 8B.20) range in age from middle Llandovery (*kentuckyensis* Zone) to late Wenlock – early Ludlow (upper *sagitta* Zone) and are 520 and 600 m thick respectively on Devon and southern Ellesmere islands (Fig. 8B.34; Mayr, 1974; Sodero and Hobson, 1979; Thorsteinsson and Mayr, 1987; Packard and Mayr, unpub. ms.). On Devon Island the middle member consists of laminated and massive dolostone; the upper member of shallowing-upward hemicycles in which grainstones composed of mud lumps etc. are succeeded by light coloured dolosiltite. The units were deposited on a broad, shallow-rimmed shelf, the outer part of which probably included sabkha islands (i.e. supratidal, semi-arid to arid settings). Initially the shelf interior was of more open marine character with abundant stenohaline biota (that had a low tolerance for variation in salinity) and extensive stromatoporoid patch reefs. With time, more restricted

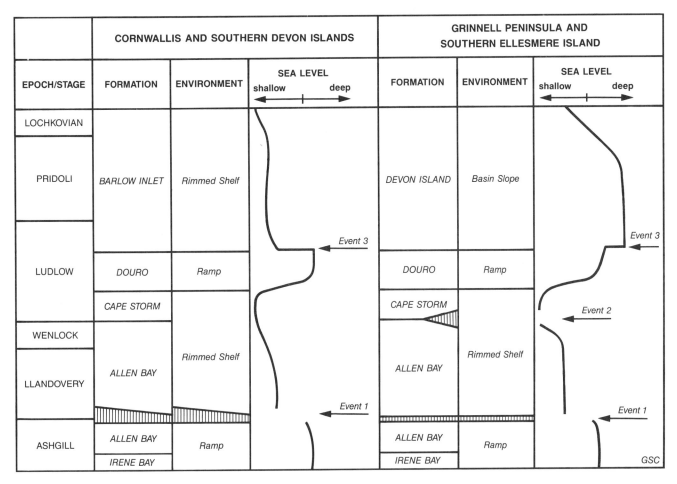

Figure 8B.19. Silurian sea level changes, Cornwallis Island to southern Devon Island.

conditions prevailed (upper member) and the shelf interior was occupied by tidal flats, channels, and protected lagoons.

A facies equivalent of the middle and upper part of the Allen Bay Formation that includes gypsum-anhydrite in Somerset Island (Mayr, 1978; Reinson, 1978; Miall and Kerr, 1980; Stewart, 1987) and solution breccias in Baffin Island (Trettin, 1969a) has been mapped as the **Cape Crauford Formation.** On Somerset Island the unit is about 579 m thick and consists mainly of dolostone (including stromatolitic rocks and intraformational conglomerate) with lesser amounts of dolomitic limestone and thinly interstratified gypsum-anhydrite. The latter occurs in carbonate-evaporite cycles that have been interpreted as the product of repeated progradation along a subsiding lagoon-sabkha shoreline (Reinson, 1978). In Baffin Island both stratiform and pipe-shaped breccia bodies (all composed of carbonate rock fragments) are present in correlative dolostones and limestones. The origin of the stratiform breccias has been attributed to solution of interstratified evaporites (Trettin, 1969a); or to removal of limestone, below a disconformity within the formation, by karst processes (Nentwich and Jones, 1985).

In Foxe Basin (Trettin, 1975) only strata of middle and late Llandovery age are preserved. The lower part of the succession is 137 m thick in the Rowley well; the thickness of the remainder cannot be established because of lack of topographic relief. The strata are mainly dolomitic limestone and dolostone with trace amounts of quartzose and feldspathic sandstone, and represent subtidal and peritidal environments. Characteristic rock types are: dolomitic limestone (lime wackestone) with abundant brachiopods and corals; pelletal limestone; dolostone with coated grains; stromatolitic dolostone, and dolomitic flat-pebble conglomerate. A stratified breccia may be a vestige of evaporites. The strata are correlative mainly with the Severn River Formation and perhaps also with the Ekwan River Formation of the Hudson Platform (cf. Norford, 1971).

The **Cape Storm Formation** (Fig. 8B.20; Kerr, 1975) is a widely distributed unit, currently recognized from the Vendom Fiord area of south-central Ellesmere Island through Prince of Wales Island (Mortensen, 1985) and Boothia Peninsula to the subsurface of Victoria Island (Mayr et al., 1980). It ranges in thickness from 120 to 600 m and has been redefined (Packard and Mayr, unpub. ms.) to include strata of early Ludlow age (lower part of *siluricus* Zone) only. The formation consists mainly of silty dolosiltite and dolomitic quartz siltstone. Sedimentary structures (desiccation polygons, ripple marks, small channels, crossbedding) and the paucity of macrofossils and presence of low-domal stromatolites suggest a largely intertidal environment of deposition. On Cornwallis Island the shelf edge is characterized by massive, vuggy, reefal(?) dolostone

Figure 8B.20. Allen Bay (OS$_{AB}$) and Cape Storm (S$_{CS}$) formations at South Cape Fiord, southern Ellesmere Island.

Figure 8B.21. Douro (S$_{DO}$), Devon Island (SD$_{DI}$), and Goose Fiord (D$_{GF}$) formations near type section of Cape Storm Formation, vicinity of Muskox Fiord, southwestern Ellesmere Island.

(Sodero and Hobson, 1979) and on Grinnell Peninsula by intraclastic shoal deposits (Morrow and Kerr, 1977).

The **Douro Formation** (Fig. 8B.21, 8B.22) is 97-460 m thick and late Ludlow in age (upper part of *siluricus* Zone; Thorsteinsson, 1980). It consists of rubbly argillaceous limestone (Jones et al., 1979); mottled, dolomitic carbonate rocks; calcareous shale; and wavy bedded calcisiltite. Mud mounds, composed of micrite, lithistid sponges and less common corals and stromatoporoids, are prevalent on Griffith and Somerset islands. The shelly fauna, divisible into a brachiopod and a brachiopod-coral biofacies, is dominated by a single genus of the brachiopod, *Atrypoidea* (Narbonne and Dixon, 1982). The Douro Formation represents deposits of a carbonate ramp extending from shallow to deep subtidal. The rubbly limestones grade into basin slope lime mudstones and shales with no apparent intervening high-energy facies belts (e.g. ooid shoals or reefs). The mud mounds grew on deeper parts of the ramp.

The **Barlow Inlet Formation** (Fig. 8B.22) is currently mapped on Cornwallis, Devon, and Griffith islands (Thorsteinsson, 1980; Packard, 1985), and equivalent strata undoubtedly are contained within the undivided Allen Bay – Read Bay succession that outcrops through central southern Ellesmere Island and occurs in the subsurface of Victoria, Banks and Melville islands. It is about 1300 m thick on Cornwallis Island and 750 m on Devon Island and ranges in age from late Ludlow (*latialata* Zone) to earliest Devonian (earliest Lochkovian, *hesperius* Zone). The lower member, about 65 m thick on central eastern Cornwallis Island, consists mainly of mudrock with fine grained subarkose at the top. It is interpreted as a prodelta and delta-front beach deposit, derived from the Boothia Uplift to the south (see below). The upper member consists of a variety of carbonate rocks, mainly lime mudstone, wackestone, and mottled dolomitic limestone, that represent a rimmed carbonate shelf complex comprising tidal flat, restricted shelf lagoon, back-barrier, shelf edge and foreslope environments, the latter two environments being reefal (Fig. 8B.23). Although exposures of the shelf edge are generally dolomitized, both true barrier reefs and extensive crinoidal shoal deposits with local stromatoporoid-coral

Figure 8B.22. View to north along Cape Hotham escarpment, central east coast of Cornwallis Island. Resistant carbonates along crest of escarpment are platform margin facies of the upper member of the Barlow Inlet Formation (SDBI2); talus below covers prodeltaic mudstone of the lower member (SDBI1); subdued terrain to west represents Duro Formation (SDO).

buildups have been identified. The shelf interior is characterized by shallowing-upward cycles.

Middle carbonate unit (P6b)

The middle carbonate unit ranges in age from latest Silurian (*latialata* Zone) to late Pragian (*kindlei* Zone) and comprises the Goose Fiord and Sutherland River formations, which to some extent are correlative; the correlative Sophia Lake Formation is discussed below in the context of the Boothia Uplift. The unit is separated from the lower carbonate unit by a tongue of deep water sediments (Devon Island Formation) in part of the area and is overlain by syntectonic clastic sediments of Prince Alfred or Vendom Fiord formations. In some areas, however, it has conformable contacts with the lower and upper carbonate units.

Both Goose Fiord and Sutherland River formations are composed of silty, thin to thick bedded dolostone with shallow water features (ripple crosslamination, flat lamination, birdseye structures, and flat-pebble conglomerates), dolomitic sandstone and rare fossiliferous beds. The sediments were deposited in restricted intertidal to shallow subtidal settings. Detrital impurities in the Sutherland River Formation are inferred to have been derived from the Boothia Uplift (Morrow and Kerr, 1977).

Conversely, sandstone in the upper part of the Goose Fiord Formation on southern Ellesmere Island was likely derived from an easterly source (Inglefield Uplift).

Upper carbonate unit (P6c)

The upper carbonate unit comprises carbonate sediments of Emsian age assigned to the Disappointment Bay and Blue Fiord formations and to an unnamed new formation, previously included in the latter; lower Middle Devonian (Eifelian) strata assigned to the Blue Fiord Formation on Devon Island (Grinnell Peninsula) and Bathurst Island are not considered here. The **Disappointment Bay Formation,** 170-200 m thick and early-middle Emsian in age (*dehiscens* to *inversus* Zones, Uyeno, unpub. ms.), has been mapped on Cornwallis, eastern Bathurst, northwestern Devon and Lowther and Young islands (Thorsteinsson and Kerr, 1968; Kerr, 1974; Thorsteinsson, 1980) and somewhat similar strata (so far generally assigned to the Blue Fiord Formation) occur in central and southwestern Ellesmere Island (Smith, 1984). On Cornwallis Island, it lies with angular unconformity on the bevelled structures of the Cornwallis Fold Belt (Boothia Uplift) and consists of a basal chert-pebble conglomerate, a middle unit of dolostone and dolomitic quartz sandstone and an upper unit mainly of

Figure 8B.23. Sponge mud mound complex of the Barlow Inlet Formation at Depot Point, Cornwallis Island (maximum exposed stratigraphic thickness 120 m).

dolostone. It probably was deposited in shallow marine, somewhat restricted environments.

The unnamed new formation, probably early Emsian in age, lies disconformably between Goose Fiord and Blue Fiord formations in southwesternmost Ellesmere Island, North Kent Island, and northeasternmost Devon Island (Goodbody et al., 1988; Prosh, 1988). About 60 m thick, it consists of predominantly peritidal sediments, chiefly silty and sandy dolostone, in part gypsiferous or solution-brecciated, with some dolomitic pebble conglomerate and limestone.

Lower Devonian carbonate rocks of the Blue Fiord Formation are widely exposed in central and southern Ellesmere Island; rocks assigned to this formation also occur in northeasternmost Devon Island, in Melville, Victoria, and Banks islands, and the Princess Royal Islands (Thorsteinsson and Tozer, 1962; Tozer and Thorsteinsson, 1964; Miall, 1976; Goodbody et al., 1988; Prosh et al., 1988). In southwestern Ellesmere Island, where the formation overlies a tongue of deep water and transitional sediments (Cape Phillips and Eids formations, Chapter 8C), it ranges in age from earliest to latest Emsian (*dehiscens* to *serotinus* Zones; Uyeno and Klapper, 1980), i.e. is partly correlative with the Disappointment Bay Formation; in central Ellesmere Island, where it overlies clastic and carbonate sediments derived from the Inglefield Uplift (Vendom Fiord Formation, see below), it is restricted to the late Emsian (Trettin, 1978). The thickness of the unit increases toward the shelf margin, attaining more than 1300 m in southwestern Ellesmere Island (Smith, 1984). The unit consists of a wide spectrum of carbonate rocks (e.g. lime mudstones, wackestones, grainstones, boundstones, and dolostones) that include a considerable variety of fossils, notably corals, brachiopods, stromatoporoids, trilobites, and bryozoans (cf. Smith, 1984). The depositional environments represented by these strata extend from marginal parts of the deep water basin, across the shelf, to intertidal settings, in which dolostone is predominant and variegated clastic sediments are present. Crinoid banks and coral buildups are present at the shelf margin, coral and stromatoporoid buildups on the adjacent slope.

Upper Silurian – Lower Devonian sediments related to the Boothia Uplift

This section briefly describes those formations deposited in shallow marine to nonmarine environments that contain significant amounts of terrigenous clastic material derived from the Boothia Uplift (Fig. 8B.35-8B.37); structural aspects are discussed in Chapter 12C. The stratigraphic record demonstrates that the movements began earlier in the region south of Barrow Strait than in the region to the north. There also are some lateral differences within these two regions.

In the area south of Barrow Strait syntectonic deposition on both flanks of the Boothia Uplift began in the late Ludlow (Somerset Island Formation in the east, lower member of Peel Sound Formation in the west). The lower member of the Barlow Inlet Formation to the north (see above) also was derived from this source. That the crystalline basement forming the core of the uplift was breached in the late Ludlow is suggested by a change in the nature of the detrital quartz in the Somerset Island Formation (from mature monocrystalline to stressed metamorphic; P. Mortensen, pers. comm., 1985) and the occurrence of fresh oligoclase and microcline in the subarkosic beach facies of the lower member of the Barlow Inlet Formation (Packard, 1985).

Movements north of Barrow Strait are heralded by the early to middle Lochkovian Sophia Lake Formation, which contains a pebble conglomerate on Baillie Hamilton Island. Rugged relief is indicated by the late Lochkovian Snowblind Bay and Prince Alfred formations.

The youngest syntectonic deposits probably are unnamed sandstones and conglomerates of Pragian – early Emsian age on Cornwallis Island (see below).

South of Barrow Strait

The clastic wedges flanking the eastern and western sides of the Boothia Uplift differ in stratigraphy (cf. Thorsteinsson, 1980), and correlations across the uplift are problematic for the Pridoli and Lower Devonian because of sparse fossil control. In spite of these difficulties, however, it is apparent that the shoreline generally was closer to the western margin of the uplift than to its eastern margin; this asymmetry (Miall and Gibling, 1977) is important for the structural interpretation of the uplift (Chapter 12C).

Syntectonic sediments on the **eastern flank** occur mainly on Somerset Island, but also on Boothia Peninsula (Thorsteinsson and Tozer in Fortier et al., 1963, p.117-129; Blackadar and Christie, 1963; Miall and Gibling, 1977; Miall et al., 1978; Miall and Kerr, 1977; Thorsteinsson, 1980; Miall, 1983; Stewart, 1987). On Somerset Island, the uppermost Ludlow **Somerset Island Formation** is 150 to 400+ m thick and divisible into two members. The lower member consists mainly of grey and buff plane-laminated dolostone and limestone with some mottled limestone and dolostone. Invertebrates, mainly ostracodes and gastropods, are sparse but vertebrates, small domal stromatolites and oncolites locally are abundant. Lithology and fauna indicate cyclical deposition on intertidal to shallow subtidal mudflats. The upper member, in addition to the same rock types, contains variegated terrigenous dolosiltite, quartzose siltstone and mudstone with common desiccation features and some gypsum nodules and halite casts. These strata demonstrate eastward progradation of intertidal carbonate mud flats, followed by the fringe of an alluvial plain.

The overlying **Peel Sound Formation** has a minimum thickness of about 600 m on Somerset Island. There, a lower sandstone-siltstone member is succeeded by two conglomeratic and sandy members, in turn unconformably overlain by an upper conglomerate member. Primary structures and facies changes indicate deposition mainly by eastward-flowing braided rivers emerging from alluvial fans. On the eastern flank, as on the western, changes in clast composition demonstrate gradual stripping of lower Paleozoic and Proterozoic carbonates and quartzites, and increasing exposure of the gneissic basement. Beds in the lowermost part of the formation have yielded fish of probable early Pridoli age; no other diagnostic fossils have been found in this region.

On the **western flank** of the uplift, strata correlative with the combined Somerset and Peel Sound formations of Somerset Island have previously all been assigned to the Peel Sound Formation (Miall, 1970a, b; Thorsteinsson, 1980), although equivalents of the Somerset Island Formation now are recognized there (Mortensen, 1985; Mortensen and Jones, 1986). The lower member of the Peel Sound Formation (of previous usage) is up to 450 m thick, ranges in age from late Ludlow to early Pridoli, and consists of interbedded limestone, siltstone, sandstone, and conglomerate with some red beds. The fossil content of carbonates and sandstones (heterostracan fish, invertebrates, conodonts) indicates predominantly marine environments.

The upper member has a minimum thickness of about 300 m. It changes from cobble- and boulder-grade conglomerate, deposited on alluvial fans, through conglomeratic and sandy braided stream deposits into nearshore marine clastic and carbonate sediments over distances of only a few tens of kilometres. The most distant facies has yielded fish and conodonts of probable late Lochkovian age at one locality.

As a consequence of poor age control, the continuity of the diastrophism and related deposition of the Peel Sound Formation is not fully resolved. Unconformities within the Peel Sound Formation have been regarded as local (Miall, 1983) or regional in extent (Thorsteinsson, 1980); recent fossil collections favour the latter interpretation (Thorsteinsson, pers. comm., 1985).

Farther west on Prince of Wales and Russell islands, the Peel Sound Formation grades into a succession that is dominated by carbonates (mainly dolostone, minor limestone) but also contains an upward increasing proportion of siliciclastic material, mainly siltstone with minor sandstone and shale, known as the **Drake Bay Formation** (Mayr, 1978; Thorsteinsson, 1980). It is about 1550 m thick and ranges in age from late Ludlow to late Pragian; its top is not preserved.

North of Barrow Strait

In this region, sedimentation on the western flank of the uplift is represented by the Bathurst Island and Stewart Bay formations of Bathurst Island, discussed below in the context of deep water sediments. Sedimentation on the uplift itself and on its eastern flank is represented by three major units occurring mainly on Cornwallis Island and in part also on Baillie Hamilton and western Devon islands: (1) the Sophia Lake Formation; (2) the Snowblind Bay Formation; and (3) the Prince Alfred Formation, and related(?) unnamed sandstones and conglomerates of Cornwallis Island.

On Cornwallis, Devon, and Baillie Hamilton islands, significant amounts of terrigenous detritus, deposited in shallow water settings, are first encountered in the lower to middle Lochkovian **Sophia Lake Formation** (Thorsteinsson, 1980; Gibling and Narbonne, 1977). The formation is characterized by erosionally-based carbonate-clastic cycles, reflecting repeated transitions from subtidal to supratidal conditions. The terrigenous material is mainly of silt to sand grade, but a pebble conglomerate occurs on Baillie Hamilton Island. In addition there are local evaporites. The thickness of the unit increases from about 290 m on western Devon, through 580 m on Cornwallis, to 980 m on Baillie Hamilton Island. Remarkably similar to the upper member of the Somerset Island Formation (although younger), this formation probably also was deposited in peritidal settings that received clastic sediments from the Boothia Uplift.

The conformably overlying **Snowblind Bay Formation** is 578 m thick and probably late Lochkovian and(?) younger in age (Thorsteinsson, 1980; Muir and Rust, 1982). It is composed of an coarsening-upward succession of mudrock, sandstone and conglomerate that represents a coastal alluvial fan prograding into shallow marine environments. The conglomerate clasts are composed mainly of limestone and dolostone with lesser amounts of chert and other sediments, and were derived from northerly and northwesterly sources.

The **Prince Alfred Formation** of northwestern Devon Island lies on the Boothia Uplift with an angular unconformity that passes into a disconformity toward the east. From west to east the 6 to 318 m thick formation consists of conglomerate, sandstone, and silty dolostone facies that represent alluvial fan and marginal marine environments respectively (Morrow and Kerr, 1977). The unit has yielded no fossils, but regional mapping suggests that it is Lochkovian-Pragian in age and coeval with a hiatus between the Sutherland River Formation and the unnamed new formation that underlies the Blue Fiord (see above). It has tentatively been correlated with unnamed occurrences of unfossiliferous clastic sediments — red sandstone, and siltstone and cobble- to boulder-grade conglomerate — on Cornwallis Island and environs (Thorsteinsson, 1988), that lie with angular unconformity on a folded succession (including the Snowblind Bay and older formations) and are, in turn, unconformably overlain by the post-tectonic Disappointment Bay Formation of mid-Emsian age.

Lower Devonian red beds related to Inglefield Uplift and(?) other sources

Variegated Lower Devonian sediments in west-central Ellesmere Island with a large proportion of red beds have been assigned to a lower unit of mostly siliciclastic sediments (Red Canyon River Formation) and an upper unit of detrital and primary carbonates, siliciclastics and minor evaporites (Vendom Fiord Formation). The two are separated by an unconformity.

The **Red Canyon River Formation** is a 1764 m thick succession of siltstone and very fine grained silty sandstone with minor amounts of gypsiferous sandstone and sandy and silty limestone and dolostone (Trettin, 1978). Scanty fossil evidence indicates a late Lochkovian – early Pragian age. The formation occurs mainly on the south side of Cañon Fiord where it conformably overlies the Eids Formation and is overlain by the Vendom Fiord Formation with angular unconformity. It was deposited in progressively shallower, subtidal to supratidal environments. Sandstone and siltstone contain about one third detrital calcite and dolomite.

A unit of unfossiliferous red dolostone with minor sandstone and siltstone, unconformably underlying the Vendom Fiord Formation east of Vendom Fiord, is reminiscent of the Red Canyon River Formation. It is about 60 m thick and its base is not exposed (Roblesky, 1979).

The provenance of the Red Canyon River Formation is difficult to establish, because of its limited distribution and lack of fluvial sediments that might yield diagnostic paleocurrent directions. Several alternatives have to be considered. First, the formation is comparable in mineral composition, grain size and order of thickness to adjacent flysch of the deep water basin (Danish River Formation), suggesting that it was derived either from the same northeasterly sources as the latter, or from uplifted parts of the deep water basin. Alternatively, it may have been derived from cratonic sources — either the Inglefield Uplift or, because of its fine grain size, from more remote southeasterly sources.

The **Vendom Fiord Formation** (Kerr, 1967b; Trettin, 1978; Roblesky, 1979) underlies an arcuate south-trending belt between Cañon Fiord and south of Makinson Inlet. It overlies strata of Read Bay Group, Red Canyon River Formation or Eids Formation and is overlain by the Blue Fiord Formation. Limited fossil evidence indicates that the unit is mainly early Emsian and locally also late Pragian in age. The lower boundary is an angular unconformity at some localities and a disconformity at others, but is a conformable contact in the westernmost exposures. Measured thicknesses range from 57 to 551 m and decrease from east to west at given latitudes.

A massive, clast-supported carbonate cobble and pebble conglomerate with minor amounts of sandstone and siltstone (Fig. 8B.24), up to 66 m thick, marks the base of the formation in the eastern parts of its outcrop area. It probably represents a prograding alluvial fan. In some areas it is followed by quartzose and arkosic sandstones and siltstones, reminiscent of braided stream deposits in semiarid and arid regions. The largest part of the formation consists of impure dolostone and dolomitic sandstone and siltstone with minor gypsum-anhydrite, that collectively are of supratidal to shallow subtidal, restricted origin. Only a small proportion of strata in the uppermost part, including fossiliferous limestone, originated in unrestricted subtidal environments.

The basal conglomerate of the Vendom Fiord Formation, which contains early Paleozoic fossils, clearly was derived from adjacent parts of the Arctic Platform (Inglefield Uplift), and overlying arkosic sandstones probably came from the Precambrian crystalline basement in that area.

Figure 8B.24. Interbedded massive limestone conglomerate and crosslaminated red siltstone and sandstone in basal Vendom Fiord Formation south of Cañon Fiord.

Upper Ordovician to lower Middle Devonian carbonate buildups isolated from the main shelf

Progressive downwarping affected outer parts of the shelf in Late Ordovician to Early Devonian time (Chapter 8C). While much of this area was covered first by graptolitic mudrock and redeposited carbonates (Cape Phillips Formation and related units), and subsequently by flysch (Danish River, Bathurst Island, Stuart Bay formations) or transitional sediments (Eids Formation), carbonate production kept pace with the subsidence in some areas so that isolated buildups developed that are surrounded by deep water sediments.

The buildups have an overall age range from Late Ordovician to early Middle Devonian (Eifelian) and extend along the shelf margin from Melville Island to central Ellesmere Island. Four major occurrences, shown on the facies map for the Wenlock (Fig. 8B.34), are named for the Raglan Range of Melville Island (R); the Bent Horn and Towson Point wells of Cameron and Melville islands (BHTP); Hoved, a small island in Baumann Fiord (HI); and Cañon Fiord (CF), respectively. Comparable buildups in Greenland include the Hauge Bjerge and Pentamerus Bjerge formations of Washington Land (Fig. 8B.34), and the late Llandovery Samuelsen Høj Formation of Peary Land.

The Raglan buildup, known from surface exposures and two exploration wells, probably separated deep water mudrocks of the Ibbett Bay and Cape Phillips formations on the northwest and southeast respectively from Late

Ordovician to early Pragian time. Its orientation and full extent are uncertain.

The Bent Horn – Towson Point buildup (BHTP) is exposed in northeastern Melville Island and is also known from drilling and seismic investigations, especially on Cameron Island, where it contains an oil pool. The buildup existed from Late Ordovician until early or middle Eifelian time, when it was drowned and overlain by basinal mudrocks of the Cape de Bray Formation. Facies relationships in the uppermost part (late Emsian to middle Eifelian; assigned to the Blue Fiord Formation) are shown in Figure 8B.25, based on Uyeno and Mayr (1979) and Mayr (1980). The main lithofacies recognized are basin, deep reef flank, open shelf, upper reef flank, reefal margin, and bank interior. The buildup attained a thickness of 608 m. It has been variously interpreted as a single large atoll (Daae and Rutgers, 1975; Smith, 1984) with a maximum diameter of 160 km, or as several smaller, adjacent atolls (Embry and Klovan, 1976; Stuart Smith and Wennekers, 1977).

The Hoved Island reef or buildup (HI), of Silurian age and assigned to the undifferentiated Allen Bay Formation and Read Bay Group, is represented by about 135+ m of dolostone that overlie basinal limestone (Mayr, 1974; Poey, 1982). A predominantly coralline fauna is preserved in the outer part of the reef.

A buildup east of Cañon Fiord (CF), Ellesmere Island (Trettin, 1979; Hurst and Kerr, 1982) was flanked on the northwest by deep basinal sediments of the Hazen and Danish River formations and on the south by sediments of a shorter lived and probably shallower basin assigned to the Cape Phillips Formation. The buildup became separated from the main shelf by early Llandovery time and existed until Pridoli time, when it was drowned and covered by the Cape Phillips Formation after having attained a height of perhaps more than 1400 m. Thickness relationships indicate that it is not a horst or "miogeanticlinal ridge" but rather a part of the original outer shelf in which carbonate production kept pace with rapid subsidence. The lateral extent of the buildup is uncertain because it is covered by strata of the Sverdrup Basin on the southwest and by extensive icefields on the northeast. A similar buildup, or a northeastern extension of the Cañon Fiord buildup, probably separated Wenlock strata of the Danish River and Cape Phillips formations on Judge Daly Promontory, but is not preserved.

Smaller buildups or reefs are present in other parts of Ellesmere Island. In the southwesternmost part of the island, the clinothem (submarine slope), represented by the lower member of the Devon Island Formation, contains relatively small (maximum length 0.5 km, thickness 150 m), discrete, dolomitized buildups (Fig. 8B.26; Packard and Mayr, unpub. ms.). Five such buildups of latest Silurian – earliest Devonian age have been mapped. They are composed of variable amounts of coral framestone, bafflestone, crinoid rudstone, and a significant amount of dolomitic fibrous cement (botryoidal cements of submarine(?) origin).

A major, isolated downslope buildup of early Zlichovian (earliest Emsian) age has been identified and described by Smith (1984; Smith and Stearn, 1987) from the Blue Fiord Formation of southeastern Bjorne Peninsula in southwestern Ellesmere Island. The structure is more than 10 km long, 2 km wide and 100 m thick. It appears to be an atoll with an organic framework reef margin. Six growth stages, partly related to sea level changes, are recognized.

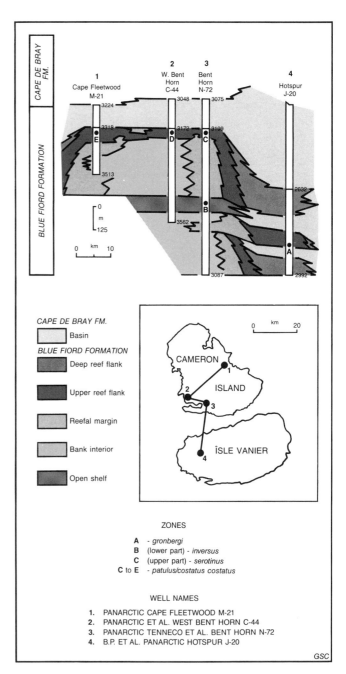

Figure 8B.25. Facies interpretation and conodont zonation (by T.T. Uyeno) of Lower Devonian carbonate buildup on Cameron Island (Bent Horn oil field).

Other similar, if smaller, isolated buildups occur in the formation elsewhere in southwestern Ellesmere Island.

Summary

Carbonate sediments of shallow marine to supratidal origin make up by far the largest proportion of the Lower Cambrian to Lower Devonian succession discussed. Second in abundance are coarse to fine clastic sediments, mostly derived from the cratonic margin and interior (except for

Figure 8B.26. Downslope dolomitic buildup in the Devon Island Formation (S$_{DI}$), overlying the Douro Formation (S$_{DO}$), southwesternmost Ellesmere Island; stratigraphic height of buildup approximately 150 m.

Upper Silurian – Lower Devonian clastics derived from the Boothia Uplift, which extends from the Arctic Platform far into the mobile belt). Evaporites (or related solution breccias) occur intermittently throughout the Middle Cambrian to Middle Devonian record, but are quantitatively significant only in the Early and Middle Ordovician. The depositional record is compatible with the results of paleomagnetic studies indicating that the region occupied equatorial latitudes in early and middle Paleozoic time (Irving, 1974).

The patterns of sedimentation have been shaped by two major factors, changes in apparent sea level and changes in shelf architecture.

(1) Changes in apparent sea level, related to eustatic sea level changes and/or subtle vertical movements of the continent, have resulted in major, unconformity-bounded transgressive-regressive cycles. The most important unconformities of this type occur within the Lower Cambrian and the upper Middle Cambrian; suspected unconformities between the Lower and Middle Cambrian and at the Ordovician-Silurian boundary are difficult to establish on a regional scale. From Early Cambrian to Late Ordovician time, each transgression was more extensive than the previous one, so that the unconformable base of the Phanerozoic deposits becomes progressively younger toward the cratonic interior.

(2) The most important changes affecting the shelf architecture were:

(a) Intermittent development of carbonate buildups at the margin of the deep water basin, which transformed a slightly inclined, distally steepened ramp into a rimmed shelf. Widespread evaporite deposition in the shelf interior during parts of late Early and early Middle Ordovician time probably was caused by the development of a rather continuous rim. The factors that governed the growth of the buildups are not obvious from the stratigraphic record; wind and current directions that controlled the supply of nutrients may have played a role.

(b) Accelerated tectonic subsidence of the outer shelf, resulting in the invasion of deep water sediments in some areas and the development of isolated carbonate buildups in others. This process was important from Late Ordovician time onward.

(c) Late Silurian – Early Devonian movements of the Boothia Uplift.

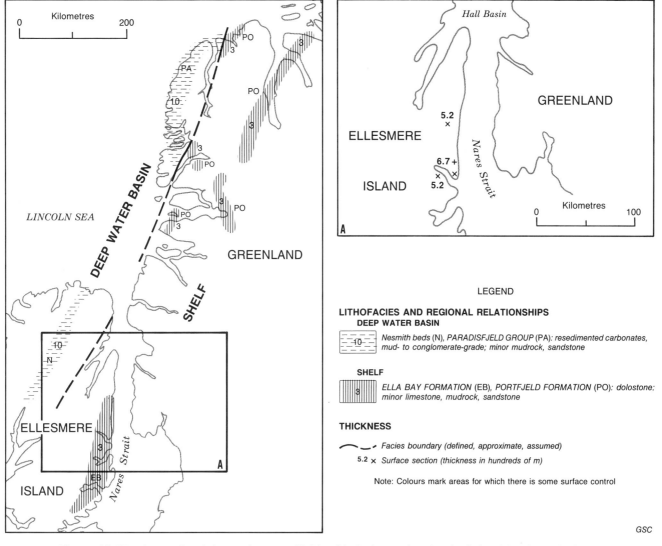

Figure 8B.27. Lower Cambrian carbonates (P1b) – lithofacies and regional relationships. Inset A shows thickness. (Pearya and Clements Markham Fold Belt have been omitted from Fig. 8B.27 to 8B.37 because of uncertainty about their position at given times.)

(d) Early Devonian movements of the Inglefield Uplift.

The depositional history of the shelf and of the adjacent sedimentary subprovince of the deep water basin in the Arctic Islands and North Greenland is portrayed in Figures 8B.27 to 8B.37, a series of lithofacies maps for selected time intervals that have reasonable paleontological control. The local stratigraphic nomenclature is indicated by letter codes. Thickness data for the time-rock units in the Arctic Islands are shown on the lithofacies maps or on separate isopach maps. These data are relatively sparse because many of the time-rock units chosen do not correspond to established rock units (formations etc.).

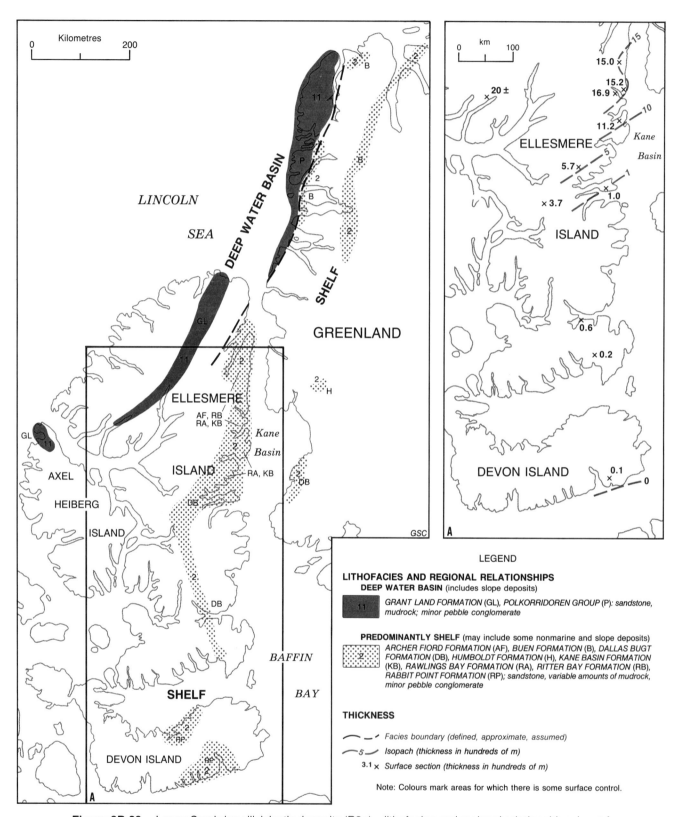

Figure 8B.28. Lower Cambrian siliciclastic deposits (P2a) – lithofacies and regional relationships. Inset A shows thickness.

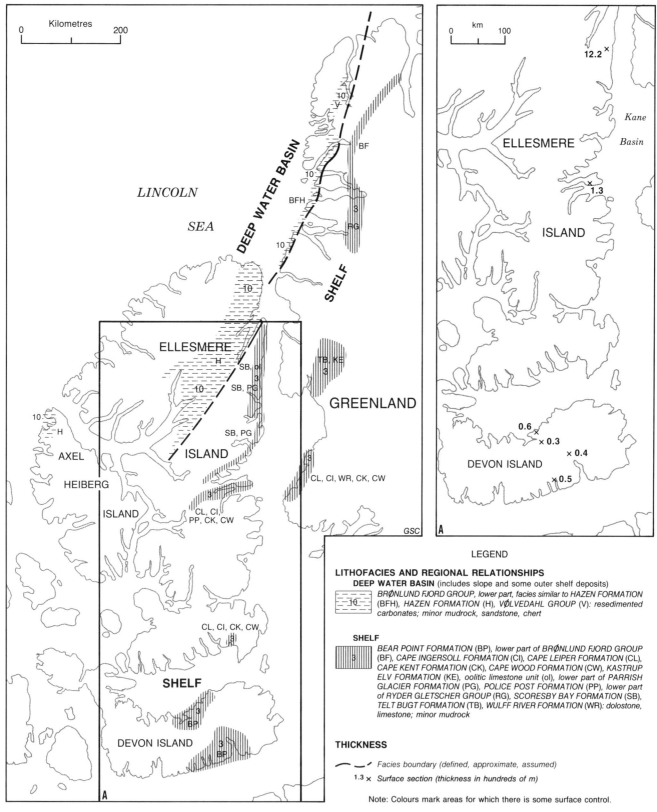

Figure 8B.29. Lower – Middle Cambrian carbonates (P2b) – lithofacies and regional relationships. Inset A shows thickness.

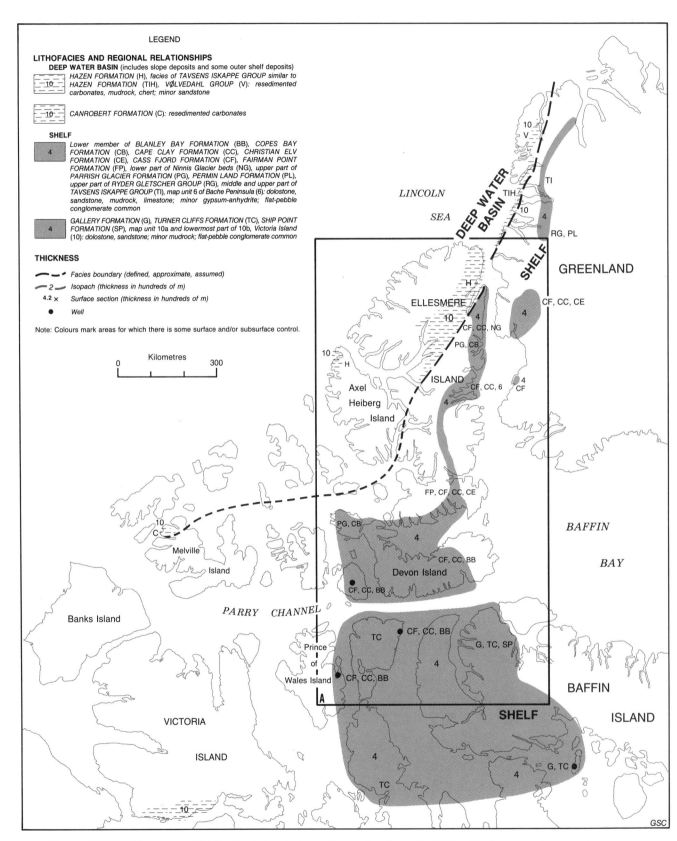

Figure 8B.30. Uppermost Middle Cambrian to lower Tremadoc deposits (P3) – lithofacies and regional relationships. Inset A shows thickness.

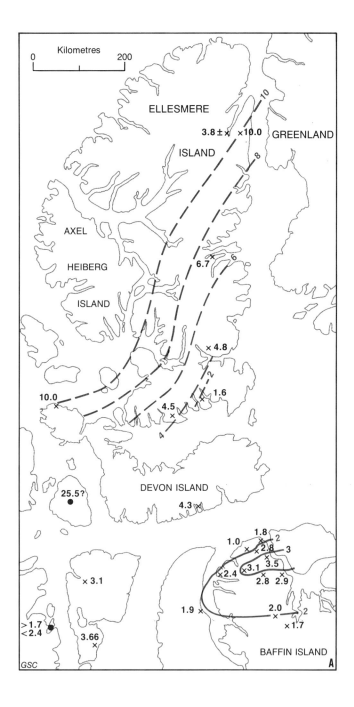

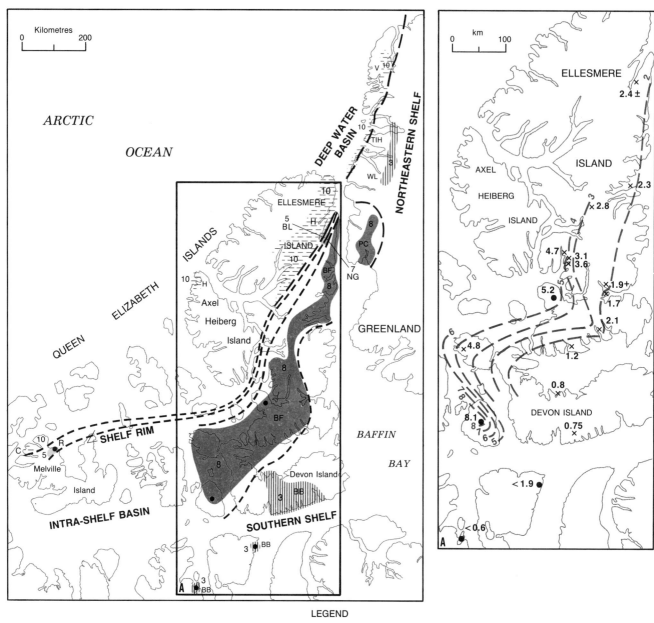

Figure 8B.31. Middle Tremadoc deposits (P4a) – lithofacies and regional relationships. Inset A shows thickness.

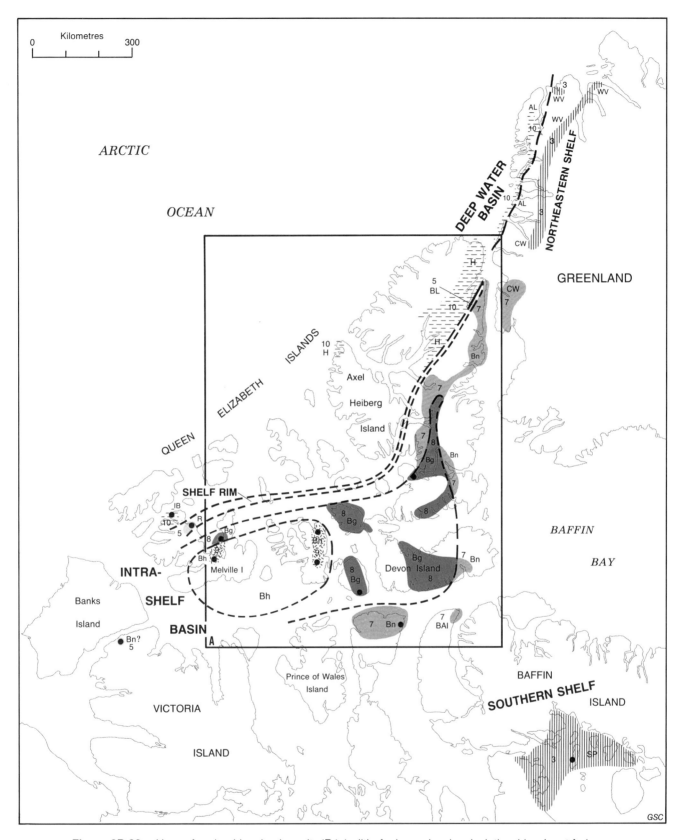

Figure 8B.32. Upper Arenig – Llanvirn deposits (P4c) – lithofacies and regional relationships. Inset A shows thickness.

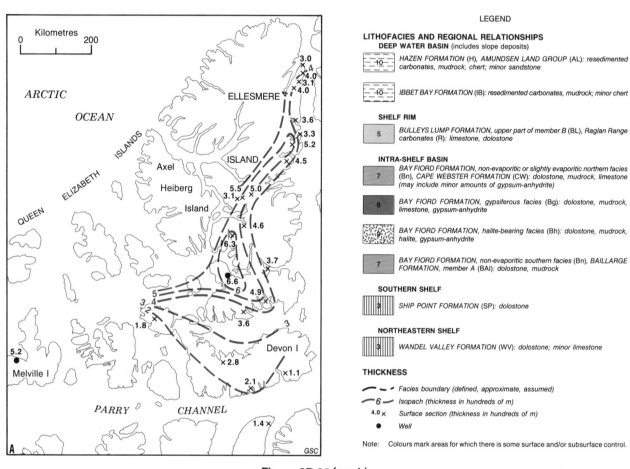

Figure 8B.32 (cont.)

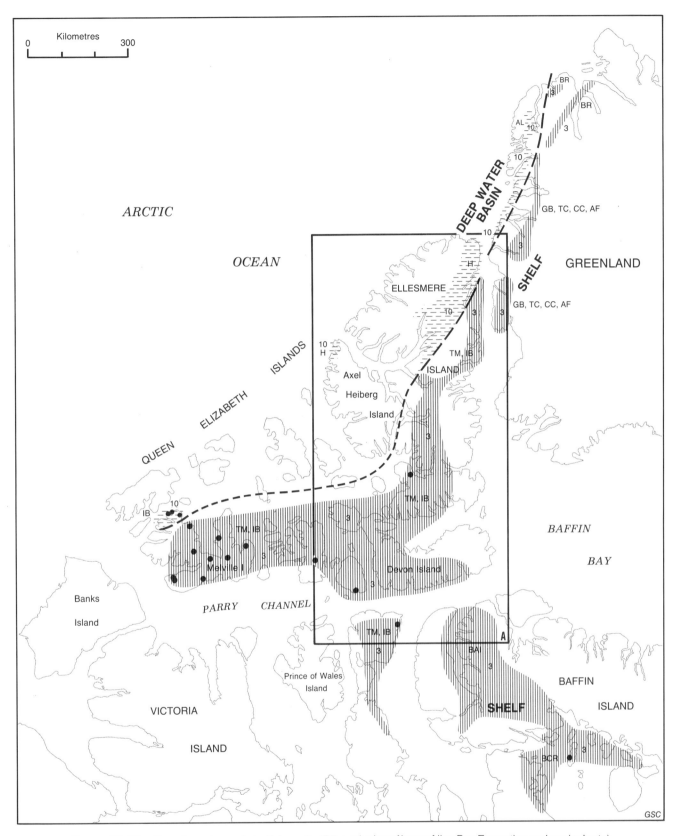

Figure 8B.33. Caradoc – lower Ashgill deposits (P5, exclusive of lower Allen Bay Formation and equivalents) – lithofacies and regional relationships. Inset A shows thickness.

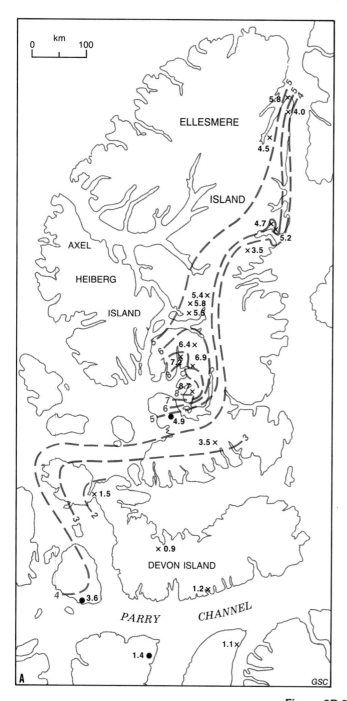

Figure 8B.33 (cont.)

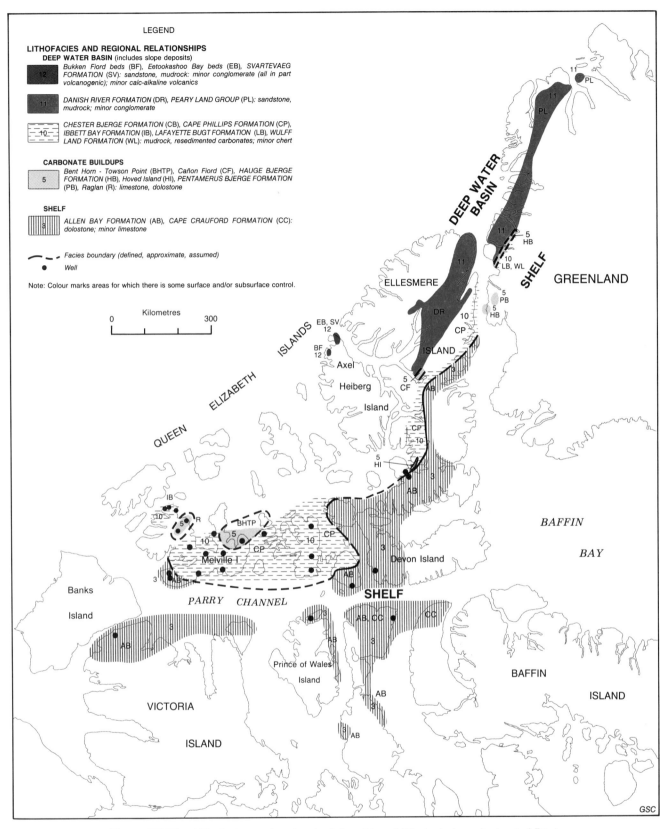

Figure 8B.34. Lithofacies and regional relationships of Wenlock deposits (part of P6a).

CHAPTER 8

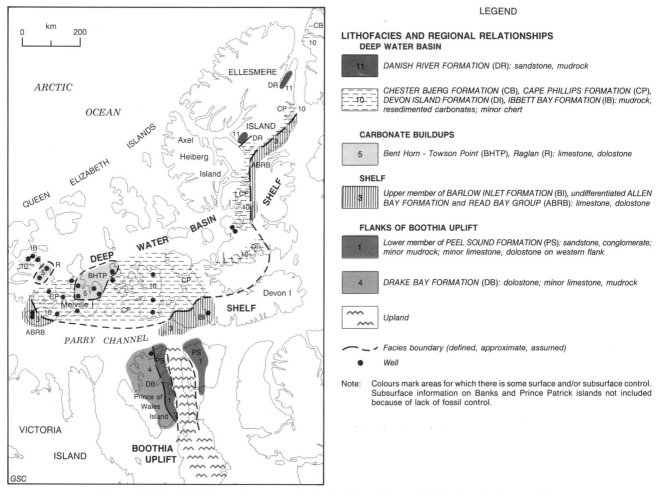

Figure 8B.35. Lithofacies and regional relationships of lower Pridoli deposits (part of P6a):

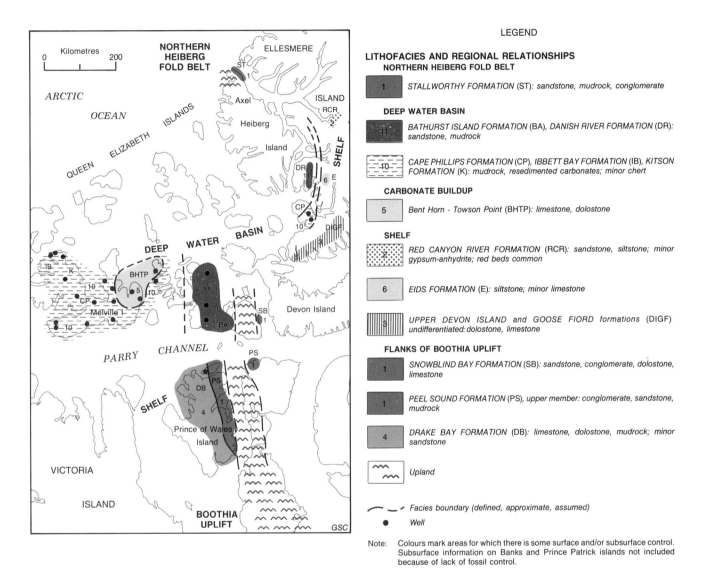

Figure 8B.36. Lithofacies and regional relationships of upper Lochkovian deposits (part of P6b).

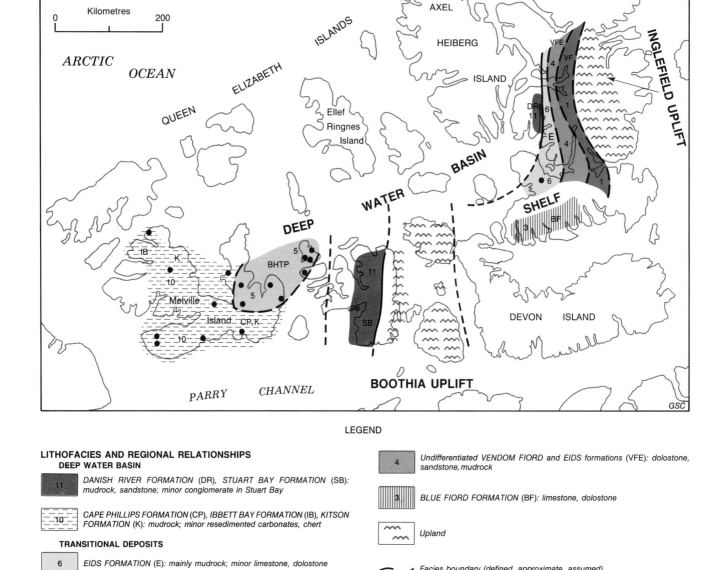

Figure 8B.37. Lithofacies of lowermost Emsian deposits (prior to the deposition of the Disappointment Bay Formation on the Boothia Uplift) (part of P6c).

C. FRANKLINIAN DEEP WATER BASIN

Introduction

Deep water conditions existed in northwestern parts of the Arctic Islands from Early Cambrian to Early Devonian time and persisted into the early Middle Devonian in the southwestern part (northwestern Melville Island). The Franklinian deep water basin is divisible into an exclusively sedimentary subprovince, exposed intermittently from northeastern Greenland to northwestern Melville Island, and a sedimentary and volcanic subprovince, exposed only in northern Ellesmere and Axel Heiberg islands (Fig. 8C.1), which also includes older and younger deposits that are of shallow marine to nonmarine origin.

The summaries for northern Ellesmere Island and Axel Heiberg Island are based partly on published accounts (Trettin, 1969b, 1971, 1978, 1979, 1987; Trettin et al., 1979), and partly on unpublished manuscripts resulting from extensive field and laboratory investigations since 1975; only additional sources will be quoted.

Sedimentary subprovince

Introduction

In the Arctic Archipelago, the sedimentary subprovince is intermittently exposed in an area extending from eastern Ellesmere Island through northwesternmost Devon Island, northern Cornwallis Island, and Bathurst Island to northwestern Melville Island. It is divisible into a northwestern and northern "axial" belt that lay beyond the Early Cambrian to Middle Ordovician shelf margin, and a southeastern and southern "marginal" belt that overlies Middle Ordovician or younger shelf deposits (Fig. 8B.2-8B.4, 8C.1-8C.2, 8C.14, 8C.26). The term "axial" implies that the deep water basin was bordered on the northwest by a bathymetric or topographic high. This configuration is demonstrable for the Silurian and Early Devonian on the basis of paleocurrent determinations (Fig. 8C.17), but not for the preceding time. The most complete record of the deep water basin is exposed in the axial region of North Greenland and northern Ellesmere Island, where five first-order rock units are distinguished that are informally designated B1 to B5 (B for basin). Units B3 and B4 are further divisible into three or four major subunits each. The resulting ten units, listed in Figure 8C.2 and Table 8C.1, represent distinct phases in the depositional and tectonic history of the region (see below, Summary and interpretation). Their relationship with formal rock units is shown in Figure 8C.2 and further explained below; their relationship with units of the Franklinian Shelf is apparent in Figures 8B.2-8B.4 and 8B.27-8B.37. A brief descriptive summary of these units, and of related units in the sedimentary-volcanic subprovince, will be followed by an interpretative outline of the depositional history. Two deep water formations of Middle Devonian age, the Blackley (B4d) and the Cape de Bray (B5), are discussed in Chapter 10.

Lower Cambrian carbonates, northern Ellesmere Island (B1)

The Nesmith beds are exposed in three narrow, elongate belts in the northwestern part of the Hazen Fold Belt (Fig. 8C.3). They underlie the Grant Land Formation with a conformable, gradational contact and their base is not exposed. The thickness of the exposed part of the unit has not been determined because of poor exposure and complex structure.

The Nesmith beds consist mainly of impure resedimented carbonates ranging in grade from calcilutite through calcarenite to pebble conglomerate with lesser amounts of calcareous sandstone, calcareous mudrock, and mudrock. The carbonate rock types represented by the clasts include oolitic and oncolitic grainstones. Calcarenite and calcilutite commonly contain 10-30% of siliciclastic impurities, largely quartz with a few percent of feldspar, muscovite and chlorite. At Henrietta Nesmith Glacier (Fig. 8C.3, locality HNG), the uppermost part of the formation is characterized by turbidites that commonly consist of relatively thick A-divisions (Ta) alone, which show an upward decrease both in the maximum grain size of the carbonate particles and in quartz content (Fig. 8C.4). If transported by channelized turbidity currents, these strata probably were deposited in the mid-fan environment (cf. Mutti, 1979); if by sheet flows, in lower slope or base-of-slope settings (cf. McIlraith, 1978).

The Nesmith beds have not yielded any fossils but are comparable in lithology and stratigraphic position to the Paradisfjeld Group of Peary Land (Chapter 7, stage 2B), dated as Early Cambrian on the basis of *Chancelloria* and stratigraphic relationships. The correlative unit on the Franklinian Shelf is the Ella Bay Formation, which also contains quartz sand and oncoids (see Chapter 8B, P1b). Upper parts may have been derived from uplifted portions of the Ella Bay Formation, which is truncated by a disconformity on southwestern Judge Daly Promontory, and lower parts from unconsolidated submarine sediments of that unit.

Lower Cambrian silicilastic sediments, northern Ellesmere Island (B2)

The Grant Land Formation is exposed mainly in the northwestern Hazen Fold Belt, and also in the Northern Heiberg Fold Belt and the southwestern part of the Clements Markham Fold Belt (Fig. 8C.3). The unit is intensely deformed and so far has been studied in reconnaissance fashion only. The following discussion of its stratigraphy is based mainly on stratigraphic sections at Hare Fiord (Fig. 8C.3, localities HFW, HFE), Tanquary Fiord (TF), Rollrock Lake (RL) and Henrietta Nesmith Glacier (HNG).

The Grant Land Formation overlies the Nesmith beds with conformable contact and is overlain by the Hazen Formation with an abrupt contact that may indicate a submarine disconformity. Its thickness is greater than 1.5 km and perhaps in the order of 2 km.

The Grant Land Formation consists mainly of quartzite and slate or phyllite with small amounts of pebble conglomerate and intraformational conglomerate. It is divisible into three members, informally referred to as members A, B and C.

Member A, the lowest unit, is about 70 m thick northeast of Tanquary Fiord. Composed mainly of sandstone and mudrock that are partly calcareous, it is characterized by turbidites consisting of division A (Ta) alone, of divisions A and B (Tab), or of divisions A to C (Ta-c), with intercalated mudrocks and nongraded, flat-laminated sandstones. The mineral composition of the calcareous sandstones is

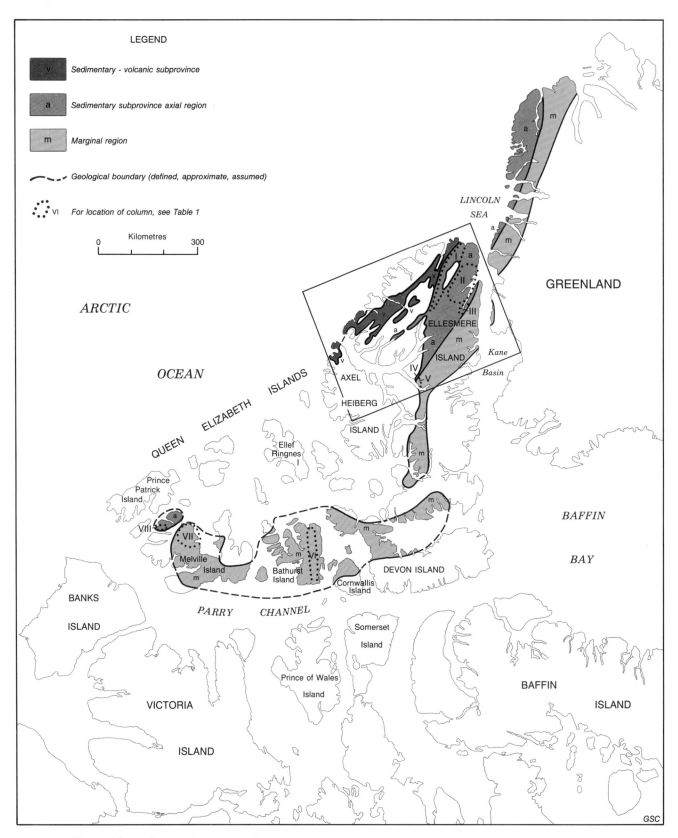

Figure 8C.1. Subdivisions of the Cambrian-Devonian deep water basin, Arctic Archipelago and North Greenland, and index for Figures 8C.3 and 8C.7. Generalized distribution areas include outcrop and subcrop.

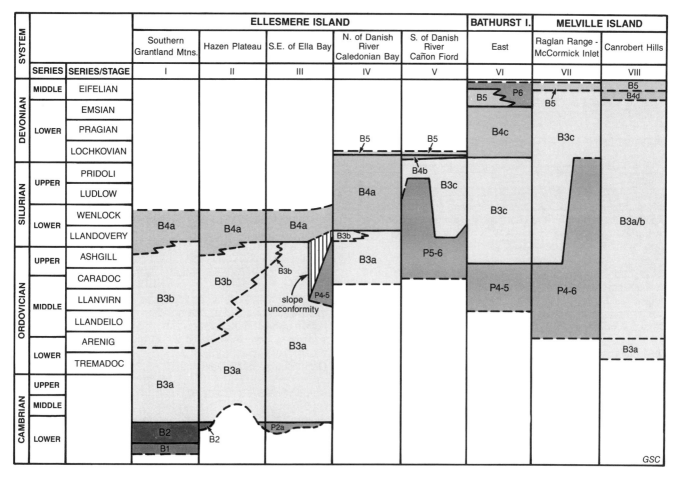

Figure 8C.2. Age relationships of deep water units.

comparable to that of the Nesmith beds. Individual turbidities show an upward decrease in maximum grain size and carbonate content (Fig. 8C.5).

Member B includes most of the formation. Bouma sequences, comparable to those of member A, occur in the lowermost part. The bulk of the member consists of alternating units that are dominated by sandstone or mudrock and separated by sharp, flat contacts. The sandstones of the sandstone-dominated units are mainly medium to coarse grained. Nongraded, structureless beds, up to about 7 m thick (Fig. 8C.6b), are most characteristic, but flat-laminated, crosslaminated, and rare trough crossbedded strata (Fig. 8C.6c) also are present. Upward-fining units, 0.15-10 m thick and generally not of Bouma-type, make up more than one quarter of a detailed section. Characteristic of the sandstone composition in this unit is a high feldspar content (mean value of 30 analyzed specimens is 17% albite, 4% K-feldspar) and the presence of a secondary matrix of intergrown muscovite, chlorite, quartz and feldspar. The conglomerates are massive or trough crossbedded and partly lenticular. Rip-up clasts of mudrock (Fig. 8C.6d) occur in nearly all rock types.

The mudrock-dominated units are grey or consist of alternating units of greenish-grey and purplish-red layers that commonly are extensive. Chlorite (with minor muscovite), pseudomorphous after glauconite, is common in the upper part of the member.

Member C, probably no more than a few hundred metres thick, consists mainly of mudrock with an upward decreasing proportion of sandstone that forms structureless beds up to several metres thick, and thinner, flat-laminated or crosslaminated units. The sandstones differ from those in member B by a finer grain size and larger quartz content. The mudrocks are similar to those in member B but contain larger proportions of altered glauconite.

The thick, structureless sandstones and conglomerates of the Grant Land Formation probably were deposited by sandy debris flows or turbidity currents of high density that were capable of eroding and transporting large fragments of mudrock. The upward-fining sequences are interpreted as the fill of channels that gradually were abandoned. Most channels must have been so wide and shallow that they are not recognizable in outcrop, but small channels, metres wide and less than a metre deep, are exposed locally. The mudrock-dominated units do not seem to be facies equivalents of the sand-dominated units and probably were deposited from "nepheloid" (i.e. mud-charged) layers during intervals when there was no sediment gravity flow activity, perhaps because of a high stand of sea level. The scarcity of Bouma sequences, which appear to be limited to the

Table 8C.1. Age relationships of major rock units, sedi-mentary subprovince of deep water basin. (For location see Figure 8C.1)

B5: mainly mudrock, minor sandstone and carbonates of transitional (submarine fan to outer shelf) aspect; restricted to marginal belt where it overlies units B3c, B4b, or B4c; Early to Middle Devonian, diachronous

- Eids Formation of Ellesmere and Bathurst islands

- Cape de Bray Formation, (lower part), northwestern Melville Island

B4: sandstone, mudrock, and minor conglomerate derived mainly from orogenic sources (flysch); minor shelf-derived carbonates; highly diachronous with overall age range from Late Ordovician to Middle Devonian

B4a: mudrock, sandstone, minor conglomerate and shelf-derived carbonates; axial region of North Greenland and northern Ellesmere Island where it overlies B3b or locally B3a; diachronous, with overall age range from Late Ordovician to Early Devonian

- Merqujôq, Lauge Koch, and Nordkronen formations with tongues of Wulff Land Formation, Greenland

- Danish River Formation of northern Ellesmere Island

B4b: sandstone, mudrock, minor shelf-derived carbonates; marginal region of northern Greenland and central Ellesmere Island where it overlies B3c or shelf carbonates; diachronous with overall age range from late Early Silurian to Early Devonian

- Lauge Koch Formation of Greenland

- Danish River Formation of central Ellesmere Island

B4c: sandstone, mudrock, minor shelf-derived carbonates; marginal region of Bathurst and Cornwallis islands (western flank of Boothia Uplift) where it overlies B3c; Early Devonian

- Bathurst Island and Stuart Bay formations of Bathurst Island

B4d: sandstone, mudrock; overlies B3b in axial region of northwestern Melville Island; late Early to Middle Devonian

- Blackley Formation

B3: variable proportions of shelf-derived carbonates, graptolitic mudrock, and radiolarian chert — with or without secondary chert and minor amounts of sandstone; highly diachronous with overall age range from latest Early Cambrian to Lower Devonian

B3a: carbonates predominant, radiolarian chert subordinate; restricted to axial region; latest Early Cambrian to Late Ordovician / Early Silurian; upper boundary diachronous

- Vølvedahl Group of Greenland

- carbonate member of Hazen Formation, northern Ellesmere Island

- Canrobert Formation and parts of Ibbett Bay Formation, northwestern Melville Island

B3b: mainly graptolitic mudrock and chert, carbonates subordinate or absent, sandstone rare or absent; restricted to northern/northwestern part of axial region; highly diachronous with overall age range from Early Ordovician to Early Devonian

- Amundsen Land Group of Peary Land, Greenland

- chert member of Hazen Formation, northern Ellesmere Island

- parts of Ibbett Bay Formation, northwestern Melville Island

B3c: mainly shelf-derived carbonates and graptolitic mudrock, minor radiolarian chert and sandstone; confined to marginal region where it overlies Ordovician or Silurian shelf carbonates; separated from main basin by carbonate buildups in some areas; highly diachronous, with overall age range from Late Ordovician to Early Devonian

- most of Lafayette Bugt and Wulff Land formations, Greenland

- Cape Phillips Formation of Ellesmere, Cornwallis, Bathurst, and Melville islands

- Devon Island Formation of northwestern Devon and southwestern Ellesmere islands

- Kitson Formation of northwestern Melville Island

B2: craton-derived, fine to coarse siliciclastic sediments; exposures confined to North Greenland and northern Ellesmere Island; middle to late Early Cambrian

- Polkorridoren Group of Greenland

- Grant Land Formation of southern Grantland Mountains, northern Ellesmere Island

B1: mainly shelf-derived carbonates, minor siliciclastic sediments; exposures confined to North Greenland and northern Ellesmere Island; Early Cambrian

- Paradisfjeld Group of Greenland

- Nesmith beds of southern Grantland Mountains, northern Ellesmere Island

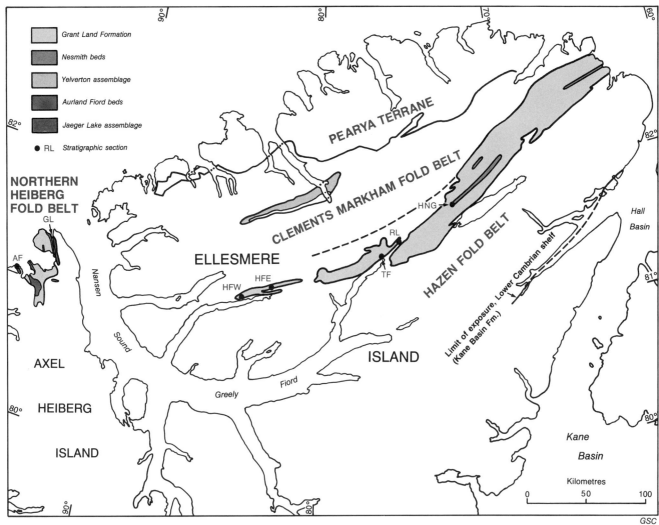

Figure 8C.3. Generalized outcrop areas of Lower Cambrian sedimentary units, northern Ellesmere and Axel Heiberg islands; also shown are units of uncertain Proterozoic and/or early Paleozoic age. (Section designations: AF = Aurland Fiord; GL = Greenstone Lake; HFE = Hare Fiord east; HFW = Hare Fiord west; HNG = Henrietta Nesmith Glacier; RL = Rollrock Lake; TF = Tanquary Fiord.)

lowermost part of the formation, indicates that turbidity currents of average or low density played only a minor role in the deposition of members B and C. These units, therefore, either represent inner fan environments and channels of the mid-fan environment; or, more likely, a variant of the "low transport efficiency fan" of Mutti (1979) or "inefficient" fan of Homewood and Caron (1982).

The carbonate fragments in member A, as those in the underlying Nesmith beds, probably were derived from carbonates of the Shelf Province (Chapter 8B), especially the Ella Bay Formation, which has a disconformity at the top. Facies relationships with the Ellesmere Group suggest derivation of the silicates from the craton. U-Pb analysis of four zircon fractions from a coarse grained, feldspar-rich sandstone near Tanquary Fiord yielded early Proterozoic or older ages in excess of 2200 Ma (Trettin et al., 1987).

The unfossiliferous Grant Land Formation is correlated with the Polkorridoren Group of North Greenland and the Lower Cambrian Ellesmere Group of the Franklinian Shelf on the basis of lithostratigraphic relationships.

Lower Cambrian to Lower Silurian carbonates, mudrock, and chert, northern Ellesmere Island (B3a, B3b)

The Hazen Formation, a condensed succession of resedimented carbonates, impure, mainly fine grained clastic sediments, and chert (Fig. 8C.7, 8C.8) overlies the upper Lower Cambrian Kane Basin Formation of the Ellesmere Group of the Shelf Province (see Chapter 8B) with conformable, gradational contact near the head of Archer Fiord (Fig. 8C.9). The late Early Cambrian trilobites *Olenellus* sp. and *Wanneria*(?) sp. have been found in the basal part of the Hazen Formation in that area. Farther northwest, the Hazen overlies the Grant Land Formation with an abrupt contact that may indicate a submarine

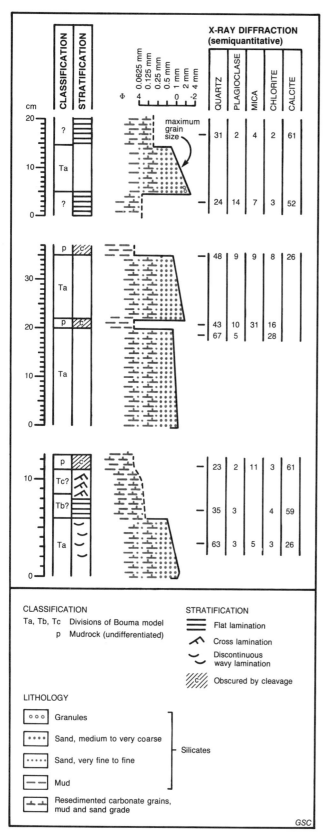

Figure 8C.4. Nesmith beds, stratigraphy of typical turbidites at Henrietta Nesmith Glacier.

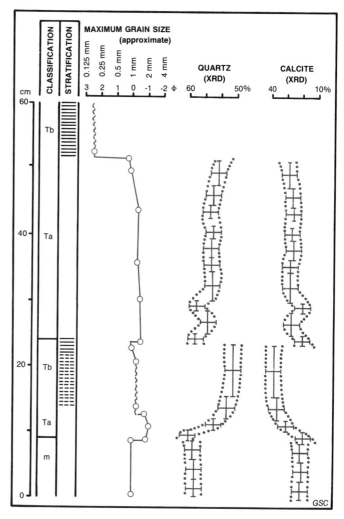

Figure 8C.5. Turbidites in member A of Grant Land Formation: variations in grain size and mineral composition.

disconformity. On the southeast it grades into carbonate units of the Franklinian Shelf. It is overlain, mostly abruptly, by flysch of the Danish River Formation. Fossils from the uppermost Hazen Formation are latest Ordovician in age at section SPB (Fig. 8C.7, 8C.8); Llandovery (older than latest Llandovery) at locality F1; middle Llandovery at locality F2, and late Llandovery at section CF1.

The thickness of the Hazen Formation is greatest at its southeastern facies boundary but cannot be measured there; a restored palinspastic cross-section suggests that more than 1 km of strata are present (Fig. 8C.26). It thins markedly to the northwest, decreasing through 632 m at the head of Ella Bay (Fig. 8C.7 and 8C.8, section EB1) to about 250 m at Bent Glacier (section BG).

The Hazen Formation is divisible into two units that are of regional extent, a lower unit dominated by resedimented carbonates (B3a — carbonate member) and an upper unit characterized by radiolarian chert (B3b — chert member). The contact between them is highly diachronous, becoming older along the line of the stratigraphic cross-section (Fig. 8C.8). The age of this

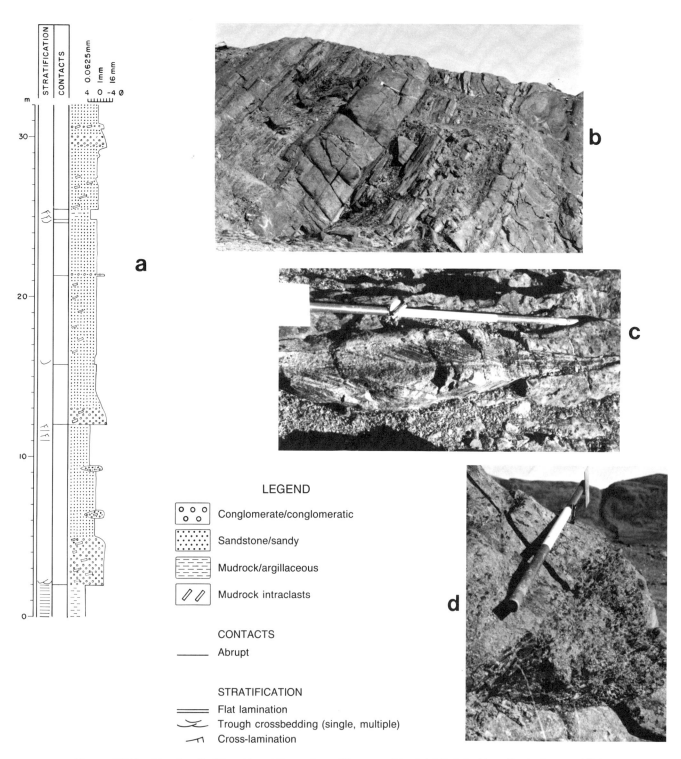

Figure 8C.6. Member B of Grant Land Formation at Tanquary Fiord. (a) Detailed log of typical upward-fining sequences. (b) Typical exposures; 1.5 m staff on massive sandstone bed. (c) Trough crossbedding. (d) Trough filled with rip-up clasts of mudrock.

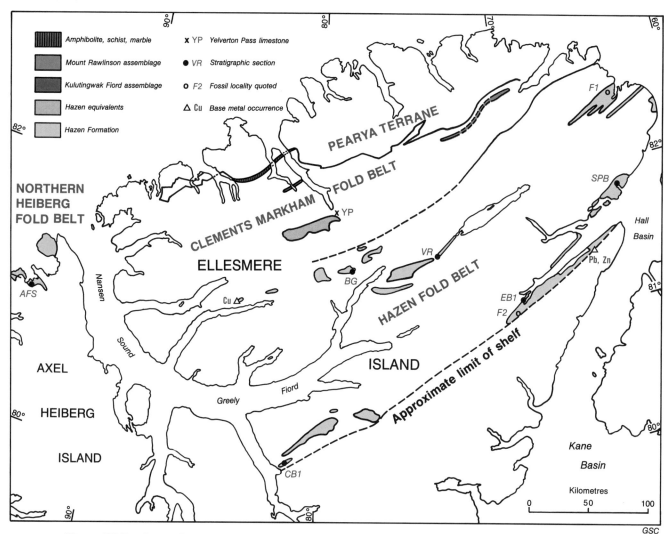

Figure 8C.7. Generalized outcrop map of Hazen Formation and correlative units, and index for Figure 8C.8. (Section designations: BG = Bent Glacier; CB1 = Caledonian Bay no. 1; EB1 = Ella Bay no. 1; SPB = St. Patrick Bay; VR = Very River; AFS = Aurland Fiord South. Llandovery fossils occur at top of formation at localities F1 and F2.)

boundary is middle Llandovery at Cañon Fiord (section CF1), Ashgill or slightly older at Ella Bay (section EB1), and Arenig at St. Patrick Bay (section SPB). In some areas the two members are difficult to distinguish because of extensive replacement by secondary chert.

The **carbonate member** can be described in terms of three lithofacies that are distinguished in the stratigraphic sections of Figure 8C.8 but have not been mapped. The southeasternmost facies (A) consists of thinly interlaminated calcilutite and calcareous and dolomitic mudrock only. It occurs adjacent to the facies boundary with the shelf sediments and is exposed only in the area southeast of Ella Bay and Archer Fiord.

The central facies (B) consists mainly of resedimented carbonates ranging in grade from calcilutite to boulder conglomerate, with thinly interstratified fine grained clastic sediments and very small proportions of radiolarian chert. It is well developed at sections CF1 and EB1.

The northwestern facies (C), characteristic of sections SPB, VR, and BG, consists of thinly interstratified carbonates and predominantly fine grained clastic sediments with small proportions of radiolarian chert. The carbonates are mainly calcilutite although calcarenite and rare pebble conglomerate also are present.

The **chert member** consists mostly of thinly interstratified radiolarian chert and sparsely graptolitic claystone, but also includes a carbonate conglomerate at Cañon Fiord (section CF1) and north of the head of Ella Bay. The age range of the member, and the proportion it forms of the entire formation, both decrease toward the southeast. At Ella Bay the member wedges out before reaching the facies contact with the shelf carbonates.

The rock types of the Hazen Formation can be assigned to eight genetic assemblages, described in the order of decreasing proximity to the southeastern facies boundary.

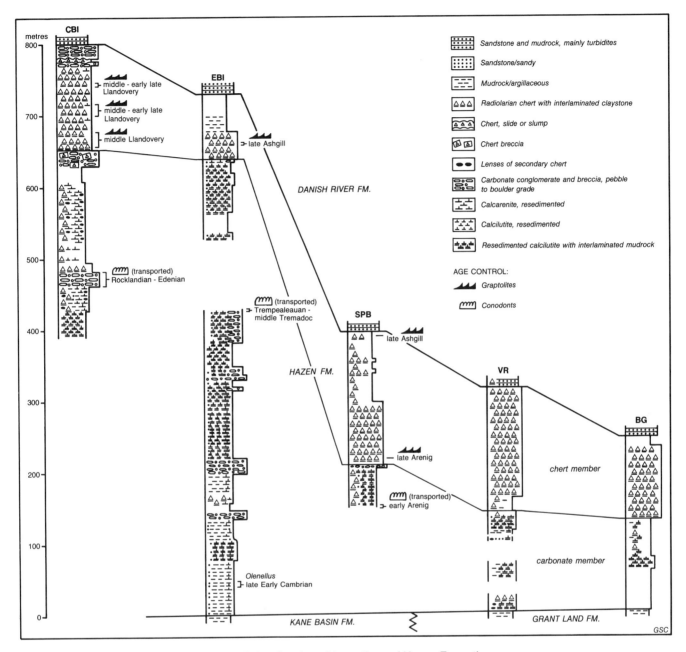

Figure 8C.8. Stratigraphic sections of Hazen Formation.

1. **Thinly interlaminated impure calcilutite and calcareous and dolomitic mudrock** constitute facies A exclusively, but also occur in facies B and C. The rocks are characterized by thin, flat or undulating lamination with some crosslamination, generally of microscopic scale. The transitional setting and indigenous(?) trilobites (*Glossopleura* associated with *Ogygopsis*) suggest an upper slope environment of deposition for facies A, but the laminites have a wider depositional range; they probably were deposited from nepheloid layers generated by storms in the adjacent shelf environment.

2. **Disrupted laminites and related breccias and conglomerates,** composed of disc-shaped fragments, bedding parallel or inclined to bedding, that can be metres in length (Fig. 8C.10e), are common in facies B. Measured thicknesses range from 0.35 to 31 m. Two distinctive units in the Ella Bay area have a minimum palinspastic extent of 16 x 10 km (parallel and normal to strike, respectively).

3. Only one example of a **shelf-derived submarine slide** is known, which occurs in the southeastern part of facies B at Ella Bay. It consists of a block of massive dolostone, about 30 m thick and tens of metres or more in strike length, that grades laterally into carbonate conglomerate.

4. **Submarine conglomerates with benthonic faunas** have been observed only at Caledonian Bay in the

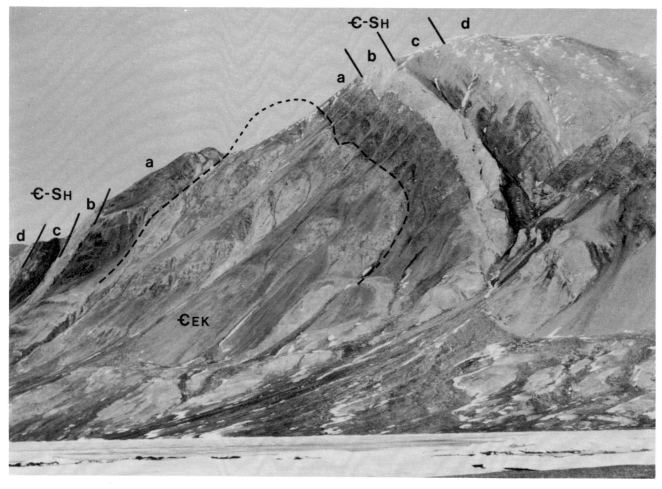

Figure 8C.9. Contact between Kane Basin Formation (ЄEK) and Hazen Formation (Є-SH) southeast of Ella Bay, Archer Fiord; units a to d are subdivisions of the lower part of the carbonate member, and unit b is carbonate conglomerate of Figure 8C.10e.

carbonate member (facies B) and in the chert member. They are matrix-supported, massive and disorganized, and represent subaqueous debris flows derived from the adjacent shelf margin. Their thickness ranges from 18 to 27 m, and one has a strike length of more than 14 km.

5. **Graded calcarenite-calcilutite** beds are common only in the Caledonian Bay area (Fig. 8C.10e), where they are associated with the fossiliferous conglomerates mentioned. They probably represent turbidites that developed from subaqueous debris flows.

6. **Thinly interstratified impure calcilutite and calcarenite and calcareous and dolomitic sandstone and mudrock,** showing flat lamination and some crosslamination (Fig. 8C.10c), are common in facies B and C. Most of these deposits probably are distal turbidites beginning with Tb, Tc, or Td, although graded bedding is scarce or very subtle.

7. **Radiolarian chert and interlaminated claystone** occur mainly in the chert member, and only in smaller proportion in facies B and C of the carbonate member. The chert forms laminae and beds that are 0.5-20 cm thick and separated by laminae of claystone (Fig. 8C.10a). The radiolarians usually are preserved as inclusion-free spherules or ellipsoids, one to a few tens of micrometres in diameter, that rarely show skeletal details (Fig. 8C.10b). The rocks are medium to predominantly dark grey and contain sparse graptolites.

8. **Secondary chert** is most common as lenses that may be centimetres or decimetres in length, but in some areas replaces large parts of the formation.

In summary, the Hazen Formation consists mainly of resedimented carbonates that were generated on the adjacent Franklinian Shelf and carried into the basin by a variety of processes such as sliding, sediment gravity flow, and settling from storm-generated nepheloid layers. Associated with the carbonates are subordinate clastic sediments of clay to fine sand grade that passed across the shelf and were involved in the same processes. The third component are radiolarian chert beds and associated claystones of pelagic origin. The proportion of the clastic and carbonate sediments and the thickness of the formation decrease toward the northwest, i.e. away from the source of these sediments, whereas the proportion of chert decreases in the opposite direction. The proportion of carbonate and

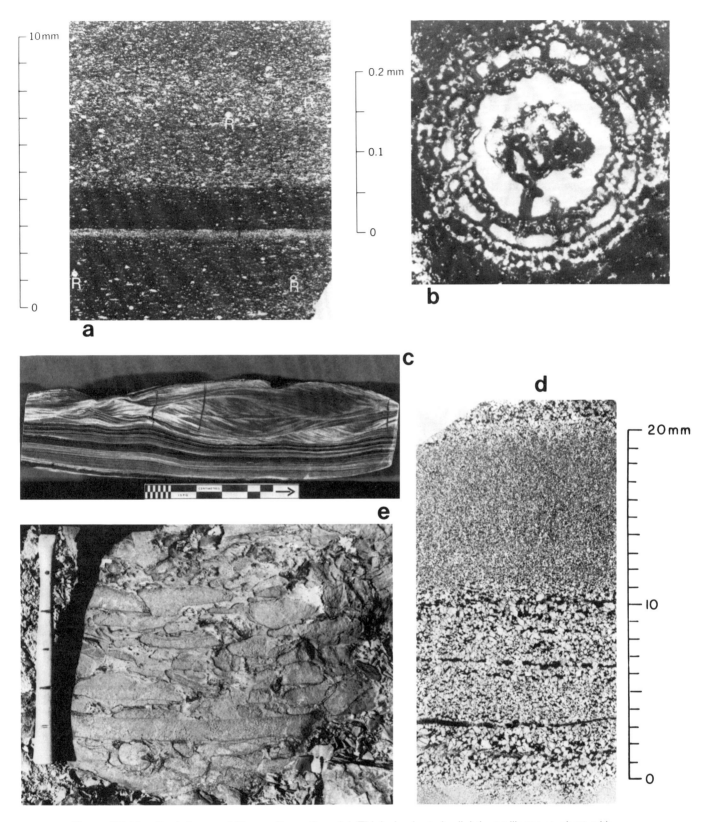

Figure 8C.10. Rock types of Hazen Formation. (a) Thinly laminated, slightly argillaceous chert with radiolarians (R). (b) Closeup of radiolarian (*Flustrella* sp.). (c) Calcareous mudrock showing flat lamination and small-scale crosslamination; (d) Graded calcarenite. (e) Carbonate conglomerate.

clastic sediments versus radiolarian chert also decreases upward within the formation, indicating that conditions became increasingly unfavourable for processes of resedimentation during the Middle and Late Ordovician. In the Shelf Province, this time interval is marked by an extensive transgression, coupled with progressive tectonic subsidence of the outer shelf from Late Ordovician time onward, which resulted in extensive and relatively deeply submerged shelves (see Chapter 8B).

Lower Ordovician to Lower Devonian carbonates, mudrock and chert, northwestern Melville Island (B3a, B3b)

Deep water carbonate sediments, mudrock and chert, comparable to the Hazen Formation, are included in the Canrobert and Ibbett Bay formations. Established by Tozer and Thorsteinsson (1964), these units are presently being restudied in more detail, but only preliminary data are available at this time (Robson, 1985; Goodbody and Christie, unpub. ms.). The Canrobert Formation corresponds to unit B3a; the Ibbett Bay Formation is transitional between B3a and B3b.

The **Canrobert Formation** has a minimum thickness of about 460 m and its base is not exposed. The top is dated as late Arenig on the basis of graptolites; the base of the exposed part as Tremadoc on the basis of trilobites (Fig. 8C.11).

Three lithofacies can be distinguished, as follows.

1. The most abundant lithofacies consists of calcilutite and related dolostone with flat lamination and small-scale crosslamination. These rocks probably were deposited from storm-generated nepheloid layers or turbidity currents of low density and velocity. Common pinching and swelling and boudinage structures are attributed to extension in a slope environment, local folds to slumping (Fig. 8C.12b).

2. Fairly abundant clast-supported conglomerates or breccias (Fig. 8C.12a), composed mainly of laminated carbonate rocks, and to a lesser extent of mudrock and chert, commonly are 0.5-4 m thick. The phenoclasts, ranging from pebble to boulder grade, are disc-shaped and frequently show imbricate structure. A complete spectrum of textural types, ranging from slightly displaced and rotated boudins to chaotic breccias and locally to upward fining breccias or conglomerates, is present. These textures demonstrate the transition from creep or sliding to subaqueous debris flows and turbidity currents and are characteristic of slope and proximal basin environments.

3. Thinly interstratified graptolitic mudrock and chert are confined to the upper 6-16 m. These strata are of basinal origin and more characteristic of the Ibbett Bay than the Canrobert Formation but are separated from the former by a carbonate conglomerate.

The **Ibbett Bay Formation** is about 1200 m thick and ranges in age from latest Early Ordovician (late Arenig) probably to earliest Middle Devonian (early Eifelian). It consists mainly of dark grey, commonly graptolitic mudrock with lesser proportions of thin bedded dark grey chert and argillaceous to sandy dolostone and minor amounts of carbonate breccia, similar to those in the Canrobert Formation (Fig. 8C.12e). The formation is divisible into five major units mappable throughout the Canrobert Hills (Fig. 8C.11). Slump structures of microscopic to large scale are common in all but the lowest unit; channelling was noted in unit 4 (Fig. 8C.12d). Generally, the Ibbett Bay Formation is characteristic of moderately distal basin environments.

In summary, the Canrobert and Ibbett Bay formations, combined, represent successive slope, proximal basin, and distal basin environments. They are comparable to the most proximal parts of the Hazen Formation in overall lithology and thickness, but not in age range and stratigraphy.

Upper Ordovician to Lower Devonian carbonates, mudrock and chert, central Ellesmere Island to eastern Melville Island (B3c)

Unit B3c consists of resedimented carbonates, predominantly of fine grain size, but ranging up to boulder grade; graptolitic mudrock, generally calcareous and/or dolomitic, with small proportions of radiolarian chert; and local occurrences of very fine grained sandstone. These sediments generally are dark grey and laminated, except for the carbonate conglomerates, which are light grey and commonly massive. They overlie shelf carbonates of different ages and are highly diachronous with an overall age range from Late Ordovician (Ashgill) to Early Devonian (Pragian). Most of the strata have been assigned to the Cape Phillips Formation. The name, Devon Island Formation, has been used for strata of Late Silurian – earliest Devonian age on northwestern Devon and southwestern Ellesmere islands; and the name, Kitson Formation, for strata of Early to early Middle Devonian age in the Kitson River area of northwestern Melville Island.

The predominant flat lamination and fine grain size suggest that most of these sediments were deposited from nepheloid layers, generated by storms in the adjacent shelf region. A relatively small proportion of coarser grained carbonate sediments was deposited by subaqueous debris flows and turbidity currents caused by slope failure.

Upper Ordovician to Lower Devonian predominantly siliciclastic sediments, northern and central Ellesmere Island (B4a, B4b)

The Danish River Formation covers all of the axial region and much of the marginal region in northern and central Ellesmere Island respectively (Fig. 8C.13). The stratigraphy in most of the axial region, which coincides largely with the Hazen Plateau, is uncertain because of complex folds and faults. The stratigraphy in the southwestern part of the axial region, however, is well represented by the type section, located north of Danish River on the east side of Cañon Fiord (Fig. 8C.14, section CF1), close to the Ordovician shelf margin. There the formation overlies unit B3b (chert member of Hazen Formation), is 2.74-2.8 km thick, and divisible into three members, informally referred to as members A, B, and C (Fig. 8C.14).

Member A, 1057 m thick, consists mainly of interstratified sandstone and mudrock with minor amounts of redeposited carbonate sediments, and ranges in age from late Llandovery probably to Ludlow. The base of the formation, however, appears to become older toward the northeast and may be Late Ordovician at St. Patrick Bay, Archer Fiord (see Fig. 8C.8, section SPB).

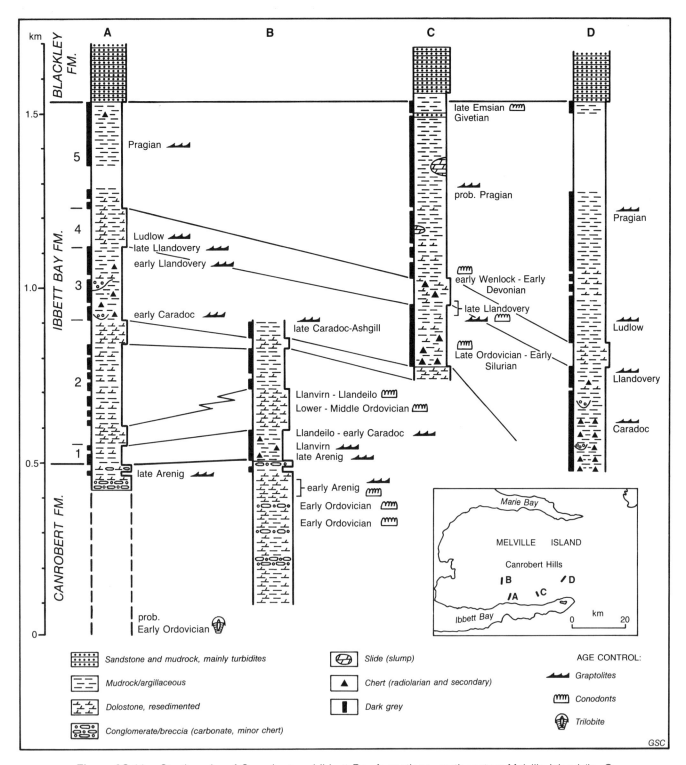

Figure 8C.11. Stratigraphy of Canrobert and Ibbett Bay formations, northwestern Melville Island (by Q. Goodbody).

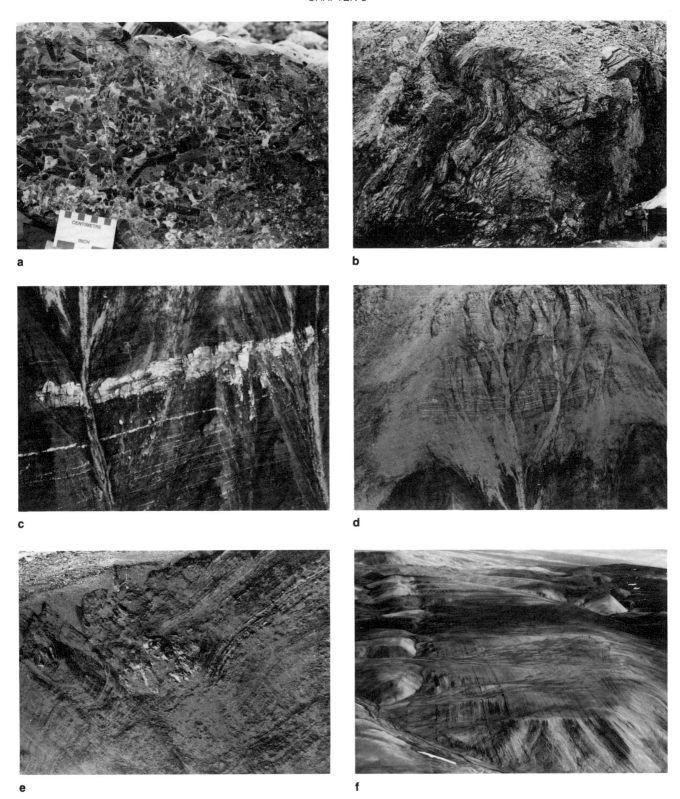

Figure 8C.12. Cambrian to Devonian deep water sediments in the Canrobert Hills, northwestern Melville Island (by Q. Goodbody). (a) Breccia in the Canrobert Formation; clasts of variably siliceous dolostone (light and dark grey) in matrix of silty dolostone. (b) Slump folds in thin bedded, argillaceous dolostones of Canrobert Formation. (c) Slumped dolostone in dark grey shale of unit 3 of the Ibbett Bay Formation. Stratigraphic interval shown is about 50 m thick. (d) Ibbett Bay Formation, contact between black shale of unit 3 and cherty dolostone of unit 4. Note lateral accretion surfaces in unit 4, due to migration of submarine channel. Stratigraphic interval shown is about 50 m thick. (e) Ibbett Bay Formation, unit 5; blocks of slumped dolostone in dark grey shale. (f) Contact between black shale of Ibbett Bay Formation, unit 5 (left, resistant) and siliciclastic turbidites of Blackley Formation (right, recessive, striped).

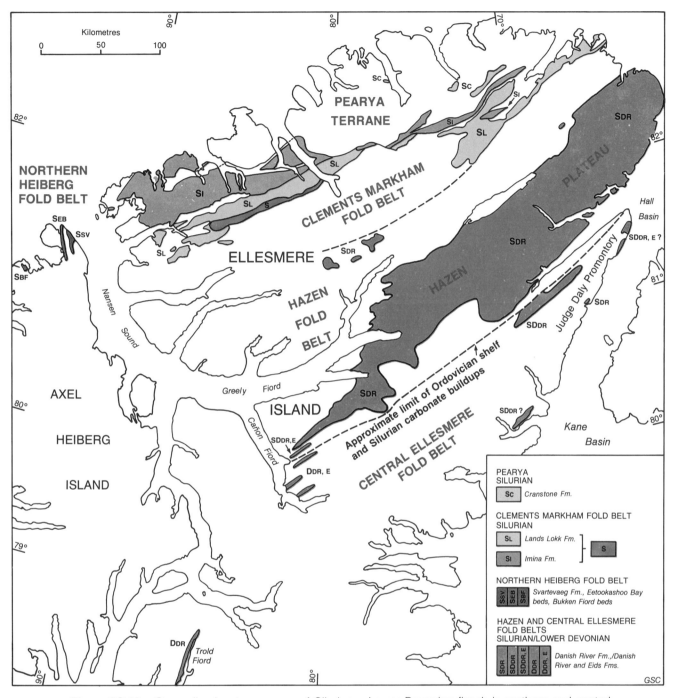

Figure 8C.13. Generalized outcrop areas of Silurian – Lower Devonian flysch in northern and central Ellesmere Island.

Member B, also known as the **Caledonian Bay Conglomerate Member,** is 41-101 m thick and consists of terrigenous conglomerate with lesser amounts of sandstone and mudrock. It probably is Ludlow in age and seems to be restricted to this area.

Member C, similar in lithology to member A, is 1637 m thick at the type section and ranges in age from Ludlow to early Lochkovian (earliest Devonian).

Directly south of the type section and farther south, only the upper part of member C is represented. There, member C lies on the Cape Phillips Formation (B3c), in turn overlying Silurian carbonate rocks. Member C, together with the Cape Phillips and Eids formations (B5), thus represents the marginal belt in this area. Member C in that region contains strata of early Lochkovian (earliest Devonian) age only, but includes strata of Pragian and early Emsian age, about 133 km to the southwest, at Trold Fiord.

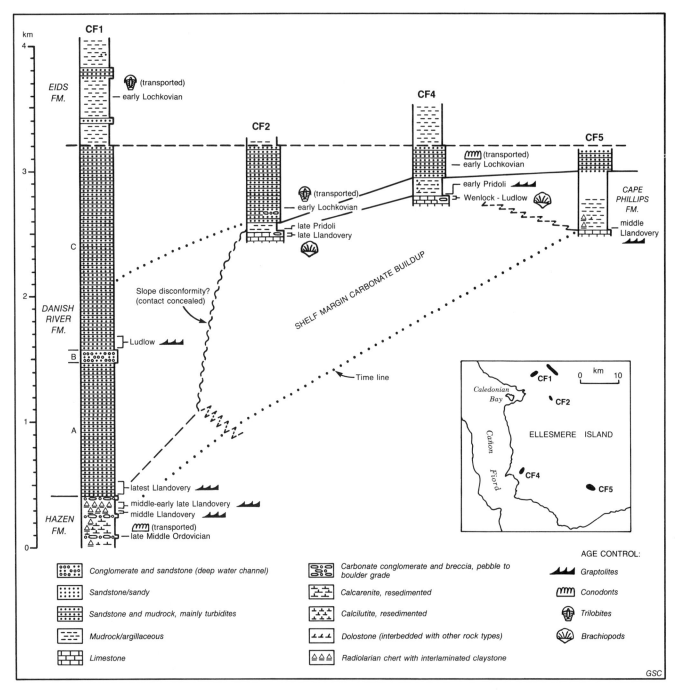

Figure 8C.14. Stratigraphy of the Danish River Formation east of Cañon Fiord.

Southeast of Ella Bay, the Danish River Formation lies directly on shelf carbonates of the lower Allen Bay Formation (Fig. 8C.26), but farther southeast on Judge Daly Promontory, it rests on the Cape Phillips Formation.

Sandstone and mudrock of the Danish River Formation (both in the axial and marginal belts) can be assigned to the following genetic assemblages:

1. Bouma sequences (Fig. 8C.15c, d) deposited from "classical" turbidity currents, a few centimetres to 1.5 m, commonly decimetres thick (abundant);

2. massive, mostly nongraded sandstones, up to 9 m thick and commonly containing mudrock clasts (Fig. 8C.15d, e, f), probably deposited from turbidity currents of high density (common);

3. nongraded, flat-laminated or crosslaminated sandstones, probably deposited by contour currents (rare);

4. undifferentiated mudrock units, up to about 60 m thick, deposited by turbidity currents, contour currents, and nepheloid layers (common).

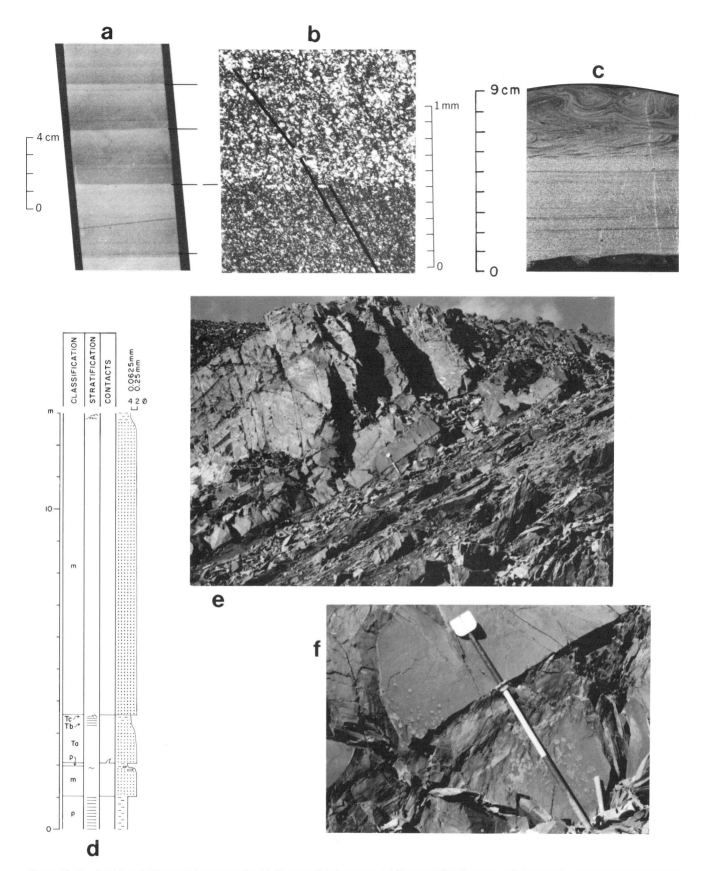

Figure 8C.15. Turbidites of different thickness in the Danish River and Eids formations. (a) Siltstone of Eids Formation at Caledonian Bay, showing plane lamination and repetitive graded bedding; upward increase in darkness is due to increase in carbon content. (b) Photomicrograph showing abrupt basal contact of graded layer; dark lines indicate slaty cleavage. (c) Turbidite in Danish River Formation, member C, composed mainly of divisions B and C (Tb, Tc) of the Bouma model. (d) Graphic log of turbidites and massive sandstone bed near base of Danish River Formation on southwestern Judge Daly Promontory; for legend see Fig. 8C.24. (e) Photograph of same section; massive sandstone bed, 9.2 m thick, forms cliff. (f) Abrupt lower contact of massive bed (1 m above base of 1.5 m long measuring staff).

Extensive X-ray diffraction and thin section studies reveal a surprising compositional homogeneity. The silicate fraction of the sandstones consists mainly of quartz (mean content about 52%) with lesser proportions of feldspar (albite 8%, K-feldspar 4%), phyllosilicates (muscovite, chlorite and biotite combined 4%), lowgrade metamorphic rock fragments (3%), and chert (1%). The balance (about 28%) consists of carbonate grains that are intricately mixed with the silicates.

Paleocurrent determinations, made on sole marks of sandstones (mainly flute marks, [Fig. 8C.16], grooves and ridges, and tool marks) indicate predominantly strike-parallel transport from northeast to southwest (Fig. 8C.17), but with two significant exceptions. Transverse transport in predominantly southeastward directions is apparent from more than 600 (out of a total of 2600) determinations in the Hazen Fold Belt (Fig. 8C.17, inset A). The beds here are not well dated, but tentatively inferred to be Llandovery – Wenlock in age. Transverse transport in westward directions is evident in Pragian – lower Emsian strata at Trold Fiord (inset D). The significance of these determinations is discussed in the concluding section of this chapter.

Terrigenous conglomerate, as far as known, is restricted to Member B at Caledonian Bay, exposed along strike for 13 km. It is mainly of pebble grade but includes terrigenous cobbles and boulders up to 60 cm in diameter, as well as large rip-up clasts and slumped blocks of sandstone. The conglomerate occurs in massive, lenticular bodies and probably was deposited in a submarine channel. The phenoclasts consist mainly of chert and to a lesser extent of quartzite and carbonate rocks.

The resedimented carbonates range from mud to boulder grade, and show evidence of deposition by turbidity currents and subaqueous debris flows. Restricted to the southeastern margin of the basin, they obviously were derived from the adjacent shelf. They differ from the carbonate particles in the sandstones and mudrocks discussed above by the common occurrence of skeletal material.

Lower Devonian, predominantly siliciclastic sediments, Bathurst Island (B4c)

Unit B4c comprises the Lochkovian-Pragian Bathurst Island Formation and the Emsian Stuart Bay Formation (Kerr, 1974). The former is up to 1862 m thick and consists mainly of turbidites of mudrock and very fine grained sandstone with minor amounts of limestone. The Stuart Bay Formation overlies the Bathurst Island unconformably near the Boothia Uplift, but conformably farther west. It is lithologically similar to the Bathurst Island Formation but separated from it by a basal pebble conglomerate. Slide blocks of richly fossiliferous carbonate strata, up to 10 m across, were derived from bioherms on the western flank of the Boothia Uplift (Polan and Stearn, 1984) and most other sediments in the two formations probably also came from the uplift.

Lower-Middle Devonian transitional sediments, Ellesmere and Bathurst islands (B5)

The transitional sediments are assigned to the Eids Formation, which lies between deep water formations (Danish River or Cape Phillips formations in central Ellesmere Island, Stuart Bay Formation in Bathurst Island) and shelf deposits (Red Canyon River, Vendom Fiord or Blue Fiord formations in Ellesmere Island, Blue Fiord Formation in Bathurst Island). As presently mapped, the Eids Formation is early Lochkovian to early Emsian in age and diachronous in central Ellesmere Island, and late Emsian – early Eifelian in Bathurst Island (Kerr, 1974).

On Ellesmere Island, where the unit is more than 1.15 km thick locally, it consists mainly of flat-laminated silty mudrock with rather uniform mineral composition, containing about one quarter to one third calcite and dolomite. The colour — hues of olive, green or grey — is lighter than that of the Cape Phillips Formation (unit B3c), reflecting slightly more oxidizing conditions. Rare tentaculitids, conodonts, and scelodonts are the only indigenous fossils. Associated other rock types indicate a spectrum of depositional environments. Submarine fan environments evidently existed at Caledonian Bay, Cañon Fiord, where sandy and silty turbidites occur (Fig. 8C.14, section CF1; Fig. 8C. 15a, 15b). Farther south in west-central Ellesmere Island, subtidal shelf conditions are indicated by richly fossiliferous, thin bedded impure limestones — mainly skeletal wackestone and lime mudstone, with smaller proportions of skeletal grainstone and packstone, and coralline boundstone.

On Bathurst Island (Kerr, 1974), the formation is between 330 and 600 m thick and typically composed of thin bedded calcareous mudrock and argillaceous limestone. An eastern facies, transitional with the Disappointment Bay Formation, consists of thin bedded impure limestone and calcareous mudrock.

Sedimentary-volcanic subprovince
Introduction

The sedimentary-volcanic subprovince is exposed in the Clements Markham Fold Belt of northern Ellesmere Island

Figure 8C.16. Large flute marks in the Danish River Formation (lower part of member C) at Caledonian Bay. Transport was from upper left to lower right. Four foot (1.2 m) staff for scale.

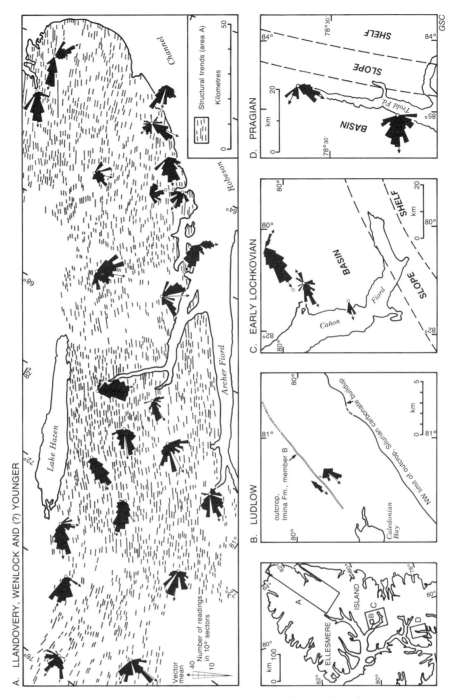

Figure 8C.17. Paleocurrent directions in the Danish River Formation.

and the Northern Heiberg Fold Belt of Axel Heiberg Island. The two belts differ in structural trend and aspects of their stratigraphy, and therefore are treated separately although they have some units in common.

Clements Markham Fold Belt

Proterozoic and/or lower Paleozoic volcanic and sedimentary rocks

The possibly oldest exposed strata in the Clements Markham Fold Belt are basic metavolcanics and associated sills and metasediments assigned to the Yelverton assemblage. The unit underlies a 93 km long and up to 8 km wide area extending from south of Phillips Inlet to the head of Yelverton Inlet (Fig. 8C.3). The Yelverton assemblage has been thrust over the Lower Cambrian Grant Land Formation on the south (Fig. 8C.18) and is in fault contact with Silurian flysch (Lands Lokk Formation) and Carboniferous strata in the north. Stratigraphic top and bottom are not preserved. The total thickness of the unit has not been established because of its structural complexity, but is estimated to be in excess of 1 km; about 464 m of strata have been measured in detail.

Sedimentary units range in thickness from a few to perhaps a few hundred metres. They are composed of weakly to distinctly metamorphosed limestone and dolostone (or marble); argillaceous sediments metamorphosed to slate, phyllite or schist; and lesser proportions of chert. Lime mudstone is predominant among the carbonate rocks but oolitic grainstone also has been observed. The sedimentary strata appear to be barren of fossils (including conodonts and palynomorphs).

The volcanic rocks, forming units up to several hundred metres thick, comprise tuffs and flows (some with pillow structure; Fig. 8C.19) and have sills associated with them. Greenschist-grade metamorphism is incomplete in much of the unit, as remnants of clinopyroxene commonly are preserved. Stable trace elements indicate that most rocks are in the boundary field of andesite and subalkaline basalt (Fig. 8C.20a). The highly problematic age and origin (cf.

Figure 8C.19. Pillow basalt in Yelverton assemblage.

Fig. 8C.21) of the assemblage are discussed below (under Summary and interpretation), along with that of the related(?) Jaeger Lake assemblage of northern Axel Heiberg Island.

Lower Cambrian siliciclastic sediments (B2)

Strata comparable to the Grant Land Formation of Ellesmere Island and the Polkorridoren Group of North Greenland constitute a thrust sheet south of Phillips Inlet (Fig. 8C.3) that lies upon upper Paleozoic strata and, in turn, is overthrust by the Yelverton assemblage (Fig. 8C.18). The formation consists mainly of quartzite and grey or green, rarely purplish-red slate or phyllite with minor amounts of fine pebble conglomerate.

Lower Paleozoic chert, deep water carbonates, and mudrock (equivalents of B3a, B3b)

Strata comparable, to some extent, to the Hazen Formation of Ellesmere Island and the combined Vølvedahl and Amundsen Land groups of northeastern Greenland are exposed in a belt of nunataks, 50 km long and up to 10 km wide, which is southwest of the head of Yelverton Inlet (Fig. 8C.7, "Hazen equivalents"). They consist mainly of secondary chert with relicts of flat-laminated mudrock, calcilutite, and calcareous turbidites of sand to pebble grade. The relatively coarse grade of the turbidites, which also contain medium grained quartz sand, shows that they were probably not derived from the carbonate shelf of central Ellesmere Island, and this poses a tectonic problem (Chapter 12B). Outcrops of green and purple chert in this belt are tentatively interpreted as chertified mudrocks of the underlying uppermost part of the Grant Land Formation (or Polkorridoren Group).

Upper Middle Ordovician – Lower Silurian volcanics, carbonates, and chert

The informal name, Mount Rawlinson assemblage (unit MR, Fig. 8C.7), has been applied to a fault slice of intercalated volcanic and sedimentary rocks in the northeastern part of

Figure 8C.18. Thrust fault placing Yelverton assemblage (dark) upon Lower Cambrian Grant Land Formation (light).

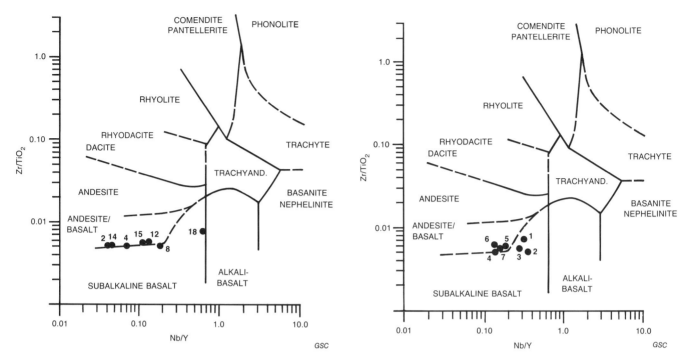

Figure 8C.20. Stable trace element ratios and rock classification (after Winchester and Floyd, 1977, Fig. 6) for: (a) Yelverton assemblage, (b) Jaeger Lake assemblage.

the Clements Markham Fold Belt that lies structurally between Lands Lokk and Imina formations. The volcanic rocks are intensely altered tuffs and less abundant flows, classified as andesite, rhyodacite/dacite and trachyandesite on the basis of stable trace elements (Fig. 8C.22). The sediments include impure marble and radiolarian chert. A U-Pb analysis on zircon from a rhyolite in the lower part of the unit yielded an age of 454.7 + 9.7/-4.6 Ma (Trettin et al., 1987), Llandeilo according to current time scales (e.g. Snelling, 1987). Volcanic and carbonate rocks of comparable lithology (Fig. 8C.22, triangles), occurring 120 km to the southwest (Fig. 8C.7), are referred to as the Kulutingwak Fiord assemblage. Conodonts from a pelmatozoan limestone in this unit are Early Silurian, probably early or middle Llandovery in age. The chemical composition and predominantly pyroclastic nature of the volcanics in both assemblages suggest that they originated in an arc that may have been affected by extension. If the two assemblages represent the same arc, then it had a southwesterly extent of no less than 200 km and an overall age range from Llandeilo (or older) to Llandovery.

An isolated occurrence of a 100-200 m thick, richly fossiliferous but recrystallized limestone of middle or late Llandovery age, exposed 39 km southeast of the Kulutingwak Fiord assemblage at Yelverton Pass (unit YP in Fig. 8C.7), may represent a carbonate buildup developed upon the arc, but its base is not exposed.

Silurian clastic sediments and volcanics (equivalents of B4a)

Imina Formation

The Imina Formation (unit Si in Fig. 8C.13), a thick(?) succession of interbedded sandstone, mudrock and minor fine grained pebble conglomerate, seems to lie stratigraphically between the Kulutingwak Fiord assemblage and the Lands Lokk Formation, but the contact with the former is faulted. The sandstone is very fine to very coarse grained and occurs as massive beds or in Bouma sequences, together with mudrock and conglomerate. Combined X-ray diffraction and point count analyses indicate compositions very similar to those of the Danish River

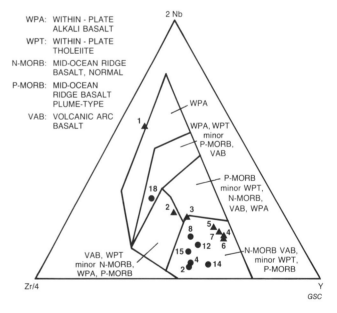

Figure 8C.21. Tectonic environments of mafic rocks in Yelverton and Jaeger Lake assemblages according to stable trace elements (after Meschede, 1986, Fig. 1).

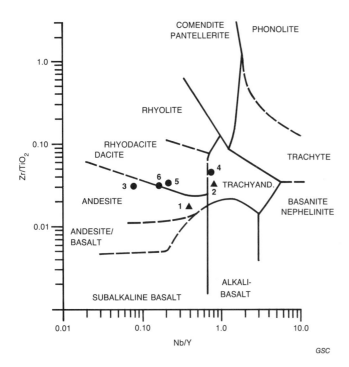

Figure 8C.22. Stable trace element ratios of Mount Rawlinson assemblage (circles) and Kulutingwak Fiord assemblage (triangles); classification after Winchester and Floyd, 1977.

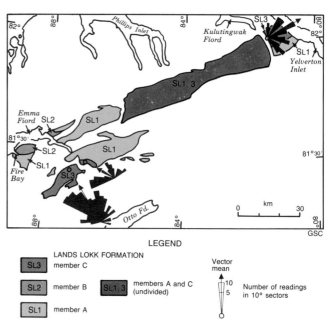

Figure 8C.23. Members of the Lands Lokk Formation in northwestern Ellesmere Island and paleocurrent determinations.

Formation (see above). Silicates, mainly quartz (mean content about 51%) with lesser amounts of feldspar (albite 13%, K-feldspar 3%), phyllosilicates and related metamorphic fragments (10%), and chert (3%) are intricately mixed with carbonate grains (20%). The thickness of the formation has not been determined because of tight folding and complex faulting. Generally of sub-greenschist grade, it is metamorphosed to greenschist or amphibolite grade near Phillips and Yelverton inlets (see Chapter 12B). If the unit indeed overlies the Kulutingwak Fiord assemblage (the latter may represent a different fault slice) its base can be no older than early Llandovery. The top probably is late Llandovery in age on the basis of poorly preserved graptolites.

Lands Lokk Formation

The Lands Lokk Formation underlies much of the Clements Markham Fold Belt (unit SL in Fig. 8C.13). In the western part of the fold belt it is divisible into three members, informally named members A, B and C (Fig. 8C.23). Member B is restricted to the westernmost part; members A and C have not been distinguished in the eastern half of the belt.

Member A probably is more than 1.5 km thick and consists mainly of medium to dark grey slaty mudrock with minor proportions of submarine tuff and tuffaceous sediments of siliceous to intermediate composition. Graptolites indicate an age range from late Llandovery or Wenlock to early Ludlow. The unit probably was deposited in basin plain environments.

Member B consists of perhaps 300 m of volcanic-derived sediments and volcanics with some carbonate olistostromes. The volcanics consist chiefly of tuff and agglomerate with lesser amounts of flow rocks. They are of siliceous to intermediate composition and conform mineralogically (but not texturally) with the greenschist facies. Two analyzed specimens of flow and tuff are classified as calc-alkaline rhyolite and calc-alkaline dacite, respectively, on the basis of major-element composition. The volcanic-derived sediments, of similar composition, range in size grade from mudrock to conglomerate and probably were deposited by sediment gravity flows. The volcanics and volcanic-derived sediments diminish toward the northeast so that member B grades into the upper part of member A less than 10 km northeast of the head of Emma Fiord. These relationships suggest that at least part of the member is early Ludlow in age[1].

Member C is composed of mudrock, quartzite, and minor pebble conglomerate with flysch-like primary structures. It overlies member A in several areas but graptolite collections indicate an age range from late Llandovery and/or Wenlock to early Ludlow so that much of the unit is a facies equivalent of member A.

Three facies are tentatively distinguished within the member. The **southwestern facies,** which is more than 1 km thick in the area southeast of Emma Fiord, is characterized by the presence of pebble conglomerate and coarse grained quartzite, in addition to finer grained quartzite and mudrock. Two sedimentary assemblages are distinguished in a roughly 800 m thick section south of the

---

[1] Conodonts from the olistostromes, obtained in 1988, are early Middle Ordovician (Whiterockian) in age (G.S. Nowlan, pers. comm., 1989). Member B and some adjacent strata of member A are now interpreted as a faultblock of Middle Ordovician rocks, broadly comparable in age and lithology to the Mount Rawlinson assemblage of northeastern Ellesmere Island.

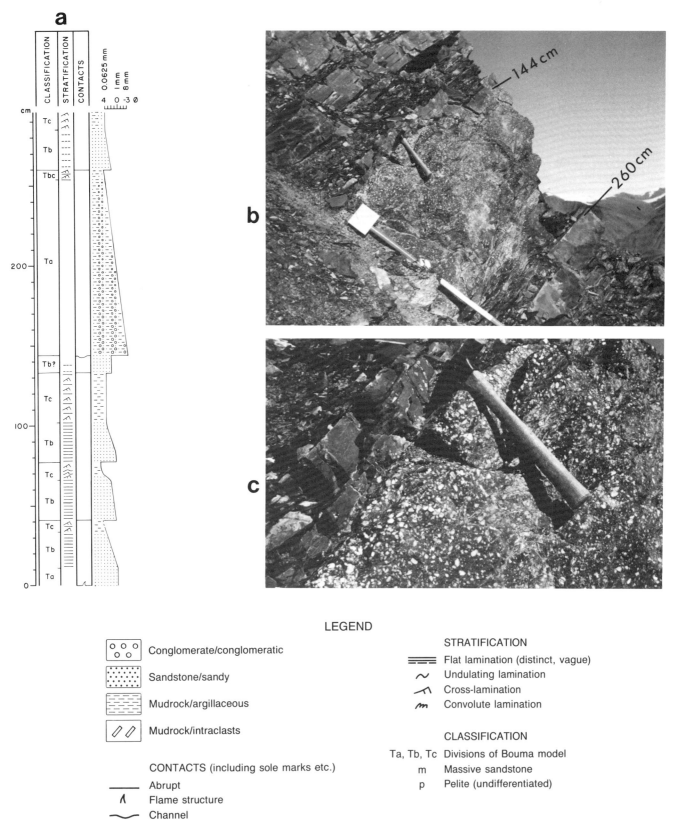

Figure 8C.24. Lands Lokk Formation, member C, southeast of Phillips Inlet. (a) Detailed log of turbidites. (b) Graded pebble conglomerate with underlying and overlying sedimentation units (section is upside down). (c) Close-up of base of pebble conglomerate.

head of Emma Fiord: (1) thinly interstratified mudrock and sandstone, commonly forming Bouma sequences, a few centimetres to a few decimetres in thickness; and (2) units composed of sandstone, 0.8-13.5 m thick, with or without small proportions of mudrock and fine pebble conglomerate. The sandstones occur in Bouma sequences, commonly decimetres in thickness, and as thick structureless beds that show an upward decrease in grain size only in the uppermost part. Individual sandstone beds commonly have undulating parting surfaces at the base that indicate minor channels, probably within broader channels of the mid-fan environment. Comparable assemblages are present southeast of Phillips Inlet (Fig. 8C.24). The **central facies** is poorly known but seems to consist of mudrock and quartzite only. In the **northeastern facies** minor amounts of green and red mudrock and very small amounts of carbonate-pebble conglomerate (one composed of resedimented oncoids) are present in addition to the predominant mudrock and very fine or fine grained quartzite.

The conglomerates and quartzites are composed mainly of quartz, quartzite and chert (commonly with "ghosts" of radiolarians), which together make up about 80-90% of the rock volume, the proportion of chert increasing with grain size; the remainder consists of feldspar, mica, chlorite, metamorphic rock fragments, and carbonates. Paleocurrent determinations show northwestward transport in the area south and southeast of Emma Fiord, and northeastward transport in the area southwest of Yelverton Inlet (Fig. 8C.23).

Regional relationships

The Imina Formation is similar in mineral composition both to the Danish River Formation of Ellesmere Island, and to the Merqujôq Formation of North Greenland (cf. Chapter 7), but coarser grained than the former. Its age (middle-late Llandovery) is closer to the age span of the Merqujôq Formation (Late Ordovician – Llandovery) than to that of the Danish River Formation (Late Ordovician or Early Silurian to Lower Devonian). The Lands Lokk Formation is comparable to the combined Lauge Koch and Nordkronen formations of northeastern Greenland with respect to composition (quartz and chert abundant, carbonate absent or subordinate), relatively coarse grade, and age (late Llandovery or Wenlock to Ludlow), but markedly different from the bulk of the Danish River Formation, except for member B at Caledonian Bay. The implications of this relationship are considered in Chapter 12B in the context of Late Silurian(?) sinistral strike slip.

Upper Silurian carbonate and siliciclastic sediments

Two different carbonate units of Late Silurian age are present in the northeastern part of the Clements Markham Fold Belt (Fig. 12B.2), that appear to be correlative with the upper part of the Marvin Formation of the Pearya Terrane (Chapter 9).

1. The **Markham River beds,** exposed southwest of Clements Markham Inlet overlie dark grey mudrock of the

Figure 8C.25. Unconformity between Eetookashoo Bay beds (SE) and cliff-forming basal conglomerate of Stallworthy Formation (Ds) on northern Axel Heiberg Island; view to the northeast. The angular discordance appears to be small because the exposures are subparallel with strike.

Lands Lokk Formation with seemingly abrupt, probably disconformable contact. The unit has a minimum thickness of 166 m; its top is not exposed. It has yielded ostracodes of unspecified late Llandovery to late Ludlow age, but directly underlying strata of the Lands Lokk Formation have yielded palynomorphs of probable Ludlow age so that the Markham River beds are probably confined to the Ludlow. The Markham River beds consist mainly of a variety of dolostones, including "quiet-water oolite" (cf. Flügel, 1982, p. 146) and bacterial(?) oncolite with lesser proportions of cherty and dolomitic spiculite, mudrock, sandstone, and conglomerate. Seven conglomerate beds, 0.2-2 m thick, are mainly of pebble grade but include cobbles up to 10 cm long in the lowermost unit. The clasts are composed of siliceous materials such as quartzite, vein quartz, chert, and chalcedony and of dolomitic rocks with sponge spicules and "quiet-water ooids". It appears that the Markham River beds were deposited in shallow marine to peritidal settings adjacent to an uplifted region where similar carbonate rocks and quartzite were eroded. Disconformities probably are present within the formation, especially in its lowermost part.

2. The **Piper Pass beds** form a 47 m long, narrow outcrop belt in the southeasternmost part of the Clements Markham Fold Belt that is in fault contact with the Lands Lokk and Grant Land formations. The unit consists mainly of medium dark grey, commonly laminated lime mudstone, but also includes smaller proportions of calcareous mudrock, dolostone and skeletal lime wackestone and packstone with a diverse benthonic fauna. The lime mudstones and mudrocks are interpreted as basinal deposits and the packstones and wackestones as subaqueous debris flows, derived from a nearby shelf margin. The strata have yielded conodonts of late Ludlow (Zone of *latialata*) and slightly older age.

Northern Heiberg Fold Belt

Proterozoic and/or lower Paleozoic volcanic and sedimentary rocks

The Jaeger Lake assemblage, possibly the oldest exposed unit in the Northern Heiberg Fold Belt, occurs as small fault blocks and as sheets that have been thrust on the Lower Cambrian Grant Land Formation. A 425 m thick partial section at Greenstone Lake (Fig. 8C.3, loc. GL) comprises nearly equal proportions of volcanic and carbonate rocks, occurring in three units each. The volcanic rocks are flows and tuffs of mafic composition (Fig. 8C.20b), metamorphosed in the greenschist facies, and pervasively altered by calcite and quartz. Age and origin of this enigmatic unit are discussed below (under Summary and interpretation), together with that of the Yelverton assemblage.

Lower Cambrian and/or older carbonates (B1?)

This unit, informally named Aurland Fiord beds, consists of carbonate strata that also occur as fault blocks and as thrust sheets, lying on the Grant Land and Hazen formations. The unit has a minimum thickness of 1.3 m southwest of Aurland Fiord (Fig. 8C.3, loc. AF), and consists mainly of flat-laminated dolostone and related intraformational conglomerate with small proportions of nonlaminated dolostone; microcrystalline limestone; dolomitic grainstone, pisolite, and solution(?) breccia; and slate. The Aurland Fiord beds have yielded no fossils but are tentatively regarded as earliest Cambrian (or older) in age because they have been thrust over the Grant Land Formation. If they are Early Cambrian in age and underlie the Grant Land Formation stratigraphically, they probably represent a facies intermediate in water depth between the shelf sediments of the Ella Bay Formation of central Ellesmere Island and the correlative deep water carbonate sediments of the southern Grantland Mountains (Nesmith beds; B1).

Lower Cambrian siliciclastic sediments (B2)

Quartzite and multicoloured phyllite and slate with minor amounts of fine pebble conglomerate, widely exposed in northern Axel Heiberg Island, can be assigned to the Grant Land Formation (unit B2) on the basis of macroscopic and microscopic characteristics. The thickness and internal stratigraphy of the unit are unknown because of its structural complexity.

Lower Cambrian – Lower Silurian carbonate, mudrock, and chert (B3a, B3b)

In a syncline south of Aurland Fiord (Fig. 8C.7, loc. AFS), the Grant Land Formation is overlain by more than 60 m of carbonate and minor mudrock that must represent the basal part of the carbonate member of the Hazen Formation if the concealed contact between the two units is not faulted. The carbonate rocks include: (1) laminated, variably argillaceous dolostone; (2) graded dolosiltite and dolarenite; and (3) dolomitic intraformational conglomerate that originated by sliding of carbonate boudins in a submarine slope environment. Thin bedded dark grey chert, exposed in several, generally small areas, probably represents the chert member of the Hazen Formation. No fossils have been found in the Hazen Formation of Axel Heiberg Island.

Silurian clastic sediments and volcanics (equivalents of B4a)

Three different facies belts are represented by the Silurian strata of northern Axel Heiberg Island, referred to as the Svartevaeg Formation, Eetookashoo Bay beds, and Bukken Fiord beds (Fig. 8C.13, units SSV, SEB, SBF, the last two previously assigned to the Lands Lokk Formation (Trettin, 1969b).

Eetookashoo Bay beds

The Eetookashoo Bay beds are overthrust on the southwest by the Aurland Fiord beds and unconformably overlain on the northeast by the Stallworthy Formation (Fig. 8C.25). The unit consists mainly of medium to dark grey slaty mudrock, and lesser proportions of lithic and tuffaceous sandstone and tuff, with at least one carbonate conglomerate of pebble to cobble grade. The volcanic fraction in the tuffs and sandstones comprises rock fragments of intermediate to siliceous composition, feldspar, chlorite, and volcanic quartz; the sedimentary fraction of the sandstones consists mainly of quartz, chert with "ghosts" of radiolarians, and carbonate fragments. The presence of graded bedding indicates that these strata were deposited by sediment gravity flows (including turbidites); the carbonate conglomerate appears to be a subaqueous debris flow. Benthonic fossils in the carbonate clasts are of Llandovery or Wenlock, perhaps late Llandovery, age. Graptolites near the top of the unit are probably late Wenlock in age.

Bukken Fiord beds

This facies is exposed in three small outcrop areas southwest of Aurland Fiord, where it is in fault contact with the Hazen Formation. It consists mainly of sandstone with lesser amounts of slate and small proportions of pebble conglomerate. Sandstone and conglomerate occur both as thick massive beds, and, associated with mudrock, in Bouma sequences. The sandstone is composed of volcanic material of intermediate to siliceous composition and of quartz and chert, the latter reminiscent of the Hazen Formation with respect to lamination and "ghosts" of radiolarians. No fossils have been found in this area.

Svartevaeg Formation

The Svartevaeg Formation overlies the Stallworthy Formation in a structural sense. The concealed boundary between these two units, earlier interpreted as a normal stratigraphic contact, is now regarded as a thrust fault. The Svartevaeg Formation is overlain by the Viséan (upper Lower Carboniferous) Emma Fiord Formation with angular unconformity.

The Svartevaeg Formation consists largely of sediment gravity flows that are composed mainly of volcanogenic (tuffaceous?) sandstone with lesser amounts of siltstone and breccia or conglomerate; tuff and volcanic flows are a minor component. The proportion of volcanic rocks and mudrock versus sandstone is greater in the lower part (member A) than in the upper (member B). The sandstones consist mostly of volcanic rock fragments of siliceous to mafic, predominantly intermediate (andesitic) composition with lesser proportions of feldspar, carbonate, and chlorite. A chemically analyzed tuff from member A is classified as calc-alkaline andesite, and a flow from the upper part of member B as calc-alkaline basalt, both on the basis of major elements. The plagioclase in the sedimentary rocks studied and in the andesitic tuff mentioned (but not in the basalt) has been altered to albite to a large extent.

The formation contains several limestone breccias with benthonic fossils. The most diagnostic faunule, from a large slide block of limestone about 60 m above the base of the exposures, is late Llandovery or Wenlock in age.

Lower Devonian siliciclastic sediments

The Stallworthy Formation, divided into three informal members, comprises roughly 4 km of predominantly clastic sediments that overlie the Eetookashoo Bay beds with angular unconformity (Fig. 8C.25). Its stratigraphic top is not preserved.

Member A, 0-900 m thick, has an 11 m thick massive chert-pebble conglomerate at the base, which is overlain by about 66 m of interbedded pebble conglomerate and sandstone. The balance consists of predominantly red weathering mudrock and minor sandstone.

Member B is about 75-350 m thick and consists of interstratified conglomerate or breccia and sandstone with lesser amounts of mudrock. The conglomerates and breccias are mainly of pebble grade but include cobbles. Medium-scale crossbedding is common in the sandstones.

Member C, roughly 2450 m thick, is composed mainly of red weathering, laminated to thick bedded, coarse grained mudrock, alternating with finer grained greenish grey and dark grey weathering mudrock. Associated pebble conglomerate and partly crossbedded sandstone, tuffaceous sediments, and argillaceous limestone are minor components. The unit contains brachiopods, gastropods, pelecypods and rare ostracoderms.

The conglomerates and breccias of the Stallworthy Formation consist mainly of chert, and the sandstones of variable proportions of quartz and chert with very small amounts of feldspar, muscovite and argillaceous rock clasts. Carbonate clasts and volcanic materials (volcanic rocks fragments, feldspar, chlorite) are present only in some sandstone and conglomerate beds in the upper part of member C.

The Stallworthy Formation is interpreted as a syntectonic to post-tectonic unit derived from an upland to the west-southwest in which chert (Hazen Formation) and quartzose and feldspathic sandstone (Grant Land Formation) were exposed. Members A and B were deposited in alluvial fan and braided river environments; member C was laid down in brackish delta environments.

The only age-diagnostic fossils found so far are ostracoderms from a limestone about 1 km below the top of the formation. They are late Lochkovian or early Pragian (mid-Early Devonian) in age (R. Thorsteinsson and H.P. Schultze, pers. comm., 1985).

Summary and interpretation

Enigmatic volcanic-sedimentary assemblages of Proterozoic or early Paleozoic age

Intercalated volcanic and sedimentary strata that have been thrust over the Grant Land Formation, both in the Northern Heiberg Fold Belt (Jaeger Lake assemblage) and in the western part of the Clements Markham Fold Belt (Yelverton assemblage), may represent the oldest units of the deep water basin, but are highly problematic with respect to age and origin. Structural relationships would imply an earliest Cambrian or Late(?) Proterozoic age if these assemblages belonged to the same stratigraphic succession as the Grant Land Formation, but an early Paleozoic age is possible if they are exotic. The total absence of fossils may indicate an age older than Silurian.

The reliability of chemical analyses for the interpretation of such highly altered rocks is questionable, but the fact that the Zr/TiO_2 versus Nb/Y ratios (Fig. 8C.20a, b) of all samples cluster in the fields of basaltic andesite and subalkaline basalt may indicate that these trace elements have not been seriously affected by alteration. Ratios of Zr, Nb, and Y (Fig. 8C.21) permit three different settings — volcanic arc, within-plate, and mid-ocean ridge — but associated shallow marine carbonate rocks are incompatible with the mid-ocean ridge environment.

Combining all available information, the two most probable interpretations for the Yelverton and Jaeger Lake assemblages are: either an island arc suite of Late Proterozoic and/or early Paleozoic age, possibly exotic with respect to the Grant Land Formation; or a rift-related suite of Late Proterozoic or earliest Cambrian age, possibly related to the Upper Proterozoic Natkusiak Formation of Victoria Island (Chapter 6). The first interpretation is more compatible with the bulk of the chemical analyses (the few deviating analyses would indicate rifting episodes in the development of the arc or arcs); the second would simplify the regional stratigraphy and tectonic history.

Cambrian to Middle Devonian depositional phases of the sedimentary subprovince

Five major phases, with some subdivisions, are distinguished in the development of the sedimentary subprovince of the deep water basin, that correspond to the first-order lithological units described above. Comparable deposits occur in the sedimentary-volcanic subprovince, but other rock types, such as volcanics and shallow marine and nonmarine sediments also are present that indicate a different tectonic history for that region. Regional facies patterns are shown in Figures 8B.27-8B.37. and tectonic developments during this interval are discussed in Chapters 4, 12, and 21.

Phase BI — Early Cambrian: resedimentation of carbonates

The stratigraphic record of the first phase is restricted mainly to the sedimentary subprovince of North Greenland and northern Ellesmere Island, although poorly dated carbonate rocks in northern Axel Heiberg Island may also be related to it. The record indicates redeposition of shelf-derived carbonate and minor siliciclastic sediments in deep water settings (Fig. 8B.27). The strata probably are correlative with shelf carbonates in central Ellesmere Island and North Greenland (Ella Bay Formation and Paradisfjeld Group; cf. Chapter 8B and Chapter 7, stage 2).

Phase BII — middle to late Early Cambrian: deposition of craton-derived siliciclastic sediments

Deposits of this phase are exposed in North Greenland, and in northern Ellesmere and Axel Heiberg islands where they occur in both subprovinces. They consist of roughly 2 km of variably feldspathic quartzite, mudrock, and minor pebble conglomerate, deposited in submarine fan environments. The sediments were derived from the cratonic interior and are correlative with a cratonward-tapering wedge of siliciclastic sediments in the Shelf Province (Fig. 8B.28; Chapter 8B, Ellesmere Group and correlative units [P2a]; Chapter 7, stage 3).

Phase BIII — late Early Cambrian to Early Devonian (diachronous): starved-basin type sedimentation

Sedimentary subprovince. Deposits of this phase (Fig. 8B.30-8B.37), exposed throughout the deep water region of North Greenland and the Arctic Archipelago, consist of variable proportions of: 1. fine grained siliciclastic sediments (predominantly mudrock with minor proportions of very fine grained sandstone) of mainly cratonic provenance; 2. shelf-derived carbonates; and 3. radiolarian and secondary chert. The thickness of the sediments decreases markedly, and the proportion of radiolarian chert increases, with distance from the shelf margin. Evidently, "starved-basin" conditions existed in the northern part of the sedimentary subprovince where the entire uppermost Lower Cambrian to lowermost Silurian succession (units 3a and 3b combined) is only 250 m thick — 5% of the thickness of correlative shelf margin deposits.

The base of these deposits ascends in age toward the southeast and south because of the diachronous downwarping of outer parts of the shelf, mainly in Late Ordovician to Llandovery time. The top ascends in age both along the axis of the deep water basin in a southwestward direction, and perpendicular to it in a cratonward direction (see Phase BIV).

Sedimentary-volcanic subprovince. Units B3a and B3b (Hazen Formation) are recognized in northern Axel Heiberg Island. A comparable, largely chertified unit in northwestern Ellesmere Island differs from strata of the Hazen Formation farther south in Ellesmere Island (sedimentary subprovince) by the presence of coarse grained quartz and of carbonate pebbles and appears to be "out of place". Uncertain, also, is the significance of arc-type volcanics with associated chert and carbonate sediments in northeastern and northwestern Ellesmere Island.

Phase BIV — Late Ordovician to Middle Devonian (diachronous): flysch deposition

Sedimentary subprovince. This phase is represented by extensive deposits of sandstone, mudrock and minor terrigenous conglomerate with small proportions of shelf-derived carbonates, intermittently exposed from northeastern Greenland to northwestern Melville Island (Fig. 8B.34-8B.37). Primary structures and lithological associations indicate transportation by sediment gravity flows, predominantly turbidity currents, in submarine fan and basin plain environments. These synorogenic deposits can be designated as flysch (cf. Hsü, 1970).

In the axial region of the basin, the flysch prograded in a southwestward direction, as the age of its base decreases from Late Ordovician or Early Silurian in northeastern Greenland to late Early Devonian in northwestern Melville Island. This conclusion is supported by the predominantly strike-parallel, southwesterly paleocurrent directions recorded in Greenland and Ellesmere Island. The base of the flysch also becomes younger in southeasterly directions, owing to the progressive downwarping of the shelf, mentioned above. The downwarping probably was caused by rapid sedimentary loading of the basin floor.

At least five different sources contributed to the Silurian – Lower Devonian flysch of the Arctic Islands:

1. The Caledonian Mountains at the northeastern termination of the deep water basin off northeastern Greenland constituted a major source, evident from the predominantly southwesterly paleocurrent directions in North Greenland and Ellesmere Island. It is apparent from Late Ordovician or Early Silurian to Early Devonian time and probably accounted for the bulk of the Danish River Formation (B4a, B4b). The composition of this formation indicates that the source terrain consisted of relatively

weakly metamorphosed (greenschist-grade) quartzose and argillaceous metasediments with no less than 25% carbonate rocks. The high proportion of carbonate material indicates rapid mechanical erosion and little chemical weathering (i.e. an arid climate) in the source area.

2. Southeastward paleocurrent directions indicate that a source of similar composition lay somewhere northwest of the Hazen Plateau during Llandovery and(?) Wenlock times (unit B4a). It was previously equated with Pearya, but the assumption that Pearya had attained its present position by that time no longer is tenable (Chapters 9, 12B). The relatively fine grade of these sediments (very fine grained sandstone and mudrock) sugests that the source was distant.

3. A much closer source must have supplied the pebble and cobble conglomerate of member B of the Danish River Formation at Caledonian Bay. The chert, which is predominant, was derived from strata either of the deep water basin (Hazen Formation or Amundsen Land Group) or Pearya (successions II and III); the quartzite and marble are petrographically reminiscent of succession II of Pearya.

4. Westward paleocurrent directions in the Danish River Formation (B4b) indicate that a source of siliciclastic and carbonate sediments existed east of Trold Fiord in Pragian – early Emsian time; it is tentatively equated with the Inglefield Uplift (see Chapters 8B, 12D).

5. The Bathurst Island and Stuart Bay formations of Bathurst Island (B4c) evidently were derived from the Boothia Uplift (Chapters 8B, 12C).

6. Regional relationships and limited petrographic evidence suggest that the Blackley Formation of northwestern Melville Island (unit B4d) was derived from Pearya and adjacent parts of the deep water basin, but paleocurrent determinations are lacking.

Sedimentary-volcanic subprovince. The Silurian flysch of northern Ellesmere Island (Imina and Lands Lokk formations) appears to be more closely related to correlative sediments in northeastern Greenland (Merqujôq, Lauge Koch, and Nordkronen formations) than to sediments farther south in Ellesmere Island (Danish River Formation). This (together with stratigraphic and structural evidence given in Chapters 9 and 12B) suggests that these deposits originally were located somewhere north of eastern Greenland and that their present position is due to sinistral strike slip. Northerly paleocurrent directions are common on the east side of northern Greenland (Chapter 7), and the seemingly anomalous northwestward paleocurrent directions at Emma Fiord (Fig. 8C.23) may have been inherited from that region, but horizontal rotation may also have occurred during the inferred strike slip.

The Silurian sediments in northern Axel Heiberg Island and in the Emma Fiord area of northwestern Ellesmere Island differ from other Silurian flysch in the region by their content of volcanogenic material and association with volcanic rocks. The calc-alkaline composition of this material indicates proximity to an island arc, probably located in the present Nansen Sound region, as the largest amount of these materials occurs in a southwestward transported thrust sheet on the northeastern side of Axel Heiberg Island (Svartevaeg Formation). The tectonic significance of this relationship is discussed in Chapter 12B.

Deep water sedimentation in the sedimentary-volcanic subprovince was terminated by Late Silurian tectonism, the youngest preserved flysch being late Wenlock in age in the Northern Heiberg Fold Belt, and Ludlow in the Clements Markham Fold Belt. The unconformably overlying nonmarine and shallow marine deposits of Late Silurian and Early Devonian age will not be considered in this summary.

Phase BV: Early-Middle Devonian (diachronous): deposition of transitional mudrocks

The last phase in the development of the sedimentary subprovince was characterized by deposition of mudrock in transitional, upper fan or slope to outer shelf settings. These sediments are preserved only in southeastern parts of the marginal belt (Chapter 10).

Development of the shelf-basin transition, central Ellesmere Island

In the Arctic Archipelago, shelf-basin relationships are apparent only in northern parts of central Ellesmere Island. The late Early Cambrian to Early Silurian record, exposed on southwestern Judge Daly Promontory (Fig. 8C.26, 8C.27), can be summarized as follows.

Late Early Cambrian to early Middle Ordovician. The deep water basin, represented by the carbonate member of the Hazen Formation (unit B3b), extended to the southeastern side of Archer Fiord and is separated from exposures of the shelf facies (Scoresby Bay, Cass Fjord, Cape Clay formations) by a gap of a few kilometres only. The shelf evidently was bordered by a moderately inclined slope that was a site of both deposition and sliding.

Early Middle Ordovician to Early Silurian (early or middle Llandovery). The absence of the Hazen Formation over the slope, and the presence of a hiatus between carbonate units of the shelf margin and the Silurian Danish River Formation (unit B4a), indicates a steep bypass margin. A submarine relief of 1-1.5 m and a relatively steep slope can be inferred from the geometry of the unconformity. The development of this margin is attributed to subsidence of both shelf and basin, combined with slow sedimentation in the basin and rapid sedimentation on the shelf. The architecture of the shelf margin was characterized by a carbonate buildup (Bulleys Lump Formation) in the late Early to early Middle Ordovician, and by a distally inclined ramp (Thumb Mountain, Irene Bay and lower Allen Bay formations) from late Middle Ordovician to earliest Silurian.

Silurian. Deposition of sediment gravity flows of northeasterly provenance (unit B4a) in the basin probably began in the middle Llandovery in this area. The rate of deposition exceeded the rate of subsidence so that the sediment gravity flows filled the basin, accumulating laterally against the marginal carbonate escarpment rather than vertically upon it. On the adjacent shelf, carbonate sedimentation ceased in about middle Llandovery time and was followed by flysch sedimentation (unit B4b), probably after an interval of nondeposition of uncertain length. The subsidence of the outer shelf during this interval probably was caused by loading of the basin with flysch sediments.

At Cañon Fiord, on the other hand, upward growth of a carbonate buildup at the original shelf margin kept pace with the subsidence until Ludlow (early Late Silurian) time. It was covered by a thin unit of graptolitic carbonates and mudrocks of latest Silurian (Pridoli) age (Cape Phillips

Formation — unit B3c), before it was overstepped by the flysch (unit B4b) in earliest Devonian time (Fig. 8C.14).

ACKNOWLEDGMENTS

M.P. Smith (University of Cambridge) helped to clarify the age relationships between Ordovician shelf deposits in the Arctic Islands and Greenland. Q. Goodbody (then a Research Fellow at the Geological Survey of Canada, Calgary) contributed illustrations and an unpublished manuscript on Ordovician to Devonian deep water sediments in northwestern Melville Island. Critical reading of the manuscript by A.K. Higgins is gratefully acknowledged.

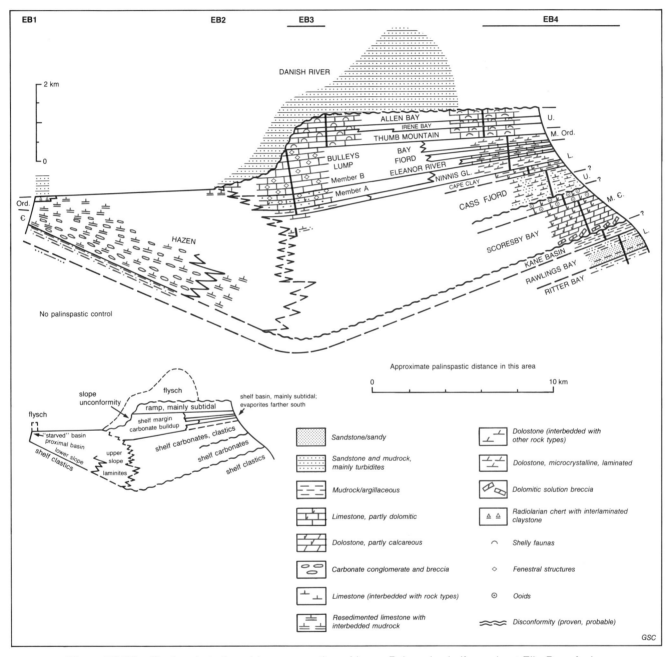

Figure 8C.26. Restored stratigraphic cross-section of lower Paleozoic shelf margin at Ella Bay, Archer Fiord, central eastern Ellesmere Island.

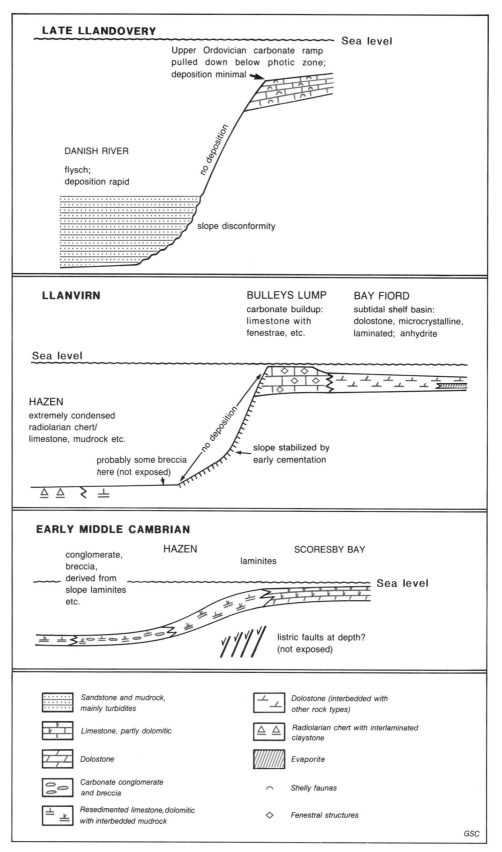

Figure 8C.27. Development of the shelf-basin transition at Ella Bay.

REFERENCES

Barnes, C.R.
1974: Ordovician conodont biostratigraphy of the Canadian Arctic; in Proceedings of the Symposium on the Geology of the Canadian Arctic, J.D. Aitken and D.J. Glass (ed.); Geological Association of Canada and Canadian Society of Petroleum Geologists, p. 221-240.

Barnes, C.R., Norford, B.S., and Skevington, D.
1981: The Ordovician System in Canada, correlation chart and explanatory notes; International Union of Geological Sciences, Publication no. 8, 27 p.

Blackadar, R.G. and Christie, R.L.
1963: Geological reconnaissance, Boothia Peninsula, and Somerset, King William, and Prince of Wales Islands, District of Franklin; Geological Survey of Canada, Paper 63-19, 15 p.

Bolton, T.E., Sanford, B.V., Copeland, M.J., Barnes, C.R., and Rigby, J.K.
1977: Geology of Ordovician rocks, Melville Peninsula and region, southeastern District of Franklin; Geological Survey of Canada, Bulletin 269, 137 p.

Brunton, F.R.
1986: Depositional framework of Lower Cambrian carbonates from northeast Ellesmere Island, Arctic Canada; unpublished B.Sc. thesis, Department of Geology, Laurentian University, Sudbury, Ontario, 38 p.

Christie, R.L.
1967: Bache Peninsula, Ellesmere Island, Arctic Archipelago; Geological Survey of Canada, Memoir 347, 63 p.
1973: Three new lower Paleozoic formations of the Boothia Peninsula region, Canadian Arctic Archipelago; Geological Survey of Canada, Paper 73-10, 31 p.

Daae, H.D. and Rutgers, A.T.C.
1975: Geological history of the Northwest Passage; Bulletin of Canadian Petroleum Geology, v. 23, p. 84-108.

Embry, A.F. and Klovan, J.R.
1976: The Middle-Upper Devonian clastic wedge of the Franklinian geosyncline; Bulletin of Canadian Petroleum Geology, v. 24, p. 485-639.

Ethington, R.L. and Clark, D.L.
1971: Lower Ordovician conodonts in North America; in Symposium on Conodont Biostratigraphy, W.C. Sweet and S.M. Bergstrom (ed.); Geological Society of America, Memoir 127, p. 63-82.
1982: Lower and Middle Ordovician conodonts from the Ibex area, western Millard County, Utah; Brigham Young University Geology Studies, v. 28, Part 2, 155 p.

Ethington, R.L., Engel, K.M., and Elliott, K.L.
1987: An abrupt change in conodont faunas in the Lower Ordovician of the Midcontinent Province; in Palaeobiology of Conodonts, R.J. Aldridge (ed.); Ellis Horwood Limited, Chichester, England, p. 111-127.

Ethington, R.L. and Repetski, J.E.
1984: Paleobiographic distribution of Early Ordovician conodonts in central and western United States; in Conodont Biofacies and Provincialism, D.L. Clark (ed.); Geological Society of America, Special Paper 196, p. 89-101.

Flügel, E.
1982: Microfacies Analysis of Limestones; Springer-Verlag, Berlin, Heidelberg, New York, 633 p.

Fortier, Y.O., Blackadar, R.G., Glenister, B.F., Greiner, H.R., McLaren, D.J., McMillan, N.J., Norris, A.W., Roots, E.F., Souther, J.G., Thorsteinsson, R., and Tozer, E.T.
1963: Geology of the north-central part of the Arctic Archipelago, Northwest Territories (Operation Franklin); Geological Survey of Canada, Memoir 320, 671 p.

Fox, F.G.
1985: Structural geology of the Parry Islands Fold Belt; Bulletin of Canadian Petroleum Geology, v. 33, p. 306-340.

Fritz, W.H.
1981: Cambrian biostratigraphy, southern Canadian Rocky Mountains, Alberta and British Columbia; in Second International Symposium on the Cambrian System, Guidebook for Field Trip 2, M.E. Taylor (ed.); Denver, Colorado, p. 29-33.

Gibling, M.R. and Narbonne, G.M.
1977: Siluro-Devonian sedimentation on Somerset and Cornwallis Islands, Arctic Canada; Bulletin of Canadian Petroleum Geology, v. 25, p. 1145-1156.

Goodbody, Q., Uyeno, T.T., and McGregor, D.C.
1988: The Devonian sequence on Grinnell Peninsula and in the region of Arthur Fiord, Devon Island, Arctic Archipelago; in Current Research, Part D, Geological Survey of Canada, Paper 88-1D, p. 75-82.

Henriksen, N. and Peel, J.S.
1976: Cambrian – early Ordovician stratigraphy in south-western Washington Land, western North Greenland; Grønlands Geologiske Undersøgelse, Report no. 80, p. 17-23.

Homewood, P. and Caron, C.
1982: Flysch of the western Alps; in Mountain Building Processes, K.J. Hsü, (ed.); Academic Press, London, p. 157-168.

Hurst, J.M. and Kerr, J.W.
1982: Upper Ordovician and Silurian facies patterns in eastern Ellesmere Island and western North Greenland and their bearing on the Nares Strait lineament; in Nares Strait and the Drift of Greenland: a Conflict in Plate Tectonics, P.R. Dawes and J.W. Kerr (ed.); Meddelelser om Grønland, Geoscience 8, p. 137-145.

Hsü, K.J.
1970: The meaning of the word flysch — a short historical search; in Flysch Sedimentology in North America, J. Lajoie (ed.); The Geological Association of Canada, Special Paper 7, p. 1-11.

Irving, E.
1974: Latitude variation of the Canadian Arctic Archipelago during the Phanerozoic; in Proceedings of the Symposium on the Geology of the Canadian Arctic, J.D. Aitken and D.J. Glass (ed.); Geological Association of Canada and Canadian Society of Petroleum Geologists, p. 1-3.

Jones, B.
1974: Facies and faunal aspects of the Silurian Read Bay Formation of northern Somerset Island, District of Franklin; unpublished Ph.D. thesis, Department of Geology, University of Ottawa, Ottawa, Ontario, 448 p.

Jones, B., Oldershaw, A.E., and Narbonne, G.M.
1979: Nature and origin of rubbly limestone in the Upper Silurian Read Bay Formation of Arctic Canada; Sedimentary Geology, v. 24, p. 227-252.

Kerr, J.W.
1967a: Stratigraphy of central and eastern Ellesmere Island, Arctic Canada. Part I. Proterozoic and Cambrian; Geological Survey of Canada, Paper 67-27, 63 p.
1967b: Vendom Fiord Formation — a new red-bed unit of probable early Middle Devonian (Eifelian) age, Ellesmere Island, Arctic Canada; Geological Survey of Canada, Paper 67-43, 8 p.
1968: Stratigraphy of central and eastern Ellesmere Island, Arctic Canada. Part II. Ordovician; Geological Survey of Canada, Paper 67-27, 92 p.
1974: Geology of the Bathurst Island group and Byam Martin Island, Arctic Canada (Operation Bathurst Island); Geological Survey of Canada, Memoir 378, 152 p.
1975: Cape Storm Formation — a new Silurian unit in the Canadian Arctic; Bulletin of Canadian Petroleum Geology, v. 23, p. 67-83.
1976: Stratigraphy of central and eastern Ellesmere Island, Arctic Canada. Part III. Upper Ordovician (Richmondian), Silurian and Devonian; Geological Survey of Canada, Bulletin 260, 55 p.
1977: Cornwallis Fold Belt and the mechanism of basement uplift; Canadian Journal of Earth Sciences, v. 14, p. 1374-1401.

Klapper, G.
1977: Lower and Middle Devonian conodont sequence in central Nevada, with contributions by D.B. Johnson; in Western North America: Devonian, M.A. Murphy, W.B.N. Berry, and C.A. Sandberg (ed.); University of California, Riverside Campus Museum Contributions, no. 4, p. 33-54.

Klapper, G. and Murphy, M.A.
1975: Silurian – Lower Devonian conodont sequence in the Roberts Mountains Formation of central Nevada; University of California Publications in Geological Sciences, v. 3 (1974), 62 p.

Kurtz, V.E., McNair, A.H., and Wales, D.B.
1952: Stratigraphy of the Dundas Harbour area, Devon Island, Arctic Archipelago; American Journal of Science, v. 250, p. 636-655.

Landing, E., Barnes, C.R., and Stevens, R.K.
1987: Tempo of earliest graptolite faunal succession: conodont-based correlations from the Ordovician of Quebec; Canadian Journal of Earth Sciences, v. 23, p. 1928-1949.

Lemon, R.R.H. and Blackadar, R.G.
1963: Admiralty Inlet area, Baffin Island, District of Franklin; Geological Survey of Canada, Memoir 328, 84 p.

Mayr, U.
1974: Lithologies and depositional environments of the Allen Bay – Read Bay formations (Ordovician-Silurian) on Svendsen Peninsula, central Ellesmere Island; in Proceedings of the Symposium on the geology of the Canadian Arctic, J.D. Aitken and D.J. Glass (ed.); Geological Association of Canada and Canadian Society of Petroleum Geologists, p. 143-157.

1978: Stratigraphy and correlation of lower Paleozoic formations, subsurface of Cornwallis, Devon, Somerset, and Russell islands, Canadian Arctic Archipelago; Geological Survey of Canada, Bulletin 276, 55 p.

1980: Stratigraphy and correlation of lower Paleozoic formations, subsurface of Bathurst Island and adjacent smaller islands, Canadian Arctic Archipelago; Geological Survey of Canada, Bulletin 306, 52 p.

Mayr, U., Uyeno, T.T., and Barnes, C.R.
1978: Subsurface stratigraphy, conodont zonation, and organic metamorphism of the lower Paleozoic succession, Bjorne Peninsula, Ellesmere Island, District of Franklin; in Current Research, Part A, Geological Survey of Canada, Paper 78-1A, p. 393-398.

Mayr, U., Uyeno, T.T., Tipnis, R.S., and Barnes, C.R.
1980: Subsurface stratigraphy and conodont zonation of the lower Paleozoic succession, Arctic Platform, southern Arctic Archipelago; in Current Research, Part A, Geological Survey of Canada, Paper 80-1A, p. 209-215.

McClure, H.A.
1978: Early Paleozoic glaciation in Arabia; Paleogeography, Paleoclimatology, Paleoecology; v. 25, p. 315-326.

McCracken, A.D. and Barnes, C.R.
1981: Conodont biostratigraphy across the Ordovician-Silurian boundary, Ellis Bay Formation, Anticosti Island, Québec; in Subcommission on Silurian Stratigraphy, Ordovician-Silurian Boundary Working Group, Field Meeting, Anticosti-Gaspé, Québec, v. II: Stratigraphy and Paleontology, P.J. Lespérance, (ed.); International Union of Geological Sciences, p. 61-69.

McGill, P.
1974: The stratigraphy and structure of the Vendom Fiord area; Bulletin of Canadian Petroleum Geology, v. 22, p. 361-386.

McIlreath, I.A.
1978: Facies Models – 13. Carbonate Slopes; Geoscience Canada, v. 5, no. 4, p. 189-199.

Meschede, M.
1986: A method of discriminating between different types of mid-ocean ridge basalts and continental tholeiites with the Nb-Zr-Y diagram; Chemical Geology, v. 56, p. 207-218.

Miall, A.D.
1970a: Continental marine transition in the Devonian of Prince of Wales Island, Northwest Territories; Canadian Journal of Earth Sciences, v. 7, p. 125-144.

1970b: Devonian alluvial fans, Prince of Wales Island, Arctic Canada; Journal of Sedimentary Petrology, v. 40, p. 556-571.

1976: Proterozoic and Paleozoic geology of Banks Island, Arctic Canada; Geological Survey of Canada, Bulletin 258, 77 p.

1983: Stratigraphy and tectonics of the Peel Sound Formation, Somerset and Prince of Wales islands: discussion; in Current Research, Part A; Geological Survey of Canada, Paper 83-1A, p. 493-495.

Miall, A.D. and Gibling, M.R.
1977: The Siluro-Devonian clastic wedge of Somerset Island, Arctic Canada, and some regional paleogeographic implications; Sedimentary Geology, v. 21, p. 85-127.

Miall, A.D. and Kerr, J.W.
1977: Phanerozoic stratigraphy and sedimentology of Somerset Island and northeastern Boothia Peninsula; in Report of Activities, Part A, Geological Survey of Canada, Paper 77-1A, p. 99-106.

1980: Cambrian to Upper Silurian stratigraphy, Somerset Island and northeastern Boothia Peninsula, District of Franklin, N.W.T.; Geological Survey of Canada, Bulletin 315, 43 p.

Miall, A.D., Kerr, J.W., and Gibling, M.R.
1978: The Somerset Island Formation: an upper Silurian to ?Lower Devonian intertidal/supratidal succession, Boothia Uplift region, Arctic Canada; Canadian Journal of Earth Sciences, v. 15, p. 181-189.

Miller, J.F.
1980: Taxonomic revisions of some Upper Cambrian and Lower Ordovician conodonts with comments on their evolution; The University of Kansas, Paleontological Contributions, Paper 99, 44 p.

Mirza, K.
1976: Late Ordovician to Late Silurian stratigraphy and conodont biostratigraphy of the eastern Canadian Arctic Islands; unpublished M.Sc. thesis, Department of Earth Sciences, University of Waterloo, Waterloo, Ontario, 302 p.

Mortensen, P.S.
1985: Stratigraphy and sedimentology of the Upper Silurian strata on eastern Prince of Wales Island, Arctic Canada; unpublished Ph.D. thesis, Department of Geology, University of Alberta, Edmonton, Alberta, 335 p.

Mortensen, P.S. and Jones, B.
1986: The role of contemporaneous faulting on Late Silurian sedimentation in the eastern M'Clintock Basin, Prince of Wales Island, Arctic Canada; Canadian Journal of Earth Sciences, v. 23, p. 1401-1411.

Morrow, D.W. and Kerr, J.W.
1977: Stratigraphy and sedimentology of lower Paleozoic formations near Prince Alfred Bay, Devon Island; Geological Survey of Canada, Bulletin 254, 122 p.

1986: Geology of Grinnell Peninsula and the Prince Alfred Bay area, Devon Island, District of Franklin, N.W.T.; Geological Survey of Canada, Open File 1325, 45 p.

Mossop, G.D.
1979: The evaporites of the Ordovician Baumann Fiord Formation, Ellesmere Island, Arctic Canada; Geological Survey of Canada, Bulletin 298, 52 p.

Muir, I.D. and Rust, B.R.
1982: Sedimentology of a Lower Devonian coastal alluvial fan complex: the Snowblind Bay Formation of Cornwallis Island, Northwest Territories, Canada; Bulletin of Canadian Petroleum Geology, v. 30, p. 245-263.

Mutti, E.
1979: Turbidites et cônes sous-marins profonds; in Sédimentation détritique (fluviatile, littorale, et marine), P. Homewood (ed.); Institut de Géologie de l'Université de Fribourg, Imprimerie Claraz SA, Fribourg, Switzerland, p. 353-419.

Narbonne, G.M.
1981: Stratigraphy, reef development and trace fossils of the Upper Silurian Douro Formation in the southeastern Canadian Arctic Islands; unpublished Ph.D. thesis, University of Ottawa, Ottawa, Ontario, 259 p.

Narbonne, G.M. and Dixon, O.
1982: Physical correlation and depositional environments of Upper Silurian rubbly limestone facies in the Canadian Arctic Islands; in Arctic Geology and Geophysics, A.F. Embry and H.R. Balkwill (ed.); Canadian Society of Petroleum Geologists, Memoir 8, p. 135-146.

Nentwich, F.W. and Jones, B.
1985: Geology of northwestern Brodeur Peninsula, Baffin Island, N.W.T.; in Contributions to the Geology of the Northwest Territories, Volume 2, J.A. Brophy (ed.); Canada, Department of Indian and Northern Affairs, p. 71-84.

Norford, B.S.
1971: Silurian stratigraphy of northern Manitoba; in Geoscience Studies in Manitoba, A.C. Turnock, (ed.); The Geological Association of Canada, Special Paper 9, p. 199-206.

Nowlan, G.S.
1976: Late Cambrian to Late Ordovician conodont evolution and biostratigraphy of the Franklinian Miogeosyncline, eastern Canadian Arctic Islands; unpublished Ph.D. thesis, Department of Biology, University of Waterloo, Waterloo, Ontario, 591 p.

1981: Late Ordovician – Early Silurian conodont biostratigraphy of the Gaspé Peninsula — a preliminary report; in Subcommission on Silurian Stratigraphy, Ordovician-Silurian Boundary Working Group, Field Meeting, Anticosti-Gaspé, Québec, 1981, v. II: Stratigraphy and Paleontology, P.J. Lespérance (ed.); International Union of Geological Sciences, p. 257-291.

1985: Late Cambrian and Early Ordovician conodonts from the Franklinian Miogeosyncline, Canadian Arctic Islands; Journal of Paleontology, v. 59, p. 96-122.

Packard, J.J.
1985: The Upper Silurian Barlow Inlet Formation, Cornwallis Island, Arctic Canada; unpublished Ph.D. thesis, Department of Geology, University of Ottawa, Ottawa, Ontario, 744 p.

Packard, J.J. and Mayr, U.
1982: Cambrian and Ordovician stratigraphy of southern Ellesmere Island, District of Franklin; in Current Research, Part A, Geological Survey of Canada Paper 82-1A, p. 67-74.

Palmer, A.R., Cowie, J.W., and Eby, R.G.
1981: A Late Cambrian (Dresbachian, *Creciphalus* Zone) fauna from the Boothia Peninsula, District of Franklin; in Contributions to Canadian Paleontology; Geological Survey of Canada, Bulletin 300, p. 1-6.

Palmer, A.R. and Peel, J.S.
1981: Dresbachian trilobites and stratigraphy of the Cass Fjord Formation, western North Greenland; Grønlands Geologiske Undersøgelse, Bulletin 141, 46 p.

Peel, J.S. and Christie, R.L.
1982: Cambrian-Ordovician platform stratigraphy: correlations around Kane Basin; in Nares Strait and the Drift of Greenland: a Conflict

in Plate Tectonics, J.W. Kerr and P.R. Dawes (ed.); Meddelelser om Grønland, Geoscience 8, p. 117-135.

Poey, J.L.
1982: Preliminary report on Upper Ordovician to Upper Silurian carbonates, Baumann Fiord area, southwestern Ellesmere Island, District of Franklin; in Current Research, Part A, Geological Survey of Canada, Paper 82-1A, p. 75-77.

Polan, K.P. and Stearn, C.W.
1984: The allochthonous origin of the reefal facies of the Stuart Bay Formation (Early Devonian), Bathurst Island, Arctic Canada; Canadian Journal of Earth Sciences, v. 21, p. 657-668.

Prosh, E.C., Lesack, K.A., and Mayr, U.
1988: Devonian stratigraphy of northwestern Devon Island, District of Franklin; in Current Research, Part D, Geological Survey of Canada, Paper 88-1D, p. 1-10.

Reinson, G.E.
1978: Carbonate-evaporite cycles in the Silurian rocks of Somerset Island, Arctic Canada; Geological Survey of Canada, Paper 76-10, 13 p.

Repetski, J.E.
1982: Conodonts from El Paso Group (Lower Ordovician) of westernmost Texas and southern New Mexico; New Mexico Bureau of Mines and Mineral Resources, Memoir 40, 121 p.

Roblesky, R.F.
1979: Upper Silurian (late Ludlovian) to Upper Devonian (Frasnian?) stratigraphy and depositional history of the Vendom Fiord region, Ellesmere Island, Arctic Canada; unpublished M.Sc. thesis, Department of Geology and Geophysics, University of Calgary, Calgary, Alberta, 230 p.

Robson, M.J.
1985: Lower Paleozoic stratigraphy of northwestern Melville Island, District of Franklin; in Current Research, Part B, Geological Survey of Canada, Paper 85-1B, p. 281-284.

Ross, R.J., Jr. et al.
1982: The Ordovician System in the United States, correlation chart and explanatory notes; International Union of Geological Sciences, Publication no. 12, 73 p.

Smith, G.P.
1984: Stratigraphy and paleontology of the Lower Devonian Sequence, southwest Ellesmere Island, Canadian Arctic Archipelago; unpublished Ph.D. thesis, Department of Geological Sciences, McGill University, Montreal, Quebec, 513 p.

Smith, G.P. and Stearn, C.W.
1987: Anatomy and evolution of a Lower Devonian reef complex, Ellesmere Island, Arctic Canada; Bulletin of Canadian Petroleum Geology, v. 35, p. 251-262.

Sodero, D.E. and Hobson, J.P., Jr.
1979: Depositional facies of lower Paleozoic Allen Bay carbonate rocks and contiguous shelf and basin strata, Cornwallis and Griffith islands, Northwest Territories, Canada; The American Association of Petroleum Geologists Bulletin, v. 63, p. 1059-1091.

Snelling, N.J.
1987: Measurement of geological time and the geological time scale; Modern Geology, v. 11, p. 365-374.

Stewart, W.D.
1987: Late Proterozoic to early Tertiary stratigraphy of Somerset Island and northern Boothia Peninsula, District of Franklin, N.W.T.; Geological Survey of Canada, Paper 83-26, 78 p.

Stuart Smith, J.H. and Wennekers, J.H.N.
1977: Geology and hydrocarbon discoveries of Canadian Arctic Islands; The American Association of Petroleum Geologists Bulletin, v. 61, p. 1-27.

Sweet, W.C.
1979: Late Ordovician conodonts and biostratigraphy of the western Midcontinent Province; Brigham Young University Geology Studies, v. 26, Part 3, p. 45-86.
1984: Graphic correlation of upper Middle and Upper Ordovician rocks, North American Midcontinent Province, U.S.A; in Aspects of the Ordovician System, D.L. Bruton (ed.); Palaeontological Contributions from the University of Oslo, no. 295, Universitetsforlaget, Oslo, p. 23-25.

Sweet, W.C., Ethington, R.L., and Barnes, C.R.
1971: North American Middle and Upper Ordovician conodont faunas; in Symposium on Conodont Biostratigraphy, W.C. Sweet and S.M. Bergstrom (ed.); Geological Society of America, Memoir 127, p. 163-193.

Thorsteinsson, R.
1958: Cornwallis and Little Cornwallis Islands, District of Franklin, Northwest Territories; Geological Survey of Canada, Memoir 294, 134 p.
1963: Copes Bay; in Geology of the north-central part of the Arctic Archipelago, Northwest Territories, Y.O. Fortier et al.; Geological Survey of Canada, Memoir 320, p. 386-395.
1980: Part I. Contributions to stratigraphy (with contributions by T.T. Uyeno); in Stratigraphy and Conodonts of Upper Silurian and Lower Devonian Rocks in the Environs of the Boothia Uplift, Canadian Arctic Archipelago; Geological Survey of Canada, Bulletin 282, p. 1-38.
1988: Geology of Cornwallis Island and neighbouring smaller islands, Canadian Arctic Archipelago, District of Franklin, Northwest Territories; Geological Survey of Canada, Map 1626A (scale 1: 250 000)

Thorsteinsson, R. and Kerr, J.W.
1968: Cornwallis Island and adjacent smaller islands, Canadian Arctic Archipelago; Geological Survey of Canada, Paper 67-64, 16 p.

Thorsteinsson, R. and Mayr, U.
1987: The sedimentary rocks of Devon Island, Canadian Arctic Archipelago; Geological Survey of Canada, Memoir 411, 182 p.

Thorsteinsson, R. and Tozer, E.T.
1962: Banks, Victoria, and Stefansson Islands, Arctic Archipelago; Geological Survey of Canada, Memoir 330, 85 p.
1970: Geology of the Arctic Archipelago; in Geology and Economic Minerals of Canada, R.J.W. Douglas (ed.); Geological Survey of Canada, Economic Geology Report no.1, p. 547-590.

Tozer, E.T. and Thorsteinsson, R.
1964: Western Queen Elizabeth Islands, Arctic Archipelago; Geological Survey of Canada, Memoir 332, 242 p.

Trettin, H.P.
1969a: Lower Paleozoic sediments of northwestern Baffin Island; Geological Survey of Canada, Bulletin 157, 70 p.
1969b: Pre-Mississippian geology of northern Axel Heiberg and northwestern Ellesmere islands, Arctic Archipelago; Geological Survey of Canada, Bulletin 171, 82 p.
1971: Geology of lower Paleozoic formations, Hazen Plateau and southern Grant Land Mountains, Ellesmere Island, Arctic Archipelago; Geological Survey of Canada, Bulletin 203, 134 p.
1975: Investigations of lower Paleozoic geology, Foxe Basin, northeastern Melville Peninsula, and parts of northwestern and central Baffin Island; Geological Survey of Canada, Bulletin 251, 177 p.
1978: Devonian stratigraphy, west-central Ellesmere Island; Geological Survey of Canada, Bulletin 302, 119 p.
1979: Middle Ordovician to Lower Devonian deep-water succession at southeastern margin of Hazen Trough, Cañon Fiord, Ellesmere Island; Geological Survey of Canada, Bulletin 272, 84 p.
1987: Investigations of Paleozoic geology, northern Axel Heiberg and northwestern Ellesmere islands; in Current Research, Part A, Geological Survey of Canada, Paper 87-1A, P. 357-367.

Trettin, H.P., Barnes, C.R., Kerr, J.W., Norford, B.S., Pedder, A.E.H., Riva, J., Tipnis, R.S, and Uyeno, T.
1979: Progress in lower Paleozoic stratigraphy, northern Ellesmere Island, District of Franklin; in Current Research, Part B, Geological Survey of Canada, Paper 79-1B, p. 269-279.

Trettin, H.P., Parrish, R., and Loveridge, W.D.
1987: U-Pb age determinations on Proterozoic to Devonian rocks from northern Ellesmere Island, Arctic Canada; Canadian Journal of Earth Sciences, v. 24, p. 246-256.

Uyeno, T.T.
1980: Systematic study of conodonts; Part II in Stratigraphy and Conodonts of Upper Silurian and Lower Devonian Rocks in the Environs of the Boothia Uplift, Canadian Arctic Archipelago, Geological Survey of Canada, Bulletin 292, p. 39-75.
1990: Conodonts and conodont biostratigraphy of Upper Ordovician through Middle Devonian rocks of southwestern Ellesmere Island and northwestern Devon Island, Canadian Arctic Archipelago (with contributions by U. Mayr and R.A. Roblesky); Geological Survey of Canada, Bulletin 401.

Uyeno, T.T. and Klapper, G.
1980: Summary of conodont biostratigraphy of the Blue Fiord and Bird Fiord formations (Lower-Middle Devonian) at the type and adjacent areas, southwestern Ellesmere Island, Canadian Arctic Archipelago; in Current Research, Part C, Geological Survey of Canada, Paper 80-1C, p. 81-93.

Uyeno, T.T. and Mayr, U.
1979: Lithofacies interpretation and conodont biostratigraphy of the Blue Fiord Formation in the subsurface of Cameron and Vanier islands, Canadian Arctic Archipelago; in Current Research, Part A; Geological Survey of Canada, Paper 79-1A, p. 233-240.

Warren, J.K. and Kendall, C.G.St.C.
1985: Comparison of sequences formed in marine sabkha (subaerial) and salina (subaqueous) settings — modern and ancient; The American Association of Petroleum Geologists Bulletin, v. 69, p. 1013-1023.

Winchester, J.A. and Floyd, P.A.
1977: Geochemical discrimination of different magma series and their products using immobile elements; Chemical Geology, v. 20, p. 325-343.

Authors' addresses

G.D.F. Long
Department of Geology
Laurentian University
Ramsay Lake Road,
Sudbury, Ontario
P3E 2C6

U. Mayr
Geological Survey of Canada
3303-33rd St. N.W.
Calgary, Alberta
T2L 2A7

J.J. Packard
Esso Resources Canada Limited
Box 2480, Station M
Calgary, Alberta
T2P 3M9

H.P. Trettin
Geological Survey of Canada
3303-33rd St. N.W.
Calgary, Alberta
T2L 2A7

Printed in Canada

Chapter 9
THE PROTEROZOIC TO LATE SILURIAN RECORD OF PEARYA

Introduction

Upper Middle Proterozoic (Neohelikian) granitoid basement (succession I)

Upper Proterozoic (Hadrynian) – Lower Ordovician metasediments and metavolcanics (sucession II)

Lower-Middle Ordovician(?) metavolcanics and metasediments (sucession III; Maskell Inlet assemblage)

Plutonic rocks related to succession III

M'Clintock Orogeny

Middle Ordovician to Upper Silurian sedimentary and volcanic rocks (succession IV)

Tectonic interpretation of successions I-IV

Comparison with Franklinian and Caledonian mobile belts

References

Chapter 9

THE PROTEROZOIC TO LATE SILURIAN RECORD OF PEARYA

H.P. Trettin

INTRODUCTION

A geological province in northernmost Ellesmere Island, named Pearya by Schuchert (1923) and interpreted as the relict of a Precambrian borderland, is now regarded as an exotic continental fragment with an internal suture, i.e. as a composite terrane (Chapter 4). This chapter outlines its late Middle Proterozoic to Late Silurian record and briefly compares it with the records of the Franklinian and Caledonian mobile belts. Special emphasis is placed on Ordovician tectonic and plutonic events, which are important for the evaluation of Pearya's relationships with these two regions.

Establishment of this history has been difficult because of sparse age control for the pre-Caradoc stratigraphic record, the effects of five orogenies, and partial ice cover. On first approach, Pearya is divisible into four major successions that differ in age and overall lithology (Fig. 9.2; for geographic names throughout this chapter see Fig. 9.1). Succession I includes sedimentary and(?) volcanic strata of uncertain age that have been deformed, metamorphosed in the amphibolite facies, and intruded by granitic plutons in late Middle Proterozoic time (1.0-1.1 Ga). Succession II, ranging in age from Late Proterozoic to Early Ordovician, consists mainly of "miogeoclinal" or "platformal" sediments (carbonates, quartzite, mudrock) with lesser proportions of mafic and siliceous volcanics, diamictite, and chert. Its concealed contact with succession I is tentatively interpreted as an angular unconformity. Succession III includes arc-type and(?) ocean floor volcanics, mudrock, chert, and carbonates, and has associated with it fault slices of ultramafic-mafic complexes of late Early Ordovician (Arenig) age. The faulted contact between these rocks and succession II is unconformably overlain by succession IV: 7-8 km of sedimentary and volcanic strata of late Middle Ordovician (Caradoc) to Late Silurian age. The angular unconformity at the base of succession IV represents the early Middle Ordovician M'Clintock Orogeny, a collision that was accompanied by metamorphism of subgreenschist to amphibolite grade and by the emplacement of granitic plutons of different kinds. Successions I to III occur in nine separate belts that have been named. Succession IV overlaps two or three belts underlain by successions II and III. To avoid confusion, the term terrane will be restricted to Pearya as a whole although at least one belt (Bromley Island) or its components (succession III and Thores Suite) could be so classified.

Early geological investigations in this area were carried out by Blackadar (1954), Christie (1957, 1964), Trettin (1969a, b; 1971) and Frisch (1974). The present summary is based on a systematic re-investigation in progress since 1975 (Trettin, 1981; Mayr et al., 1982; Trettin, 1987a, b; Trettin and Frisch, 1987); only additional sources will be quoted.

UPPER MIDDLE PROTEROZOIC GRANITOID BASEMENT (SUCCESSION I)

Succession I (Fig. 9.2) consists of granitoid gneiss and lesser proportions of amphibolite, schist, marble and quartzite. It occurs in three major outcrop areas with different structural setting and trends, referred to as the Cape Columbia, Mitchell Point and Deuchars Glacier belts, and as small fault blocks within succession II (Empire Belt).

Description of belts

The **Cape Columbia Belt** is about 40 km long in a westerly direction and up to 3 km wide. Its contact with the Mount Disraeli Belt to the south (succession II) probably is faulted. The gneissic foliation strikes westerly and is vertical or dips steeply north or south. The most abundant rock type is banded garnet-biotite gneiss with two generations of feldspar porphyroblasts; garnetiferous hornblende gneiss and related amphibolite also are common. The persistent compositional layering of the gneisses probably reflects bedding, but the nature of the protoliths is uncertain. Chemical characteristics of the amphibolites suggest that they are meta-igneous rocks. The gneisses are cut by numerous granitic and pegmatitic dykes. Less common are schist, quartzite and marble. The original metamorphism is of amphibolite grade, but retrograde metamorphism and cataclasis are ubiquitous.

The **Mitchell Point Belt** proper is about 110 km long and has a maximum width of about 30 km. It is in fault contact with Upper Proterozoic to Cambrian rocks of the Empire and Cape Alfred Ernest belts, and with Ordovician-Silurian strata of the deep water basin (Clements Markham Fold Belt). The belt consists mainly of granitoid gneiss with lesser proportions of amphibolite and clastic metasediments (Fig. 9.3, 9.4). Foliation and internal lithological contacts are parallel with the regional southwesterly orientation; dip directions are variable. Cataclasis and retrograde metamorphism are common, especially at the faulted southeastern contact.

Trettin, H.P.
1991: The Proterozoic to Late Silurian record of Pearya; Chapter 9 in Geology of the Innuitian Orogen and Arctic Platform of Canada and Greenland, H.P. Trettin (ed.); Geological Survey of Canada, Geology of Canada, no. 3; (also Geological Society of America, The Geology of North America, v. E).

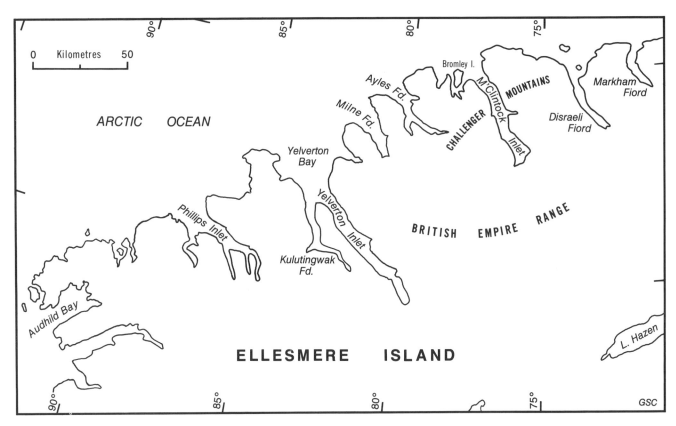

Figure 9.1. Index for geographic names.

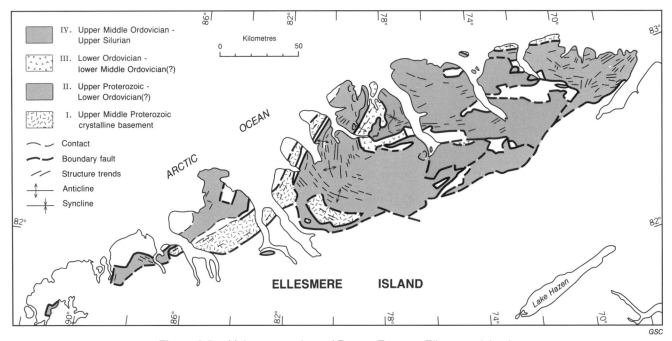

Figure 9.2. Major successions of Pearya Terrane, Ellesmere Island.

Figure 9.3. Mitchell Point Belt east of Kulutingwak Fiord, view to the east.

Figure 9.4. Granitoid gneiss of Mitchell Point Belt west of Kulutingwak Fiord, view to the west.

A relatively large fault block of gneiss also occurs on a peninsula in Phillips Inlet, and small fault blocks of gneiss are present northeast of Ayles Fiord. Both occurrences are aligned with the Mitchell Point Belt proper but are separated from it by fiords and younger strata. This relationship suggests that the Mitchell Point Belt proper is continuous with these outliers in the subsurface.

Deuchars Glacier Belt is a roughly 30 km long, up to 12 km wide, crescent-shaped body that is in fault contact with surrounding Upper Proterozoic – Cambrian strata of the Empire Belt. Foliation and schistosity are arcuate in westerly to northwesterly directions. This belt consists mostly of granitic rocks, smaller proportions of schist, and very small amounts of amphibolite. The granitic rocks, mainly granite and granodiorite according to their present composition, are mostly medium- or coarse-grained. Some are massive and others somewhat gneissic owing to cataclasis and concentration of mica in discontinuous, undulating S-planes. They are comparable to S-type granites by mineral and peraluminous chemical composition. The schists differ from the gneisses by finer grain size, well defined lamination and higher quartz to feldspar and plagioclase to K-feldspar ratios.

Age of metamorphism and granitic plutonism

A nine-point Rb-Sr isochron indicates an age of 1060 ± 18 Ma (recalculated from Sinha and Frisch, 1976) for the metamorphism of the Cape Columbia Belt. It coincides, within the confidence limits stated, with a 1037 +25/-20 Ma U-Pb zircon age (upper intercept of discordant concordia plot) on granitoid gneiss of plutonic aspect from the Deuchars Glacier Belt (Trettin et al., 1987). The presence of an upper

Middle Proterozoic granitoid basement also is indicated by upper intercept ages of inherited zircon from the Ordovician Cape Richards plutonic complex (1300 ± 300 Ma), the Devonian Cape Woods pluton (1450 ± 400 Ma), and the Cretaceous Hansen Point volcanics (1150 ± 300 Ma). The Mitchell Point Belt is tentatively correlated with the Cape Columbia and Deuchars Glacier belts on the basis of lithology, although it has yielded somewhat younger Rb-Sr ages of 726 ± 12 Ma and 802 ± 19 Ma (recalculated from Sinha and Frisch, 1975).

UPPER PROTEROZOIC – LOWER ORDOVICIAN METASEDIMENTS AND METAVOLCANICS (SUCCESSION II)

Succession II (Fig. 9.5) comprises variably metamorphosed sedimentary and subordinate volcanic rocks of five different types: (1) a predominant suite of original shelf sediments such as limestone, dolostone, quartzite and mudrock; (2) diamictite and related greywacke and mudrock that probably are glaciogenic; (3) mafic volcanics, analyzed samples of which are subalkaline basalt; (4) siliceous volcanics; and (5) chert. The mafic and siliceous volcanics have not been observed together but may represent a bimodal suite of rift origin, if related. The metamorphic grade of the succession ranges from subgreenschist to amphibolite facies and is mainly of greenschist facies.

The base of succession II is not known to be exposed and its relationship with succession I is problematic. A small inlier of succession I northeast of Ayles Fiord is overlain by succession II, but succession I seems to have pierced through this cover to some extent (Frisch, 1974, Fig. 12A); a similar relationship may exist at Deuchars Glacier. Elsewhere the contacts between the two successions are interpreted as original strike slip faults that also had large vertical components of movement, partly acquired during later compressive or extensional deformations. Considerations of a more general kind can be summarized as follows. (1) The abundance of shelf carbonates and quartzites characterizes succession II as part of an essentially "platformal" or "miogeoclinal" suite deposited on a continental basement. (2) Available evidence indicates that succession II is younger than the 1.0-1.1 Ga orogeny that transformed succession I into a crystalline basement. (3) The close spatial association of the two successions suggests an original basement-cover relationship. (4) The unconformity between crystalline basement and supracrustal succession may have become a detachment surface during later deformations because of the difference in mechanical properties above and below it.

Succession II occurs in six separate belts with different structural trends, referred to as the Mount Disraeli, Disraeli Glacier, Empire (for British Empire Range), Kulutingwak (for Kulutingwak Fiord), Cape Alfred Ernest, and Audhild Bay belts (Fig. 9.2). Within these belts the rocks have been mapped mainly by lithology, although preliminary stratigraphic units have been established in some areas. The stratigraphy of the succession is very difficult to establish because of lack of fossils and extreme structural complexity — most contacts are faulted. It is likely that several units occur in more than one belt, but the only deposits for which this is obvious are the diamictites and related greywackes and mudrocks. They are comparable to Upper Proterozoic ("Varangian") diamictites both in the northern Cordillera and in the North Atlantic region (cf. Hambrey, 1983) and are tentatively correlated with them. Those in the Cordillera and in the Atlantic region usually consist of two sets of glacial deposits that are separated by a unit of nonglacial origin, but this has not been established for Pearya. Important also is a less deformed unit, informally referred to as the Milne Fiord assemblage, which contains isotopically dated volcanics of earliest Ordovician age (503.2 +7.8/1.7 Ma, Tremadoc or Arenig; Trettin et al., 1987) in the middle part. It seems to be preserved in one area only (Fig. 9.4) and probably occurs at the top of succession II.

These observations permit tentative assignment of at least some of the rocks to four major units: (A) Upper Proterozoic strata underlying diamictites; (B) Upper Proterozoic diamictites and associated sediments and volcanics; (C) Upper Proterozoic and Cambrian strata between diamictites and Milne Fiord assemblage; and (D) the Milne Fiord assemblage (lowermost Ordovician and [?] Upper Cambrian).

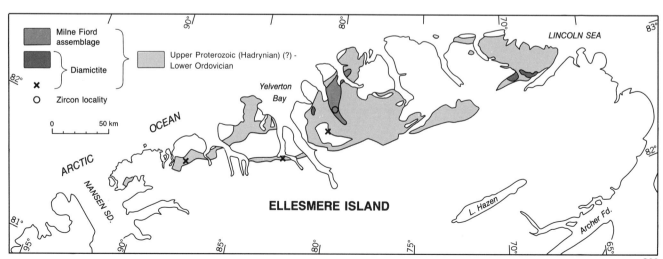

Figure 9.5. Components of succession II.

Unit A includes: (1) very fine- to very coarse-grained quartzite and interstratified phyllite that have been thrust over diamictites west of Yelverton Bay (Cape Alfred Ernest Belt); and (2) schist, phyllite, quartzite and carbonates that appear to underlie diamictites in the Mount Disraeli Belt.

Unit B consists mainly of diamictite and related greywacke and mudrock, but also includes some carbonate rocks, chert and volcanics. Exposures in the southeastern part of the Mount Disraeli Belt appear to be hundreds of metres thick.

The diamictite has an abundant muddy matrix in which are embedded smaller fractions of sand and pebbles with or without rare cobbles and boulders (Fig. 9.6). The greywacke and related sandy mudrock differ from the diamictite only by the absence of gravel-grade material. The phenoclasts, discoid to equant in shape and generally somewhat rounded, are unsorted and have random orientations. The grains or phenoclasts in diamictite and greywacke represent intrabasinal sediments (mainly carbonate rocks, also clastic rocks), extrabasinal sediments (rounded quartz, metamorphics etc.), and volcanic materials. The diamictites are similar in texture to the subaqueous debris flows found, for example, in flysch, but the diamictite-bearing unit as a whole differs from flysch by its largely massive character. On the other hand, the rocks seem to be comparable to the Upper Proterozoic glaciogenic deposits mentioned, in spite of the fact that conclusive evidence for a glaciogenic origin, such as a striated basement, striated phenoclasts or dropstone conglomerate, has not been found.

A metamorphosed volcanic flow associated with diamictite of the Mount Disraeli Belt is classified as subalkaline basalt on the basis of stable trace elements (Zr/TiO_2 vs. Nb/Y).

Unit C makes up the bulk of succession II but its internal stratigraphy is uncertain. It consists mainly of calcareous and dolomitic shelf carbonates, mudrock and quartzite with small proportions of volcanics.

Unit D, the Milne Fiord assemblage (Fig. 9.7), contains in ascending order: (1) 150-200 m of calcareous and dolomitic marble; (2) about 900 m of phyllite with lesser amounts of impure dolostone and argillaceous limestone, and small amounts of volcanics, mainly siliceous tuff; and (3) 540+ m of very fine to medium grained, mostly massive quartzite.

LOWER-MIDDLE ORDOVICIAN(?) METAVOLCANICS AND METASEDIMENTS (SUCCESSION III, MASKELL INLET ASSEMBLAGE)

Succession III, which constitutes the Bromley Island Belt, underlies an L-shaped area of discontinuous outcrops (Fig. 9.2, 9.9). It consists mainly of volcanic and sedimentary rocks assigned to the Maskell Inlet assemblage. Closely associated with these strata are the ultramafic to granitic rocks of the Thores Suite discussed below. This succession is in fault contact with succession II (Empire Belt) and overlain by upper Middle Ordovician (early Caradoc, Blackriveran) strata of succession IV with angular unconformity. Succession III is tentatively assigned to the late Early – early Middle Ordovician on the basis of its stratigraphic and tectonic relationships.

The thickness and stratigraphy of the Maskell Inlet assemblage also are uncertain because of its complex structure. Metamorphism is mainly of subgreenschist facies but attains greenschist and locally amphibolite grade in the vicinity of the M'Clintock West and M'Clintock East bodies of the Thores Suite (see below). The sediments consist of limestone, chert, and mudrock with lesser proportions of sandstone and dolostone. The most common limestone type is an argillaceous and slightly sandy lime mudstone, but calcarenite is also present. The chert beds, light to dark grey and partly laminated, contain structures interpreted as poorly preserved radiolarians and sponge spicules. Tuffs are predominant over flows in the volcanic suite. Ten out of 11 samples analyzed for stable trace elements (Zr, Ti, Nb, Y) are in the compositional fields of andesite or undifferentiated andesite/basalt; one is a subalkaline basalt (Fig. 9.8). The tuffaceous character and andesitic to basaltic composition of the rocks suggest that most are of island arc origin. However, the mafic rocks fall into the fields of undifferentiated volcanic arc basalt and normal mid-ocean ridge basalt according to a Nb-Zr-Y plot of Meschede (1986, Fig. 2); thus the presence of ocean floor rocks cannot be excluded.

Figure 9.6. Diamictite of Empire Belt; light coloured phenoclasts are carbonate rocks.

Figure 9.7. Divisions 1 (ЄO$_{M1}$) and 2 (ЄO$_{M2}$) of Milne Fiord assemblage northeast of lower Ayles Fiord, view to the northwest.

PLUTONIC ROCKS RELATED TO SUCCESSION III

Ultramafic-mafic complexes (Thores Suite)

The Thores Suite of ultramafic-mafic complexes comprises the M'Clintock West, M'Clintock East, Thores River, Bromley Island, and Ootah Bay bodies (Fig. 9.9).

Structural Setting

The **M'Clintock West body,** about 13 km long and up to 6.8 km wide, is nonconformably overlain on the east side by Middle Pennsylvanian clastic sediments; its western contact with metavolcanics of succession III is covered by ice (Fig. 9.10). On the north and south it is bounded by linear faults that also terminate the adjacent metavolcanics. The southern fault, which probably dips steeply to the north, is bordered on the south by greenschist-grade metasediments of the Empire Belt (succession II). The northern fault is bordered on the north by Upper Ordovician sediments of succession IV; its dip is unknown.

The **M'Clintock East body** is 3.8 km long and 1.2 km wide, and has been thrust southward upon metavolcanics and minor metasediments of the Maskell Inlet assemblage, which in turn have been thrust upon Upper Ordovician sediments that probably lie unconformably on the Empire Belt. It consists of two (or more) thrust slices of plutonic rocks that locally are separated by thin slices of metasediments. On the north side the body is separated from Carbonifereous sediments by a steeply-dipping fault.

The 2.5 km long, narrow **Thores River body** also is close to the northern border of the Empire Belt, from which it is separated, however, by a fault block of Carboniferous strata. On the north it is overthrust by schist of the Maskell Inlet assemblage.

The **Bromley Island body,** about 2 km long and up to 1 km wide, is separated by steeply-dipping faults from the surrounding Maskell Inlet assemblage. Plutonic rocks and Maskell Inlet strata both are locally unconformably overlain by nearly horizontal carbonate strata of member B of the upper Middle Ordovician Cape Discovery Formation.

The small **Ootah Bay body,** about 1 km long and up to 170 m wide, occurs in a structurally complex area where it is in fault contact with the Middle and Upper Ordovician Cape Discovery, M'Clintock, and Taconite River formations.

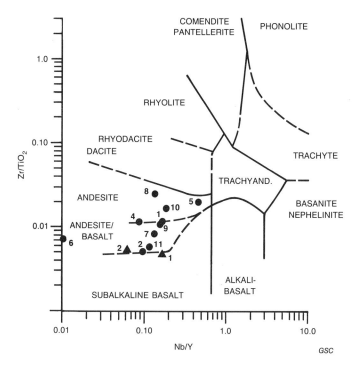

Figure 9.8. Maskell Inlet assemblage, classification of analyzed specimens according to stable trace elements; based on Winchester and Floyd, 1977, Fig. 6. (Triangle indicates specimens from outcrop east of M'Clintock East body).

Lithology

The Thores Suite consists mainly of an ultramafic-mafic association, which in some bodies is intruded by a mafic to predominantly felsic association (Table 9.1). In the **M'Clintock West body** the former is composed mainly of serpentinite with local occurrences of gabbro, wehrlite and clinopyroxenite (Frisch, 1974). The mafic-felsic association occurs mainly as sheets or groups of sheets that are up to 10 m thick and up to several hundred metres long. It comprises clinopyroxene-hornblende diorite; hornblende diorite, in part pegmatitic with large hornblende crystals; monzodiorite; quartz monzodiorite; leucocratic tonalite (trondhjemite); and granodiorite.

In the **M'Clintock East body,** the ultramafic-mafic association consists of layered cumulates that either are monomineralic clinopyroxenite or are composed of various mixtures of clinopyroxene and hornblende with or without lesser proportions of plagioclase and rare olivine and phlogopitic biotite; these rocks are classified as wehrlite, olivine clinopyroxenite, hornblende clinopyroxenite, clinopyroxene hornblendite, plagioclase-clinopyroxene hornblendite, gabbro, or hornblende gabbro. The mafic-felsic suite is represented by thin sheets of leucocratic rocks, including leuco-monzodiorite, that are separated by serpentinite.

In the **Thores River body** the ultramafic-mafic suite includes clinopyroxenite, and the mafic-felsic suite pegmatitic hornblende diorite, and sheared and altered monzodiorite and quartz diorite.

The **Bromley Island body** consists of gabbro and serpentinite with remnants of clinopyroxene and phlogopitic biotite, and the **Ootah Bay body** of variably serpentinized wehrlite.

Mode of origin

The Thores Suite differs significantly from layered mafic intrusions (such as the Bushveld, Skaergaard or Stillwater intrusions) by the general absence of orthopyroxene and the high proportion of ultramafic rocks in the M'Clintock West body, but has significant features in common with ophiolites and Alaskan-type intrusions, i.e. plutons associated with ensimatic arcs.

The standard ophiolite consists of five major units (Coleman, 1977). The lowest unit, peridotite tectonite, may be present in the M'Clintock West body but is not identifiable there because of pervasive serpentinization — a common feature of ancient ophiolites and their counterparts in modern oceans. The second and third units, cumulate complexes and leucocratic associates (or plagiogranite), are well represented — the lack of orthopyroxene, and the presence of small amounts of K-feldspar notwithstanding. The absence of the upper two units of the standard ophiolite, sheeted dykes and pillow basalt, may be due to faulting or may be an original feature, as is the case in parts of the Mediterranean region where serpentinite and gabbro are directly overlain by sediments (Lemoine et al., 1987).

Important properties in common with Alaskan-type intrusions are the (apparent) absence of orthopyroxene and the common occurrence of hornblende in a variety of rock types, including pegmatitic hornblende diorite; important differences are the absence of concentric zoning and the presence of felsic differentiates.

Apart from the lithological features, one of the most important characteristics of ophiolites is the fact that they usually occur in thrust sheets associated with suture zones. This concept is applicable to the M'Clintock Inlet Suite (see below) if one considers its stratigraphic-structural setting and broader regional relationships (see below).

Age

U-Pb determinations on zircon from a felsic intrusion in the M'Clintock West body (Fig. 9.10) gave an apparent age of 481 +7/-6 Ma (Trettin et al., 1982), Arenig according to all recent time scales (e.g. Harland et al., 1982; McKerrow et al., 1985; Odin, 1985). This determination is in agreement with the observation that the presumably cogenetic Bromley Island body is unconformably overlain by the Cape Discovery Formation of late Middle Ordovician (early Caradoc) age.

Granitic intrusion (Markham Fiord pluton)

This 16 km long, up to 2.5 km wide pluton is in fault contact with Upper Proterozoic or Cambrian strata of the Mount Disraeli Belt, with Ordovician rocks of succession IV, and with upper Paleozoic sediments (Fig. 9.9). As a result of the faulting, the rocks commonly are shattered and altered. The pluton consists of fine- to predominantly medium-grained felsic rocks, mainly granodiorite, quartz monzodiorite, and quartz diorite; and of minor amounts of intermediate and mafic rocks, including fine grained gabbro. The original mafic silicates are mainly hornblende and biotite, the latter commonly altered to chlorite, but small

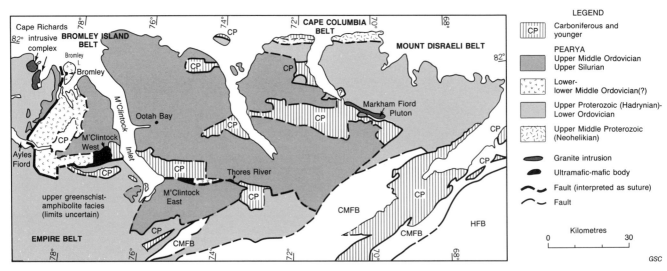

Figure 9.9. Features of M'Clintock Orogen. CMFB = Clements Markham Fold Belt; HFB = Hazen Fold Belt.

Figure 9.10. Setting of M'Clintock West body (LOum), view to the northwest (oblique air photograph T404R-15). This body is predominantly ultramafic in composition but has associated with it felsic differentiates (g). On the left (south) it is in fault contact with upper Proterozoic – Cambrian marble (Hl∈c) and schist (Hl∈s) and on the right (north) with Middle Ordovician to Silurian formations (OMI = Maskell Inlet assemblage; OCD = Cape Discovery Formation; OCD1 = Cape Discovery Formation, member A; OA = Ayles Formation; OMC = M'Clintock Formation; OTR = Taconite River Formation; OZC = Zebra Cliffs Formation; ODG = Disraeli Glacier beds; SCR = Cranstone Formation). Its western contact with the Maskell Inlet assemblage (OMI) is covered by a glacier. The upper Proterozoic or Cambrian to Silurian rocks all are unconformably overlain by Upper Carboniferous (C) and Quaternary (Q) sediments.

Table 9.1. Rock types of the Thores Suite.

ROCK TYPES / OCCURRENCES	M'CLINTOCK WEST	M'CLINTOCK EAST	THORES RIVER	BROMLEY ISLAND	OOTAH BAY
Serpentinite	X	X		X	X
Wehrlite	X	X		X	X
Olivine clinopyroxenite		X			
Clinopyroxenite	X	X	X		
Hornblende clinopyroxenite		X			
Clinopyroxene hornblendite		X			
Plagioclase-clinopyroxene hornblendite		X			
Gabbro	X	X		X	
Hornblende gabbro		X			
Clinopyroxene-hornblende diorite	X		X		
Hornblende diorite	X				
Monzodiorite	X	X	X		
Quartz monzodiorite	X				
Tonalite (leucocratic = trondjhemite)	X				
Granodiorite	X				

GSC

amounts of clinopyroxene also occur. Chemical analyses of three relatively fresh samples indicate metaluminous compositions.

U-Pb determinations on zircon gave a slightly discordant upper intercept age of 462 ± 11 Ma, Llandeilo (Harland et al., 1982; Palmer, 1983) or Llanvirn (McKerrow et al., 1985; Odin, 1985). The chemical and mineralogical composition of the pluton, combined with the lack of inherited Precambrian zircon, suggests an "I-type" intrusion, derived from igneous rocks of the mantle.

M'CLINTOCK OROGENY
Stratigraphic-structural relationships

Relationships between successions II, III and IV indicate an orogeny that was accompanied by metamorphism. This event probably affected all of Pearya, but conclusive evidence is restricted to the M'Clintock Inlet region. The evidence consists primarily of an angular unconformity at the base of succession IV, exposed at several localities in the vicinity of M'Clintock Inlet. Beneath the unconformity are strata of the Empire Belt (succession II) in the west and south (Fig. 9.11, 9.12), and strata of succession III (Maskell Inlet assemblage) and ultramafic-mafic rocks (Bromley Island body) in the east and north. The oldest strata above the unconformity are early Caradoc (Blackriveran) in age (see below) and the youngest rocks beneath it perhaps late Early Ordovician (Bromley Island body) or slightly younger (Maskell Inlet assemblage, assumed age).

The structural effects of this event are difficult to distinguish from those of later deformations, but complex folds and faults are apparent beneath the unconformity at several localities (e.g. Fig. 9.11). The most important structure recognized is a major fault between the Bromley Island and Empire belts. The fact that it places an arc-type suite plus ultramafic rocks against a miogeoclinal suite suggests that it may be a suture (see Tectonic Interpretation). The fault now is L-shaped in plan, one arm trending north and the other east, and the strata of Bromley Island and Empire belts are parallel with it, but this configuration probably is the result of Late Silurian horizontal bending (Chapter 12B); the original outline probably was straight or slightly arcuate in a westerly direction (in terms of present geographic directions). The original attitude of the fault also is uncertain because of later deformations that have involved Upper Ordovician strata. If these later deformations have been influenced by Middle Ordovician structures, then the main fault (and inferred suture) probably was a south-directed thrust fault.

Metamorphism

As succession IV is nearly unmetamorphosed, the metamorphism of successions III and II must have resulted from an older event — almost certainly the M'Clintock Orogeny. A K-Ar (muscovite) age of 452 ± 8 Ma (Stevens et al., 1982), on schist of the Empire Belt west of upper M'Clintock Inlet (Fig. 9.11), is interpreted as a cooling age related to this event.

The metamorphism of Empire and Bromley Island belts in this area was Barrovian in nature and ranges from subgreenschist to amphibolite grade. Subgreenschist-grade metamorphism is displayed by those volcanics of the Maskell Inlet assemblage that contain remnants of clinopyroxene in addition to albite, chlorite, etc. The greenschist facies, characterized by albite, chlorite, muscovite and/or biotite and locally by chloritoid, is widely developed in the Empire Belt and also in strata of the Maskell Inlet assemblage associated with the M'Clintock West and East ultramafic bodies. The amphibolite facies, characterized by hornblende, staurolite, kyanite, garnet and rare sillimanite is widely developed in pelites of the Empire Belt southeast and southwest of M'Clintock Inlet. It also is locally developed in metavolcanics of the Bromley Island Belt (succession III) associated with the M'Clintock West ultramafic body and in garnetiferous schist associated with the M'Clintock East body.

Granitic intrusions

With the possible exception of the Markham Fiord pluton discussed above, there is no radiometric evidence for syntectonic granitic plutonism associated with the M'Clintock Orogeny. However, field evidence suggests that a small syntectonic granitic body north of upper Ayles Fiord may may have been intruded during this event, and if so, others probably are present. The intrusion is a structurally concordant sheet of medium grained granodiorite that occurs in the axial region of a small recumbent fold formed in amphibolite-grade pelitic schist. In thin section, the granodiorite shows sigmoidal folds that must have developed during its crystallization.

The Cape Richards intrusive complex (Frisch, 1974), composed of quartz monzonite, granodiorite, and syenite with alkalic affinity, also is early Middle Ordovician (Llandeilo) in age on the basis of a U-Pb determination (463 ± 5 Ma; Trettin et al., 1987). Its unmetamorphosed state and slightly alkaline composition, and the cauldron-like outline of its two outcrop areas, suggest that it is a post-tectonic intrusion, emplaced during an extensional regime.

Figure 9.11. Exposures west of the head of M'Clintock Inlet. Complexly folded Upper Proterozoic – Cambrian marble (uP€c) and schist (uP€s) are overlain with angular unconformity by the Upper Ordovician Taconite River (OTR) and Zebra Cliffs (Ozc) formations.

MIDDLE ORDOVICIAN TO UPPER SILURIAN SEDIMENTARY AND VOLCANIC ROCKS (SUCCESSION IV)

Succession IV, identical with the Challenger Mountains Supergroup, represents a successor basin that developed upon the Middle Ordovician orogen and overlaps the Bromley Island Belt and marginal parts of Empire and(?) Mount Disraeli belts (Fig. 9.2, 9.9), but contacts with the latter are faulted or concealed. The succession is more than 7 km thick and consists of sedimentary and volcanic strata assigned to four major units, the first and second (Egingwah Bay and Harley Ridge groups) separated by an unconformity (Fig. 9.14).

Middle-Upper Ordovician carbonates, clastics, and volcanics

The Egingwah Bay Group, which is bounded by unconformities at base and top, comprises a lower unit of carbonate and clastic sediments with minor volcanics (Cape Discovery Formation), a middle, predominantly volcanic unit (M'Clintock Formation), and an upper carbonate unit (Ayles Formation).

Sedimentary and volcanic unit

The Cape Discovery Formation (about 1 km thick) has been divided into four members. **Member A** unconformably overlaps Bromley Island and Empire belts. It ranges in thickness from perhaps a few metres to more than 144 m and is divisible into three genetic units. The lower unit, probably nonmarine in origin, consists of massive, upward-fining conglomerate of boulder to pebble grade (Fig. 9.13a) and sandstone. It was derived mainly from the Empire Belt (phyllite) and Maskell Inlet assemblage (chert, siliceous volcanics), and to a lesser extent also from the Thores Suite (ultramafic-mafic clasts; trace amounts of chromite — Fig. 9.13c). The middle unit comprises calcareous and dolomitic sandstones of shallow marine origin that contain benthonic fossils, including *Gonioceras* sp. of early Caradoc (Blackriveran) age. The upper unit is characterized by basic to siliceous, partly alkaline tuffs and flows that probably indicate an extensional regime.

Member B consists of 107 m or more of shelf carbonate sediments (dolostone, skeletal lime wackestone, pelletal packstone, oncolitic grainstone etc.) with minor amounts of calcareous sandstone and mudrock. Macrofossils and conodonts combined indicate a late Caradoc age.

Member C, 307 m in thickness, is made up of laminated, argillaceous to sandy limestone and related intraformational flat-pebble conglomerate. This unfossiliferous, red weathering unit was deposited in a very shallow marine or peritidal environment.

Member D, 468 m thick, consists mainly of reddish sandstone derived from contemporaneous(?) siliceous volcanics with minor amounts of dolostone and rhyodacite/dacite in the lower part.

Figure 9.12. Exposures east of the head of M'Clintock Inlet. Upper Proterozoic – Cambrian schist and marble (uPЄsc) are overlain with angular unconformity by the Upper Ordovician Zebra Cliffs Formation (Ozc), in turn overlain with low angular discordance by the Upper Ordovician Lorimer Ridge Formation (OLR).

Volcanic unit

The M'Clintock Formation, which overlies the Cape Discovery with an abrupt but conformable contact, is 1.4 km or more, possibly 3 km, thick. The formation consists mostly of volcanic flows and tuffs with minor proportions of volcanic-derived clastic sediments and limestone. Most analyzed samples (Fig. 9.15) are in the boundary field between andesite and subalkaline basalt, but siliceous rocks (rhyolite, dacite) and alkali basalt also are present. Apart from the alkali basalt, the rocks are suggestive of a volcanic arc. A lime mudstone about 85 m below the top of the formation yielded *Paleofavosites*? sp. of Late Ordovician age.

Carbonate unit

The Ayles Formation, which conformably overlies the M'Clintock, has a minimum thickness of about 600 m. It consists of dolostone, calcareous dolostone and dolomitic limestone, mainly lime mudstone. Rare fossils, mainly corals, indicate a Late Ordovician, probably late Ashgill (Richmondian) age.

Upper Ordovician clastic and carbonate sediments

The Upper Ordovician (upper Ashgill, Richmondian) Harley Ridge Group consists of closely associated clastic and carbonate sediments, clastics predominating in the lower and upper parts (Taconite River and Lorimer Ridge formations) and carbonates in the middle part (Zebra Cliffs Formation).

The base of the group is unconformable. In the southernmost part of the Challenger Basin, i.e. south of the ultramafic outcrops (M'Clintock West, M'Clintock East, and Thores River bodies), the Harley Ridge Group lies on metamorphic rocks of the Empire Belt; the Egingwah Group, if deposited there originally, was eroded prior to the deposition of the Harley Ridge Group. West of M'Clintock Inlet, the Empire Belt is overlain by a thin Taconite River Formation (Fig. 9.11) and east of the inlet by the Zebra Cliffs Formation (Fig. 9.12), a relationship that suggests encroachment on a paleotopographic high. In the remainder of the basin, the Taconite River Formation lies disconformably on different parts of the Egingwah Group (Ayles or M'Clintock formations).

Lower, predominantly clastic unit

The Taconite River Formation attains a thickness of more than 600 m. It is divisible into three facies: (1) a subordinate nonmarine facies of boulder to pebble conglomerate and sandstone, occurring mainly at the base but also in the upper part of the formation; (2) a predominant shallow marine facies of variegated sandstone and mudrock; and (3)

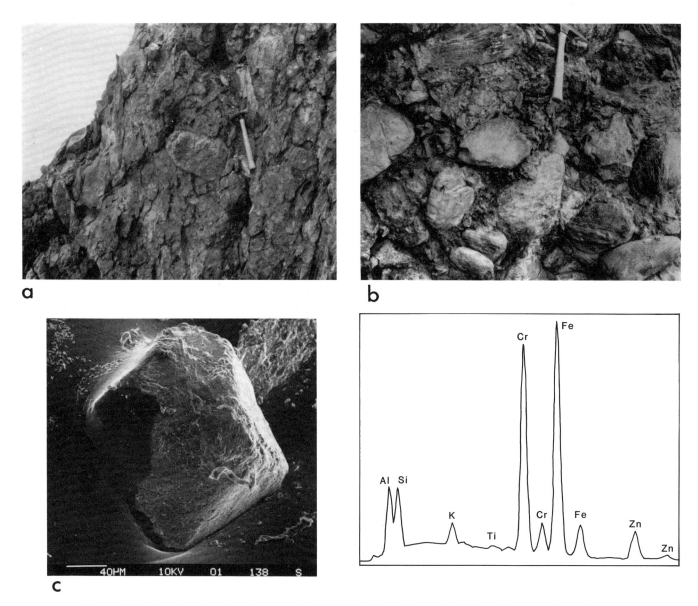

Figure 9.13. Basal sediments of succession IV. (a) Basal conglomerate of Cape Discovery Formation on unnamed peninsula east of Bromley Island. (b) Basal conglomerate of Taconite River Formation west of head of M'Clintock Inlet. (c) SEM micrograph and semiquantitative spectral analysis of chromite from basal Cape Discovery Formation.

a subordinate shallow marine facies of impure, commonly fossiliferous lime mudstone and skeletal lime wackestone. At the head of M'Clintock Inlet, the basal conglomerate consists mainly of marble derived from the underlying Empire Belt (Fig. 9.13b). The sandstones consist chiefly of quartz and limestone fragments with smaller proportions of feldspar, schist, muscovite, biotite, chlorite, opaques and sporadic volcanic rock fragments.

Middle, predominantly carbonate unit

The Zebra Cliffs Formation comprises about 300-600 m of carbonate rocks with a small proportion of siliciclastic sediments. Characteristic are thick successions of thinly stratified limestone with minor amounts of intercalated mudrock and dolostone; less common are thick bedded to massive limestones and dolostones, some of which represent small reefs. The predominant limestone types are skeletal lime wackestone and lime mudstone deposited in quiet subtidal shelf environments, but oncolitic-skeletal grainstone, indicative of high-energy environments, also is present. The carbonate rocks contain a rich and varied benthonic fauna of Late Ordovician (Late Ashgill, Richmondian) age that is dominated by corals but also includes brachiopods, trilobites, cephalopods, gastropods, bryozoans, echinoderms, ostracodes, dasycladacean algae, and rare graptolites. It has diagnostic elements in common with faunas in Siberia (*Sibiriolites sibiricus* Sokolov),

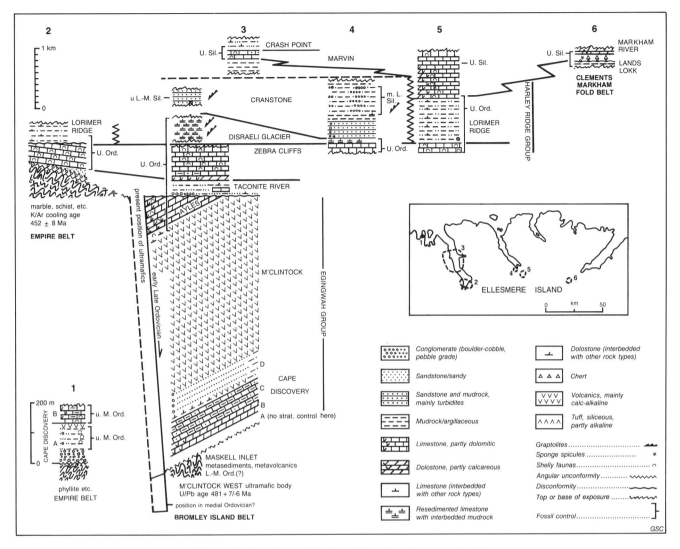

Figure 9.14. Generalized stratigraphy of succession IV.

northern Greenland (*Troedssonites conspiratus* (Troedsson)), and the Arctic Platform (*Armenoceras michaudae* Bolton, *Palaeophyllum raduguini* Nelson var.).

Units of variegated siliciclastic sediments, up to about 40 m thick, occur locally in the lower and middle part of the formation. They consist mainly of mudrock and very fine- to fine-grained sandstone, but include coarser grained sandstone and pebble conglomerate at one locality. Bioturbation, rare fossils, and close association with fossiliferous limestones indicate a shallow marine origin.

Upper, predominantly clastic unit

The Lorimer Ridge Formation overlies the Zebra Cliffs Formation with a low-angle unconformity on Harley Ridge, southeast of the head of M'Clintock Inlet (Fig. 9.12); elsewhere the contact probably is conformable. It is 280-770 m thick and composed chiefly of mudrock and sandstone with lesser amounts of limestone and rare pebble conglomerate. Grey and green beds alternate with red beds, the latter conspicuous in the eastern outcrop areas.

Mudrock and sandstone both are chiefly composed of quartz with lesser proportions of carbonate minerals (calcite, minor dolomite), feldspar, mica and chlorite. The limestone units, commonly one to a few metres thick, consist of argillaceous and sandy lime mudstone and skeletal lime wackestone. They contain corals, including the Late Ordovician *Sibiriolites sibiricus*, brachiopods, gastropods, bryozoans, stromatoporoids, echinoderms, and ostracodes. The formation was deposited in shallow marine, oxygenated environments.

Upper Ordovician – Silurian deep water succession

Discontinuous outcrops of Upper Ordovician to Middle Silurian deep water sediments occur in a narrow belt that is 103 km long in a westerly direction and locally up to 8 km wide (Fig. 9.16). The deep water succession overlies the Zebra Cliffs Formation and comprises an Upper Ordovician unit of resedimented carbonates and argillaceous sediments (Disraeli Glacier beds) that probably is correlative with the

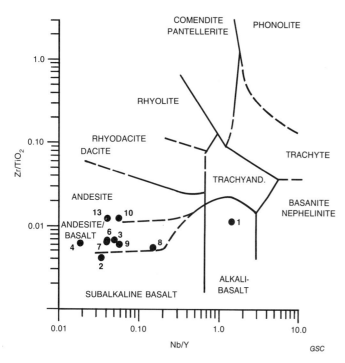

Figure 9.15. M'Clintock Formation: classification of specimens according to stable trace elements (based on Winchester and Floyd, 1977, Fig. 6).

Lorimer Ridge Formation, and a Silurian (Llandovery-Wenlock) unit of flysch (Cranstone Formation) that is correlative with lower parts of the Marvin Formation; the top of the Cranstone Formation is not preserved. The present distribution of these units suggests that the deep water belt was flanked by shelves both on the north and south (Fig. 9.16), but the original configuration is difficult to reconstruct because of scarcity of outcrop and structural uncertainties. The two deep water units both are divisible into a relatively coarse eastern facies and a relatively fine western facies.

Carbonate-mudrock unit

The eastern facies of the Disraeli Glacier beds has been recognized only southwest of Disraeli Fiord where it is 56 m thick and has yielded conodonts of Late Ordovician age. It consists mainly of laminated lime wackestone and argillaceous lime mudstone, both with calcareous spongae, but also includes a lime packstone bed interpreted as a submarine slide. The western facies has a minimum thickness of 400 m and consists of interbedded calcareous and dolomitic mudrock and argillaceous lime mudstone that both contain graptolites of the Late Ordovician *Dicellograptus complanatus ornatus* or *Pacificograptus pacificus* Zones.

Conglomerate-sandstone-mudrock unit

The following summary of the **eastern facies** of the Cranstone Formation is based on exposures west of lower Disraeli Glacier where it overlies the Disraeli Glacier beds with abrupt but probably conformable contact; a 27 km long belt in the ice fields east of Disraeli Fiord is known only from brief helicopter landings. The formation is about 1 km thick and composed of alternating units of mudrock, conglomerate, and sandstone and very small amounts of resedimented carbonates.

Conglomeratic units, 0.01-20 m thick, are mostly of pebble grade but contain cobbles and boulders up to 40 cm in diameter. In some units the phenoclasts are embedded in a muddy or sandy matrix, in others they support each other. Most form thick or massive beds that are bounded by flat surfaces and show no internal structure, but graded beds, concave foresets, and lenticular bodies also are present. Pebbles consist of chert; vein quartz; metaquartzite; volcanic, hypabyssal, and granitic rocks; limestone; dolostone; calcareous sandstone; and dolomitic siltstone. Cobbles and boulders are composed mostly of limestone and dolostone; some contain favositid corals of Upper Ordovician or Silurian age.

The sandstones can be divided into structureless beds, Bouma sequences, and flatbedded, lenticular and crossbedded units associated with conglomerates. Bouma sequences in the lower part of the formation consist mainly of A + B divisions (Fig. 9.17a, 9.17b). The sandstones are composed of quartz (40%), feldspar (13%, mainly sodic plagioclase), carbonates (12%, limestone, dolostone, fossil fragments), chert (10%), metamorphic rock fragments (10%, phyllite, schist), quartzite (8%), volcanic and hypabyssal rock fragments of predominantly siliceous composition (4%), chlorite (1%) and trace amounts of muscovite, biotite, mudrock and granitoid rocks. These data indicate that the eastern facies of the Cranstone Formation is markedly richer in siliceous detritus and poorer in carbonate detritus than correlative turbidites of the adjacent Clements Markham Fold Belt (Chapter 8C, Imina Formation).

The mudrocks are medium dark grey on fresh surfaces and commonly laminated and calcareous and dolomitic. They occur as discrete major units, tens of metres or more in thickness, that contain sparse graptolites and very rare sponges, and as laminae or thin beds associated with sandstones and conglomerates.

The deep water origin of the formation is apparent from its graptolitic fauna and the presence of Bouma sequences. Facies relationships suggest that it was deposited in two separate submarine channels (Fig. 9.16). The siliciclastic sediments and part of the carbonate sediments were derived from terrestrial sources perhaps to the northeast, but other carbonate materials, especially outsized and fossiliferous phenoclasts, probably came from flanking marine strata of the Marvin Formation.

Graptolites from about 400 m and 820-860 m above the base of the formation all are middle Llandovery in age. The lower few hundred metres may be early Llandovery and(?) latest Ordovician in age.

The **western facies** overlies the Disraeli Glacier beds with abrupt and probably faulted contact. It consists mainly of interbedded sandstone and mudrock with very small amounts of sandy pebble conglomerate and has a minimum thickness of about 230 m. Sandstone and mudrock commonly occur in Bouma sequences. The sandstones, very fine to very coarse and commonly medium grained, are composed mainly of quartz with variable proportions of carbonates, feldspar, and volcanic rock fragments, small amounts of muscovite, chlorite, biotite, and metamorphic rock

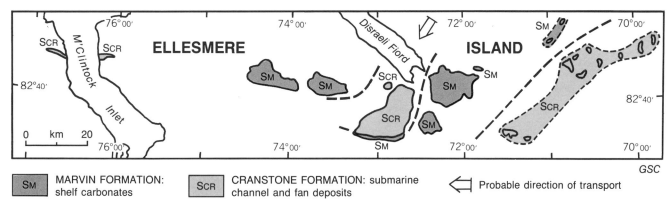

Figure 9.16. Setting of Cranstone Formation.

fragments, and very rare fossil fragments. A conglomerate in the lowermost part of the formation has pebbles of limestone, fossil fragments and chert. These strata probably represent a more distal facies, perhaps a submarine fan that received small amounts of skeletal material from an adjacent shelf. Beds in the lower 100 m or so yielded a graptolite of late Llandovery or Wenlock age (*Monograptus* aff. *M. priodon*).

Uppermost Ordovician to Upper Silurian carbonate and clastic sediments

The fourth major unit of succession IV consists mainly of carbonate sediments and subordinate clastic sediments assigned to the Marvin Formation; an overlying predominantly clastic unit (Crash Point beds) is preserved at one locality only.

Lower, predominantly carbonate unit

In the vicinity of Disraeli Fiord the Marvin Formation overlies the Lorimer Ridge Formation, ranges in age from latest Ordovician to Late Silurian (Ludlow), and has a preserved thickness of about 750 m; the top is eroded. East of M'Clintock Inlet, a tongue of the upper part of the formation lies stratigraphically between an unnamed mudrock unit and the Crash Point beds. The Marvin Formation consists mainly of limestone (wackestone, packstone, grainstone) but is extensively replaced by dolomite in some areas; it also includes minor amounts of calcareous sandstone east of M'Clintock Inlet. The inferred depositional environments range from deep subtidal shelf to intertidal settings with a considerable range of energy conditions. The age range of the formation is inferred mainly from conodonts but it also includes well preserved macrofossils, especially corals and brachiopods. The upper part of the formation probably is correlative with two carbonate units of the Clements Markham Fold Belt, the Piper Pass beds, which have yielded conodonts of middle to late Ludlow age, and the Markham River beds of probable Ludlow age (Chapter 8C).

Upper clastic and carbonate unit

The Crash Point beds are preserved only east of M'Clintock Inlet where they overlie the Marvin Formation with conformable contact. The unit has a minimum thickness of 100-200 m; the upper part is too disturbed for measurement and the top not preserved. It is composed of sandstone, mudrock and lesser amounts of limestone, that can be assigned to two facies: a low-energy subtidal facies, comprising flat-laminated or bioturbated mudrock and very fine grained sandstone and related impure pelletal packstone; and a high-energy shallow water facies comprising fine- to medium-grained calcareous sandstone and sandy skeletal packstone that show flat stratification and medium-scale trough crossbedding and contain relatively thick pelmatozoan stems and ossicles. The Crash Point beds have yielded conodonts of Late Silurian (Ludlow and/or Pridoli) age.

TECTONIC INTERPRETATION OF SUCCESSIONS I-IV

Succession I evidently represents a Middle Proterozoic or older sedimentary and(?) volcanic assemblage that was affected by a 1.0-1.1 Ga orogeny.

Succession II, presumed to overlie succession I with angular unconformity, probably originated in rift-related and passive margin settings, the former suggested by mafic and siliceous volcanics, and the latter by thick units of shelf carbonates, quartzite and mudrock.

The Thores Suite is tentatively interpreted as basement of an oceanic basin that formed in Early Ordovician time. It is speculated that contraction of the basin resulted first in the development of a subduction-related volcanic arc (Maskell Inlet assemblage) and subsequently in a mid-Ordovician collision between the arc and the Empire Belt (M'Clintock Orogeny; Fig. 9.18). The suture between these two belts is unconformably overlapped by strata of the upper Middle Ordovician to Upper Silurian Challenger Basin.

The northeastern limit of the Bromley Island Belt is concealed by strata of the Challenger Basin that are in fault contact with the Upper Proterozoic – Cambrian Mount Disraeli Belt on the north and east. The relationship between the Bromley Island and Mount Disraeli belts thus is unknown. However, the fact that the eastern part of the Challenger Basin shows similar structural trends as the adjacent Mount Disraeli Belt may indicate that it is underlain by the latter at depth. If this is the case, then the Bromley

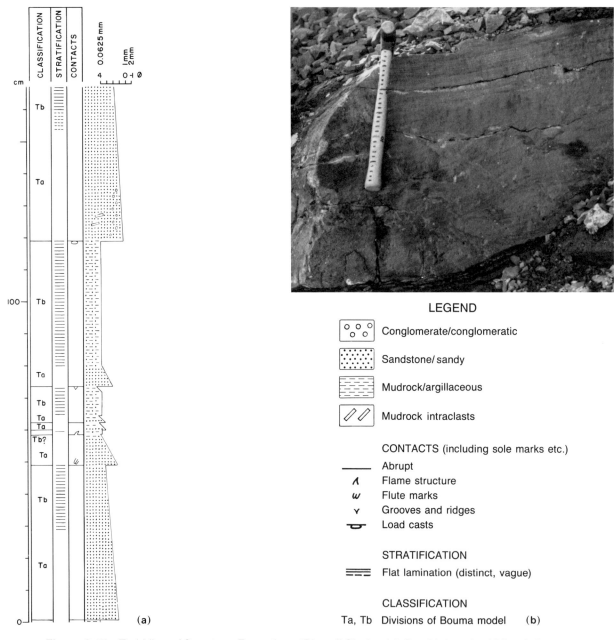

Figure 9.17. Turbidites of Cranstone Formation at Disraeli Glacier. (a) Graphic log of turbidites in lower part of formation. (b) Turbidite in upper part of section shown in (a).

Island Belt probably collided not only with the Empire Belt but also with the Mount Disraeli Belt prior to the development of the Challenger Basin.

The late Middle to early Late Ordovician history of the Challenger Basin is divisible into four phases that demonstrate continued tectonic instability: (1) initial rifting, characterized by coarse clastic sedimentation and felsic, in part alkaline, volcanism, and subsequent thermal(?) subsidence; (2) subduction, evident from andesitic-basaltic volcanism; (3) carbonate sedimentation; and (4) uplift and block faulting. Horizontal plate motions evidently continued during this interval but the nature of these motions is not clear. The subsequent, Late Ordovician to Late Silurian, record of the Challenger Basin was characterized by renewed rifting and thermal(?) subsidence without volcanism or deformation. In some areas deposition kept pace with this subsidence so that shallow marine environments persisted; in others, deposition was slower so that deep water environments developed.

COMPARISON WITH FRANKLINIAN AND CALEDONIAN MOBILE BELTS

The most obvious counterpart for **succession I** is the Grenville-Sveconorwegian Orogen of the North Atlantic region (Fig. 9.19; cf. Zwart and Dornsiepen, 1978; and

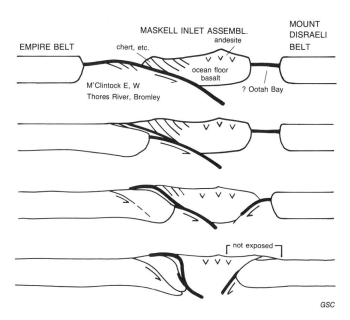

Figure 9.18. Speculative interpretation of M'Clintock Orogeny (adapted from Dewey, 1977, Fig. 3).

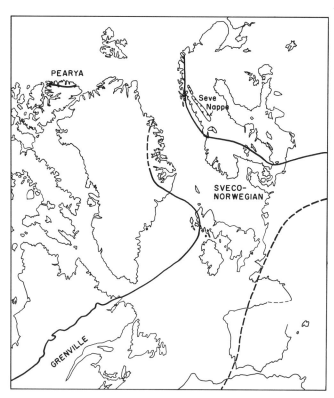

Figure 9.19. Generalized area of Grenville-age basement in North Atlantic region. Modified from Zwart and Dornsiepen, 1978. Base, indicating configuration prior to Cretaceous drift, after Le Pichon et al., 1977. Some authors question the presence of Grenville-age basement in central East Greenland.

Williams, 1984), which also is late Middle Proterozoic in age; age determinations for east-central Greenland are given by Steiger et al., 1979; Hansen et al., 1978; and Peucat et al., 1985; relevant age determinations for Scandinavia include Griffin et al., 1978; Reymer et al., 1980; and Wielens et al., 1981. The basement of the Franklinian mobile belt, on the other hand, is Early Proterozoic – Archean in age (Chapter 6).

Succession II is comparable to the Upper Proterozoic – lower Paleozoic succession of Svalbard in that it consists of metamorphosed shelf sediments as well as diamictite units and siliceous to mafic volcanics (Harland, 1978), but its stratigraphy is too poorly known to permit specific correlations. Comparison with the Franklinian mobile belt is inconclusive because the latest Proterozoic – earliest Cambrian record is not exposed.

Lower Ordovician ultramafic-mafic complexes. The U-Pb zircon age on a leucocratic differentiate of the M'Clintock West body is surprisingly close to ages obtained by the same method for ophiolites in Newfoundland and Scotland (Fig. 9.20). The total age range of the Newfoundland ophiolites as redetermined by U-Pb zircon ages is 477.5 +2.6/-2.0 Ma to 493.9 +2.5/-1.9 Ma (Dunning and Krogh, 1985). Closely comparable to the M'Clintock determination is an age of 481.4 +4.0/-1.9 Ma on the Annieopsquotch Complex (Fig. 9.20, loc. A). A trondhjemite from the Ballantrae complex in Scotland (Fig. 9.20, loc. B) has a U-Pb zircon age of 483 ± 4 Ma (Bluck et al., 1980).

Lower Paleozoic ophiolites and associated felsic intrusions are widely distributed in the Caledonian Orogen of Norway (Furnes et al., 1985). Recent U-Pb age determinations on zircon (Dunning and Pedersen, 1988) have established at least three generations of ophiolites with the age ranges: 497 ± 2 to 489 ± 3 Ma, 470 +9/-5 Ma, and 443 ± 3 Ma. Agewise, the M'Clintock West body lies between the first and second generations of Norwegian ophiolites and is close to arc-related tonalites in southwestern Norway with ages of 485 ± 2 and 482 +6/-4 Ma.

The **M'Clintock Orogeny** is comparable in age, and probably also in nature, to the Taconian Orogeny of eastern North America, best known from Newfoundland (Williams and Hatcher, 1983). There, it undeniably consisted of obduction of Lower Ordovician ophiolites onto the eastern margin of the North American continent in Arenig to late Llandeilo time. The Grampian Orogeny of the British Isles, more difficult to decipher, has been correlated with the Taconian (Dewey and Shackleton, 1984).

The Ordovician tectonic history of the Scandinavian Caledonides is difficult to establish because of extensive thrust faulting in the Silurian and Devonian, and concepts are in a state of flux as new radiometric, petrological and stratigraphic data are being evaluated (e.g. Roberts and Gee, 1985; Thon, 1985; Dallmeyer and Gee, 1986; Dunning and Pedersen, 1988). It now appears that the different generations of ophiolites represent short-lived marginal basins that closed at different times in the late Early – Middle Ordovician interval, prior to continent – continent collision in the Silurian (Dunning and Pedersen, 1988). The M'Clintock Orogeny fits such a scenario in general, but has not yet been dated precisely enough to permit correlation with specific events in Norway.

An angular unconformity between Proterozoic and lower Caradoc strata provides proof for a pre-Caradoc

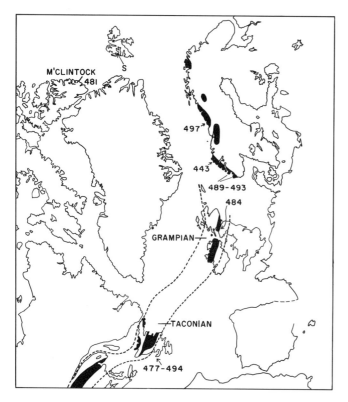

Figure 9.20. Generalized distribution of ophiolitic rocks in the North Atlantic region, and Taconian and Grampian orogens (adapted from Williams, 1984, Fig. 3 and 7; base as in Fig. 9.19). Numbers indicate generalized ranges of U-Pb (zircon) ages on ophiolites (Bluck et al., 1980 [Scotland]; Dunning and Krogh, 1985 [Newfoundland]; Dunning and Pedersen, 1988 [Norway]). S: area of Arenig(?) high-pressure metamorphism in Svalbard.

deformation in central western Spitsbergen (Ohta et al., 1986; Armstrong et al., 1986; Fig. 9.20, loc. C). High-pressure metamorphism of the Proterozoic succession, indicative of a subduction regime, has been dated at 479 ± 14 and 484 ± 14 Ma (recalculated from Horsfield, 1972).

In the Franklinian mobile belt, on the other hand, there is no angular unconformity indicative of a mid-Ordovician deformation. Furthermore, clastic sediments derived from northerly sources do not appear in the deep water basin of the Arctic Islands before latest Ordovician or Early Silurian time.

Collectively, all this evidence strongly suggests that Pearya was part of the Caledonides at least until Late Ordovician time.

Provenance studies on flysch in the Franklinian deep water basin indicate that the terminal orogeny of the northern Caledonides began in Late Ordovician or Early Silurian time (Chapters 7, 8). The Pearya Terrane was not involved in the initial deformation, but must have been close to the orogen in the Early Silurian, when pebble- and cobble-grade conglomerate accumulated in submarine channels in eastern parts of the Challenger Basin. The Lower Silurian flysch of the Challenger Basin is comparable to coeval flysch in the Franklinian deep water basin in general, although there are differences in detrital composition. In spite of analogies between the deep water facies of the Challenger Basin and the Franklinian deep water basin, it is unlikely that accretion of Pearya to the Franklinian mobile belt occurred before the Late Silurian because that process was accompanied by intense deformation and no orogeny is apparent in Pearya between latest Ordovician and early Late Silurian time.

REFERENCES

Armstrong, H.A., Nakrem, H.A., and Ohta, Y.
1986: Ordovician conodonts from the Bulltinden Formation, Motalafjella, central-western Spitsbergen; Polar Research, v. 4 n.s., p. 17-23.

Blackadar, R.G.
1954: Geological reconnaissance of the north coast of Ellesmere Island, Arctic Archipelago, Northwest Territories; Geological Survey of Canada, Paper 53-10, 22 p.

Bluck, B.J., Halliday, A.N., Aftalion, M., and Macintyre, R.M.
1980: Age and origin of Ballantrae ophiolite and its significance to the Caledonian Orogeny and Ordovician time scale; Geology, v. 8, p. 492-495.

Christie, R.L.
1957: Geological reconnaissance of the north coast of Ellesmere Island, District of Franklin, Northwest Territories; Geological Survey of Canada, Paper 56-9, 40 p.

1964: Geological reconnaissance of northeastern Ellesmere Island, District of Franklin (120, 340, parts of); Geological Survey of Canada, Memoir 331, 79 p.

Coleman, R.G.
1977: Ophiolites — Ancient Oceanic Lithosphere?; Springer-Verlag, Berlin, 229 p.

Dallmeyer, R.D. and Gee, D.G.
1986: $^{40}Ar/^{39}Ar$ mineral dates from retrogressed eclogites within the Baltoscandian miogeocline: Implications for a polyphase Caledonian orogenic evolution; Geological Society of America Bulletin, v. 97, p. 26-34.

Dewey, J.F.
1977: Suture zone complexities: a review; Tectonophysics, v. 40, p. 53-67.

Dewey, J.F. and Shackleton, M.
1984: A model for the evolution of the Grampian tract in the early Caledonides and Appalachians; Nature, v. 312, p. 115-121.

Dunning, G.R. and Krogh, T.E.
1985: Geochronology of ophiolites of the Newfoundland Appalachians; Canadian Journal of Earth Sciences, v. 22, p. 1659-1670.

Dunning, G.R. and Pedersen, R.B.
1988: U/Pb ages of ophiolites and arc-related plutons of the Norwegian Caledonides: implications for the development of Iapetus; Contributions to Mineralogy and Petrology, v. 98, p. 13-23.

Frisch, T.
1974: Metamorphic and plutonic rocks of northernmost Ellesmere Island, Canadian Arctic Archipelago; Geological Survey of Canada, Bulletin 229, 87 p.

Furnes, H., Ryan, P.D., Grenne, T., Roberts, D., Sturt, B.A., and Prestvik, T.
1985: Geological and geochemical classification of the ophiolitic fragments in the Scandinavian Caledonides; in The Caledonide Orogen — Scandinavia and Related Areas, D. Gee and B.A. Sturt (ed.); John Wiley and Sons, Chichester, England, p. 657-669.

Griffin, W.L., Taylor, P.N., Hakkinen, J.W., Heier, K.S., Iden, I.K., Krogh, E.J., Malm, O., Olsen, K.I., Ormaasen, D.E., and Tveten, E.
1978: Archaean and Proterozoic crustal evolution in Lofoten-Vesterålen, N Norway; Journal of the Geological Society of London, v. 135, p. 629-647.

Hambrey, M.J.
1983: Correlation of Late Proterozoic tillites in the North Atlantic region and Europe; Geological Magazine, v. 120, p. 209-232.

Hansen, B.T., Higgins, A.K., and Bär, M.T.
1978: Rb-Sr and U-Pb age patterns in polymetamorphic sediments from the southern part of the East Greenland Caledonides; Bulletin of the Geological Society of Denmark, v. 27, p. 55-62.

Harland, W.B.
1978: The Caledonides of Svalbard; in Caledonian-Appalachian Orogen of the North Atlantic Region, E.T. Tozer and P.E. Schenk, (ed.); Geological Survey of Canada, Paper 78-13, p. 3-11.

Harland, W.B., Cox, A.V., Llewellyn, P.G., Pickton, C.A.G., Smith, A.G., and Walters, A.R.
1982: A geologic time scale; Cambridge University Press, Cambridge, United Kingdom, 131 p.

Horsfield, W.T.
1972: Glaucophane schists of Caledonian age from Spitsbergen; Geological Magazine, v. 109, p. 29-36.

Lemoine, M., Tricart, P., and Boillot, G.
1987: Ultramafic and gabbroic ocean floor of the Ligurian Tethys (Alps, Corsica, Apennines): in search of a genetic model; Geology, v. 15, p. 622-625.

Le Pichon, X., Sibuet, J.C., and Francheteau, J.
1977: The fit of the continents around the North Atlantic Ocean; Tectonophysics, v. 38, p. 169-209.

Mayr, U., Trettin, H.P., and Embry, A.F.
1982: Preliminary geological map and notes, Clements Markham Inlet and Robeson Channel map-areas, District of Franklin (NTS 120E, F, G); Geological Survey of Canada, Open File 833, 32 p.

McKerrow, W.S., Lambert, R.St.J., and Cocks, L.R.M.
1985: The Ordovician, Silurian and Devonian periods; in The Chronology of the Geological Record, N.J. Snelling (ed.); The Geological Society, Memoir 10, p. 81-88.

Meschede, M.
1986: A method of discriminating between different types of mid-ocean ridge basalts and continental tholeiites with the Nb-Zr-Y diagram; Chemical Geology, v. 56, p. 207-218.

Odin, G.S.
1985: Remarks on the numerical scale of Ordovician to Devonian times; in The Chronology of the Geological Record, N.J. Snelling (ed.); The Geological Society, Memoir 10, p. 93-98.

Ohta, Y., Hirajima, T., and Hiroi, Y.
1986: Caledonian high-pressure metamorphism in central western Spitsbergen; in Blueschists and Eclogites, B.W. Evans and E.H. Brown (ed.); Geological Society of America, Memoir 164, p. 205-216.

Palmer, A.R.
1983: The Decade of North American Geology 1983 time scale; Geology, v. 11, p. 503-504.

Peucat, J.J., Tisserant, D., and Clauer, N.
1985: Resistance of zircons to U-Pb resetting in a prograde metamorphic sequence of Caledonian age in East Greenland; Canadian Journal of Earth Sciences, v. 22, p. 330-338.

Reymer, A.P.S., Boelrijk, N.A.I.M., Hebeda, E.H., Priem, H.N.A., Verdurmen, E.A.Th., and Verschure, R.H.
1980: A note on Rb-Sr whole-rock ages in the Seve Nappe of the central Scandinavian Caledonides; Norsk Geologisk Tidsskrift, v. 60, p. 139-147.

Roberts, D. and Gee, D.G.
1985: An introduction to the structures of the Scandinavian Caledonides; in The Caledonide Orogen — Scandinavia and Related Areas, D.G. Gee and B.A. Sturt (ed.); John Wiley and Sons, Chichester, England, p. 55-68.

Schuchert, C.
1923: Sites and nature of the North American geosynclines; Bulletin of the Geological Society of America, v. 34, p. 151-229.

Sinha, A.K. and Frisch, T.
1975: Whole-rock Rb/Sr ages of metamorphic rocks from northern Ellesmere Island, Canadian Arctic Archipelago. I. The gneiss terrain between Ayles Fiord and Yelverton Inlet; Canadian Journal of Earth Sciences, v. 12, p. 90-94.
1976: Whole-rock Rb/Sr and zircon U/Pb ages of metamorphic rocks from northern Ellesmere Island, Canadian Arctic Archipelago. II. The Cape Columbia Complex; Canadian Journal of Earth Sciences, v. 13, p. 774-780.

Steiger, R.H., Hansen, B.T., Schuler, C., Bär, M.T., and Henriksen, N.
1979: Polyorogenic nature of southern Caledonian fold belt in East Greenland: an isotopic age study; Journal of Geology, v. 87, p. 475-495.

Stevens, R.D., Delabio, R.N., and Lachance, G.R.
1982: Age determinations and geological studies, K-Ar isotopic ages, Report 16; Geological Survey of Canada, Paper 82-2, 56 p.

Thon, A.
1985: Late Ordovician and Early Silurian cover sequences to the west Norwegian ophiolite fragments: stratigraphy and structural evolution; in The Caledonide Orogen — Scandinavia and related areas, D.G. Gee and B.A. Sturt (ed.); John Wiley and Sons, Chichester, England, p. 407-415.

Trettin, H.P.
1969a: Pre-Mississippian geology of northern Axel Heiberg and northwestern Ellesmere islands, Arctic Archipelago; Geological Survey of Canada, Bulletin 171, 82 p.
1969b: Geology of Ordovician to Pennsylvanian rocks, M'Clintock Inlet, north coast of Ellesmere Island, Canadian Arctic Archipelago; Geological Survey of Canada, Bulletin 183, 93 p.
1971: Reconnaissance of lower Paleozoic geology, Phillips Inlet region, north coast of Ellesmere Island, District of Franklin; Geological Survey of Canada, Paper 71-12, 29 p.
1981: Geology of Precambrian to Devonian rocks, M'Clintock Inlet area, District of Franklin (NTS 340E, H) — preliminary geological map and notes; Geological Survey of Canada, Open File 759, 26 p.
1987a: Pearya: a composite terrane with Caledonian affinities in northern Ellesmere Island; Canadian Journal of Earth Sciences, v. 24, p. 224-245.
1987b: Pre-Carboniferous geology, M'Clintock Inlet map-area, northern Ellesmere Island, interim report and map (340E, H); Geological Survey of Canada, Open File 1652, 210 p.

Trettin, H.P. and Frisch, T.
1987: Bedrock geology, Yelverton Inlet map-area, northern Ellesmere Island, interim report and map (340F, 560D); Geological Survey of Canada, Open File 1651, 98 p.

Trettin, H.P., Loveridge, W.D., and Sullivan, R.W.
1982: U-Pb ages on zircon from the M'Clintock West massif and the Markham Fiord pluton, northernmost Ellesmere Island; in Current Research, Part C, Geological Survey of Canada, Paper 82-1C, p. 161-166.

Trettin, H.P., Parrish, R., and Loveridge, W.D.
1987: U-Pb age determinations of Proterozoic to Devonian rocks from northern Ellesmere Island, Arctic Canada; Canadian Journal of Earth Sciences, v. 24, p. 246-256.

Wielens, J.B.W., Andriessen, P.A.M., Boelrijk, N.A.I.M., Hebeda, E.H., Priem, H.N.A., Verdurmen, E.A.T., and Verschure, R.H.
1981: Isotope geochronology in the high-grade metamorphic Precambrian of southwestern Norway: new data and reinterpretations; Norges geologiske undersøkelse, v. 359, p. 1-30.

Williams, H.
1984: Miogeoclines and suspect terranes of the Caledonian-Appalachian Orogen: tectonic patterns in the North Atlantic region; Canadian Journal of Earth Sciences, v. 21, p. 887-901.

Williams, H. and Hatcher, R.D., Jr.
1983: Appalachian suspect terranes; in Contributions to the Tectonics and Geophysics of Mountain Chains, R.D. Hatcher, Jr., H. Williams, and I. Zietz (ed.); Geological Society of America, Memoir 158, p. 33-53.

Winchester, J.A. and Floyd, P.A.
1977: Geochemical discrimination of different magma series and their products using immobile elements; Chemical Geology, v. 20, p. 325-343.

Zwart, H.J. and Dornsiepen, U.F.
1978: The tectonic framework of central and western Europe; Geologie en Mijnbouw, v. 57, p. 627-654.

Author's address

H.P. Trettin
Geological Survey of Canada
3303-33rd Street N.W.
Calgary, Alberta
T2L 2A7

CHAPTER 9

Printed in Canada

Chapter 10
MIDDLE-UPPER DEVONIAN CLASTIC WEDGE OF THE ARCTIC ISLANDS

Introduction

Stratigraphy, sedimentology and depositional history

External factors affecting the clastic wedge

References

Chapter 10

MIDDLE-UPPER DEVONIAN CLASTIC WEDGE OF THE ARCTIC ISLANDS

Ashton F. Embry

INTRODUCTION

Carbonate deposition dominated the Franklinian miogeocline from Late Cambrian until earliest Middle Devonian. Following a transgression in early Eifelian (within the *costatus-costatus* conodont Zone), quartzose clastics replaced carbonates as the dominant sediment type and, from that time until Early Carboniferous, clastic sedimentation was widespread across the Franklinian miogeocline. During this interval an enormous clastic wedge prograded southwestward, heralding the advance of Ellesmerian deformation.

Middle-Upper Devonian clastic sediments are widely preserved and are most widespread in the western Arctic, where they occur over much of Bathurst, Melville, Prince Patrick and Banks islands (Fig. 10.1). In the eastern Arctic the deposits occur mainly in a broad synclinorium which stretches from central Ellesmere Island to eastern Grinnell Peninsula. Isolated occurrences are present on northern Ellesmere Island in the Yelverton Pass region and in Tertiary grabens on Cornwallis Island (Fig. 10.1, Fig. 4 [in pocket]). Forty-two wells have penetrated the strata and numerous surface sections are described in the literature (Fig. 10.1, Fig. 1 [in pocket]). The maximum preserved thickness of the clastic wedge is about 4000 m, although thermal maturation levels of strata within and directly below the wedge suggest that original thicknesses may have been nearly twice this figure in some areas.

Regional mapping studies carried out by the Geological Survey of Canada in the 1950s and 1960s established a general stratigraphic framework for these clastic sediments (McLaren, 1963; Thorsteinsson and Tozer, 1962; Tozer and Thorsteinsson, 1964; Kerr, 1974). Embry and Klovan (1976) reviewed all previous work up to 1975 and presented a revised stratigraphic nomenclatural system for the clastic wedge as well as interpretations on the depositional history of the strata. Chi and Hills (1976) provided the necessary palynological data for Embry and Klovan's (1976) study. Figure 10.2 illustrates the general stratigraphy and sedimentology of the clastic wedge as determined by Embry and Klovan (1976).

In the past decade few studies have appeared on the clastic wedge. Trettin (1978) and Roblesky (1979) describe the lower portion of the wedge on central Ellesmere Island, between Vendom and Cañon fiords. Mayr (1980) describes four subsurface sections of the deposits on Cameron and Vanier islands (northwest of Bathurst Island) and his report also contains biostratigraphic data in an appendix. McGregor and Camfield (1982) provide detailed palynological data from the Middle Devonian portion of the wedge in a section on northeastern Melville Island. Of special note are two detailed sedimentological studies which are presently nearing completion, Goodbody (1985) on the Bird Fiord Formation and Rice (1982, 1985, 1987) on the Okse Bay Group of southwestern Ellesmere Island. Studies on Melville Island are in the early stages (Harrison et al., 1985).

This summary is based mainly on Embry and Klovan (1976) with the results of the above quoted studies also included. All available well data and unpublished Geological Survey of Canada paleontological reports on the strata have also been reviewed and incorporated.

STRATIGRAPHY, SEDIMENTOLOGY AND DEPOSITIONAL HISTORY
Introduction

The time-stratigraphic framework for the Middle-Upper Devonian clastic wedge is illustrated on the correlation chart in the pocket (Fig. 4) and on Figure 10.3. The biostratigraphic data used in constructing this framework are from Chi and Hills (1976) and subsequent palynological data, which occur in both published and unpublished reports. The only notable change to the ages assigned by Chi and Hills (1976) to their megaspore zones is McGregor and Camfield's (1982) conclusion that the *macromanifestus* Zone of Chi and Hills (1976) is mainly Eifelian in age rather than early to middle(?) Givetian.

The clastic wedge can be conveniently divided into three stratigraphic intervals on the basis of the occurrence of two widespread lithological changes within the wedge. These changes occur at the base of the Beverley Inlet Formation and its equivalents (earliest Frasnian) and at the base of the Parry Islands Formation (late Frasnian) (Fig. 10.2, 10.3). Both these changes probably coincide with regional unconformities. In the following section each of these intervals is described and interpreted in regards to stratigraphy, sedimentology, provenance and depositional history.

Early Eifelian – earliest Frasnian

This stratigraphic interval consists of five formations and four facies, which represent a broad spectrum of depositional environments ranging from deep marine to fluvial plain (Fig. 10.2). The upper portion of black shale and chert-

---

Embry, A.F.
1991: Middle-Upper Devonian clastic wedge of the Arctic Islands; Chapter 10 *in* Geology of the Innuitian Orogen and Arctic Platform of Canada and Greenland, H.P. Trettin (ed.); Geological Survey of Canada, Geology of Canada, no. 3; (*also* Geological Society of America, The Geology of North America, v. E).

CHAPTER 10

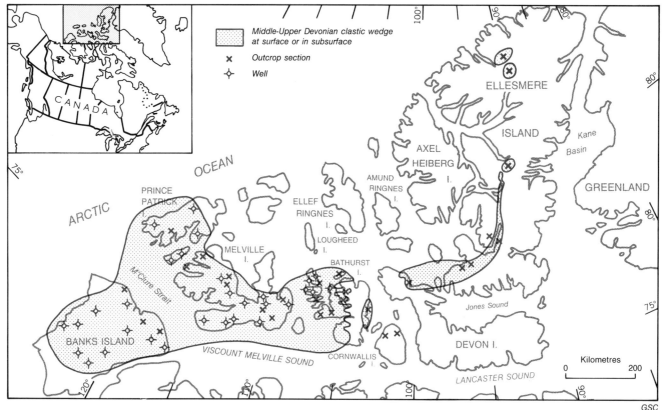

Figure 10.1. Distribution of Middle-Upper Devonian clastic strata and control points. (Special Transverse Mercator Projection with variable scale; this also applies to other regional maps in this chapter.)

dominant formations which underlie the clastics (Fig. 10.2, 10.3) also probably fall within this time-stratigraphic interval on the basis of regional stratigraphic relationships, but biostratigraphic data are not available to evaluate this interpretation. Overall, the interval is characterized by quartz-rich sediments representing a progradational sedimentary succession.

The Blackley Formation is the lowermost clastic unit in the far west and it conformably overlies black, bituminous shale and chert of starved basin origin. The Blackley consists of rhythmically interbedded very fine- to fine-grained sandstone, siltstone and shale; turbidites, representing A to D divisions of the Bouma model, are common. The lithologies and stratigraphic position of this formation indicate a deep marine environment with turbidity currents being the main method of sediment transport. The Blackley Formation is up to 700 m thick and occurs only on western Melville, Prince Patrick and Banks islands (see Chapter 8C).

The Cape de Bray Formation conformably overlies the Blackley and consists almost entirely of grey shale and siltstone with minor amounts of very fine grained sandstone. The rock types and stratigraphic position of the formation and the identification of clinoforms on seismic sections all indicate a marine slope environment of deposition. The Cape de Bray is up to 1025 m thick and thins dramatically to 100 m over carbonate buildups (Blue Fiord Formation). The Cape de Bray is much more widespread than the Blackley and occurs over Bathurst, Melville, Prince Patrick and Banks islands. Where the Blackley is not present, the Cape de Bray overlies black shale with conformable contact, or shallow water carbonates of the Blue Fiord Formation with an unconformable contact.

The Cape de Bray is gradationally overlain by either the Bird Fiord or the Weatherall Formation. These two formations are lithologically similar and are characterized by repetitive, coarsening-upward cycles of shale, siltstone and very fine- to fine-grained sandstone, which are up to 50 m thick (Fig. 10.4). These sediments were deposited on a marine shelf that was under deltaic influence, and the cycles most likely are due to shifting deltaic centres. Fine grained, white sandstones with crossbedding and horizontal lamination of delta front to strand plain origin occur at the top of cycles in the upper part of these formations (Fig. 10.5). The entire Bird Fiord Formation on Bathurst Island and Grinnell Peninsula and the upper portion of the formation on southern Ellesmere Island are included in this interval. The formation reaches its maximum thickness in the Bathurst Island area where it is about 800 m thick. To the west, on Melville, Banks and Prince Patrick islands, marine shelf strata of this interval are assigned to the Weatherall Formation which is up to 1500 m thick. The Bird Fiord Formation differs from the Weatherall Formation by the common presence of brachiopod-rich, arenaceous limestones and calcareous sandstones. Pelecypods are the dominant faunal element of the Weatherall Formation.

In the eastern Arctic, the Strathcona Fiord Formation is laterally equivalent to the Bird Fiord Formation or overlies it. The formation consists mainly of fining-upward

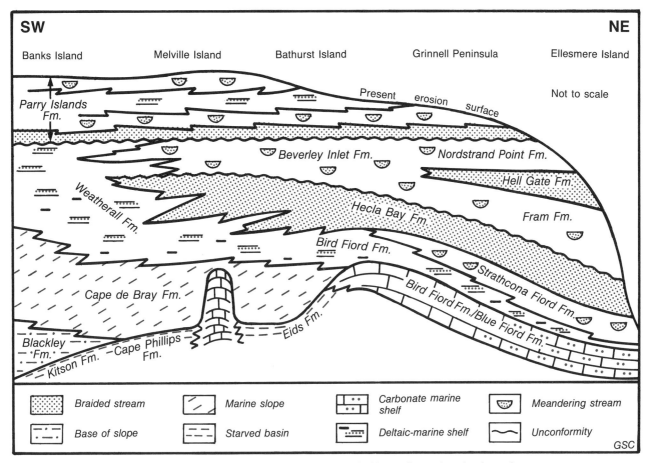

Figure 10.2. Schematic cross-section, Middle-Upper Devonian clastic wedge.

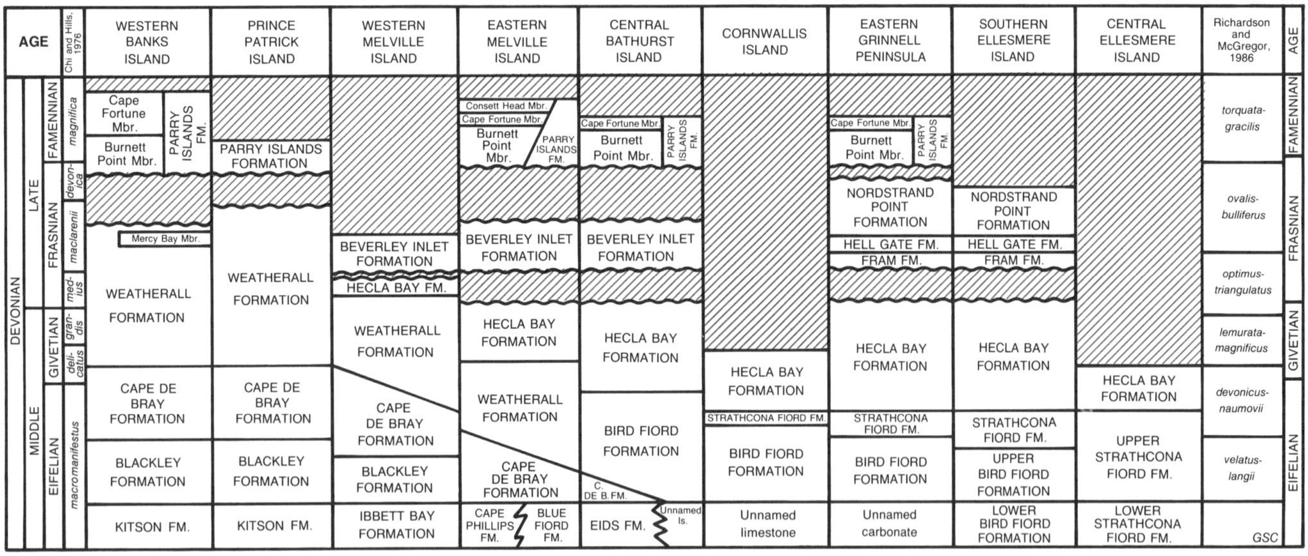

Figure 10.3. Correlation chart, Middle-Upper Devonian formations, Canadian Arctic Islands.

Figure 10.4. Coarsening-upward cycles of Weatherall Formation, western Melville Island.

Figure 10.5. White, fine grained sandstone with horizontal bedding, Bird Fiord Formation, Helena Island.

cycles of very fine- to medium-grained sandstone and mudrock and rare chert-pebble conglomerate, coloured in hues of green, brown and red, that are of meandering stream origin. Also present are thin limestones of brackish water origin that contain fish fragments, charophyte oogonia, and ostracodes. The Eifelian part of the formation is up to 500 m thick in central Ellesmere Island, but thins southwestward and pinches out on Grinnell Peninsula.

The Strathcona Fiord Formation is conformably overlain by the Hecla Bay Formation, which extends for over 1000 km from central Ellesmere Island to western Melville Island (Fig. 10.2, 10.3). The Hecla Bay consists mainly of fine- to medium-grained quartzose sandstone, which is usually crossbedded and occurs in units up to 100 m thick (Fig. 10.6). Sandstone units commonly have scoured basal contacts and are separated by thin intervals of carbonaceous siltstone, shale and rare coal. The formation is mainly of braided stream origin. In the western Arctic, where the Hecla Bay Formation conformably overlies either the Bird Fiord or the Weatherall Formation, basal strata of the Hecla Bay Formation include strata of delta front to lower deltaic plain origin. Also in this area, intervals of sandstone, siltstone and shale of marine shelf origin penetrate the formation.

The main stratigraphic relationships of the formations of this interval are depicted in Figure 10.7. The Strathcona Fiord Formation is laterally equivalent to the Bird Fiord Formation and thins westward, due to facies change, to marine shelf strata. The Hecla Bay Formation extends much farther westward than the Strathcona Fiord Formation and changes facies westward to deltaic-marine strata of the Bird Fiord and Weatherall formations. Two major tongues of deltaic-marine strata can be recognized penetrating the Hecla Bay Formation, and the formation is completely replaced by Weatherall strata in the southwest (Banks and Prince Patrick islands) (Fig. 10.8). The Bird Fiord and Weatherall grade westward into the Cape de Bray slope shales, and the Cape de Bray together with the Blackley Formation represent the fill of marginal parts of the original deep water basin.

The main source areas of the clastic sediments of this interval were the Precambrian Shield of Greenland and the western slopes of the Caledonian Mountains. This interpretation is based on the highly quartzose nature of the sediments, ubiquitous west-southwest directed paleocurrents, and nonmarine to marine facies change to the southwest. Uplifts in northern Ellesmere Island and possibly in northernmost Greenland also contributed sediment, but this same area was subsidiary to the eastern sources. Goodbody (1985) has demonstrated that the Canadian Shield to the southeast also contributed quartzose sediment.

The early Eifelian to earliest Frasnian interval records a major, southwestward progradation of a thick clastic succession across the Franklinian miogeocline. Clastic influx to the miogeocline greatly increased in early Eifelian following a transgression, and a deltaic plain, characterized by meandering streams (Strathcona Fiord Formation), prograded southwestward over marine shelf sediment (Bird Fiord Formation). As these depositional systems reached the western Arctic slope, mud and silt (Cape de Bray) began to fill deep marine areas. In the far west, turbidites, perhaps derived from both northern parts of the Franklinian mobile

Figure 10.6. Crossbedded, fine grained sandstone of Hecla Bay Formation, eastern Melville Island.

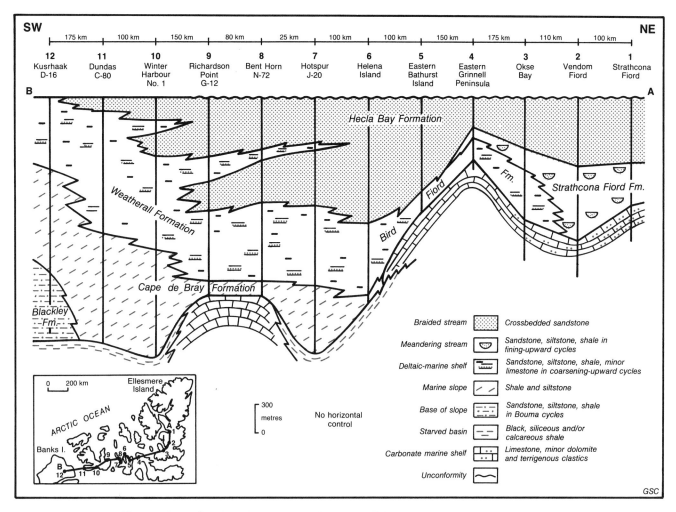

Figure 10.7. Stratigraphic cross-section, lower Eifelian – lowermost Frasnian strata.

belt and the Canadian Shield, were deposited on the basin floor (Fig. 10.9). The climate from early to late Eifelian was "savanna", that is temperate to tropical with seasonal rainfall, as evidenced by red shales and caliche deposits in overbank strata.

In late Eifelian the climate became much more humid and the water and sediment supply to the coastal plain and shelf greatly increased. The fluvial system was transformed into a braided one and plant debris accumulated in overbank areas (Hecla Bay Formation). The deltaic-coastal plain continued to expand westward over marine shelf deposits (Weatherall Formation), and marine slope deposits (Cape de Bray Formation) infilled deep marine areas in the west. By earliest Frasnian time the deltaic plain had reached southern and western Melville Island, and the marine shelf extended over Prince Patrick and Banks islands (Fig. 10.10).

Earliest Frasnian – late Frasnian

The early-late Frasnian interval consists of five formations with three facies, which represent coastal plain and shallow marine environments (Fig. 10.2). The contact with the underlying interval is abrupt and is marked by the sudden appearance of chert and rock fragments in the sandstones, a higher proportion of overbank deposits in the fluvial sediments, and the presence of red and green shales in the overbank strata. The contact of the basal strata of this interval with the underlying Hecla Bay Formation is interpreted to be a regional unconformity over most of the area on the basis of (a) the abrupt lithological changes across the contact, (b) the presence of a kaolinite-rich shale unit (soil?) at the contact at some localities, and (c) palynological data that indicate the uppermost Hecla Bay Formation is earliest Frasnian on southern and western Melville Island but is no younger than Givetian to the east (Chi and Hills, 1976).

In the eastern Arctic, from central Ellesmere Island to eastern Grinnell Peninsula, the Fram, Hell Gate and Nordstrand Point formations comprise the interval and, combined, reach a maximum thickness of 2500 m. The Fram Formation is up to 1125 m thick and consists of repetitive coarsening-upward cycles of very fine- to medium-grained sandstone, siltstone and shale which are 5 to 10 m thick (Fig. 10.11). The argillaceous units are often red and green in the lower portion but become more carbonaceous and include rare coal seams in the upper portion. The strata are interpreted to represent meandering stream deposits.

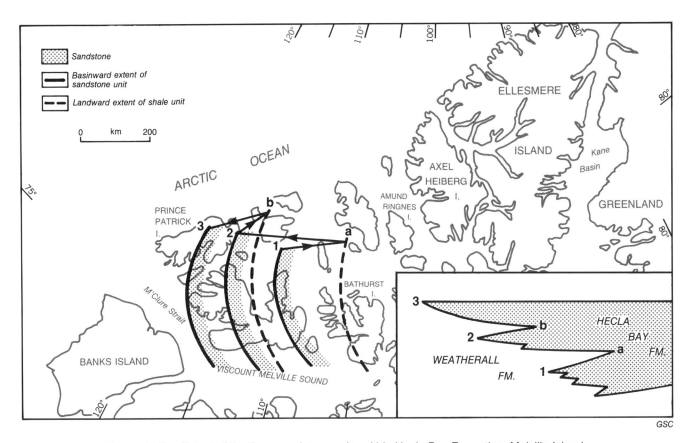

Figure 10.8. Extent of the three sandstone units within Hecla Bay Formation, Melville Island.

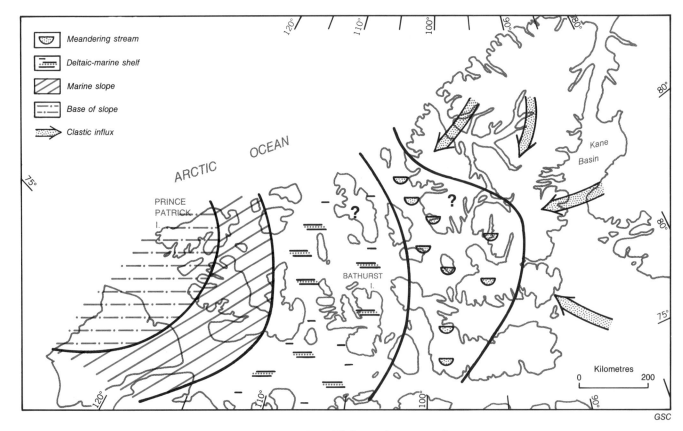

Figure 10.9. Late Eifelian paleogeography.

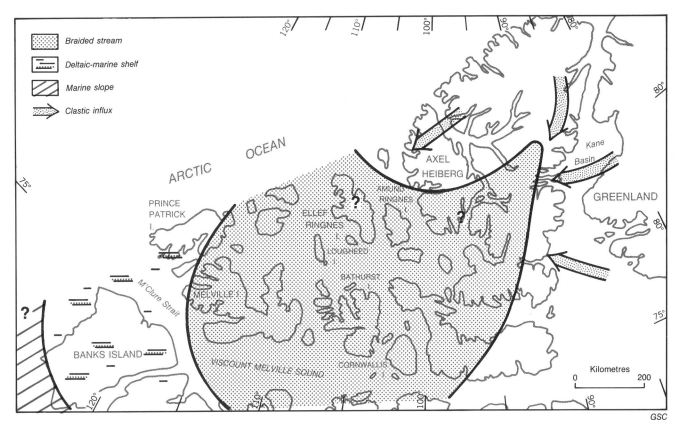

Figure 10.10. Latest Givetian – earliest Frasnian paleogeography.

Figure 10.11. Fining-upward cycles in Fram Formation, southern Ellesmere Island.

Figure 10.12. Coral-stromatoporoid bioherm of Mercy Bay Member, Weatherall Formation, Banks Island. Off-reef strata to right.

The overlying Hell Gate Formation, which is up to 650 m thick, consists mainly of fine- to medium-grained, crossbedded sandstone with thin intervals of carbonaceous siltstone and shale. These strata are of braided stream origin. The Nordstrand Point Formation, which is up to 725 m thick, is similar to the Fram Formation and consists of coarsening-upward cycles of meandering stream origin.

In the Yelverton Pass region of northern Ellesmere Island, a thick succession of Upper Devonian sediments, assigned to the Okse Bay Formation, is exposed in two areas that are surrounded by unconformably overlying Carboniferous and younger strata (Mayr et al., 1982; Fig. 12E.2); the base of the succession is not exposed. The beds are markedly less deformed than the Silurian and older rocks in northern Ellesmere Island. It is presumed that they occupy a graben (or two separate grabens), but the bounding faults are not exposed. Two units are distinguished. The lower one is 280-320+ m thick and is tentatively placed in the early-late Frasnian interval on the basis of sparse palynological data and lithological similarity. It consists of interbedded very fine- to coarse-grained, pebbly sandstone, siltstone and shale arranged in fining-upward cycles of meandering stream origin. Argillaceous units are usually maroon although dark grey beds, rich in carbonaceous matter, occur near the top of the unit. The sandstones are lithic arenites and contain chert and detrital carbonate as well as quartz. Paleocurrent data indicate derivation of the sediment from the southwest (Fig. 12E.1).

In the western Arctic, the strata of this interval are placed in the Beverley Inlet Formation on Bathurst and Melville islands and in the upper Weatherall Formation on Banks and Prince Patrick islands. The Beverley Inlet Formation is up to 777 m thick and consists mainly of very fine- to medium-grained sandstone, siltstone, shale and rare coal of meandering stream origin. Shale and siltstone dominate, with sandstone content and grain size as well as formation thickness increasing to the northeast. On southern and western Melville Island, units of brachiopod-bearing siltstone and shale are present within the formation. Farther south and west, the Beverley Inlet is replaced by marine shelf strata of the upper Weatherall Formation.

The upper portion of the Weatherall Formation on Banks and Prince Patrick islands consists of coarsening-upward cycles of very fine- to fine-grained sandstone, siltstone and shale with thin brachiopod coquina often capping the cycles. These strata are of marine shelf origin. On Banks Island, a 200 m limestone unit, the Mercy Bay Member (Embry and Klovan, 1971), occurs near the top of the Weatherall Formation (Fig. 10.3) and consists mainly of coral and stromatoporoid bioherms and biostromes (Fig. 10.12).

Figure 10.13 illustrates the main stratigraphic relationships of the early-late Frasnian strata. Fluvial sediments are mainly of meandering stream origin, with braided stream deposits in the northeast and east. These fluvial deposits change facies to marine shelf strata to the southwest and very little coastal plain progradation is apparent during this interval. The units thicken markedly to the northeast and east.

The main source areas for the deposits of this interval were again the northern parts of the Franklinian mobile belt and the Precambrian Shield and Caledonian Mountains to the east. However, in contrast to the previous interval, the northern mountains contributed substantially more sediment to the basin, as evidenced by the common occurrence of chert and rock fragments and by more southwardly directed paleocurrents. Chert content increases westward, suggesting the southward flowing streams from the northern mountains joined westward flowing streams which originated on the Precambrian Shield (Fig. 10.14).

Widespread uplift occurred in latest Givetian – earliest Frasnian. This was followed by a marked increase in subsidence rates especially in areas closest to the source areas, a notable increase in sediment supply from the northern mountains, a change in climate from very humid to savanna, and possibly by graben initiation within the uplifted northern mountains. All of these changes taken together indicate that major tectonic activity occurred to the north near the Givetian-Frasnian boundary.

The fluvial system on the coastal plain responded to the drier climate by changing from a braided pattern (Hecla Bay Formation) to a meandering one (Fram, Beverley Inlet formations). The high rates of subsidence resulted in an initial transgression followed by only minor coastal plain progradation during this interval, because most of the incoming sediment remained within the fast-subsiding coastal plain area. These thick fluvial sediments included braided stream deposits in areas close to the source areas (Fig. 10.15). A transgressive event in mid-Frasnian is recorded by the establishment and growth of a coral-stromatoporoid reef tract on the marine shelf in the northeast Banks Island area. Widespread uplift in late Frasnian terminated this interval.

Late Frasnian – middle Famennian

This interval contains the youngest portion of the preserved clastic wedge, and the strata occur in the cores of synclines in widely separated areas between eastern Grinnell Peninsula and northern Banks Island. All of these deposits are assigned to a single formation, the Parry Islands, which has three members, Burnett Point, Cape Fortune and Consett Head. Three facies, representing braided stream, meandering stream and marine shelf environments, are recognized. The upper part of the Okse Bay Formation in Yelverton Pass is tentatively correlated with the Parry Islands Formation on the basis of lithological similarlity and stratigraphic position.

The basal contact of the Parry Islands Formation is everywhere abrupt with coarse grained, pebbly sandstone of fluvial origin overlying much finer grained, coastal plain or marine shelf strata. Embry and Klovan (1976, p. 587) interpreted this contact as a regional unconformity.

The lower member of the Parry Islands Formation, the Burnett Point, is up to 720 m thick, with thicknesses decreasing to the southwest. The lower portion of the member consists mainly of braided stream deposits characterized by fine- to coarse-grained, pebbly, crossbedded sandstone (Fig. 10.16). The upper portion contains fining-upward cycles of very fine- to medium-grained sandstone, siltstone, shale and coal (meandering stream deposits). Overall sandstone content and grain size decrease to the southwest and upward in the section.

The middle member, the Cape Fortune, consists mainly of very fine- to fine-grained sandstone, siltstone and shale with brachiopods and crinoids. The member is up to 250 m thick and was deposited in a marine shelf environment.

The upper member, the Consett Head, is present only on eastern Melville Island and Byam Martin Island and outcrop is poor. White, fine- to medium-grained sandstone is the predominant lithology, with lesser amounts of siltstone, shale and coal. These strata originated on a fluvial-deltaic plain.

Sandstone compositions in the Parry Islands Formation are distinguished by high chert and rock fragment content, which usually exceeds the quartz content. Chert content appears to decrease eastward.

The upper portion of the Okse Bay Formation, preserved in grabens(?) in the Yelverton Pass area, consists mainly of fine- to coarse-grained pebbly sandstone with only minor

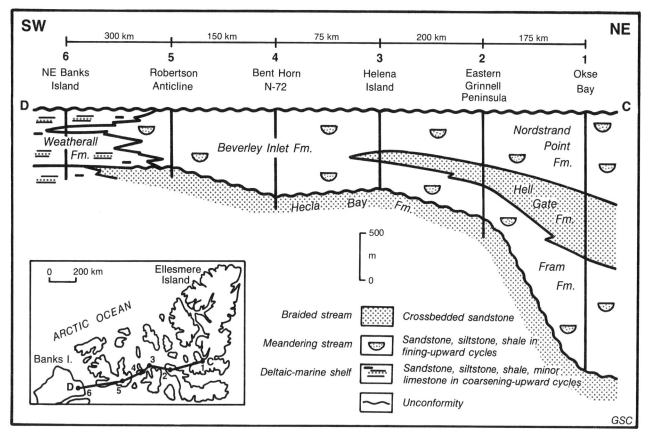

Figure 10.13. Stratigraphic cross-section, lower-middle Frasnian strata.

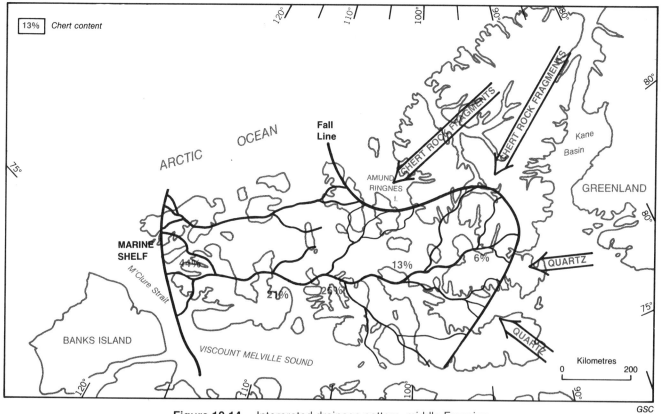

Figure 10.14. Interpreted drainage pattern, middle Frasnian.

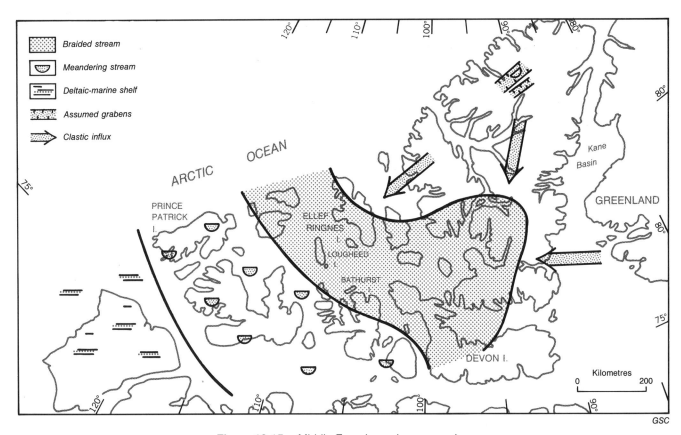

Figure 10.15. Middle Frasnian paleogeography.

Figure 10.16. White sandstones of Parry Islands Formation unconformably overlying Weatherall Formation, northeast Banks Island.

siltstone and shale. The strata are up to 1000 m thick and are braided stream deposits.

Stratigraphic relationships within this interval are poorly understood because of its poor preservation. The Burnett Point Member thins and becomes finer grained toward the southwest (Fig. 10.17), and the upper contact may become older in that direction. The Cape Fortune Member is widespread indicating that much of the coastal plain was drowned in early Famennian. The member is very thick on Banks Island and contains lower deltaic plain strata which are in part equivalent to the Consett Head Member of Melville Island.

The high chert and rock fragment content of the strata, in conjunction with southwesterly paleocurrents and overall grain size increase to the northeast, indicates that the main source area was in the northern part of the Franklinian mobile belt. The increase in quartz content to the east suggests that the shield areas to the east also contributed sediment but were secondary to the northern mountains. The sediments in the Yelverton Pass area were derived from uplifted Silurian and older strata to the east-southeast (U. Mayr, pers. comm., 1985, and Chapter 12E).

Much of the Franklinian Shelf was uplifted in late Frasnian, probably in response to tectonic movements to the north. The area then began to subside and the shelf was gradually transgressed from the southwest, so that marine shelf conditions prevailed throughout the basin by early Famennian time (Cape Fortune Member). Following this, rate of supply again exceeded the subsidence rate and the coastal plain built out toward the southwest (Consett Head Member) (Fig. 10.18).

Late Famennian – Tournaisian

Strata of late Famennian to Tournaisian age have not been identified in the Arctic Islands, so that the geological history of the Arctic during this time interval is rather speculative.

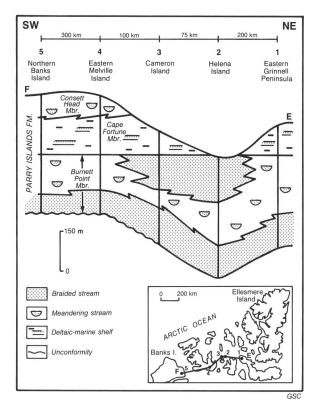

Figure 10.17. Stratigraphic cross-section of upper Frasnian – middle Famennian strata.

The occurrence of reworked late Famennian to Tournaisian spores in Upper Carboniferous and younger units of the Sverdrup Basin (unpublished palynological reports) suggests that sediments of this age originally were present but have subsequently been eroded. Rough estimates of the thickness of these strata in various areas can be made by using thermal maturation data from the uppermost beds of the clastic wedge, and from strata immediately underlying the wedge, to determine maximum depths of burial.

On eastern Melville Island a coal unit from very near the top of the clastic succession (Consett Head Member) had a vitrinite reflectance of 0.68 indicating that it is thermally mature (F.A. Goodarzi, pers. comm., 1985). Assuming normal heat flow values, this level of maturity would be achieved at a depth of about 2500-3000 m. Such burial could have been achieved only by late Famennian to Early Carboniferous sedimentation, because the area was emergent during most of late Carboniferous to Tertiary time. The strata directly underlying the clastic wedge in this area are overmature (Powell, 1978), which indicates a maximum burial depth, i.e. maximum thickness of the clastic wedge, of 7000 to 8000 m, thus confirming the estimate for eroded strata.

To the east, on southern Ellesmere Island, the carbonate strata underlying the clastics are mature. As the preserved thickness of the clastic wedge is similar to that of eastern Melville Island, this suggests that upper Famennian to Tournaisian strata were thinner here (1000-1500 m). On western Banks Islands the strata underlying the clastic wedge are highly thermally altered (graphite present [Crain, 1977]), indicating that the clastic wedge reached its maximum thickness in this area (8000 – 10 000 m). Only 4000 m of clastic strata are preserved, and thus upper Famennian to Tournaisian strata may have been up to 6000 m thick in this area.

Some insight into the composition of upper Famennian to Tournaisian strata in the eastern Melville Island area can be gained from compositional data of the Upper Carboniferous Canyon Fiord Formation, which comprises the initial deposits of the Sverdrup Basin in this area. The strata consist mainly of fine- to coarse-grained, chert-rich sandstones and chert-pebble conglomerates. The highly fractured chert pebbles show evidence of having been originally derived from deformed strata. However, paleocurrent data and facies analysis demonstrate that Canyon Fiord sediments were derived mainly from the south where no deformed, chert-rich strata occur. The derivation of the chert pebbles is best explained by reworking of pebbles from upper Famennian – Lower Carboniferous pebbly sandstones and conglomerates, which were initially derived from uplifted and deformed cherty strata to the north (Fig. 10.19). Overall, it is concluded that a thick succession of upper Famennian – Tournaisian coarse clastic strata of fluvial to alluvial fan origin derived from the northern mountains were the final deposits of the foreland basin. The depocentre for these strata occurred in the western Banks Island area.

The clastic wedge was uplifted and the proximal portion deformed in the final phase of Ellesmerian deformation in the Tournaisian. The strata have been undergoing erosion for most of the time following this event, except during transgressive intervals in Cretaceous time when shallow seas covered much of the Arctic Islands.

EXTERNAL FACTORS AFFECTING THE CLASTIC WEDGE

The primary external factors were tectonics and climate, which together governed the sediment supply.

Tectonics

Orogenic activity occurred in the northern parts of the mobile belt from Late Silurian time onward, but until Early Devonian time the sediments derived from this region (and from the northern Caledonides) were trapped in the deep water basin. The present foreland (or molasse) basin came into existence when the northeastern part of the deep water basin had been filled, or was uplifted, so that the clastic sediments now had access to the Franklinian Shelf. At the same time, epeirogeny affected extensive areas on the southeastern side of the basin. The subsidence in the basin probably was caused by tectonic loading of its northwestern margin combined with sedimentary loading of the basin itself.

The three distinct stratigraphic intervals which comprise the overall clastic wedge are interpreted by Embry (1988) to reflect major tectonic episodes in the adjacent orogenic belt. Each interval was initiated by the widespread occurrence of accelerated subsidence related to crustal shortening and tectonic loading in the Franklinian mobile belt. The unconformity which caps the intervals may be the result of crustal uplift related to the partial removal of the tectonic load by erosion or to an expanding forebulge.

The tectonic record, discussed in the preceding parts of this chapter, may be summarized as follows.

Early Eifelian to late Givetian: Sediments were derived mainly from the northeast and southeast; the contribution from northern Ellesmere Island and environs was small. During this time the foreland basin filled from northeast to southwest, roughly parallel to its axis (Fig. 10.20).

Early and middle Frasnian: An abrupt increase in clastic supply from the northern part of the Franklinian mobile belt indicates tectonic activity in that area. Subsidence in the foreland basin generally increased toward the northeast, and most sediments were deposited in the coastal plain adjacent to the source areas while the shoreline remained relatively stable.

Late Frasnian – early Famennian: Uplift of much of the foreland basin in late Frasnian was followed by increasing subsidence, which by early Famennian time exceeded sediment input, so that regional transgression occurred. The sediments, supplied mainly by the northern mountains, are markedly coarser than previously. Basin fill occurred in a more southward direction perpendicular to the northern mountains but parallel to the extension of the mountain chain to the south (Fig. 10.20).

Late Famennian – early Tournaisian: Inferred thick and coarse clastic sediments indicate continuing uplift in the orogenic belt and the migration of the depocentre to the southwest. The final phase of tectonism and southwestward migration of the mountain front probably occurred in the Tournaisian and resulted in uplift of much of the foreland basin and deformation of proximal areas. The northern mountains were affected by extensional tectonics in the Viséan, and became an area of subsidence and deposition (Sverdrup Basin) rather than one of uplift and supply.

Climate

The foreland basin and its source areas were located in an equatorial zone during Middle and Late Devonian (Embry and Klovan, 1976; Heckel and Witzke, 1979) and consequently were influenced by a generally humid climate. This climatic regime was favourable for the production and transportation of clastic sediments, and this in part may explain the enormous size of the clastic wedge. For comparison it is about ten times the size of the Middle-Upper Devonian "Catskill" clastic wedge, which was deposited under a savanna climate in a foredeep basin west of the Appalachian Mountains.

The sedimentary record, indicates three semi-cycles, progressing from savanna to very humid (Fig. 10.21). The relatively arid climate at the beginning of each semi-cycle probably was caused by uplifts in the northern mountains as evidenced by the coincidence of the abrupt climate changes with tectonic episodes.

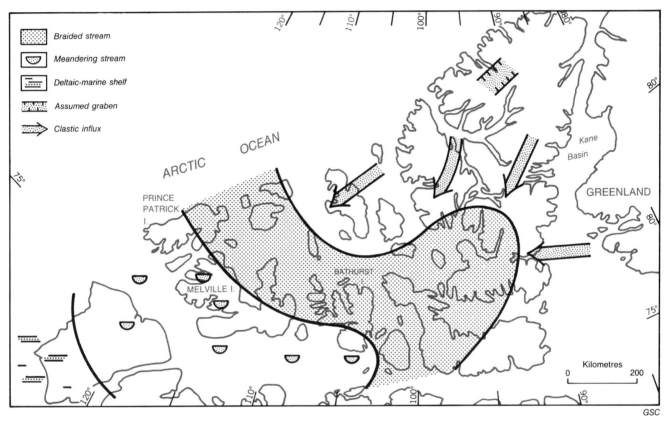

Figure 10.18. Early Famennian paleogeography.

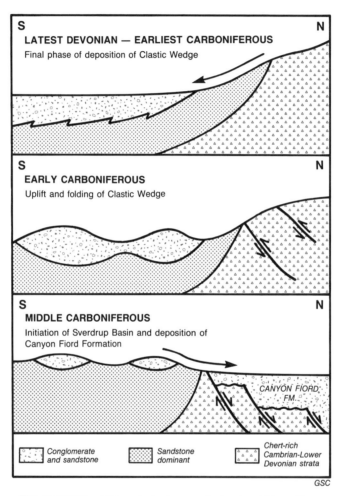

Figure 10.19. Interpreted source of chert pebbles, Upper Carboniferous Canyon Fiord Formation.

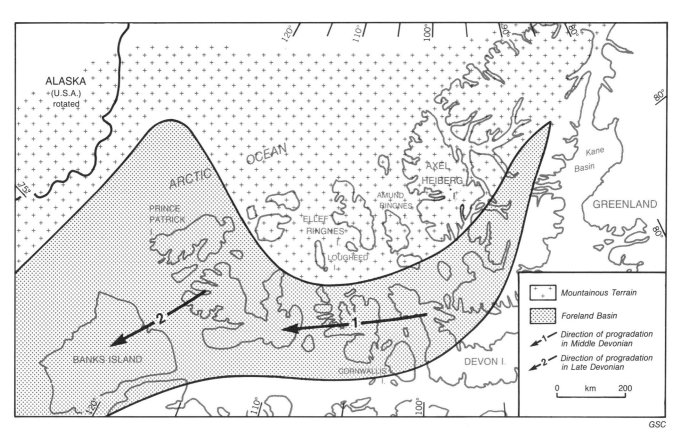

Figure 10.20. Interpreted basin fill pattern for Middle-Upper Devonian foreland basin. Plate reconstruction from Harland et al. (1984).

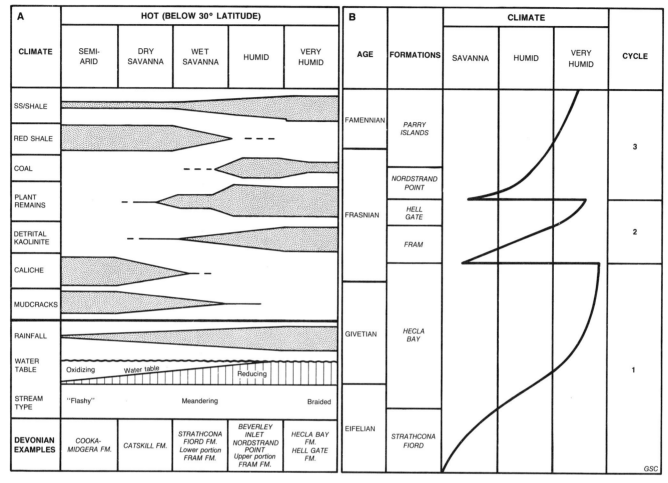

Figure 10.21. (a) Climatic and lithological variations in Middle-Upper Devonian fluvial strata, Arctic Islands. (b) Climatic changes, Middle-Upper Devonian.

REFERENCES

Chi, B.I. and Hills, L.V.
1976: Biostratigraphy and taxonomy of Devonian megaspores, Arctic Canada; Bulletin of Canadian Petroleum Geology, v. 24, p. 640-818.

Crain, E.R.
1977: Log interpretation in the High Arctic; in Transactions of the CWLS Sixth Formation Evaluation Symposium; Canadian Well Logging Society, p. S1-S32.

Embry, A.F.
1988: Middle-Upper Devonian sedimentation in the Canadian Arctic Islands and the Ellesmerian Orogeny; in Devonian of the World, N.J. McMillan, A.F. Embry and D.J. Glass (ed.); Canadian Society of Petroleum Geologists, Memoir 14, v. 2, p. 15-28.

Embry, A.F. and Klovan, J.E.
1971: A Late Devonian reef tract on northeastern Banks Island, Northwest Territories; Bulletin of Canadian Petroleum Geology, v. 19, p. 730-781.
1976: The Middle-Upper Devonian clastic wedge of the Franklinian Geosyncline; Bulletin of Canadian Petroleum Geology, v. 24, p. 485-639.

Goodbody, Q.H.
1985: Stratigraphy, sedimentology and paleontology of the Bird Fiord Formation, Canadian Arctic Archipelago; unpublished Ph.D. thesis, University of Alberta, Edmonton, Alberta, 368 p.

Harland, W.B., Gaskell, B.A., Heafford, A.P., Lind, E.K., and Perkins, P.J.
1984: Outline of Arctic post-Silurian continental displacements; in Petroleum Geology of the North European Margin, A.M. Spencer (ed.); Norwegian Petroleum Society, p. 137-148.

Harrison, J.C., Goodbody, Q.H., and Christie, R.L.
1985: Stratigraphic and structural studies on Melville Island, District of Franklin; in Current Research, Part A, Geological Survey of Canada, Paper 85-1A, p. 629-637.

Heckel, P.H. and Witzke, B.J.
1979: Devonian world paleogeography determined from distribution of carbonates and related lithic paleoclimatic indicators; in The Devonian System, M.R. House et al. (ed.); The Paleontological Association of London, Special Papers in Paleontology, no. 23, p. 99-123.

Kerr, J.W.
1974: Geology of Bathurst Island Group and Byam Martin Island, Arctic Canada (Operation Bathurst Island); Geological Survey of Canada, Memoir 378, 152 p.

Mayr, U.
1980: Stratigraphy and correlation of Lower Paleozoic formations, subsurface of Bathurst Island and adjacent smaller islands, Canadian Arctic Archipelago; Geological Survey of Canada, Bulletin 306, 50 p.

McGregor, D.C. and Camfield, M.
1982: Middle Devonian miospores from the Cape de Bray, Weatherall and Hecla Bay formations of northeastern Melville Island, Canadian Arctic; Geological Survey of Canada, Bulletin 348, 105 p.

McLaren, D.J.
1963: Goose Fiord to Bjorne Peninsula; in Geology of the north-central part of the Arctic Archipelago, Northwest Territories (Operation Franklin), Y.O. Fortier et al.; Geological Survey of Canada, Memoir 320, p. 310-338.

Powell, T.G.
1978: An assessment of the hydrocarbon source rock potential of the Canadian Arctic Islands; Geological Survey of Canada, Paper 78-12, 82 p.

Rice, R.J.
1982: Sedimentology of the Late Middle and Upper Devonian Okse Bay Group, southwestern Ellesmere Island: a progress report; in Current Research, Part A, Geological Survey of Canada, Paper 82-1A, p. 79-82.

1985: Sedimentology of the Middle and Upper Devonian Okse Bay Group, southwestern Ellesmere Island and North Kent Island, Arctic Archipelago; Geological Association of Canada, Program with Abstracts, Joint Annual Meeting, p. A51.

1987: The sedimentology and petrology of the Okse Bay Group (Middle and Upper Devonian) on S.W. Ellesmere Island and North Kent Island in the Canadian Arctic Archipelago; unpublished Ph.D. thesis, McMaster University, Hamilton, Ontario, 769 p.

Richardson, J.B. and McGregor, D.C.
1986: Silurian and Devonian spore zones of the Old Red Sandstone Continent and adjacent regions; Geological Survey of Canada, Bulletin 364, 79 p.

Roblesky, R.F.
1979: Upper Silurian (Late Ludlovian) to Upper Devonian (Frasnian?) stratigraphy and depositional history of the Vendom Fiord region, Ellesmere Island, Canadian Arctic; unpublished M.Sc. thesis, Department of Geology and Geophysics, University of Calgary, Calgary, Alberta, 230 p.

Thorsteinsson, R. and Tozer, E.T.
1962: Banks, Victoria and Stefansson Islands, District of Franklin, Northwest Territories; Geological Survey of Canada, Memoir 330, 85 p.

Tozer, E.T. and Thorsteinsson, R.
1964: Western Queen Elizabeth Islands, Arctic Archipelago; Geological Survey of Canada, Memoir 332, 242 p.

Trettin, H.P.
1978: Devonian stratigraphy, west-central Ellesmere Island, Arctic Archipelago; Geological Survey of Canada, Bulletin 302, 119 p.

Author's address

Ashton F. Embry
Geological Survey of Canada
3303-33rd Street N.W.
Calgary, Alberta
T2L 2A7

CHAPTER 10

Printed in Canada

Chapter 11
DEVONIAN – EARLY CARBONIFEROUS DEFORMATION AND METAMORPHISM, NORTH GREENLAND

A. Deformation – *N.J. Soper and A.K. Higgins*

B. Metamorphism – *A.K. Higgins and N.J. Soper*

References

Chapter 11

DEVONIAN – EARLY CARBONIFEROUS DEFORMATION AND METAMORPHISM, NORTH GREENLAND

A. DEFORMATION

N.J. Soper and A.K. Higgins

INTRODUCTION

As outlined above (Chapters 7, 8), the sedimentary subprovince of the Franklinian deep water basin (Hazen Trough) extends eastward from the Canadian Arctic Islands (Chapter 8C) into northern Greenland (Chapter 7), and the pulses of deformation which affected the Canadian sector (Hazen Fold Belt) in Late Silurian or Devonian to Early Carboniferous time (Chapter 12) likewise extended into Greenland producing an east-west striking mobile zone, known traditionally as the North Greenland fold belt. The fold belt has an exposed width of up to 100 km on the north coast of Greenland and is flanked to the south by the weakly deformed lower Paleozoic platform sequence described in Chapter 7.

Figure 11.1 shows that several distinct tectonic zones can be recognized in the fold belt, and Figure 11.2 illustrates how these are spatially related to the geometry of the deep water basin (Hazen Trough). A southern fold and thrust zone coincides with a region which was transitional between the platform and trough for much of the Cambrian and Ordovician. It is bounded to the south by the Navarana Fjord lineament, which represents the position to which the platform margin had retreated by Early Silurian time. The structure of the southern zone is of thin skinned fold and thrust type, with an approximately east-west strike and southerly vergence.

A northern orthotectonic zone is developed on the site of the trough proper with its thick fill of dominantly Lower Cambrian turbidites. It forms the mountainous regions of Johannes V. Jensen Land, Nansen Land, and adjacent islands along the north coast. The structural style may be termed orthotectonic, with multiphase deformation and northwardly increasing metamorphism. Folds are upright in the south of this zone, but northward become progressively tighter and more recumbent, with northerly vergence (see cross-section, Fig. 3A in pocket).

Between the northern and southern zones is a tract across which the vergence changes. This divergence zone also coincides with a profound change in the stratigraphic level of rocks exposed at the present surface. To the north, in the orthotectonic zone on the site of the trough, Lower Cambrian rocks are exposed (except locally in a plunge depression in central Johannes V. Jensen Land), whereas to the south, across the fold and thrust zone and adjacent parts of the platform, Lower Silurian rocks prevail, except where older strata are brought to the surface in anticlinal cores and thrust sheets. As shown below, the divergence zone is interpreted as the root zone of the southerly directed thin skinned thrusts. Along it, the deformed trough sequence was displaced upward and to the south, over the platform margin.

In southern Nansen Land, the divergence zone occupies a poorly exposed through valley cut in Lower Cambrian mudrocks (Frigg Fjord mudstones of Polkorridoren Group). To the west, this tract is obscured by the outer part of J.P. Koch Fjord (Fig. 1 and 2, in pocket) but eastward, at Kap Bopa, are exposed some of the complex structures which are developed immediately south of the root zone (see cross-section, Fig. 3A in pocket). Farther east, in Amundsen Land (Fig. 1, 2), the divergence zone widens and is characterized by imbricate thrusts, which repeat the strata above the Frigg Fjord mudstones and probably root into that horizon. The imbricate thrusts have curvilinear traces, with "anomalous" vergence directions between west and south. Still farther east, along the outer part of Frederick E. Hyde Fjord (Fig. 1, 2), the divergence zone narrows again and coincides approximately with the much later (Tertiary) Harder Fjord fault zone.

The age of deformation is delimited by strata of middle Wenlock and Late Carboniferous age, respectively, in the northern orthotectonic zone; and by strata of latest Silurian (late Pridoli) and Quaternary age, respectively, in the southern fold and thrust zone. In the Arctic Islands, the term Ellesmerian Orogeny *sensu stricto* is applied to the terminal deformation of the Franklinian mobile belt in Late Devonian – Early Carboniferous time (Chapters 4, 12). This extensive event almost certainly affected the North Greenland fold belt. In addition, a deformation of Late Silurian – Early Devonian age has been inferred for the Clements Markham Fold Belt and(?) the northwestern Hazen Fold Belt (Chapter 12B), and a deformation of Givetian and/or early Frasnian age for the latter. The

---

Soper, N.J. and Higgins, A.K.
1991: Deformation (Devonian – Early Carboniferous deformation and metamorphism, North Greenland); in Chapter 11 of Geology of the Innuitian Orogen and Arctic Platform of Canada and Greenland, H.P. Trettin (ed.); Geological Survey of Canada, Geology of Canada, no. 3; (also Geological Society of America, The Geology of North America, v. E).

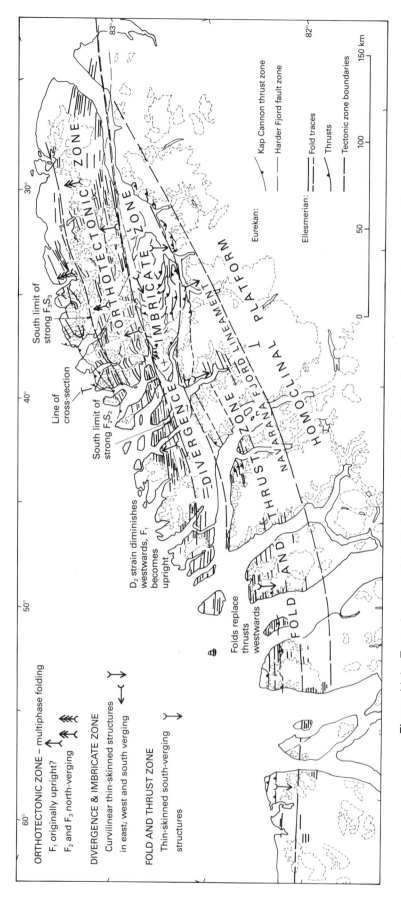

Figure 11.1. Tectonic zones of the North Greenland fold belt. For place names used in the text, see Figure 11.2. Cross-section in pocket (Fig. 3A).

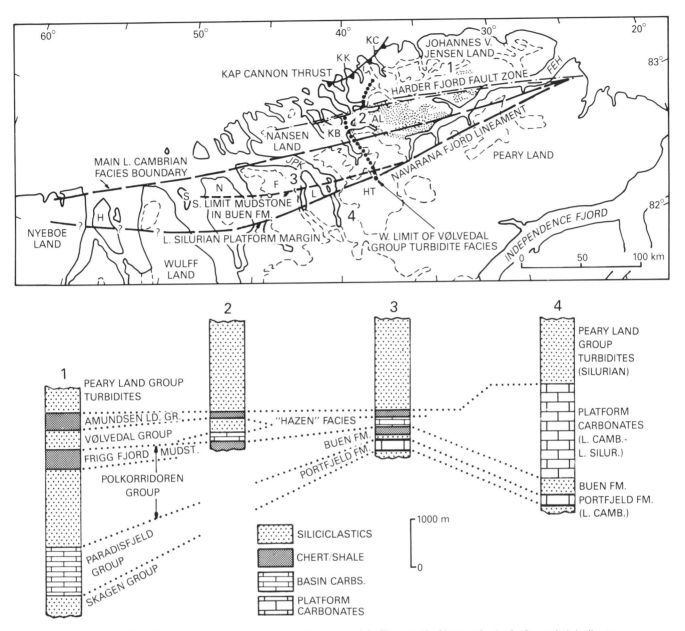

Figure 11.2. Stratigraphic control of tectonic zones (cf. Fig. 11.1). Numerals 1, 2, 3, and 4 indicate approximate location of columnar sections below map. For further discussion of facies boundaries see Chapter 7. AL = Amundsen Land; F = Freuchen Land; FEH = Frederick E. Hyde Fjord; H = Hendrik Ø; HT = Hans Tavens Iskappe; JPK = J.P. Koch Fjord; KB = Kap Bopa; KC = Kap Cannon; KK = Kap Kane; L = Lauge Koch Land; N = Nares Land; S = Stephenson Ø.

evidence for the Givetian-Frasnian movements consists of syntectonic clastic sediments preserved in parts of the Clements Markham Fold Belt (Chapters 10, 12E). The problem, whether or not the orthotectonic zone of the North Greenland fold belt also has been affected by these movements, cannot be answered because of the absence of Devonian strata in this area.

THE SOUTHERN, THIN SKINNED, FOLD AND THRUST ZONE

This description is based mainly on the central part of the zone, between Wulff Land and western Peary Land, and represents the state of knowledge as of 1985 (Soper and Higgins, 1985). The sequence is thinner than that of the platform to the south or the trough to the north, less than 2 km of Cambro-Ordovician sediments being exposed below the Silurian turbidites. It takes the form of a layered sequence, with strong competence contrasts between the weak mudrock horizons (Frigg Fjord mudstones and equivalent Buen shales) and the more rigid Portfjeld Formation carbonates and debris-flow breccias in the Cambrian – Lower Silurian starved slope/basin sequence (equivalent to the Hazen Formation of northern Ellesmere Island). Some of the units of intermediate competence are strongly laminated, for example the cherts and dolomite mudstones of the starved slope/basin sequence.

The deformation is of classic fold and thrust style, with flexural-slip folds located above thrust ramps. Depth-to-décollement calculations indicate major detachments along the weak horizons mentioned above, with local thrust flats along shaly units in the Silurian turbidites and the possibility of an even deeper detachment level (some 2.5 km below sea level south of J.P. Koch Fjord), which may represent the top of the crystalline basement.

Figure 3A is a north-south section across the whole fold belt between Kap Kane in the north and Hans Tavens Iskappe in the south (see Fig. 11.1 for section line). The southern limit of deformation is located a few kilometres north of the Lower Silurian platform margin at a "terminal syncline" where the gently north-dipping turbidites turn up steeply. To the north are a pair of anticlines situated where the basal detachment in the Buen Formation ramps up through the Cambro-Ordovician sequence, probably at the facies change from starved slope/basin sequence rocks to platform carbonates. The terminal syncline can be traced westward, stepping northward en echelon. Its position may provide the best indication of the platform margin in mid-Early Cambrian to earliest Silurian time, in areas where the older rocks are obscured by a continuous cover of Silurian turbidites.

In the north are examples of medium and small scale south-facing asymmetric folds in the Silurian turbidites, which pass down into thrusts related to the same basal detachment, now in Frigg Fjord mudstones. Total displacement along this sector is no more than 7.5 km. A deeper detachment may be present, below the Portfjeld Formation, to account for the general southward tilt of the thin skinned zone seen in the cross-section.

The cross-section also illustrates the complex structure at Kap Bopa, at the southern edge of the divergence zone. This structure restores to give a displacement of 13 km and involves both back-thrusting and refolding. An interpretation envisages southward thrusting toward the platform, with deep detachments locking as blind thrusts beneath the southward-thickening Cambro-Ordovician sequence, followed by a further southwardly propagating wave of compression, which resulted in backthrusting and refolding at Kap Bopa. The latter deformation is thought to relate to the D2 structures of the northern orthotectonic zone, and thus Kap Bopa may mark the southern limit of F2 refolding.

North of Kap Bopa, the thrusts root in the divergence zone, marked by an 8 km wide tract of largely unexposed Frigg Fjord mudstone. The mudstones are probably imbricated, but the displacement need not exceed a few kilometres. Total southward displacement on the fundamental thrust, integrated across the whole thin skinned zone, is estimated at 20-25 km. This detachment must continue at a shallow level (within Lower Cambrian strata) northward beneath the orthotectonic zone, accommodating the shortening due to F1 folding in that region (Soper and Higgins, 1987).

To the west, similar fold and thrust structures occur in inner J.P. Koch Fjord, Freuchen Land, Nares Land, Stephenson Ø, and Wulff Land (Fig. 11.1, 11.3, also Fig. 1, 2). The zone narrows westward and folds become more important than thrusts in that direction. Fold traces lie en echelon and thrusts transfer displacement from one fold to another. In northern Nyeboe Land, the structure is dominated by a single south-facing monocline with a vertical limb some 5 km in width. The fold exposes Lower Cambrian strata to the north, Silurian to the south and thus has the same effect as the divergence zone, stepping the stratigraphy down to the south. As indicated in Figure 11.2, the Early Cambrian and Early Silurian positions of the platform margin converge in this area, and the effect has been to concentrate the deformation into a single large structure. The vertical limb can be traced eastward to inner J.P. Koch Fjord as a narrowing steep belt, but the individual folds associated with it are discontinuous.

THE NORTHERN ORTHOTECTONIC ZONE

This zone is developed in the thick clastic sequence of the eastern part of the trough, and siliciclastic and calcareous turbidites of the Lower Cambrian Polkorridoren and Paradisfjeld groups are exposed almost exclusively over the area. Small outliers indicate that a thick Silurian turbidite sequence also probably covered the whole tract, and another turbidite fan system of Cambrian age (Vølvedal Group) is present in the east (Fig. 11.2). The whole exposed sequence, including Skagen Group clastics at the base, exceeded 7 km.

In the **eastern part** of the orthotectonic zone, in Johannes V. Jensen Land, structures referable to three tectonic episodes have been recognized (Dawes and Soper, 1973, 1979; Higgins et al., 1985). Major folds of all three episodes are effectively coaxial and trend approximately east-west with small plunges (Fig. 11.1). F1 folds are dominant in the southern part of the zone, immediately north of the Harder Fjord fault zone. They are upright, with no cleavage, and are developed in rocks which show little recrystallization. To the north, F2 folds become superimposed on F1. These are consistently overturned to the north and develop a south-dipping axial planar fabric (S2) of pressure solution type which cuts obliquely across F1 axial planes. D2 strain intensifies northward and imposes a northward vergence on F1 (Fig. 11.3). The metamorphic grade also rises northward and S2 becomes penetrative in many lithologies. Ultimately F1 and F2 folds become isoclinal, indistinguishable from each other, and strongly north vergent. Near the north coast, where the metamorphic grade rises to low amphibolite facies, third generation folds are superimposed on the S2 schistosity and compositional layering. They also verge north and carry an S3 crenulation fabric, which dips more steeply south than S2.

In the **western part** of the orthotectonic zone, in Nansen Land and adjacent islands, D1 has produced a spectacular series of F1 folds (Fig. 11.3), but D2 strain is low and diminishes westward (Friderichsen and Bengaard, 1985). Along the line of section, which is located centrally, the F1 folds are north vergent but not strongly so, and S2 is weak. In the extreme west of Nansen Land, F1 folds are effectively upright and D2 structures absent. It is evident that the curious feature seen in Johannes V. Jensen Land — the northward overturning of structures away from the platform — is a local feature related to the strong development of the later deformation episodes in that region.

Figure 11.3. Examples of structures in the southern fold and thrust zone: (a) imbricate thrusts repeating contact between the starved slope/basin sequence (Ɛ-Sh) and Peary Land Group (Sm), inner J.P. Koch Fjord; (b) trains of northerly overturned F1 folds east of Nansen Land, outlined by the contact between the Paradisfjeld Group (Ɛpar, lighter coloured) and Polkorridoren Group (Ɛpol).

THE DIVERGENCE AND IMBRICATE ZONE

The divergence zone widens eastward into Amundsen Land where a distinctive tectonic style is developed in strata above the Frigg Fjord mudstones (Pedersen, 1980, 1986). The structure is characterized by imbricate thrusts, which root in the mudstones and have an anomalous curvilinear trend, convex to the southwest (Fig. 11.1). The pattern may be due in part to refolding of the thrusts, but it is certainly also an original feature and presumably indicates displacement toward the west and southwest. In the southern part of the area, the thrusts swing toward parallelism with the east-west structures of the fold and thrust region described above, and have southerly vergence.

Interpretations have been made of this difficult area, which involve speculations about its exotic nature, relating it to the East Greenland Caledonides and northern Ellesmere Island (Håkansson and Pedersen, 1982; Pedersen, 1986; see also Chapter 16). However, the stratigraphy is that of the surrounding region and therefore not exotic. The structure should be interpreted in the context of the regional compression direction, which is north-south. Points to note are that the anomalous structures are located above the divergence zone, implying considerable shortening of the section at depth, and that there is a strong ductility contrast between the Frigg Fjord mudstones and overlying Vølvedal Group turbidites, suggesting that imbrication was initiated at that boundary. Our preferred interpretation involves some 20 km of north-south shortening against the abrupt Lower Silurian platform margin (Navarana Fjord lineament), which trends obliquely west-southwest in this region (Fig. 11.1, 11.2). This compression at depth produced intense imbrication of the Vølvedal Group turbidites, and the obliquity of the platform margin induced a westward component of displacement. The west vergent structures are completely detached from those beneath, at the level of the Frigg Fjord mudstones.

DISCUSSION

Little is known of the geotectonic context of the North Greenland fold belt and a synthesis cannot yet be attempted. Some facts are available. The presence of gneiss xenoliths in dykes and volcanic rocks of Johannes V. Jensen Land demonstrates that the deep water basin was floored by continental crust in Late Cretaceous time. This probably indicates that the southern part of the deep water succcession was ensialic (Higgins, 1986; Parsons, 1981). The supracrustal shortening appears to be too small to support the alternative hypothesis that it was thrust onto sialic basement during the Late Silurian or Devonian. No studies have been devoted specifically to the subsidence history, but it appears to relate well, in the simplest sense, to a McKenzie (1978) lithosphere stretching model, with a period of rapid subsidence and thick clastic sedimentation in the Early Cambrian followed by a long period of basin expansion until Late Llandovery (Early Silurian) time, when both basin and adjacent platform were swamped by turbidites derived from the rising East Greenland Caledonides (Hurst et al., 1983). No evidence of a northern margin to the basin is seen in Greenland, and subduction-related volcanic and plutonic rocks are absent here, as in the stratigraphically related part of adjacent northern Ellesmere Island (sedimentary subprovince of deep water basin or Hazen Fold Belt).

Subsequent north-south compression of the trough, during the Ellesmerian Orogeny (and possibly earlier events), was accomplished in three southwardly propagating waves. The first produced upright buckle folds in the trough sediments, a detachment at approximately the position of the Lower Cambrian platform margin and the southern, thin skinned, fold and thrust zone. The latter locked just before reaching the Lower Silurian platform margin. A second wave of deformation produced the D2 structures of the orthotectonic zone whose strong northward vergence indicates southward underthrusting. This wave propagated as far south as the divergence zone, for which it is responsible, and it produced the backthrusts at Kap Bopa and refolding of the Amundsen Land imbricate structures. A further phase of southward underthrusting produced the D3 structures of northernmost Johannes V. Jensen Land. The cause of this underthrusting is unknown, but its effects are limited to the eastern part of the North Greenland fold belt.

B. METAMORPHISM

A.K. Higgins and N.J. Soper

INTRODUCTION

The Late Silurian or Devonian to Early Carboniferous deformation in the North Greenland fold belt was accompanied by metamorphism, and the metamorphic grade increases northward in line with the northward increase in intensity of deformation. Metamorphic effects are confined to the sediments of the lower Paleozoic deep water trough. In the southernmost outcrops of the trough sediments, the rocks are non-metamorphic and preserve unmodified detrital textures, whereas in the extreme north, metamorphic grade reaches low amphibolite facies and the rocks are totally recrystallized. These metamorphic variations have previously been described in terms of tectonic-metamorphic zones by Dawes and Soper (1973, 1979), Dawes (1976), and Higgins et al. (1985); this account and Figure 11.4 are based on studies of new rock collections. The progressive metamorphic changes permit distinction of a sequence of approximately east-west trending metamorphic zones (Fig. 11.4), which roughly run parallel to the tectonic zones.

ZONE A

The southernmost zone coincides with the southern part of the fold and thrust zone (Fig. 11.1) and the unfolded platform strata to the south. Textures in thin section are detrital and almost entirely unmodified by deformation or recrystallization. However, organic geochemical analyses related to hydrocarbon source rock studies (Christiansen et al., 1985; Chapter 12B) show an increase in maturity northward, accompanying an increase in the thermal alteration index; the northern boundary of the zone corresponds approximately to the limit between thermally mature and thermally overmature rocks. Studies of the thermal alteration of conodonts (J.S. Peel, pers. comm., 1985) also indicate northward increases in temperature across the zone. No investigations have been made to determine whether the zeolite or prehnite-pumpellyite facies can be identified in zone A; nor have illite crystallinity studies been undertaken.

ZONE B

This zone corresponds to the northern part of the thin skinned fold and thrust zone, and most of the divergence and imbricate zone of the fold belt (Fig. 11.1). The southern boundary of zone B is defined by the appearance of cleavage, which is of spaced pressure-solution type in silty lithologies but often penetrative in pelitic rocks. In thin section, textures are essentially detrital, although there are some signs of recrystallization of the matrix, and in pelitic rocks prograde sericitic material is frequently seen aligned parallel to the cleavage. Organic geochemical analyses related to hydrocarbon source rock studies indicate that the rocks of this zone are thermally overmature (Christiansen et al., 1985).

Within the main outcrop of zone B around J.P. Koch Fjord there occurs an elongate zone characterized by a higher degree of recrystallization, more appropriate to zone C (Fig. 11.4). There, prograde chlorite co-exists with muscovite in pelitic rocks, and there is a widespread development of chloritoid. The anomalous area appears to coincide with locally higher temperatures developed in the divergence zone where thin skinned thrusts reach the surface. This implies that chloritoid was developed in the footwall of the main thrust due to emplacement of relatively hot hanging-wall rocks on top.

ZONE C

In this zone the argillaceous matrix of psammitic rocks is generally completely recrystallized. Detrital textures are still discernible, although modified by deformation or partial recrystallization. Chlorite co-exists with muscovite, and chloritoid is conspicuous in appropriate pelitic lithologies. The main outcrop of zone C coincides with the southern part of the orthotectonic zone of the fold belt (Fig. 11.1).

ZONE D

Zone D (Fig. 11.4) is marked by the appearance of biotite, which co-exists with muscovite and/or chlorite. Textures of nearly all rocks show strong deformation and complete recrystallization, and only occasionally are traces of detrital textures still discernible.

ZONE E

A relatively small area adjacent to the north coast of Johannes V. Jensen Land contains metamorphic mineral assemblages which are generally characteristic of the cordierite-amphibolite subfacies. Staurolite and cordierite are present in appropriate lithologies in some parts of this zone, andalusite is occasionally present, and kyanite-bearing assemblages have been recorded from two localities. Textural evidence indicates that the thermal peak, some 500-525°C at 3-4 kbar, was attained during and after the D2 deformation. A cover of some 12-15 km above the present erosion level is inferred; in addition to tectonic thickening of the basin sequence during D1 and D2, it seems necessary to involve a northward stratigraphic thickening of both the Cambrian Polkorridoren Group and the Silurian Peary Land Group. The latter, along with any representatives of the Cambrian – Lower Silurian starved basin sequence (= Hazen Formation of Ellesmere Island), have been entirely removed from the northern zones by post-Ellesmerian, pre-Late Carboniferous erosion. The absence of syn- or post-

---

Higgins, A.K. and Soper, N.J.
1991: Metamorphism (Devonian – Early Carboniferous deformation and metamorphism, North Greenland); in Chapter 11 of Geology of the Innuitian Orogen and Arctic Platform of Canada and Greenland, H.P. Trettin (ed.); Geological Survey of Canada, Geology of Canada, no. 3; (also Geological Society of America, The Geology of North America, v. E).

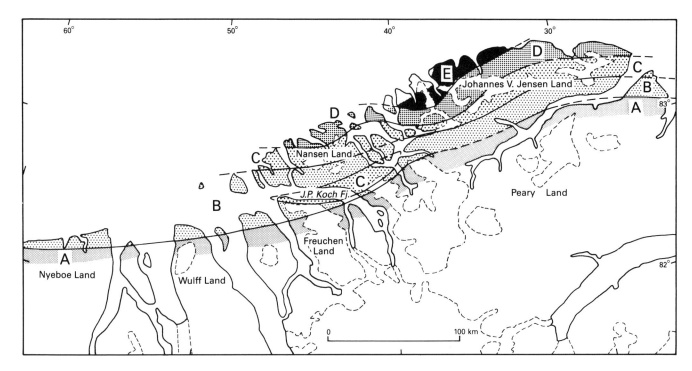

Figure 11.4. Metamorphic zones in the North Greenland fold belt.

orogenic igneous rocks makes it difficult to account for the relatively high thermal gradient required to produce the observed assemblages. Indeed, the complete absence of evidence for a magmatic arc is one of the puzzling features of the Late Silurian or Devonian to Early Carboniferous orogeny in North Greenland. The northward tectonic vergence, which is characteristic of the orthotectonic part of the fold belt (see above), implies southward underthrusting, but this seems not to have been associated with oceanic subduction. Perhaps the sedimentary subprovince of the Franklinian deep water trough remained ensialic throughout its history, despite major subsidence episodes in the Early Cambrian and Silurian.

Retrogressive assemblages are widely developed in zone E. Some of this retrogression is clearly associated with D3. An important retrogressive episode, accompanied by high shear strains, is also associated with the Tertiary (Eurekan) Kap Cannon thrust zone, which truncates zone E and part of zone D (Chapter 16). Post-Ellesmerian dolerite dykes are converted to greenstones within 1-2 km of the thrust zone, while greenschist-facies mylonites are developed in the hanging wall of the main thrust.

REFERENCES

Christiansen, F.G., Nør-Hansen, H., Rolle, F., and Wrang, P.
1985: Preliminary analysis of the hydrocarbon source rock potential of central and western North Greenland; Grønlands Geologiske Undersøgelse, Report no. 126, p. 117-128.

Dawes, P.R.
1976: Precambrian to Tertiary of northern Greenland; in Geology of Greenland, A. Escher and W.S. Watt, (ed.); The Geological Survey of Greenland, Copenhagen, p. 248-303.

Dawes, P.R. and Soper, N.J.
1973: Pre-Quaternary history of Greenland; in Arctic Geology, M.G. Pitcher (ed.); American Association of Petroleum Geologists, Memoir 19, p. 117-134.

1979: Structural and stratigraphic framework of the North Greenland fold belt in Johannes V. Jensen Land, Peary Land; Grønlands Geologiske Undersøgelse, Report no. 93, 40 p.

Friderichsen, J.D. and Bengaard, H-J.
1985: The North Greenland fold belt in eastern Nansen Land; Grønlands Geologiske Undersøgelse, Report no. 126, p. 69-78.

Håkansson, E. and Pedersen, S.A.S.
1982: Late Paleozoic to Tertiary tectonic evolution of the continental margin in North Greenland; in Arctic Geology and Geophysics, A.F. Embry and H.R. Balkwill (ed.); Canadian Society of Petroleum Geologists, Memoir 8, p. 331-348.

Higgins, A.K.
1986: Geology of central and eastern North Greenland; Grønlands Geologiske Undersøgelse, Report no. 126, p. 37-54.

Higgins, A.K., Soper, N.J., and Friderichsen, J.D.
1985: North Greenland fold belt in eastern North Greenland; in The Caledonide Orogen: Scandinavia and Related Areas, D.G. Gee and B.A. Sturt (ed.); John Wiley and Sons, Chichester, England, p. 1017-1029.

Hurst, J.M., McKerrow, W.S., Soper, N.J., and Surlyk, F.
1983: The relationship between Caledonian nappe tectonics and Silurian turbidite deposition in North Greenland; Journal of the Geological Society, London, v. 140, p. 123-131.

McKenzie, D.
1978: Some remarks on the development of sedimentary basins; Earth and Planetary Science Letters, v. 40, p. 25-32.

Parsons, I.
1981: Volcanic centres between Frigg Fjord and Midtkap, eastern North Greenland; Grønlands Geologiske Undersøgelse, Report no. 106, p. 69-75.

Pedersen, S.A.S.
1980: Regional geology and thrust fault tectonics in the southern part of the North Greenland fold belt, north Peary Land; Grønlands Geologiske Undersøgelse, Report no. 99, p. 79-87.

1986: A transverse, thin-skinned, thrust-fault belt in the Paleozoic North Greenland Fold Belt; Geological Society of America Bulletin, v. 97, p. 1442-1455.

Soper, N.J. and Higgins, A.K.
1985: Thin-skinned structures at the trough-platform transition in North Greenland; Grønlands Geologiske Undersøgelse, Report no. 126, p. 87-94.

1987: A shallow detachment beneath the North Greenland fold belt: implications for sedimentation and tectonics; Geological Magazine, v. 124, p. 441-450.

Authors' addresses

A.K. Higgins
Geological Survey of Greenland
Øster Voldgade 10
DK-1350, Copenhagen K
Denmark

N.J. Soper
Department of Earth Sciences
University of Leeds
Leeds LS2 9JT
U.K.

CHAPTER 11

Printed in Canada

Chapter 12

SILURIAN – EARLY CARBONIFEROUS DEFORMATIONAL PHASES AND ASSOCIATED METAMORPHISM AND PLUTONISM, ARCTIC ISLANDS

A. Introduction — *H.P. Trettin*

B. Late Silurian – Early Devonian deformation, metamorphism, and granitic plutonism, northern Ellesmere and Axel Heiberg islands – *H.P. Trettin*

C. Late Silurian – Early Devonian deformation of the Boothia Uplift — *A.V. Okulitch, J.J. Packard, and A.I. Zolnai*

D. Early Devonian movements of the Inglefield Uplift — *G.P. Smith and A.V. Okulitch*

E. Middle Devonian to Early Carboniferous deformations, northern Ellesmere and Axel Heiberg islands — *H.P. Trettin*

F. Late Devonian – Early Carboniferous deformation of the Central Ellesmere and Jones Sound fold belts — *A.V. Okulitch*

G. Late Devonian – Early Carboniferous deformation of the Parry Islands and Canrobert Hills fold belts, Bathurst and Melville islands — *J.C. Harrison, F.G. Fox and A.V. Okulitch*

H. Late Devonian – Early Carboniferous deformation, Prince Patrick and Banks islands — *J.C. Harrison and T.A. Brent*

I. Summary — *H.P. Trettin*

References

Chapter 12

SILURIAN – EARLY CARBONIFEROUS DEFORMATIONAL PHASES AND ASSOCIATED METAMORPHISM AND PLUTONISM, ARCTIC ISLANDS

A. INTRODUCTION

H.P. Trettin

In the Arctic Islands, intermittent and localized deformational phases of Late Silurian to Frasnian age were followed by the Ellesmerian Orogeny *sensu stricto* of latest Devonian (Famennian) to Early Carboniferous age (Thorsteinsson and Tozer, 1970, p. 566), which affected the entire mobile belt. Regional metamorphism of greenschist and higher grades, and granitic plutonism were restricted to parts of northern Ellesmere and Axel Heiberg islands.

The structural events and associated phases of metamorphism and granitic plutonism will be described in historical order, and, within given phases, from the oceanic side toward the craton. The most important conclusions are summarized at the end of the chapter. For an overview of the orogen, see Figures 12A.1 and 12I.1, and for geographic names, Figure 1 (in pocket).

Trettin, H.P.
1991: Introduction (Silurian – Early Carboniferous deformational phases and associated metamorphism and plutonism, Arctic Islands); in Chapter 12 of Geology of the Innuitian Orogen and Arctic Platform of Canada and Greenland, H.P. Trettin (ed.); Geological Survey of Canada, Geology of Canada, no. 3; (also Geological Society of America, The Geology of North America, v. E).

B. LATE SILURIAN – EARLY DEVONIAN DEFORMATION, METAMORPHISM, AND GRANITIC PLUTONISM, NORTHERN ELLESMERE AND AXEL HEIBERG ISLANDS

H.P. Trettin

DEFORMATION IN NORTHERN ELLESMERE ISLAND

The most important event in the structural history of northern Ellesmere Island was the accretion of Pearya, which resulted in intensive deformation in Pearya and the Clements Markham Fold Belt, and possibly also in adjacent parts of the Hazen and Northern Heiberg fold belts (Fig. 12B.1). Preliminary geological maps for this region have been published by Trettin and Mayr (1981), Mayr et al. (1982a, b), Trettin (1982, 1987b), Trettin et al. (1982), and Trettin and Frisch (1987); other relevant references are quoted in Chapters 8 and 9.

Age of accretion of Pearya and of associated deformations

Available evidence strongly suggests that Pearya is an exotic terrane that was part of the Caledonides, at least

---

Trettin, H.P.
1991: Late Silurian – Early Devonian deformation, metamorphism, and granitic plutonism, northern Ellesmere and Axel Heiberg islands; in Chapter 12 of Geology of the Innuitian Orogen and Arctic Platform of Canada and Greenland, H.P. Trettin (ed.); Geological Survey of Canada, Geology of Canada, no. 3; (also Geological Society of America, The Geology of North America, v. E).

CHAPTER 12

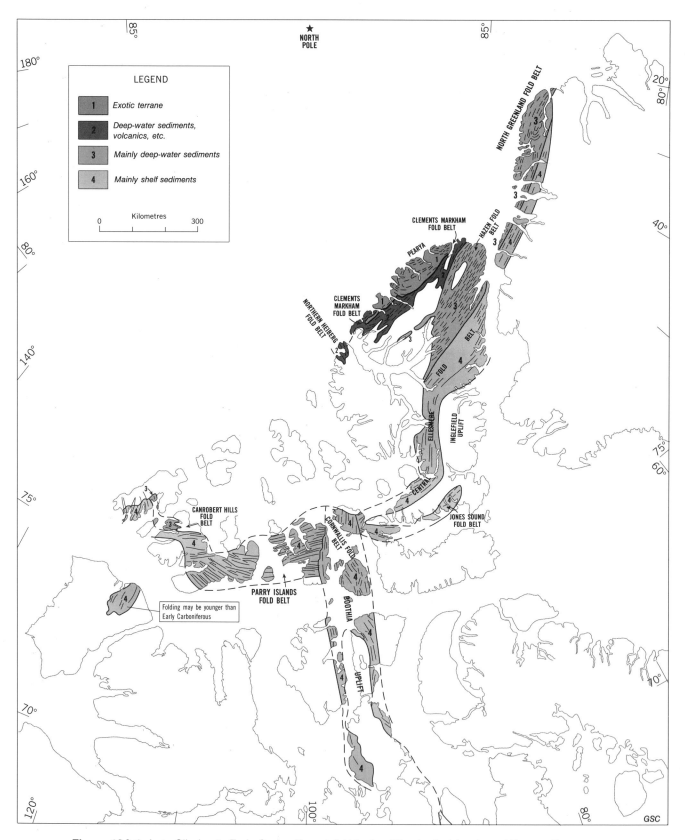

Figure 12A.1. Late Silurian to Early Carboniferous fold belts of the Arctic Islands and Pearya Terrane.

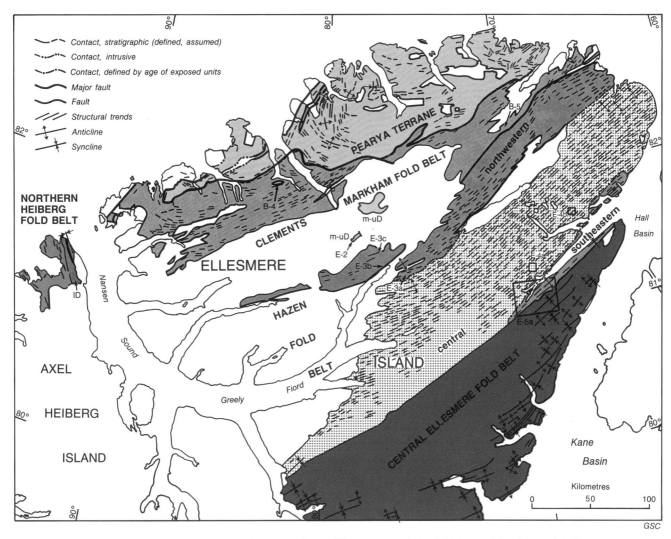

Figure 12B.1. Structural subdivisions, northern Ellesmere and Axel Heiberg islands, and index for illustrations in Chapters 12B and 12E (labelled B-4, E-3c etc.). lD: Lower Devonian, m-uD: Middle-Upper Devonian.

until late Middle Ordovician time (Chapter 9). The Ashgill-Wenlock deep water facies of Pearya are broadly comparable to those of the Franklinian deep water basin, but the flysch in the two regions differs in detrital composition.

The oldest known sedimentary unit in the Franklinian mobile belt possibly derived from Pearya is Member B of the Danish River Formation, a submarine channel deposit of Ludlow age that is exposed at Caledonian Bay, in the southwestern part of the deep water basin, near the Ordovician-Silurian shelf margin (Chapter 8C). The phenoclasts of this conglomeratic unit consist mainly of radiolarian chert, presumably derived from northern parts of the deep water basin, but also of quartzite and marble, reminiscent of succession II of Pearya. Ranging up to cobble grade, these sediments are too coarse to have been derived from the Caledonides, and consequently may herald the approach (but not necessarily the accretion) of Pearya. The same may apply to the Nordkronen Formation of North Greenland (Chapter 7), which is comparable to Member B of the Danish River Formation in clast composition, size grade, and age. The southwestward longitudinal progradation of that unit has been taken as evidence for derivation from the Caledonides on the northeast (Hurst and Surlyk, 1982; Surlyk and Hurst, 1984) but may, alternatively, record the southwestward advance of Pearya.

Two different carbonate units in the northeastern part of the Clements Markham Fold Belt are comparable, in some respects, to the upper part of the Marvin Formation of Pearya (Chapters 8C, 9). The Piper Pass beds, which have yielded conodonts of middle to late Ludlow age, are in fault contact with the Lands Lokk Formation (Wenlock to lower Ludlow); the Markham River beds of unspecified Ludlow age appear to overlie the Lands Lokk Formation with an unconformable contact. If these two units and the Marvin Formation represent diverse facies of an originally coherent carbonate body, then together, they would represent an overlap assemblage that dates the accretion of Pearya, but this is not certain. Nevertheless, conglomerates within the Markham River beds, and perhaps also the presumed unconformity at its base, are signs of a tectonic instability that may mark the beginning of the deformation associated with the accretion of Pearya.

An upper limit for the possible age range of that deformation is defined by Givetian or Frasnian spores from exposures of the Okse Bay Formation in the Clements Markham Fold Belt. In contrast to other parts of the Arctic Islands, the Okse Bay Formation (elsewhere Group) in this area is far less deformed than the Cambrian to Silurian units (see below, Chapter 12E), which suggests an intervening deformation. The Okse Bay Formation is separated from Upper Carboniferous and younger strata by an angular unconformity but structures below the unconformity indicate tilting rather than folding prior to Carboniferous deposition. Unfortunately, the base of the Okse Bay Formation, and the entire succession between it and the Upper Silurian units of the Clements Markham Fold Belt are concealed so that the stratigraphic position of the suspected angular unconformity is uncertain: rather than at the base of the Okse Bay Formation, for example, it may occur beneath (postulated) Lower Devonian clastic sediments, as in the Northern Heiberg Fold Belt (see below).

A somewhat older upper age limit for the deformation may be given by the 390 ± 10 Ma U-Pb (sphene) age of the Cape Woods Pluton if the latter indeed is a post-tectonic intrusion that resulted from crustal thickening during the accretion event (see below).

Interpretation of structures in Pearya

Understanding of Late Silurian – Early Carboniferous events in Pearya is exceedingly difficult and necessarily speculative because of the limited distribution of succession IV (upper Middle Ordovician to Upper Silurian); the absence of Devonian and Lower Carboniferous strata; and the bewildering complexity of structures. It appears, however, that the most important structures developed during the last stages of transport and the subsequent accretion to the Clements Markham Fold Belt. Evidence and considerations that have led to the present model of transport and amalgamation can be summarized as seven points.

(1) Two basic modes (or end members) for the emplacement of exotic terranes are known, strike-normal collision and strike slip. Collision usually results in broad thrust belts, in which slices of the exotic terrane may be associated with ophiolites and slivers of the host terrane. Strike slip, on the other hand, normally results in well defined, linear boundaries, with or without associated narrow thrust belts and pull-apart basins. From northeasternmost Ellesmere Island to Yelverton Inlet, the southeastern boundary of Pearya generally is linear and well defined and thus fits the second category. The only pre-Devonian units that seem to transgress it are the carbonate rocks comparable to the upper part of the Marvin Formation, and this is attributed to sedimentary overlap rather than to thrust faulting. This southeastern boundary, known as the Mount Rawlinson Fault in the northeast and the M'Clintock Glacier Fault in the southwest, is offset by some cross faults that probably developed during late stages of the emplacement of Pearya or at later times (Fig. 12B.2).

(2) At its southwestern end, the M'Clintock Glacier Fault merges with the arcuate, southwestward-directed Yelverton Thrust. This implies that the M'Clintock Glacier Fault had southwestward (sinistral) motion.

(3) The Yelverton Thrust is subparallel with arcuate trends in the southwestern part of Empire, Deuchars Glacier and Bromley Island belts and the Challenger Basin, that differ from the east-west trends predominant in the central part of Pearya as well as from the northeast-southwest trends dominant in the rest of northern Ellesmere Island (Fig. 12B.1), and require an explanation. The curvature in this region is neither a primary feature, such as an ancient shore line, because interference structures are seen locally; nor does it represent a southwest-plunging major fold — the region, which includes a fault slice of succession I of Pearya (Deuchars Glacier Belt), is far too large and complex. Because this curvature occurs at the southwestern end of a southwestward moving block it probably was caused by an obstruction — most likely the gneissic Mitchell Point Belt, which has a more southerly trend.

(4) This leads to the concept that Pearya was emplaced, not as a single block, but as number of fault slices, that interfered with each other. One consists of the Mitchell Point Belt and another of the combined Empire, Deuchars Glacier, Bromley Island, Mount Disraeli, and Disraeli Glacier belts (here collectively termed "Southeast Pearya"), which probably were amalgamated during the mid-Ordovician M'Clintock Orogeny, prior to deposition of the overlapping Challenger Basin. It is here proposed that the curvature of trends in the southwestern part of Southeast Pearya developed, as a result of obstruction and drag, when Southeast Pearya was squeezed between the Mitchell Point Belt and the Clements Markham Fold Belt (Fig. 12B.3). This probably occurred in the last stage of strike slip, shortly before Southeast Pearya became locked into its present position. If established in the Late Silurian, the curvature must have been enhanced by the later northwest-southeast compression that produced the Hazen and Central Ellesmere fold belts in Devonian to early Carboniferous time.

The remainder of Pearya either was attached to the Mitchell Point Belt or, more likely, consisted of one or two additional major slices. One of these is the gneissic Cape Columbia Belt, which either is contiguous with the comparable Mitchell Point Belt in the offshore area or constitutes a separate slice. The other consists of the Cape Alfred Ernest Belt, and possibly also of the Audhild Bay Belt, which is aligned with it and comparable in lithology. The Cape Alfred Ernest Belt is separated on the southeast from the Mitchell Point Belt by the Mitchell Point Fault. The Cape Alfred Ernest and Audhild Bay belts now are separated by younger strata, and their relationships have been obscured by later movements. The original northwestern boundary of the Cape Alfred Ernest Belt is not exposed.

A belt of variably metamorphosed sedimentary and volcanic strata directly south and southeast of the Petersen Bay Fault (the Petersen Bay assemblage of Trettin and Frisch, 1987), is tentatively included in the Clements Markham Fold Belt although it may be part of Pearya. If the first-mentioned interpretation is correct, then a fault sliver of the Clements Markham Fold Belt lies between the Mitchell Point Belt and Southeast Pearya in the area northeast and southwest of Milne Fiord.

(5) The present model implies that the Mount Rawlinson, M'Clintock Glacier, Yelverton, Parr Bay, Petersen Bay and Mitchell Point faults (Fig. 12B.2) all

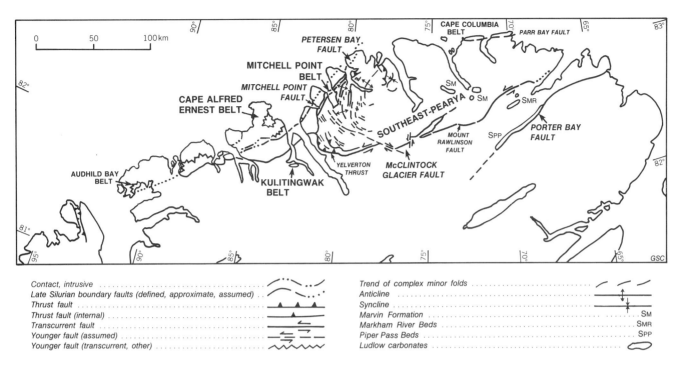

Figure 12B.2. Suggested fault slices of Pearya.

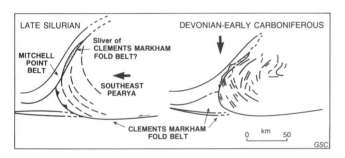

Figure 12B.3. Suggested explanation for curvature of structural trends in Southeast Pearya. Shaded arrows show direction of overall tectonic transport and compression.

originated as strike-slip faults in the Late Silurian. The Porter Bay Fault, which bounds the Clements Markham Fold Belt on the southeast, also was initiated as a sinistral strike-slip fault at this time, if the Clements Markham Fold Belt was dragged along by Pearya to some extent. Wherever Carboniferous or younger strata are exposed in the vicinity of these faults, it is apparent that they have also been active in post-Carboniferous (presumably Cretaceous-Tertiary) time. This applies to the Mount Rawlinson, M'Clintock Glacier, Mitchell Point, and Porter Bay faults. The post-Carboniferous motions clearly had vertical components, but strike slip cannot be ruled out in general, and is likely for the Porter Bay Fault. The present distribution of Ludlow carbonate units in Pearya (Marvin Formation) and in the Clements Markham Fold Belt (Markham River and Piper Pass beds), however, suggests that the net displacement between Pearya and the Clements Markham Fold Belt was less than 200 km.

(6) The present strike-slip model does not exclude the possibility that remnants of oceanic crust have been trapped between Pearya and the Clements Markham Fold Belt and that thrust faulting has occurred, especially in the region southwest of the Yelverton Thrust and southwest and south of Phillips Inlet. Both features are evident at the southern boundary of the Kulitingwak Belt (Fig. 12B.3) where strata of the Clements Markham Fold Belt are in fault contact with strata of Pearya and with serpentinite that has relict textures suggestive of both fractured igneous source rock and derived talus and stratified sediments of conglomerate to siltstone grade (Fig. 12B.4). Comparable Recent deposits occur in oceanic fracture zones where mantle is exposed on submarine ridges (LaGabrielle and Auzende, 1982).

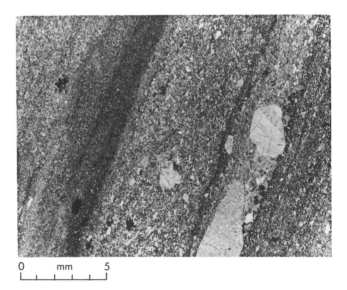

Figure 12B.4. Photomicrograph of laminated and size-graded serpentine sandstone and siltstone; ordinary light.

Clements Markham Fold Belt and(?) adjacent Parts of Hazen Fold Belt

The first and most intense phase of deformation of the Clements Markham Fold Belt, which postdates the Lands Lokk Formation and the Markham River ans Piper Pass beds and predates the Okse Bay Formation, is characterized by tight and complex folding (Fig. 12B.5), comparable to the folding of the Danish River Formation in the central Hazen Fold belt (Fig. 12E.4). If Pearya indeed was emplaced by sinistral strike slip, then the Clements Markham Fold Belt probably participated in this motion to a lesser extent. This hypothesis may explain apparent stratigraphic relationships with formations in northeastern Greenland, and incongruities with the Hazen Fold Belt (Chapter 8C).

Figure 12B.5. Crest of anticline formed in sandstone and slate of Lands Lokk Formation, Clements Markham River, northeastern part of Clements Markham Fold Belt.

DEFORMATION IN NORTHERN AXEL HEIBERG ISLAND

In northern Axel Heiberg Island, a Late Silurian – Early Devonian deformation is indicated by the angular unconformity between the Eetookashoo Bay beds and the Stallworthy Formation (Fig. 8C.27). The age of the event is constrained by late Wenlock graptolites directly below the unconformity and by mid-Early Devonian (late Lochkovian – Pragian) ostracoderms about 2 km above it. Differences in structural style between the two units suggest that it was a compressive deformation, characterized by tight and complex folding and thrust faulting, but at present it is impossible to distinguish its effects from those of a later, post-Early Devonian pre-Viséan deformation (see below, Chapter 12E).

The central part of the Northern Heiberg Fold Belt is a structural high, characterized by extensive exposures of Lower Cambrian and older(?) rocks, that is flanked by Silurian strata on the southwest and Silurian and Devonian strata on the northeast (Fig. 12E.6). It is likely that this high developed during the Late Silurian – Early Devonian deformation because the Stallworthy Formation appears to have been derived from the Hazen and Grant Land formations to the southwest. Complex fault zones on the margins of the high, which probably developed during the Late Silurian – Early Devonian deformation, were modified by later events (see below, Chapter 12E).

Tectonic Interpretation. The structural trends of the Northern Heiberg Fold Belt are almost perpendicular to those in the Clements Markham Fold Belt but conform with northerly to northwesterly structural trends prevalent in the central part of the Sverdrup Basin and the Boothia Uplift. The northerly trends in the sedimentary cover of the Boothia Uplift clearly were inherited from the Precambrian crystalline basement, and the same may apply to the entire region, including the Northern Heiberg Fold Belt. Counterclockwise rotation during the Eurekan Orogeny probably was less than 10° in this area (Chapter 17).

Reactivation of these inferred basement trends must have been caused by stress, oriented in a southwesterly or westerly direction. The fact that the deformation in northern Axel Heiberg Island was approximately coeval with that of the Boothia Uplift (Chapter 12C) and aligned with it, may indicate that both were caused by the same stress system. On the other hand, the approximate coincidence of this deformation with the emplacement of Pearya may imply a relationship with that event. It is possible, for example, that the compression was caused by a southwestward motion of northern Ellesmere Island, in which northern Axel Heiberg Island did not participate because of the northwesterly "grain" of its crystalline basement. In applying this hypothesis, however, it must be borne in mind that palinspastic reconstruction would place northern Ellesmere Island farther northwest with respect to northern Axel Heiberg Island than it is now because of the different orientation of structural trends. Reliable palinspastic reconstruction is impossible at present because of lack of information about the total (Silurian to Oligocene) crustal shortening in the Central Ellesmere and Hazen fold belts. If the total shortening was more than about 18%, parts of the Hazen Fold Belt must have lain adjacent to northern Axel Heiberg Island, and must have been involved in the sinistral motion — according to this hypothesis.

Another alternative explanation would be compression resulting from (unknown) events in the region west and northwest of Axel Heiberg Island.

REGIONAL METAMORPHISM

Amphibolite-grade metamorphism is restricted to a narrow elongate belt on the southeast side of the Petersen Bay Fault, which is bordered on the northwest by gneiss of the Mitchell Point Belt (Fig. 9.1, 12B.2, 12B.6). It is displayed mainly by a unit of amphibolite, schist and marble that is no older than Cambrian because it contains pelmatozoan columnals. In the adjacent Imina Formation, the metamorphism decreases from upper to lower greenschist facies, but the latter is difficult to distinguish from lower grades of metamorphism because the formation contains muscovite, chlorite, and albite of detrital origin; only the sporadic preservation of detrital biotite in these rocks indicates that the metamorphism was incomplete.

Elsewhere, regional metamorphism of muscovite-chlorite-albite grade is attained locally in the Jaeger Lake and Yelverton assemblages of the Northern Heiberg and Clements Markham fold belts (Upper Proterozoic and/or lower Paleozoic), and in the Grant Land Formation (Lower Cambrian) on the southeast side of the Porter Bay Fault — the extent of this belt has not yet been mapped.

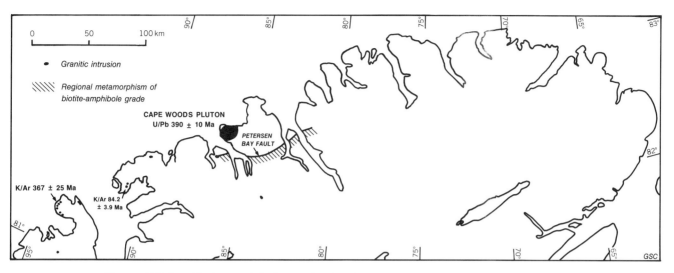

Figure 12B.6. Biotite-amphibolite grade of regional metamorphism, and granitic plutons.

The regional metamorphism has not been dated isotopically and the observations made so far indicate no more than that it is post-Wenlock and pre-Viséan in age. However, the synkinematic phase of this metamorphism probably was associated with the most intensive deformation, which is inferred to have occurred in latest Silurian – Early Devonian time.

GRANITIC PLUTONISM

Granitic plutonism, as far as is known, is restricted to northwestern Ellesmere and northern Axel Heiberg islands (Fig. 12B.6).

A pluton at Cape Woods (Frisch, 1974), at the entrance of Phillips Inlet, 166 km^2 in area, is the largest known Phanerozoic granitic body in the Arctic Islands. It intrudes carbonate and clastic sediments of assumed Late Proterozoic – Cambrian age (succession II of Pearya, Cape Alfred Ernest Belt) that are generally in the greenschist facies, but attain the amphibolite facies in the aureole of the pluton. The typical intrusive rock is a medium grained biotite quartz monzonite or granodiorite with abundant phenocrysts of K-feldspar. Aplite and pegmatite veins and segregations are common. The bulk of the intrusion is undeformed, but crush texture and foliation occur at its margin.

U-Pb determinations on sphene (Trettin et al., 1987) indicate that the pluton had cooled to a temperature between 600 and 500°C at 390 ± 10 Ma, (Emsian-Eifelian boundary according to Snelling, 1987). It contains abundant inherited zircon with an upper intercept age of about 1360 ± 200 Ma, suggesting that the bulk of the magma was derived from the crystalline basement of Pearya. The melting of the basement is thought to have been caused by crustal thickening during the Late Silurian – Early Devonian phase of accretion and deformation. The structural discordance between the pluton and trends in the Cape Alfred Ernest Belt suggest that it is post-tectonic.

Four small bodies of tonalite, granodiorite and quartz diorite, intruding the Lower Cambrian Grant Land Formation on the west side of northern Axel Heiberg Island, may be extensions of a continuous body at depth. One yielded a K-Ar (biotite) age of 367 ± 25 Ma, taken as a minimum age[1].

Other undated granitic intrusions in the region include: a quartz diorite – tonalite pluton in fault contact with the Stallworthy Formation in northeasternmost Axel Heiberg Island; a linear belt of porphyritic felsitic dykes[2] and plugs south of Rens Fiord; and five small intrusions of tonalite, quartz diorite, monzonite etc. in northwestern Ellesmere Island that occur in the Clements Markham Fold Belt and in succession II of Pearya (Cape Alfred Ernest and Audhild Bay belts). Probably most of these intrusions are Devonian in age but some may be Cretaceous (cf. Trettin and Parrish, 1987).

---

[1] A U-Pb (zircon) determination on a sample from the same locality has yielded an age of 360 ± 3 Ma (R. Parrish, pers. comm., 1988), close to the Devonian-Carboniferous boundary. This pluton therefore is related to the Ellesmerian Orogeny, discussed in Chapter 12E.

[2] A U-Pb (zircon) determination on a dyke has given an age of 365.7 ± 0.5 Ma (R. Parrish, pers. comm., 1988). It falls into the Famennian and confirms that granitic plutonism occured in northern Axel Heiberg Island during the Ellesmerian Orogeny (cf. Chapter 12E).

CHAPTER 12

C. LATE SILURIAN – EARLY DEVONIAN DEFORMATION OF THE BOOTHIA UPLIFT

A.V. Okulitch, J.J. Packard, and A.I. Zolnai

INTRODUCTION

The Boothia Uplift is a major positive cratonic and supracrustal structural feature, formed mainly in the Late Silurian to Early Devonian, that extends due north from Boothia Peninsula nearly 1000 km to Grinnell Peninsula (northwestern Devon Island; Fig. 12C.1). The uplift is composed of two parts. A lower structural level, with an exposed width of 80 km, contains faulted, poly-deformed Early Proterozoic or older metamorphic rocks of the Canadian Shield and outcrops over the southern two thirds of the uplift. An upper structural level consists of Middle Proterozoic, and Cambrian to Lower Devonian strata of the Arctic Platform and Franklinian Shelf, which have been folded and faulted to form the Cornwallis Fold Belt. This upper level mantles the crystalline core and is preserved in the northern one-third of the uplift over an exposed width of 150 km.

Study of structures of the uplift has been predominantly at a reconnaissance scale, and their geometry is thereby not fully understood. In contrast, stratigraphic, sedimentological and paleontological investigations, although incomplete, have provided data of high resolution that document the timing, location, and magnitude of vertical components of tectonic pulses in considerable detail (e.g. Thorsteinsson and Uyeno, 1980).

Such studies have indicated that the uplift is an asymmetric basement block, bounded on the west by steeply to perhaps moderately east-dipping reverse faults, and on the east by normal and reverse faults (Stewart and Kerr, 1984), and segmented by northeast- to northwest-trending normal and reverse faults. Arctic Platform strata directly west of the basement are steeply overturned, tightly folded, and overthrust. To the east, they form west-verging folds and east-facing monoclines cut by steep normal and reverse faults. Within the thicker succession of the Franklinian Shelf to the north, structures at the surface are broad open synclines with tighter intervening anticlines and numerous steep faults of normal, and less commonly, reverse displacements. Although several weak pulses of uplift are documented in the tectonically sensitive platformal succession from the Middle Proterozoic to the Late Devonian, the primary pulse that produced most of the structures and structural relief of the uplift occurred during the Late Silurian to Early Devonian epochs. To clearly understand events of this age, it is necessary to include the earlier and subsequent tectonic history in this review.

EARLY HISTORY OF THE BOOTHIA UPLIFT

The crystalline core of the uplift is composed of Early Proterozoic or older metasedimentary and metavolcanic gneissic units and granitoids that underwent polyphase deformation and uniform high grade metamorphism prior to the Middle Proterozoic (ca. 1600-1700 Ma; Blackadar, 1967; Brown et al., 1969). Predominant structural trends are north-south, changing to northeast-southwest at the southern end of Boothia Peninsula to parallel those typical of the adjacent part of the Canadian Shield. Diabase dykes, faults, and topographic lineaments generally trend northwest and northeast. In northern Boothia Peninsula and southern Somerset Island, northerly structural trends combine with east-west lineaments and faults to produce a reticulate pattern.

Unconformable deposition of sandstone of the Middle Proterozoic Aston Formation on the crystalline core was followed by intrusion (ca. 1240 Ma) of dykes and sills related to the Mackenzie dyke swarm (Jones and Fahrig, 1978). An interval of uplift and erosion preceded deposition of dolomite of the Hunting Formation of Middle or Late Proterozoic age (ca. 1200-700 Ma; Dixon, 1974). Minor folding and faulting preceded intrusion of Franklinian diabase dykes and sills during the Late Proterozoic (ca. 775-640 Ma). Possibly coeval disconformities in Middle and Upper Proterozoic successions elsewhere in the Arctic Archipelago (Young, 1981; Jackson and Iannelli, 1981) suggest that these minor pulses are not restricted to the Boothia Uplift, and are probably unrelated to its structural genesis (see Chapter 6). Middle Proterozoic rifting on northern Baffin Island formed several northwest-trending basins and uplifts. Both the Aston and Hunting formations were probably deposited in the western part of the Borden Basin (Fig. 12C.5). The adjacent Steensby and Devon highs (Fig. 12C.5) may have extended to Somerset Island, and tectonism affecting the Aston and Hunting formations may be related to them (Jackson and Iannelli, 1981, Fig. 16.35). These structures were subsequently incorporated into the younger Boothia Uplift.

The Paleozoic history of the uplift began with unconformable deposition of Upper Cambrian clastic sediments and continued with largely uninterrupted sedimentation of a predominantly carbonate succession prior to the Late Silurian Cape Storm Formation (Miall and Kerr, 1980; Thorsteinsson and Uyeno, 1980; Stewart, 1987). Three disconformities within this platformal to miogeoclinal sequence (Dixon 1974, 1978; U. Mayr, pers. comm.,1985) are regional in extent and apparently reflect tectonic pulses and/or eustatic falls in sea level not restricted to the vicinity of the uplift (see Correlation Chart, Fig. 4 in pocket).

Okulitch, A.V., Packard, J.J., and Zolnai, A.I.
1991: Late Silurian – Early Devonian deformation of the Boothia Uplift; in Chapter 12 of Geology of the Innuitian Orogen and Arctic Platform of Canada and Greenland, H.P. Trettin (ed.); Geological Survey of Canada, Geology of Canada, no. 3; (also Geological Society of America, The Geology of North America, v. E).

SILURIAN – EARLY DEVONIAN TECTONISM AND FORMATION OF THE BOOTHIA UPLIFT

Kerr (1977) defined the Cornwallis Disturbance as a series of four pulses of uplift with the following ages: 1. late Llandovery – early Wenlock (weak); 2. Gedinnian (strong); 3. mid-Emsian (moderate); 4. mid-Frasnian (moderate). More recently, the disturbance has been described as a single, more or less continuous event of Late Silurian (latest Ludlow) to Early Devonian (Pragian or early Emsian) age, with a different history in the regions north and south of Barrow Strait (Okulitch at al., 1986).

Kerr's pulses 2 and 3 fall into the main interval of tectonic activity and subsequent peneplanation discussed below. The occurence of a mid-Frasnian Uplift (pulse 4) is not supported by present stratigraphic information (Chapter 8B; R. Thorsteinsson, pers. comm., 1986). Pulse 1, also dismissed now, requires discussion. It was inferred by Kerr (1975) from a probable angular unconformity at the base of the Cape Storm Formation on eastern Grinnell Peninsula, the northeastern flank of the future Boothia Uplift. (The formation was then considered to range in age from late Llandovery to Ludlow but has subsequently been restricted to the middle Ludlow.) The significance of this unconformity is uncertain. A disconformity — caused either by a eustatic drop in sea level or by a very weak tectonic uplift — occurs at this level all along the cratonic margin but has not been reported from other parts of the Boothia Uplift. Furthermore, there is no sedimentological evidence for positive movements of the Boothia Uplift at this time; the detrital quartz and feldspar of the Cape Storm Formation — generally of silt grade and subordinate to nondetrital carbonate sediments — more likely were derived from the remote Canadian Shield than from a nearby source such as the Boothia Uplift. However, the variations in depositional patterns and unit thickness of the formation on eastern Prince of Wales Island reported by Mortensen (1985) may have been caused by local faulting.

Late Silurian – Early Devonian sedimentation on the Boothia Uplift and its environs has been discussed in Chapter 8B and only the two most important tectonic implications are repeated here.

(1) Two tectonic phases (not identical with Kerr's pulses 2 and 3) are apparent (Thorsteinsson, 1980). During the first phase of Late Silurian (latest Ludlow and Pridoli) age, only the southern part of the uplift, including Somerset Island, eastern Prince of Wales Island, and Boothia Peninsula was emergent and produced detrital sediments. The northern submerged part was also tectonically unstable but to a lesser degree (Packard, 1985). The transition, in the latest Ludlow – early Pridoli, from submarine ramp through prodelta and delta front to rimmed shelf conditions (event 3 of Fig. 8B.19; Douro Formation and lower part of Barlow Inlet Formation), for example, coincides with the initiation of the Boothia Uplift. Depositional cycles in the upper member of the Barlow Inlet Formation suggest episodic (every $10^4 - 10^5$ years) submergence of blocks bounded by northerly- and easterly-trending faults. These small (in the order of 10 m) downward movements are thought to be reciprocal to uplifts farther south. Olistostromes of shelf limestone in the Cape Phillips Formation of northern Cornwallis Island (Thorsteinsson, pers. comm., 1986) probably indicate earthquake activity at the shelf margin,

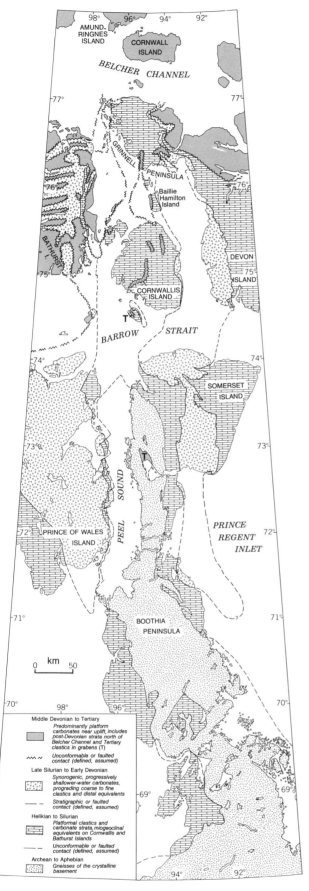

Figure 12C.1. Tectonostratigraphic elements and major geographic features of the Boothia Uplift.

but they are not restricted to the environs of the future Boothia Uplift.

During the second phase, of Early Devonian age, the entire uplift became emergent and produced coarse clastic sediments with a large proportion of conglomerate. The movements probably ceased in Pragian time and peneplanation was completed prior to the deposition of the Disappointment Bay Formation in middle Emsian time.

(2) Facies relationships in the Peel Sound Formation of Somerset and Prince of Wales islands indicate that the uplift was asymmetrical (Miall and Gibling, 1978). The west side of the uplift appears to have been bounded by steep slopes adjacent to which sand and gravel were deposited in coastal fan complexes. On the east side, finer sediments were derived from gentle slopes and transported onto a broad alluvial plain and tidal flat.

LATE DEVONIAN AND TERTIARY EVENTS AFFECTING THE BOOTHIA UPLIFT

Compression during the Late Devonian – Early Carboniferous Ellesmerian Orogeny produced the Parry Islands Fold Belt west of the uplift on Bathurst Island and resulted in dramatic interference structures along its east coast (Temple, 1965; Kerr 1974). Although the uplift acted as a buttress, pre-existing north-south folds in its cover were tightened, and unconformably overlying Lower and Middle Devonian strata were folded. Cross folds of lesser amplitude developed on Cornwallis Island and Grinnell Peninsula. Folding and south-directed thrust faulting east of the Peninsula (Morrow and Kerr, 1977) may also have occurred at this time. Effects of the Ellesmerian Orogeny were not felt south of Barrow Strait.

Tertiary faulting of the uplift produced grabens, the largest of which is on Somerset Island, preserving Tertiary (Eureka Sound Formation) and older strata. Some older faults of the uplift and adjacent areas may have been reactivated at this time. Peel Sound may be a fault-controlled topographic depression of Tertiary age (Kerr, 1977). The uplift remains a locus of weak seismic activity up to the present (Chapter 5C).

TECTONIC MODELS

The Boothia Uplift has long been regarded as a simple basement "horst" of possible isostatic origin (Walcott, 1970), which produced faults and drape folds in its cover strata (Kerr and Christie, 1965; Kerr, 1977). The uplift was compared to models of vertical block uplifts of the Wyoming Province (Stearns, 1971) by Kerr (1977). Subsequent revision of Wyoming uplift models from vertical to thrust-faulted blocks (Smithson et al., 1978, 1979) and the asymmetry of clastic wedges adjacent to the Boothia Uplift led Miall (1983a) to revise analogies between these uplifts. He suggested that the Boothia Uplift is: "... a deep-seated, east-dipping thrust block." This suggestion gains support from geophysical (gravity) data (Berkhout, 1973) and reinterpretation of structures shown by Stewart and Kerr (1984).

Gravity data suggest the presence of low-density sediments below basement in Peel Sound (Fig. 12C.4). In regions of low relief, fault attitudes are uncertain and faults shown with west side down normal slip (Stewart and Kerr, 1984) may be interpreted as west-directed, east-dipping reverse or thrust faults. This interpretation is augmented by the presence of west-verging folds and faults on eastern Prince of Wales Island (Christie et al., 1966), as well as thrust faults west of the uplift and possible basement wedges along its eastern margin on Boothia Peninsula. Structures shown in Figures 12C.2 and 12C.3 reflect such a thrust model for the uplift. Cover strata are intimately involved with the basement in southern parts of the uplift. To the north, where basement is not exposed, structures in the cover show marked asymmetry only at the leading edge of the uplift on eastern Bathurst Island. On Cornwallis Island and Grinnell Peninsula, evidence for west-directed thrusting is meagre. Differences in style of cover deformation between northern and southern parts of the uplift can likely be attributed to exposure of different structural levels, and to the presence of a thick succession containing ductile shale and evaporite units in the north in contrast to thinner platformal carbonates in the south. The analogy between the Boothia Uplift and structures of the Wyoming Province (Gries, 1983) is enhanced by comparing stratigraphic columns in terms of their mechanical behaviour (cf. Cook, 1983). Ductile strata (in the Cornwallis Fold Belt, evaporite of the Ordovician Baumann Fiord Formation and, possibly, evaporite-bearing siltstone units within the Cambro-Ordovician Cass Fjord Formation) lie above competent units and their crystalline basement, and below thick carbonate formations of the upper cover succession. Their presence provides a basis for interpretation of structures using a thick skinned tectonic model.

Structures of the Cornwallis Fold Belt can be used to infer the approximate loci of hypothetical deep-seated basement faults (Fig. 12C.3), which are also suggested by step-wise changes in uplift amounts from west to east. Amounts of uplift were estimated by determining the thickness of stratigraphic sections missing in the vicinity of the uplift. These estimates yielded minimum figures where the basement is exposed and where an unknown thickness of basement has been eroded. Approximate contours indicate that uplift increased from south to north but likely did not exceed 4-5 km. North of Barrow Strait the influence of Late Devonian folding must be accounted for; thicknesses of missing sections exceed 7-9 km in places, but amounts missing beneath post-uplift platformal strata (Disappointment Bay Formation) are 2-4 km (R. Thorsteinsson, pers. comm.). Uplift contours suggest a series of basement sheets whose upper surfaces dip east beneath folded and faulted cover. Estimates of horizontal displacement that might have accompanied uplift depend upon the dip of reverse faults. Berkhout's (1973) gravity model suggests dips of 5-15°, implying about 30 km of crustal shortening. Steeper attitudes (25-30°), derived from analogy with Wyoming structures, would result in lower estimates.

Changes in level of cover stratigraphy and basement-uplift contours (Fig. 12C.1, 12C.3) suggest the presence of faults (Fig. 12C.3) in Barrow Strait. These hypothetical features are postulated to account for reverse-fault relationships observed on eastern Prince of Wales Island and inferred at the northwestern end of "exposed" basement, and normal-fault relationships inferred between northern Somerset and southeastern Cornwallis islands. Rotational backsliding on thrust faults may have produced such features. Barrow Strait also marks a change in the possible

SILURIAN – EARLY CARBONIFEROUS DEFORMATIONAL PHASES AND ASSOCIATED METAMORPHISM AND PLUTONISM, ARCTIC ISLANDS

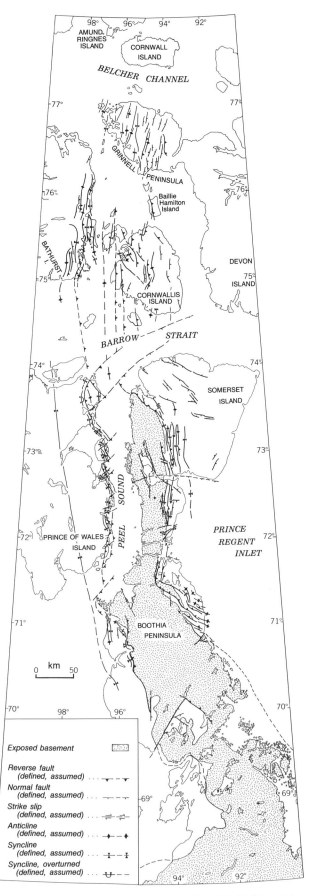

Figure 12C.2. Cover structures of the Boothia Uplift and the Cornwallis Fold Belt. Data from Christie et al. (1966), Kerr (1977), Stewart and Kerr (1984), Morrow and Kerr (1986), and Thorsteinsson (1988).

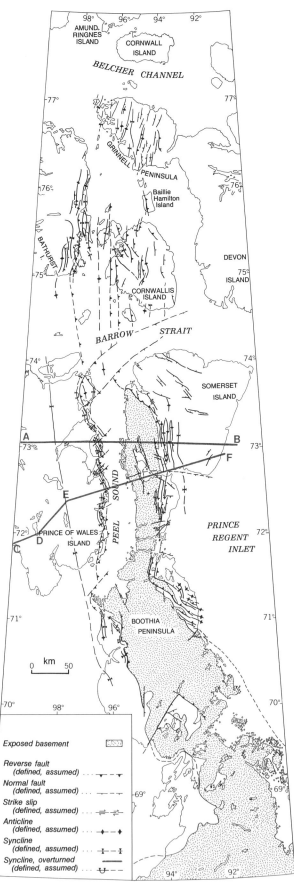

Figure 12C.3. Basement structures of the Boothia Uplift. Data sources as for Figure 12C.2.

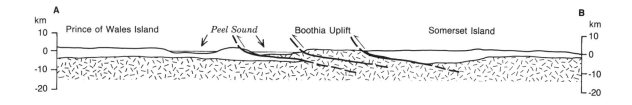
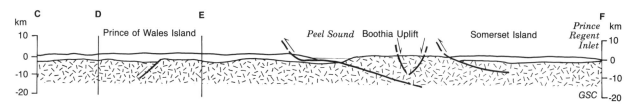

Figure 12C.4. Cross-sections of the Boothia Uplift on Somerset and Prince of Wales islands based on gravity data and models of Berkhout (1973). Altered to eliminate vertical exaggeration (horizontal and vertical scales equal), and some details of geography and geology added. Locations of cross-sections are on Figure 12C.3.

basement structure of the uplift, from a large single mass (perhaps with internal faults and a few subordinate slivers) in the south, to a broad array of fault-bounded crystalline sheets in the north. Significant (>30 km) transcurrent fault motions of either sense, such as proposed by Drummond (1973) and Hamilton (1983), do not seem to have occurred within the strait during or after the formation of the Boothia Uplift.

The uplift disappears beyond the north coast of Grinnell Peninsula beneath Carboniferous to Mesozoic strata of the Sverdrup Basin, presumably truncated by east-west normal faults in Belcher Channel. Some residual control of Tertiary folds by foundered parts of the uplift may be manifested by the Cornwall Arch on Cornwall and Amund Ringnes islands (Fig. 17.1), but the palinspastically restored position of this structure is uncertain and the juxtaposition may be fortuitous.

REGIONAL IMPLICATIONS OF BOOTHIA UPLIFT TECTONISM

Tectonic activity that formed the uplift occurred at about the same time as the closing pulses of the Ordovician to Early Devonian Caledonian Orogeny of northwestern Europe and northeastern Greenland (Soper and Hutton, 1984; Powell and Phillips, 1985). Sanford et al. (1985) suggested that compressional stresses generated at the margins of Laurentia during Caledonian orogenesis may have caused rejuvenated uplift of selected fault-bounded basement segments of the cratonic interior. Crustal accommodation, including some tilting of fault blocks, may have preceded and indeed presaged the more intense deformation of the Boothia Uplift. Although closure of Iapetus Ocean between Baltica and Laurentia was completed by Wenlock time, late Caledonian sinistral transpression continued to form compressive structures up to the Emsian. In the model of Soper and Hutton (1984, Fig. 4), vectors of north-northwest plate convergence within the Caledonian Orogen are nearly orthogonal to the Boothia Uplift on a polar projection (Fig. 12C.5).

Uplifts appear to have formed where structural trends in the basement were northerly and presumably favoured response to gentle west-directed compression. Areas of basement with other trends apparently did not react to such stresses. Spatial relationships among tectonically active areas are not obvious however, since structural trends, although northerly, have divergent trends in a crustal plate reconstruction on a polar projection (Fig. 12C.5).

SILURIAN – EARLY CARBONIFEROUS DEFORMATIONAL PHASES AND ASSOCIATED METAMORPHISM AND PLUTONISM, ARCTIC ISLANDS

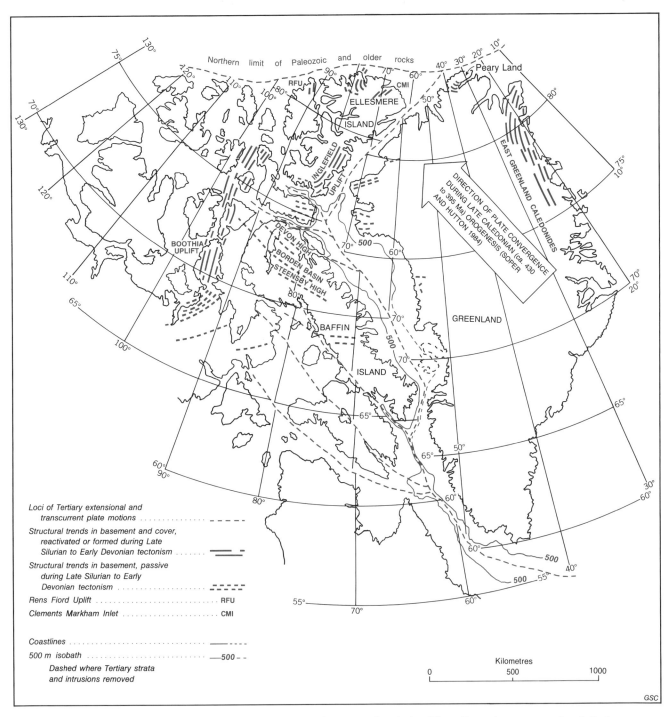

Figure 12C.5. Structural trends in basement and cover active during Siluro-Devonian orogenesis, plotted on a polar projection restored to show possible crustal plate configurations in the Early Devonian. Restoration follows that of Okulitch and Mayr (1984), utilizing mechanisms suggested by Rice and Shade (1982), Peirce (1982), and Menzies (1982) for Late Cretaceous and Tertiary plate motions. Restoration of relatively minor Devono-Carboniferous crustal shortening of the Ellesmerian Orogeny has also been made.

CHAPTER 12

D. EARLY DEVONIAN MOVEMENTS OF THE INGLEFIELD UPLIFT

G.P. Smith and A.V. Okulitch

INTRODUCTION

The Inglefield Uplift (Smith and Roblesky, 1984) is an Early Devonian positive feature located in southeastern Ellesmere Island (Fig. 12D.1) that is defined as the source of fluvial clastic sediments of that age. The name is derived from the Inglefield Mountains on the east coast of Ellesmere Island, which were part of this tectonically active landmass. Lower Paleozoic strata and crystalline rocks of the Canadian Shield comprised this land area, but neither the structures bounding the uplift nor the nature of the tectonism that produced it are known. The Inglefield Uplift is a new term that replaces the Bache Peninsula Arch of former usage (Kerr, 1967a).

PREVIOUS STUDIES

An uplifted land area in this region was previously recognized by Kerr (1967a, p. 485): "The Bache Peninsula Arch, a tectonic feature extending from Greenland to Ellesmere Island ... is a narrow positive basement feature that influenced the depositional pattern of overlying sediments of Early Cambrian to Late Devonian age, and has been delineated from stratigraphic and structural data." Kerr (1967b) also reported the "widespread emergence of central Ellesmere Island in latest Early to early Middle Devonian time". Kerr (1967c; Fig. 12D.1) illustrated the arch as a narrow, east-west trending tectonic feature extending from the Bache Peninsula region of Ellesmere Island across Nares Strait to Greenland, but did not include in it the Precambrian Shield to the south. Trettin et al. (1972, p. 126) and Kerr (1976, p. 15) also referred to an uplifted area on east-central Ellesmere Island during the Early Devonian. They ascribed the clastic deposits of the Vendom Fiord Formation to this emergent area. Trettin (1978) referred to this uplifted region as the southeasterly or cratonic Bache Peninsula Arch. Roblesky (1979) followed this nomenclature. Miall (1983b, p. 227) used stratigraphic and structural data to shed doubt on the existence of the Bache Peninsula Arch and contended that the arch "probably did not cross the strait [Nares Strait] but is a north-south feature extending south from Bache Peninsula". Supporting this concept is the regional trend of Precambrian structures in the basement of southeastern Ellesmere Island from the south coast up to and including Bache Peninsula (Taylor, 1981, Fig. 2; Frisch, 1983). Smith and Roblesky (1984) proposed a new term, the Inglefield Uplift, to avoid nomenclatural confusion. The northern portion of the uplift contains the region previously referred to as the Bache Peninsula Arch.

Smith, G.P. and Okulitch, A.V.
1991: Early Devonian movements of the Inglefield Uplift; in Chapter 12 of Geology of the Innuitian Orogen and Arctic Platform of Canada and Greenland, H.P. Trettin (ed.); Geological Survey of Canada, Geology of Canada, no. 3; (also Geological Society of America, The Geology of North America, v. E).

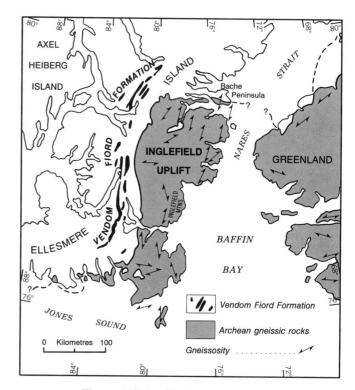

Figure 12D.1. The Inglefield Uplift.

TECTONIC HISTORY

The Precambrian history of rocks that comprise the uplift is poorly known because of extensive ice cover. The crystalline rocks are of Archean age and complexly deformed and intruded (Frisch, 1983; Chapter 6). Regional tectonic trends are predominantly northerly, swinging to southwesterly toward the south coast of Ellesmere Island. Deposition of the Proterozoic Thule Group and Rensselaer Bay Formation unconformably on the craton initiated platformal sedimentation that continued without major interruption until the Lochkovian to late Pragian Goose Fiord Formation (Packard and Mayr, 1982; Mayr, 1982; Mayr and Okulitch, 1984; Chapter 8B).

Evidence for the rise of the Inglefield Uplift is based primarily on the unconformable deposition of clastic sediments of the Vendom Fiord Formation (Kerr, 1967b). The formation ranges in age from late Pragian to late Zlichovian. It contains a basal dolostone conglomerate, with a matrix containing quartz and minor feldspar and micas, overlain by a complex interbedded sequence of quartzose sandstone, siltstone, limestone, and gypsum and anhydrite (Trettin, 1978; Roblesky, 1979). These sediments were interpreted to have been deposited in an alluvial fan environment that passed into braided streams and playa lakes to the west (Roblesky, 1979) and to shallow marine conditions still farther west (Trettin, 1978). A fining-upward megasequence implies the presence of coeval fault-controlled relief. The primary proximal source of clastics in the Vendom Fiord Formation is clearly eastern parts of the lower Paleozoic succession of the Arctic Platform (cf. Fig. 8B.37). Fine, subangular matrix grains of quartz and feldspar may have been introduced from areas still farther east where cratonal cover may have been thin and readily breached, exposing the craton. Current directions (Roblesky, 1979) indicate sources to the northeast, east and southeast.

No structures that can be directly linked to the rise of the Inglefield Uplift are known. Structural trends in the basement are northerly and several normal faults parallel to this trend lie between nunataks of the craton and outcrops of the Vendom Fiord Formation (Thorsteinsson et al., 1972; Okulitch, 1982). Some of these structures may have been active during the late Pragian to Zlichovian. Numerous northerly- and east-west trending faults on Bache Peninsula (Kerr, 1973a) may mark the northernmost expression of the uplift. North of the peninsula, any structures possibly associated with the uplift are over-ridden by Tertiary thrust sheets. The Inglefield Uplift is similar to the Boothia Uplift in that it formed at about the same time and was likely controlled by reactivation of northerly-trending basement structures. During the late Zlichovian to late Dalejan, upward movement of the Inglefield Uplift ceased, and the sea encroached over its western margin, depositing dolostone of the Disappointment Bay (formerly Blue Fiord) Formation. Central and eastern parts of the Inglefield Uplift probably remained tectonically active and formed one of several sources of clastic sediments during deposition of the late Dalejan to late Frasnian Bird Fiord Formation and Okse Bay Group.

E. MIDDLE DEVONIAN TO EARLY CARBONIFEROUS DEFORMATIONS, NORTHERN ELLESMERE AND AXEL HEIBERG ISLANDS

H.P. Trettin

NORTHERN ELLESMERE ISLAND

Two major pulses of deformation occurred in northern Ellesmere Island during the Middle Devonian to Early Carboniferous interval; the first preceded and/or accompanied the deposition of the Okse Bay Formation (Givetian(?)-Frasnian), and the second followed it (Late Devonian–Early Carboniferous Ellesmerian Orogeny *sensu stricto*).

Givetian(?)-Frasnian uplift of northwestern Hazen Fold Belt

Paleocurrent determinations and petrographic studies (U. Mayr, pers. comm., 1986) indicate that the Okse Bay Formation in the Clements Markham Fold Belt (Givetian(?) and Frasnian) was derived from chert-rich source rocks to the southeast that included the Hazen Formation. They probably were located in the northwestern part of the Hazen Fold Belt where the basal units of the Sverdrup Basin lie directly on the tightly folded strata of Grant Land Formation and Nesmith beds (Lower Cambrian), the Danish River and Hazen formations having been removed by intervening erosion; farther southeast the Hazen Formation is preserved (Fig. 12E.1). The complex folding of these strata either preceded this uplift (having formed during the Late Silurian accretion event; Chapter 12B) or accompanied it.

The strata in the northwestern part of the Hazen Fold Belt can be assigned to four structural units with contrasting mechanical properties and structural style. The lowest unit, comprising the Nesmith beds and member A of the Grant Land Formation, consists of structurally incompetent mudrock and thinly stratified carbonate rocks. Exposures at Rollrock Lake and Henrietta Glacier (Fig. 8C.3, locs. RL and HNG) show complex minor folds that commonly are isoclinal. Farther northeast this structural unit occurs in three narrow elongate belts with linear, probably faulted boundaries. Exposure is too poor to decipher the internal structure of these belts, but the outward vergence of folds in the adjacent member B of the Grant Land Formation suggests that the Nesmith beds form fan-shaped anticlinoria.

Member B of the Grant Land Formation constitutes the lowest competent unit exposed. It forms complex, but relatively large slip folds that are held up by massive sandstone or conglomerate units (Fig. 12E.3b).

A second incompetent unit comprises member C of the Grant Land Formation (Fig. 12E.3c) and the carbonate member of the Hazen Formation. Complex minor folds, generally overturned or recumbent, are characteristic of these thinly stratified carbonate and mudrock units. In some areas the folds are so disrupted by minor faults that the structure is chaotic (Fig. 12E.3a).

The overlying, rather massive chert member of the Hazen Formation constitutes a second competent unit that forms well defined, doubly-plunging folds, up to 12 km long. Directly overlying strata of the Danish River Formation conform with the Hazen Formation (Fig. 12E.3a), but at higher levels complex minor folds and faults become predominant.

The cause of the uplift is not apparent, but southwestward strike-slip motion, for example on the Porter Bay Fault Zone, would provide a plausible explanation for: (1) the uplift of this source area; (2) the preservation of the

---

Trettin, H.P.
1991: Middle Devonian to Early Carboniferous deformations, northern Ellesmere and Axel Heiberg islands; in Chapter 12 of Geology of the Innuitian Orogen and Arctic Platform of Canada and Greenland, H.P. Trettin (ed.); Geological Survey of Canada, Geology of Canada, no. 3; (also Geological Society of America, The Geology of North America, v. E).

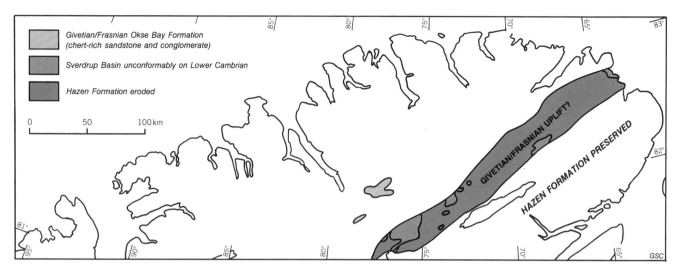

Figure 12E.1. Outcrop areas of Givetian/Frasnian Okse Bay Formation and inferred source area.

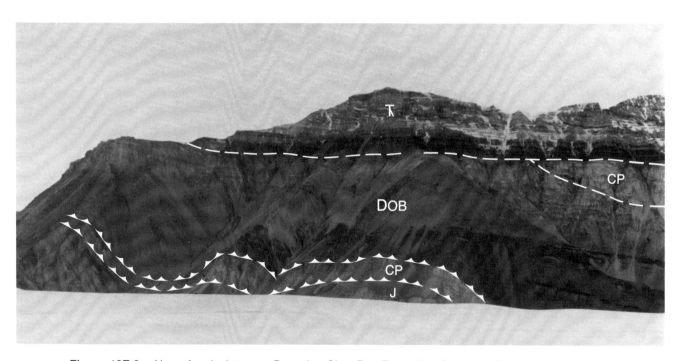

Figure 12E.2. Unconformity between Devonian Okse Bay Formation (DOB) and Sverdrup Basin (CP = Carboniferous-Permian; T [with small slash] = Triassic units) in Yelverton Pass, about 38 km northwest of head of Tanquary Fiord, looking northwest. The Okse Bay Formation has been thrust over upper Paleozoic strata (CP), in turn thrust over Jurassic (J) strata. Relief about 300 m. (Geology from unpublished work by U. Mayr).

SILURIAN – EARLY CARBONIFEROUS DEFORMATIONAL PHASES AND ASSOCIATED METAMORPHISM AND PLUTONISM, ARCTIC ISLANDS

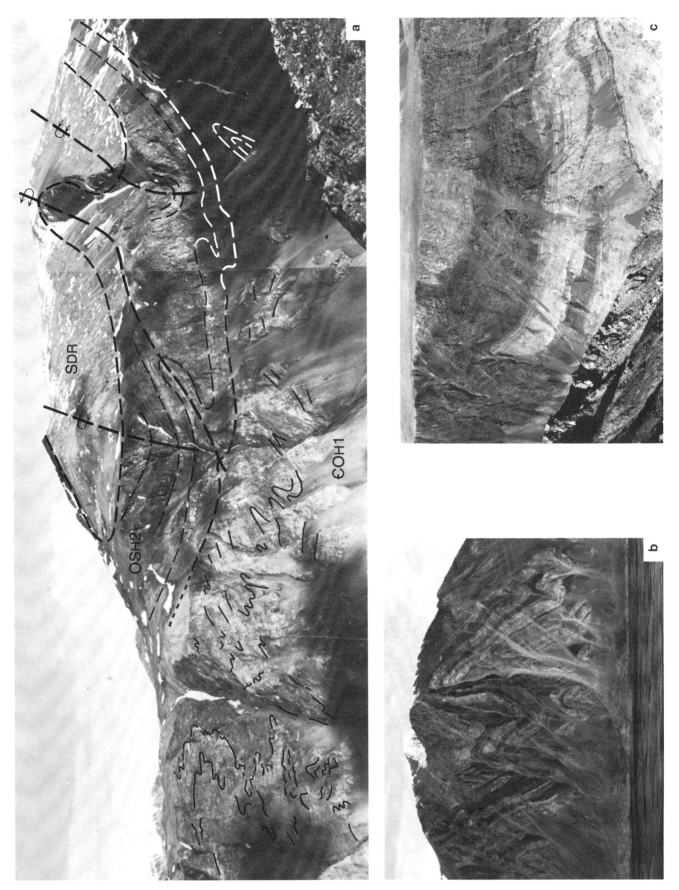

Figure 12E.3. Structures in northwestern part of Hazen Fold Belt. (a) Structures 14 km ESE of head of Tanquary Fiord, looking east. The carbonate member of the Hazen Formation (ϵO_{H1}) shows chaotic folds and faults. Major overturned folds are formed by the competent chert member (OS_{H2}), which controls structures in the overlying Danish River Formation (S_{DR}). Relief about 700 m. (b) Complex folds and faults in member B of the Grant Land Formation north of head of Tanquary Fiord. Relief about 800 m. (c) Recumbent anticline formed in uppermost strata of Grant Land Formation (light grey) and basal Hazen formation (dark grey); about 12 km northwest of head of Tanquary Fiord.

311

clastic sediments in a downfaulted pull-apart(?) basin; and (3) renewed deformation in northern Axel Heiberg Island (see below).

Late Devonian – Early Carboniferous deformation (Ellesmerian Orogeny *sensu stricto*)

Clements Markham Fold Belt, Pearya, and northwestern Hazen Fold Belt

The extensive deformation of latest Devonian – Early Carboniferous age that produced the Central Ellesmere and Parry Islands fold belts probably also affected northern Ellesmere Island, but is difficult to distinguish there from earlier deformations because of the scarcity of Upper Devonian outcrops. To date, only two such outcrop areas are known, occurring in the Yelverton Pass area of the Clements Markham Fold Belt (Fig. 12B.1). There, the Okse Bay Formation is unconformably overlain by upper Paleozoic and Triassic strata of the Sverdrup Basin (Fig. 12E.2). Relationships at the unconformity indicate that the Okse Bay Formation was tilted, but not folded, prior to the deposition of the Upper Carboniferous strata.

The two areas are too small to be representative for such a large region, but it is reasonable to assume that Pearya, the Clements Markham Fold Belt, and perhaps also the northwestern part of the Hazen Fold Belt were stabilized by the preceding events to such an extent that tight folding no longer was possible. However, the northwest-southeast compression that formed the Central Ellesmere Fold Belt, may have caused thrust faulting throughout the region and enhancement of the arcuate trends in southwestern Pearya (Fig. 12B.3).

Central and southeastern parts of Hazen Fold Belt

The age of deformation of this region, bracketed by strata of earliest Devonian and early Late Carboniferous age, respectively, is poorly constrained. However, absence of sedimentological evidence for uplift in Early or Middle Devonian time suggests that it is no older than Late Devonian.

The central Hazen Fold Belt is underlain mostly by the Danish River Formation with doubly-plunging anticlines of the Hazen Formation, up to 17 km long, in the northeastern part. The Danish River Formation forms complicated minor folds that approach the chevron type, typical of many flysch successions (Ramsay, 1974). Innumerable minor faults are associated with these folds, but major faults have not been detected (Fig. 12E.4).

In the **southeastern Hazen Fold Belt**, surface exposures range in age from late Early Cambrian to Silurian. Near Archer Fiord and Ella Bay, the deep water succession (Hazen and Danish River formations) overlies clastic units of the Lower Cambrian shelf succession (Ellesmere Group, Kane Basin and Rawlings Bay formations). Farther southeast it abuts laterally against younger shelf sediments with an unconformable contact (Fig. 8C.26). The Lower Cambrian units are exposed in a large, southwesterly-plunging anticlinorium. Second-order en echelon folds are well developed on the flanks of the anticlinorium on the northwest side of Ella Bay, the head of Archer Fiord.

The structures adjacent to the Ordovician shelf margin southeast of upper Archer Fiord are shown in the cross-sections of Figure 12E.5. There the northwesternmost strata of the Cambrian-Ordovician shelf succession form the broad Ninnis Glacier Syncline. The contact with the Hazen

Figure 12E.4. Vertical air photographs of typical folds, Danish River Formation, central Hazen Fold Belt. (a) At Murray Lake (air photo A16609-115; Canada, Department of Energy, Mines and Resources); relief about 700-800 m. (b) At Heintzelman Lake (A16609-44); relief about 600 m.

Formation, not well exposed, appears to be a steeply-dipping thrust fault at this locality but does not seem to be faulted everywhere. It probably coincides with the facies change between shelf carbonates and slope deposits of Middle to Late Cambrian age and was localized by it. Directly northwest of the fault, slaty laminite of the outer shelf — upper slope facies of the Hazen Formation displays very complex minor folds; the overall structure in this part of the cross-section is uncertain because of the absence of stratigraphic markers. Farther northwest, proximal basin deposits in the lower part of the Hazen Formation and underlying strata of the Kane Basin Formation form folds that are smaller and more complicated than those in the shelf carbonates (half-wavelength about 0.5-1.5 km), but larger and simpler than those in the slope facies. These folds probably are controlled by thick and massive carbonate conglomerate beds of proximal basin facies in the lower part of the Hazen Formation. A northwestward increase in the structural complexity of the Hazen Formation, apparent at the head of Ella Bay, may be due to the thinning of these beds in that direction.

NORTHERN AXEL HEIBERG ISLAND

A second phase of compression occurred in the Northern Heiberg Fold Belt some time after the deposition of the Lower Devonian Stallworthy Formation and before the deposition of the Lower Carboniferous (Viséan) Emma Fiord Formation. The only structure that can be attributed to this event with some confidence is the inferred southwestward-directed thrust fault that places the Svartevaeg Formation upon the Stallworthy (Svartevaeg Thrust of Fig. 12E.6; cf. Trettin, 1987a). Although at least some of the faults bounding the Rens Fiord Uplift probably date back to the Late Silurian – Early Devonian deformation, it is likely that they were reactivated or modified during the Middle Devonian – Early Carboniferous deformation. Faults on the northeastern side of the Rens Fiord Uplift place the Jaeger Lake assemblage and Aurland Fiord beds on Lower Cambrian to Silurian strata. They include the northeastward directed Eetookashoo Bay Thrust and the southwestward directed Greenstone Lake and Jaeger Lake thrusts. It is uncertain whether the Eetookashoo Bay Thrust is overturned or whether it is an older structure, truncated at depth by the Greenstone Lake Thrust. The southeastward projections of these faults, under Quaternary cover, do not seem to break the Sverdrup Basin succession, indicating that they all are pre-Viséan in age.

On the southwestern flank of the Rens Fiord Uplift, a large, subhorizontal thrust sheet of the Aurland Fiord beds, locally with a subsidiary sheet of the Grant Land Formation at the base, lies on generally steeply-dipping strata of the Grant Land Formation and locally of the Hazen Formation. The vergence of folds beneath the Aurland Fiord Thrust suggests northeastward motion, and it is possible that this thrust is rooted in a steeply-dipping fault that bounds vertical fault blocks of the Aurland Fiord beds in the area west of Aurland Fiord. This thrust fault had a minimum horizontal displacement of 9 km (cross-section C-C' in Fig. 12E.6). The Aurland Fiord thrust sheet and underlying strata both are cut by mafic dykes presumably of Late Cretaceous or older age, suggesting that the fault is older than the Eurekan Orogeny. However, this has to be confirmed by isotopic dating as post-Eurekan dykes are present in parts of central Ellesmere Island (Chapter 17).

The cause of the post-Early Devonian pre-Viséan deformation of northern Axel Heiberg Island is unknown, but renewed southwestward motion of the Hazen and Clements Markham fold belts is a possible explanation.

CHAPTER 12

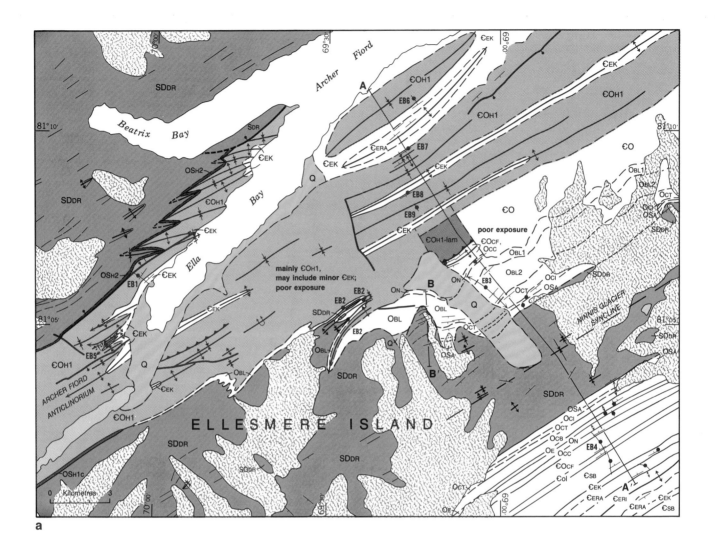

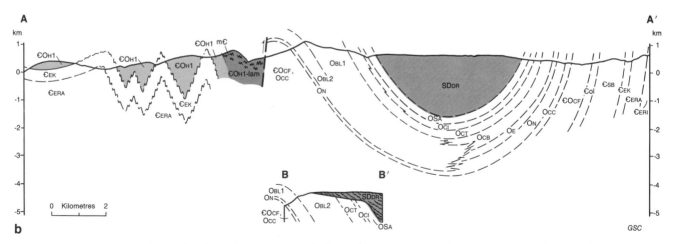

Figure 12E.5. Transition from Hazen Fold Belt to Central Ellesmere Fold Belt. The boundary between the two belts coincides with the southeastern limit of the Hazen Formation. (a) Geological map and index for structural cross-sections and for composite stratigraphic cross-section (Fig. 8C.26, sections EB1 to EB4). (b) Structural cross-sections A-A' and B-B'.

SILURIAN – EARLY CARBONIFEROUS DEFORMATIONAL PHASES AND ASSOCIATED METAMORPHISM
AND PLUTONISM, ARCTIC ISLANDS

LEGEND

QUATERNARY

Q

DEEP WATER SUCCESSION

SILURIAN, SILURIAN AND (?)LOWER DEVONIAN

S$_{DR}$
SD$_{DR}$ — DANISH RIVER FORMATION

UPPER ORDOVICIAN TO LOWER SILURIAN

OS$_{H2}$ — HAZEN FORMATION (ϵO$_{H1}$-lam – OS$_{H2}$)
Chert member: (Upper Ordovician and ?Lower Silurian)

OS$_{H1c}$ — Carbonate member: upper limestone unit
(Upper Ordovician? and Lower Silurian)

ϵO$_{H1}$ — Carbonate member: undifferentiated (upper
Lower Cambrian to Upper Ordovician)

ϵO$_{H1}$-lam — Carbonate member: laminate facies (full age range
uncertain, but includes mϵ, Middle Cambrian fossils)

UNDIFFERENTIATED

CAMBRIAN AND ORDOVICIAN

ϵO — CASS FJORD, CAPE CLAY, and HAZEN formations,
and Ninnis Glacier beds

SHELF MARGIN CARBONATE BUILDUP

UPPER LOWER AND LOWER MIDDLE ORDOVICIAN

O$_{BL2}$ — BULLEYS LUMP FORMATION (O$_{BL1}$-O$_{BL2}$)
Member B

O$_{BL1}$ — Member A

Geological boundary (defined, approximate,
projected through ice or overburden)
Anticline, syncline (arrow indicates plunge)
Anticline (overturned)
Fault (unspecified)
Normal fault (defined, approximate)
Thrust fault (defined, approximate)
Bedding (strike and dip direction; general trend)

SHELF SUCCESSION

UPPER ORDOVICIAN AND LOWER SILURIAN

OS$_A$ — ALLEN BAY FORMATION

ORDOVICIAN

CORNWALLIS GROUP (O$_{CB}$-O$_{CI}$)

O$_{CI}$ — IRENE BAY FORMATION

O$_{CT}$ — THUMB MOUNTAIN FORMATION

O$_{CB}$ — BAY FIORD FORMATION

O$_E$ — ELEANOR RIVER FORMATION

O$_N$ — Ninnis Glacier beds

O$_{CC}$ — CAPE CLAY FORMATION

MIDDLE CAMBRIAN TO LOWER ORDOVICIAN

O$\epsilon$$_{CF}$ — CASS FJORD FORMATION

LOWER AND MIDDLE CAMBRIAN

$\epsilon$$_{OL}$ — Oolitic limestone unit

$\epsilon$$_{SB}$ — SCORESBY BAY FORMATION

ELLESMERE GROUP ($\epsilon$$_{ERI}$-$\epsilon$$_{EK}$)

$\epsilon$$_{EK}$ — KANE BASIN FORMATION

$\epsilon$$_{ERA}$ — RAWLINGS BAY FORMATION

$\epsilon$$_{ERI}$ — RITTER BAY FORMATION

Structural cross-section .... A⊢——⊣A'
Stratigraphic section .... • EB1 •
Glacier
Lake

GSC

SILURIAN – EARLY CARBONIFEROUS DEFORMATIONAL PHASES AND ASSOCIATED METAMORPHISM AND PLUTONISM, ARCTIC ISLANDS

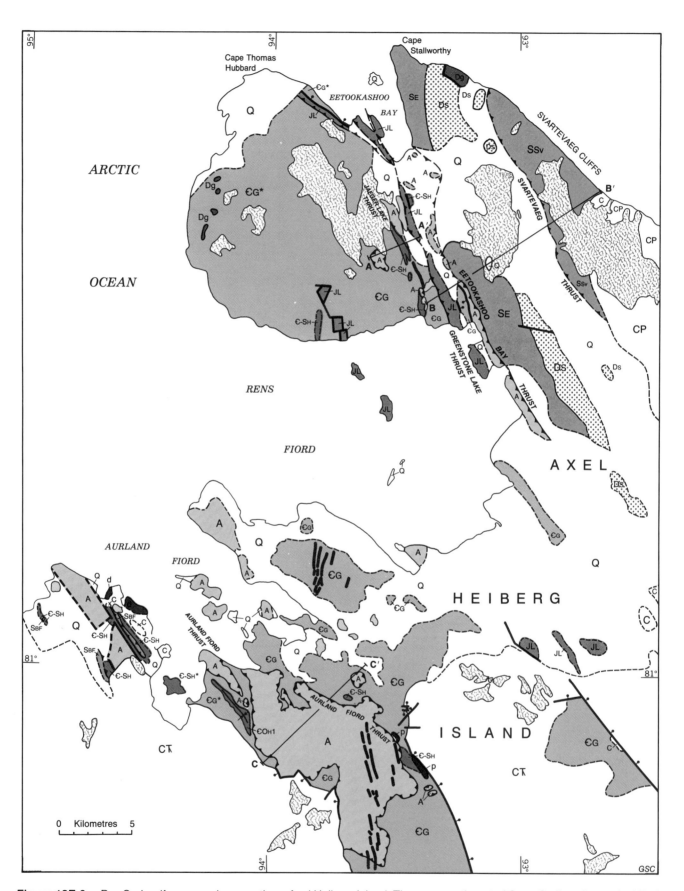

Figure 12E.6. Pre-Carboniferous geology, northern Axel Heiberg Island. The area southeast of Cape Stallworthy marked Dg is underlain mainly by metasedimentary rocks with minor granitic intrusions. Two new U-Pb zircon ages have recently been established: 360 ± 3 Ma (earliest Tournaisian) for map unit Dg southwest of Cape Thomas Hubbard, and 365.7 ± 0.5 Ma (Famennian) for map unit p southeast of Rens Fiord (R. Parrish, pers. comm., Sept. 1989).

SILURIAN – EARLY CARBONIFEROUS DEFORMATIONAL PHASES AND ASSOCIATED METAMORPHISM AND PLUTONISM, ARCTIC ISLANDS

DIAGRAMMATIC CROSS-SECTIONS

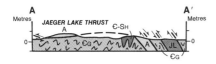

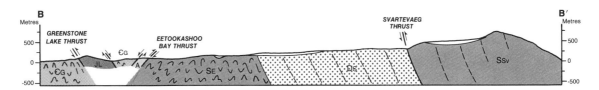

LEGEND

SEDIMENTARY AND VOLCANIC ROCKS

QUATERNARY
- Q   Sand, gravel, mud

CARBONIFEROUS, CARBONIFEROUS AND PERMIAN, CARBONIFEROUS TO TRIASSIC
- C
- CP
- CTR

LOWER DEVONIAN
- Ds   STALLWORTHY FORMATION: sandstone, mudrock, conglomerate

SILURIAN
- Ssv   SVARTEVAEG FORMATION: volcanogenic sandstone; minor mudrock, tuff, volcanic flows, conglomerate, breccia
- SE   Eetookashoo Bay beds: mudrock; minor sandstone, volcanogenic sandstone, tuff, conglomerate
- SBF   Bukken Fiord beds: volcanogenic sandstone, mudrock; minor conglomerate

LOWER CAMBRIAN TO LOWER SILURIAN
- Є-SH   HAZEN FORMATION: chert, slate, resedimented carbonates
- Є-SH*   May include Grant Land Formation and/or Bukken Fiord beds
- ЄOH1   Carbonate member (Cambrian-Ordovician)

LOWER CAMBRIAN
- ЄG   GRANT LAND FORMATION: sandstone, slate, phyllite
- ЄG*   May include Hazen Formation etc.

UPPER PROTEROZOIC AND/OR LOWER PALEOZOIC
- A   Aurland Fiord beds (lower Lower Cambrian?): dolostone, minor limestone
- JL   Jaeger Lake assemblage: metamorphosed basaltic flows and tuff, dolostone, limestone

Note: stratigraphic order uncertain

INTRUSIVE ROCKS

CRETACEOUS AND/OR OLDER
- Diabase dykes

DEVONIAN AND/OR CRETACEOUS
- d   Diorite, gabbro
- p   Felsite porphyry

UPPER DEVONIAN AND/OR OLDER
- Dg   Granodiorite, tonalite, quartz diorite

- Geological boundary (defined, approximate)
- Fault (exposed, projected through overburden)
- Thrust fault (exposed, projected through overburden)
- Normal fault
- Syncline
- Glacier
- Lake

GSC

CHAPTER 12

F. LATE DEVONIAN – EARLY CARBONIFEROUS DEFORMATION OF THE CENTRAL ELLESMERE AND JONES SOUND FOLD BELTS

A.V. Okulitch

CENTRAL ELLESMERE FOLD BELT

Introduction

The Central Ellesmere Fold Belt is an "S"-shaped belt of cratonally-directed folds and thrust faults that extends from the eastern margin of the Boothia Uplift on Devon Island eastward across southern Ellesmere Island. At Baumann Fiord, tectonic trends change to north-northeast; these persist north to Cañon Fiord, where they curve again to become easterly to northeasterly (Fig. 12F.1). Continuations of the fold belt lie across Kennedy and Robeson channels on northwestern Greenland (Chapter 11).

The age of deformation of the fold belt is bracketed by strata of Famennian and Viséan ages (Parry Islands and Emma Fiord formations; cf. Chapters 10, 11). Throughout southern and central Ellesmere Island, Ellesmerian structures were reactivated and deformed by the Tertiary Eurekan Orogeny. In central Ellesmere Island, where structures of both orogenies are of similar trend and compressional in nature, their age remains uncertain except in the few areas where relatively undisturbed strata of Carboniferous to Tertiary ages lie unconformably on deformed Devonian and older units, and in areas where Tertiary strata are deformed.

Because of differing trends and somewhat variable structural styles, the fold belt will be described in three segments: a southern one from northwest Devon Island to Baumann Fiord, a central segment along Vendom Fiord to Cañon Fiord, and a northern segment from Cañon Fiord to Robeson Channel.

Southern segment

In this segment, Ellesmerian structures are east-trending on Devon Island and curve gently to become north-trending east of Baumann Fiord. Exposed northern and northwestern limits of the fold belt are at outcrops of unconformably overlying Carboniferous strata on northern Grinnell and central Bjorne peninsulas. Beneath the unconformity, folds are upright with horizontal axes, large, open, and concentric in competent carbonate units, and smaller (less than 1 km in wavelength) in units such as the siltstone of the Lower Devonian Eids Formation (Fig. 12F.1, cross-section A-B; Fig. 3B [in pocket], section E-F). The southeasternmost major structure in this segment is the Schei Syncline, which extends from Grinnell Peninsula to south of Vendom Fiord.

Central segment

Throughout the central segment north of Baumann Fiord are north-northeast trending, upright to eastward-overturned folds, east-verging thrust faults, and normal faults, which formed during the Ellesmerian and Eurekan orogenies. Ellesmerian structures have likely been reactivated during the latter event, thus in most instances the ages of structures cannot be determined.

Demonstrable Ellesmerian structures outcrop in the region of Trold Fiord, where gently folded Carboniferous strata unconformably overlie folded carbonates of early Paleozoic age (Thorsteinsson, 1972b; Fig. 12F.1, cross-section C-D, west end). Two broad, open synclines west of and within Vendom Fiord (Fig. 12F.1, section C-D; Fig. 3B, section C-D) can be reasonably inferred to be extensions of Ellesmerian folds east of Eids and Blue fiords and of the Schei Syncline. North of Bay Fiord, Carboniferous to Mesozoic strata overlie folded Devonian and older units with gentle unconformity, but are themselves folded and faulted along trends subparallel to Ellesmerian structures. A few examples of reactivated Ellesmerian structures have been mapped (Thorsteinsson, 1972a). Most folds are concentric; however, disharmonic folds are evident in the evaporitic Baumann Fiord and Bay Fiord formations. Most faults are concealed in valleys and their attitudes are constrained only by slopes of adjacent ridges. Some faults shown as thrusts may be normal faults.

Northern segment

This segment extends generally northeast from Cañon Fiord under the Agassiz Ice Cap to Judge Daly Promontory. Easterly trends south of the ice cap may be the result of warping about the northern end of the cratonic Inglefield Uplift (Chapter 12D) and/or the superimposition of major, southerly-verging Eurekan thrust faults. Such Tertiary features are common throughout this segment and few unequivocal Ellesmerian structures are known. On the basis of structural style, tight to open, upright folds and associated southeast-verging thrust faults can be inferred to be of Ellesmerian age. Paleozoic strata throughout this region have been affected by this tectonism, except near the craton. This narrow zone of relatively undeformed platformal strata can be observed near the east end of Bay Fiord and in the Bache Peninsula area (Thorsteinsson, 1972c; Kerr, 1973a, b). The intensity of deformation increases to the northwest toward the Hazen Fold Belt (Chapter 12E, Fig. 12E.6).

Folds are concentric, controlled by competent carbonate units, with subhorizontal fold axes up to 50 km long. Anticlines are generally tight, whereas synclines are broad (Higgins et al. 1982; Fig. 12F.1, cross-sections E-F, G-H). On Judge Daly Promontory recognizable Ellesmerian folds are monoclines, tight anticlines, some of which are overturned to the northwest (Higgins et al., 1982; de Vries

---

Okulitch, A.V.
1991: Late Devonian – Early Carboniferous deformation of the Central Ellesmere and Jones Sound fold belts; in Chapter 12 of Geology of the Innuitian Orogen and Arctic Platform of Canada and Greenland, H.P. Trettin (ed.); Geological Survey of Canada, Geology of Canada, no. 3; (also Geological Society of America, The Geology of North America, v. E).

SILURIAN – EARLY CARBONIFEROUS DEFORMATIONAL PHASES AND ASSOCIATED METAMORPHISM AND PLUTONISM, ARCTIC ISLANDS

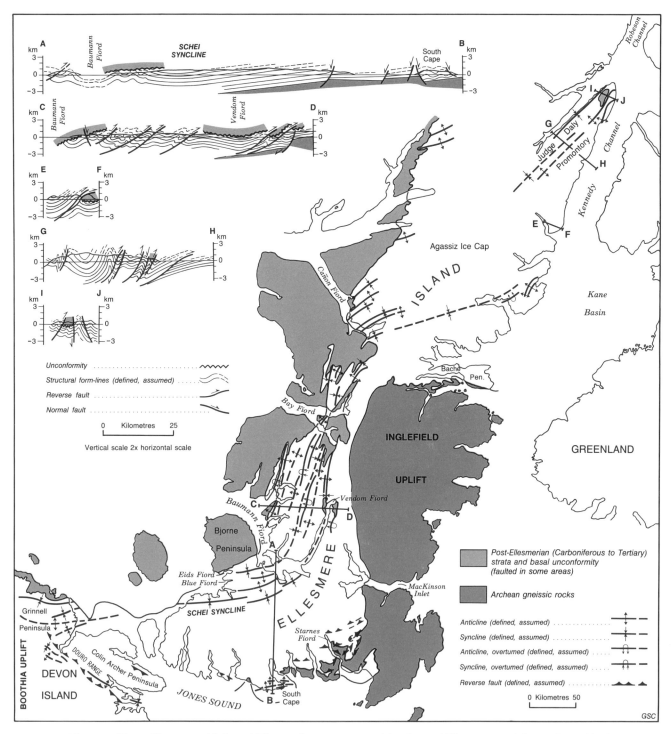

Figure 12F.1. Known and inferred Ellesmerian structures of the Central Ellesmere and Jones Sound fold belts. Tertiary structures and those of uncertain age are not shown except on the cross-sections. Data, listed from southwest to northeast, from: Morrow and Kerr (1977), Mayr and Okulitch (1984), Okulitch, (1982), Kerr and Thorsteinsson (1972), Thorsteinsson (1972a, b, c), Kerr (1973a, b), Thorsteinsson and Kerr (1972), Kerr (1973c), Thorsteinsson and Trettin (1971) and de Vries (1974). The northwestern part of cross-section G-H coincides with cross-section A-A' in Figure 12E.5.

1974; Fig. 12F.1, cross-section I-J), and the broad Ninnis Glacier Syncline (Fig. 12E.6).

Summary

The overall geometry of the Central Ellesmere Fold Belt is not fully understood, primarily because of pervasive overprinting during the Eurekan Orogeny. No estimates can be made of the amount of crustal shortening (although it is unlikely to exceed 50 km). As in the Parry Islands Fold Belt, the deformation is thought to be thin skinned with décollement within evaporitic strata (Bay Fiord and Baumann Fiord formations) and near the crystalline basement (Trettin and Balkwill, 1979; de Vries, 1974). Models of crustal plate tectonism that may have produced the fold belt remain highly speculative.

JONES SOUND FOLD BELT

The Jones Sound Fold Belt is an arcuate belt of folds and associated reverse and normal faults of small displacement that diverges from the Central Ellesmere Fold Belt on Grinnell Peninsula, northwestern Devon Island, to extend southeast and east into Jones Sound, and along the coast of Ellesmere Island to the Canadian Shield at Makinson Inlet. The west end of the fold belt abuts against the Boothia Uplift and interferes with it. Structural trends are northwesterly near the uplift and westerly on Devon and much of Ellesmere Island, becoming southwesterly toward Makinson Inlet. Over much of its extent, the Jones Sound Fold Belt is separated from the Central Ellesmere Fold Belt by a 70 km wide, block-faulted homocline (the southeastern limb of the Schei Syncline), that dips gently northwest (sections A-B of Fig. 12F.1, E-F-G of Fig. 3B).

On Devon Island, southwest-verging thrust faults and overturned folds occur in the Douro Range of Grinnell Peninsula (Morrow and Kerr, 1977, 1986). Upright folds and minor north-verging thrust faults on Colin Archer Peninsula may continue eastward under Jones Sound to South Cape on Ellesmere Island. Farther east on Ellesmere Island, there are south- to southeast-verging reverse faults that have been interpreted to involve the Precambrian crystalline basement (Mayr and Okulitch, 1984).

This eastern end of the fold belt is complex (Fig. 3B, sections J-K and L-M): faults and folds of probable Ellesmerian age have been cut by flat detachment faults developed within Ordovician evaporitic layers, and by three sets of faults of latest Cretaceous – Tertiary age, related to the opening of Baffin Bay (Chapter 17). Given such relationships, neither the geometry nor the history of the inferred Ellesmerian structures is well understood. Reverse faults do not appear to be controlled by obvious zones of weakness in the Paleozoic succession, but may be the result of reactivation of structures in the crystalline basement, here trending approximately east-west. Magnetic anomalies (McGrath and Fraser, 1981) suggest that the basement trends are northwest-southeast in Jones Sound and adjacent parts of Devon Island. Basement trends are not apparent in the magnetic pattern of northwestern Devon Island (except for easternmost Colin Archer Peninsula), which has a thicker sedimentary cover and probably was depressed below the Curie point (cf. Chapter 5D). However, the parallelism of the structures in this area with the anomalies in Jones Sound and central Devon Island suggests that the westernmost part of the Jones Sound Fold Belt also is basement-controlled. The absence of Ellesmerian structures in the Arctic Platform between the Jones Sound Fold Belt and the Schei Syncline remains unexplained, although some parallels may be suggested with the Wyoming Province of the western United States. There too, a belt of basement-controlled faults and folds occurs cratonward of a more extensive belt of thin skinned folds and thrust faults.

The youngest unit known to be involved in deformation of the fold belt is the Upper Devonian Nordstrand Point Formation of the Okse Bay Group. On Colin Archer Peninsula, remnants of the Upper Carboniferous Canyon Fiord Formation are preserved in a Tertiary(?) graben within the fold belt. Strata within the graben are tilted, and evaporitic units are drag-folded along the margins of the graben. Flanking the graben are extensively folded and thrust-faulted Ordovician to Devonian strata. Tertiary strata of the Eureka Sound Group preserved in other grabens to the south are tilted but otherwise undeformed. Such relationships indicate that the Jones Sound Fold Belt most likely formed in latest Devonian – Early Carboniferous time, i.e. during the Ellesmerian Orogeny.

G. LATE DEVONIAN – EARLY CARBONIFEROUS DEFORMATION OF THE PARRY ISLANDS AND CANROBERT HILLS FOLD BELTS, BATHURST AND MELVILLE ISLANDS

J.C. Harrison, F.G. Fox, and A.V. Okulitch

INTRODUCTION

This chapter discusses the Late Devonian – Early Carboniferous (Ellesmerian) deformation on Melville, Bathurst and adjacent smaller islands, which, together with Cornwallis Island, constitute the Parry Islands (for geographic names throughout this subchapter, see Fig. 1, in pocket). The strata involved in this deformation form part of three depositional realms: the Franklinian Shelf (or miogeocline), the sedimentary subprovince of the deep water basin, and the Middle-Late Devonian foreland basin (cf. Chapters 4, 8, 10). Structurally, they were previously all included in the Parry Islands Fold Belt (as redefined by Fortier, 1957), but the strata deposited in the axial region of the deep water basin (cf. Chapter 8C, Fig. 8C.1 and Table 8C.1), which are exposed in northwestern Melville Island, are now assigned to the Canrobert Hills Fold Belt (Harrison et al., 1988; Harrison and Bally, in press). The new subdivision is analogous to that on Ellesmere Island, the Parry Islands Fold Belt corresponding to the Central Ellesmere Fold Belt, and the Canrobert Hills Fold Belt to the southeastern part of the Hazen Fold Belt.

The Parry Islands Fold Belt is delimited on eastern Bathurst Island by the north-trending Cornwallis Fold Belt. Southern limits are gradational; the fold belt passes into relatively undeformed Cambrian to Devonian strata of the Arctic Platform. West of Melville Island, the Parry Islands Fold Belt is covered by Jurassic and Cretaceous strata of the Eglinton Graben but continues, with change of structural trend, on southeastern Prince Patrick Island, beyond which it is covered by Neogene strata of the Arctic Coastal Plain (Harrison et al., 1988; Chapter 12H, below).

In the Parry Islands, the Parry Islands Fold Belt is divisible into two segments. The eastern segment, exposed on Bathurst and eastern Melville islands, is characterized by a thin skinned detachment style of deformation. In contrast, deformation in the western segment, exposed on central and western Melville Island, involves most if not all of the lower and middle Paleozoic succession. The existence of and depth to a (postulated) basal decoupling surface in the western segment remain uncertain.

The Canrobert Hills Fold Belt (Fig. 12A.1 and 3C, in pocket) is exposed on northwestern Melville Island and continues to the west and north beneath younger cover rocks of Eglinton Graben and Sverdrup Basin, respectively. Like the Hazen Fold Belt, it consists mainly of basinal sediments (resedimented carbonates, mudrock, chert, and flysch) and was involved in the Ellesmerian Orogeny and younger tectonic events. Various scales of plunging and asymmetric chevron folding typify the style of deformation in the Canrobert Hills Fold Belt in contrast to the long-wavelength, upright, and cylindrical style of folding typical of many areas of the Parry Islands Fold Belt.

The age of the Ellesmerian deformation is constrained by strata of Famennian (latest Devonian) and Moscovian (early Late Carboniferous) ages in the Parry Islands Fold Belt, and by strata of Givetian (late Middle Devonian) and Moscovian ages in the Canrobert Hills Fold Belt (Fig. 12G.1). However, broader regional considerations indicate that the orogeny terminated before the Viséan (late Early Carboniferous), the age of the oldest known sediments of the Sverdrup Basin in other parts of the Arctic Islands (Chapter 13).

The eastern segment of the Parry Islands Fold Belt is, structurally, the best known part of the Franklinian mobile belt. Systematic mapping and regional stratigraphic studies (Fortier et al., 1963; Tozer and Thorsteinsson, 1964; Kerr, 1974) were followed or accompanied by drilling, reflection seismic surveys, and additional surface investigations (Fox, 1983, 1985; Texaco Resources Canada Ltd., 1983; Harrison and Bally, in press) and remapping (Harrison et al., 1985; Christie, 1986; Fig. 3D, in pocket). The exposed portion of the Canrobert Hills Fold Belt has not been investigated geophysically.

The Canrobert Hills and Parry Islands fold belts are important for the understanding of the styles of Late Devonian to Early Carboniferous deformation in the Canadian Arctic Islands because, in contrast to the eastern Arctic, the superimposed effects of the Eurekan Orogeny are slight.

DEPOSITIONAL FRAMEWORK AND ITS INFLUENCE ON STRUCTURAL STYLE

The lower Paleozoic depositional framework of the Canrobert Hills and Parry Islands fold belts is discussed in Chapter 8; for an overview the reader is referred to the stratigraphic cross-sections, Figures 8B.4 and 12G.2, and to the regional lithofacies maps, Figures 8B.30 to 8B.37. The pre-Ordovician record is not exposed. Reflection seismic records from eastern Melville Island reveal a pronounced angular unconformity 3.5 to 4.5 seconds (8 to 11 km) below the surface that may be close to the base of the Phanerozoic section. A comparable stratigraphy is exposed on Victoria Island where undeformed Cambrian strata lie unconformably on either gently folded sedimentary and volcanic rocks of the Upper Proterozoic Shaler Group or on more highly deformed strata of the Lower Proterozoic Goulburn Group (Thorsteinsson and Tozer, 1962; Campbell, 1981; Chapter 6). Seismic records suggest that the interval of laterally continuous reflectors,

---

Harrison, J.C., Fox, F.G., and Okulitch, A.V.
1991: Late Devonian – Early Carboniferous deformation of the Parry Islands and Canrobert Hills fold belts, Bathurst and Melville islands; in Chapter 12 of Geology of the Innuitian Orogen and Arctic Platform of Canada and Greenland, H.P. Trettin (ed.); Geological Survey of Canada, Geology of Canada, no. 3; (also Geological Society of America, The Geology of North America, v. E).

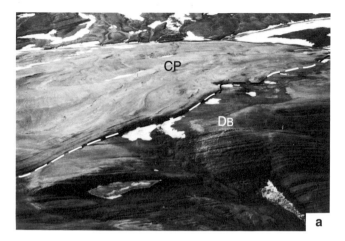 

Figure 12G.1. The angular unconformity separating Moscovian and younger strata of the Canyon Fiord Formation (CP) from: (a) peneplained Middle Devonian Blackley Formation (DB) in Canrobert Hills Fold Belt; (b) basinal Ordovician to Devonian Cape Phillips Formation (ODCP) near McCormick Inlet.

corresponding to the Cambrian to Lower Ordovician section above the unconformity but below the deepest well penetrations, may represent 4 to 6 kilometres of platform sediments.

In the Canrobert Hills Fold Belt, a pre-existing deep water basin received resedimented carbonates, mudrock, and chert from late Early Ordovician (Arenig) to early Middle Devonian (early Eifelian) time (Canrobert and Ibbett Bay formations). The basin then filled with submarine-fan and argillaceous slope deposits in the Middle Devonian (Blackley and Cape de Bray formations).

In the eastern segment of the Parry Islands Fold Belt, the known stratigraphic record begins above the base of the Eleanor River Formation, a Lower to Middle Ordovician shelf carbonate unit. In the Middle Ordovician, bedded salt, gypsum-anhdrite, and carbonates of the Bay Fiord Formation were deposited in an extensive intra-shelf basin that was separated from the deep water basin of the Canrobert Hills Fold Belt by a carbonate barrier, exposed in the western segment of the Parry Islands Fold Belt. Following a period of widespread platform carbonate deposition (upper Bay Fiord and Thumb Mountain formations), large areas of the Parry Islands Fold Belt subsided below shelf level and formed part of the marginal belt of the deep water basin from Late Ordovician to early Middle Devonian time. This area received mainly condensed successions of fine grained clastic sediments, redeposited carbonates, and minor chert (Cape Phillips, Kitson, and lower Eids formations), while Lower Devonian submarine-fan and argillaceous slope deposits accumulated near the Boothia Uplift (Bathurst Island and Stuart Bay formations). Offshore carbonate buildups (the upper part of which are referred to as the "Blue Fiord Formation") persisted in two areas until Early to early Middle Devonian time.

In the Middle and Late Devonian, the Parry Islands formed part of a foreland basin that received a thick and extensive succession of shallow marine and nonmarine siliciclastic sediments (Bird Fiord, Weatherall and Hecla Bay formations, Griper Bay Subgroup, etc.; cf. Chapter 10).

The extent and structural style of deformation in the Canrobert and Parry Islands fold belts is governed by the variations in ductility of units. In the eastern segment of the Parry Islands Fold Belt, the evaporitic lower member of the Bay Fiord Formation and, to a lesser extent, the argillaceous Eids and Cape de Bray formations, have responded to stress by flow. In contrast, pre-Bay Fiord carbonates, various Caradoc to Eifelian carbonates and basinal facies rocks, and Eifelian to Famennian coarse clastic formations have responded by buckling and faulting at various scales. Significant departures from these patterns of deformation exist where, due to changes in depositional environment, Ordovician evaporites pass into carbonate strata beneath the western segment of the Parry Islands Fold Belt, and elsewhere where Middle Devonian argillaceous strata are thin or absent above the "Blue Fiord Formation". Structural style is also different in the Canrobert Hills Fold Belt, where all units display varying degrees of ductile behaviour.

STRUCTURE
Eastern segment of Parry Islands Fold Belt

The Late Devonian – Early Carboniferous Parry Islands Fold Belt is bordered on the east by the Cornwallis Fold Belt, which originated in the Early Devonian, but was reactivated during the Ellesmerian Orogeny. The intersection area is attended by considerable structural complexity, caused by discrete episodes of deformation, the 90° difference in tectonic trend, and the thin skinned detachment style of the Parry Islands structures, in contrast to the basement-involved style of the Cornwallis structures. Temple (1965; and in Kerr, 1974, p. 74-83) suggested that the southward translation of rocks required by the east-trending folds was accomplished along a wrench fault within the Cornwallis Fold Belt, which permitted the westernmost Cornwallis anticline to move southward also and to be mildly deformed. However, it is important to realize that some of these north-trending structures have refolded westerly-trending Parry Islands folds, thrusts, and detachment surfaces, and are, therefore, in part younger than the latter (see Fig. 12G.3).

West of the Cornwallis Fold Belt on Bathurst Island, all of the folds plunge southwest to Austin Channel (between Bathurst and Byam Martin islands). Between Bathurst and Melville islands, a plunge depression provides anticlines

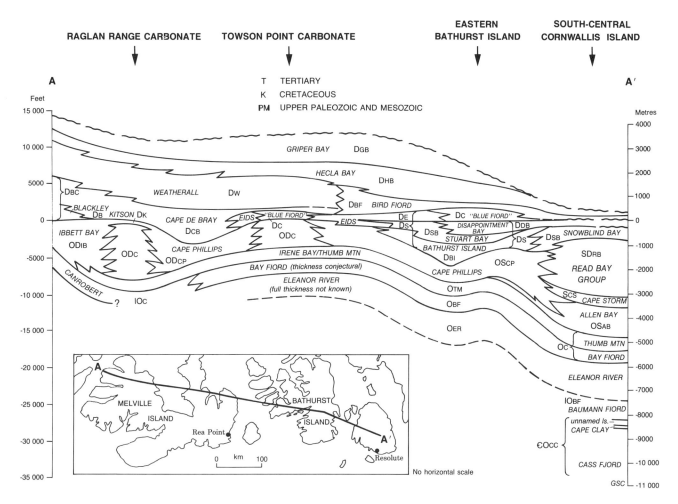

Figure 12G.2. Generalized stratigraphic cross-section showing stratigraphic relationships of Devonian and older rocks on Cornwallis, Bathurst, and Melville islands.

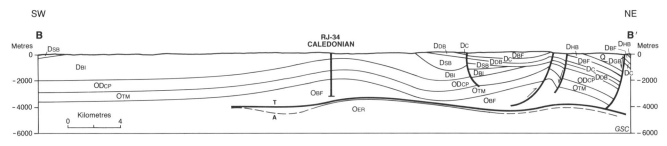

Figure 12G.3. Structural cross-section of Caledonian River Anticline and adjacent structures. A: footwall of fault has moved away from viewer. T: hanging wall towards viewer. (Location on Fig. 3C, in pocket; for legend see Fig. 12G.2).

of eastern Melville Island some easterly plunge. There is also a gradual change in tectonic trend from N60-70°E on Bathurst Island to due east on eastern Melville Island. To the west, the trend swings gradually to N75°W. The eastern segment of the fold belt ends along a line between Liddon Gulf and Hecla and Griper Bay in the vicinity of which the basement-involved structures of the western segment of the fold belt have obliquely cross folded the shallow detachment structures, creating a dome-and-basin interference pattern at the surface.

On Bathurst Island, the folds display gradually diminishing amplitude from north to south. The change is not evident on eastern Melville Island where the southern limit of folding is concealed under Viscount Melville Sound but is obvious farther west on Dundas Peninsula. This indicates that the effects of the Ellesmerian Orogeny did not extend all the way across the region occupied by the deep water basin from Late Ordovician to early Middle Devonian time (Fig. 8B.34-8B.37).

On western Bathurst Island and on the eastern half of Melville Island, the exposed anticlines are up to 150 km long and narrow (Fig. 12G.4). Axial zones, which commonly expose Middle Devonian beds, dip 50 to 90° north and south. Flank dips range from 30 to 45°. The folds are regularly spaced, with wavelengths of 12 to 15 km and amplitudes less than 3.5 km. The folds are only slightly asymmetric with an equal representation of northward and southward apparent vergence. Synclines are also very long and are on average 2.4 times the width of the anticlines (measured between dip inflection points near the surface). Departures from this scale of deformation are observed in the Bent Horn (Fig. 12G.5) and Towson Point (Fig. 12G.6) structures.

These structures have been influenced by a thickened competent section, a consequence of the continuous deposition of Upper Ordovician to Middle Devonian carbonates, and a thinned section of ductile mudrocks at the base of the Middle-Upper Devonian clastic wedge. These structures have also been greatly modified by extensional and wrench-tectonic events associated with the initiation and evolution of the Sverdrup Basin.

Faults exposed at the surface that are parallel or subparallel to the axial trace of folds are a common feature of the fold belt. They have a strike length of 10 to 40 km and stratigraphic throws ranging from a few hundred metres to several kilometres. Although the attitudes of these faults

Figure 12G.4. Oblique aerial photograph of the Parry Islands Fold Belt looking east from central Melville Island. Byam Martin Island is visible at top right. D_{PI} = Parry Islands Fm.; D_{BI} = Beverley Inlet Fm.; D_{HB} = Hecla Bay Fm.; D_W = Weatherall Fm. (T414L-45)

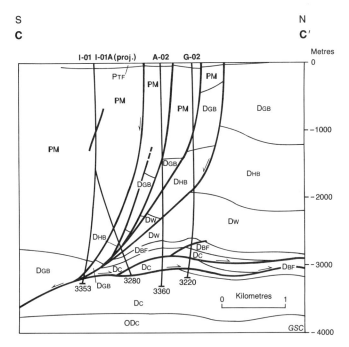

Figure 12G.5. Cross-section of the Bent Horn structure, Cameron Island. (Location on Fig. 3C, in pocket; for legend see Fig. 12G.2).

are generally unknown, it is likely that they display a reverse sense of slip and that they were formed through brittle failure of the evolving anticlines.

The role of the Bay Fiord evaporites in the development of the distinctive style of deformation on eastern Melville Island was first suggested by Tozer and Thorsteinsson (1964). Kerr (1964) noted the presence of two areas of gypsiferous rocks in northeast Bathurst Island and suggested that the intrusive nature of the gypsum is more probably the result of folding than the cause of it. Workum (1965) suggested that salt in the Caledonian River J-34 well "may be the zone of décollement beneath the Parry Islands fold belt". Kerr (1974) assigned the salt and anhydrite in this well to the Middle Ordovician Bay Fiord Formation, suggested that the Eleanor River and Baumann Fiord formations should be present below the salt, and related development of the Parry Islands folds to salt mobility. He also noted that salt had migrated to anticlinal axes and intruded diapirically in some structures.

Seismic studies and further deep drilling have confirmed these observations and have also shown that, at depth, thrust faulting played an important role in fold development. The maximum shortening indicated by these data is at least 13-16 km (8-10%) on Bathurst Island, and roughly 23-27 km (10-12%) on Melville Island. The regional cross-section of the fold belt on eastern Melville Island shown in Figure 3D, together with the cross-section of the Beverley Inlet and Robertson Point anticlines (Fig. 12G.7), incorporates many of the features indicated by the subsurface data. The bulk of the deformation occurs above a subhorizontal decoupling level above the Eleanor River Formation, and north of the limit of bedded salt within the overlying Bay Fiord Formation (left side of Fig. 3D and Fig. 12G.8). There is an equal number of forward- and backward-verging thrusts and folds, subsurface structures extend farther onto the platform than surface folds, and shortening between the two zones of ductile flow is more than twice that observed in the fold belt above the upper ductile zone at the surface.

The diversity of structures and the apparent discrepancy in the shortening (see cross-section, Fig. 3D) at the two structural levels can both be explained by a kinematic model discussed in detail by Harrison and Bally (in press).

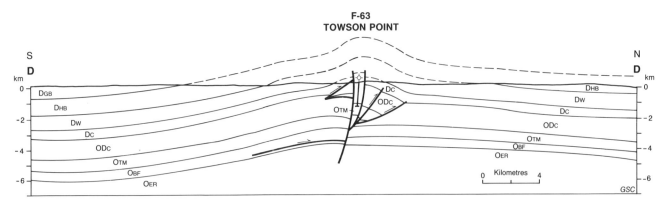

Figure 12G.6. Cross-section of the Towson Point Anticline, northeastern Melville Island. (Location on Fig. 3C, in pocket; for legend see Fig. 12G.2).

CHAPTER 12

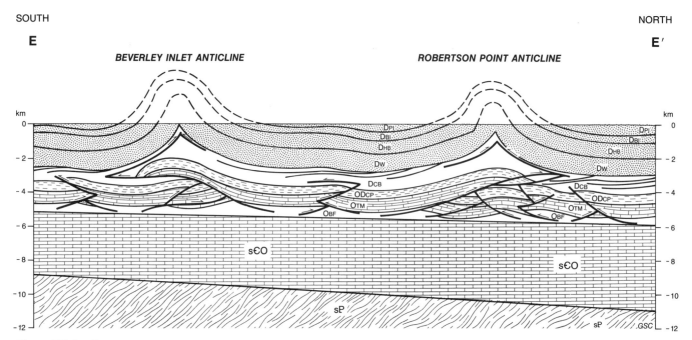

Figure 12G.7. Structural cross-section of Beverley Inlet and Robertson Point anticlines, eastern Melville Island. (Location on Fig. 3C, in pocket; for legend see Fig.12G.2 and 12G.4. sP = Proterozoic seismic unit; s∈O = Cambrian – Lower Ordovician seismic unit.) Cretaceous strata at surface are less than 10 m thick and have been ignored in this cross-section.

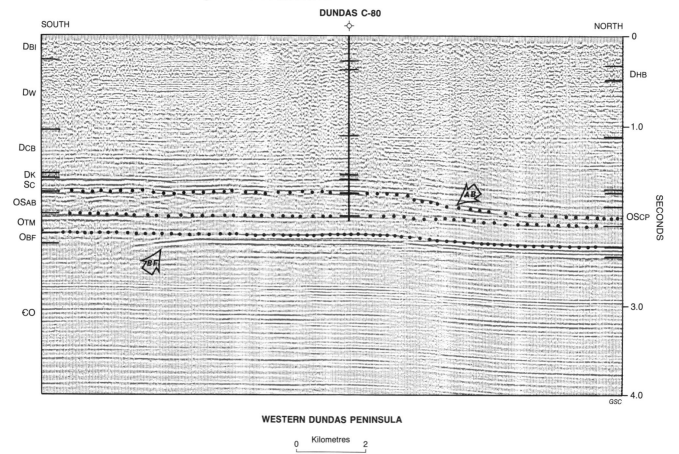

Figure 12G.8. Portion of reflection seismic section (Panarctic Oils Limited line 1168) showing the edge of the evaporitic facies of the Bay Fiord Formation, indicated by the abrupt disappearance of the related reflector (arrow labelled "BF"). A similar change in seismic character of the lower Bay Fiord Formation westward on Melville Island is paralleled by a progressive replacement of bedded halite by anhydrite and dolostone, inferred from outcrops and three wells. Also shown is the edge of a Silurian carbonate buildup in the Allen Bay Formation (arrow labelled "AB"). (For legend see Fig.12G.2-12G.4).

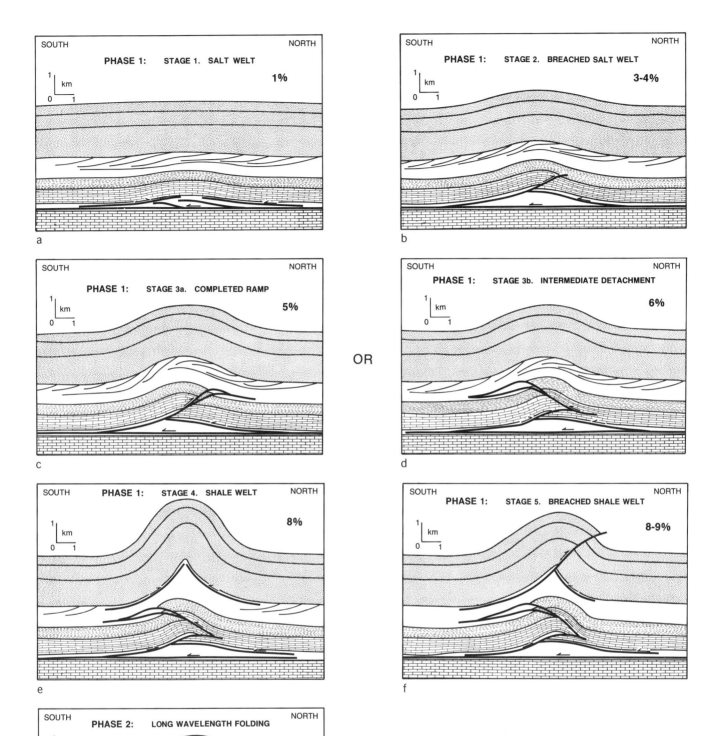

Figure 12G.9. Phases and stages in the evolution of anticlines in the Parry Islands Fold Belt, eastern Melville Island. Apparent shortening (in per cent) is indicated in the upper right corner of each diagram. 3a and 3b are two alternative stages that could develop simultaneously during anticline evolution. (For explanation of patterns see Fig. 12G.7).

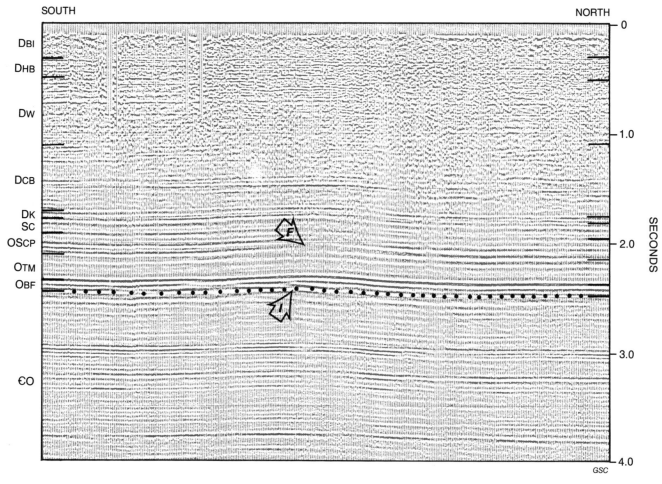

Figure 12G.10. Portion of reflection seismic line (Panarctic Oils Limited line 1171) over Cape Providence Anticline (see also Fig. 3D, in pocket). This represents the seismic expression of the embryonic stage (phase 1, stage 1 of Fig. 12G.9) of deformation, characterized by the development of a salt welt in the lower Bay Fiord Formation (O$_{BF}$). Note the imbricated weak reflector in the salt welt (arrow labelled "I"), implying southward transport of rock units in the welt. Although the strong reflectors above the welt appear to be undeformed, a reflector near the top of the Thumb Mountain interval (O$_{TM}$) appears to be offset over the crest of the welt (arrow labelled "F"). Amplitude of structures decreases in and above the Cape de Bray (D$_{CB}$) interval, and there appears to be no displacement or folding of the upper Hecla Bay (D$_{HB}$) reflector (for legend for Fig.12G.10-12G.14 see Fig.12G.2-12G.4).

In general, the folds in the inner part of the fold belt appear to have formed earlier and evolved farther than the more embryonic external structures, although there is considerable overlap in the "lifespans" of adjacent folds (Fig. 12G.9). Anticlines were initiated through buckling of the most competent section between the two ductile zones. This was accompanied by the lateral migration of evaporites in the lower ductile zone to form a linear salt welt, and may have coincided with tectonic compaction of the siliciclastic strata above the upper ductile zone (Fig. 12G.9a, 12G.10). Through progressive shortening, the competent carbonate-dominated section failed in a brittle fashion which led to the propagation of thrusts. Low-amplitude buckling was initiated in the siliciclastic strata (Fig. 12G.9b-12G.9d, 12G.11-12G.13).

An important observation to arise from the cross-section is that thrusts emerging from the lower ductile zone are older than those higher in the section. A key factor in the relative dating of thrusts is the resistance to sliding of various detachment surfaces, and the tendency for competent strata to buckle and fail at an early stage while other loosely compacted strata may initially absorb the shortening through lateral compaction and porosity loss. This contrasting response to contractional stresses can also explain the apparent differences in shortening at the two structural levels.

The more complicated folds consist of pop-up structures (formed where slip is transferred along strike from forward-verging to backward-verging thrusts), thrust splays, multiple breached salt welts, and/or faults that zigzag up section, utilizing a number of intermediate detachment horizons to reverse the sense of slip (Fig. 12G.14, 12G.15). At an advanced stage of anticlinal evolution, argillaceous strata in the upper ductile zone migrated into the axial region of

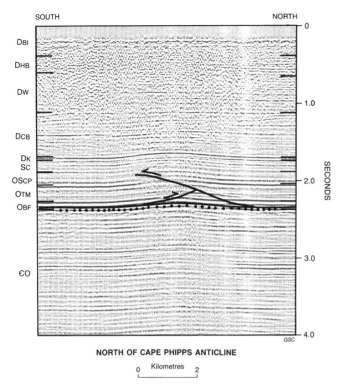

Figure 12G.11. Portion of reflection seismic section (line 1171) over an unnamed structure north of Cape Phipps Anticline (see Fig. 3D, in pocket). This is the seismic expression of the second stage of deformation (stage 2, Fig. 12G.9). Faulting now appears through most of the competent beds above the shale welt although the thrust ramp appears to die out upsection into an anticline short of the upper ductile zone.

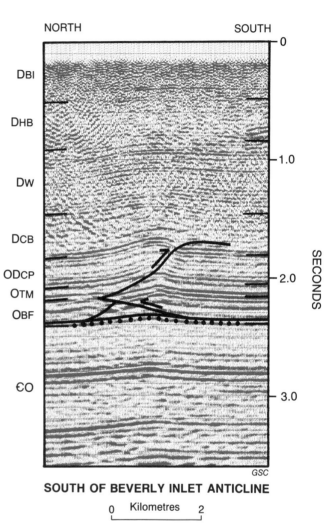

Figure 12G.13. Portion of time-migrated reflection seismic line (Panarctic Oils Limited line 1660), stage 3b of Figure 12G.9. Wedging has created a triangle-zone geometry in the seismic image of this unnamed anticline south of Beverly Inlet Anticline (Fig. 12G.7). Thrust vergence reversal occurs near the base of the Thumb Mountain Formation (OTM).

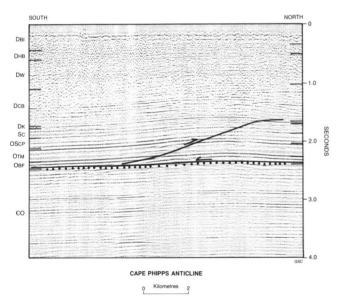

Figure 12G.12. Portion of reflection seismic line (1171) over Cape Phipps Anticline (see also Fig. 3D, in pocket). This represents the seismic expression of the completed ramp stage of fold development (stage 3a, Fig. 12G.9). Diffractions mask somewhat the location of fault-terminated reflectors, and the sigmoid nature of the thrust ramp would tend to be accentuated by depth conversion.

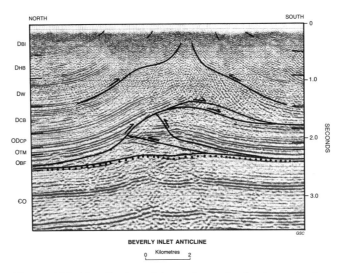

Figure 12G.14. Portion of time-migrated reflection seismic line (1660) over Beverley Inlet Anticline (Fig. 12G.7). This is the seismic image of the stage of fold development when surface fold amplitude increases dramatically due to diapiric growth of a shale welt (stage 4, Fig. 12G.9) in the Cape de Bray section perched vertically above the salt welt.

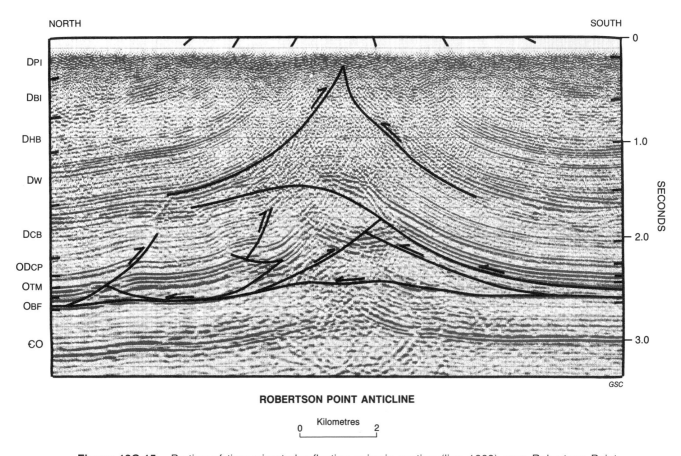

Figure 12G.15. Portion of time-migrated reflection seismic section (line 1660) over Robertson Point Anticline (Fig. 12G.7). This structure represents the stage of fold development when demonstrable shortening exceeds 10%. In this instance, the simple surface anticline and near-surface cuspate shale welt mask a complex array of salt welts and major and minor thrusts of variable vergence and variable level of intermediate detachment.

the anticline to form a second welt, stacked vertically above the deeper salt welt. This was coeval with a rapid growth in the amplitude of the surface anticline (Fig. 12G.9e, 12G.14, 12G.15). Eventually this shallow anticline failed in a brittle fashion and thrusts emerging from the upper ductile zone were propagated up through the overlying siliciclastic strata (Fig. 12G.9f).

Western segment of Parry Islands Fold Belt

Folds of the eastern Parry Islands Fold Belt gradually die away west of a line between Liddon Gulf and Hecla and Griper Bay. Associated with this transition is the lateral facies change of the lower member of the Bay Fiord Formation from a halite-dominated facies through an anhydrite-dominated to a carbonate-dominated facies (Balkwill and Fox, 1982). These changes eliminate the lower ductile zone that greatly influenced the formation of buckle folds and thrusts in the overlying strata. Instead, shortening in the western segment of the Parry Islands Fold Belt has been accomplished by long-wavelength (40 km) folding, and by reverse faulting of much if not all of the Cambrian to Devonian section. Ductile flow of the Cape de Bray Formation is still evident and many significant thrusts fail to offset the overlying siliciclastic strata. The long-wavelength structures (that extend onto the line of the cross-section shown in Fig. 3D) trend east-northeast and are younger than the west-northwest trending structures above the lower Bay Fiord Formation (Phase 2 of folding of Fig. 12G.9g). As a result, there is a region on central Melville Island where the two phases of folding have produced a pattern of dome-and-basin interference structures at the surface.

To the southwest on Melville Island, long-wavelength folds pass laterally into an area of steep faults that strike variably N45°E through N90°E to N70°W, the Blue Hills Fault Belt of Tozer and Thorsteinsson (1964). Knowledge of this area is limited by a lack of geophysical data and by the absence of exposed stratigraphic section; in consequence, the attitude, timing of motion, and sense of slip of the exposed faults is unknown. The faults are, however, associated with the monoclinal flexure and doming of surface strata, which suggests a component of contractional slip. Other faults in this area, striking N5°E to N20°E, are spatially and, probably, genetically related to the evolution of the Eglinton Graben (Middle Jurassic to Late(?) Cretaceous).

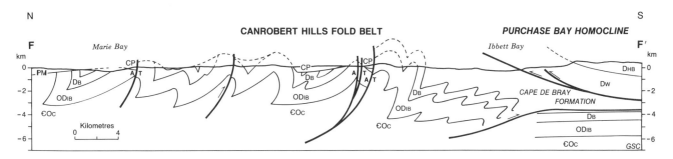

Figure 12G.16. Diagrammatic structural cross-section of Canrobert Hills Fold Belt, northwestern Melville Island. For location see Figure 3C, in pocket. Stratigraphic units are: PM = Permian and Mesozoic strata; CP = Carboniferous and Permian Canyon Fiord Fm.; D_{HB} = Hecla Bay Fm.; D_W = Weatherall Fm.; D_B = Blackley Fm.; OD_{IB} = Ibbett Bay Fm.; ϵOc = Cambrian and Lower Ordovician strata (undivided). A = strike slip away from viewer; T = strike slip toward viewer.

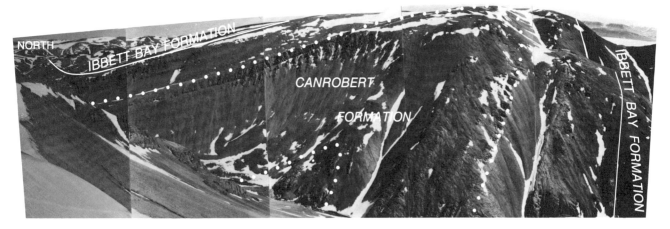

Figure 12G.17. Panoramic view of an overturned anticline in Canrobert Hills Fold Belt. Wavelength of this structure is approximately 12 km. Dotted lines indicate structural markers within Canrobert Formation.

Canrobert Hills Fold Belt

Deformation in the Canrobert Hills Fold Belt is typified by various scales of chevron folding, and two distinct tectonic events. The earlier deformation affects Givetian and older strata and may be penecontemporaneous with the folding and thrusting in the Parry Islands Fold Belt. The younger deformation (Melvillian Disturbance), affecting lower Lower Permian, Upper Carboniferous, and older strata is discussed elsewhere in this volume (Chapter 13).

Tectonic elements (shown in Fig. 3C) define an arcuate fold belt that trends N70°E on the east side through N90°E to N65°W on the west side. Folds are tight, upright, asymmetric, and locally overturned, south-verging structures, that plunge to the east on the southeast side of the fold belt and plunge west along the western limit of the exposed fold belt adjacent to Kellett Strait. Folding is associated with the development of an axial planar cleavage evident in the axial region of the larger folds, and a weak, pre-Carboniferous low-grade metamorphism. Other features typical of the fold belt include sigmoidal carbonate veins up to 11 m wide, several reverse faults with limited displacement, northerly-striking tear faults, and various younger normal and dextral strike-slip faults. The schematic cross-section of the Canrobert Hills Fold Belt shown in Figure 12G.16 emphasizes the styles of folding and known mechanisms of shortening based on the available surface information. The role of thrust faulting in the subsurface, and the depth to the assumed basal decoupling level, remain uncertain.

The scale of the chevron folding varies with stratigraphic position. The largest folds have been formed in the Canrobert and Ibbett Bay formations and possess an average wavelength of 8 km (Fig. 12G.17). Folds in the Blackley Formation are up to 2-3 km in wavelength and are parasitically distributed over the larger underlying structures (Fig. 12G.18). Deformation in the overlying Cape de Bray Formation is characterized by apparent ductile flow and outcrop-scale, tight to isoclinal folds and minor thrusts. These small structures, like most contractional elements in the Canrobert Hills Fold Belt, verge to the south. However, there is a plane in the upper Cape de Bray or lower Weatherall formations, above which mesoscopic folds and thrusts display an opposite (northerly) sense of displacement. This plane (shown on Fig. 2G.16, 12G.19) is interpreted to be the roof thrust of a triangle zone above which Devonian siliciclastic strata have been passively raised into a south-facing homoclinal flexure (Purchase Bay Homocline), and below which the southern part of the Canrobert Hills Fold Belt has been inserted as a tectonic wedge. The preserved width of this triangle zone is equal to the width of Purchase Bay Homocline, approximately 30 km.

The westward continuation of the Canrobert Hills Fold Belt is masked at the surface by younger strata of Eglinton Graben and Sverdrup Basin, and by the waters of the inter-island channels. Reflection seismic data collected from Eglinton and northern Prince Patrick islands confirm the continuation of the belt of late Paleozoic deformation without lateral offset (Chapter 12H).

Estimated shortening in the Canrobert Hills Fold Belt is approximately 30%.

Figure 12G.18. Parasitic folds developed in the Blackley Formation in the Canrobert Hills Fold Belt. Note the minor tear faults that offset the folds at high oblique angles to the dominant structural trend.

SILURIAN – EARLY CARBONIFEROUS DEFORMATIONAL PHASES AND ASSOCIATED METAMORPHISM AND PLUTONISM, ARCTIC ISLANDS

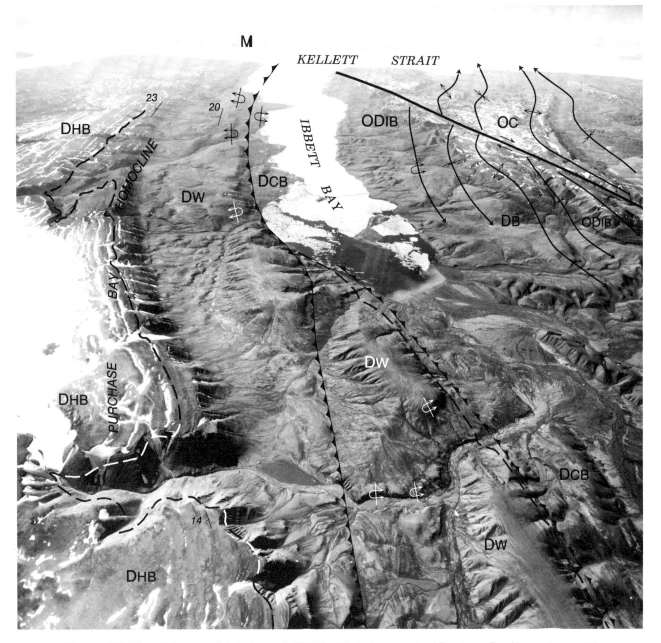

Figure 12G.19. Oblique aerial photograph (T414L-45), looking west, of Purchase Bay Homocline and the southern limit of Canrobert Hills Fold Belt (background, right side). Stratigraphic units are the same as those in the caption of Figure 12G.16 but also include Oc = Canrobert Fm.; DCB = Cape de Bray Fm. and M = Mesozoic strata of Eglinton Graben (exposed on Eglinton Island; centre, background). Vergence of asymmetric and overturned major and minor folds aids in defining the location of the northward-vergent backthrust that is traceable across the photo.

CHAPTER 12

H. LATE DEVONIAN – EARLY CARBONIFEROUS DEFORMATION, PRINCE PATRICK AND BANKS ISLANDS

J.C. Harrison and T.A. Brent

It has long been known that deformed strata of Devonian age are exposed in southwestern Prince Patrick and northeastern Banks islands, but their relationship with the Ellesmerian Orogen in the Parry Islands, and the trend of the orogen west of these islands, have remained problematic. Published surface observations and unpublished, industry-acquired, seismic reflection data relevant to these problems are summarized in this subchapter.

PRINCE PATRICK ISLAND

On Prince Patrick Island, Tozer (1956) first recognized an anticline in Devonian strata at the head of Mould Bay. An angular unconformity was later identified by Tozer and Thorsteinsson (1964) beneath Jurassic strata east of the Mould Bay Weather Station. However, the fold structures typical of the Canrobert Hills area did not obviously continue into the Devonian outcrop belt of southern Prince Patrick Island. To explain the termination of the fold belt, Thorsteinsson and Tozer (1970) suggested that a pre-Mesozoic dextral wrench fault may have caused a northward offset of the deformed belt west of Melville Island.

Field data collected in 1987 (Harrison, et al., 1988) indicate that pre-Mesozoic folding and thrusting was active throughout Devonian outcrop areas of southeastern Prince Patrick Island. In addition, reflection seismic data indicate a wide distribution of related structures throughout the subsurface of Eglinton and Prince Patrick islands and the inter-island channels west of Melville Island (Fig. 12H. 1). The surface structures include an angular unconformity (generally less than 30° but locally up to 90°) beneath Lower Cretaceous strata west of Mould Bay, beneath Jurassic strata between Green and Mould bays, and beneath Upper Triassic strata in the vicinity of Intrepid Inlet. A series of seven anticlines and six intervening synclines are developed in Middle and Upper Devonian strata. The folds are open, upright and gently-plunging structures that have an average wavelength of 15 km, an amplitude of up to 1.5 km, and a trend that varies from N20°W near Dyer Bay, through N20°E west of Mould Bay, and back to N45°W near Green Bay. The folded terrane has been shattered by a complex array of faults that strike at various oblique angles to the axial plane of the folds, displace Mesozoic and older strata, and thus appear to be unrelated to fold development. Anticlines are narrow, cuspate structures that imply a shallow decoupling level, probably in the Cape de Bray Formation (Middle Devonian). The contractional aspect of the surface structures is also indicated by the development of outcrop-scale minor folds and thrusts, and a penetrative slaty cleavage that is evident in the axial regions of the anticlines. An important feature of the surface geology is a south-facing homocline that has exposed the upper part of the Cape de Bray Formation north of Green Bay and east of Intrepid Inlet.

Seismic data also reveal an extensive sub-Mesozoic angular unconformity and several, long-wavelength, low-amplitude contractional structures involving pre-Carboniferous strata beneath southern Eglinton and southern Prince Patrick islands. Several of the surface-mappable folds and one thrust fault, north of Dyer and Mould bays, can also be identified on reflection seismic lines that image structures of comparable trend in the pre-Carboniferous succession beneath the Arctic Coastal Plain. Although surface observations suggest a detachment in the Cape de Bray Formation, some folds in the Devonian clastic units also involve the underlying Silurian and older basinal mudrock and/or carbonate succession. The ultimate depth to detachment of these folds (if a detachment exists) is unknown.

Continuity of surface and subsurface tectonic elements in the Prince Patrick Island area is also implied by the distribution of south-facing homoclinal features throughout the region shown on Figure 12H.1. The Purchase Bay Homocline of western Melville Island is also evident to the west on seismic images of the sub-Mesozoic interval beneath central Eglinton Island. In addition, the surface homocline near Green Bay appears to continue in the subsurface northwest of Intrepid Inlet. It is reasonable to conclude that these similar and spatially-related tectonic elements represent a single, continuous, homoclinal flexure (of sigmoidal shape in plan view) that is genetically related to the uplift, in pre-Triassic time, of the northeastern part of the map area.

Wherever the Paleozoic succession is exposed or geophysically imaged in this uplifted area, the intensity of deformation and degree of shortening is substantially higher than in structurally lower areas southwest of the homoclinal flexure. Westerly-plunging folds of the Canrobert Hills Fold Belt (northwestern Melville Island) can be individually correlated with seismically-defined, westerly-plunging folds of similar wavelength beneath northern Eglinton Island. Close-folded, doubly-plunging anticlines and northeasterly-vergent thrust faults of unknown displacement are also recognized beneath Permian and younger strata on the seismic records of subsurface northeastern Prince Patrick Island. These structures, presumably, represent the continuation of the Canrobert Hills Fold Belt toward the northwest.

The plan view of this intensely deformed belt is one that defines a structural re-entrant centred beneath Eglinton Island, paired with the southeastern part of a northeasterly-trending salient beneath southwestern Prince Patrick Island. The full extent of this salient to the northwest is unknown.

Harrison, J.C. and Brent, T.A.
1991: Late Devonian – Early Carboniferous deformation, Prince Patrick and Banks islands; in Chapter 12 of Geology of the Innuitian Orogen and Arctic Platform of Canada and Greenland, H.P. Trettin (ed.); Geological Survey of Canada, Geology of Canada, no. 3; (also Geological Society of America, The Geology of North America, v. E).

SILURIAN – EARLY CARBONIFEROUS DEFORMATIONAL PHASES AND ASSOCIATED METAMORPHISM AND PLUTONISM, ARCTIC ISLANDS

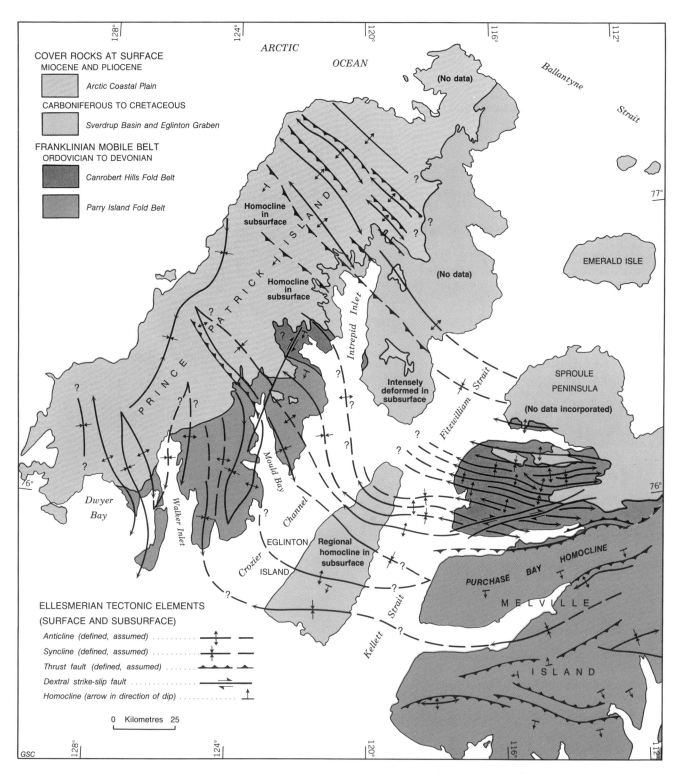

Figure 12H.1. Paleozoic tectonic elements of Prince Patrick and adjacent islands. Structural features shown in areas of the exposed Franklinian mobile belt have been defined by field observation. The remaining structures have been compiled from reflection-seismic records that image the top of the lower Paleozoic carbonate succession beneath younger cover rocks of the Sverdrup Basin, Eglinton Graben, and the Arctic Coastal Plain. Question marks have been added where the interpolation or extrapolation of tectonic elements is not confirmed by geophysical data.

Well and seismic data may also provide an explanation for the change in structural trends west of the Parry Islands. These data indicate that an environment of platform carbonate deposition may have existed throughout the early Paleozoic in the Prince Patrick Island area. These rocks could have had a buttressing effect on folds and thrust imbricates propagating southward through the relatively incompetent strata of the Canrobert Hills Fold Belt of northwestern Melville Island.

A major, but somewhat anomalous, feature on the subsurface data of western Prince Patrick Island, is a doubly-plunging N30°E-trending pre-Jurassic(?) syncline that can be mapped over a distance of 140 km. The orientation of this syncline departs from the dominant northwesterly tectonic fabric by as much as 70 to 90°. It is conceivable that this second structural trend has been generated by a separate phase of Paleozoic folding. Similar cross fold axes have been identified by Dailly and Plauchut (1986) in the subsurface of Banks Island, and may also exist in the Canrobert Hills area, where Ellesmerian folds have been refolded during the Early Permian Melvillian Disturbance (Thorsteinsson and Tozer, 1970).

NORTHEASTERN BANKS ISLAND

Folds on northern Banks Island have been mapped and described by Thorsteinsson and Tozer (1962), Klovan and Embry (1971), and Miall (1979). They are formed in strata of Middle to Late Devonian age (Weatherall and Parry Islands formations). The dominant feature is a broad syncline, which trends about N10°E (see Fig. 2, in pocket). The eastern limb of the syncline, which dips about 2°, passes into an extensive homocline. Smaller folds, spaced 5-8 km apart, occur on the western limb of the syncline. Dips on the flanks of these folds average 5° and never exceed 10°. Farther west, an extensive, but gentle, northerly-trending syncline has been detected in the subsurface by seismic surveys, and four other low amplitude folds trending about N50°E are revealed by the time-structure map constructed above the Devonian carbonate succession (Dailly and Plauchut, 1986). The fact that no comparable folds occur in unconformably overlying Cretaceous strata indicates a pre-Cretaceous age of deformation. The structural trends are similar to those on southwestern Prince Patrick Island, and are also parallel to aeromagnetic anomalies and some structural trends on the adjacent mainland.

DISCUSSION

The combined surface and subsurface information presented here clearly indicates structural continuity of the pre-Carboniferous rocks in western Melville Island with those in Prince Patrick Island, i.e. absence of a major transcurrent fault east of Melville Island. The Canrobert Hills Fold Belt is a well defined and continuous belt of intense northwesterly-trending deformation that envelops the southeastern and northeastern margins of a salient. Part of this buttress underlies southern Prince Patrick Island.

The most important remaining problems are the continuation of the Ellesmerian Orogen in the offshore region and its relationship with structures of similar age in the North Yukon and northern Alaska. Three different hypotheses can be formulated.

Bally (in communication with Kerr, 1982) has suggested that the predominantly westerly trend of the orogen in the Parry Islands may continue to the west, beneath younger sediments, and may thus be truncated by the modern (i.e. Cretaceous) continental margin of the Canadian Arctic Islands west of Prince Patrick Island. Such a configuration would support his view, and that of Oldow et al. (1987), that the reappearance of the Ellesmerian deformed belt in the Richardson, British and Barn mountains of the northern Yukon (Bell, 1974) may be due to a Mesozoic sinistral offset of the Paleozoic fold belt by a proposed northeasterly-striking megashear, close to the present continental margin and parallel with it, that was active during the creation of the Canada Basin.

Alternatively, the fold belt may change direction west of the Parry Islands Fold Belt and may be linked to the Yukon portion of the Franklinian mobile belt by a belt of northeasterly-trending deformation located beneath younger cover rocks of the continental shelf and slope, west of Banks and Prince Patrick islands. This solution may be supported by the image of pre-Jurassic thrusts on crustal seismic-reflection data near Campbell Uplift east of the Mackenzie Delta (Cook et al., 1987), and by the northeasterly-striking folds on Banks Island, if both are Late Devonian – Early Carboniferous (Ellesmerian) age. These structures could be taken as evidence supporting the oroclinal model for the creation of the Canada Basin (Carey, 1958).

A third possibilty is that the continuation of the fold belt has a westerly trend, that it is truncated by the continental margin, and that an oroclinal restoration of the Brooks Range – northern Alaska region would create a pre-existing link with the deformed lower Paleozoic strata documented from well penetrations near Point Barrow, Alaska (Carter and Laufield, 1975) and from seismic reflection data recorded over the Barrow Arch (Grantz and May, 1983).

It is difficult to evaluate these hypotheses because of the lack of subsurface information from the continental shelf west of Prince Patrick and Banks islands. However, the fact that the structural trends of the Ellesmerian Orogen meet the continental shelf at high angles without a noticeable turn to the west would seem to argue against the second hypothesis. The only possible evidence for such a turn are the northeasterly-trending folds on Banks Island, but their age is poorly constrained and may well be younger than the Ellesmerian Orogeny (cf. Chapter 13, Melvillian Disturbance).

SILURIAN – EARLY CARBONIFEROUS DEFORMATIONAL PHASES AND ASSOCIATED METAMORPHISM AND PLUTONISM, ARCTIC ISLANDS

I. SUMMARY

H.P. Trettin

The Late Silurian to Early Carboniferous deformational history of the Arctic Islands can be summarized in terms of six major events (Fig. 12I.1).

(1) The accretion of Pearya to the Clements Markham Fold Belt was one of the most significant events in the structural history of northern Ellesmere Island. Available information suggests that it was due to Late Silurian sinistral transpression, but the age and mechanism of the accretion require further studies. Regional metamorphism, locally up to amphibolite grade, and the emplacement of a basement-derived, post-tectonic(?) granitic pluton with a late Early to Middle Devonian U-Pb (zircon) age seem to be related to this event.

(2) The structural trends of the Northern Heiberg Fold Belt, subparallel with those of the Boothia Uplift, probably were inherited from the Precambrian basement. Reactivation of these trends by Late Silurian – Early Devonian thrust faulting and folding either was caused by

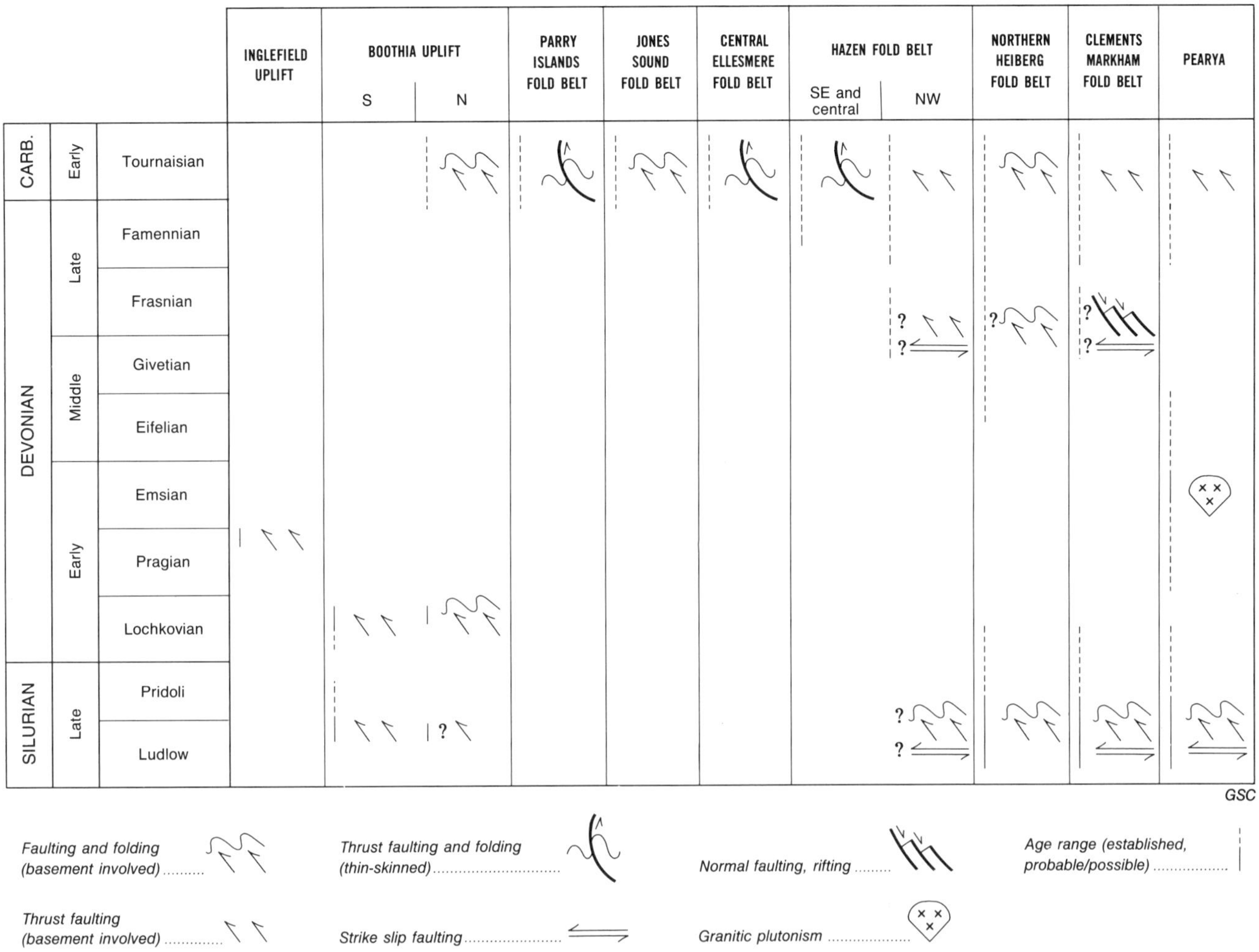

Figure 12I.1. Correlation of Late Silurian to Early Carboniferous tectonic events. Vertical scale is absolute time scale of Palmer, 1983.

Trettin, H.P.
1991: Summary (Silurian – Early Carboniferous deformational phases and associated metamorphism and plutonism, Arctic Islands; *in* Chapter 12 of Geology of the Innuitian Orogen and Arctic Platform of Canada and Greenland, H.P. Trettin (ed.); Geological Survey of Canada, Geology of Canada, no. 3; (also Geological Society of America, The Geology of North America, v. E).

southwestward motion of northern Ellesmere Island or by the same regional stress that caused Late Silurian movements of the Boothia Uplift.

(3) Westward-directed thrust faulting occurred on the western flank of (and likely throughout) the Boothia Uplift in Late Silurian (late Ludlow) and Early Devonian (Lochkovian) time, but the Silurian movements were weak in the area north of Barrow Strait. The thrust faults involve the crystalline Precambrian basement and are controlled by its structural "grain". The compression is thought to have been related to convergence in the Caledonian mobile belt, which has a comparable meridional orientation.

(4) Middle to late Early Devonian movements of the Inglefield Uplift, known only from the Pragian – early Emsian sedimentary record of central and southern Ellesmere Island, may be comparable in character and origin to those of the Boothia Uplift.

(5) A Givetian(?)-Frasnian uplift of the northwestern Hazen Fold Belt, perhaps accompanied by downdrop of a part of the Clements Markham Fold Belt, is tentatively attributed to renewed sinistral strike slip, and correlated with a second, poorly dated episode of compressional deformation in northern Axel Heiberg Island.

(6) Extensive compressive deformation, known as the Ellesmerian Orogeny *sensu stricto*, occurred throughout the Franklinian mobile belt after the deposition of the Okse Bay Formation/Group in latest Devonian – Early Carboniferous time. In northernmost Ellesmere Island (Pearya, Clements Markham Fold Belt and(?) adjacent parts of Hazen Fold Belt) this compression probably resulted mainly in faulting because of partial stabilization by earlier events; farther south it caused thin skinned folding and thrust faulting in the central and southeastern parts of the Hazen Fold Belt, and in the Central Ellesmere, Canrobert Hills, and Parry Islands fold belts. The evolution and structural style of this deformation is best known from seismic sections of the Parry Islands Fold Belt, where most of the shortening occurs above the base of Ordovician evaporites.

In the northern Boothia Uplift — the structural trends of which are at high angles but not perpendicular to the regional stress — basement faults were reactivated, and complex interference structures were formed in the sedimentary cover, especially at its intersection with the Parry Islands Fold Belt.

The origin of the Ellesmerian Orogeny has not been discussed in the preceding sections because it is not apparent from the exposures on land. Convergence of North America with another plate, continental or oceanic, in the present Arctic Ocean region, is the most probable cause, but that plate has not yet been identified with any assurance, and the postulated suture, if it exists, must be hidden in the present offshore area. The convergence may have been orthogonal or oblique, and, in the latter case, would have been accompanied by strike slip, but pure strike slip without a component of convergence could not have caused such extensive deformation.

REFERENCES

Balkwill, H.R. and Fox, F.G.
1982: Incipient rift zone, western Sverdrup Basin, Arctic Canada; in Arctic Geology and Geophysics, A.F. Embry and H.R. Balkwill (ed.); Canadian Society of Petroleum Geologists, Memoir 8, p. 171-187.

Bell, J.S.
1974: Late-Paleozoic orogeny in the northern Yukon; in Proceedings of the Symposium on the Geology of the Canadian Arctic, J.D. Aitken and D.J. Glass (ed.); Geological Association of Canada and Canadian Society of Petroleum Geologists, p. 23-38.

Berkhout, A.W.J.
1973: Gravity in the Prince of Wales, Somerset and northern Baffin islands region; in Proceedings of the Symposium on the Geology of the Canadian Arctic, J.D. Aitken and D.J. Glass (ed.); Geological Association of Canada and Canadian Society of Petroleum Geologists, p. 63-80.

Blackadar, R.G.
1967: Precambrian geology of Boothia Peninsula, Somerset Island, and Prince of Wales Island, District of Franklin; Geological Survey of Canada, Bulletin 151, 62 p.

Brown, R.L., Dalziel, I.W.D., and Rust, B.R.
1969: The structure, metamorphism, and development of the Boothia Arch, Arctic Canada; Canadian Journal of Earth Sciences, v. 6, p. 525-543.

Campbell, F.H.A.
1981: Stratigraphy and tectono-depositional relationships of the Proterozoic rocks of the Hadley Bay area, northern Victoria Island, District of Franklin; in Current Research, Part A, Geological Survey of Canada, Paper 81-1A, p. 15-22.

Carey, S.W.
1958: The tectonic approach to continental drift; in Continental Drift. A symposium, S.W. Carey (convener); Geology Department, University of Tasmania, Hobart, p. 177-358.

Carter, C. and Laufield, S.
1975: Ordovician and Silurian fossils in well cores from North Slope of Alaska; The American Association of Petroleum Geologists Bulletin, v. 59, p. 457-464.

Christie, R.L.
1986: The Melville Project, 1984-85: progress report; in Current Research, Part A, Geological Survey of Canada, Paper 86-1A, p. 795-799.

Christie, R.L., Thorsteinsson, R., and Kerr, J.W.
1966: Prince of Wales Island; Geological Survey of Canada, Open File 66, 6 maps at 1:125 000 scale.

Cook, D.G.
1983: The northern Franklin Mountains, N.W.T., Canada: a scale model for the Wyoming Province; in Rocky Mountains Foreland Basins and Uplifts, J.D. Lowell (ed.); Rocky Mountain Association of Geologists, p. 315-338.

Cook, F.A., Coflin, K.C., Lane, L.S., Dietrich, J.R., and Dixon, J.
1987: Structure of the southeast margin of the Beaufort-Mackenzie basin, Arctic Canada, from crustal seismic-reflection data; Geology, v. 15, p. 931-935.

Dailly, G. et Plauchut, B.
1986: Relations entre l'archipel arctique canadien et la North Slope d'Alaska; Bulletin de la Société Géologique de la France, 8e série, v. 2, p.703-710.

de Vries, C.D.S.
1974: Structure; in Geology and Hydrocarbon Prospects of P. and N. G. Permit Nos. A6846 to A6886 (excluding Permit No. A6855), northeastern Ellesmere Island, Canadian Arctic Islands, written and compiled by U. Mayr and J.H. Stuart Smith, J.C. Sproule and Associates Ltd., Calgary, Alberta, p. 28-34 (unpublished).

Dixon, J.
1974: Revised stratigraphy of the Hunting Formation (Proterozoic), Somerset Island, N.W.T.; Canadian Journal of Earth Sciences, v. 11, p. 635-642.
1978: A Middle Ordovician unconformity in the Arctic Lowlands: a discussion; Bulletin of Canadian Petroleum Geology, v. 26, p. 208-217.

Drummond, K.J.
1973: Canadian Arctic Islands; in Future Petroleum Provinces of Canada — their Geology and Potential, R.G. McCrossan (ed.); Canadian Society of Petroleum Geologists, Memoir 1, p. 442-472.

Fortier, Y.O.
1957: The Arctic Archipelago; in Geology and Economic Minerals of Canada, C.H. Stockwell (ed.); Geological Survey of Canada, Economic Geology Series no. 1 (fourth edition), p. 393-442.

Fortier, Y.O., Blackadar, R.G., Glenister, B.F., Greiner, H.R., McLaren, D.J., McMillan, N.J., Norris, A.W., Roots, E.F., Souther, J.G., Thorsteinsson, R., and Tozer, E.T.
1963: Geology of the north-central part of the Arctic Archipelago, Northwest Territories (Operation Franklin); Geological Survey of Canada, Memoir 320, 671 p.

Fox, F.G.
1983: Structure sections across Parry Islands fold belt and Vesey Hamilton salt wall, Arctic Archipelago, Canada; in Seismic Expression of Structural Styles, a Picture and Work Atlas, A.W. Bally (ed.); American Association of Petroleum Geologists, Studies in Geology, Series 15, v. 3, p. 3.4.1-54 to 3.4.1-72.
1985: Structural geology of the Parry Islands Fold Belt; Bulletin of Canadian Petroleum Geology, v. 33, p. 306-340.

Frisch, T.
1974: Metamorphic and plutonic rocks of northernmost Ellesmere Island, Canadian Arctic Archipelago; Geological Survey of Canada, Bulletin 229, 87 p.
1983: Reconnaissance geology of the Precambrian Shield of Ellesmere, Devon and Coburg islands, Arctic Archipelago: a preliminary account; Geological Survey of Canada, Paper 82-10, 11 p.

Grantz, A. and May, S.D.
1983: Rifting history and structural development of the continental margin north of Alaska; in Studies in Continental Margin Geology, J.S. Watkins and C.L. Drake (ed.); American Association of Petroleum Geologists, Memoir 34, p. 77-100.

Gries, R.
1983: North-south compression of Rocky Mountains foreland structures; in Rocky Mountains Foreland Basins and Uplifts, J.D. Lowell (ed.); Rocky Mountain Association of Geologists, p. 9-32.

Hamilton, W.
1983: Cretaceous and Cenozoic history of the northern continents; Annals of the Missouri Botanical Garden, v. 70, p. 440-458.

Harrison, J.C. and Bally, A.W.
in press: Cross-sections of the Parry Islands Fold Belt on Melville Island, Canadian Arctic Islands; Bulletin of Canadian Petroleum Geology.

Harrison, J.C., Embry, A.F., and Poulton, T.P.
1988: Field observations on the structural and depositional history of Prince Patrick Island and adjacent areas, Canadian Arctic Islands; in Current Research, Part D, Geological Survey of Canada, Paper 88-1D, p. 41-49.

Harrison, J.C., Goodbody, Q.H., and Christie, R.L.
1985: Stratigraphic and structural studies on Melville Island, District of Franklin, in Current Research, Part A, Geological Survey of Canada, Paper 85-1A, p. 629-637.

Higgins, A.K., Mayr, U., and Soper, N.J.
1982: Fold belts and metamorphic zones of northern Ellesmere Island and North Greenland; in Nares Strait and the Drift of Greenland: a Conflict in Plate Tectonics, P.R. Dawes and J.W. Kerr (ed.); Meddelelser om Grønland, Geoscience 8, p. 159-166.

Hurst, J.M. and Surlyk, F.
1982: Stratigraphy of the Silurian turbidite sequence of North Greenland; Grønlands Geologiske Undersøgelse, Bulletin no. 145, 121 p.

Jackson, G.D. and Ianelli, T.R.
1981: Rift-related cyclic sedimentation in the Neohelikian Borden Basin, northern Baffin Island; in Proterozoic Basins of Canada, F.H.A. Campbell (ed.); Geological Survey of Canada, Paper 81-10, p. 269-302.

Jones, D.L. and Fahrig, W.F.
1978: Paleomagnetism and age of the Aston dykes and Savage Point sills of the Boothia Uplift, Canada; Canadian Journal of Earth Sciences, v. 10, p. 1605-1612.

Kerr, J.W.
1964: Bathurst Island; in Summary of activities: field, 1963, S.E. Jenness (comp.); Geological Survey of Canada, Paper 64-1, p. 4.
1967a: Nares submarine rift valley and the relative rotation of north Greenland; Bulletin of Canadian Petroleum Geology, v. 15, p. 483-520.
1967b: Vendom Fiord Formation — a new red-bed unit of probable early Middle Devonian (Eifelian) age, Ellesmere Island, Arctic Canada; Geological Survey of Canada, Paper 67-43, 8 p.
1967c: Devonian of the Franklinian Miogeosyncline and adjacent Central Stable Region, Arctic Canada; in International Symposium on the Devonian System, D.H. Oswald (ed.); Alberta Society of Petroleum Geologists, Calgary, Alberta, v. 1, p. 677-692.
1973a: Geology, Sawyer Bay, District of Franklin, scale 1:250 000; Geological Survey of Canada, Map 1357A.
1973b: Geology, Dobbin Bay, District of Franklin, scale 1:250 000; Geological Survey of Canada, Map 1358A.
1973c: Geology, Kennedy Channel and Lady Franklin Bay, District of Franklin, scale 1:250 000; Geological Survey of Canada, Map 1359A.
1974: Geology of the Bathurst Island group and Byam Martin Island, Arctic Canada (Operation Bathurst Island); Geological Survey of Canada, Memoir 378, 152 p.
1975: Cape Storm Formation — a new Silurian unit in the Canadian Arctic; Bulletin of Canadian Petroleum Geology, v. 23, p. 67-83.
1976: Stratigraphy of central and eastern Ellesmere Island, Arctic Canada. Part III. Upper Ordovician (Richmondian), Silurian and Devonian; Geological Survey of Canada, Bulletin 260, 55 p.
1977: Cornwallis Fold Belt and the mechanism of basement uplift; Canadian Journal of Earth Sciences, v. 14, p. 1374-1401.
1982: Evolution of sedimentary basins in the Canadian Arctic; in Evolution of Sedimentary Basins, P. Kent (ed.); Philosophical Transactions of the Royal Society of London, Series A, v. 305, p. 193-205.

Kerr, J.W. and Christie, R.L.
1965: Tectonic history of Boothia Uplift and Cornwallis Fold Belt, Arctic Canada; Bulletin of the American Association of Petroleum Geologists, v. 49, p. 905-926.

Kerr, J.W. and Thorsteinsson, R.
1972: Geology, Baumann Fiord, District of Franklin, scale 1:250 000; Geological Survey of Canada, Map 1312A.

Klovan, J.E. and Embry, A.F., III.
1971: Upper Devonian stratigraphy, northeastern Banks Island, N.W.T.; Bulletin of Canadian Petroleum Geology, v. 19, p. 705-729.

LaGabrielle, Y. and Auzende, J-M.
1982: Active in situ disaggregation of oceanic crust and mantle on Gorringe Bank: analogy with ophiolite massives; Nature, v. 297, p. 490-493.

Mayr, U.
1982: Lithostratigraphy (Ordovician to Devonian) of the Grise Fiord area, Ellesmere Island, District of Franklin; in Current Research, Part A, Geological Survey of Canada, Paper 82-1A, p. 63-66.

Mayr, U. and Okulitch, A.V.
1984: Geological maps (1:125 000) of North Kent Island and southern Ellesmere Island; Geological Survey of Canada, Open File 1036.

Mayr, U., Trettin, H.P., and Embry, A.F.
1982a: Preliminary geological map and notes, Clements Markham Inlet and Robeson Channel map-areas, District of Franklin (NTS 120E, F, G); Geological Survey of Canada, Open File 833, 40 p.
1982b: Preliminary geological map and notes, part of Tanquary Fiord map-area, District of Franklin (NTS 340D); Geological Survey of Canada, Open File 835, 32 p.

McGrath, P.H. and Fraser, I.
1981: Magnetic anomaly map of Arctic Canada, scale 1:3.5 million; Geological Survey of Canada, Map 1512A.

Menzies, A.W.
1982: Crustal history and basin development of Baffin Bay; in Nares Strait and the Drift of Greenland: a Conflict in Plate Tectonics, P.R. Dawes and J.W. Kerr (ed.); Meddelelser om Grønland, Geoscience 8, p. 295-312.

Miall, A.D.
1979: Mesozoic and Tertiary geology of Banks Island, Arctic Canada: the history of an unstable cratonic margin; Geological Survey of Canada, Memoir 387, 235 p.
1983a: Stratigraphy and tectonics of the Peel Sound Formation, Somerset and Prince of Wales islands: discussion; in Current Research, Part A, Geological Survey of Canada, Paper 83-1A, p. 493-495.
1983b: The Nares Strait problem: a re-evaluation of the geological evidence in terms of a diffuse oblique-slip plate boundary between Greenland and the Canadian Arctic Islands; Tectonophysics, v. 100, p. 227-239.

Miall, A.D. and Gibling, M.R.
1978: The Siluro-Devonian clastic wedge of Somerset Island, Arctic Canada, and some regional paleogeographic implications; Sedimentary Geology, v. 21, p. 85-127.

Miall, A.D. and Kerr, J.W.
1980: Cambrian to Upper Silurian stratigraphy, Somerset Island and northeastern Boothia Peninsula, District of Franklin, N.W.T.; Geological Survey of Canada, Bulletin 315, 43 p.

Morrow, D.W. and Kerr, J.W.
1977: Stratigraphy and sedimentology of lower Paleozoic formations near Prince Alfred Bay, Devon Island; Geological Survey of Canada, Bulletin 254, 122 p.
1986: Geology of Grinnell Peninsula and the Prince Alfred Bay area, Devon Island, District of Franklin, N.W.T.; Geological Survey of Canada, Open File 1325, 45 p.

Mortensen, P.S.
1985: Stratigraphy and sedimentology of the Upper Silurian strata on eastern Prince of Wales Island, Arctic Canada; unpublished Ph.D. thesis, University of Alberta, Edmonton, Alberta, 335 p.

Oldow, J.S., Ave Lallement, H.G., Julian, F.E., and Seidensticker, C.M.
1987: Ellesmerian(?) and Brookian deformation in the Franklin Mountains, northeastern Brooks Range, Alaska, and its bearing on the origin of the Canada Basin; Geology, v. 15, p. 37-41.

Okulitch, A.V.
1982: Preliminary structure sections, southern Ellesmere Island, District of Franklin; in Current Research, Part A, Geological Survey of Canada, Paper 82-1A, p. 55-62.

Okulitch, A.V. and Mayr, U.
1984: Compressive and extensional tectonics near dextral continental transform faults within the Arctic Platform, southeastern Ellesmere Island, Canada; Geological Society of America, Abstracts with Programs, v. 16, p. 613.

Okulitch, A.V., Packard, J.J., and Zolnai, A.I.
1986: Evolution of the Boothia Uplift, Arctic Canada; Canadian Journal of Earth Sciences, v. 23, p. 350-358.

Packard, J.J.
1985: The Upper Silurian Barlow Inlet Formation, Cornwallis Island, Arctic Canada; unpublished Ph.D. thesis, Department of Geology, University of Ottawa, Ottawa, Ontario, 744 p.

Packard, J.J. and Mayr, U.
1982: Cambrian and Ordovician stratigraphy of southern Ellesmere Island, District of Franklin; in Current Research, Part A, Geological Survey of Canada, Paper 82-1A, p. 67-74.

Palmer, A.R.
1983: The Decade of North American Geology 1983 time scale; Geology, v. 11, p. 503-504.

Peirce, J.W.
1982: The evolution of the Nares Strait lineament and its relation to the Eurekan orogeny; in Nares Strait and the Drift of Greenland: a Conflict in Plate Tectonics, P.R. Dawes and J.W. Kerr (ed.); Meddelelser om Grønland, Geoscience 8, p. 237-251.

Powell, D. and Phillips, W.E.A.
1985: Time and deformation in the Caledonide Orogen of Britain and Ireland; in The Nature and Timing of Orogenic Activity in the Caledonian Rocks of the British Isles, A.L. Harris (ed.); The Geological Society, Memoir 9, p. 17-40.

Ramsay, J.G.
1974: Development of chevron folds; Geological Society of America Bulletin, v. 85, p. 1741-1754.

Rice, P.D. and Shade, B.D.
1982: Reflection seismic interpretation and seafloor spreading history of Baffin Bay; in Arctic Geology and Geophysics, A.F. Embry and H.R. Balkwill (ed.); Canadian Society of Petroleum Geologists, Memoir 8, p. 245-267.

Roblesky, R.F.
1979: Upper Silurian (late Ludlovian) to Upper Devonian (Frasnian?) stratigraphy and depositional history of the Vendom Fiord region, Ellesmere Island, Arctic Canada; unpublished M.Sc. thesis, Department of Geology and Geophysics, University of Calgary, Calgary, Alberta, 230 p.

Sanford, B.V., Thomson, F.J., and McFall, G.H.
1985: Plate tectonics — a possible controlling mechanism in the development of hydrocarbon traps in southwestern Ontario; Bulletin of Canadian Petroleum Geology, v. 33, p. 52-71.

Smith, G.P. and Roblesky, R.F.
1984: The Inglefield Uplift: a new tectonic unit in the Canadian Arctic Islands; Canadian Society of Petroleum Geologists, National Convention, Program and Abstracts, p. 171-172.

Smithson, S.B., Brewer, J.A., Kaufman, S., and Oliver, J.
1979: Structure of the Laramide Wind River Uplift, Wyoming, from COCORP deep reflection data and from gravity data; Journal of Geophysical Research, v. 84, p. 5955-5972.

Smithson, S.B., Brewer, J.A., Kaufman, S., Oliver, J., and Hurich, C.
1978: Nature of the Wind River thrust, Wyoming, from COCORP deep reflection and gravity data; Geology, v. 6, p. 648-652.

Snelling, N.J.
1987: Measurement of geological time and the geological time scale; Modern Geology, v. 11, p. 365-374.

Soper, N.J. and Hutton, H.W.
1984: Late Caledonian sinistral displacements in Britain: implications for a three-plate collision model; Tectonics, v. 3, p. 781-794.

Stearns, D.W.
1971: Mechanism of drape folding in the Wyoming Province; in Twenty-Third Annual Field Conference Guidebook; Wyoming Geological Association, p. 125-143.

Stewart, W.D.
1987: Late Proterozoic to early Tertiary stratigraphy of Somerset Island and northern Boothia Peninsula, District of Franklin, N.W.T.; Geological Survey of Canada, Paper 83-26, 78 p.

Stewart, W.D. and Kerr, J.W.
1984: Geology of Somerset Island north, Somerset Island south, Boothia Peninsula north, and central Boothia Peninsula, scale 1:250 000; Geological Survey of Canada, Maps 1595A, 1596A, 1597A, and 1598A.

Surlyk, F. and Hurst, J.M.
1984: The evolution of the early Paleozoic deep-water basin of North Greenland; Geological Society of America Bulletin, v. 95, p. 131-154.

Taylor, F.C.
1981: Precambrian geology of the Canadian North Atlantic borderlands; in Geology of the North Atlantic Borderlands, J.W. Kerr and A.J. Ferguson (ed.); Canadian Society of Petroleum Geologists, Memoir 7, p. 11-30.

Temple, P.G.
1965: Geology of Bathurst Island Group, District of Franklin, Northwest Territories; unpublished Ph.D. thesis, Princeton University, Princeton, New Jersey (University Microfilms 66-1, 332 Ann Arbor, Michigan), 186 p.

Texaco Canada Resources Ltd.
1983: Melville Island, Northwest Territories, Canada Line no. 7; in Seismic Expression of Structural Styles, a Picture and Work Atlas, A.W. Bally (ed.); American Association of Petroleum Geologists, Studies in Geology, Series 15, p. 3.4.1-73 to 3.4.1-78.

Thorsteinsson, R.
1972a: Geology, Cañon Fiord, District of Franklin, scale 1:250 000; Geological Survey of Canada, Map 1308A.
1972b: Geology, Eureka Sound, District of Franklin, scale 1:250 000; Geological Survey of Canada, Map 1300A.
1972c: Geology, Strathcona Fiord, District of Franklin, scale 1:250 000; Geological Survey of Canada, Map 1307A.
1980: Part I. Contributions to stratigraphy (with contributions by T.T. Uyeno); in Stratigraphy and conodonts of Upper Silurian and Lower Devonian rocks in the environs of the Boothia Uplift, Canadian Arctic Archipelago, Geological Survey of Canada, Bulletin 282, p. 1-38.
1988: Geology of Cornwallis Island and neighbouring smaller islands, Canadian Arctic Archipelago, District of Franklin, Northwest Territories, scale 1: 250 000; Geological Survey of Canada, Map 1626A.

Thorsteinsson, R. and Kerr, J.W.
1972: Geology, Greeley Fiord East, District of Franklin, scale 1:250 000; Geological Survey of Canada, Map 1348A.

Thorsteinsson, R., Kerr, J.W., and Tozer, E.T.
1972: Geology, Strathcona Fiord, District of Franklin, scale 1:250 000; Geological Survey of Canada, Map 1307A.

Thorsteinsson, R. and Tozer, E.T.
1962: Banks, Victoria, and Stefansson Islands, Arctic Archipelago; Geological Survey of Canada, Memoir 330, 85 p.
1970: Geology of the Arctic Archipelago; in Geology and Economic Minerals of Canada, R.J.W. Douglas (ed.); Geological Survey of Canada, Economic Geology Report no. 1, p. 547-590.

Thorsteinsson, R. and Trettin, H.P.
1971: Geology, Tanquary Fiord, District of Franklin, scale 1:250 000; Geological Survey of Canada, Map 1306A.

Thorsteinsson, R. and Uyeno, T.
1980: Stratigraphy and conodonts of Upper Silurian and Lower Devonian rocks in the environs of the Boothia Uplift, Canadian Arctic Archipelago; Geological Survey of Canada, Bulletin 292, 75 p.

Tozer, E.T.
1956: Geological reconnaissance, Prince Patrick, Eglinton, and western Melville Islands, Arctic Archipelago, Northwest Territories; Geological Survey of Canada, Paper 55-5, 32 p.

Tozer, E.T. and Thorsteinsson, R.
1964: Western Queen Elizabeth Islands, Arctic Archipelago; Geological Survey of Canada, Memoir 332, 242 p.

Trettin, H.P.
1978: Devonian stratigraphy, west-central Ellesmere Island; Geological Survey of Canada, Bulletin 302, 119 p.
1981: Geology of Precambrian to Devonian rocks, M'Clintock Inlet area, District of Franklin (NTS 340E, H) — preliminary geological map and notes; Geological Survey of Canada, Open File 759, 26 p.
1982: Lower Paleozoic geology in parts of Greeley Fiord East, Greeley Fiord West and Cañon Fiord map-areas, District of Franklin (NTS 340A, B, 49H) — preliminary geological map and notes; Geological Survey of Canada, Open File 836, 7 p.
1987a: Investigations of Paleozoic geology, northern Axel Heiberg and northwestern Ellesmere islands; in Current Research, Part A, Geological Survey of Canada, Paper 87-1A, p. 357-367.
1987b: Pre-Carboniferous geology, M'Clintock Inlet map-area, northern Ellesmere Island, interim report and map (340E, H); Geological Survey of Canada, Open File 1652, 210 p.

Trettin, H.P. and Balkwill, H.R.
1979: Contributions to the tectonic history of the Innuitian Province, Arctic Canada; Canadian Journal of Earth Sciences, v. 16, p. 748-769.

Trettin, H.P., De Laurier, I., Frisch, T.O., Law, L.K., Niblett, E.R., Sobczak, L.W., Weber, J.R., and Witham, K.
1972: The Innuitian Province; in Variations in Tectonic Styles in Canada, R.A. Price and R.J.W. Douglas (ed.); Geological Association of Canada, Special Paper 11, p. 83-179.

Trettin, H.P. and Frisch, T.
1987: Bedrock geology, Yelverton Inlet map-area, northern Ellesmere Island, interim report and map (340F, 560D); Geological Survey of Canada, Open File 1651, 98 p.

Trettin, H.P. and Mayr, U.
1981: Preliminary geological map and notes, parts of Otto Fiord and Cape Stallworthy map-areas, District of Franklin (NTS 340C, 560D); Geological Survey of Canada, Open File 757, 10 p.

Trettin, H.P., Mayr, U., Embry, A.F., and Christie, R.L.
1982: Preliminary geological map and notes, part of Lady Franklin Bay map-area, District of Franklin; Geological Survey of Canada, Open File 834, 31 p.

Trettin, H.P. and Parrish, R.
1987: Late Cretaceous bimodal magmatism, northern Ellesmere Island: isotopic age and origin; Canadian Journal of Earth Sciences, v. 24, p. 257-265.

Trettin, H.P., Parrish, R., and Loveridge, W.D.
1987: U-Pb age determinations of Proterozoic to Devonian rocks from northern Ellesmere Island, Arctic Canada; Canadian Journal of Earth Sciences, v. 24, p. 246-256.

Walcott, R.I.
1970: An isostatic origin for basement uplifts; Canadian Journal of Earth Sciences, v. 7, p. 931-937.

Workum, R.H.
1965: Lower Paleozoic salt, Canadian Arctic Islands; Bulletin of Canadian Petroleum Geology, v. 13, p. 181-191.

Young, G.M.
1981: The Amundsen Embayment, Northwest Territories; relevance to the Upper Proterozoic evolution of North America; in Proterozoic Basins of Canada, F.H.A. Campbell (ed.); Geological Survey of Canada, Paper 81-10, p. 203-218.

Authors' addresses

T.A. Brent
Geological Survey of Canada
3303-33rd Street N.W.
Calgary, Alberta
T2L 2A7

F.G. Fox
Formerly with Panarctic Oils Ltd.
Deceased 13 February, 1990

J.C. Harrison
Geological Survey of Canada
3303-33rd Street N.W.
Calgary, Alberta
T2L 2A7

A.V. Okulitch
Geological Survey of Canada
3303-33rd Street N.W.
Calgary, Alberta
T2L 2A7

J.J. Packard
Esso Resources Canada Limited
Box 2480, Station M
Calgary, Alberta
T2P 3M9

G.P. Smith
Shell Canada Ltd.
Box 100, Station M
Calgary, Alberta
T2P 2H5

H.P. Trettin
Geological Survey of Canada
3303-33rd Street N.W.
Calgary, Alberta
T2L 2A7

A.I. Zolnai
36, Cedardale Mews SW
Calgary, Alberta
T2W 5G4

CHAPTER 12

Printed in Canada

Chapter 13
CARBONIFEROUS AND PERMIAN HISTORY OF THE SVERDRUP BASIN, ARCTIC ISLANDS

Introduction

Tectonic development of the Sverdrup Basin during the late Paleozoic

Stratigraphy and lithofacies

Organic mounds and reefs

References

Chapter 13

CARBONIFEROUS AND PERMIAN HISTORY OF THE SVERDRUP BASIN, ARCTIC ISLANDS

G.R. Davies and W.W. Nassichuk

INTRODUCTION

Upper Paleozoic carbonates and evaporites with associated basal redbeds, marine shales and minor chert and basic volcanic rocks form the initial fill of the Sverdrup Basin. These rocks are exposed in spectacular fiord cliff faces on northwestern Ellesmere and northern Axel Heiberg islands, and in more subdued terrain on southern Ellesmere, northwestern Devon, northern Melville, Cameron, and Helena islands (Fig. 13.1; for geographic names and regional geology see Fig. 1 and 2 [in pocket]). They also have been penetrated, totally or in part, by 35 exploratory wells, most of which are in the western sector of the basin, on and between Melville and Cameron islands.

Carboniferous evaporites deposited in the centre of the basin have generated halite-cored diapirs and other linear intrusive bodies that influenced the structural development of the basin, deformation of the Mesozoic section, and localization of oil and gas accumulations. Similarity with the upper Paleozoic section on Svalbard and northern Greenland extends the regional implications of upper Paleozoic strata in the Sverdrup Basin in terms of circum-Arctic tectonics and paleoclimates.

TECTONIC DEVELOPMENT OF THE SVERDRUP BASIN DURING THE LATE PALEOZOIC

Subsidence

The Ellesmerian Orogeny (Chapters 10, 12) probably extended into the earliest Carboniferous (Tournaisian), but appears to have terminated before the deposition of the Viséan Emma Fiord Formation. Too little of this formation is preserved to obtain a clear picture of tectonic conditions just prior to the subsidence of the Sverdrup Basin, which commenced in the Namurian and continued to the Late Cretaceous. Structures formed during the initial phase of subsidence, and geophysical evidence for crustal thinning below the axis of the basin (Forsyth et al., 1979; Chapter 5B), indicate that the subsidence was due to extension; the ultimate cause of the extension is unknown.

The basic architecture of the basin was created by rifting in the Carboniferous. On the northwestern side of the basin (Fig. 13.1, 13.2), fault zones originated on the flank of a major "sill" known as the Sverdrup Rim, which separates the basin from the Arctic Ocean region (cf. Chapter 5B). The upper Paleozoic – Mesozoic succession over the rim is thinner than that in the basin and contains a larger proportion of shallow water sediments and unconformities (Thorsteinsson and Tozer, 1960, 1970; Hobson, 1962; Meneley et al., 1975; Balkwill, 1978). On the southern and southeastern flanks of the basin, horst-and-graben structures and down-to-the-basin normal faults with associated nonmarine conglomerates and sandstones characterize the initial rifting phase (Fig. 13.3; and G. Varney, pers. comm., 1984). Drilling and seismic investigations on Cameron Island have revealed listric growth faults related to these syntectonic rocks (F.G. Fox, pers. comm., 1984). In the axial region of the basin, subsidence must have exceeded deposition, because the sedimentary record indicates a gradual progression from nonmarine to deep water environments.

In Permian time, subsidence was slower and its mechanism is less clearly understood. Seismic records suggest that it was accomplished by both flexuring of the crust and subtle growth faulting (H.R. Balkwill, pers. comm., 1985).

The Melvillian Disturbance

The name "Melvillian Disturbance" was given by Thorsteinsson and Tozer (1970) to an episode of folding and faulting between the deposition of Upper Carboniferous and Upper Permian strata observed at two localities at the margin of the Sverdrup Basin. The effects are best displayed on northwestern Melville Island (Tozer and Thorsteinsson, 1964) where the Upper Carboniferous Canyon Fiord Formation was folded along the same westerly trend as unconformably underlying Devonian and older strata. The overlying Upper Permian (Guadalupian, Wordian) Trold Fiord Formation, which is separated from the Canyon Fiord by an angular unconformity, is not folded at the Melville Island locality. The Canyon Fiord and Trold Fiord formations again are separated by an angular unconformity at the head of Trold Fiord on Ellesmere Island, where the Upper Carboniferous strata were faulted against Ordovician limestones prior to the deposition of the Upper Permian rocks (Thorsteinsson and Tozer, 1970).

The occurrence of folding within a regional regime of subsidence, and the wide separation of the only two areas known to have been affected by this event, posed a major tectonic problem. Subsequent subsurface studies in other

---

Davies, G.R. and Nassichuk, W.W.
1991: Carboniferous and Permian history of the Sverdrup Basin, Arctic Islands; Chapter 13 in Geology of the Innuitian Orogen and Arctic Platform of Canada and Greenland, H.P. Trettin (ed.); Geological Survey of Canada, Geology of Canada, no. 3; (also Geological Society of America, The Geology of North America, v. E).

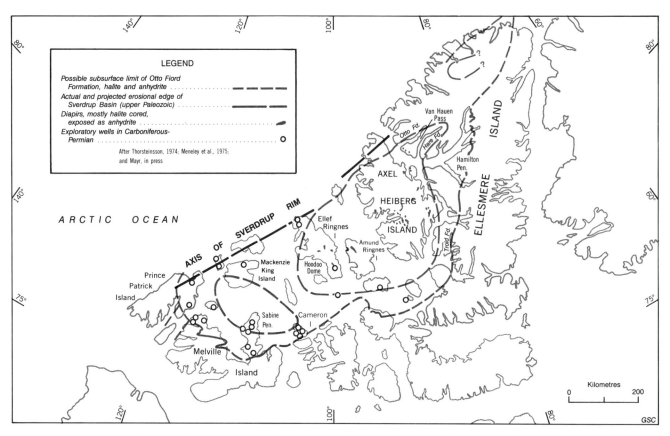

Figure 13.1. Sverdrup Basin, evaporite depocentres, exposed diapirs, and locations of wells intersecting Carboniferous-Permian section (after Thorsteinsson, 1974; Meneley et al., 1975; and Mayr, in press).

parts of the basin, and recent fieldwork on Melville Island, have thrown some light on these questions. Meneley and others (1975), using seismic and well data, demonstrated significant thinning of Upper Carboniferous (Pennsylvanian) deposits over horsts of the Sverdrup Rim beneath Brock and Ellef Ringnes islands (Fig. 13.2). They inferred uplift and erosion in Late Carboniferous – Early Permian time, but the data can be interpreted alternatively as indicating growth faulting during the deposition of the Carboniferous to Lower Permian shelf carbonates of the Nansen Formation. In any case, these observations showed: (1) that the Melvillian movements were more extensive than apparent from the surface exposures; and (2) that they are compatible with the overall regime of regional extension inferred for that time interval. Meneley and others (1975) also suggested that the Sverdrup Rim was breached by crosscutting grabens at this time.

The results of more detailed structural studies on northwestern Melville Island by Harrison (Harrison et al., 1985) are compatible with these conclusions and provide a clue to the possible origin of the extension and deformation. These studies confirm the presence of broad, gently deformed folds in the Carboniferous rocks described by Tozer and Thorsteinsson, but also demonstrate that the axes of the folds are associated with the termination of fault segments. Other Melvillian structures observed by Harrison are east-northeast striking normal faults and easterly and northeasterly striking oblique faults, some of which may have had dextral strike-slip motion (see below). Where the Canyon Fiord Formation lies unconformably on shale of the Cape de Bray Formation, the unconformity commonly is marked by a moderately- to gently-dipping normal fault that probably is listric at depth. Harrison concluded that the deformation style of the Carboniferous strata on Melville Island is representative of horst-and-graben tectonics, the Melvillian folds being due to drape over normal-faulted blocks in the lower Paleozoic "basement". Erosional truncation of such structures prior to Late Permian time would account for the angular unconformity found locally at the base of the Trold Fiord Formation. Overall, the Melvillian tectonic assemblage closely resembles that associated with the creation of pull-apart basins between diverging and converging dextral strike-slip fault arrays (cf. Crowell, 1974; Harrison, pers. comm., 1985).

A regional unconformity separating Carboniferous from Permian strata occurs in the eastern Cordillera and in the Brooks Range of Alaska (Nassichuk and Bamber, 1978), but deformations comparable to the Melvillian Disturbance have not been recognized in these regions.

Salt tectonics

Diapirs and other salt-cored intrusive structures in the Sverdrup Basin are displaced sedimentary evaporites of the Carboniferous (Upper Mississippian to Middle Pennsylvanian) Otto Fiord Formation (Thorsteinsson, 1974;

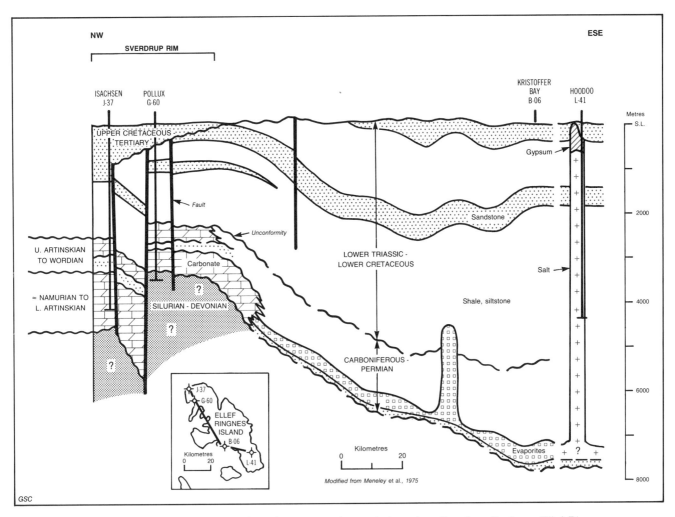

Figure 13.2. Schematic structural section across the central-western Sverdrup Basin on Ellef Ringnes Island, illustrating the horst-and-graben structure of the Sverdrup Rim with overlying Nansen-equivalent shelf carbonates, Otto Fiord halite facies in the basin centre, and salt-cored diapirs exemplified by the Hoodoo L-41 well (modified from Meneley et al., 1975).

Davies, 1975b; Nassichuk and Davies, 1980). Diapirs exposed at the present erosional surface form spectacular concentric domes of anhydrite (with enclosed blocks of limestone, dolomite and gabbro) preserved by the extremely arid Arctic climate (Fig. 13.4). Present distribution of halokinetic structures (structures formed by salt movements) in the basin (Fig. 13.1) suggests two evaporite depocentres or sub-basins — one in the southwestern sector of the basin, north of Sabine Pennisula, Melville Island (Barrow Basin of Meneley et al., 1975, or Barrow segment of Balkwill, 1978), and a larger depocentre extending from the Ringnes islands to northwestern Ellesmere Island (the Axel Heiberg Basin of Meneley et al., 1975, or Axel Heiberg segment of Balkwill, 1978). These two evaporite segments (Fig. 13.1) apparently were separated by a "Central Basin Platform" on which accumulated more or less contemporaneous carbonates (Meneley et al., 1975).

In the southern part of the basin, halokinetic structures range from ovate non-piercement anticlines closer to the basin margin, to circular stocks and large salt piercement structures farther basinward (Balkwill, 1978). In the Ringnes-Heiberg segment, halokinetic structures include large ovate anticlines, sinuous evaporite piercement "walls", and linear evaporite bodies in the axes of tight anticlines and along fault surfaces, and also large near-circular intrusive stocks.

Some structures are due to compression during the Eurekan Orogeny, but most are due to upward flow of salt caused by sediment loading through Mesozoic time (Balkwill, 1978; Chapters 14, 17).

Vertical movement of some of the diapirs in the Sverdrup Basin may be as much as 10 000 m. The Panarctic Dome et al. Hoodoo L-41 well, which was drilled off the exposed margin of Hoodoo Dome in central Ellef Ringnes Island (Fig. 13.3), penetrated 280 m of anhydrite and limestone, commencing at a depth of 337 m. Deeper in the well, from 620 to 4280 m, a steeply-dipping Otto Fiord halite section was encountered, giving a diapiric halite thickness in the

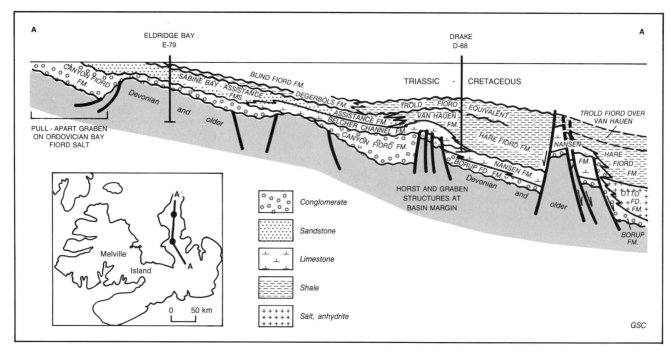

Figure 13.3. Diagrammatic south to north cross-section through Sabine Peninsula of Melville Island, illustrating the pull-apart horst-and-graben structures forming the southern margin of the Sverdrup Basin, with localized syntectonic conglomerates of the Canyon Fiord Formation, overlain by younger Carboniferous and Permian siliciclastics and carbonates (modified from Panarctic Oils Ltd., G. Varney, pers. comm., 1985).

well of 3660 m and confirming the presence of a salt core in these diapiric structures (Davies, 1975b).

Volcanics and intrusions

The earliest phase of volcanism in the Sverdrup Basin is represented by the Audhild Formation of Kleybolte Peninsula, northwesternmost Ellesmere Island (Thorsteinsson, 1974; also Fig. 1, 2 in pocket) and by a local volcanic member in the lowermost part of the Borup Fiord Formation in northern Axel Heiberg Island (Trettin, 1988). The Audhild Formation lies stratigraphically between the Borup Fiord and Nansen formations and is probably Namurian (late Early Carboniferous) in age. It consists of up to 561 m of dark coloured, basaltic and spilitic flows and lesser amounts of pyroclastic rocks. The volcanic member of the Borup Fiord Formation is about 32 m thick and composed of spilitized amygdaloidal flows, identified as alkali basalt on the basis of stable trace elements. A limestone about 35 m above the volcanics has yielded conodonts of early Namurian (Arnsbergian) age. The member is either correlative with the lowermost part of the Audhild Formation or slightly older.

A second phase of late Paleozoic volcanism is represented by the Lower Permian Esayoo Formation, exposed on northwestern Ellesmere and northern Axel Heiberg islands (Thorsteinsson, 1974). The unit overlies the Nansen Formation, and is overlain variably by the Trold Fiord, van Hauen or Degerböls formations. It reaches a maximum measured thickness of 297 m, and is composed of dark coloured basaltic volcanic flows and agglomerates.

Unmetamorphosed dykes and sills are abundant in the Paleozoic succession of the Arctic Islands, especially in northern Ellesmere and Axel Heiberg islands. It is reasonable to assume that at least some are late Paleozoic in age although the few that have been dated are Mesozoic.

STRATIGRAPHY AND LITHOFACIES

The regional stratigraphic framework for the Lower Carboniferous to Permian succession in the Sverdrup Basin (Fig. 13.5) was established by Thorsteinsson (1974). The oldest sedimentary formation in the Sverdrup Basin is the Viséan Emma Fiord Formation, a nonmarine marlstone unit, which is overlain by an Upper Carboniferous (Namurian) redbed conglomeratic unit, the Borup Fiord Formation. The latter marks the onset of major faulting and crustal extension leading to marine transgression in the Sverdrup Basin. Subsequent Carboniferous to Upper Permian rocks are dominantly marine in character.

Viséan lithofacies

The Viséan Emma Fiord Formation is exposed on erosional (post-Ellesmerian) topography above lower Paleozoic rocks on Grinnell Peninsula of northwestern Devon Island, on northeastern Axel Heiberg Island, and on northwestern Ellesmere Island. The formation consists of up to 300 m of marlstone ("oil shale"), shale, siltstone, lensoid sandstone, fine conglomerate, and coal seams, with some nonmarine algal and oolitic (lacustrine) limestones on Grinnell Peninsula (Davies and Nassichuk, 1988). The "oil shale" component of the Emma Fiord Formation is characterized

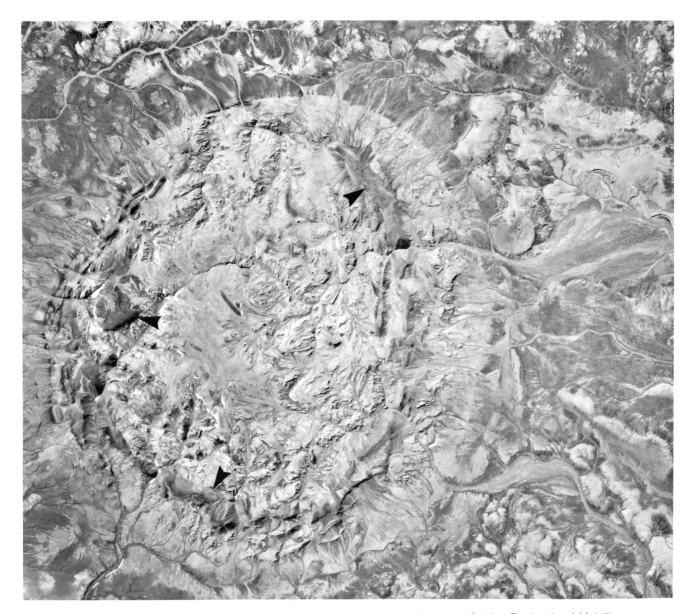

Figure 13.4. Vertical aerial photograph of Barrow Dome on north-central Sabine Peninsula of Melville Island. The dome is about 6 km in diameter, and is composed of light coloured anhydrite with a dark coloured concentric mafic intrusion (arrows). Although halite is not exposed in the diapir, other evidence in the Sverdrup Basin indicates that the structure is cored by salt. Note the radial drainage pattern connecting outward with dendritic drainage on the Mesozoic siliciclastic host rocks. Air photo A16764-19.

by high alginite content in the liptinite fraction (coalified remnants of resinous plants), which forms up to 95 percent of the total organic matter (Goodarzi et al., 1987). The Emma Fiord Formation is interpreted to have been deposited in a density and/or chemically stratified lake environment.

From a paleoclimatic standpoint, the Emma Fiord Formation records at least a seasonally-pluvial climate, unlike that inferred for the younger Carboniferous to Lower Permian carbonate-evaporite sequence. For this reason, and also because the coarse syntectonic redbed conglomerates of the Borup Fiord or Canyon Fiord formations overlie this unit, the Emma Fiord may be considered a precursor of the Sverdrup Basin sequence, recording peripheral and lacustrine deposition over an undulatory Ellesmerian eroded surface prior to, or contemporaneously with, the onset of the major extension that created the Sverdrup Basin *sensu stricto*.

On a larger regional scale, the Emma Fiord Formation shows a striking similarity in setting, age and lithology to part of the nonmarine coal-bearing rocks of the Billefjorden Group of Vestspitzbergen (Steel and Worsley, 1984). Other stratigraphic, faunal and sedimentological similarities to the Emma Fiord (and overlying syntectonic conglomerates) occur in the U.S.S.R., northeastern Greenland, northern

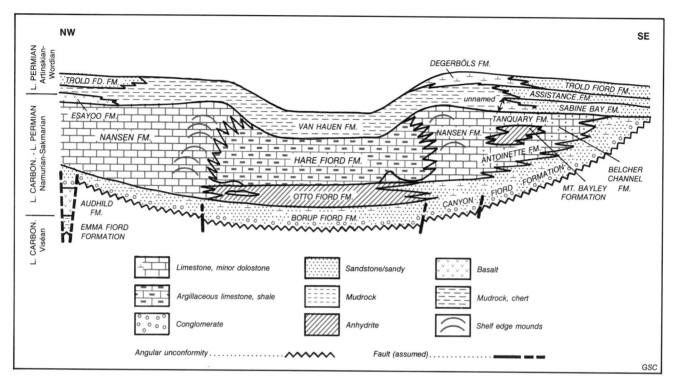

Figure 13.5. Schematic lithofacies section for the Carboniferous and Permian of the northeastern Sverdrup Basin, based on exposures on Ellesmere and Axel Heiberg islands (modified from Thorsteinsson, 1974; and Mayr, in press).

Alaska, northern and southeastern Yukon, northeastern British Columbia, and southwestern District of Mackenzie (Nassichuk and Davies, 1980, p. 9). It also is closely comparable in composition and depositional aspects to the upper Tournaisian Albert Formation of New Brunswick, a lacustrine marlstone "oil shale" unit.

Namurian to Artinskian lithofacies

Thorsteinsson (1974) recognized that the Namurian to Artinskian succession of the Sverdrup Basin could be divided into a series of five facies belts roughly paralleling the depositional and structural axis of the basin. Moving westward from the eastern edge of the basin, these facies belts (Fig. 13.5, 13.6) and stratigraphic units are:

1. **The Marginal Clastic Belt** — a narrow but conspicuous belt of redbed sandstone and conglomerate of Early Pennsylvanian (Bashkirian) to Early Permian age, adjacent to the basin margin of west-central Ellesmere Island, and assigned to the Canyon Fiord Formation. These rocks are syntectonic alluvial deposits shed off active fault scarps and horsts created by extension and rifting of the Sverdrup Basin. They are highly variable in thickness because of their syntectonic origin, but are estimated to be about 1700 m thick near Cañon Fiord on Ellesmere Island. The phenoclasts are composed of varying proportions of carbonate rocks, sandstone, siltstone, and chert that reflect lithological variations in the pre-Carboniferous source terrains.

2. **The Marginal Clastic and Carbonate Belt** — a facies belt lying basinward of the marginal clastic belt and in other areas along the margin of the basin, marked by development of limestone interbeds within the redbed sandstone and conglomerates of the mainly Moscovian (Middle Pennsylvanian) Canyon Fiord Formation. The Canyon Fiord in this facies belt is overlain by the Belcher Channel Formation, a sequence of cyclical carbonate and clastic strata ranging in age from latest Carboniferous (Late Pennsylvanian) to Early Permian (Artinskian). This unit reaches a maximum outcrop thickness of about 580 m on Hamilton Peninsula of Ellesmere Island, and shows a general increase in frequency and thickness of limestone units toward the top of the formation, locally culminating in organic mounds (see section on Organic Mounds and Reefs).

In the vicinity of Cañon Fiord and Tanquary Fiord on central Ellesmere Island, the Canyon Fiord – Belcher Channel couplet becomes four formations. There, the Canyon Fiord Formation is overlain successively by the mainly Upper Carboniferous Antoinette Formation (limestone), the Lower Permian (Asselian) Mount Bayley Formation (anhydrite), and the Lower Permian (Sakmarian) Tanquary Formation (limestone) (Fig. 13.7). All three units are lateral equivalents of the Belcher Channel Formation and also of the upper part of the Nansen Formation of the Northwestern Carbonate Belt.

The Antoinette Formation consists mainly of thinly bedded grey limestone with subsidiary siltstone and thin beds of gypsum/anhydrite. Some limestone beds contain abundant fusulinacean foraminifers. Maximum measured thickness of this unit, again in the area of Hamilton Peninsula, is about 810 m.

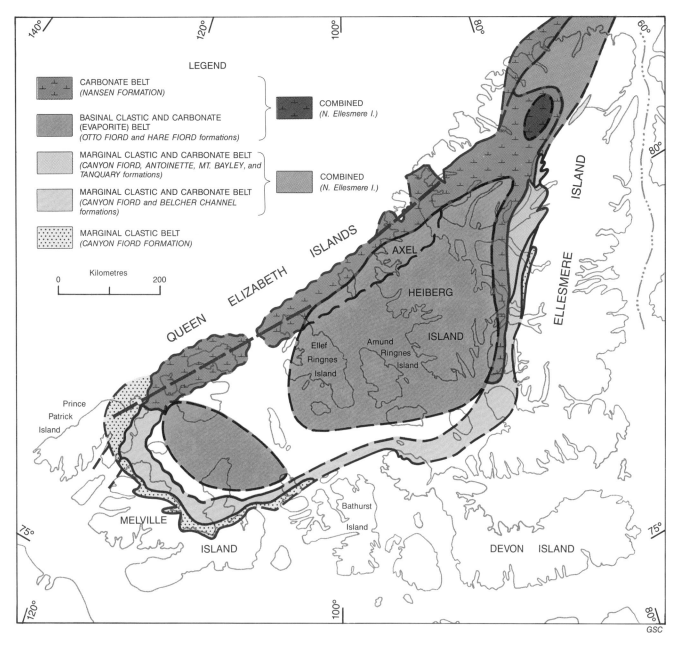

Figure 13.6. Generalized Namurian to Lower Artinskian lithofacies of the Sverdrup Basin, based on the facies belts of Thorsteinsson (1974) in the northeastern outcrop region of Ellesmere and Axel Heiberg islands, and on exploratory well data (Fig. 13.1) for the southwestern section.

The Mount Bayley Formation, which overlies the Antoinette, is areally restricted and is composed of cyclically interbedded anhydrite, siltstone and silty limestone, locally intersected by mafic sills (Fig. 13.7). It reaches a maximum measured thickness of 250 m on Hamilton Peninsula.

The overlying Tanquary Formation is characterized by multiple cycles of siltstone, sandstone and limestone. The latter are rich in fusulinaceans and solitary corals. Maximum measured thickness for this unit is about 835 m near Cañon Fiord.

3. **Southeastern Carbonate Belt** — the strata of this belt are included in the Namurian to Artinskian Nansen Formation, composed mainly of cyclic carbonates that represent shallow marine shelf deposits and shelf edge buildups. In his original definition of this belt, Thorsteinsson (1974) extended it from Bjorne Peninsula northward to Borup Fiord on Ellesmere Island, where it merges with the Northwestern Carbonate Belt (Fig. 13.6). Nassichuk (1975), however, suggested that rocks in the southern part of the belt near Blind Fiord, north of Bjorne Peninsula, might

Figure 13.7. Type locality of the Asselian Mount Bayley Formation (MB) on the northwest side of Greely Fiord, Ellesmere Island, illustrating the three-fold division of the shelf sequence with the underlying upper Carboniferous Antoinette Formation (An) and overlying Sakmarian Tanquary Fiord Formation (TF). The Mount Bayley Formation is composed dominantly of interbedded anhydrite, silty limestone and siltstone, and is about 170 m thick at this location. The evaporites are intruded by several dark coloured mafic sills (arrows).

better be placed in the Belcher Channel Formation (underlain by Canyon Fiord redbeds), implying that the Belcher Channel in that area grades basinward into the deeper water Hare Fiord Formation of the Basinal Clastic and Evaporite Belt. B. Beauchamp (pers. comm., 1986), on the other hand, is of the opinion that the Belcher Channel Formation grades basinward into the Nansen Formation at Blind Fiord. The Nansen Formation is more fully described under the heading "Northwestern Carbonate Belt".

4. **Basinal Clastic and Evaporite Belt** — this facies belt delineates the axial region of subsidence of the Sverdrup rift basin or trough. It is marked by a vertical succession of marine and nonmarine conglomeratic redbeds of the Namurian Borup Fiord Formation, overlain by cyclic halite, anhydrite, marine limestones and minor sandstones of the Namurian (Upper Mississippian) to Moscovian (Middle Pennsylvanian) Otto Fiord Formation, and in turn by shales and basinal limestones of the Moscovian to Lower Permian Hare Fiord Formation (Fig. 13.8). Depositionally and lithologically, the Borup Fiord Formation is similar to the Canyon Fiord Formation, characterized by syntectonic redbeds eroded from fault scarps. The phenoclasts, commonly composed of quartzite, mudrock or chert, are comparable in composition to the lower Paleozoic rocks of this region. The onset of the first major marine transgression into the Sverdrup Basin is recorded by shallow, marginal marine carbonates (including oolitic grainstones) interbedded in the upper part of the Borup Fiord redbed sequence in various parts of the basin. The Borup Fiord Formation has a maximum measured thickness of about 395 m near the head of Hare Fiord on Ellesmere Island.

The Otto Fiord Formation is the major basin-centre evaporitic unit of the Sverdrup Basin, forming part of the initial marine transgressive infill of the developing rift trough. In the type area at van Hauen Pass on northwestern Ellesmere Island, the Otto Fiord Formation ranges from Early Pennsylvanian (Morrowan) to Middle Pennsylvanian age, and is composed of 410 m of rhythmically interbedded

Figure 13.8. Cyclically bedded anhydrite and limestone of the Otto Fiord Formation (OF) overlain by dark coloured argillaceous limestones and shales of the Hare Fiord Formation (HF) on the north side of the head of Hare Fiord, Ellesmere Island. The Otto Fiord anhydrite section at this locality is 365 m thick. The Hare Fiord Formation in this view is overlain by tongues of limestone of the Nansen Formation (N).

marine limestone and bedded anhydrite (Fig. 13.9), with one horizon marked by multiple organic mounds formed by beresellid and kamaenid algae with pervasive submarine cements. Nassichuk (1975) pointed out that the base of the type section is faulted and that, elsewhere in the basin, faunas of Late Mississippian (Chesterian) age occur in the lower Otto Fiord Formation.

The evaporites in the Otto Fiord Formation are interpreted to be of hypersaline subaqueous origin, rather than of sabkha type, on the basis of various criteria, including: position in the axial region of the Sverdrup Basin, seaward of shallow marine limestones; presence of marine biota in interbedded limestones; bathymetric relief (to more than 300 m, see below) on interbedded marine limestones; thickness of anhydrite units (typically more than 10 and up to 60 m); partial preservation of laterally continuous thin bedding and parallel lamination; preservation of subaqueous gypsum pseudomorphs; and absence of nonmarine indicators (Davies and Nassichuk, 1975; Nassichuk and Davies, 1980). East of the type section on Ellesmere Island, the limestone-anhydrite cycles of the Otto Fiord are capped by thick channelized and crossbedded sandstones (Fig. 13.10), probably tongues of marine deltaic sediment derived from the eastern margin of the basin, and contiguous with fluvial rocks of the Canyon Fiord Formation (Nassichuk and Davies, 1980, Fig. 32).

The Otto Fiord Formation is buried below younger Paleozoic and Mesozoic sediments in the central and southern part of the basin, where it is the source of numerous diapiric structures. Halite is known to occur in the subsurface in axial regions of the basin and probably was deposited in the Barrow and Axel Heiberg sub-basins, which may be separated from each other by a carbonate-dominated platform (Fig. 13.1).

The Hare Fiord Formation, when observed distant from abrupt facies transitions with the Nansen Formation, is composed of parallel, thin to medium bedded cherty limestones, chert, siltstone and shale. Crinoids and brachiopods are the most common biotic components. Debris flows and carbonate turbidites are common closer to the

Figure.13.9. Cyclically bedded anhydrite and limestone of the Otto Fiord Formation (OF) overlain by dark coloured basinal rocks of the Hare Fiord Formation (HF) near van Hauen Pass, Ellesmere Island. The Otto Fiord section is 410 m thick at this locality. Note the three small beresellid-algal mounds (arrows) of Bashkirian age which occur at one limestone horizon.

transition zone into the Nansen Formation (Davies, 1977c), but many of the other finer grained beds in the Hare Fiord may be distal turbidites. The maximum recorded outcrop thickness is 1250 m.

In the Blue Mountains of northern Ellesmere Island, and at several other localities, large carbonate reefs constructed by fenestellid bryozoans and submarine cements occur at the base of the Hare Fiord section (Fig. 13.11). These reefs, informally referred to as the "Tellevak" limestone (Bonham-Carter, 1966), are of early Moscovian (Atokan) age (also see section on Organic mounds and reefs).

Along the southwestern side of Otto Fiord, on Ellesmere Island, large truncation structures are preserved in steep and inaccessible cliffs in the Hare Fiord Formation (Davies, 1977c). These are interpreted to be large listric slide structures infilled by younger Hare Fiord sediments, and probably triggered by earthquake shock, generated by faulting along the Sverdrup Rim during basin subsidence.

5. **Northwestern Carbonate Belt** — represented by the Namurian (Upper Mississippian) to Artinskian (Lower Permian) shelf and shelf edge Nansen Formation, exposed over a large area of northern Axel Heiberg and northern Ellesmere islands (Plate 13.1, Fig. 13.12). It also is projected into the subsurface along the Sverdrup Rim (Fig. 13.2), across the postulated central platform between the two evaporite depocentres, and possibly along the southern flanks of the basin.

In gross character, the Nansen Formation is composed of up to 2370 m of rhythmically bedded or cyclic shelf carbonates of shallow water aspect. In mid-shelf position, closer to the transition into the Mount Bayley evaporite section on central Ellesmere Island, the Nansen may be selectively or totally dolomitized (Nassichuk and Davies, 1980). Many of the cyclic beds in the mid-shelf part of the Nansen Formation are capped by oolitic grainstones. Other grain components in the Nansen rocks include a wide variety of crinoids, brachiopods, bryozoans, fusulinacean and other foraminifers, algae, micritized grains, and other shell fragments and grains of shallow water origin.

Closer to or at the edge of the Nansen shelf, lenticular organic mounds (reefs), constructed of phylloid algae and the enigmatic organism *Palaeoaplysina* (Davies and Nassichuk, 1973) and stabilized by syndepositional submarine cements, may be present (see section on Organic mounds and reefs). Thick beds of oolitic sand and coarse fusulinacean grainstones also are common in the shelf edge setting.

Along the northwestern cliff face of Hare Fiord on northwestern Ellesmere Island, spectacular facies transitions between Otto Fiord evaporites, Nansen shelf carbonates, and basin-fill Hare Fiord rocks are exposed (Fig. 13.13). In some localities, the entire Lower Carboniferous to Lower Permian section is completely exposed. Tongues of Nansen carbonate stabilized by submarine cements (preserving paleo-slopes exceeding 35°) may be traced as they plunge downward and roll out as thin

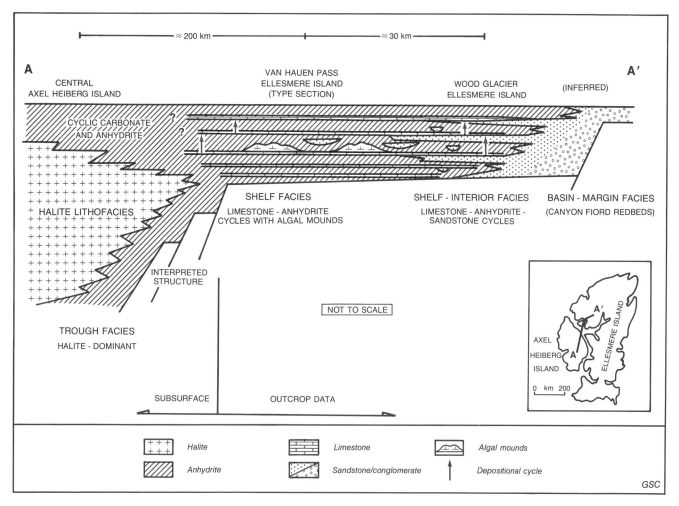

Figure.13.10. Schematic lithofacies section of the Namurian to Moscovian Otto Fiord Formation on Ellesmere and Axel Heiberg islands (based on data from Nassichuk and Davies, 1980).

debris flows and carbonate turbidites interbedded with shales of the Hare Fiord Formation, or, lower in the section, between anhydrite units of the upper Otto Fiord Formation (Davies, 1977a). Vertical relief on some of these steeply-dipping Nansen shelf transitions, from the original upper shelf surface to the paleo-seafloor, commonly is well in excess of 300 m, and may be 500 m or more. Drawdown of sea level during evaporitic maxima in Otto Fiord time would have exposed the surrounding shelf carbonate edges. Overall, these facies relationships demonstrate that, at least for Otto Fiord and lower Hare Fiord rocks, sedimentation was contemporaneous on carbonate shelves and basin floor. On a more specific scale, sedimentation may have shown a reciprocal relationship (Wilson, 1967), with periods of more active shelf sedimentation alternating with more active basinal sedimentation.

Artinskian-Wordian lithofacies

Overlying the major Namurian-Artinskian facies belts is a sequence of younger Permian rocks (Fig. 13.5), which were deposited in more widely transgressive seas and were not restricted to the same facies belts (Fig. 13.14). These younger sediments are dominated by siliciclastic rocks, although carbonates and cherts are still significant components. Starved, cool water shelf environments with glauconite, phosphate, and sponge spicules characterize several of the younger Permian units.

The oldest of these units is the Sabine Bay Formation, of Artinskian to Roadian age. This unit is exposed intermittently along the southern and eastern margin of the basin, overlying the Belcher Channel Formation and commonly bounded by disconformities. This unit is of fluvio-deltaic to marginal-marine origin, and is composed of quartzose sandstone with interbeds of conglomerate and coal, and minor limestone. It reaches a maximum thickness in outcrop of 194 m on Hamilton Peninsula of Ellesmere Island, where the presence of brachiopods, bryozoans and crinoids, as well as minor glauconite, attest to a marine depositional setting.

The Sabine Bay sandstone is overlain conformably by the uppermost Lower Permian (Roadian) Assistance Formation, composed commonly of friable to unconsolidated sandstone, often glauconitic, carbonaceous, and calcareous, with variable amounts of siltstone and concretionary siliceous limestone. The Assistance Formation reaches a

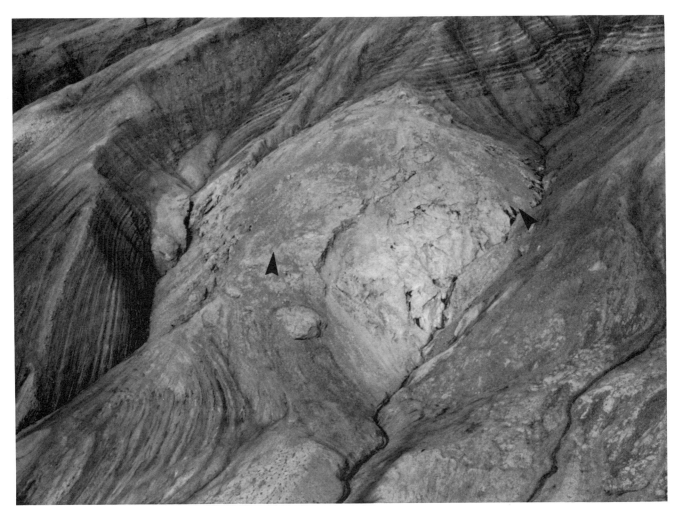

Figure 13.11. Oblique aerial photograph of a Moscovian bryozoan mound (arrows) enclosed in and overlain by well bedded shales, siltstones and argillaceous limestones of the Hare Fiord Formation at Stepanow Creek on the west side of Hare Fiord, Ellesmere Island. The mound is underlain by evaporites of the Otto Fiord Formation, projected from a nearby section. The mound is constructed by fenestellid bryozoans with extensive submarine cements and "stromatactoid" fabrics; it is identical to the "Waulsortian" type of mounds in the Blue Mountains of Ellesmere Island (Fig. 13.17A).

maximum thickness of 400 m in the Sawtooth Range of central Ellesmere Island. Thin chert-pebble conglomerate beds commonly are present at the base and higher in the unit. Brachiopods and bryozoans are common wherever the formation occurs, but ammonoids and shark (elasmobranch) remains also have been recovered from a few localities on Devon Island, Melville Island and Ellesmere Island (Nassichuk, 1970; Nassichuk and Spinosa, 1970). Trace fossils, including *Physonemas*, *Spirophyton*, and *Zoophycos*, are characteristic of more argillaceous facies where they occur on bedding planes.

The Assistance Formation grades basinward into a deeper water equivalent, the van Hauen Formation, composed mainly of dark coloured shale, siltstone and spicular chert (Fig. 13.15) and bounded by regional disconformities. With the exception of sponge spicules and scattered brachiopods, fossils are extremely rare in the van Hauen, but Nassichuk (1975) reported a few significant

ammonoid species from the formation, including *Daubichites fortieri* (Harker) and *Sverdrupites harkeri* Ruzhencev, which are also known from the Assistance Formation. Thorsteinsson (1974) identified two informal members in the van Hauen Formation — a lower recessive unit of mainly interbedded shale and siltstone, and an upper member of chert and siltstone, with siliceous sponge spicules common in the chert. The formation has a maximum outcrop thickness of 690 m near Blind Fiord on Ellesmere Island.

On southwestern Ellesmere Island, and on northern Melville Island, an unnamed formation composed in part of bioclastic limestone was initially recognized by Nassichuk (1975), and Nassichuk and Wilde (1977), but important stratigraphic refinement has resulted from recently completed fieldwork by B. Beauchamp (pers. comm., 1985). The unnamed unit, included for convenience in the Assistance Formation by Thorsteinsson (1974), is mainly of Artinskian age but some strata near the top, including

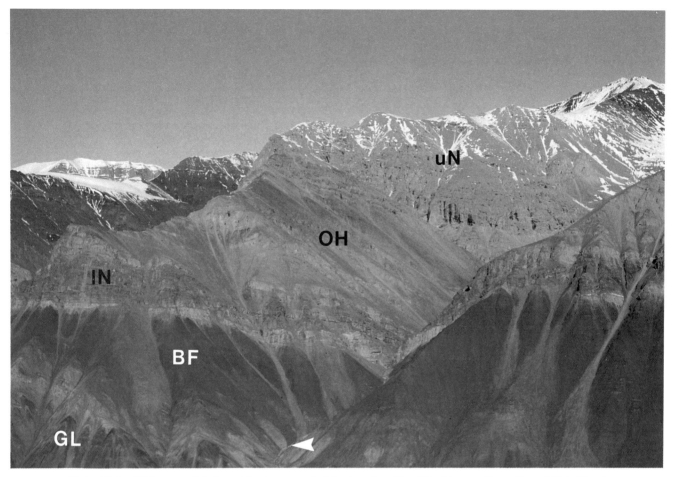

Plate 13.1. Oblique aerial view of the lower part of the type locality of the Nansen Formation at Girty Creek (arrow) on the north side of the head of Hare Fiord, Ellesmere Island. Red conglomerates and finer siliciclastics and limestones of the Borup Fiord Formation (BF) overlie the Lower Cambrian Grant Land Formation (GL) with angular unconformity at the base of the section (arrow), and are in turn overlain by lenticular (reefal?) carbonates of the lower Nansen (lN). The recessive weathering section above the carbonates marks a tongue of Otto Fiord evaporites and Hare Fiord cherty limestones and shales (OH). This section in turn is overlain by massive cliff-forming limestones of the upper Nansen Formation (uN).

strata at Great Bear Cape that Thorsteinsson (1974) referred to the Degerböls Formation, may be of latest Early Permian (Roadian) age, or possibly even slightly younger. Indeed, "Unit A" of Nassichuk (1965), which directly overlies the Assistance Formation on Sabine Peninsula, is also equivalent to the uppermost part of the unnamed formation. The unnamed formation overlies the top of the Belcher Channel Formation. In the Blind Fiord area of Ellesmere Island, it is both overlain by the upper van Hauen Formation, and grades laterally into the lower part of the van Hauen Formation (Fig. 13.16). This relationship indicates that the shoaling-up carbonate cycle represented by the unnamed formation marks the edge of a carbonate shelf sequence that grades laterally and basinward into thinner shales and cherts of the lower van Hauen Formation.

The youngest Permian formations in the Sverdrup Basin are the lowermost Upper Permian (Wordian) Trold Fiord Formation and the Degerböls Formation (Fig. 13.5). The Trold Fiord Formation is composed of glauconitic sandstone and siltstone with some shale, chert-pebble conglomerate, and minor limestone. Large ramose bryozoans and brachiopods are common in some units (Thorsteinsson, 1974), with crinoids and benthonic foraminifers as less common components. On Ellesmere Island, very fine- to fine-grained Trold Fiord sandstones contain up to 35% glauconite peloids; many of the latter are partly to completely replaced by iron oxides, which, on weathering, impart the characteristic red colour to the Trold Fiord Formation. The formation also contains units of spiculite composed of monaxon sponge spicules replaced, infilled, and cemented by chalcedonic silica; these units probably represent transitional intertongues of Degerböls-like facies. The Trold Fiord is found in eastern and southern regions of the basin where it overlies older Permian, Carboniferous, and in some areas, lower Paleozoic rocks of the Franklinian mobile belt; thus it clearly is transgressive in its distribution. The Trold Fiord has a maximum thickness, in the Sawtooth Range on Ellesmere Island, of about 315 m. The Degerböls Formation

Figure 13.12. Exposure of Otto Fiord (OF), Hare Fiord (HF) and Nansen (N) rocks along the north side of the head of Hare Fiord on Ellesmere Island, with Hare Glacier in foreground. The length of outcrop at the glacier contact is about 3 km. Note the steep bedding planes in the Nansen limestone (arrows) marking submarine-cemented depositional dip of the shelf carbonates prograding into and over the basinal Hare Fiord rocks.

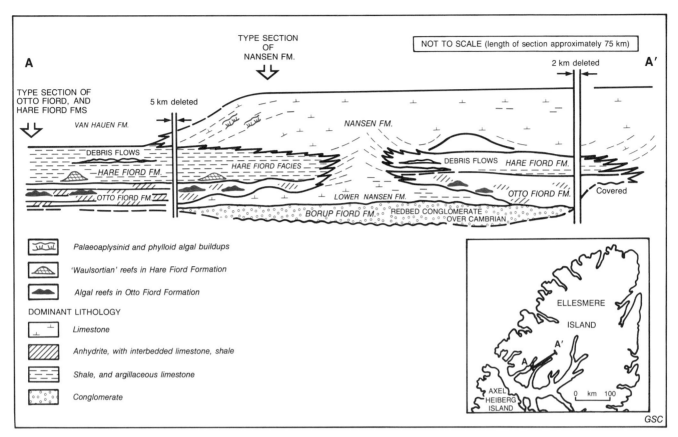

Figure 13.13. Schematic reconstruction of facies relationships between Borup Fiord, Otto Fiord, Hare Fiord and Nansen formations along the northern cliff face of Hare Fiord, Ellesmere Island. The sketch is based on vertical air-photos, field photographs and observations, but is (artificially) shortened to incorporate all of the major facies changes within a 60 km section.

is the basinal equivalent to the Trold Fiord, and is characterized by skeletal limestones composed of brachiopods, bryozoans, and crinoids. Lithological variants include grey-blue spiculitic chert and cherty limestone. The unit has a maximum measured thickness of about 300 m in the Krieger Mountains of northern Ellesmere Island.

Subsurface facies, southwestern Sverdrup Basin

About 35 of the 170 or so exploratory wells that have been drilled in the Sverdrup Basin (Fig. 13.1) have penetrated Carboniferous and/or Permian rocks. Most of the wells were drilled on Melville, Cameron and adjacent islands in the southwestern part of the Sverdrup Basin, but several were drilled in the channels between these islands from artificial "ice islands".

At the present time, the outcrop stratigraphic terminology that was established by Thorsteinsson (1974) for surface exposures in the northern and eastern sectors of the Sverdrup Basin is applied to the subsurface without significant modification. In this summary section, formational identifications and thicknesses are tentative and subsurface thicknesses have not been corrected for drillhole deviation.

Because of the location of many of the wells along the southern margin of the basin, the upper Paleozoic section encountered by the wells is biased toward the younger Permian (upper Artinskian–Wordian) siliciclastic sequence and the Canyon Fiord syntectonic redbeds. The latter record the erosion of active fault scarps delineating the structural margin of the Sverdrup Basin (Fig. 13.3). The Canyon Fiord Formation in this basin-margin tectonic setting is highly variable in thickness; on Cameron Island, for example, in the closely-spaced wells of the Bent Horn field, the Canyon Fiord redbeds vary in thickness from less than 30 to about 2000 m over a distance of less than 1 km. In other wells along the southern margin of the basin, the Canyon Fiord Formation is commonly 600 to 1000 m thick.

Other wells drilled on Prince Patrick, Brock, Mackenzie King and Ellef Ringnes islands along the trend of the Sverdrup Rim, encountered a more complete but attenuated Carboniferous to Permian stratigraphic section. In this structural setting, elevated horsts forming the rim are overlain by some 500 to 600 m of shelf limestones of the Nansen Formation (Fig. 13.2), much thinner than the thickest Nansen (2370 m) recorded from exposures on Ellesmere and Axel Heiberg islands.

Since all of the wells drilled in the central parts of the basin in the vicinity of Melville and adjacent islands had only shallower exploratory objectives, none encountered

CHAPTER 13

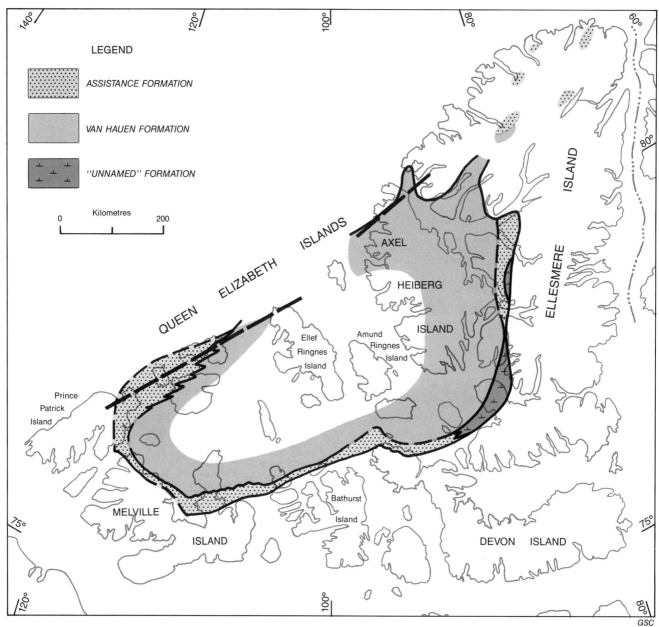

Figure 13.14. Distribution of uppermost Lower Permian (Roadian) lithofacies (Assistance and van Hauen formations) in the Sverdrup Basin, based on outcrop data from Thorsteinsson (1974) and Mayr (in press); subsurface data from exploratory wells in the southwestern sector. In the area between Bjorne Peninsula and Trold Fiord, Ellesmere Island, the distribution of an "unnamed formation" that is partly equivalent in age to the Assistance and van Hauen formations, but also including clastic and carbonate strata as old as late Sakmarian or early Artinskian, is also shown. Very heavy broken line marks axis of Sverdrup Rim.

the Carboniferous Otto Fiord evaporites in the Basinal Clastic and Evaporite Belt. As noted earlier, however, the Hoodoo L-41 well, drilled off the flank of the Hoodoo diapir on southern Ellef Ringnes Island, penetrated 3944 m of intrusive halite and anhydrite of the Otto Fiord Formation, and was still in halite when abandoned (Fig. 13.2).

The greatest contrast in thickness of stratigraphic units between the outcrop belts of Ellesmere and Axel Heiberg islands in the northeast, and the subsurface of the south and west of the Sverdrup Basin, occurs in the upper Artinskian to Wordian succession. From the youngest downward, these units are:

1. The Degerböls Formation, ranging in thickness in seven wells along the southern margin and the Sverdrup Rim from 360 to 1200 m under Brock Island, about four times the maximum measured thickness of 300 m in the northeastern outcrop belt.

Figure 13.15. Sandstones of the Roadian Assistance Formation (A) overlain by cyclically interbedded shale, siltstone and limestone of the Upper Permian (Wordian) Trold Fiord Formation (TF) in the Sawtooth Mountains near Notch Lake, Ellesmere Island. Throughout the Sverdrup Basin, the upper Artinskian to Wordian section is dominated by siliciclastics, in contrast to the carbonate-dominant Namurian to lower Artinskian sequence.

2. The Trold Fiord Formation, encountered in 23 wells, commonly in thicknesses exceeding 200 m, reaches a maximum thickness of about 800 m in the subsurface of Emerald Island, and slightly more than 600 m under central Sabine Peninsula of Melville Island and under northern Prince Patrick Island.

3. The van Hauen Formation, encountered in at least 11 wells, varies widely in thickness from less than 100 to a tentative 890 m in the Panarctic et al. Cornwall O-30 well on Cornwall Island, close to the axis of the Sverdrup Basin, and to about 1200 m in the Bent Horn field area on Cameron Island. The latter thickness is approximately twice the measured thickness of the van Hauen Formation exposed on Ellesmere and Axel Heiberg islands.

4. The Assistance Formation, encountered in nine of the southern and western wells, is generally less than 200 m thick.

5. The Sabine Bay Formation is recognized in only about three wells in the southern and western Sverdrup Basin, and in two of the wells on the southern end of Sabine Peninsula, Melville Island. The formation is 70 to 80 m thick at these localities.

In a number of wells drilled in the southern and western Sverdrup Basin, the Artinskian to Wordian sequence is represented mainly by shales or very argillaceous fine siltstones that are not readily correlated with the existing stratigraphic units erected on outcrop sections. No attempt has been made in this paper to assign new formational names to these subsurface sections; instead, they have been correlated on biostratigraphic criteria with the existing units.

In summary, the Artinskian and younger Permian clastic section above the Belcher Channel and equivalent formations in the southern and western part of the Sverdrup Basin generally is thicker than surface exposures on

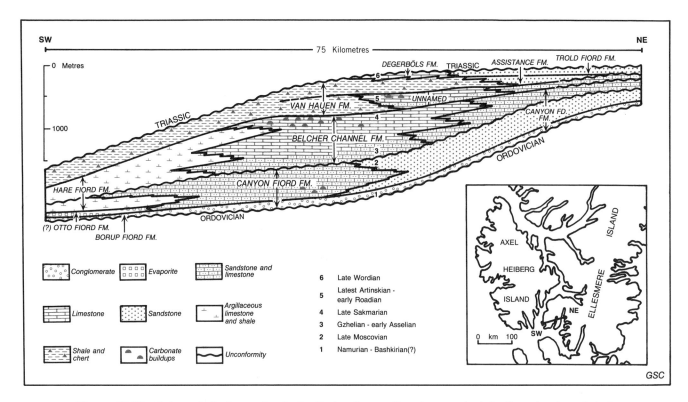

Figure 13.16. Schematic facies section for the Carboniferous-Permian marginal clastic and carbonate belt, southwestern Ellesmere Island, demonstrating the complex facies relationships between siliciclastics and carbonates, multiple unconformities, relationship of "unnamed" unit, and presence of carbonate mounds at three stratigraphic horizons (B. Beauchamp, pers. comm., 1985).

Ellesmere and Axel Heiberg islands. Down-to-the-basin faulting and development of horst-and-graben structures paralleling the southern margin of the basin and the Sverdrup Rim, plus the evidence for reactivation of faulting during the Melvillian Disturbance in the south and west, might reasonably account for the increase in thickness of the younger Permian section. The lack of a broad, shallow water Carboniferous to Lower Permian limestone platform extending well offshore along the southern margin, in contrast to the broad, regional distribution of the Nansen Formation in the northeastern part of the basin, is both a response to the active extensional tectonics and siliciclastic input in the south, and a factor allowing greater infill by younger Permian sediments after the Melvillian Disturbance. It is also consistent with a half-graben extensional model, with siliciclastics in the graben to the south and carbonates on the ramp to the north.

The Permian-Triassic boundary

The Permian-Triassic boundary in the Sverdrup Basin is marked by a regional unconformity (disconformity), below which all post-Wordian Permian rocks are missing (Nassichuk et al., 1973; Thorsteinsson, 1974). Throughout most of the Sverdrup Basin, Triassic rocks rest directly on the Permian, but locally, (in parts of western Ellesmere Island for example) Permian rocks have been removed by erosion and Triassic rocks rest on Carboniferous strata (Thorsteinsson, 1974). The Triassic section overlying the upper Paleozoic rocks consists of red weathering quartzose sandstone of the Bjorne Formation near the margin of the basin and a sequence of dark siltstone, shale and fine grained sandstone of the Blind Fiord Formation deposited in deeper water nearer the axis of the basin. Tozer (1969) correlated the *Otoceras concavum* Zone at the base of the Blind Fiord Formation with the *Otoceras woodwardi* Zone of the Himalayas. The latter zone effectively defines the base of the Triassic System.

Nondeposition and/or erosion of uppermost Permian rocks indicates basin-wide regression of the latest Permian sea. Two stages in the global time-stratigraphic scale for the Permian, the Dzhulfian and Changhsingian, are missing throughout the Sverdrup Basin and indeed throughout North America, with a 10 to 12 million year time gap. The lack of a major tectonic event throughout North America, Greenland and Svalbard at this time is indicated by the invariable structural conformity of the Triassic on Permian rocks.

Inter-regional correlations

The Carboniferous and Permian succession in the Sverdrup Basin shows striking similarities to the stratigraphic succession on Svalbard and northeastern Greenland. Perhaps the most striking similarity is with Svalbard, where the Carboniferous to Permian succession overlies unconformably a sequence of deformed and eroded Devonian and older rocks equivalent to those deposited in the Arctic

Islands. The upper Paleozoic succession is divisible into three sequences (Steel and Worsley, 1984), roughly correlative with the three time-stratigraphic sequences outlined for the Sverdrup Basin in this chapter. The main sequences on Svalbard are:
1. The Billefjorden Group of fluvial and alluvial clastics with associated coals, mainly of Tournaisian(?) to Viséan and possibly Namurian age, partly correlative with the nonmarine carbonaceous rocks of the Viséan Emma Fiord Formation in the Sverdrup Basin;
2. The Gipsdalen Group of limestone, dolomite and evaporite with localized clastics at block-faulted margins, of Bashkirian to Artinskian age, correlative with the Otto Fiord, Hare Fiord and Nansen formations, and with the Canyon Fiord redbeds in the Sverdrup Basin;
3. The Tempelfjorden Group of silicified marine clastics and associated limestones of Artinskinian to Late Permian age, correlative with the Sabine Bay to Degerböls sequence of the Sverdrup Basin.

From these similarities, it is clear that the Carboniferous and Permian sequence in the Sverdrup Basin and on Svalbard were influenced by the same tectonic and climatic conditions. Both were the product of sedimentation into sublinear graben-and-horst topography imposed on deformed Devonian and older rocks. Initially, during Viséan time, climatic conditions were humid, but later they became increasingly arid, favouring deposition of redbeds, evaporites, and shallow marine carbonates. Late Permian climatic amelioration in tectonically less active depressions in Svalbard and in the Sverdrup Basin resulted in accumulation of terrigenous clastics with extensive silica mineralization and cementation.

ORGANIC MOUNDS AND REEFS

Five distinct types of organically-constructed buildups or reefs occur within the upper Paleozoic section of the Sverdrup Basin (Fig. 13.17, 13.18). Each is characterized by the dominance of a specific organism, or group of organisms (Davies and Nassichuk, 1986; Beauchamp, 1987), and also by the development (in the majority of mounds) by pervasive syndepositional submarine cement (Davies, 1977b). Five types of mounds are recognized (Fig. 13.18).

1. Lowermost Upper Carboniferous (Bashkirian) tubular algal reefs

The type section of the Otto Fiord Formation on Ellesmere Island contains at least four algal reefs (Fig. 13.9) developed at one horizon within a carbonate cycle (Nassichuk and Davies, 1980; Fig. 13.8, 13.16). The mounds are up to 35 m thick and about 350 m long in outcrop dimension. Internally, the core of the mound is composed of a basal, open-marine crinoid wackestone unit, overlain by a more restricted (metahaline?) organic boundstone network of beresellid and donezellid algae cemented by submarine cements (Nassichuk and Davies, 1980; Fig. 13.20). Reconstruction of mound development (Nassichuk and Davies, 1980, Fig. 21) reveals a complex series of marine carbonate and hypersaline evaporitic events contributing to the evolution and burial of the mounds.

Other algal mounds of this type and of similar age occur within the Otto Fiord sequence at several horizons exposed along the northwestern cliff face of Hare Fiord on northwestern Ellesmere Island (Fig. 13.13). At least one of these mounds is built off the flank of an erosional high on redbeds in the upper part of the Borup Fiord Formation (Nassichuk and Davies, 1980, Fig. 28).

2. Upper Carboniferous (Moscovian) fenestellid bryozoan reefs

Large individual and coalesced reefs, at least 300 m thick and possibly as thick as 550 m, are common near the base of the Hare Fiord Formation on northwestern Ellesmere Island (Fig. 13.11, 13.17A). These reefs were formed by accumulation of fenestellid bryozoans and entrapped sediments stabilized by pervasive submarine cements (Davies, 1977b). They are buried by argillaceous and cherty limestones and shales of the basinal Hare Fiord Formation, which locally develop classic fining-upward graded crinoidal and intraclastic turbidite beds off the steep flanks of the reefs.

In internal fabrics and composition, the Moscovian bryozoan reefs of Ellesmere Island are identical to the classical Waulsortian (Lower Carboniferous) bryozoan reefs of Ireland, Belgium and elsewhere (Davies, 1975a). Stromatactis-like primary cavity systems, now filled by submarine and later burial cements (Fig. 13.17B), are dominant features of those groups of bryozoan reefs (see also James, 1983, Fig. 147-150).

3. Uppermost Carboniferous (Gzhelian) or Lower Permian (Asselian) phylloid algal mounds

Lenticular carbonate buildups and mounds constructed by the accumulation of fragmented plates of phylloid algae (Wray, 1968) are common at the Nansen shelf edge on northwestern Ellesmere Island (Fig. 13.17D). Once again, the algal plates commonly are cemented by submarine cements, some of which were originally precipitated as multi-generation isopachous magnesian calcite cement and others as botryoidal aragonite (similar to Quaternary submarine cements of the British Honduras barrier reefs; James, 1983).

Molds of fragmented phylloid algae have been recorded in Lower Permian (Asselian) rocks in cores from several wells close to the southern margin of the Sverdrup Basin. Elsewhere, particularly near the head of Hare Fiord on northern Ellesmere Island, they occur near the shelf edge in strata of the Nansen Formation, tentatively considered to be latest Carboniferous and earliest Permian (Asselian) in age. The occurrence of phylloid algae near the southern margin of the basin suggests that phylloid algal buildups may be present at the Nansen or Belcher Channel shelf edges in the subsurface of the southern (and southeastern) margin of the basin. Similar phylloid algal buildups, deposited in a shelf-margin setting in the Paradox Basin of the southern United States, have extensive primary and secondary porosity and form productive reservoirs (Choquette, 1983). In the Sverdrup Basin, however, the porosity of the inferred mounds may have been reduced by massive submarine syndepositional cementation.

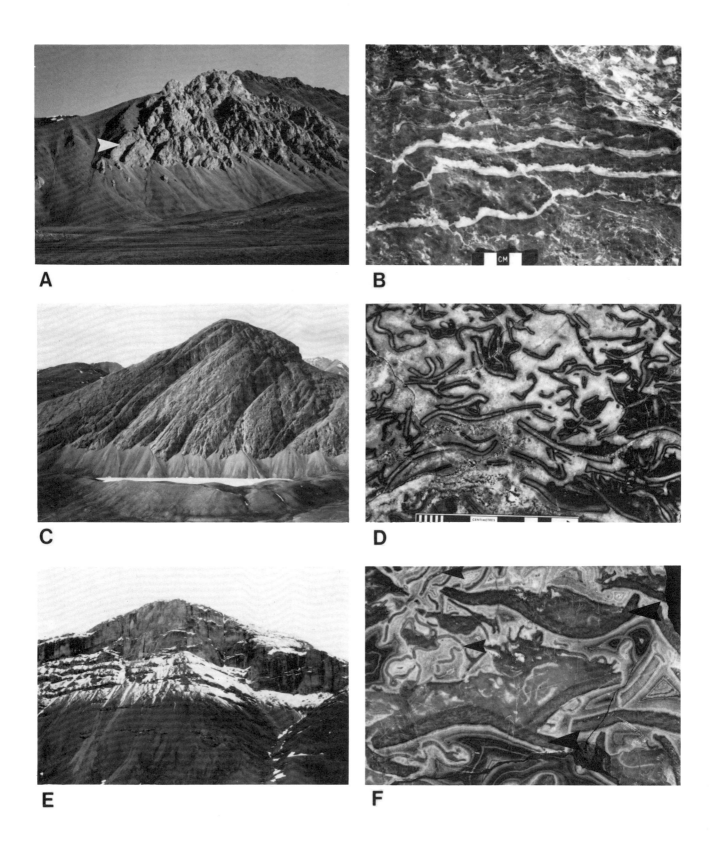

4. Lower Permian (Asselian) *Palaeoaplysina* reefs

The Nansen Formation on northwestern Axel Heiberg Island and northwestern Ellesmere Island, and the Belcher Channel Formation on southern Ellesmere Island, contain lenticular buildups constructed of fragmented plates of the enigmatic organism *Palaeoaplysina* (Fig. 13.17E, 13.17F). This organism was first recognized in Canada in small mounds in the northern Yukon Territory (Davies, 1971), but has since been discovered in Idaho, in the Yukon Territory, throughout the Sverdrup Basin, and in Greenland and Spitsbergen.

In an earlier description of *Palaeoaplysina* from the Sverdrup Basin (Davies and Nassichuk, 1973), this organism was tentatively identified as a hydrozoan, but a genetic relationship with sponges, or even algae, cannot be discounted entirely.

In its most massive development, buildups of *Palaeoaplysina* form cyclic limestone (less commonly dolomite) units, 10 to 30 m thick, with lenticular morphology, usually at or near the top of the Nansen or Belcher Channel formations. Buildups close to or at the Nansen shelf edge often show early submarine cements and multiple stages of

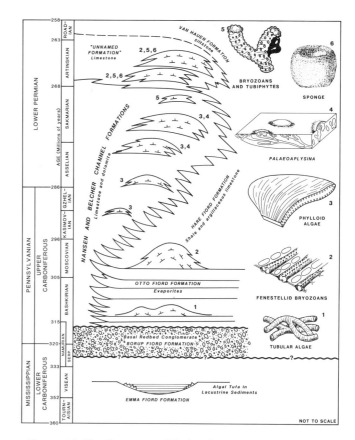

Figure 13.17. Carbonate reefs and fabrics, Sverdrup Basin.

A. Moscovian bryozoan reef of the "Tellevak" limestone unit enclosed by the Hare Fiord Formation at the southern end of the Blue Mountains on Ellesmere Island. Note the steep flank beds of the mound (arrow) cemented by pervasive submarine cements. The mound is about 300 m thick.

B. "Laminar stromatactoid" fabric in bryozoan reef (Fig. 13.17A) supported by fenestellid bryozoans and cemented by multiple generations of isopachous submarine cements. These fabrics are identical to the "Waulsortian" facies of Ireland and other areas, but are Late rather than Early Carboniferous in age.

C. Exposure of lenticular bedded shelf carbonates of the Nansen Formation, Ellesmere Island. Closer to the shelf edge, the Nansen contains organic buildups of phylloid algae and *Palaeoaplysina*.

D. Polished block of phylloid algal fabric from a buildup in the Nansen Formation, with ponded internal sediment and pervasive submarine cements. The phylloid algae and mound fabrics are similar to those found in west Texas, New Mexico, the Uinta Basin, and the Urals in the Soviet Union.

E. Stacked Lower Permian *Palaeoaplysina* mounds in the upper Belcher Channel Formation near Blind Fiord, Ellesmere Island. The mounds are about 150 m in total thickness (B. Beauchamp, pers. comm., 1985).

F. Polished block of *Palaeoaplysina* reef fabric from the head of Hare Fiord, showing thick fragmented sections of *Palaeoaplysina* plates (large arrows) with ponded sediment admixed with thin plates of phylloid algae (small arrows), and cemented by pervasive submarine cements. *Palaeoaplysina* is an organism of uncertain taxonomic affinity, and is a major mound-builder in the Sverdrup Basin, Spitsbergen and the Urals in the Soviet Union.

Figure 13.18. Summary of Carboniferous to Lower Permian organic mound- or reef-development in the Sverdrup Basin, illustrating the stratigraphic occurrences of the various types of organic buildups (modified from Davies and Nassichuk, 1986).

internal sedimentation. Those on northwestern Axel Heiberg Island, however, show the effect of extensive leaching and recrystallization, suggesting possible subaerial exposure and/or freshwater diagenesis; similar uplift and diagenesis may have played a role in dolomitization of *Palaeoaplysina* buildups on Bjorne Peninsula on southern Ellesmere Island (Davies and Nassichuk, 1973; Beauchamp, 1987).

Near the top of the type section of the Nansen Formation on northwestern Ellesmere Island, fragmented plates of *Palaeoaplysina*, exceeding 1 m in length and 2 to 4 m in thickness, are intermixed with coarse fusulinacean grainstones. It is not certain whether *Palaeoaplysina* grew as a subhorizontal encrusting plate on the substrate, or as free-standing fronds with some type of holdfast. Where both *Palaeoaplysina* and phylloid algal buildups are present, the stratigraphic and facies relationships suggest that phylloid algae grew below, and thus slightly deeper than, the more robust and clearly much larger *Palaeoaplysina* plates.

Several wells penetrating the Lower Permian section close to the southern margin of the Sverdrup Basin in the vicinity of Sabine Peninsula encountered a few fragments of *Palaeoaplysina* plates. Dispersed plates and abraded fragments of *Palaeoaplysina* also have been found in outcrops of inner shelf units of the Belcher Channel Formation on Grinnell Peninsula, Devon Island, and on other areas of southern Ellesmere Island. These occurrences suggest the possibility of shelf edge buildups of *Palaeoaplysina* in Nansen or Belcher Channel sequences in the subsurface of the southern Sverdrup Basin. Again, the chances of such occurrences of *Palaeoaplysina* mounds having preserved or secondary porosity may be downgraded by pervasive submarine cements, but possibly enhanced by subaerial exposure and leaching and/or dolomitization, as demonstrated by outcrops of similar mounds on southern Ellesmere Island.

5. Lower Permian (Artinskian) sponge-bryozoan-*Tubiphytes* reefs

Beauchamp (1987) described lenticular mound carbonates in the upper but not uppermost part of an unnamed formation exposed in the Blind Fiord area of southwestern Ellesmere Island (Fig. 13.16, 13.18). This unit forms a shallowing-upward cycle overlying *Palaeoaplysina* and phylloid algal buildups in the upper Belcher Channel Formation. The upper part of the unnamed formation is composed generally of thick bedded bioclastic limestones, but locally includes lens-shaped buildups of probable Artinskian age more than 15 m thick. These mounds are characterized by early cemented assemblages of fenestrate bryozoans with minor amounts of the enigmatic encrusting organism *Tubiphytes*. The sequence may be considered as another type of Lower Permian organically-controlled buildup in the Sverdrup Basin.

In several wells along the southern margin of the Sverdrup Basin, centred around Sabine Peninsula on Melville Island, Permian cuttings and cores are characterized by a diverse biota of brachiopods, bryozoans and crinoids but commonly lacking fusulinaceans, while often containing scattered remains of *Tubiphytes*. Nassichuk (1965) designated a sequence of limestones above the Assistance Formation on Sabine Peninsula, Melville Island as "Unit A". Those limestones, of probable Roadian age, are most likely equivalent to the uppermost part of the "unnamed formation" at Blind Fiord and Great Bear Cape, Ellesmere Island. To date, mounds of this age have not been identified from surface strata on Melville Island, but the presence of *Tubiphytes* in subsurface cores and cuttings may indicate a comparable environment of deposition.

REFERENCES

Balkwill, H.R.
1978: Evolution of Sverdrup Basin, Arctic Canada; The American Association of Petroleum Geologists Bulletin, v. 62, p. 1004-1028.

Bonham-Carter, G.F.
1966: The geology of the Pennsylvanian sequence of the Blue Mountains, northern Ellesmere Island; unpublished Ph.D. thesis, Department of Geology, University of Toronto, Toronto, Ontario, 172 p.

Beauchamp, B.
1987: Stratigraphy and facies analysis of the Upper Carboniferous to Lower Permian Canyon Fiord, Belcher Channel and Nansen formations, southwestern Ellesmere Island; unpublished Ph.D. thesis, Department of Geology and Geophysics, University of Calgary, Calgary, Alberta, 370 p.

Choquette, P.W.
1983: Platy algal reef mounds, Paradox Basin; in Carbonate Depositional Environments, P.A. Scholle, D.G. Bebout, and C.H. Moore (ed.); American Association of Petroleum Geologists, Memoir 33, p. 454-462.

Crowell, J.C.
1974: Origin of late Cenozoic basins in southern California; in Tectonics and Sedimentation, W.R. Dickinson (ed.); Society of Economic Paleontologists and Mineralogists, Special Publication 22, p. 190-204.

Davies, G.R.
1971: A Permian hydrozoan mound, Yukon Territory; Canadian Journal of Earth Sciences, v. 8, p. 973-988.
1975a: Upper Paleozoic carbonates and evaporites in the Sverdrup Basin, Canadian Arctic Archipelago; in Current Research, Part B, Geological Survey of Canada, Paper 75-1B, p. 209-214.
1975b: Hoodoo L-41: diapiric halite facies of the Otto Fiord Formation in the Sverdrup Basin, Arctic Archipelago; in Current Research, Part C, Geological Survey of Canada, Paper 75-1C, p. 23-39.
1977a: Carbonate to anhydrite facies relationships in the Otto Fiord Formation (Mississippian-Pennsylvanian), Canadian Arctic Archipelago; in Reefs and Evaporites — Concepts and Depositional Models, J.H. Fisher (ed.); The American Association of Petroleum Geologists, Studies in Geology no. 5, p. 145-167.
1977b: Former magnesium calcite and aragonite submarine cements in upper Paleozoic reefs of the Canadian Arctic: a summary; Geology, v. 5, p. 11-15.
1977c: Turbidites, debris sheets and truncation structures in upper Paleozoic deep-water carbonates of the Sverdrup Basin, Arctic Archipelago; in Deep-Water Carbonate Environments, H.E. Cook and P. Enos (ed.); Society of Economic Paleontologists and Mineralogists, Special Publication 25, p. 221-248.

Davies, G.R. and Nassichuk, W.W.
1973: The hydrozoan *Palaeoaplysina* from the upper Paleozoic of Ellesmere Island, Arctic Canada; Journal of Paleontology, v. 47, p. 251-265.
1975: Subaqueous evaporites of the Carboniferous Otto Fiord Formation, Canadian Arctic Archipelago — a summary; Geology, v. 3, p. 273-278.
1986: Ancient reefs in the high Arctic; Energy, Mines and Resources Canada, Geos, v. 15, no. 4, p. 1-5.
1988: An Early Carboniferous (Viséan) lacustrine oil shale in Canadian Arctic Archipelago; The American Association of Petroleum Geologists Bulletin, v. 72, p. 8-20.

Forsyth, D.A., Mair, J.A., and Fraser, I.
1979: Crustal structure of the central Sverdrup Basin; Canadian Journal of Earth Sciences, v. 16, p. 1581-1598.

Goodarzi, F., Davies, G.R., Nassichuk, W.W., and Snowdon, L.R.
1987: Organic petrology and Rock-Eval analysis of Lower Carboniferous Emma Fiord Formation in the Sverdrup Basin, Canadian Arctic Archipelago; Marine and Petroleum Geology, v. 4, p. 132-145.

Harrison, J.C., Goodbody, Q.H., and Christie, R.L.
1985: Stratigraphical and structural studies on Melville Island, District of Franklin; in Current Research, Part A, Geological Survey of Canada, Paper 85-1A, p. 629-637.

Hobson, G.D.
1962: Seismic exporation in the Canadian Arctic Islands; Geophysics, v. 27, p. 253-272.

James, N.P.
1983: Reef environment; in Carbonate Depositional Environments, P.A. Scholle, D.G. Bebout, and C.H. Moore (ed.); American Association of Petroleum Geologists, Memoir 33, p. 345-453.

Mayr, U.
in press: Reconnaissance and preliminary interpretation of Upper Devonian to Permian stratigraphy of northeastern Ellesmere Island, Canadian Arctic Archipelago; Geological Survey of Canada, Paper, 90-25.

Meneley, R.A., Henao, D., and Merritt, R.K.
1975: The northwest margin of the Sverdrup Basin; in Canada's Continental Margin and Offshore Exploration, C.J. Yorath, E.R. Parker, and D.J. Glass (ed.); Canadian Society of Petroleum Geologists, Memoir 4, p. 531-544.

Nassichuk, W.W.
1965: Pennsylvanian and Permian rocks in the Parry Islands Group, Canadian Arctic Archipelago; in Report of Activities, Field, 1964; Geological Survey of Canada, Paper 65-1, p. 9-12.
1970: Permian ammonoids from Devon and Melville Islands, Canadian Arctic Archipelago, Journal of Paleontology, v. 44, no. 1, p. 77-97.
1975: The stratigraphic significance of Permian ammonoids on Ellesmere Island; in Report of Activities, Part B, Geological Survey of Canada, Paper 75-1B, p. 277-283.

Nassichuk, W.W. and Bamber, E.W.
1978: Middle Pennsylvanian biostratigraphy, eastern Cordillera and Arctic Islands, Canada — a summary; in Western and Arctic Canadian Biostratigraphy, C.R. Stelck and B.E.D. Chatterton (ed.); Geological Association of Canada, Special Paper 18, p. 395-413.

Nassichuk, W.W. and Davies, G.R.
1980: Stratigraphy and sedimentology of the Otto Fiord Formation — a major Mississippian-Pennsylvanian evaporite of subaqueous origin in the Canadian Arctic Archipelago; Geological Survey of Canada, Memoir 286, 87 p.

Nassichuk, W.W. and Spinosa, C.
1970: Helicoprion sp., a Permian elasmobranch from Ellesmere Island, Canadian Arctic; Journal of Paleontology, v. 44, no. 6, p. 1130-1132.

Nassichuk, W.W., Thorsteinsson, R., and Tozer, E.T.
1973: Permian-Triassic boundary in the Canadian Arctic Archipelago; in The Permian and Triassic Systems and their Mutual Boundary, A. Logan and L.V. Hills (ed.); Canadian Society of Petroleum Geologists, Memoir 2, p. 286-293.

Nassichuk, W.W. and Wilde, G.L.
1977: Permian fusulinaceans and stratigraphy at Blind Fiord, southwestern Ellesmere Island; Geological Survey of Canada, Bulletin 268, 59 p.

Steel, R.J. and Worsley, D.
1984: Svalbard's post-Caledonian strata — an atlas of sedimentational patterns and paleogeographical evolution; in Petroleum Geology of the North European Margin, A.M. Spencer et al. (ed.); Norwegian Petroleum Society, p. 109-135.

Thorsteinsson, R.
1974: Carboniferous and Permian stratigraphy of Axel Heiberg Island and western Ellesmere Island, Canadian Arctic Archipelago; Geological Survey of Canada, Bulletin 224, 115 p.

Thorsteinsson, R. and Tozer, E.T.
1960: Summary account of structural history of the Canadian Arctic Archipelago since Precambrian time; Geological Survey of Canada, Paper 60-7, 23 p.
1970: Geology of the Arctic Archipelago; in Geology and Economic Minerals of Canada; R.J.W. Douglas (ed.); Geological Survey of Canada, Economic Geology Report no. 1, p. 546-590.

Tozer, E.T.
1969: A standard for Triassic time; Geological Survey of Canada, Bulletin 156, 103 p.

Tozer, E.T. and Thorsteinsson, R.
1964: Western Queen Elizabeth Islands, Arctic Archipelago; Geological Survey of Canada, Memoir 332, 242 p.

Trettin, H.P.
1988: Early Namurian (or older) alkali basalt in the Borup Fiord Formation, northern Axel Heiberg Island, Arctic Canada; in Current Research, Part D, Geological Survey of Canada, Paper 88-1D, p. 21-26.

Wilson, J.L.
1967: Cyclic and reciprocal sedimentation in Virgilian strata of southern New Mexico; Geological Society of America Bulletin, v. 78, p. 805-818.

Wray, J.L.
1968: Late Paleozoic phylloid algal limestones in the United States; 23rd International Geological Congress, Prague, Czechoslovakia, Proceedings, v. 8, p. 113-119.

Authors' addresses

G.R. Davies
Graham Davies Geological Consultants Ltd.
1, 2835-19th Street N.E.
Calgary, Alberta
T2E 7A2

W.W. Nassichuk

Geological Survey of Canada
3303-33rd Street N.W.
Calgary, Alberta
T2L 2A7

CHAPTER 13

Printed in Canada

Chapter 14
MESOZOIC HISTORY OF THE ARCTIC ISLANDS

Introduction

Stratigraphy, sedimentation and depositional history

Factors affecting Mesozoic stratigraphy

Summary

References

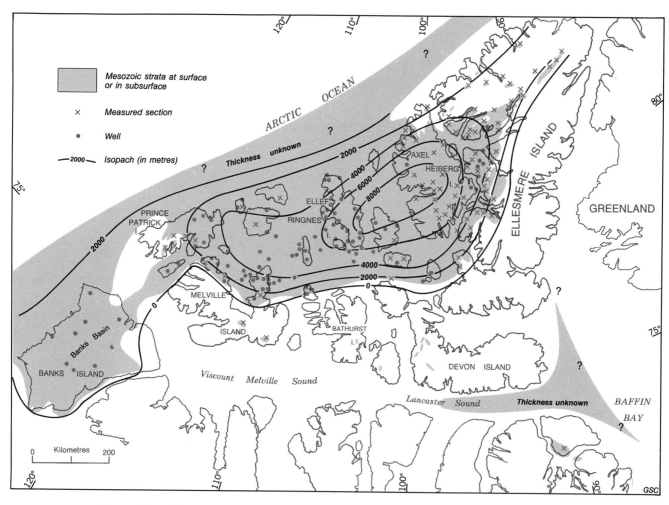

Figure 14.1. Distribution and isopach map of Mesozoic strata, Canadian Arctic Archipelago.

Chapter 14

MESOZOIC HISTORY OF THE ARCTIC ISLANDS

Ashton F. Embry

INTRODUCTION

Mesozoic strata are widespread in the Canadian Arctic Islands and occur in diverse tectonic-stratigraphic settings (Fig. 14.1). The Sverdrup Basin, which was a major depocentre in the Arctic Islands from Carboniferous to early Tertiary, contains the thickest and most complete Mesozoic succession in the region. In the central portion of the basin Triassic to Cretaceous deposits are up to 9 km thick (Fig. 14.1; for well sites and geographic names see Fig. 1, in pocket). Mesozoic rocks in eastern Sverdrup Basin were folded and faulted by regional compression in early Tertiary, and excellent exposures occur in mountainous terrain. To the west, the structures and terrain have much lower relief and outcrop is mainly Cretaceous or younger in age.

Banks Basin on Banks Island (Fig. 4.3) contains a gently-dipping, 1200 m succession of Upper Jurassic to uppermost Cretaceous strata. Scattered outliers of flat-lying to tilted Cretaceous strata occur in the Franklinian mobile belt, Arctic Platform and Canadian Shield geological provinces. These outliers are areally restricted and thin, and commonly lie in grabens. Cretaceous strata are also interpreted to occur beneath thick Tertiary deposits along the continental shelf northwest of the Arctic Islands, in eastern Lancaster Sound, and on the continental shelf east of Baffin Island. The nature and thickness of these offshore and deeply buried strata are unknown owing to a lack of data.

The Mesozoic succession in the Arctic Islands consists almost entirely of clastic sediments. In Sverdrup Basin, sandstone units occur mainly on the basin margins with shale-siltstone units predominating in the basin centre (Fig. 14.2). A few thick sandstone-dominant units extend across the basin. The Upper Jurassic to Cretaceous strata lying south of the Sverdrup Basin can be related to the stratigraphic units within the basin and represent sediments deposited during, and directly following, major transgresssions (Fig. 14.2).

Previous work

The description and mapping of Mesozoic strata in the Arctic Islands has been carried out principally by the Geological Survey of Canada. Maps at a scale of 1:500 000 or better are presently available for almost all known Mesozoic occurrences. Published observations on the Mesozoic strata were first made by explorers during the nineteenth century and first half of the twentieth century. The most systematic of such descriptions was made by Troelsen (1950). Formal study of the Mesozoic strata was begun by E.T. Tozer of the Geological Survey of Canada in 1954. He carried out fieldwork on most of the Arctic Islands over a span of 10 years, usually in collaboration with R. Thorsteinsson (Tozer 1956, 1961a, b, 1963a, b, c, d, 1967; Thorsteinsson and Tozer, 1962; Tozer and Thorsteinsson, 1964). During this time, observations on Mesozoic strata were also made by Heywood (1957), Souther (1963), McMillan (1963), Greiner (1963), Glenister (1963), Fricker (1963) and Christie (1964). These early studies resulted in a solid stratigraphic framework which was summarized by Tozer (1970a).

From the late 1960s to the mid-1970s a succession of officers of the Geological Survey of Canada studied Mesozoic strata, and resulting publications include Trettin and Hills (1966), Nassichuk and Christie (1969), Petryk (1969), Stott (1969), Roy (1972, 1973, 1974), Nassichuk and Roy (1975), Wilson (1976), Rahmani and Hopkins (1977), Rahmani (1978), Rahmani and Tan (1978), Miall (1975, 1979), Miall et al. (1980), Balkwill and Hopkins (1976), Balkwill and Roy (1977), Balkwill et al. (1977), Balkwill and Fox (1982), Balkwill et al. (1982) and Balkwill (1983).

Petroleum companies became interested in the Mesozoic strata of the Arctic following the early descriptions of the Geological Survey. Surface studies were conducted by numerous companies in the 1960s and early 1970s, but the only publication from this work is that of Plauchut and Jutard (1976), who ably described the Mesozoic succession of Banks and Eglinton islands. Field reports, which vary greatly in quality, were submitted by many companies to the federal government, and most are available from the government.

The surface studies of government and industry indicated that the Mesozoic strata of the Sverdrup Basin have good petroleum potential. The first well to penetrate Mesozoic formations was drilled in 1969 and more than 100 wells have been drilled in Mesozoic strata since that time. Subsurface studies of Mesozoic strata are not common. They include those by Meneley et al. (1975), Reinson (1975), Crain (1977), Henao-Londoño (1977), Meneley (1977), and Douglas and Oliver (1979). Subsurface data are also found in Balkwill's regional reports quoted above.

Recent summaries of Arctic geology which contain substantial sections on Mesozoic stratigraphy and depositional history include Plauchut (1971), Stuart Smith and Wennekers (1977), Balkwill (1978), Hea et al. (1980), Kerr (1981), Rayer (1981), and Balkwill et al. (1983).

Embry, A.F.
1991: Mesozoic history of the Arctic Islands; Chapter 14 in Geology of the Innuitian Orogen and Arctic Platform of Canada and Greenland, H.P. Trettin (ed.); Geological Survey of Canada, Geology of Canada, no. 3; (also Geological Society of America, The Geology of North America, v. E).

Table 14.1. Guide fossils and biostratigraphic zones.

TRIASSIC			JURASSIC					CRETACEOUS			
AGE		Ammonites and Pelecypods (Tozer, 1970)	AGE		Ammonites and Pelecypods (Frebold, 1970)	Foraminifera (Wall, 1983)	Palynology (Davies, 1983; Suneby, 1984)	AGE		Ammonites and Pelecypods (Jeletzky, 1970)	Foraminifera (Wall, 1983)	Palynology (Davies, 1983, unpublished)
LATE	Norian	*Monotis ochotica* *Neohimavatites* spp.	LATE	Tithonian	*Buchia fischeriana* *Buchia piochii*	*Arenoturrispirillina jeletzkyi*	L	LATE	Maastrichtian			*Integricorpus* sp. 1
							K					
				Kimmeridgian	*Buchia mosquensis* *Buchia concentrica*	*Glomospirella* sp. 174	J		Campanian	*Inoceramus lobatus*	*Verneuilinoides bearpawensis*	
	Carnian	*Halobia zitteli* *Sirenites nanseni*			*Amoeboceras* sp.		I					*Chatangiella chetensis*
				Oxfordian	*Cardioceras* sp.	*Ammodiscus thomsi*			Santonian		*Dorothia smokyensis*	
							H					
MIDDLE	Ladinian	*Nathorstites mcconnelli* *Ptychites nanuk*	MIDDLE	Callovian	*Cadoceras* sp.		G		Coniacian			
	Anisian	*Anagymnotoceras varium* *Lenotropites caurus*		Bathonian	*Arcticoceras ishmae* *Arctocephalites* sp.		F		Turonian	*Scaphites delicatulus*		
							E					*Isabelidinium magnum*
				Bajocian	*Cranocephalites vulgaris* *Arkelloceras tozeri*				Cenomanian	*Inoceramus pictus*		
EARLY	Spathian	*Keyserlingites subrobustus* *Olenikites pilatiens*		Aalenian	*Pseudolioceras mclintocki* *Lioceras opalium*	*Ammodiscus* cf. *asper*	D	EARLY	Albian	*Neogastroplites* sp. *Cleoniceras canadense* *Arcthoplites belli*	*Verneuilinoides borealis*	*Diplofusa gearlensis*
							C				*Quadrimorphina albertensis*	
	Smithian	*Wasatchites tardus* *Euflemingites romunderi*	EARLY	Toarcian	*Pernoceras polare* *Dactylioceras commune* *Harpoceras* sp.	*Flabellammina* sp. 1	B		Aptian		*Verneuilinoides neocominensis*	
							A		Barremian			
	Dienerian	*Vavilovites sverdrupi* *Proptychites candidus*		Pliensbachian	*Amaltheus stokesi*		MD biozone		Hauterivian			
				Sinemurian	*Echioceras arcticum* *Arietites* sp.				Valanginian	*Buchia inflatia*	*Uvigerinammina* sp. 1	Q P O
	Griesbachian	*Ophiceras commune* *Otoceras boreale*		Hettangian	unknown		AM biozone		Berriasian	*Buchia okensis*		N M

GSC

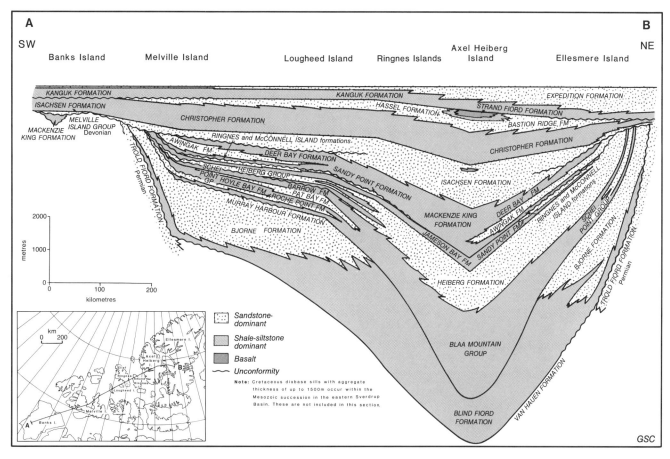

Figure 14.2. Schematic stratigraphic cross-section across Banks Basin and Sverdrup Basin.

Published biostratigraphic data of the Mesozoic of the Arctic Islands are reasonably abundant and include Tozer (1961a, 1965, 1967, 1970b), Frebold (1958, 1960, 1961, 1964, 1970, 1975), Jeletzky (1966, 1970, 1980, 1981, 1984), Logan (1967) and Kemper (1975, 1977) on macrofossils; Hopkins and Balkwill (1973), Johnson and Hills (1973), Doerenkamp et al. (1976), Pocock (1976), Felix (1975), Felix and Burbridge (1977), Fisher (1979), Fisher and Bujak (1975), Bujak and Fisher (1976), Tan (1979), Tan and Hills (1978), Davies (1983) and Suneby (1984) on palynomorphs; and Wall (1983) and Souaya (1976) on foraminfera. Table 14.1 summarizes various biostratigraphic zonations and guide fossils for Mesozoic strata of the Arctic Islands.

Present work

The author began a study of the Mesozoic strata of the Arctic Islands in 1977, with the aims of the project being to determine 1) regional stratigraphic relationships within the Mesozoic succession, 2) environments of deposition, 3) the geological history of the Arctic Islands during the Mesozoic, 4) the economic potential of the Mesozoic strata. Numerous surface sections and all available subsurface data have been studied, and these control points combined with the data of previous workers, have provided an excellent regional network for stratigraphic and sedimentological studies (Fig. 14.1).

To accomplish a detailed stratigraphic analysis of the Mesozoic strata, transgressive events, which are readily recognizable, very widespread, and common throughout the Mesozoic column, have been used to subdivide the succession. Thirty-one regional transgressions have been recognized, allowing subdivision of the Mesozoic Era into 30 transgressive-regressive cycles (T-R cycles), or depositional sequences in the terminology of Vail et al. (1977).

The T-R cycles (Fig. 14.3) are the basic building blocks for the Mesozoic succession. On the basin margins, subaerial unconformities form the cycle boundaries. A thin, transgressive unit of glauconitic sandstone or sandy limestone generally overlies the basal unconformity. A submarine unconformity commonly occurs on top of the transgressive unit. The transgressive deposits thin and eventually disappear basinward, and submarine unconformities form the cycle boundaries in the basin centre (Fig. 14.3). The cycle consists mainly of a regressive succession of marine shale and siltstone which coarsens upward into shallow marine and deltaic sandstones. The sandstones and subaerial unconformities disappear basinward as shown on Figure 14.3.

The cycles are the product of the interplay of varying rates of subsidence, sediment supply and eustatic sea level change. Boundaries between the cycles can be recognized in most sections and wells, but occasionally they are masked by the effects of high rates of sedimentation or subsidence.

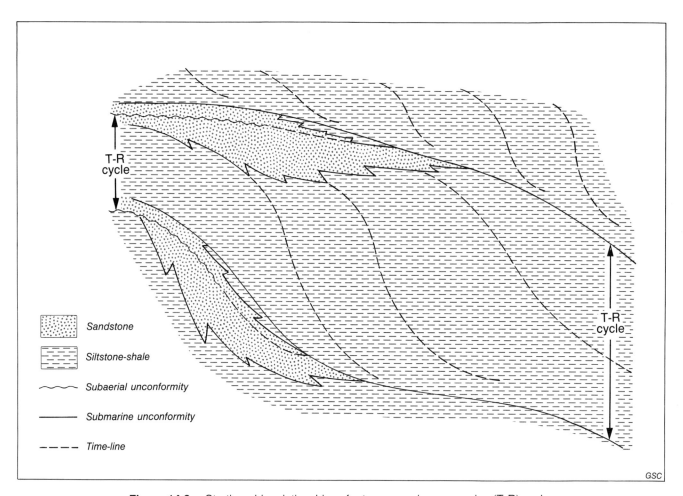

Figure 14.3. Stratigraphic relationships of a transgressive-regressive (T-R) cycle.

However, their stratigraphic position can usually be approximated in such cases for the purposes of isopach, facies and paleogeographic mapping. Biostratigraphic data are invaluable for dating the cycles and for their correlation throughout the basin.

The existence of 30 T-R cycles within the Mesozoic succession has resulted in a complex lithostratigraphic mosaic which does not lend itself to a simple stratigraphic nomenclature system. The present stratigraphic nomenclature for the Mesozoic is found on the correlation chart (Fig. 4, in pocket) and is illustrated on Figure 14.4. Most cycle boundaries are marked by solid lines and all major sandstone-dominant and shale-siltstone units have been assigned formal stratigraphic names. This nomenclature system is a combination of names proposed during the early phase of study, twenty to twenty-five years ago, and names which have been recently proposed on the basis of surface and subsurface studies (Embry, 1983a, b; 1984a, b, c; 1985a, b; 1986a, b; 1988).

Triassic-Valanginian tectonic setting

An isopach map (Fig. 14.5) of Triassic to Valanginian (Lower Cretaceous) strata of the Arctic Islands portrays the general tectonic setting for this time interval. This tectonic setting was established in mid-Carboniferous when the Sverdrup Basin originated by rifting (Balkwill, 1978). In general, the trends of the Sverdrup extensional faults, which were active between mid-Carboniferous and Early Permian, coincide with the Devonian – Early Carboniferous structural trends. The Sverdrup Basin was the main depocentre in the Arctic Islands during deposition of Triassic – lowest Cretaceous sediments. A thick, upper Paleozoic succession underlies Mesozoic strata (Davies and Nassichuk, Chapter 13). Of special note is the presence of a thick, widespread halite unit (Otto Fiord Formation) near the base of the upper Paleozoic succession. Halokinetic salt structures derived from this unit affected Mesozoic sedimentation over much of the basin.

The basin is centred in the Amund Ringnes Island – central Axel Heiberg Island area and is elongate in a northeast-southwest direction. It should be noted that the northeastern portion of the Sverdrup Basin was compressed perpendicular to its axis during the early Tertiary (Eurekan Orogeny) and therefore was wider in the Mesozoic than is illustrated on the non-palinspastic maps in this paper. Structural studies in this area are not sufficiently detailed to estimate the amount of shortening across the Sverdrup Basin, but preliminary estimates suggest that at least 100 km of shortening took place (U. Mayr, pers. comm.,

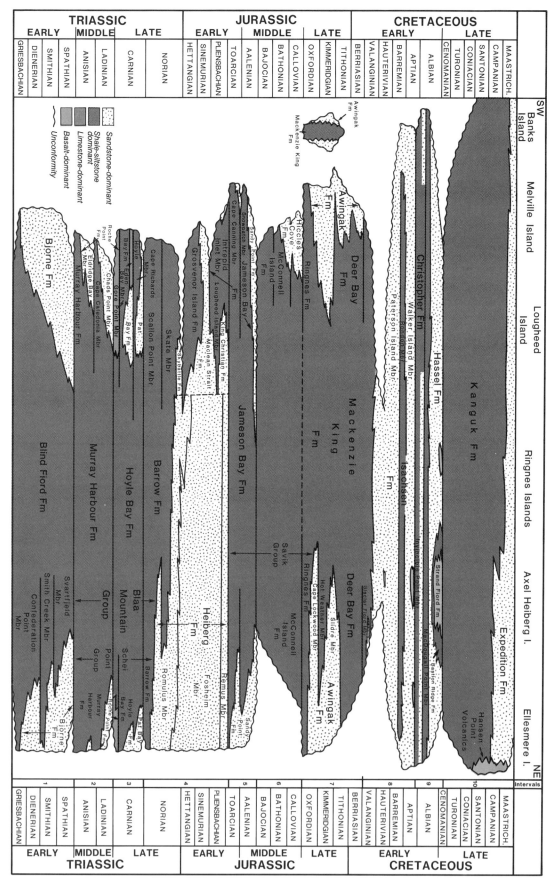

Figure 14.4. Stratigraphic nomenclature, Mesozoic strata, Arctic Islands.

1984). Low-lying landmasses were present to the east, south, and northwest of the basin. The land to the east and south consisted of a foreland fold and thrust belt of lower Paleozoic and Proterozoic carbonate and clastic rocks, displaced southeastward and southward toward the craton during mid-Paleozoic, Ellesmerian orogenesis. Lying eastward and southward from the foreland fold belt are undisturbed, flat, lower Paleozoic and Proterozoic platform cover rocks, and Precambrian crystalline basement rocks.

Concepts of nature and size of the land area to the northwest of the basin are speculative. This area is now occupied by the continental shelf and oceanic crust of the Amerasian Basin, which formed during the Cretaceous. The amounts and types of sediment derived from the northwest during the Triassic-Valanginian suggest a low-lying landmass of limited size and with a variety of sedimentary, igneous and metamorphic rock types. This landmass is interpreted to have moved counterclockwise away from the Arctic Islands in Cretaceous (Grantz et al., 1979) and now lies submerged beneath the Chukchi and East Siberian seas.

The Tanquary High (Nassichuk and Christie, 1969), a major basement arch in the northeastern part of the Sverdrup Basin, extends into the basin from the southeast (Fig. 14.5). Upper Paleozoic to Middle Triassic strata thin onto the flanks of this feature and are absent on its crest, indicating that it was exposed at this time. Limited data suggest the Tanquary High was only a minor positive structural element during the Jurassic and Cretaceous.

The southwestern portion of the Sverdrup Basin was little affected by Tertiary compression, and forms a broad platform with thicknesses considerably less than the central portion (Fig. 14.5). The axis of the basin swings markedly to the northwest in the southwestern area (again parallel to the older Devonian–Early Carboniferous structural trend) and abuts the present continental margin.

Another component of the Triassic-Valanginian tectonic setting is Banks Basin, which began to subside in Late Jurassic. Most of the basin fill is Barremian and younger, but 400 m of Upper Jurassic strata occur at the base of the succession. The basin is connected to the Sverdrup Basin through Eglinton Graben, a north-trending downwarp which formed in Middle Jurassic. Banks Basin and Eglinton Graben were flanked to the east and west by land areas and connected southward with shelf seas on the northern mainland.

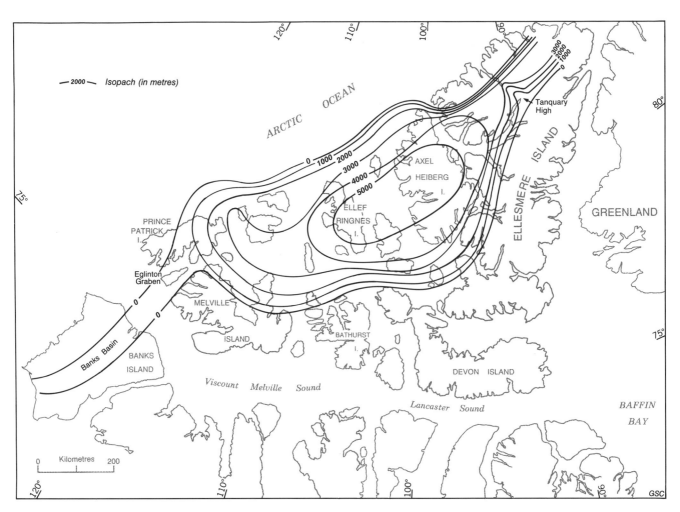

Figure 14.5. Isopach map, Triassic-Valanginian strata.

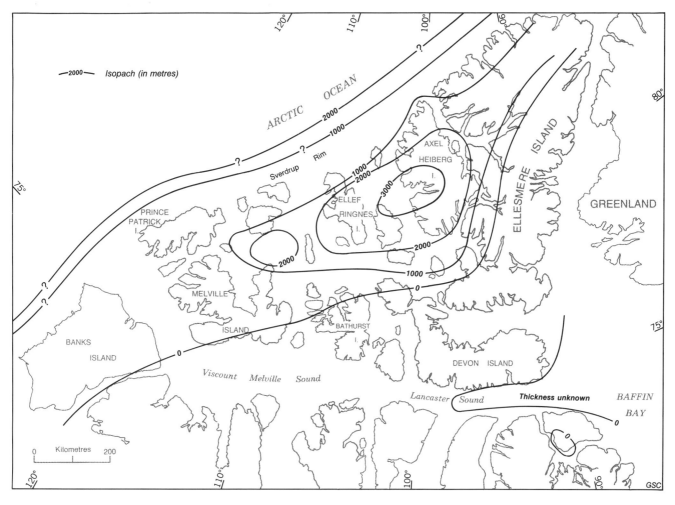

Figure 14.6. Isopach map, Hauterivian-Maastrichtian strata.

Hauterivian-Maastrichtian tectonic setting

Crustal rifting in the Valanginian-Hauterivian, which eventually led to the opening of the Amerasian Basin and Baffin Bay, changed the tectonic regime in the Arctic Islands. The isopach map of Hauterivian-Maastrichtian strata in the Arctic Islands (Fig. 14.6) illustrates the main tectonic features of this time interval. The Sverdrup Basin was still the main depocentre but had an oval shape in contrast to the dog leg of the Triassic-Valanginian interval. Banks Basin and Eglinton Graben expanded and became more prominent depocentres. Also in contrast to the preceding interval, deposition occurred on the broad platform area south and east of the Sverdrup Basin.

Two oceanic areas formed adjacent to the Arctic Islands during the Cretaceous, the Amerasian Basin in mid-Late Cretaceous, and Baffin Bay in Late Cretaceous – early Tertiary. The continental shelves of these ocean basins most likely have Cretaceous strata in the basal portion of the continental terrace wedges. The thickness and nature of such late Mesozoic deposits are unknown owing to lack of data.

The Amerasian Basin was separated from the Sverdrup and Banks basins by a broad arch named Sverdrup Rim (Balkwill, 1978; cf. Chapters 5B, 13). This positive feature is interpreted to be a rift shoulder, peripheral to the oceanic Amerasian Basin. It was intermittently exposed during the Cretaceous with most of the uplift occurring in Early Cretaceous time. Mesozoic strata were progressively truncated northwestward before being transgressed by Upper Cretaceous – Tertiary strata.

STRATIGRAPHY, SEDIMENTATION AND DEPOSITIONAL HISTORY

Thirty T-R cycles have been recognized in the Mesozoic succession of the Arctic Islands. Each cycle represents a distinct depositional interval. Sufficient data are not yet available to delineate each of these cycles on a regional basis. Accordingly, the succession has been divided into ten time-stratigraphic intervals which are bounded by prominent transgressive events. The style of sedimentation within each interval is distinctive and often contrasts sharply with that of adjacent intervals. The boundaries for these intervals are indicated on Figure 14.4. In the following sections the distribution, thickness, stratigraphy, sedimentology and paleogeography of each interval are described.

Permian-Triassic boundary

Upper Permian deposition was restricted to the Sverdrup Basin although fossils representing latest Permian time, that is, the Capitanian, Dzhulfian and Changhsingian stages have not been found. The apparent absence of these stages and the occurrence of a marked lithological break between Permian and Triassic strata at numerous localities led Tozer (1970a), Nassichuk et al. (1972), Thorsteinsson (1974), and Davies and Nassichuk (Chapter 13) to interpret the Permian-Triassic boundary as an unconformity throughout the basin. The opinions of these authors notwithstanding, the author's studies have led to a different interpretation.

Along the margins of Sverdrup Basin, where almost all the outcrops examined by previous workers are located, the Permian-Triassic contact is very abrupt and definitely appears to be unconformable as described by previous workers. Evidence in central portions of the basin suggests that the contact is conformable. Such a contact is well exposed 5 km east of Buchanan Lake on east-central Axel Heiberg Island. At that locality, the uppermost Permian strata consist of medium to dark grey, siliceous shale interpreted to be of slope to basin origin (Van Hauen Formation). Overlying Triassic strata (Blind Fiord Formation) consist of soft, medium grey shales with thin siltstone units, which exhibit characteristics of turbidite deposits (well developed Bouma sequences). The contact between the Permian shale and the Triassic shale is gradational over one half metre with the silica content gradually decreasing upward. X-ray diffraction analysis indicates that clay minerals within the Permian shales are the same as those in the Triassic shales and are also in approximately the same proportions. The only significant difference between the two rock types is silica content present as sponge spicules in the Permian strata. The relatively abrupt disappearance of this biological component at the Permian-Triassic boundary is in keeping with the major faunal extinction event at the boundary.

Similar gradational contacts between Permian cherty strata and Triassic shales also occur in a number of wells which were drilled well within the basin (e.g. Depot Point L-24; Fig. 1, in pocket). Outcropping and subsurface cherty strata of the uppermost Permian do not contain any biostratigraphically useful fossils and remain undated. The overlying Triassic shales are earliest Triassic in age (Tozer, 1970b). Until these uppermost Permian cherty strata are dated, the question of the nature of the contact will be debatable. However, in the author's opinion, the occurrence

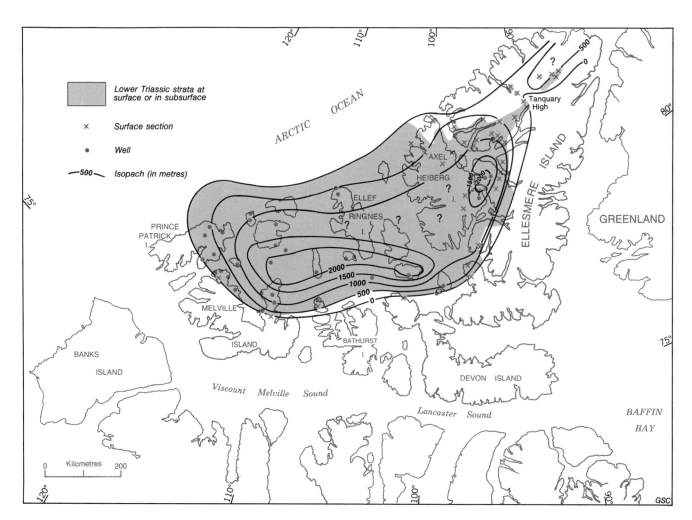

Figure 14.7. Distribution and isopach map of Lower Triassic strata.

of a gradational contact between structurally conformable, deep water deposits, favours the interpretation of a stratigraphically conformable, Permian-Triassic boundary over much of the Sverdrup Basin.

Early Triassic

Lower Triassic strata are widespread over the Sverdrup Basin except in the extreme northeast where they have been removed by Tertiary erosion (Fig. 14.7). Many surface and subsurface control points are available over most of the basin. The isopach map for the Lower Triassic (Fig. 14.7) reveals that two major depocentres were present, one centred on Sabine Peninsula, Melville Island and the other on Fosheim Peninsula, Ellesmere Island. Thicknesses exceed 2000 m in the depocentres but are usually less than 1200 m in other areas. A third depocentre may be present in the Cornwall Island area, but more data are needed to evaluate this. Lower Triassic strata thin markedly on the flanks of the Tanquary High and are absent on the crest of this feature.

Two formations comprise Lower Triassic strata, the shale-siltstone dominant Blind Fiord Formation, and the sandstone-dominant Bjorne Formation. Along the southern and eastern margins of the basin, the Blind Fiord Formation forms a thin, basal interval of medium to dark grey-green shale and siltstone of offshore shelf origin. The formation unconformably overlies Permian strata and is gradationally overlain by the Bjorne Formation. The Bjorne Formation exhibits marked thickness and facies changes along the southern and eastern margins (Fig. 14.8). The formation exceeds 1000 m in the depocentres and consists mainly of fine- to coarse-grained sandstone with minor red shale and siltstone interbeds (Fig. 14.9). The deposits are mainly of braided stream origin although delta front sediments also occur. In the eastern area, up to two shale-siltstone units of offshore marine shelf origin penetrate the Bjorne Formation.

The Bjorne Formation thins basinward because of a facies change to shale and siltstone of the Blind Fiord Formation (Fig. 14.10). The Blind Fiord Formation comprises the entire Lower Triassic succession in the central and

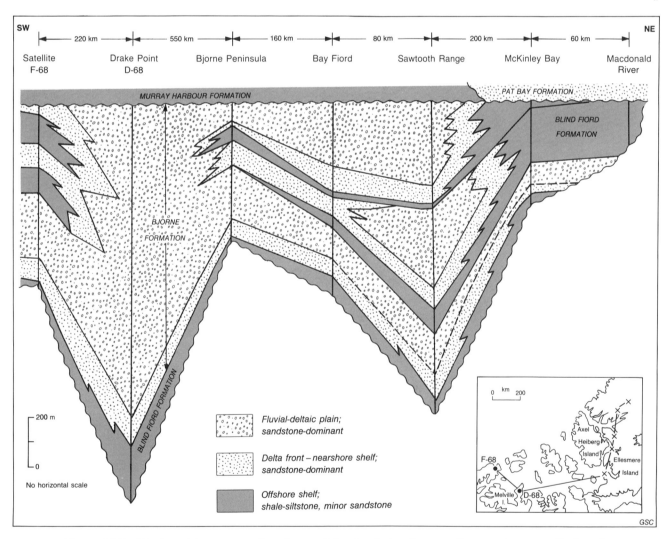

Figure 14.8. Stratigraphic cross-section, Lower Triassic strata along southern and eastern flanks of Sverdrup Basin.

Figure 14.9. Sea cliff of Bjorne Formation, consisting mainly of fine- to medium-grained crossbedded sandstone of braided stream origin. Smith Creek Member (shale-siltstone of marine shelf origin) occurs two thirds up section (SC). Black band in Smith Creek Member is a diabase sill. Northern Ellesmere Island.

northern portions of the basin. In areas where it is overlain by the Bjorne, or is close to the edge of the Bjorne, very fine- to fine-grained sandstone units of marine shelf origin (in part storm transported) occur interbedded with shale and siltstone. Farther basinward the sandstones are absent, and many of the siltstone units have characteristics of turbidites suggesting a deep water, slope to basin origin. On northwestern Ellesmere Island and northern Axel Heiberg Island three members, Confederation Point, Smith Creek and Svartfjeld, are recognized in the Blind Fiord Formation (Fig. 14.8; Embry, 1986a). Each member consists mainly of shale in the lower portion and siltstone in the upper portion (Fig. 14.11). In areas near the axis of the basin the Svartfjeld Member consists mainly of black, bituminous shale. On the northwestern margin of the basin sandstone units again appear within the Blind Fiord Formation (Fig. 14.10). To the northeast the strata onlap the Tanquary High and thin intervals of shale, siltstone and very fine grained sandstone,

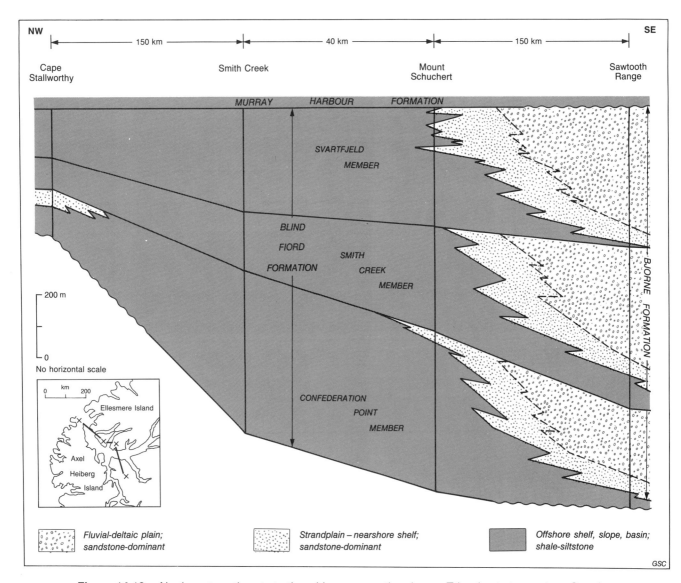

Figure 14.10. Northwest-southeast stratigraphic cross-section, Lower Triassic strata, eastern Sverdrup Basin.

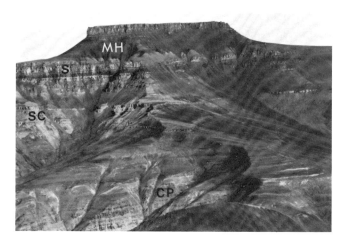

Figure 14.11. Three members of Blind Fiord Formation, northern Ellesmere Island. CP = Confederation Point; SC = Smith Creek; S = Svartfjeld. Murray Harbour Formation (MH) overlies the Blind Fiord Formation.

bounded by unconformities, surround the crest of the high (Fig. 14.8).

Three T-R cycles are recognized in the Lower Triassic strata with the contact between them placed at the base of the Smith Creek and Svartfjeld members. On the northwestern basin margin a sandstone unit occurs at the top of the lowest cycle. The basal cycle is dated as Griesbachian and Dienerian, the middle cycle as Smithian, and the upper cycle as late Smithian and Spathian.

Before discussing the Early Triassic depositional history it is important to examine briefly the basin topography that existed at the beginning of the Triassic. In Late Permian the Sverdrup Basin was ringed by shallow shelves and a deep basin occupied the central portion. In latest Permian a relative sea level fall occurred and the shelves of the basin were exposed, resulting in a widespread unconformity at the top of the Permian succession on the margins of the basin.

The shelves of Sverdrup Basin were transgressed in earliest Mesozoic time and the strandline advanced to about the present day limits of the basin. Following the

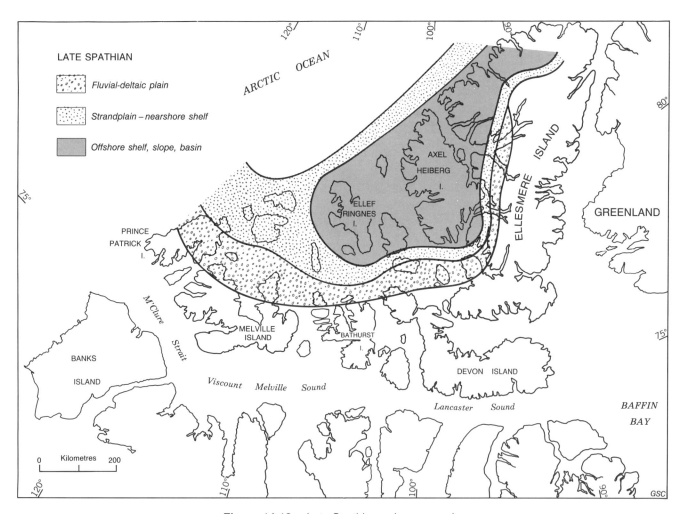

Figure 14.12. Late Spathian paleogeography.

transgression two deltaic centres, fed by rivers flowing from the east and south, were established. Sedimentation rates were high in these depocentres, and deltaic plains characterized by sandy, braided streams (Bjorne Formation) prograded into the basin. Thick, argillaceous shelf, slope and basin deposits began to fill the central basin (Blind Fiord Formation). A minor source area to the northwest of the basin also existed at this time as evidenced by the presence of a shallow shelf sandstone unit on northern Axel Heiberg Island.

Relative sea level rises occurred also in earliest and latest Smithian. The eastern delta was almost completely drowned but only the seaward edge of the western delta was transgressed. Deltaic sedimentation resumed following the transgressions and sandy deltas again prograded seaward (Bjorne Formation). They reached their maximum extent in late Spathian (Fig. 14.12). By this time the Lower Triassic succession had succeeded in filling in much of the western portion of the basin, but deep water environments still existed over the central part of the Sverdrup Basin. A sea level rise occurred at the beginning of the Middle Triassic, drowning the Early Triassic deltaic centres and initiating a new depositional regime.

Middle Triassic

Middle Triassic strata occur over much of the Sverdrup Basin except on northeastern Ellesmere Island where only a single outcrop has been found (Fig. 14.13). Outcrops are plentiful in the eastern Sverdrup Basin and well control is available for the western part of the basin. Only in the central portion of the basin is control lacking. On the basin flanks Middle Triassic strata are generally less than 300 m thick (Fig. 14.13). Thicknesses within the central portion of the basin may be considerably higher than illustrated, but no data are available for this area. Middle Triassic strata are absent over a broad area of the Tanquary High, indicating that this feature expanded considerably during this time. Of special note is that salt halokinesis was active in the basin at this time (Balkwill, 1978) and abruptly variable thickness changes are associated with syndepositional development of these features. Such local features are not displayed on Figure 14.13 and subsequent isopach maps.

In most areas along the southern and eastern flanks of the basin, Middle Triassic strata are placed in the Murray Harbour and Roche Point formations (Fig. 14.14). The Murray Harbour Formation consists of medium to very dark grey shale and siltstone with minor very fine grained

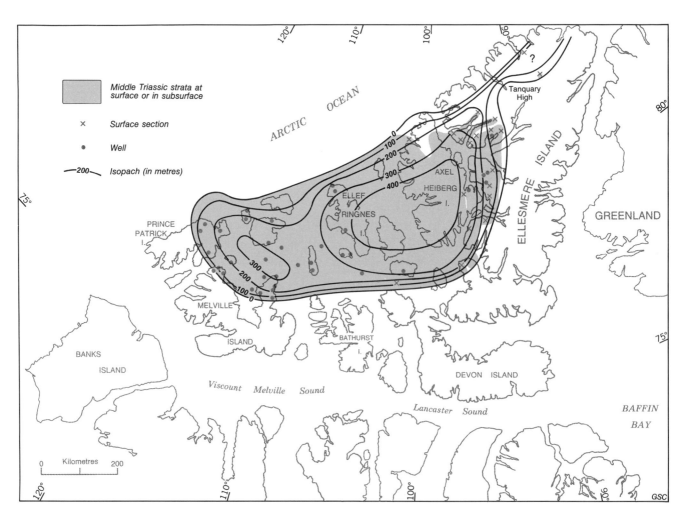

Figure 14.13. Distribution and isopach map of Middle Triassic strata.

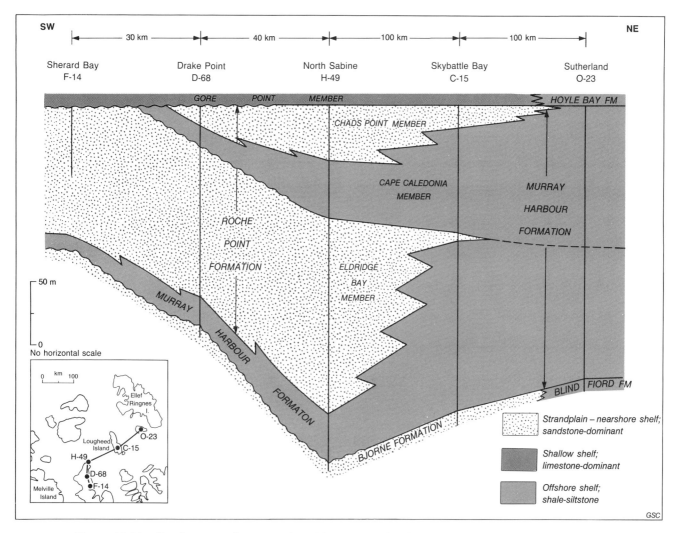

Figure 14.14. Southwest-northeast stratigraphic cross-section, Middle Triassic strata, Sverdrup Basin.

Figure 14.15. Dark, bituminous shale overlain by calcareous siltstone and silty limestone in an overall regressive succession, Murray Harbour Formation, Blind Fiord area, southern Ellesmere Island.

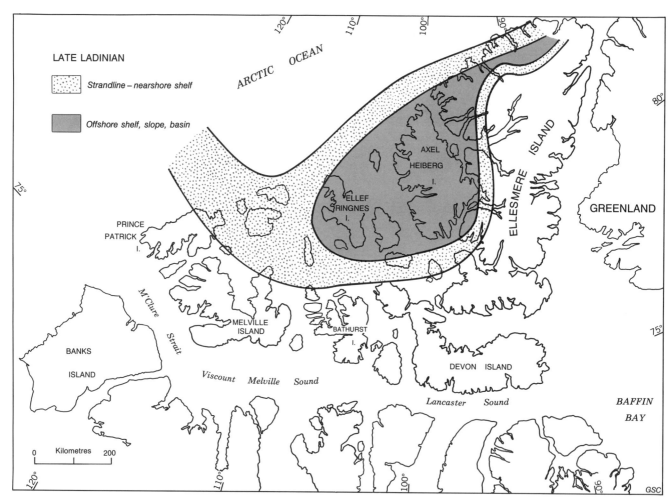

Figure 14.16. Late Ladinian paleogeography.

sandstone of offshore shelf origin. The basal contact of the Murray Harbour Formation with the underlying Bjorne Formation sandstones varies from unconformable on the basin margins to conformable farther basinward.

Roche Point Formation sandstones conformably overlie Murray Harbour Formation pelites. The Roche Point can commonly be divided into four members, three of which are Middle Triassic in age. The lowest member (Eldridge Bay) consists of interbedded very fine- to medium-grained sandstone, siltstone and shale. The lithologies are commonly arranged in coarsening-upward cycles of shallow marine shelf origin. Over the former Lower Triassic deltaic centre in the western Sverdrup Basin, fining-upward cycles with red shales and siltstone of delta plain origin occur within the Eldridge Bay Member.

The overlying Cape Caledonia Member consists of medium to dark grey shale and siltstone. Shale units commonly contain abundant organic material and the strata are of offshore marine shelf origin.

The Chads Point Member consists of very fine- to medium-grained, calcareous sandstone with minor shale, siltstone and bioclastic limestone interbeds. These strata are of shallow marine shelf origin. The contact of the Chads Point Member with the overlying Carnian strata is unconformable on the basin margins and on the flanks of the Tanquary High.

Basinward, the upper contact of the Middle Triassic strata becomes conformable (Fig. 14.14), and the sandstone units thin due to facies change to shale and siltstone. Over much of the central and northern portions of the basin, the entire Middle Triassic succession is represented by the Murray Harbour Formation. In these areas the formation consists of black, bituminous shale with abundant phosphate nodules in the lower portion and with silty lime mudstone generally capping the succession (Fig. 14.15). These strata are of offshore shelf, slope, and basin origin.

Two T-R cycles are recognized in Middle Triassic strata, with the contact between them placed at the base of the Cape Caledonia Member. This contact can often be traced basinward into the Murray Harbour Formation. The lower cycle is Anisian in age and the upper one is Ladinian.

The depositional history of the Middle Triassic began with a sea level rise which drowned the Lower Triassic deltas and caused shoreline displacement to a considerable distance landward. Following transgression, sediment supply into the Sverdrup Basin was considerably less than it had been during the Early Triassic. Sediment supply was concentrated in the western deltaic centre of the Early Triassic but the Anisian deltas were small and prograded only a short distance basinward (Eldridge Bay Member). Shallow shelf sands (Eldridge Bay Member) dominated the margins of the basin and these sediments were derived mainly from small rivers. These shelf sands prograded over deep shelf mud and silt (Murray Harbour Formation) during the Anisian.

A rise in sea level occurred in earliest Ladinian and much of the basin margin became an offshore shelf. During the Ladinian, shoreline and shallow shelf sands (Chads Point Member) prograded basinward over offshore shelf mud and silt (Cape Caledonia Member). A broad area over the Tanquary High was exposed during this regression.

Figure 14.16 illustrates the paleogeography at the close of the Ladinian when the regression was at its maximum. The shelf edge had moved basinward only slightly during the Middle Triassic due to the relatively low rate of sediment supply. A deep basin still existed in the central part of the Sverdrup Basin. Anoxic bottom conditions were present in the basin and outer portions of the shelves during the Middle Triassic, as evidenced by the high organic content preserved in these deposits.

Carnian

Carnian strata are widespread over the Sverdrup Basin. Stratigraphic control is provided mainly by surface exposures in the east and by well control in the west (Fig. 14.17). Carnian sediments have been removed from the most northeastern portion of the basin. They are less than 600 m thick on the basin flanks but thicken abruptly to over 1400 m toward the basin centre (Fig. 14.17). Control is lacking over much of the central portion of the basin and it is not known if such thicknesses occur over all the central portion of the basin. Carnian strata are generally less than 50 m thick over most of the Tanquary High and are absent on the crest of the arch.

Three formations comprise the Carnian interval, the limestone-dominant Gore Point Member of the Roche Point Formation, the sandstone-dominant Pat Bay Formation, and the shale-siltstone dominant Hoyle Bay Formation. On the southern flanks of the basin the Hoyle Bay Formation is divided into two members, Eden Bay and Cape Richards (Fig. 14.18; Embry, 1984b).

The Gore Point, the upper member of the Roche Point Formation, consists mainly of bioclastic limestone with pelecypod and crinoid fragments being the main skeletal components. The uppermost portion of the member is commonly argillaceous, and calcareous sandstone beds occur in northern localities. The Gore Point Member unconformably overlies Middle Triassic strata on the southern basin margin but the contact becomes conformable basinward (Fig. 14.18).

The Eden Bay Member of the Hoyle Bay Formation consists of interbedded dark bituminous shale, micritic limestone and siltstone with the limestone content increasing upward. Limestone content decreases northward, with calcareous siltstone interbeds becoming more common. In the extreme southwestern portion of the basin the member consists almost entirely of dark grey to black shale.

The overlying Cape Richards Member consists mainly of interbedded dark grey shale and siltstone. It generally lacks the bituminous shale and limestone interbeds which characterize the Eden Bay Member.

The Gore Point Member represents shallow shelf deposits which were laid down when clastic influx was low. The overlying Eden Bay and Cape Richards members are outer shelf deposits.

The Pat Bay Formation conformably overlies the Hoyle Bay Formation and consists mainly of very fine- to fine-grained, burrowed sandstone of shallow marine shelf origin. Thin siltstone and shale units lie within the formation and much of the formation was deposited below storm wave base. As illustrated in Figure 14.18 the Pat Bay Formation

CHAPTER 14

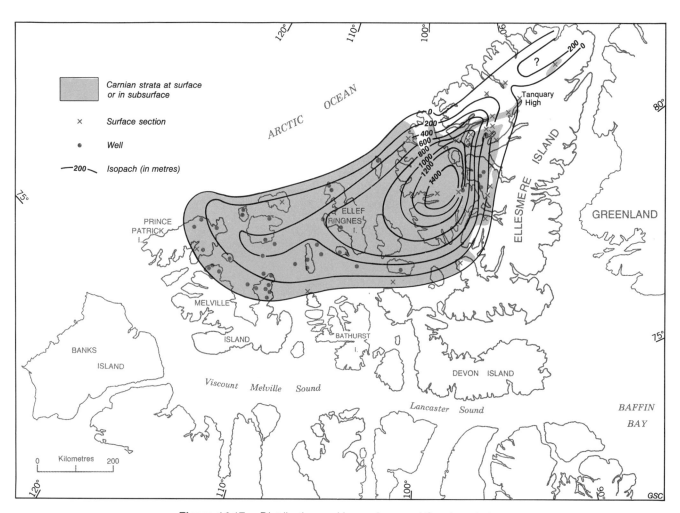

Figure 14.17. Distribution and isopach map of Carnian strata.

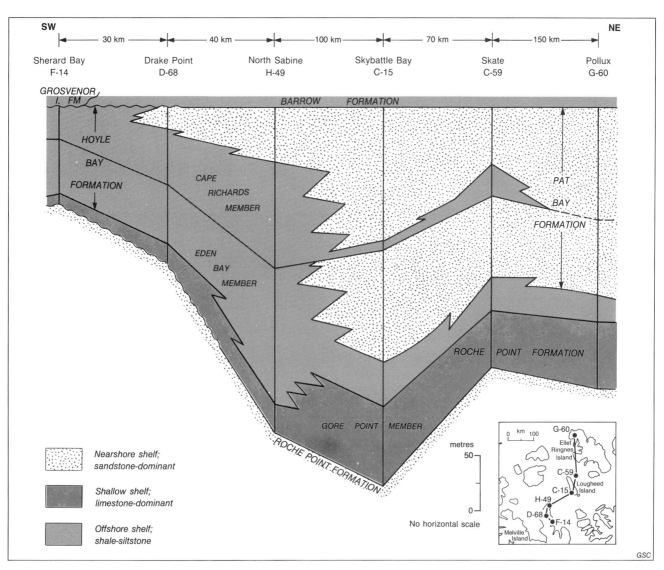

Figure 14.18. Southwest-northeast stratigraphic cross-section, Carnian strata, western Sverdrup Basin.

Three T-R cycles are recognized in the Carnian succession, with the contacts placed at the base of the Eden Bay and Cape Richards members of the Hoyle Bay Formation. The Eden Bay – Cape Richards contact can be extended into the Pat Bay Formation and occurs at the base of a shale within the formation. The cycles are dated as earliest Carnian, early Carnian, and late Carnian respectively.

Carnian deposition began with a rise in sea level which reduced clastic influx to the basin. In response to this, lime sand accumulated on the shallow shelves of the basin (Gore Point Member) with lime mud and bituminous clay being deposited farther offshore. During the subsequent regression, shallow marine sands (lower Pat Bay Formation) prograded basinward over offshore shelf and slope, silt and mud (Hoyle Bay Formation). The main areas of sediment supply were the northern and eastern margins of the basin. Lime mud and bituminuous clay accumulated on the southwestern margin of the basin in a "starved shelf" setting (Eden Bay Member).

A rise in sea level occurred in mid-Carnian and the basin margins were transgressed. In late Carnian, shallow shelf sands (upper Pay Bay Formation) again prograded basinward over offshore and slope deposits (Cape Richards Member, Hoyle Bay Formation). By the end of the Carnian, the shelf edge had moved significantly basinward owing to major infilling of the basin by slope deposits. Shallow marine sands extended over much of the western Sverdrup Basin at that time (Fig. 14.20).

Norian – earliest Toarcian

Uppermost Triassic to Lower Jurassic strata outcrop in mountainous terrain in many areas of the eastern Sverdrup Basin. To the west the strata are mainly in the subsurface and have been penetrated by numerous wells (Fig. 14.21). The strata exceed 2000 m in thickness throughout the central portion of the basin and may locally exceed 3000 m (Fig. 14.21). Thicknesses are an order of magnitude less over the western part of the basin.

The stratigraphy of this time interval is complex, as illustrated in Figure 14.4. In the eastern and central portions of the basin, two formations (Barrow and Heiberg) comprise the succession. The Barrow Formation is also recognized in the west but the Heiberg in that area is given group status and contains five component formations.

The Barrow Formation in the eastern and central Sverdrup Basin consists of medium grey, silty shale and siltstone of prodelta and marine slope origin. The strata represent the pelitic infill of the central basin. The Barrow Formation conformably overlies Carnian strata over most of the area but oversteps the Carnian strata on the basin margins and on the Tanquary High (Fig. 14.22).

Heiberg sandstones conformably overlie Barrow pelites in the eastern and central Sverdrup Basin. The formation is divided into three members, Romulus, Fosheim and Remus (Embry, 1983a). The lowest member, the Romulus, consists of coarsening-upward cycles of very fine- to fine-grained sandstone, siltstone and shale of delta front origin (Fig. 14.23).

The overlying Fosheim Member consists mainly of fine- to medium-grained sandstone with variable amounts of carbonaceous siltstone, shale and coal. Both fining-upward

Figure 14.19. Outcrop of Carnian strata, Esayoo Bay, northern Ellesmere Island. Thin transgressive limestone of Gore Point Member unconformably overlies Ladinian shale. MH = Murray Harbour Formation; GP = Gore Point Member, Roche Point Formation; HB = Hoyle Bay Formation; PB = Pat Bay Formation; B = Barrow Formation.

thickens toward the north and is equivalent to the Cape Richards Member and part of the Eden Bay Member. These stratigraphic relationships indicate that in the western Sverdrup Basin the main area of sediment supply lay to the north.

In the eastern Sverdrup Basin the Pat Bay Formation is present along both the eastern and northern basin margins, but is absent over much of the central portion of the basin where the Hoyle Bay Formation comprises the entire Carnian succession. The Pat Bay Formation consists of coarsening-upward cycles of shale, siltstone and very fine- to coarse-grained sandstone of shallow shelf origin (Fig. 14.19). The Hoyle Bay consists mainly of medium to dark grey silty shale with siltstone and micritic limestone interbeds. These strata are of outer shelf to slope origin and represent a major infill of the central basin. To the northeast, Carnian strata onlap the Tanquary High. Sandstones of the Pat Bay Formation occur near the crest of the high and rest unconformably on Lower Triassic strata.

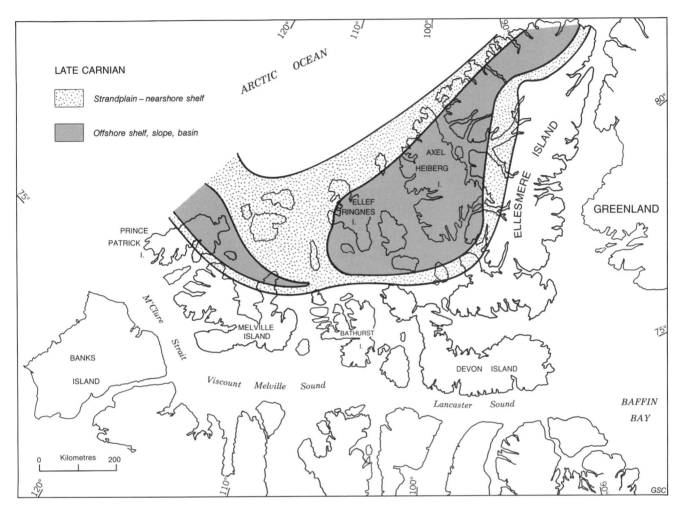

Figure 14.20. Late Carnian paleogeography.

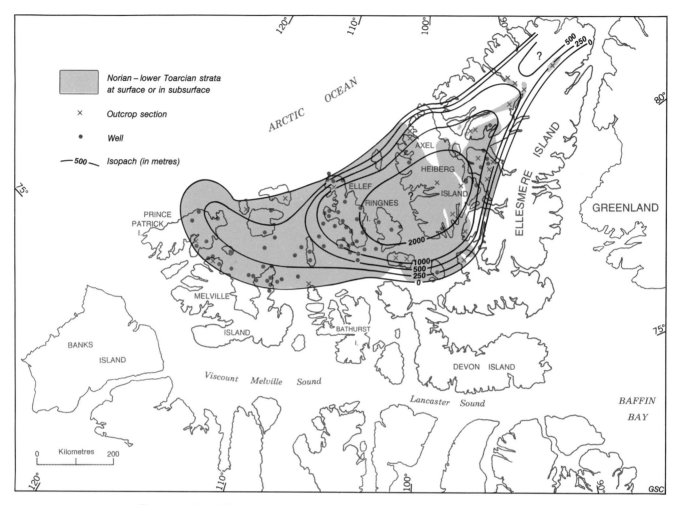

Figure 14.21. Distribution and isopach map of Norian – lower Toarcian strata.

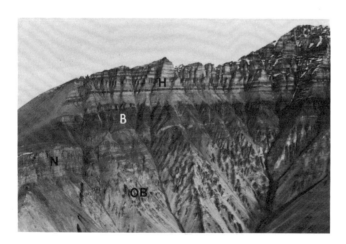

Figure 14.22. Barrow Formation (B) unconformably overlies Devonian clastics (OB), and Carboniferous carbonates of the Nansen Formation (N), and is overlain by Heiberg Formation (H), Yelverton Pass, northern Ellesmere Island.

Figure 14.23. Interbedded units of very fine grained sandstone and shale of Romulus Member, Heiberg Formation. Contact with overlying Fosheim Member at base of massive sandstone in upper right. Raanes Peninsula, southwestern Ellesmere Island.

Figure 14.24. Interbedded massive channel sandstones and recessive overbank strata, Fosheim Member, Heiberg Formation, Raanes Peninsula, southwestern Ellesmere Island.

Figure 14.25. Massive sandstone of marine shelf origin, Remus Member, Heiberg Formation (200 m). Contact with underlying Fosheim Member at arrow, Yelverton Pass, northern Ellesmere Island.

cycles of meandering stream origin and thin, coarsening-upward cycles of bay-fill origin occur in the member in the eastern Sverdrup Basin, thereby collectively representing a delta plain environment (Fig. 14.24). In the central Sverdrup Basin the Fosheim Member consists mainly of sandstone of braided stream origin which was deposited in a fluvial-deltaic plain setting.

The Remus Member caps the Heiberg Formation and consists almost entirely of fine grained sandstone (Fig. 14.25). Burrows and horizontal bedding are the only discernible sedimentary structures in the distinctive unit. Red weathering ironstone beds and hematitic sandstone are characteristic of this member. The strata are interpreted to be of shallow marine shelf origin. Figure 14.26 illustrates the interpreted stratigraphic relationships of the three members of the Heiberg Formation.

The Barrow Formation in the western Sverdrup Basin consists mainly of medium to dark grey siltstone and shale of prodelta and offshore marine shelf origin. The formation is divided into two informal members over most of the area (Scallon Point and Skate). The contact between the two is placed at a prominent gamma-ray sonic log-marker. A sandstone unit, the Jenness Member (Embry, 1984b), occurs between the Scallon Point and Skate members along the northwestern basin margin (Fig. 14.27). The Jenness consists of very fine- to medium-grained, glauconitic burrowed sandstone of marine shelf origin in the Brock Island – Mackenzie King Island area. On northern Ellef Ringnes Island the Jenness Member is thicker, contains distinct coarsening-upward cycles and is interpreted to be of delta front origin. The Jenness sandstones shale out to the southeast into the upper portion of the Scallon Point member and were derived from the northwest.

In the western Sverdrup Basin the Heiberg is raised to group status and comprises five formations: Skybattle, Grosvenor Island, Maclean Strait, Lougheed Island and King Christian (Fig. 14.27). The Skybattle, Maclean Strait and King Christian are sandstone-dominant units. They consist mainly of delta front sandstones in the Lougheed Island area with delta plain strata becoming more common eastward. All three units grade to shale northwestward (Fig. 14.27) with the King Christian Formation having the greatest westward extent. In areas where the King Christian sandstones are absent due to facies change to shale and siltstone, equivalent strata are placed in the Intrepid Inlet Member of the Jameson Bay Formation (Fig. 14.27, 14.28).

The Skybattle, Maclean Strait and King Christian formations are represented by thin beach to marine shelf sandstones along the southwestern margin of the basin, and these units all change facies to shale and siltstone northward. Landward the sandstones are capped by unconformities and each has been eroded toward the outcrop belt (Fig. 14.28). Skybattle and Maclean Strait sandstones also occur along the northwestern margin of the basin.

The Grosvenor Island and Lougheed Island formations consist mainly of grey shale and siltstone of prodelta to marine shelf origin. These units separate the three sandstone formations and extend over most of the western Sverdrup Basin. An oolitic ironstone unit commonly exists at the base of the Grosvenor Island Formation and interbeds of red shale also lie within the lower portion of the formation.

Four T-R cycles are recognized in the Norian–lowermost Toarcian strata of the western Sverdrup Basin. The ages of stratigraphic composition of each of these cycles are:

1. Early – Middle Norian — Scallon Point and Jenness members, Barrow Formation,

2. Late Norian (Rhaetian) — Skate member, Barrow Formation and Skybattle Formation,

3. Hettangian-Sinemurian — Grosvenor Island and Maclean Strain formations,

4. Pliensbachian – earliest Toarcian — Lougheed Island Formation, King Christian Formation, Intrepid Inlet Member, Jameson Bay Formation.

These cycles are not recognized in the eastern part of the basin, where high sedimentation rates prevented the transgressions from reaching this area.

The depositional history of the Norian–earliest Toarcian interval began with a sea level rise in early Norian. Following the transgression, sediment influx was concentrated on the eastern margin of the basin with lesser supply along the

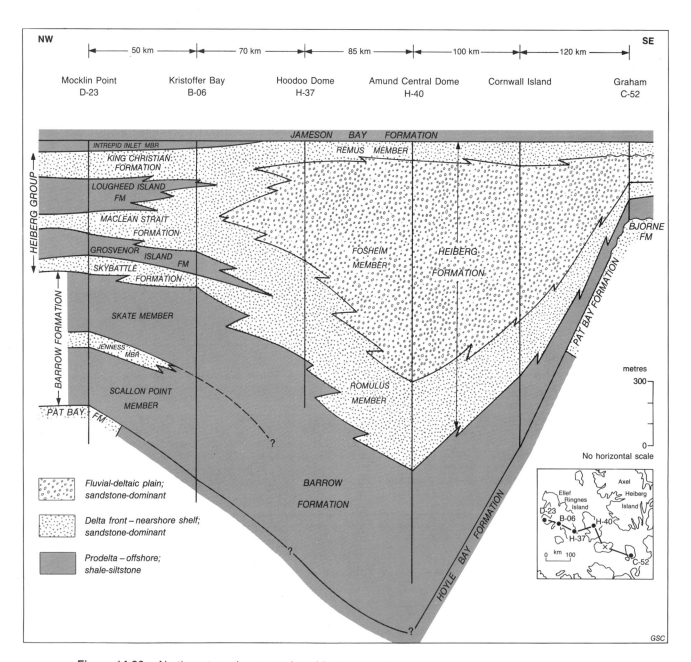

Figure 14.26. Northwest-southeast stratigraphic cross-section, Norian – lower Toarcian strata, eastern and central Sverdrup Basin.

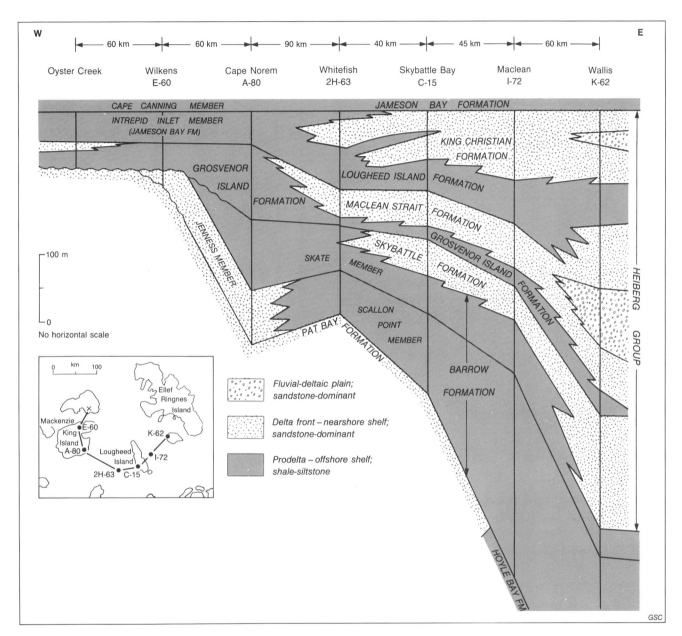

Figure 14.27. East-west stratigraphic cross-section, Norian – lower Toarcian strata, western Sverdrup Basin.

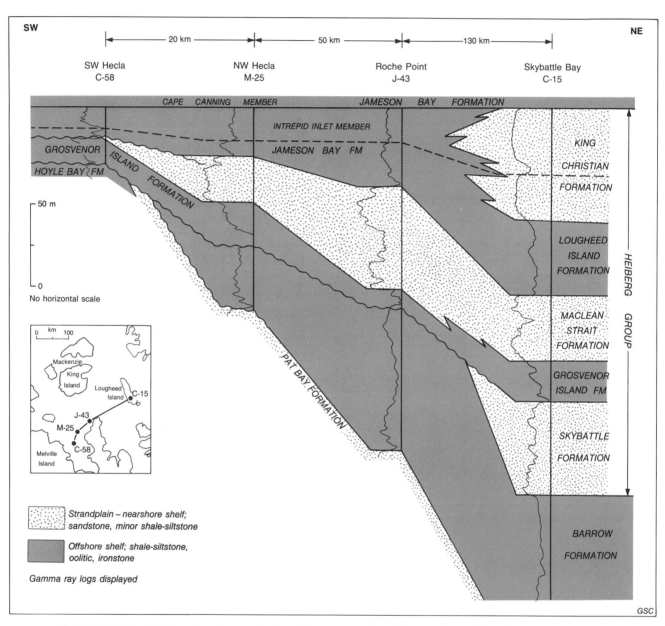

Figure 14.28. Northeast-southwest stratigraphic cross-section, Norian – lower Toarcian strata, western Sverdrup Basin.

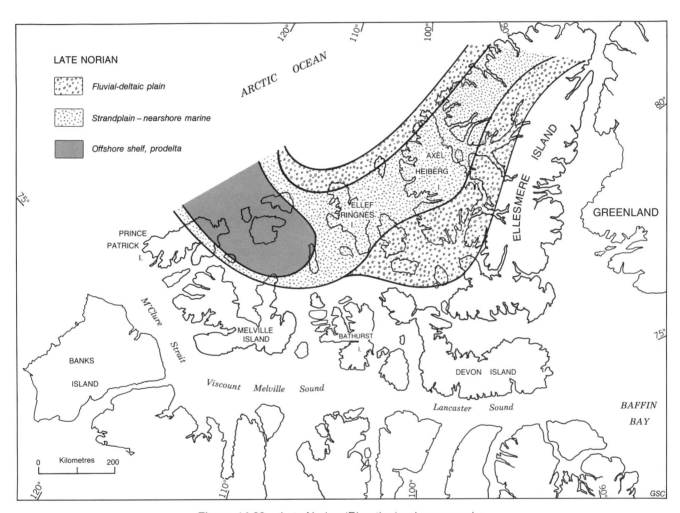

Figure 14.29. Late Norian (Rhaetian) paleogeography.

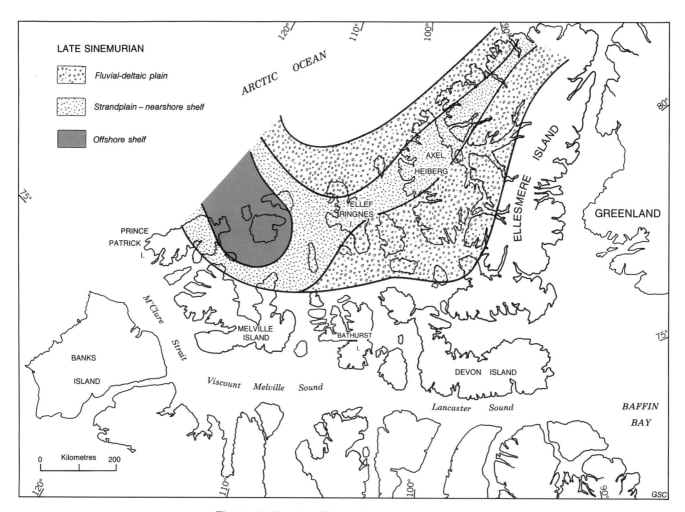

Figure 14.30. Late Sinemurian paleogeography.

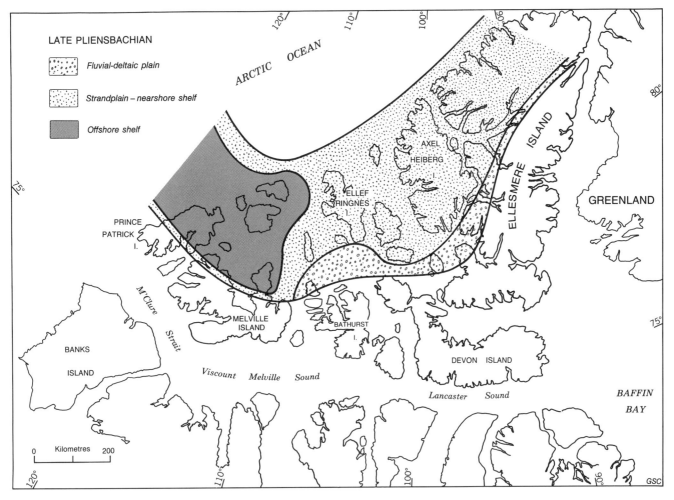

Figure 14.31. Late Pliensbachian paleogeography.

southwestern and northwestern margins. Shallow marine sandstones (Romulus Member) prograded northwestward across the shelf and slope mud, and silt (Barrow Formation) continued to infill the deep, central basin. An offshore shelf environment occupied much of the western Sverdrup Basin with narrow tracts of beach and shallow shelf sands along the basin margins.

A rise in sea level in late mid-Norian resulted in widespread transgression. Following this, sediment supply to the Sverdrup Basin greatly increased and deltaic centres were established in the eastern and central portions of the basin. Most of the sediment was derived from the adjacent Arctic Platform and Canadian and Greenland shields to the south and east. Source areas also existed to the northwest of the basin.

Owing to the high influx of sediment, heavily vegetated deltaic plains (indicated by abundant Fosheim Member coals) prograded over thick delta front sands (Romulus Member) and prodelta silts and muds (Barrow Formation). The central basin was infilled by slope sediments (Barrow Formation) as the deltaic systems prograded across the basin. In the western Sverdrup Basin, sediment influx was low and an offshore shelf receiving mud and silt (Skate Member, Barrow Formation) was present over most of the area (Fig. 14.29).

A rise in sea level in earliest Jurassic resulted in the seaward margins of the delta plains being transgressed. In late Hettangian and Sinemurian the deltas again advanced seaward and prograded westward onto the western shelf (Grosvenor Island and Maclean Strait formations). At the same time nearshore sands advanced northward along the southwestern basin margin (Fig. 14.30).

Another cycle of transgression and regression occurred in Pliensbachian, and deltas (King Christian Formation) extended slightly seaward of the underlying Sinemurian deltas. In late Pliensbachian and earliest Toarcian, sediment influx to the Sverdrup Basin was reduced and a transgression took place. During this time the extensive delta plain of the central and eastern Sverdrup Basin was gradually inundated, and beach and nearshore sand deposits (Remus Member) blanketed the area as the shoreline retreated. An offshore shelf receiving mud and silt (Intrepid Inlet Member, Jameson Bay Formation) occupied the north-central and western Sverdrup Basin at this time (Fig. 14.31).

This interval was brought to a close by a transgression in early Toarcian which resulted in an offshore shelf

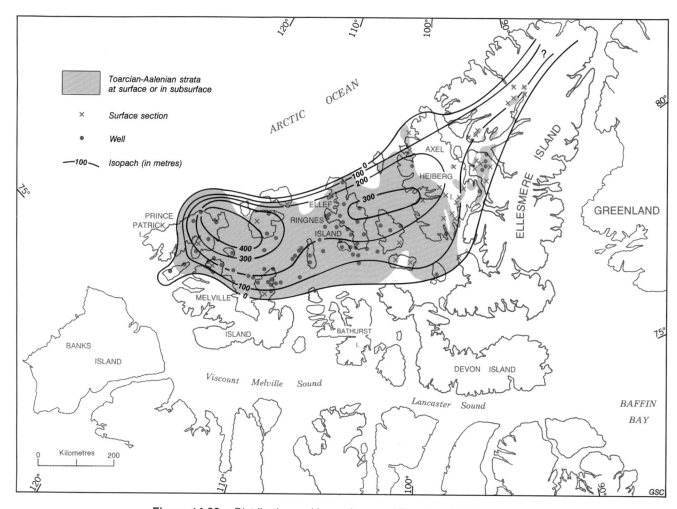

Figure 14.32. Distribution and isopach map of Toarcian-Aalenian strata.

environment occupying most of the basin. The Norian – earliest Toarcian interval was especially significant because the deep, central basin was finally infilled and the shelf-slope-basin depositional topography, which had played such a major role during the Triassic and late Paleozoic, ceased to exist.

Toarcian-Aalenian

Strata of Toarcian to Aalenian age (latest Early Jurassic to earliest Middle Jurassic) are preserved over much of the central and western parts of the Sverdrup Basin (Fig. 14.32). Within this area, numerous wells penetrate the strata, and outcrops occur along the basin margins. To the east the strata are present in synclinal areas and have been penetrated by five wells on Fosheim Peninsula (Fig. 14.32).

The deposits are relatively thin compared with the underlying interval. Two depocentres are evident on the isopach map (Fig. 14.32) with thicknesses of more than 300 m in the basin centre and in the southwestern corner of the basin. The deposits thin over the Tanquary High, indicating that this feature was still positive.

In the western Sverdrup Basin, Toarcian to Aalenian strata are included within two formations, the Jameson Bay and Sandy Point (Fig. 14.33). The Jameson Bay consists mainly of medium grey shale and siltstone of offshore shelf origin. Two members, Cape Canning and Snowpatch, are recognized within the formation. Highly burrowed, argillaceous sandstone and siltstone often occur in the uppermost portion of the Cape Canning, and are overlain by clay-rich shales of the basal Snowpatch Member. The Cape Canning Member overlies either siltstone of the Intrepid Inlet Member of the Jameson Bay Formation, or sandstone of the King Christian Formation. The contact is usually conformable, but on the basin margin Cape Canning strata rest unconformably on rocks as old as Devonian.

The Sandy Point Formation gradationally overlies the Jameson Bay Formation and consists of coarsening-upward cycles of shale, siltstone and very fine- to fine-grained sandstone. Sandstone units are often extensively burrowed and glauconitic. Cleaner, horizontally bedded sandstones occur in basin margin sections. The strata are interpreted to be of shallow shelf origin and to have been deposited mainly below wave base. The upper contact of the Sandy Point Formation with overlying Bajocian strata is unconformable on the basin margins and becomes conformable basinward (Fig. 14.33).

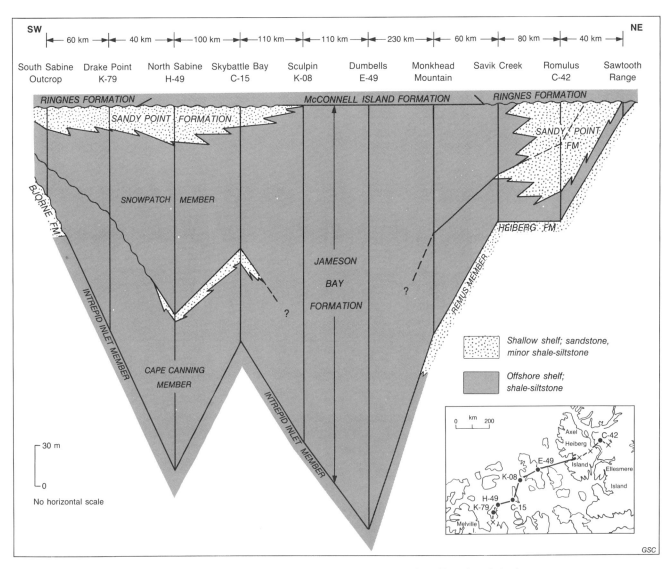

Figure 14.33. Southwest-northeast stratigraphic cross-section, Toarcian-Aalenian strata.

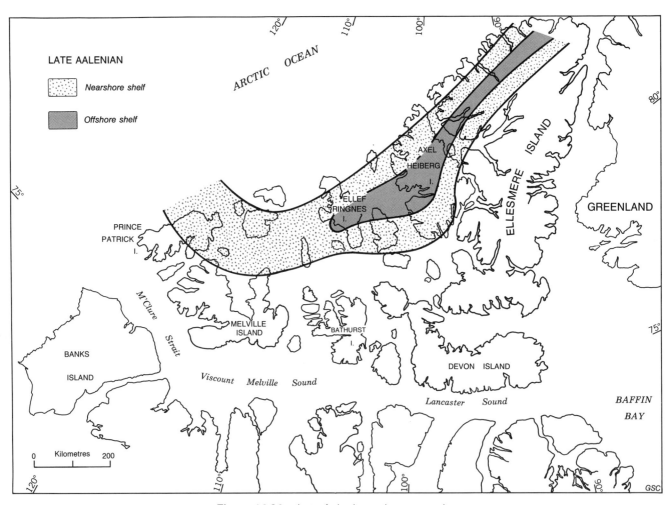

Figure 14.34. Late Aalenian paleogeography.

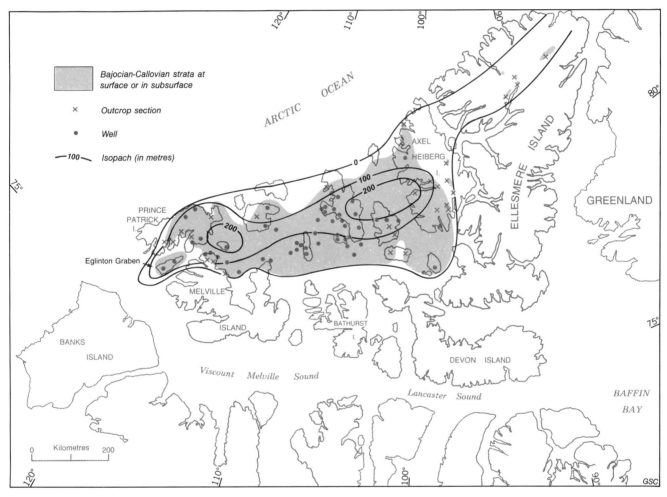

Figure 14.35. Distribution and isopach map of Bajocian-Callovian strata.

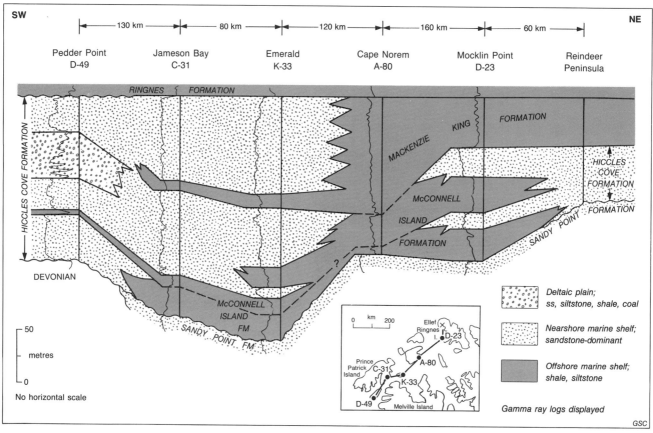

Figure 14.36. Southwest-northeast stratigraphic cross-section, Bajocian-Callovian strata, southwestern Sverdrup Basin.

In the central and eastern Sverdrup Basin the basal strata consist of very fine grained, burrowed sandstone along the basin margin. These strata are included in the Remus Member of the Heiberg Formation. The shales and siltstones of the Jameson Bay Formation are thin on the basin margin and thicken basinward. In the basin centre the Jameson Bay Formation comprises the entire Toarcian to Aalenian succession (Fig. 14.33). The Sandy Point Formation is truncated on the extreme basin margins and can be traced basinward a considerable distance (Fig. 14.33). Source areas occurred to both the southeast and northwest.

Two T-R cycles are recognized in this interval over much of the western Sverdrup Basin and at scattered localities in the eastern portion of the basin. The ages of the cycles are Toarcian and Aalenian.

The Toarcian-Aalenian interval began with a major transgression in early Toarcian, which created an offshore shelf environment over most of the basin. Shallow shelf sands continued to be deposited along the southeastern and eastern basin margins where supply was greatest. These areas were eventually drowned at the height of the transgression in Toarcian. During the subsequent regression, nearshore sand (lower Sandy Point Formation) prograded over offshore shelf mud and silt (Jameson Bay Formation) with sand influx occurring mainly in the eastern part of the Sverdrup Basin.

Following Aalenian transgression, regressive sand deposition was concentrated in the western part of the Sverdrup Basin, mainly in the Prince Patrick area, and shelf sands (Sandy Point Formation) prograded seaward. During maximum regression in latest Aalenian, a narrow seaway existed in the basin, and shelf sands almost reached its centre (Fig. 14.34). A major sea level rise terminated this interval in early Bajocian.

Bajocian-Callovian

The distribution of Bajocian-Callovian strata is similar to that of the underlying interval, with the main area of preservation stretching from western Axel Heiberg to Prince Patrick Island (Fig. 14.35). Numerous wells and surface sections are available for the strata in this area. To the east, preservation is more patchy and scattered surface and subsurface control is available.

Bajocian-Callovian deposits are relatively thin and rarely exceed 200 m (Fig. 14.35). The main depocentre was in the southwestern portion of the basin, in the northern Prince Patrick – Emerald Island area. It is also apparent that the Eglinton Graben, which lies between Melville and Prince Patrick islands, began subsiding at this time and was an integral part of the depocentre to the north.

The general stratigraphic relationships of the Bajocian-Callovian succession are illustrated in Figure 14.36. Along

Figure 14.37. McConnell Island Formation (MI) (shale-siltstone) overlying Sandy Point Formation (SP) and overlain by Ringnes Formation (R), western Axel Heiberg Island. Three T-R cycles occur within the McConnell Island Formation at this locality.

the southern margin of the basin two formations, McConnell Island (shale-siltstone) and Hiccles Cove (sandstone) encompass the interval. On the extreme basin edges only the Hiccles Cove is present and it is generally bounded by unconformities. The Hiccles Cove Formation thins and eventually disappears basinward owing to facies change to shale and siltstone of the McConnell Island Formation. In the west-central portion of the basin the McConnell Island is a member of the thick argillaceous Mackenzie King Formation which ranges in age from Bajocian to Valanginian. On the northwest margin of the basin the Bajocian-Callovian succession is represented by the McConnell Island Formation, the Hiccles Cove Formation and an unnamed shale unit which forms the basal portion of the Mackenzie King Formation (Fig. 14.36).

The McConnell Island Formation (Member) consists of medium to dark grey shale and siltstone with minor very fine grained sandstone interbeds (Fig. 14.37). Sideritic ironstone beds up to 20 cm thick occur throughout the formation and are most common in the lower portion. Over much of the basin, the McConnell Island is capped by a yellow weathering dolomite bed up to 0.5 m thick. The strata are interpreted to be of offshore shelf origin. The common occurrence of ironstone beds may indicate conditions of very slow deposition.

The Hiccles Cove Formation consists mainly of very fine- to medium-grained sandstone. The thickest development of the formation is in the northern Prince Patrick – Emerald Island area where it is up to 176 m thick. There, the lower portion of the formation consists of orange weathering, fine- to medium-grained, burrowed to massive, glauconitic sandstone. The upper portion consists of white, fine- to medium-grained, carbonaceous sandstone which is horizontally bedded to massive. Shale and siltstone units are interbedded with the sandstones and these lithologies become more common basinward (Fig. 14.36) where the Hiccles Cove consists of well defined coarsening-upward cycles. In more marginal areas thin coal seams are found within the upper portion of the formation and fining-upward cycles occur in the Eglinton Island area. The Hiccles Cove lithologies are mainly beach to shallow marine shelf in origin. The thin coal seams represent back-barrier deposits and the fining-upward cycles are likely of fluvial-deltaic plain origin. In the northern Ellef Ringnes Island area, the Hiccles Cove Formation consists of very fine- to fine-grained glauconitic sandstone of marine shelf origin.

In the Prince Patrick – Emerald Island area three T-R cycles have been delineated in the strata (Fig. 14.36). The first cycle consists of interbedded shale, siltstone and very fine grained sandstone and forms the lower portion of the McConnell Island Formation. These strata are Bajocian in age. The next cycle encompasses the upper McConnell Island and lower Hiccles Cove formations and is Bathonian in age. The uppermost cycle comprises the upper Hiccles Cove Formation and is Callovian. Over portions of the basin these three T-R cycles cannot be readily differentiated in the thin shale-siltstone succession which commonly comprises this interval. However, the three cycles are recognizable in the Ellef Ringnes Island area and in outcrop sections on western Axel Heiberg Island (Fig. 14.36, 14.37).

Bajocian-Callovian depositional history began with a sea level rise in earliest Bajocian. This event drowned the exposed margins of the basin and shorelines were extended to or beyond the present basin limits. Sediment influx into the Sverdrup Basin following the transgression was very meager and much of the basin remained a starved offshore shelf during the Bajocian. Arenaceous silts were deposited along the basin margins and vertical grain-size trends suggest a minor regression occurred in late Bajocian.

This was followed by a sea level rise in early Bathonian, causing farther starvation. By late Bathonian sediment influx to the basin had increased and the main area of deposition was along the Eglinton Island area which began to subside during this time interval. A secondary area of deposition was northern Ellef Ringnes Island, which received northerly derived sediments. Shallow shelf sands (Hiccles Cove Formation) were deposited on the basin margins with mud and silt farther offshore (McConnell Island Formation).

A sea level rise occurred in earliest Callovian, and during the following regression nearshore sands again were deposited along the basin margins with mud and silt farther

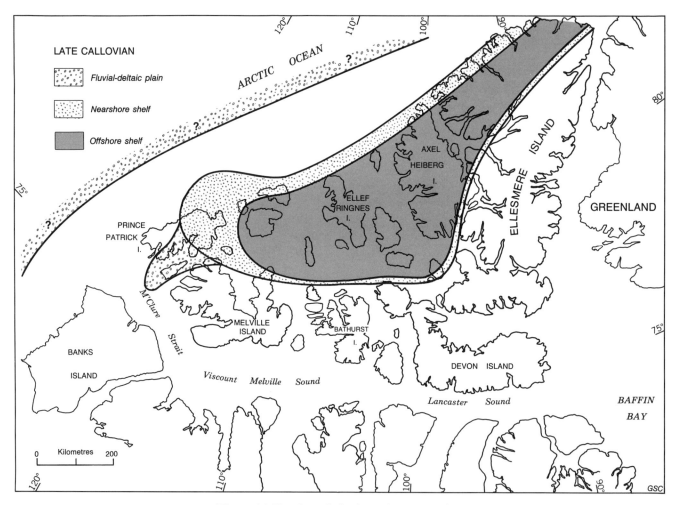

Figure 14.38. Late Callovian paleogeography.

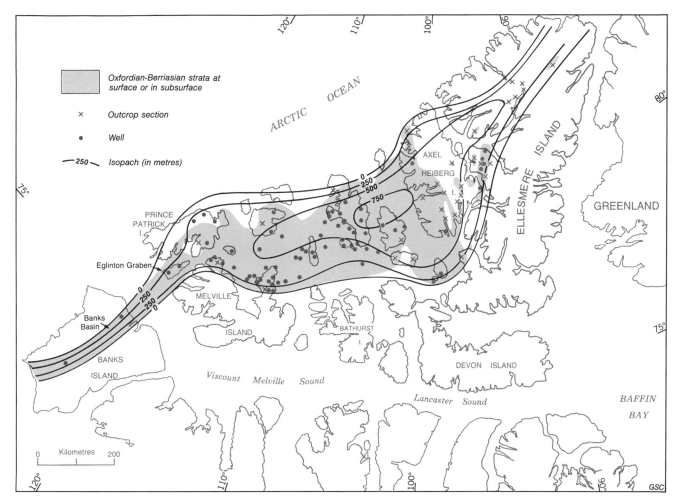

Figure 14.39. Distribution and isopach map of Oxfordian-Berriasian strata.

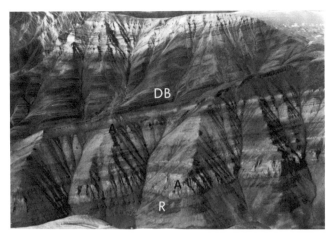

Figure 14.40. Oxfordian-Berriasian strata, western Axel Heiberg Island. R = Ringnes Formation; A = Awingak Formation; DB = Deer Bay Formation.

Figure 14.41. Coarsening-upward cycles of marine shelf origin, Awingak Formation, western Axel Heiberg Island.

offshore (Fig. 14.38). A major transgression in earliest Oxfordian terminated this depositional interval and caused widespread sediment starvation over much of the basin.

Oxfordian-Berriasian

Sediments of this time-stratigraphic interval are preserved mainly over a broad area which stretches from western Axel Heiberg to Prince Patrick Island. Subsurface control is widespread and surface sections have been measured at numerous localities. To the east the strata are preserved in scattered synclinal areas and both surface and subsurface control points are available (Fig. 14.39). Oxfordian to Berriasian sediments are also preserved in Banks Basin; two wells have penetrated the strata in this area.

The deposits exceed 400 m in thickness over most of the Sverdrup Basin and two depocentres were present (Fig. 14.39). The main one occurs in the central Axel Heiberg – Amund Ringnes Island area where thicknesses exceed 600 m. The other is in the western Sverdrup Basin and thicknesses in that area are up to 500 m. The depocentres were separated by a broad arch centred on King Christian Island. Thickness trends for these sediments in Banks Basin are speculative due to sparse control. The maximum recorded thickness is 396 m in a well (Orksut J-44) on south-central Banks Island.

Oxfordian-Berriasian strata are included in four formations. In the eastern and southern portions of the Sverdrup Basin the succession consists of the Ringnes (shale-siltstone), Awingak (sandstone), and Deer Bay (shale-siltstone) formations (Fig. 14.40). Over the northwestern portion of the basin, the Awingak Formation is not present because of a facies change, and the entire succession consists of shale and siltstone which are included within the Mackenzie King Formation. The same terminology is used for the strata in Banks Basin.

The Ringnes Formation unconformably overlies Middle Jurassic strata on the basin margins, and basinward the contact becomes conformable. The Ringnes consists mainly of dark grey to black, bituminous shale with thin siltstone interbeds which increase in abundance upward. Notably, the formation contains large, dolomitic concretions up to 4 m across. The environment of deposition of these strata is interpreted to be an offshore marine shelf which was well below storm wave base. The relatively high organic content of the shales suggests that the bottom waters were depleted in oxygen.

The Awingak Formation gradationally overlies the Ringnes Formation and for the most part consists of coarsening-upward cycles of shale, siltstone and sandstone. The cycles are up to 30 m thick and represent progradational shelf deposits (Fig. 14.41). The sandstones were deposited in a wide spectrum of environments ranging from beach to marine shelf below storm wave base. Many of the cycles are the result of migrating sand ridges on a shallow shelf. However, as discussed below, sea level variation also was a factor in the deposition of the Awingak and a few shale units can be correlated over much of the basin. A widely recognizable shale unit in the middle of the Awingak Formation in the eastern Sverdrup Basin has allowed the formation to be divided into three formal members: Cape Lockwood (sandstone), Hot Weather (shale-siltstone) and Slidre (sandstone) (Embry, 1986b).

Fining-upward cycles consisting of very fine- to medium-grained sandstone, siltstone, shale and minor coal also occur within the Awingak Formation. These deposits are of meandering stream origin and are found only on the southern margin of the basin.

The Deer Bay Formation conformably overlies the Awingak and consists of medium to dark grey, silty shale with thin siltstone interbeds. Thin, very fine grained, highly burrowed sandstones also are present within the formation. These strata are interpreted to be outer shelf deposits.

The Oxfordian-Berriasian portion of the Mackenzie King Formation consists of dark grey to black shale with interbedded siltstones. Dark, organically-rich shales tend to predominate in the lower portion. These lithologies represent offshore shelf sediments deposited well below wave base.

Figure 14.42 illustrates the Oxfordian to Berriasian stratigraphy in the western part of Sverdrup Basin. The Ringnes Formation forms a thin basal unit over the area. The overlying Awingak Formation is divided into five sandstone units separated from each other by shale-siltstone units. Most of the Awingak consists of marine shelf deposits

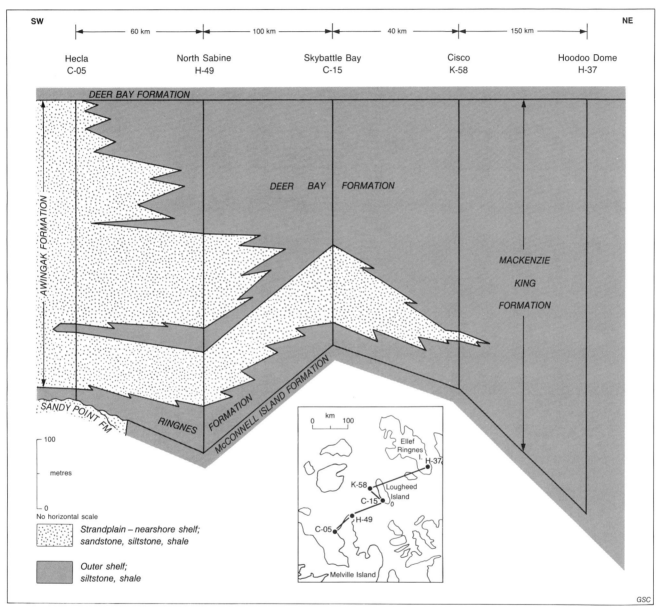

Figure 14.42. Southwest-northeast stratigraphic cross-section, Oxfordian-Berriasian strata, western Sverdrup Basin.

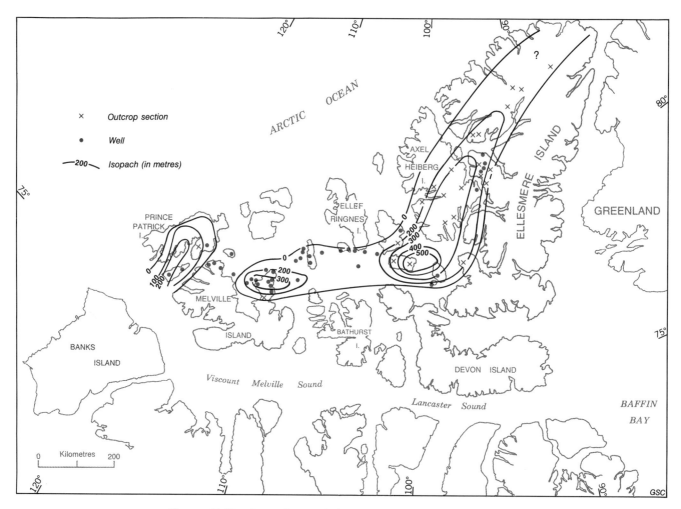

Figure 14.43. Isopach map, Awingak Formation, Sverdrup Basin.

although delta plain sediments become common on the basin margin. Each shale-sandstone couplet is a T-R cycle and five cycles have been identified in this interval. Cycle boundaries can be traced from the Awingak Formation into the Deer Bay Formation, with the contacts marked by clay-rich shales resting on siltstone or very fine grained sandstone. The ages of the cycles are early Oxfordian, late Oxfordian – early Kimmeridgian, late Kimmeridgian – early Tithonian, late Tithonian, and latest Tithonian – Berriasian. The second cycle records the maximum progradation of the Awingak Formation, and sandstones of succeeding cycles have progressively lesser extent.

In Banks Basin stratigraphic trends are poorly known because of sparse control (two wells). In the northern well the interval consists of 120 m of interbedded sandstone, siltstone and shale of the Awingak Formation. These rocks are unconformably overlain by the Isachsen Formation and represent the lower portion of the interval. The southern well contains 396 m of shale and siltstone assigned to the Mackenzie King Formation.

Figure 14.43, an isopach map of the Awingak Formation, shows that the Oxfordian-Berriasian sediments were derived mainly from the south. However, the existence of upper Tithonian to Berriasian sandstone units within the Mackenzie King Formation along the northwest margin of the basin indicates the presence of a northwestern source area which may have extended along the entire present day shelf. The main area of deposition appears to be in the southeastern re-entrant of the basin. Significant sediment influx also occurred on Sabine Peninsula and Prince Patrick Island in the western Sverdrup Basin.

Oxfordian-Berriasian depositional history began with a sea level rise in early Oxfordian and an offshore shelf environment extended onto the margins of the basin. Banks Basin was flooded by the transgression and was occupied by a north-south seaway during this time interval. Following the transgression, shallow shelf sands (lower Awingak Formation) prograded across offshore shelf mud and silt (Ringnes Formation) during the Oxfordian. Sediment influx was mainly in the southeastern part of Sverdrup Basin and maximum progradation occurred in that area. A shallow shelf with shifting sand ridges was established over much of the eastern and southern part of the basin at maximum regression (Fig. 14.44).

From Kimmeridgian to late Berriasian, three transgression-regression cycles occurred. Each cycle was initiated by a relative sea level rise, which was followed by basinward progradation of shallow shelf sands. The

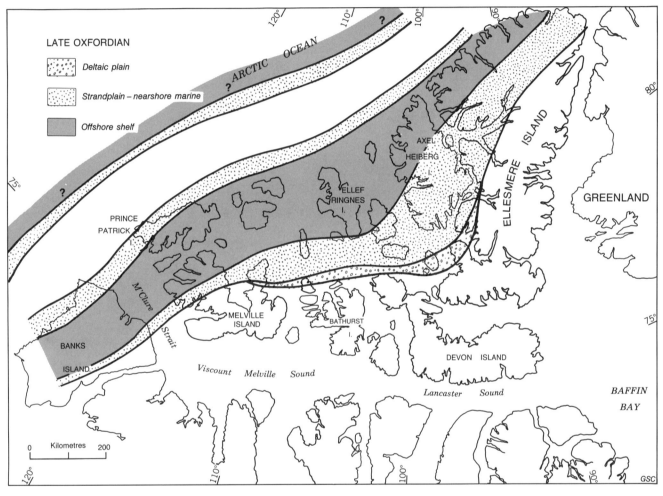

Figure 14.44. Late Oxfordian paleogeography.

CHAPTER 14

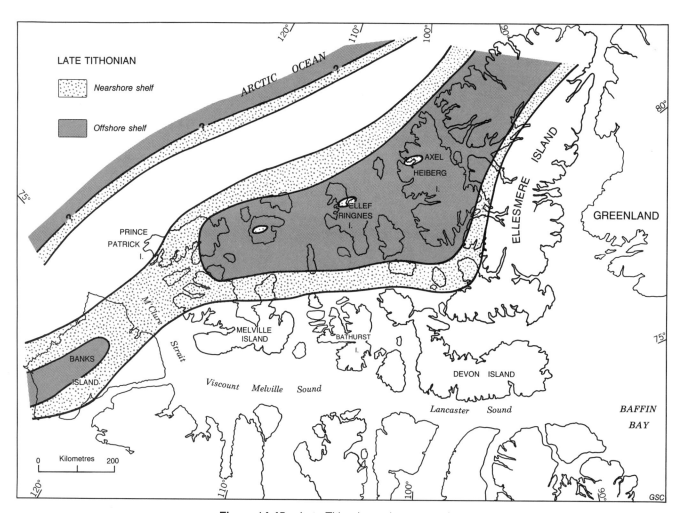

Figure 14.45. Late Tithonian paleogeography.

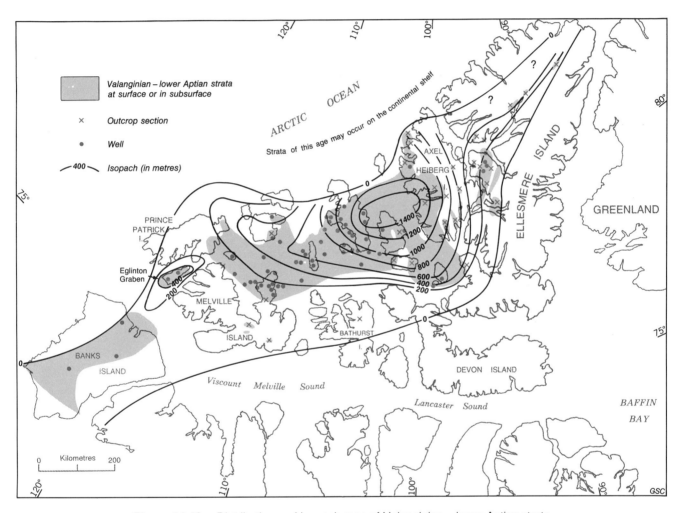

Figure 14.46. Distribution and isopach map of Valanginian – lower Aptian strata.

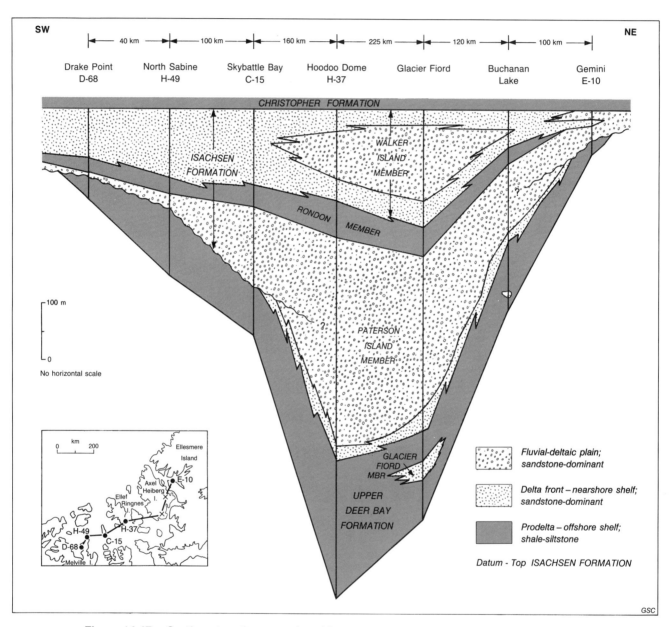

Figure 14.47. Southwest-northeast stratigraphic cross-section, Valanginian – lower Aptian strata.

Figure 14.48. Shale and siltstone of upper portion of Deer Bay Formation and overlying sandstone-dominant Isachsen Formation, western Axel Heiberg Island. PI = Paterson Island Member; R = Rondon Member.

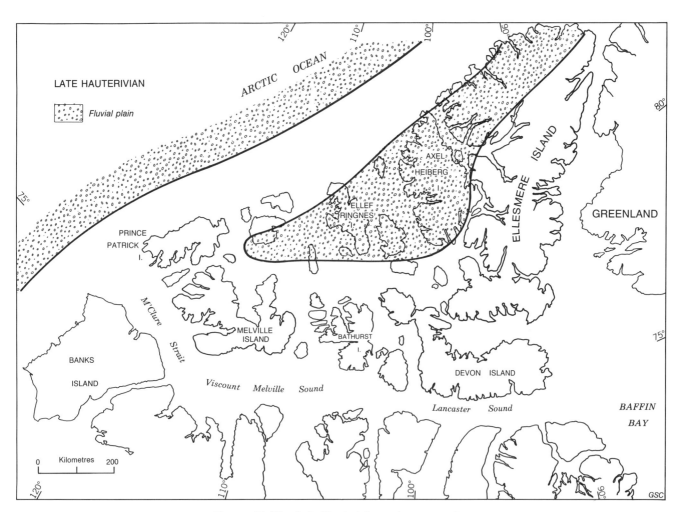

Figure 14.49. Late Hauterivian paleogeography.

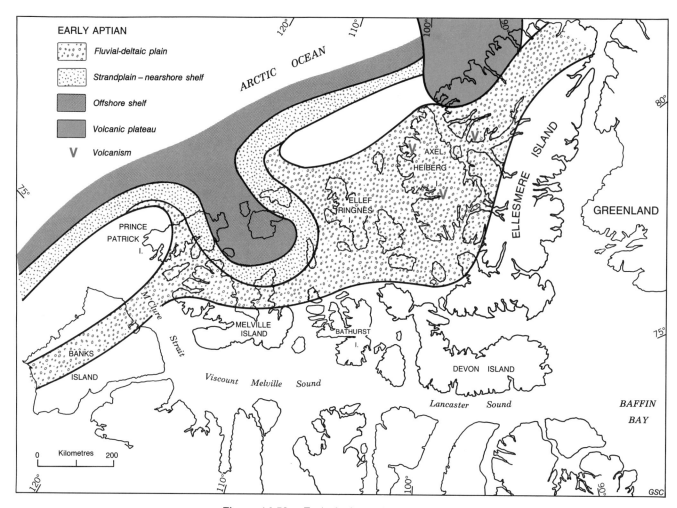

Figure 14.50. Early Aptian paleogeography.

maximum seaward extent of the sands decreased for each successive cycle, possibly indicating a gradual decline in sediment supply over that time interval. Upper Tithonian – Berriasian sands occur only on the basin margins, and the main depocentres were in the western Sverdrup Basin (Fig. 14.45). The proto-Amerasian Basin continued to develop throughout the Late Jurassic – earliest Cretaceous and it is interpreted that the basin was occupied by a seaway during this time. A significant transgression in early Valanginian initiated the next interval.

Valanginian – early Aptian

Strata of Valanginian – early Aptian age are preserved mainly in the central and western parts of Sverdrup Basin. The strata outcrop on most islands in this area and have also been penetrated by numerous wells (Fig. 14.46). In the eastern Sverdrup Basin the strata occur in widely separated synclinal areas, and surface sections have been measured at numerous localities. Also, five wells penetrate the strata in the Fosheim Peninsula area of Ellesmere Island. Outside of the Sverdrup Basin Valanginian to early Aptian strata are found in the Eglinton Graben – Banks Basin area where both surface and subsurface control is available, and on east-central Melville Island where scattered surface exposures are present (Fig. 14.46).

The main depocentre coincided with the centre of the Sverdrup Basin where the deposits exceed 1400 m (Fig. 14.46). The strata are abruptly truncated along the northwestern margin of the basin, and thickness trends in the northeastern portion of the basin are unknown due to lack of preservation of strata. A local depocentre occurred in the Eglinton Graben where the sediments are over 500 m thick.

The stratigraphic framework for the Valanginian – early Aptian interval is illustrated in Figure 14.47. The lower argillaceous portion of the succession is included within either the Deer Bay or Mackenzie King Formation. The basal contact is conformable and lies at the base of a clay-rich shale unit within one of the above-mentioned formations. On the basin margins the shale unit overlies the Awingak Formation and comprises the entire Deer Bay Formation. These argillaceous strata consist mainly of dark grey shale and siltstone with a variety of concretion types. The most distinctive of these concretions are small calcite rosettes known as "hedgehogs" or polar euhedrons (Kemper and Schmitz, 1975). The strata are interpreted to be of prodelta to offshore marine shelf origin.

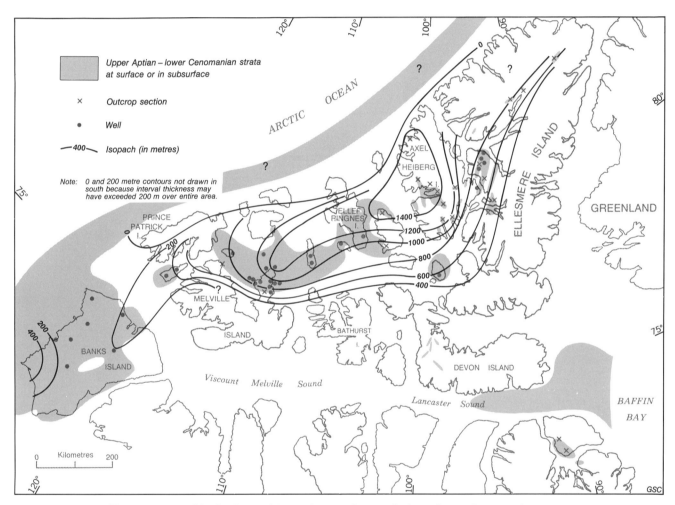

Figure 14.51. Distribution and isopach map of upper Aptian – Lower Cenomanian strata.

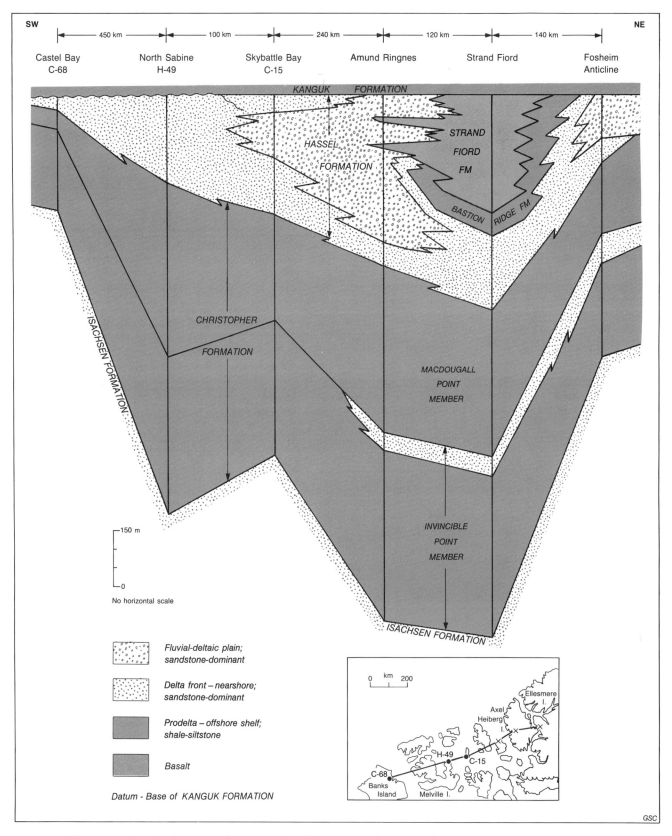

Figure 14.52. Northeast-southwest stratigraphic cross-section, upper Aptian – Lower Cenomanian strata, Banks Basin and Sverdrup Basin.

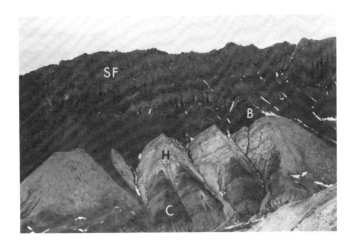

Figure 14.53. Upper Aptian to lower Cenomanian strata, western Axel Heiberg Island. C = Christopher Formation; H = Hassel Formation; B = Bastion Ridge Formation; SF = Strand Fiord Formation.

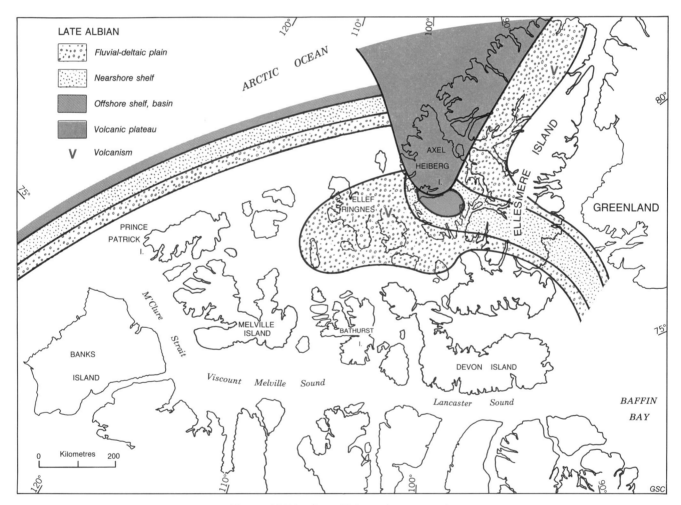

Figure 14.54. Late Albian paleogeography.

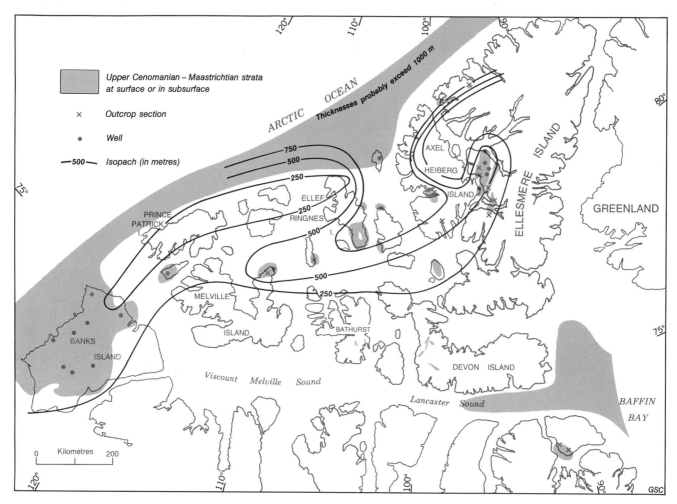

Figure 14.55. Distribution and isopach map of upper Cenomanian to Maastrichtian strata.

A sandstone unit exists within the upper part of the Deer Bay Formation in the southeastern portion of the Sverdrup Basin and is named the Glacier Fiord Member (Embry, 1985a). The member is up to 30 m thick and coarsens upward from very fine- to medium-grained. The sandstone has burrows, hummocky cross-stratification and horizontal bedding and is interpreted to be a shallow marine shelf deposit.

The overlying, arenaceous portion of the succession is assigned to the Isachsen Formation. Over much of the Sverdrup Basin the Isachsen is divided into three members, Paterson Island, Rondon and Walker Island (Fig. 14.47). In the central portion of the basin the Paterson Island Member conformably overlies shale and siltstone of either the Deer Bay Formation (Fig. 14.48) or the Mackenzie King Formation. For these areas the lower portion of the member consists of coarsening-upward cycles of delta front origin with the bulk of the member consisting of fining-upward cycles of delta plain origin. Coal seams up to 2 m thick occur within these strata. On the basin margins the Paterson Island Member unconformably overlies the Deer Bay Formation or older strata, and consists mainly of fluvial channel sandstones. Sandstones within the Paterson Island Member are commonly medium- to very coarse-grained, with quartz and quartzite pebbles as large as 30 cm being common. Basalt flows lie within the Paterson Island Member at two localities on Axel Heiberg Island (Embry and Osadetz, 1988). The Paterson Island Member is up to 880 m thick and thins dramatically onto the southern and eastern basin margins due to onlap onto the unconformity (Fig. 14.47).

The Rondon Member consists of interbedded grey shale, siltstone and very fine grained sandstone, arranged in coarsening-upward cycles. Overall, shale and siltstone predominate. The deposits are of offshore shelf origin.

The Walker Island Member conformably overlies the Rondon Member and is similar to the Paterson Island Member. The lower portion consists of coarsening-upward cycles of delta front origin, and these strata are overlain by fining-upward cycles of delta plain origin. On northern and west-central Axel Heiberg Island and northwestern Ellesmere Island, basalt flows and pyroclastics are included within the Walker Island Member (Embry and Osadetz, 1988).

There are some areas where the Isachsen Formation is not divisible into the three members. On northern and central Ellesmere, Graham Island, and in the Banks Basin area, the Rondon Member cannot be discerned and equivalents of all three members are present within an

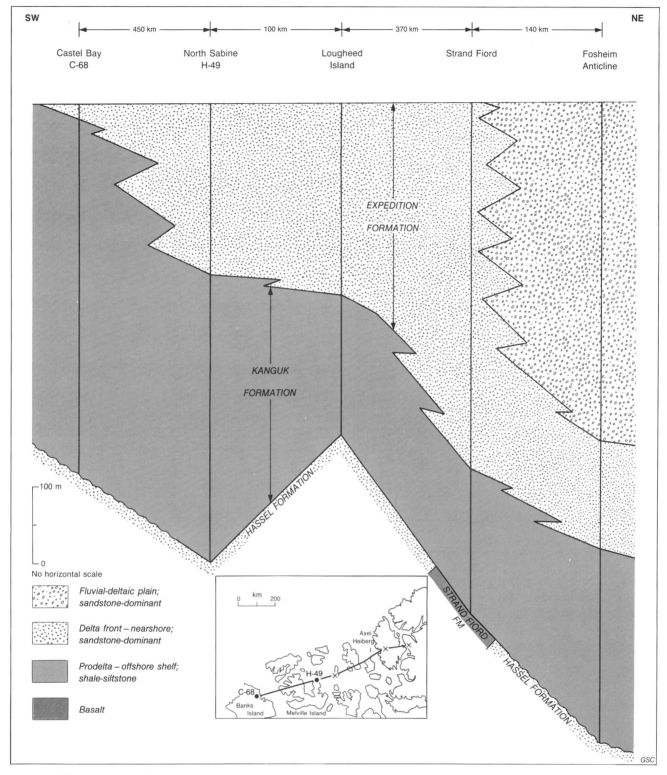

Figure 14.56. Northeast-southwest stratigraphic cross-section, upper Cenomanian – Maastrichtian strata.

Figure 14.57. Black, bituminous shale of Kanguk Formation overlying white sandstone of Hassel Formation, western Ellesmere Island.

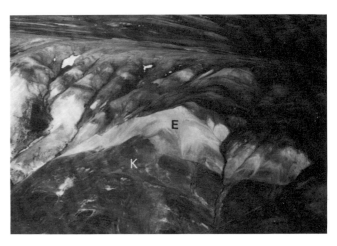

Figure 14.58. Shale and siltstone of upper portion of Kanguk Formation (K) conformably overlain by white sandstone of Expedition Formation (E), Ellesmere Island.

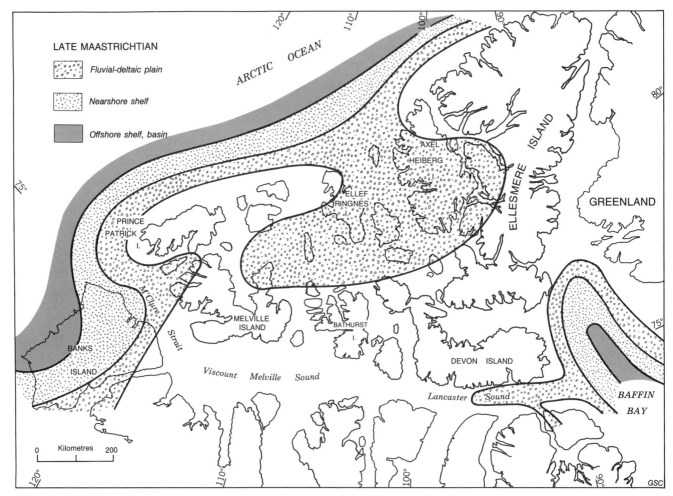

Figure 14.59. Late Maastrichtian paleogeography.

undivided, sandstone-dominant formation. In the southwestern portion of the basin (Mackenzie King and Emerald islands), the Walker Island Member grades to shale and siltstone, and equivalents of the Rondon and Walker Island members are assigned to the basal Christopher Formation.

Three T-R cycles are recognized in the Valanginian – early Aptian succession. The first cycle is of early Valanginian age and includes the Glacier Fiord Member and underlying shale and siltstone strata. The second cycle is of late Valanginian to early Barremian age and includes Deer Bay strata above the Glacier Fiord Member and the Paterson Island Member of the Isachsen Formation. The third T-R cycle is late Barremian – early Aptian in age and comprises the Rondon and Walker Island members of the Isachsen Formation. Rocks representing the first two cycles are present only within the Sverdrup Basin. The third cycle is represented in the basin and also in the Eglinton Graben – Banks Basin area and on the Arctic Platform.

Paleocurrent measurements and facies relationships indicate that the main source area lay to the south and east of Sverdrup and Banks basins. Small source areas were also present to the northwest.

This depositional interval began with a major sea level rise and transgression in earliest Valanginian. Shorelines were established south and east of the present limits of the Sverdrup Basin and an offshore shelf was widespread. During the following regression (in early Valanginian) shallow shelf sands (Glacier Fiord Member) prograded over offshore mud and silt (Deer Bay Formation) and sand supply was highest in the southeast where the sands extended into the basin.

A rise of sea level in late Valanginian again established an offshore shelf over the entire Sverdrup Basin. Following this, clastic supply increased greatly and major deltaic complexes (Paterson Island Member) prograded into the basin from the southern and eastern margins. The margins of the basin were exposed and by late Hauterivian the Sverdrup Basin was occupied by a fluvial plain surrounded by marginal uplifts (Fig. 14.49). Major rifting is interpreted to have affected the Amerasian Basin during this time and Sverdrup Rim, the arch between the two basins, underwent major uplift.

A major sea level rise and transgression occurred in mid-Barremian and much of the Sverdrup Basin was transformed into an offshore shelf (Rondon Member). Fluvial to estuarine sediments also began to accumulate in Banks Basin and on the Arctic Platform at this time. During the late Barremian and early Aptian, deltas again prograded into the eastern and central Sverdrup Basin and a fluvial deltaic plain environment (Walker Island Member) covered these portions of the basin at maximum regression (Fig. 14.50). Volcanic flows were extruded onto the delta plain in the northern Ellesmere Island – Axel Heiberg Island area and basic dykes and sills were emplaced over much of the Sverdrup Basin.

Late Aptian – early Cenomanian

Strata of this interval are preserved mainly in the western part of Sverdrup Basin and in Banks Basin. These are areas of subdued topography and poor exposure. Most wells begin within the lower portion of the succession. In the eastern part of Sverdrup Basin the strata lie in scattered synclinal areas from western Axel Heiberg Island to northeastern Ellesmere Island (Fig. 14.51). A few outliers are present south of the Sverdrup Basin on Melville, Devon and Bylot islands.

Upper Aptian – lower Cenomanian strata are thicker than 1400 m in the central part of Sverdrup Basin (Fig. 14.51). Such thicknesses also prevail on northern Axel Heiberg Island where a thick volcanic pile (800 m) caps the succession. Thicknesses for most of the northeastern portion of the Sverdrup Basin are unknown. A zero edge has been established on northwestern Ellesmere Island. In Banks Basin, thicknesses are usually less than 300 m except in the southeastern portion where the deposits are more than 400 m thick. Considerable thicknesses may be present on the shelves and slopes of the Arctic Ocean and Baffin Bay but no data are available to confirm this.

The stratigraphic framework for this interval is illustrated in Figure 14.52, a stratigraphic cross-section which extends from Banks Basin to the eastern side of Sverdrup Basin. The lower portion of the succession consists of argillaceous strata of the Christopher Formation, which rests conformably on the Isachsen Formation.

Over much of its extent the Christopher Formation is divided into two members. The lower member (Invincible Point), consists mainly of dark grey, silty shale and siltstone of offshore marine origin. Very fine- to medium-grained sandstone beds occur within the member and are most common in the uppermost portion. The sandstones are characterized by hummocky cross-stratification and burrows and are of marine shelf origin. Over most of the basin the sandstones are quartz-rich, but on northern Axel Heiberg Island they consist mainly of volcanic detritus.

Clay-rich shales of the basal Macdougall Point Member conformably overlie the Invincible Point Member. Silt content gradually increases upward in the member and thin units of very fine- to medium-grained sandstone appear in the uppermost portion. The strata are of prodelta and offshore marine shelf origin.

The sandstone-dominant Hassel Formation overlies the Christopher Formation, with the contact everywhere being conformable except at a few localities on northwestern Ellesmere Island. In most areas of the Sverdrup Basin the lowermost portion of the Hassel Formation consists of delta front, coarsening-upward cycles. The bulk of the formation consists of interbedded fine- to coarse-grained sandstone, siltstone and shale with minor coal of delta plain origin. In exception to this, the Hassel Formation of western Axel Heiberg Island and central Ellesmere Island consists entirely of coarsening-upward cycles of marine shelf origin. Fine- to medium-grained sandstone dominates the sections, but near the basin centre a thick shale-siltstone interval (90 m) occurs in the middle of the formation.

On the flanks of the Sverdrup Basin and in Banks Basin the Hassel Formation forms the uppermost portion of this interval. It is overlain by the Kanguk Formation with the contact being unconformable. In the central portion of the Sverdrup Basin two additional stratigraphic units occur in the upper portion of the interval, the Bastion Ridge and Strand Fiord formations (Fig. 14.52).

The Bastion Ridge Formation consists of dark grey, silty shale of offshore marine shelf origin and occurs on western Axel Heiberg Island. A very fine- to fine-grained, burrowed to hummocky crossbedded sandstone unit caps the formation in its southernmost occurrences.

The Strand Fiord Formation consists mainly of subaerial basalt flows and occurs on west-central and northwestern Axel Heiberg Island (Fig. 14.53). On central Axel Heiberg the Strand Fiord conformably overlies the Bastion Ridge Formation and is laterally equivalent to Bastion Ridge shale and sandstone (Ricketts et al., 1985; Embry and Osadetz, 1988). The formation thickens northward to 789+ m and rests unconformably on the Hassel Formation in the north.

Three T-R cycles occur within this interval:

1. late Aptian – early Albian: Invincible Point Member, Christopher Formation,

2. middle Albian – late Albian: Macdougall Point Member and lower Hassel Formation,

3. late Albian – early Cenomanian: Bastion Ridge Formation and Strand Fiord Formation.

The depositional history of this interval began with a major transgression in late Aptian and earliest Albian which resulted in an offshore shelf environment (Christopher Formation) over Sverdrup and Banks basins. Regression occurred during early Albian and a shallow shelf sand prograded basinward. Source areas were mainly to the east and south. A volcanic source terrane appears to have been present in the northeast.

A transgression in middle Albian re-established the offshore shelf environment over the area. This was followed by a major regression, and a delta complex (Hassel Formation) prograded into the west-central part of Sverdrup Basin during middle and late Albian.

A late Albian transgression created an offshore shelf environment in the central part of Sverdrup Basin (Bastion Ridge Formation), but deltaic and shallow shelf sands dominated the remainder of the basin. During the subsequent regression, shallow marine sands prograded basinward and an extensive volcanic pile (Strand Fiord Formation) built up in the northeastern portion of the basin (Fig. 14.54).

During this interval sediments were probably deposited in marine rift basins of the proto-Arctic Ocean. Major deltas may have prograded basinward from the uplifted Arctic Platform and Sverdrup Rim which separated Sverdrup Basin from the rift basins of the proto-Arctic Ocean. The Labrador Sea – Baffin Bay rifts were initiated during this time interval and were the site of a seaway which linked Sverdrup Basin and the Atlantic Ocean (Fig. 14.54).

Late Cenomanian – Maastrichtian

Strata of late Cenomanian to Maastrichtian age occupy a variety of geological settings but are sparsely preserved. Within the Sverdrup Basin the strata occur in numerous, relatively small, synclinal areas (Fig. 14.55). In the central and western portions of the basin the strata are mainly in outcrop, but in the east they commonly lie beneath thick Tertiary strata. To the southwest, sediments of this interval are present on southern Eglinton Island and over much of Banks Island where wells have penetrated the succession. On the Arctic Platform the strata are preserved in small grabens on Bathurst, Cornwallis, Devon, Somerset and Bylot islands. Upper Cenomanian – Maastrichtian deposits probably also are present on the continental shelves and slopes of the Arctic Ocean and Baffin Bay, and in Lancaster Sound (Hea et al., 1980).

Thickness trends are difficult to determine for this interval because of the few and widely separated control points. Thicknesses in excess of 250 m occur over the Sverdrup and Banks basins and on Bylot Island. The main depocentre appears to have been in the eastern part of the Sverdrup Basin area where thicknesses exceed 750 m (Fig. 14.55). Thick successions may also occur on the continental shelves and in the Lancaster Sound area.

The stratigraphic framework for the upper Cenomanian – Maastrichtian succession is illustrated in Figure 14.56, a stratigraphic cross-section which extends from the eastern Sverdrup Basin to Banks Basin. The Kanguk Formation comprises the argillaceous portion of the succession and it usually rests on the Hassel Formation, with the contact varying from conformable to unconformable (Fig. 14.57). On Banks Island the Kanguk Formation rests unconformably on strata ranging in age from Cretaceous to Devonian. On Meighen Island the Kanguk pelites unconformably overlie Upper Triassic strata.

The lower portion of the Kanguk Formation usually consists of black, bituminous, papery shales which represent starved, offshore shelf deposits. Bentonite beds frequently occur within these strata. The shales of the Kanguk become lighter in colour, less bituminous and siltier upward, and the upper portion of the formation consists mainly of prodelta deposits. Sandstone units, which are commonly glauconitic and burrowed, lie within marginal sections of the Kanguk and represent shallow marine shelf deposits. A 50 m thick marine sandstone unit in the upper part of the Kanguk Formation of Eglinton Island was named the Eglinton Member (Plauchut and Jutard, 1976).

The sandstone-dominant Expedition Formation of the Eureka Sound Group (Ricketts, 1986) gradationally overlies the Kanguk Formation (Fig. 14.58) and includes uppermost Cretaceous strata. The lowermost portion of the formation consists of coarsening-upward cycles of delta front origin. Overlying strata consist of fining-upward cycles of fluvial-deltaic origin, and coal seams occur in the Fosheim Peninsula area. A prominent shale unit of prodelta origin occurs low in the Expedition Formation on western Axel Heiberg Island.

In general, the contact between the Kanguk Formation and the Expedition Formation is younger westward. The Cretaceous-Tertiary boundary generally occurs within or at the top of the Expedition Formation and varies from conformable to unconformable.

On northwestern Ellesmere Island this interval is represented by a 1000 m thick succession of volcanic flows and pyroclastics with lesser amounts of sandstone, siltstone, shale and coal (Hansen Point volcanics, Trettin and Parrish, 1987). The volcanic strata are mainly basaltic in composition although rhyolite flows and dacitic and rhyolitic pyroclastics are also present (Embry and Osadetz, 1988). Clastic strata are most common in the lower portion of the succession and include both fluvial and marine shelf deposits. Dinoflagellate assemblages within a few of the shale units are very similar to those of the Kanguk Formation, which is Turonian to Campanian in age. Pollen and spore assemblages suggest the Hansen Point volcanics are as young as Maastrichtian (E. Davies, pers. comm., 1986). These strata rest unconformably on the Nansen Formation (Carboniferous-Permian).

Two T-R cycles are tentatively recognized within the Upper Cenomanian – Maastrichtian succession. The first is of late Cenomanian to early Campanian age, and the upper boundary is placed at the base of the shale tongue within the Expedition Formation and at the top of the Eglinton Member. The second cycle is late Campanian – Maastrichtian, and its top is defined by the base of the Strand Bay Formation, a Paleocene shale unit in the Eureka Sound Group (Ricketts, 1986).

The upper Cenomanian – Maastrichtian interval began with a transgression which established an offshore shelf environment over the Sverdrup Basin. Sediment supply was low, and bituminous muds and volcanic ash were the main deposits over much of the basin. Volcanism dominated the extreme northeastern portion of the basin. In early Campanian, deltas prograded southward and westward from uplifted source terranes in the northeastern portion of the Sverdrup Basin. This uplift and associated increase in sediment supply heralded a new tectonic and sedimentary regime, which lasted through to the Eocene and culminated in the final phase of the Eurekan Orogeny.

Mid-Campanian transgression drowned most of the area except for the deltaic complex in the northeast. Regression occurred during late Campanian and Maastrichtian, and deltaic sediments prograded across the Sverdrup Basin. Much of the area was either undergoing erosion or was occupied by a fluvial-deltaic plain at the close of the Mesozoic, with the sea occurring on the continental shelves and over most of Banks Basin (Fig. 14.59). The Baffin Bay rift basin continued to develop during this time interval and an arm of the rift system formed in the Lancaster Sound – Bylot Island area.

FACTORS AFFECTING MESOZOIC STRATIGRAPHY

Mesozoic stratigraphy and sedimentation were affected by a number of factors which include tectonic subsidence and uplift, magmatism, salt movement, eustatic sea level changes, sediment supply, and climatic changes. In this section each factor is briefly examined in the context of the entire Mesozoic succession and apparent trends are interpreted.

Tectonic subsidence and uplift

The Mesozoic isopach map (Fig. 14.1) provides an overall view of Mesozoic subsidence in the Arctic Islands. This subsidence is the sum of tectonic subsidence, compaction and load subsidence, and previous authors (Sweeney, 1977; Balkwill, 1978) who have discussed the subsidence history of the Sverdrup Basin have based their interpretations on

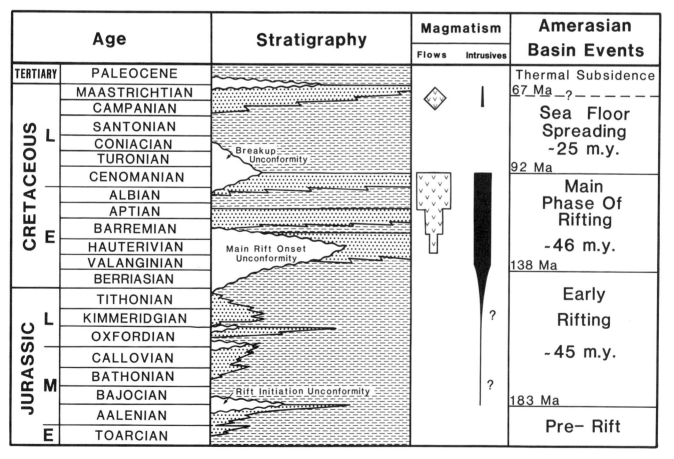

Figure 14.60. Evolution of Amerasian Basin compared with Jurassic-Cretaceous stratigraphy and volcanism of Canadian Arctic Islands.

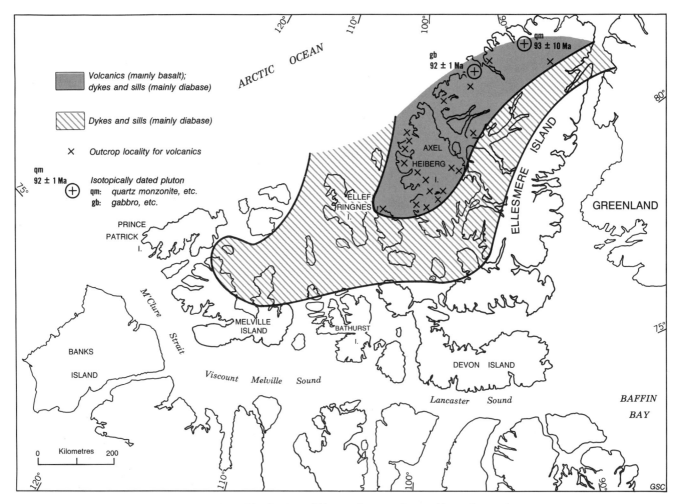

Figure 14.61. Distribution of Mesozoic extrusive and intrusive rocks.

Figure 14.62. Angular unconformity between Barrow Formation (B) and Fosheim Member, Heiberg Formation (F), eastern flank of Thompson Diapir, western Axel Heiberg Island.

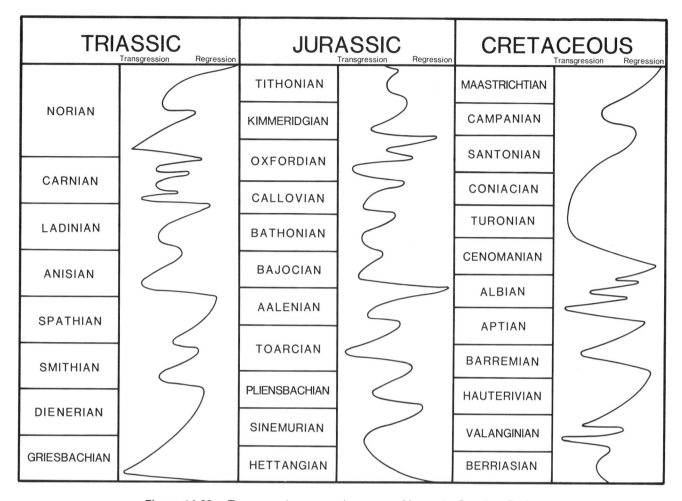

Figure 14.63. Transgressive-regressive curves, Mesozoic, Sverdrup Basin.

total subsidence trends. Recent studies of basin subsidence (e.g. Watts, 1982) emphasize that interpretations of basin evolution should be based on tectonic subsidence trends rather than total subsidence. The tectonic subsidence has been estimated for the Sverdrup Basin by Stephenson et al. (1987). The resulting subsidence curve has been interpreted as gradually decaying thermal subsidence from Triassic to earliest Cretaceous, renewed subsidence due to rifting and crustal stretching in Early Cretaceous, and thermal subsidence in Late Cretaceous. The thermal subsidence during Triassic and Jurassic was a continuation of a thermal subsidence regime which began in Early Permian and followed an interval of rifting and high subsidence in Carboniferous and Early Permian.

Tectonic uplift on the margins of the Sverdrup Basin has been identified by examining the effects of subaerial unconformities on the basin flanks. Most unconformities have a limited extent along the margins and are interpreted to be due either to eustatic sea level falls or minor tectonic events. Unconformities of latest Aalenian, late Valanginian – Hauterivian, mid-Cenomanian and late Maastrichtian age are of greater magnitude and were associated with normal faulting and/or dyke and sill intrusion. These unconformities are obviously tectonic in origin and are probably related to tectonic events in the adjacent Amerasian Basin (Fig. 14.60). The late Aalenian unconformity resulted from early rifting events related to the initiation of the Amerasian Basin. The late Valanginian – Hauterivian unconformity is interpreted to be due to the onset of a major rifting interval, which culminated in the opening of the ocean basin. The mid-Cenomanian unconformity, which is most prevalent on the northwestern margin of Sverdrup Basin, is interpreted to coincide with the onset of seafloor spreading and is therefore a cratonward representation of the "break-up unconformity" (Falvey, 1974) on the Arctic shelf. Overlying strata, which dip gently seaward over faulted Lower Cretaceous and older strata (Meneley et al., 1975), were deposited during the interval of spreading when the margins of the Amerasian Basin were undergoing thermal subsidence. The latest Maastrichtian unconformity may mark the cessation of seafloor spreading in the Amerasian Basin, although uplifts related to the early stages of the Eurekan Orogeny were present at this time and the unconformity may be tied to orogenic activity.

It must be noted that the interpreted timing of the opening of Amerasian Basin outlined above differs from previous interpretations (e.g. Grantz and May, 1983) in which seafloor spreading is interpreted to have occurred mainly in Early Cretaceous time. The timing for the opening of Amerasian Basin presented herein is based on the

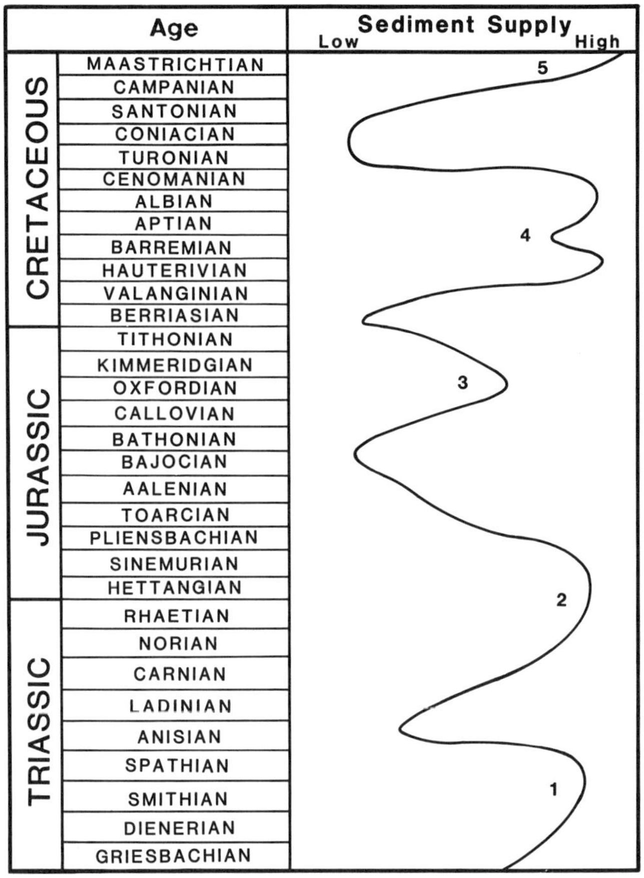

Age	Sediment Supply

Figure 14.64. Qualitative sediment supply rate curve, central Sverdrup Basin.

application of Falvey's (1974) empirical model which related the structure and stratigraphy of a continental margin to the stages of formation of the adjacent oceanic basin. The Cenomanian unconformity, which can be traced along the Arctic continental margin from Ellesmere Island to the Mackenzie Delta, compares closely with the "break-up unconformity" of Falvey (1974).

Magmatism

Mesozoic lava flows and breccias (mainly basaltic), and mafic dykes and sills are present within the Sverdrup Basin. The lavas are interbedded with strata ranging in age from Hauterivian to Maastrichtian and are restricted to the northeastern part of the basin (Fig. 14.61). The oldest flows occur in the lower portion of the Isachsen Formation and outcrop on Axel Heiberg Island. On western Axel Heiberg Island, the number of flow units in the Isachsen Formation increases northward to a maximum aggregate thickness of 230 m in the Bunde Fiord area, where the flows are interbedded with delta plain strata (Fischer, 1984). On northwestern Ellesmere Island, a single 20 m thick basalt unit is present, and it is near the top of the Isachsen Formation. Most of the basalt flows are very finely crystalline to aphanitic. Thick, extrusive basalt breccias are present on west-central Axel Heiberg Island (Ricketts, 1985).

Basalt flows also occur within the upper Albian – lower Cenomanian Hassel Formation, and have been described from northern Amund Ringnes Island (Balkwill, 1983) and northeastern Ellesmere Island (Osadetz and Moore, 1988). On Amund Ringnes Island two flow units are present and consist mainly of basalt breccia. On northeastern Ellesmere Island, two aphanitic basalt flow units, with a maximum aggregate thickness of 45 m, occur interbedded with delta plain strata of the Hassel Formation.

Lavas stratigraphically equivalent to those in the Hassel Formation comprise the Strand Fiord Formation (Souther, 1963). The Strand Fiord Formation outcrops on western Axel Heiberg Island and consists mainly of subaerial basalt flows. Volcaniclastic strata become common along the southeastern margin of the formation where it intertongues with marine shales of the Bastion Ridge Formation (Ricketts et al., 1985). The formation thickens northward from its depositional edge on southern Axel Heiberg to 789+ m on northwestern Axel Heiberg (Embry and Osadetz, 1988).

Basalt and rhyolite flows and pyroclastics assigned to the informal Hansen Point volcanics outcrop on northwestern Ellesmere Island (Fig. 14.61) and rest unconformably on upper Paleozoic strata. These are the youngest Cretaceous flows in the Sverdrup Basin and are Turonian to Maastrichtian in age. Rhyolite, northwest of Yelverton Bay, dated at 88 +20/-21 Ma, contains a high proportion of zircon inherited from an 1.15 Ga old source. This suggests that the felsic rocks in this bimodal volcanic suite (and probably also in the pluton discussed below) were in part derived from the gneissic basement of the Pearya Terrane (Trettin and Parrish, 1987).

Mafic dykes and sills have a much broader aerial extent than the lavas (Blackadar, 1964). Most of the radiometric dates obtained from these rocks indicate that they were intruded contemporaneously with the extrusion of the basalt flows (Balkwill, 1983). However, Middle to Late Jurassic ages have been obtained from a few samples suggesting intrusion may have begun as early as Bajocian (Balkwill, 1978). The amount of intrusive material increases northeastward in the Sverdrup Basin. Sills occur in strata ranging in age from Carboniferous to early Late Cretaceous; to the southwestward the sills are restricted to progressively older strata (Balkwill, 1978). Individual sills are up to 150 m thick and aggregate thicknesses can exceed 1500 m in sections on northern Axel Heiberg Island. Dykes that fed the sills and lavas can be traced for up to 75 km. Jollimore (1986) recognized two distinct dyke trends: one subparallel to the basin axis with the other having a north-south trend. He suggested that two episodes of dyke intrusion occurred in the Cretaceous.

Apart from ubiquitous dykes and sills, gabbroic and granitic plutons of Late Cretaceous age are present in northern Ellesmere Island (Trettin and Parrish, 1987). A pluton, about 28 km long, on Wootton Peninsula (Fig. 14.61; for geographic names, see Fig. 1, in pocket), has been emplaced into a northeast-trending fault in the Pearya Terrane. It is composed of a variety of gabbros but also includes felsic rocks (granite, granodiorite, quartz monzonite), and a hybrid suite. A U-Pb zircon age of 92 ± 1 Ma indicates crystallization in the Cenomanian. A smaller pluton of quartz monzodiorite, quartz monzonite, and granodiorite on Marvin Peninsula has yielded comparable K-Ar (hornblende) ages of 94.2 ± 10 Ma and 91.6 ± 9.6 Ma.

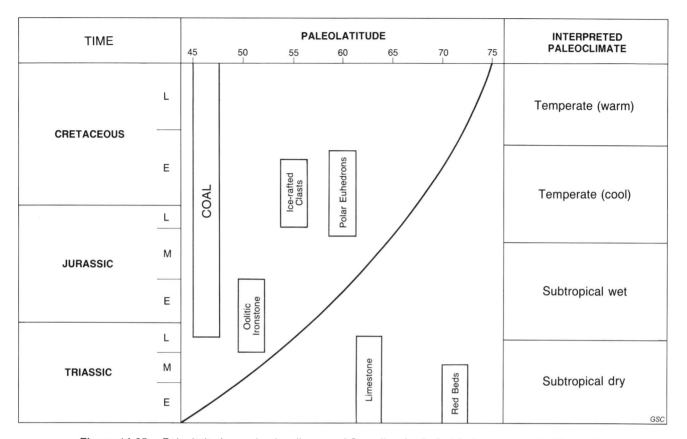

Figure 14.65. Paleolatitudes and paleoclimates of Canadian Arctic Archipelago during the Mesozoic.

Figure 14.66. Large clast of indurated very fine grained sandstone in friable, medium grained sandstone of Awingak Formation, Lake Hazen, northern Ellesmere Island. The clast is interpreted to have been transported into the marine shelf environment by floating ice. Hammer handle for scale.

Mesozoic volcanism in the Sverdrup Basin is interpreted to be related to the opening of the adjacent Amerasian Basin (Fig. 14.60). Volcanism occurred from Hauterivian to Maastrichtian and was coincident with both the main rifting and drifting phases of the Amerasian Basin. Intrusive activity apparently began earlier and included the early rifting interval. The localization of volcanism in the northeastern part of the Sverdrup Basin, in combination with the increase of intrusive activity in that direction, may indicate the presence of a hot spot in that area during Cretaceous (Embry and Osadetz, 1988). Such a hot spot may have given rise to the oceanic Alpha Ridge during the seafloor spreading (Forsyth et al., in press). The tectonic subsidence analysis discussed above indicates that rifting and crustal stretching also prevailed in the Sverdrup Basin in Early Cretaceous. Dyke and sill intrusion was also likely related to this extensional event.

Salt movement

A thick unit of salt assigned to the Otto Fiord Formation (Nassichuk and Davies, 1980) was deposited over much of the Sverdrup Basin in mid-Carboniferous. The salt was subsequently mobilized and a variety of salt structures including diapirs, ridges, salt-cored anticlines and lag domes developed within the basin. Balkwill (1978, 1983) describes these features and postulates that the salt structures began forming in Early Triassic and that pulses of growth occurred during times of high sediment supply. Stratigraphic data from outcrops, wells and seismic sections indicate that the structures were definitely growing by Middle Triassic and perhaps had begun by Early Triassic (Fox, 1983). The syndepositional structures grew intermittently during the Mesozoic. The last major pulse of mobilization took place as a result of compression during the Eurekan Orogeny in early Tertiary.

Abrupt thinning of portions of the Mesozoic succession is found on the flanks of diapirs and over the crest of salt-cored anticlines (Fox, 1983). Subaerial erosion occurred on the crest of some of the structures during intervals of accelerated upward growth (Fig. 14.62). Local thickening in rim synclines is also apparent on the flanks of some structures.

The stratigraphic effects of Sverdrup Basin salt structures usually extend out only a few kilometres from the flanks. However, in the basin centre, where there are numerous structures, the cumulative effects are major and it is often impossible to determine a "regional thickness" for a stratigraphic unit in that area.

Eustatic sea level changes

The Mesozoic succession of the Arctic Islands is divisible into 30 regional transgressive-regressive cycles which represent relative sea level changes. The cycles are closely comparable in style and magnitude to the third-order cycles or depositional sequences of Vail et al. (1977), which were interpreted to have originated mainly due to eustatic sea level changes. Embry (1988) has postulated that the Triassic cycles, which are representative of the Mesozoic cycles, were generated by an interplay of varying rates of eustatic sea level change, subsidence, and sediment supply. In this model, major transgressions coincided with rapid eustatic sea level rises and regression occurred in the intervening time intervals under conditions of eustatic sea level fall, stillstand and slow rise.

The cycles are also explainable by variations in the rates of subsidence or uplift and sediment supply without appealing to sea level variations. Certainly the significant changes in sedimentary regimes, source areas, and patterns of uplift and subsidence across many of the cycle boundaries indicate that tectonism was a key factor in the origin of the cycles. However, eustatic sea level changes are included within the model because many of the transgressions recognized in the Sverdrup Basin have also been recognized in other areas of the world (Embry, 1982, 1988; Hallam, 1978; Vail et al., 1977). Obviously many more global data are needed before the role of eustatic sea level changes can be evaluated. Figure 14.63 summarizes the major Mesozoic transgressions recorded in the Arctic Islands.

Sediment supply

Most of the sediment deposited in the Sverdrup Basin during the Mesozoic was derived from cratonic areas to the south and east of the basin. The rate of sediment supply varied markedly during the Mesozoic Era, and Figure 14.64 illustrates a qualitative sediment supply curve for the central Sverdrup Basin during this era. Five intervals of high sediment influx are apparent and these are characterized by thick intervals of prodelta to slope, argillaceous deposits and arenaceous, delta front and delta plain deposits. The intervening intervals of low sedimentation rates consist mainly of thin, marine shelf deposits.

Embry (1982) and Balkwill (1978, 1983) inferred that high rates of sedimentation reflected epeirogenic uplift of the craton. At such times, large, vigorous river systems would have been established on the craton and would have flowed northward into the Sverdrup Basin. Conversely, the intervals of low sedimentation rate would coincide with times when the craton was relatively low-lying with only small, local rivers reaching the basin. Thus the general epeirogenic history of the northern North American craton during the Mesozoic is reflected in the sediment fill of the Sverdrup Basin. Embry (1982) related the Late Triassic – Early Jurassic epeirogenic uplift of the craton to the rifting phase of the central Atlantic Ocean.

Paleoclimatology

During the Mesozoic, the Canadian Arctic Archipelago gradually shifted northward from about 40°N in earliest Triassic to about 70°N at the close of the Cretaceous (Irving and Irving, 1982). Paleoclimatic indicators within the Mesozoic succession agree with this inferred northward drift (Fig. 14.65).

Red shale and siltstone of overbank origin indicate a seasonally dry, hot, savanna-type climate (Van Houten, 1973) and such strata occur only in Lower and lower Middle Triassic strata. Overbank deposits within Upper Triassic and younger strata are generally dark grey, carbonaceous and often contain thin coal seams (Ricketts and Embry, 1984). These strata were deposited under more humid conditions.

Marine, chemical sediments which were usually deposited during times of sea level rise when clastic supply was low, and which occur at the base of most T-R cycles, also display a marked change through the Mesozoic. Shallow water limestones, which generally are indicative of low latitude, hot climates occur only in Carnian and older strata. Norian to Aalenian T-R cycles generally have oolitic ironstones at the base instead of limestones. In Bajocian and younger rocks sideritic mudstones and high concentrations of glauconite occupy this basal stratigraphic position.

Another paleoclimatic indicator is the presence of clasts up to 50 cm across which occur randomly within Upper Jurassic and Lower Cretaceous marine shale and sandstone (Fig. 14.66). These clasts are inferred to have been transported onto the marine shelf by floating ice pans which accumulated on beaches and river banks during the winter season (Embry, 1984d). Also occurring within this same interval are distinctive calcite concretions, called polar euhedrons, which probably indicate a cold water environment (Kemper and Schmitz, 1975).

The marine faunas in the Mesozoic strata also indicate that the climate became colder with time. Triassic faunas are essentially cosmopolitan in their affinities, whereas from Bajocian onward, faunas of the Sverdrup Basin are characterized by a lack of diversity.

Overall, the paleomagnetic and paleoclimatic data indicate a northward drift of the Arctic Islands from 40°N and a savanna climate in Early Triassic, through 45-60°N and a warm temperate to subtropical climate in Late Triassic and Early Jurassic, to north of 60°N and a temperate climate in Middle Jurassic and Cretaceous.

SUMMARY

Mesozoic strata in the Canadian Arctic Archipelago lie mainly in the Sverdrup Basin where the strata are up to 9000 m thick. In Banks Basin and Eglinton Graben they are up to 1500 m thick. The thicknesses of Mesozoic strata on the continental shelves bordering the Arctic Ocean and Baffin Bay are unknown. Thin, small outcrops are present on the Arctic Platform and Precambrian Shield.

The Sverdrup Basin contains a complete record of Mesozoic sedimentation. Thirty regional transgressive-regressive cycles have been recognized in the succession. These T-R cycles are the product of the interplay of subsidence, sediment supply and eustatic sea level changes. Subsidence of the Sverdrup Basin was initiated by rifting and crustal stretching in Carboniferous to Early Permian and was followed by a phase of gradually decaying thermal subsidence from late Early Permian to earliest Cretaceous. A second episode of crustal stretching with accompanying high subsidence rates and mafic volcanism occurred during Early and earliest Late Cretaceous. This rifting episode was followed by thermal subsidence over most of the basin in Late Cretaceous although the northeastern part of the basin experienced uplift at this time because of crustal compression marking the onset of the Eurekan Orogeny.

The main source areas for Sverdrup Basin sediments during the Mesozoic lay to the south and east of the basin and encompassed Paleozoic strata adjacent to the basin and Precambrian sedimentary, igneous and metamorphic rocks farther to the south and east. A source area to the northwest of the basin also supplied sediment from Triassic to early Late Cretaceous but its contribution was relatively minor compared with that of the eastern and southern sources.

Sediment supply varied markedly through the Mesozoic from very high rates of supply characterized by thick, deltaic sediments to very low rates characterized by thin, reworked marine shelf sediments. These variations are interpreted to be due to epeirogenic movements of the craton.

Relative sea level fluctuated throughout the Mesozoic, and marginal areas were exposed during times of low relative sea level. The origin of these relative sea level fluctuations is debatable and is related to either eustatic sea level changes and/or minor tectonic events.

Halokinetic salt structures, originating from deeply buried Carboniferous salt, also affected Mesozoic stratigraphy and sedimentation. A variety of structures including circular diapirs, long linear ridges, salt-cored anticlines, and lag domes deform the Mesozoic succession, and the structures started growing in Early-Middle Triassic. Marked thinning and subaerial erosion occurred over the structures during times of accelerated upward growth.

The Arctic Islands gradually moved northward during the Mesozoic from 40°N in earliest Triassic to 70°N at the end of the Cretaceous. In response to this movement the climate changed from hot and seasonally dry in Early and Middle Triassic, to warm temperate during the Late Triassic and Early Jurassic epochs, and to cool temperate from Middle Jurassic onward.

The Mesozoic succession represents a major portion of the fill of the Sverdrup Basin, and the paleogeography varied greatly throughout this time. At the close of the Paleozoic, the shelves of the Sverdrup Basin were above sea level but the central part of the basin, which had existed since mid-Carboniferous, remained under water. A major sea level rise initiated the Mesozoic Era and the shelf areas were drowned in earliest Triassic. Following this transgression, high rates of sedimentation occurred and deltaic sedimentation dominated the Early Triassic. By the end of Early Triassic, deltaic sediments had prograded into the basin but deep water outer shelf, slope and basinal environments occupied the central Sverdrup Basin.

Sediment influx was much less during Middle Triassic, and marine shelf sands prograded across outer shelf mud and silt following regional transgressions in early Anisian and early Ladinian. A transgression occurred in early Carnian and following this, sedimentation rates became relatively high, with the northwestern source area supplying considerable sediments. Carnian and early Norian shelf sandstones extend into the basin over thick, prodelta and slope, shales and siltstones. Very high rates of sedimentation prevailed from latest Triassic (Rhaetian) to Early Jurassic (Pliensbachian) and a fluvial-dominated deltaic system prograded across the eastern and central portions of the basin. The western portion of the basin received relatively little sediment and was the site of nearshore to offshore shallow shelf sedimentation.

Sediment supply waned significantly in late Pliensbachian and the deltaic complex was transgressed, leaving behind a blanket of nearshore marine sand. An offshore marine shelf environment occupied most of the Sverdrup Basin by early Toarcian. From Toarcian to early

Valanginian, sediment supply was relatively low, reaching starvation conditions in Bajocian – early Bathonian and Berriasian. Shelf sands prograded basinward during regressions with Aalenian and Oxfordian-Kimmeridgian sands being most extensive.

Sediment supply dramatically increased in late Valanginian and a fluvial-dominated deltaic system, characterized by coarse grained, pebbly deposits, prograded across the basin in late Valanginian and Hauterivian. Transgression in mid-Barremian was followed by renewed deltaic progradation in late Barremian and early Aptian. Basalt flows were extruded onto the deltaic plain in the northeastern portion of the Sverdrup Basin at this time. A major transgression began in late Aptian and much of the Arctic Archipelago was a marine shelf environment by early Albian. Shelf sands prograded basinward over silt and mud in late early Albian. This was followed by a transgression in early middle Albian and regression in middle Albian to early Cenomanian, when deltaic sediments again prograded across the Sverdrup Basin and basaltic volcanism was widespread in the northeastern part of the basin.

A major transgression began in late Cenomanian, and much of the Arctic Islands was occupied by an offshore marine shelf by Santonian with consequent low sediment input. Sedimentation rates increased in late Santonian and deltaic systems prograded southward, westward and northwestward during Campanian and Maastrichtian. Uplifts related to the early stages of the Eurekan Orogeny took place in the northeast at this time. Much of the Sverdrup Basin was a fluvial-deltaic plain at the close of the Mesozoic.

Eglinton Graben and Banks Basin form a northeast-southwest trending area of subsidence in the southwestern Arctic Archipelago. Subsidence began in Eglinton Graben in the Bajocian and in Banks Basin in the Oxfordian. The Middle Jurassic to earliest Cretaceous strata in these areas consist mainly of shelf deposits similar to those of the Sverdrup Basin. Uplift in late Valanginian to Hauterivian was followed by transgression in Barremian and the deposition of Barremian to late Albian shelf deposits. Renewed uplift occurred from latest Albian to Santonian, and the final Mesozoic deposits of this area consist of late Santonian to Maastrichtian marine shelf deposits.

REFERENCES

Balkwill, H.R.
1978: Evolution of Sverdrup Basin, Arctic Canada; The American Association of Petroleum Geologists Bulletin, v. 62, p. 1004-1028.
1983: Geology of Amund Ringnes, Cornwall and Haig Thomas islands, District of Franklin; Geological Survey of Canada, Memoir 390, 76 p.

Balkwill, H.R. and Hopkins, W.S., Jr.
1976: Cretaceous stratigraphy, Hoodoo Dome, Ellef Ringnes Island; in Report of Activities, Part B, Geological Survey of Canada, Paper 76-1B, p. 329-334.

Balkwill, H.R. and Roy, K.J.
1977: Geology, King Christian Island, District of Franklin; Geological Survey of Canada, Memoir 386, 28 p.

Balkwill, H.R., Wilson, D.G., and Wall, J.H.
1977: Ringnes Formation (Upper Jurassic), Sverdrup Basin, Canadian Arctic Archipelago; Bulletin of Canadian Petroleum Geology, v. 25, no. 6, p. 1115-1144.

Balkwill, H.R., Hopkins, W.S., Jr., and Wall, J.H.
1982: Geology of Lougheed Island and nearby small islands, District of Franklin (parts of 69C, 79D); Geological Survey of Canada, Memoir 395, 22 p.

Balkwill, H.R. and Fox, F.G.
1982: Incipient rift zone, western Sverdrup Basin, Arctic Canada; in Arctic Geology and Geophysics, A.F. Embry and H.R. Balkwill (ed.); Canadian Society of Petroleum Geologists, Memoir 8, p. 171-187.

Balkwill, H.R., Cook, D.G., Detterman, R.L., Embry, A.F., Håkansson, E., Miall, A.D., Poulton, T.P., and Young, F.G.
1983: Arctic North America and Northern Greenland; in The Phanerozoic Geology of the World, The Mesozoic, Part B, M. Moullade and A.E.M. Nairn (ed.); Elsevier, p. 1-31.

Blackadar, R.G.
1964: Basic intrusions of the Queen Elizabeth Islands, District of Franklin; Geological Survey of Canada, Bulletin 97, 36 p.

Bujak, J.P. and Fisher, M.J.
1976: Dinoflagellate cysts from the Upper Triassic of Arctic Canada; Micropaleontology, v. 22, no. 1, p. 44-70.

Christie, R.L.
1964: Geological reconnaissance of northeastern Ellesmere Island, District of Franklin; Geological Survey of Canada, Memoir 331, 79 p.

Crain, E.R.
1977: Log interpretation in the High Arctic; Transactions of the CWLS Sixth Formation Evaluation Symposium; Canadian Well Logging Society, p. S1-S32.

Davies, E.H.
1983: The dinoflagellate Oppel-zonation of the Jurassic-Lower Cretaceous sequence in the Sverdrup Basin, Arctic Canada; Geological Survey of Canada, Bulletin 359, 59 p.

Doerenkamp, A., Jardine, S., and Moreau, P.
1976: Cretaceous and Tertiary palynomorph assemblages from Banks Island and adjacent areas; Bulletin of Canadian Petroleum Geology, v. 24, no. 3, p. 372-417.

Douglas, T.R. and Oliver, T.A.
1979: Environments of deposition of the Borden Island Gas Zone in the subsurface of the Sabine Peninsula area, Melville Island, Arctic Archipelago; Bulletin of Canadian Petroleum Geology, v. 27, p. 273-313.

Embry, A.F.
1982: The Upper Triassic-Lower Jurassic Heiberg deltaic complex of the Sverdrup Basin; in Arctic Geology and Geophysics, A.F. Embry and H.R. Balkwill (ed.); Canadian Society of Petroleum Geologists, Memoir 8, p. 189-217.
1983a: Stratigraphic subdivision of the Heiberg Formation, eastern and central Sverdrup Basin, Arctic Islands; in Current Research, Part B, Geological Survey of Canada, Paper 83-1B, p. 205-213.
1983b: The Heiberg Group, western Sverdrup Basin, Arctic Islands; in Current Research, Part B, Geological Survey of Canada, Paper 83-1B, p. 381-389.
1984a: The Schei Point and Blaa Mountain groups (Middle-Upper Triassic), Sverdrup Basin, Canadian Arctic Archipelago; in Current Research, Part B, Geological Survey of Canada, Paper 84-1B, p. 327-336.
1984b: Stratigraphic subdivision of the Roche Point, Hoyle Bay and Barrow formations (Schei Point Group), western Sverdrup Basin, Arctic Islands; in Current Research, Part B, Geological Survey of Canada, Paper 84-1B, p. 275-283.
1984c: The Wilkie Point Group (Lower-Upper Jurassic), Sverdrup Basin, Arctic Islands; in Current Research, Part B, Geological Survey of Canada, Paper 84-1B, p. 299-308.
1984d: Upper Jurassic to lowermost Cretaceous stratigraphy, sedimentology and petroleum geology, Sverdrup Basin, abstract; in Program and Abstracts, CSPG-CSEG National Convention, Calgary, Alberta, June 17-20, 1984.
1985a: New stratigraphic units, Middle Jurassic to lowermost Cretaceous succession, Arctic Islands; in Current Research, Part B, Geological Survey of Canada, Paper 85-1B, p. 269-276.
1985b: Stratigraphic subdivision of the Isachsen and Christopher formations (Lower Cretaceous), Arctic Islands; in Current Reseach, Part B, Geological Survey of Canada, Paper 85-1B, p. 239-246.
1986a: Stratigraphic subdivision of the Blind Fiord and Bjorne formations (Lower Triassic), Sverdrup Basin, Arctic Islands; in Current Research, Part B, Geological Survey of Canada, Paper 86-1B, p. 329-340.
1986b: Stratigraphic subdivision of the Awingak Formation (Upper Jurassic) and revision of the Hiccles Cove Formation (Middle Jurassic), Sverdrup Basin; in Current Research, Part B, Geological Survey of Canada, Paper 86-1B, p. 341-349.
1988: Triassic sea-level changes: evidence from the Canadian Arctic Archipelago; in Sea-Level Change, an Integrated Approach, C.K. Wilgus et al. (ed.); Society of Economic Paleontologists and Mineralogists, Special Publication 42, p. 249-259.

Embry, A.F. and Osadetz, K.G.
1988: Stratigraphy and tectonic significance of Cretaceous volcanism in Queen Elizabeth Islands, Canadian Arctic Archipelago; Canadian Journal of Earth Sciences v. 25, p. 1209-1219.

Falvey, D.A.
1974: The development of continental margins in plate tectonic theory; Australian Petroleum Exploration Association Journal, v. 14, p. 95-106.

Felix, C.J.
1975: Palynological evidence for Triassic sediments on Ellef Ringnes Island, Arctic Canada; Review of Paleobotany and Palynology, v. 20, p. 109-117.

Felix, C.J. and Burbridge, P.P.
1977: A new *Ricciisporites* from the Triassic of Arctic Canada; Paleontology, v. 20, pt. 3, p. 581-587.

Fischer, B.F.G.
1984: Stratigraphy and structural geology of the region surrounding Bunde and Bukken fiords, Axel Heiberg Island, Arctic Canada; in Current Research, Part B, Geological Survey of Canada, Paper 84-1B, p. 309-314.

Fisher, M.J.
1979: The Triassic palynofloral succession in the Canadian Arctic Archipelago; American Association of Stratigraphic Palynologists, Contribution Series, no. 5B, p. 83-100.

Fisher, M.J. and Bujak, J.
1975: Upper Triassic palynofloras from Arctic Canada; Geoscience and Man, v. XI, p. 87-94.

Forsyth, D.A., Morel, P., Asudeh, I., and Green, A.G.
in press: Alpha Ridge and Iceland — Products of the same plume?; Nature.

Fox, F.G.
1983: Structure sections across Parry Islands Fold Belt and Vesey Hamilton salt wall, Arctic Archipelago, Canada; in Seismic Expression of Structural Styles, a Picture and Work Atlas, A.W. Bally (ed.); American Association of Petroleum Geologists, Studies in Geology, Series 15, v. 3, p. 3.4.1-54 to 3.4.1-72.

Frebold, H.
1958: Fauna, age and correlation of the Jurassic rocks of Prince Patrick Island; Geological Survey of Canada, Bulletin 41, 69 p.
1960: The Jurassic faunas of the Canadian Arctic: Lower Jurassic and lowermost Middle Jurassic ammonites; Geological Survey of Canada, Bulletin 59, 33 p.
1961: The Jurassic faunas of the Canadian Arctic: Middle and Upper Jurassic ammonites; Geological Survey of Canada, Bulletin 74, 43 p.
1964: The Jurassic faunas of the Canadian Arctic, Cadoceratinae; Geological Survey of Canada, Bulletin 119, 27 p.
1970: Marine Jurassic faunas; in Geology and Economic Minerals of Canada, R.J. Douglas (ed.); Geological Survey of Canada, Economic Geology Report no. 1, p. 641-648.
1975: The Jurassic faunas of the Canadian Arctic, Lower Jurassic ammonites, biostratigraphy and correlation; Geological Survey of Canada, Bulletin 243, 24 p.

Fricker, P.E.
1963: Geology of the Expedition Fiord area, west central Axel Heiberg Island, Canadian Arctic Archipelago; McGill University, Axel Heiberg Island Research Reports, Geology no. 1, 156 p.

Glenister, B.F.
1963: Localities on southern Axel Heiberg Island; in Geology of the North-Central Part of the Arctic Archipelago, Northwest Territories (Operation Franklin), Y.O. Fortier et al.; Geological Survey of Canada, Memoir 320, p. 472-480.

Grantz, A., Eittreim, S., and Dinter, D.A.
1979: Geology and tectonic development of the continental margin north of Alaska; Tectonophysics, v. 59, p. 263-291.

Grantz, A. and May, S.P.
1983: Rifting history and structural development of the continental margin north of Alaska; in Studies in Continental Margin Geology, J.S. Watkins and C.L. Drake (ed.); American Association of Petroleum Geologists, Memoir 34, p. 77-100.

Greiner, H.R.
1963: Jaeger River, eastern Cornwall Island; in Geology of the North-Central Part of the Arctic Archipelago, Northwest Territories (Operation Franklin), Y.O. Fortier et al.; Geological Survey of Canada, Memoir 320, p. 533-537.

Hallam, A.
1978: Eustatic cycles in the Jurassic; Paleogeography, Paleoclimatology, Paleoecology, v. 23, p. 1-23.

Hea, J.P., Arcuri, J., Campbell, G.R., Fraser, I., Fuglem, M.O., O'Bertos, J.J., Smith, D.R., and Zagat, M.
1980: Post-Ellesmerian basins of Arctic Canada: their depocentres, rates of sedimentation and petroleum potential; in Facts and Principles of World Oil Occurrence, A.D. Miall (ed.); Canadian Society of Petroleum Geologists, Memoir 6, p. 447-488.

Henao-Londoño, D.
1977: Correlation of producing formations in the Sverdrup Basin; Bulletin of Canadian Petroleum Geology, v. 25, p. 969-980.

Heywood, W.W.
1957: Isachsen area, Ellef Ringnes Island, District of Franklin, Northwest Territories; Geological Survey of Canada, Paper 56-8, 36 p.

Hopkins, W.S., Jr. and Balkwill, H.R.
1973: Description, palynology and paleoecology of the Hassel Formation (Cretaceous) on eastern Ellef Ringnes Island, District of Franklin; Geological Survey of Canada, Paper 72-37, 31 p.

Irving, E. and Irving, G.A.
1982: Apparent polar wander paths Carboniferous through Cenozoic and the assembly of Gondwana; Geophysical Surveys, v. 5, p. 141-188.

Jeletzky, J.A.
1966: Upper Volgian (latest Jurassic) ammonities and *Buchias* of Arctic Canada; Geological Survey of Canada, Bulletin 128, 51 p.
1970: Cretaceous macrofaunas; in Geology and Economic Minerals of Canada, R.J. Douglas (ed.); Geological Survey of Canada, Economic Geology Report no. 1, p. 649-662.
1980: New or formerly poorly known, biochronologically and paleobiogeographically important gastroplitinid and cleoniceratinid (ammonitida) taxa from Middle Albian rocks of mid-western and Arctic Canada; Geological Survey of Canada, Paper 79-22, 63 p.
1981: *Pachygrycia*, a new *Sonneratia*-like ammonite from the Lower Cretaceous (earliest Albian?) of northern Canada; Geological Survey of Canada, Paper 80-20, 25 p.
1984: Jurassic-Cretaceous boundary beds of Western and Arctic Canada and the problem of the Tithonian-Berriasian stages in the boreal realm; in Jurassic-Cretaceous Biochronology and Paleogeography of North America, G.E.G. Westerman (ed.); Geological Association of Canada, Special Paper 27, p. 175-255.

Johnson, C.D. and Hills, L.V.
1973: Microplankton zones of the Savik Formation (Jurassic), Axel Heiberg and Ellesmere islands, District of Franklin; Bulletin of Canadian Petroleum Geology, v. 21, p. 178-218.

Jollimore, W.
1986: Analyses of dyke swarms within the Sverdrup Basin, Queen Elizabeth Islands; unpublished B.Sc. thesis, Dalhousie University, Halifax, Nova Scotia, 54 p.

Kemper, E.
1975: Upper Deer Bay Formation (Berriasian-Valanginian) of Sverdrup Basin and biostratigraphy of the Arctic Valanginian; in Report of Activities, Part B, Geological Survey of Canada, Paper 75-1B, p. 245-254.
1977: Biostratigraphy of the Valanginian in Sverdrup Basin, District of Franklin; Geological Survey of Canada, Paper 76-32, 6 p.

Kemper, E. and Schmitz, H.H.
1975: Stellate nodules from the upper Deer Bay Formation (Valanginian) of Arctic Canada; in Report of Activities, Part C, Geological Survey of Canada, Paper 75-1C, p. 109-119.

Kerr, J.W.
1981: Evolution of the Canadian Arctic Islands: a transition between the Atlantic and Arctic Oceans; in The Ocean Basins and Margins, v. 5, The Arctic Ocean, A.E.M. Nairn, M. Churkin, Jr., and F.G. Stehli (ed.); Plenum Press, New York, p. 105-199.

Logan, A.
1967: Middle and Upper Triassic spiriferinid brachiopods from the Canadian Arctic Archipelago; Geological Survey of Canada, Bulletin 155, 37 p.

McMillan, N.J.
1963: Lightfort River to Wading River; in Geology of the North-Central Part of the Arctic Archipelago Northwest Territories (Operation Franklin), Y.O. Fortier et al.; Geological Survey of Canada, Memoir 320, p. 501-512.

Meneley, R.A.
1977: Exploration prospects in the Canadian Arctic Islands; Panarctic Oils Ltd., Calgary, Alberta, 16 p.

Meneley, R.A., Henao, D., and Merritt, P.K.
1975: The northwest margin of the Sverdrup Basin; in Canada's Continental Margins and Offshore Petroleum Exploration, C.J. Yorath, E.R. Parker, and D.J. Glass (ed.); Canadian Society of Petroleum Geologists, Memoir 4, p. 531-544.

Miall, A.D.
1975: Post-Paleozoic geology of Banks, Prince Patrick, and Eglinton Islands, Arctic Canada; in Canada's Continental Margins and Offshore Petroleum Exploration, C.J. Yorath, E.R. Parker, and D.J. Glass (ed.); Canadian Society of Petroleum Geologists, Memoir 4, p. 557-587.
1979: Mesozoic and Tertiary geology of Banks Island, Arctic Canada; Geological Survey of Canada, Memoir 387, 235 p.

Miall, A.D., Balkwill, H.R., and Hopkins, W.S., Jr.
1980: Cretaceous and Tertiary sediments of Eclipse Trough, Bylot Island area, Arctic Canada, and their regional setting; Geological Survey of Canada, Paper 79-23, 20 p.

Nassichuk, W.W. and Christie, R.L.
1969: Upper Paleozoic and Mesozoic stratigraphy in the Yelverton Pass region, Ellesmere Island, District of Franklin; Geological Survey of Canada, Paper 68-31, 31 p.

Nassichuk, W.W. and Roy, K.J.
1975: Mound-like carbonate rocks of Early Cretaceous (Albian) age adjacent to Hoodoo Dome, Ellef Ringnes Island, District of Franklin; in Report of Activities, Part A, Geological Survey of Canada, Paper 75-1A, p. 565-569.

Nassichuk, W.W., Thorsteinsson, R., and Tozer, E.T.
1972: Permian-Triassic boundary in the Canadian Arctic Archipelago; Bulletin of Canadian Petroleum Geology, v. 20, p. 651-658.

Nassichuk, W.W. and Davies, G.R.
1980: Stratigraphy and sedimentation of the Otto Fiord Formation; Geological Survey of Canada, Bulletin 286, 87 p.

Osadetz, K.G. and Moore, P.R.
1988: Basic volcanic rocks in the Hassel Formation (mid-Cretaceous), Ellesmere Island, District of Franklin; Geological Survey of Canada, Paper 87-21, 19 p.

Petryk, A.A.
1969: Mesozoic and Tertiary stratigraphy at Lake Hazen, northern Ellesmere Island, District of Franklin; Geological Survey of Canada, Paper 68-17, 51 p.

Plauchut, B.P.
1971: Geology of the Sverdrup Basin; Bulletin of Canadian Petroleum Geology, v. 19, p. 659-679.

Plauchut, B.P. and Jutard, G.G.
1976: Cretaceous and Tertiary stratigraphy, Banks and Eglinton Islands and Anderson Plain (N.W.T.); Bulletin of Canadian Petroleum Geology, v. 24, p. 321-371.

Pocock, S.A.J.
1976: A preliminary dinoflagellate zonation of the uppermost Jurassic and lower part of the Cretaceous, Canadian Arctic, and possible correlation in the Western Canada Basin; Geoscience and Man, v. XV, p. 101-114.

Rahmani, R.A.
1978: Fault control on sedimentation of Isachsen Formation in Sverdrup Basin; in Current Activities, Part A, Geological Survey of Canada, Paper 78-1A, p. 538-540.

Rahmani, R.A. and Hopkins, W.S., Jr.
1977: Geological and palynological interpretation of Eureka Sound Formation on Sabine Peninsula, northern Melville Island, District of Franklin; in Current Research, Part B, Geological Survey of Canada, Paper 77-1B, p. 185-189.

Rahmani, R.A. and Tan, J.T.
1978: The type section of the Lower Jurassic Borden Island Formation, Borden Island, Arctic Archipelago, Canada; in Current Activities, Part A, Geological Survey of Canada, Paper 78-1A, p. 538-540.

Rayer, F.G.
1981: Exploration prospects and future petroleum potential of the Canadian Arctic Islands; Journal of Petroleum Geology, v. 3, p. 367-412.

Reinson, G.E.
1975: Lithofacies analysis of cores from the Borden Island Formation, Drake Point, Melville Island; in Report of Activities, Part B, Geological Survey of Canada, Paper 75-1B, p. 297-301.

Ricketts, B.
1985: Volcanic breccias in the Isachsen Formation near Strand Fiord, Axel Heiberg Island, District of Franklin; in Current Research, Part A, Geological Survey of Canada, Paper 85-1A, p. 609-612.
1986: New formations in the Eureka Sound Group, Canadian Arctic Islands; in Current Research, Part B, Geological Survey of Canada, Paper 86-1B, p. 363-374.

Ricketts, B. and Embry, A.F.
1984: Summary of geology and resource potential of coal deposits in the Canadian Arctic Archipelago; Bulletin of Canadian Petroleum Geology, v. 32, p. 359-371.

Ricketts, B., Osadetz, K.G., and Embry, A.F.
1985: Volcanic style in the Strand Fiord Formation (Upper Cretaceous), Axel Heiberg Island, Canadian Arctic Archipelago; Polar Reseach, v. 3 n.s., p. 107-122.

Roy, K.J.
1972: Bjorne Formation (Lower Triassic), western Ellesmere Island; in Report of Activities, Part A, Geological Survey of Canada, Paper 72-1A, p. 224-226.
1973: Isachsen Formation, Amund Ringnes Island, District of Franklin; in Report of Activities, Part A, Geological Survey of Canada, Paper 73-1A, p. 269-273.
1974: Transport directions in the Isachsen Formation (Lower Cretaceous) Sverdrup Islands, District of Franklin; in Report of Activities, Part A, Geological Survey of Canada; Paper 74-1A, p. 351-353.

Souaya, F.J.
1976: Foraminifera of Sun-Gulf-Global Linckens Island well P-46, Arctic Archipelago, Canada; Micropaleontology, v. 22, p. 249-306.

Souther, J.G.
1963: Geological traverse across Axel Heiberg Island from Buchanan Lake to Strand Fiord; in Geology of the North-Central Part of the Arctic Archipelago, Northwest Territories (Operation Franklin), Y.O. Fortier et al.; Geological Survey of Canada, Memoir 320, p. 426-448.

Stephenson, R.A., Embry, A.F., Nakiboglu, S.M., and Hastaoglu, M.A.
1987: Rift-initiated Permian-Early Cretaceous subsidence of the Sverdrup Basin; in Sedimentary Basins and Basin-Forming Mechanisms, C. Beaumont and A. Tankard (ed.); Canadian Society of Petroleum Geologists, Memoir 12, p. 213-231.

Stott, D.F.
1969: Ellef Ringnes Island, Canadian Arctic Archipelago; Geological Survey of Canada, Paper 68-16, 38 p.

Stuart Smith, J.H. and Wennekers, J.H.N.
1977: Geology and hydrocarbon discoveries of Canadian Arctic Islands; The American Association of Petroleum Geologists Bulletin, v. 61, p. 1-27.

Suneby, L.B.
1984: Biostratigraphy of the Upper Triassic-Lower Jurassic Heiberg Formation, eastern Sverdrup Basin, Arctic Canada; unpublished M.Sc. thesis, University of Calgary, Calgary, Alberta, 245 p.

Sweeney, J.F.
1977: Subsidence of the Sverdrup Basin, Canadian Arctic Islands; Geological Society of America Bulletin, v. 88, p. 41-48.

Tan, J.T.
1979: Late Triassic-Jurassic dinoflagellate biostratigraphy, western Arctic Canada; unpublished Ph.D. thesis, University of Calgary, Calgary, Alberta, 217 p.

Tan, J.T. and Hills, L.V.
1978: Oxfordian-Kimmeridgian dinoflagellate assemblage, Ringnes Formation, Arctic Canada; in Current Research, Part C, Geological Survey of Canada, Paper 78-1C, p. 63-73.

Thorsteinsson, R.
1974: Carboniferous and Permian stratigraphy of Axel Heiberg Island and western Ellesmere Island, Canadian Arctic Archipelago; Geological Survey of Canada, Bulletin 224, 115 p.

Thorsteinsson, R. and Tozer, E.T.
1962: Banks, Victoria, and Stefansson Islands, Arctic Archipelago; Geological Survey of Canada, Memoir 330, 85 p.

Tozer, E.T.
1956: Geological reconnaissance, Prince Patrick, Eglinton, and western Melville Islands, Arctic Archipelago, Northwest Territories; Geological Survey of Canada, Paper 55-5, 32 p.
1961a: Summary account of Mesozoic and Tertiary stratigraphy, Canadian Arctic Archipelago; in Geology of the Arctic, v. 1, G.O. Raasch (ed.); Alberta Society of Petroleum Geologists, p. 381-402.
1961b: Triassic stratigraphy and faunas, Queen Elizabeth Islands, Arctic Archipelago; Geological Survey of Canada, Memoir 316, 116 p.
1963a: Mesozoic and Tertiary stratigraphy; in Geology of the North-Central Part of the Arctic Archipelago, Northwest Territories (Operation Franklin), Y.O. Fortier et al.; Geological Survey of Canada, Memoir 320, p. 74-95.
1963b: Mesozoic and Tertiary stratigraphy, western Ellesmere Island and Axel Heiberg Island, District of Franklin; Geological Survey of Canada, Paper 63-30, 38 p.
1963c: Blind Fiord (Southern Ellesmere Island, and some localities north of Bay Fiord and Graham Island); in Geology of the North-Central Part of the Arctic Archipelago, Northwest Territories (Operation Franklin), Y.O. Fortier et al.; Geological Survey of Canada, Memoir 320, p. 380-386.

1963d: Northwestern Bjorne Peninsula (Southern Ellesmere Island, and some localities north of Bay Fiord and Graham Island); in Geology of the North-Central Part of the Arctic Archipelago, northwest Territories (Operation Franklin), Y.O. Fortier et al.; Geological Survey of Canada, Memoir 320, p. 363-370.
1965: Lower Triassic stages and ammonoid zones of Arctic Canada; Geological Survey of Canada, Paper 65-12.
1967: A standard for Triassic time; Geological Survey of Canada, Bulletin 156, 103 p.
1970a: Geology of the Arctic Archipelago, Mesozoic; in Geology and Economic Minerals of Canada, R.J.W. Douglas (ed.); Geological Survey of Canada, Economic Geology Report no. 1, p. 574-583.
1970b: Marine Triassic faunas; in Geology and Economic Minerals of Canada, R.J.W. Douglas (ed.); Geological Survey of Canada, Economic Geology Report no. 1, p. 633-640.

Tozer, E.T. and Thorsteinsson, R.
1964: Western Queen Elizabeth Islands, Arctic Archipelago; Geological Survey of Canada, Memoir 332, 242 p.

Trettin, H.P. and Hills, L.V.
1966: Lower Triassic tar sands of northwestern Melville Island, Arctic Archipelago; Geological Survey of Canada, Paper 66-34, 122 p.

Trettin, H.P. and Parrish, R.
1987: Late Cretaceous bimodal magmatism, northern Ellesmere Island: isotopic age and origin; Canadian Journal of Earth Sciences, v. 24, p. 257-265.

Troelson, J.C.
1950: Contributions to the geology of northwest Greenland, Ellesmere and Axel Heiberg Islands; Meddelelser om Grønland, v. 149, no. 7, 85 p.

Vail, P.R., Mitchum, R.M. Jr., and Thompson, S., III.
1977: Seismic stratigraphy and global changes of sea level, Part 4: Global cycles of relative changes of sea level; in Seismic Stratigraphy, Applications to Hydrocarbon Exploration, C.E. Payton (ed.); American Association of Petroleum Geologists, Memoir 26, p. 83-97.

Van Houten, F.B.
1973: Origin of red beds: a review — 1961-1972; in Annual Review of Earth and Planetary Sciences, F.A. Donath (ed.); v. 1, p. 39-61.

Wall, J.H.
1983: Jurassic and Cretaceous foraminiferal biostratigraphy in the eastern Sverdrup Basin, Canadian Arctic Archipelago; Bulletin of Canadian Petroleum Geology, v. 31, p. 246-281.

Watts, A.B.
1982: Tectonic subsidence, flexure and global changes in sea level; Nature, v. 296, p. 469-474.

Wilson, D.G.
1976: Studies of Mesozoic stratigraphy, Tanquary Fiord to Yelverton Pass, northern Ellesmere Island, District of Franklin; in Report of Activities, Part A, Geological Survey of Canada, Paper 76-1A, p. 449-451.

Author's address

Ashton F. Embry
Geological Survey of Canada
3303-33rd St. N.W.
Calgary, Alberta
T2L 2A7

CHAPTER 14

Printed in Canada

Chapter 15
LATE CRETACEOUS AND TERTIARY BASIN DEVELOPMENT AND SEDIMENTATION, ARCTIC ISLANDS

Chronostratigraphic framework

Lithostratigraphic framework

Depositional history

References

Chapter 15

LATE CRETACEOUS AND TERTIARY BASIN DEVELOPMENT AND SEDIMENTATION, ARCTIC ISLANDS

A.D. Miall

CHRONOSTRATIGRAPHIC FRAMEWORK
Introduction

Macrofossils are extremely rare in Arctic Cenozoic sediments, apart from rare nondiagnostic gastropods and pelecypods, plus foraminifera, fish, scaphopods and land vertebrates recorded from the Strathcona Fiord area, Ellesmere Island (West et al., 1977, 1981). These have only a local distribution and are therefore of limited chronostratigraphic value. Macroflora, particularly pine cones, occur in Neogene sediments and some are age diagnostic. Marine fossils are rare throughout the Cenozoic section, except for rare dinoflagellates, particularly in the Banks Island area, and foraminifera on Meighen Island. However, spores and pollen are abundant throughout the Cenozoic and in the Campanian-Maastrichtian sediments of similar facies that lie conformably beneath the Cenozoic section.

Palynostratigraphic studies therefore form the main chronostratigraphic basis for the largely nonmarine or brackish water facies that comprise the bulk of the Cenozoic section of the Innuitian region (plus Campanian to Maastrichtian in the eastern Arctic).

Many local, spot age determinations have been carried out on the Upper Cretaceous and Cenozoic section. The only thorough stratigraphic investigations are those by Doerenkamp et al. (1976) in Banks Island and Rouse (1977) in Remus Creek, Ellesmere Island, for the Maastrichtian to Early Oligocene interval. These workers established zones based on comparison with contemporaneous sections in the Northern Interior Plains, Beaufort-Mackenzie Basin, British Columbia, Alaska, Siberia and elsewhere. In addition, G. Norris and M. Head (pers. comm., 1983-1985) have examined the writer's collections from Axel Heiberg and central Ellesmere islands, and G. Norris (pers. comm., 1987) examined the Eocene to Miocene section in the Meighen Island well (Crocker I-53). Hills and co-workers carried out a series of studies on Miocene to early Pliocene macroflora and microflora (Hills and Ogilvie, 1970; Hills and Fyles, 1973; Hills and Bustin, 1976). More recently, D.J. McIntyre has identified Maastrichtian to Eocene spores from various parts of the Arctic Islands (Ricketts and McIntyre, 1986, and pers. comm.). In all these studies the presence of abundant recycled Cretaceous palynomorphs has caused difficulties. This has now been identified as one of the principal problems with a chronostratigraphic study by Hickey et al. (1983) that proposed radically different age ranges for Arctic Cenozoic floras and faunas than those outlined here. Their results have now been disputed by several workers (McIntyre, 1984; Norris and Miall, 1984; Flynn et al., 1984) and are not referred to in this paper.

Few other chronostratigraphic tools are available. An attempt was made by Vinson (1981) and Hickey et al. (1983) to recognize the standard magnetostratigraphic scale in the Maastrichtian to Eocene section of Strathcona Fiord, Ellesmere Island, but the magnetic signature is weak, the data are sparse, and the results have been disputed (Kent et al., 1984; Norris and Miall, 1984; Flynn et al., 1984). A resurveying of Vinson's sections by Tauxe and Clark (1987) suggested a quite different correlation, one that can readily be accommodated to the chronostratigraphic scale adopted here.

Diagnostic fossils, mainly spore and pollen assemblages for latest Cretaceous and Tertiary time intervals, are listed below.

Campanian-Maastrichtian

Characteristic Maastrichtian palynomorphs from Ellesmere Island include species of *Wodehousia, Azonia, Expressipolis* and *Aquilapollenites* (zone A of Hickey et al., 1983). Some of these genera range lower into the Campanian, but *Wodehousia* appeared in the Maastrichtian with some species ranging into the basal Paleocene (Norris and Miall, 1984). Similar floras have been found on Ellef Ringnes Island (Felix and Burbridge, 1973). On Banks Island, the same flora occurs with marine Maastrichtian dinoflagellates in the upper part of the Kanguk Formation (Doerenkamp et al., 1976).

Paleocene-Eocene

The distribution of selected Paleocene palynomorphs in Arctic Canada is given in Table 15.1. Paleocene to mid-Eocene ranges in this table relate to studies by Rouse (1977) in the Remus Creek area, Ellesmere Island. The table is completed by determinations from Paleocene to Oligocene units in the Beaufort-Mackenzie Basin (Rouse, 1977). Comparable floras have been reported from other areas in Ellesmere Island (Hickey et al., 1983, but with age assignments revised by Norris and Miall, 1984), and from Banks Island (Doerenkamp et al., 1976). These authors all reported an abrupt change in microflora at a level that probably coincides with the Cretaceous-Tertiary boundary. Characteristic palynomorphs include *Polyvestibulopollenites verus, Paraalnipollenites confusus, P. alterniporus* and

---

Miall, A.D.
1991: Late Cretaceous and Tertiary basin development and sedimentation, Arctic Islands; Chapter 15 in Geology of the Innuitian Orogen and Arctic Platform of Canada and Greenland, H.P. Trettin (ed.); Geological Survey of Canada, Geology of Canada, no. 3; (also Geological Society of America, The Geology of North America, v. E).

Table 15.1. Distribution and zonation of selected Paleogene palynomorphs in Arctic Canada (Beaufort-Mackenzie Basin and Ellesmere Island). After Rouse (1977).

| Paleocene | | | | Eocene | | Oligocene | Age |
| Early (?) | Middle (?) | Middle (?) | Late | Early-Middle | Late | Early | |
P-1	P-2	P-3	P-4	E-1	E-2	O-1	Palynozone
xxxxxxxxxxxxxxxxxx	xx	xx					Paraalnipollenites confusus
xxxxxxxxxxxxxxxxxx	xx	xx					Triporopollenites mullensis
xxxxxxxxxxxxxxxxxxxxxxxxxxxx							Multicellaesporites A
xxxxxxxxxxxxxxxxxxxxxxxxxxxx							Dicellaesporites A
	xxxxx	xx xx					Taxodiaceapollenites A
	xxxxxxxxxxxxxxxxxxxxxxxxx						Quercoidites A
	xxxxxxxxxxxx						Momipites rotundus
		xxxxxxxxxxxxxxxxxxxxxx					Caprifoliipites A
		xxxxxx					Margocolporites cribellatus
		xxxxxx	xx				Liliacidites complexus
		xxxxxxxxxxxxxx	x x				Rhoiipites cryptoporus
		xxxxxxxxxxxxx					Tricolpites reticulatus
		xxxxxxxxxxxxxxxxxxxxxxxxxxxxxxxx					Carya veripites
		xxxxxxxxxxxxxxxxxxxxxxxxxxxxxxxx					Carya viridifluminipites
			xxxxxxxxxxxxxx				Aesculiidites A
			xxxxxxxxxxxxxx	xx x			Pesavis tagluensis
			xxxxxxxxxxxx				Nyssapollenites A
			xxxxxx				Tricolpites A
			xxxxxx	xx xx			Cupuliferoipollenites A
			xxxxxx				Subtriporopollenites A
			xxxxxxxxxxxx				Pistillipollenites mcgregorii
			xxxxxxxxxxxx				Punctodiporites A
				xxxxx			Intratriporopollenites A
				xxxxx			Lonicera - type
				xxxxxx			Holkopollenites A
				xxxx			Myricipites A
				xxxx			Caprifoliipites B
				xxxx			Diporisporites A
				xxxx			Multicellaesporites B
				xxxx			Striadiporites spp.
				xxxx			Tricolporopollenites kruschii sensu Elsik
				xxxxxxxx			cf. Acanthaceae A
				xxxxxxxxxxxx			Momipites coryloides form A
				xxxxxxxxxxxxxxxxx			Momipites coryloides form B
				xxxxxx			Rhoiipites latus
				xxxxxxxxxxxx			Tilia vescipites
				xxxxxxxxxxxx			Tilia crassipites
				xxxxxxxxxxxx			Ctenosporites wolfei
				xxxxxx			Aesculiidites B
				xxxxxxxx	x x		Multicellaesporites spp.
					xxxxxxxxxxxxxx		Juglans sp.
					xxxxxx		Tricolporopollenites A
					xxxxxxxxx		Fagus sp.
					xxxxxx		Boisduvalia clavatites
					xxxxxxx		Jussiaea sp.
						xxxxxx	Parviprojectus A

GSC

Triporopollenites mullensis. Rouse (1977) included most of these species in a basal Paleocene P-1 zone. His next youngest zone (P-2) is marked by the first appearance of *Momipites rotundus*, *Quercoidites* A, and *Taxodiaceapollenites* A.

Zone P-3 was correlated by Rouse (1977) with the lowest (T-1a) zone of Doerenkamp et al. (1976). At Remus Creek it is characterized by *Margocolporites cribellatus*, *Liliacidites complexus*, *Rhoiipites cryptoporus*, *Tricolpites reticulatus*, *Carya veripites*, *C. viridifluminipites*, *Aesculiidites* A and *Pesavis tagluensis*. On Banks Island, the characteristic palynomorphs are *Alnipollenites* sp., *Ericaceiopollenites* sp., *Ulmipollenites* sp. and *Pterocaryapollenites* sp. The marine microplankton *Epicephalopyxis indentata* occurs in one of the wells on western Banks Island.

Zone P-4 at Remus Creek was correlated with zone T-1b of Doerenkamp et al. (1976) on Banks Island by Rouse (1977). Characteristic palynomorphs that appear for the first time include *Nyssapollenites* A, *Tricolpites* A, *Cupuliferoipollenites* A, *Subtriporopollenites* A, *Punctodiporites* A and *Pistillipollenites mcgregorii*. The latter is a particularly distinctive species of late Paleocene to Middle Eocene age (Doerenkamp et al., 1976; Rouse, 1977; Norris and Miall, 1984).

The base of the Eocene is marked by the appearance of *Tilia* spp., including *T. crassipites* and *T. vescipites* (Rouse, 1977; Norris and Miall, 1984). These species, plus *P. mcgregorii*, *Diporisporites* A, *Ctenosporites wolfei*, *Intratriporopollenites* A, *Lonicera*-type, *Tricolporopollenites kruschii*, *Rhoiipites latus* and *Aesculiidites* B comprise zone E-1 of Rouse (1977). Doerenkamp et al. (1976) recorded sparse marine microplankton from Eocene strata of Banks Island, including *Wetzeliella homomorpha*, *W. symmetrica*, *Hystrichokolpoma* sp. and *Achmosphaera* cf. *alcicornu*.

A rich, diverse and well preserved microflora of Middle Eocene age has been listed by Ricketts and McIntyre (1986) from eastern Axel Heiberg Island. *Tricolporopollenites kruschii* and *Pistillipollenites mcgregorii* are the most age-diagnostic forms.

Zone E-2 (Late Eocene) of Rouse (1977) has not been recorded from the Arctic Islands, although this may reflect the incomplete state of research, as equivalent or younger beds may be present at Mokka Fiord on eastern Axel Heiberg Island. Here Sepulveda and Norris (1982) recorded the following fungal spores: *Inapertisporites circularis*, *Diporisporites harrissi*, *Brachysporisporites* sp. cf. *B. cotalis*, *Staphlosporonites delumbus* and *Plochmopeltinites* sp. These are part of a large flora present in the Middle Eocene to Lower Oligocene Richards Formation of the Beaufort-Mackenzie Basin (Norris, 1986).

Oligocene

Oligocene rocks are poorly known in the Arctic Islands. Strata of this age may be present at Mokka Fiord and Lake Hazen (Miall, 1979b, 1986), although the evidence is tentative. Regional evidence indicates that Oligocene rocks are widespread in offshore regions, but the only section that has been systematically examined is that in the Crocker I-53 well on Meighen Island. There, Miall and G. Norris (pers. comm., 1987) reported a flora very similar to that in the Beaufort-Mackenzie Basin (Norris, 1986). The middle Oligocene is well represented, including the Tetrad flora below and the *Lycopodium* flora above.

Miocene-Pliocene

Characteristic palynomorph assemblages include *Picea* sp., *Tsuga* sp., Ericaceae (tetrads), *Larix* sp., *Lycopodium* sp., *Osmunda* sp., *Alnus* sp., *Pterocarya* sp., *Betula* sp., *Salix* sp., *Acer* sp. and *Carya* sp. (Hills and Ogilvie, 1970). Macroflora elements include cones from *Picea banksii*, *Pinus* cf. *P. strobus*, *Larix* sp. and *Metasequoia alnus*, and nuts from *Juglans* cf. *J. cinerea* (Hills and Fyles, 1973).

The age-diagnostic value of these forms is in need of reassessment, in light of the work of Ricketts and McIntyre (1986) and Matthews (1987). *Picea banksii* was regarded by Hills and Bustin (1976) as indicating a Miocene-Pliocene age for the Beaufort Formation, but Ricketts and McIntyre (1986) have now recorded similar forms in beds of undoubted Eocene age. Matthews (1987) tabulated the ocurrence of 95 species of plant macrofossils in the Beaufort Formation, mainly from Banks and Meighen Islands. Among the most important are *Epipremnum* cf. *E. crassum*, *Aracispermum* sp., *Myrica* sp., *Comptonia* sp., *Betula* spp., *Alnus* spp., *Cleome* sp., *Polanisia* sp. and Lythraceae. However, he noted that the age ranges of these forms remain uncertain, concluding that the flora may be as old as mid-Miocene and probably no younger than Early Pliocene.

LITHOSTRATIGRAPHIC FRAMEWORK
Introduction

The name Eureka Sound Group was first proposed by Troelsen (1950, p. 78) for a widespread unit of sandstone, shale and lignitic coal in central Ellesmere Island and Axel Heiberg Island, which he believed to postdate the last orogeny in the area. Tozer (1963, p. 92-95) pointed out that Troelsen had not established any formations within the Eureka Sound Group and had not designated a type area or a type section. Tozer redefined the unit as a formation and stated "it seems reasonable to regard the outcrops on Fosheim Peninsula, adjacent to Eureka Sound, as typical". Thorsteinsson and Tozer (1957) had earlier demonstrated that on western Fosheim Peninsula the Eureka Sound Formation is the highest unit in a structurally conformable, folded sequence of rocks ranging in age from Permian to Tertiary, and it therefore predates the main compressive phase of the Eurekan Orogeny.

The name Eureka Sound Formation has become widely used for the last pulse of clastic sediments of the conformable Sverdrup Basin succession. The formation occurs throughout the Arctic Islands, from Banks Island in the southwest to Bylot Island in the southeast and Lake Hazen in the

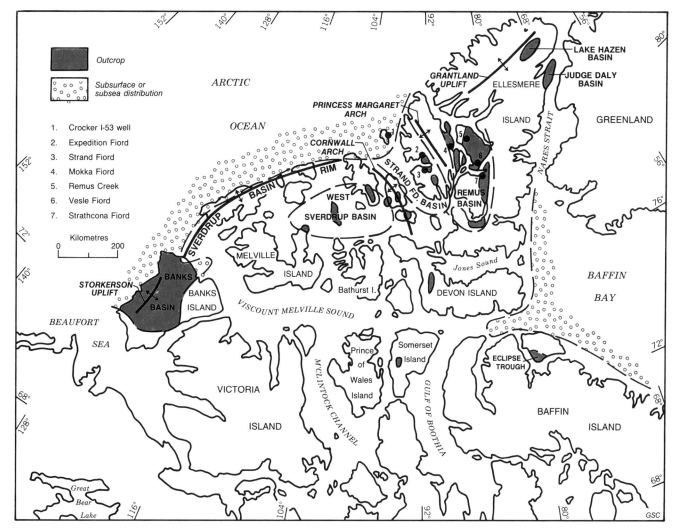

Figure 15.1. Distribution of the Eureka Sound Group in the Innuitian region.

northeast (Fig. 15.1). The formation ranges in age from Campanian or Maastrichtian to Late Eocene or earliest Oligocene, although locally only part of this age range may be represented. Recent work by H.R. Balkwill, R.M. Bustin, B.D. Ricketts and the writer has shown that the formation is highly variable in lithofacies. Nine formation-rank units can now be recognized, and the formation has been raised to group status (Miall, 1986). The formations are based on recognition of ten lithofacies assemblages, as first described by Miall (1981). Some of these assemblages are readily mappable as formations, others cannot be separated for mapping purposes and are grouped herein. The succession of formations varies from place to place within the Arctic Islands, reflecting a complex paleogeography that was continually evolving in response to contemporaneous movements of the Eurekan Orogeny. According to Miall (1986), none of the formations was ever developed simultaneously across large areas of the Arctic, unlike many of the preceding Mesozoic and Paleozoic units. This is because deposition took place within a series of isolated or semi-isolated basins separated by upwarps that acted as local sediment sources. A different approach to the lithostratigraphy, including a different set of formation names, was proposed by Ricketts (1986), based on fieldwork in Axel Heiberg Island and central Ellesmere Island. He interpreted several of the units as being Arctic-wide in distribution, reflecting regional tectonic events and sea level changes. Some of these differences in interpretation are touched on below, and more complete discussions of the different concepts are given by Miall (1988) and Ricketts (1988).

The lithofacies assemblages are summarized in Table 15.2. The new lithostratigraphic classification of Miall (1986) is given in Tables 15.3 and 15.4. Table 15.3 shows the relationship to earlier, informal stratigraphic subdivisions that have been recognized in principal basins across the Arctic.

Rocks of middle Oligocene to Early Miocene age are not exposed at the surface anywhere in the Arctic Islands. Following the main compressive phase of the Eurekan Orogeny, the Innuitian Orogen was uplifted, and large volumes of clastic detritus were shed to the continental margins of the Arctic Ocean and Baffin Bay. Sparse information from a few seismic lines and exploration wells indicates that thick wedges of upper Tertiary sediment drape the continental margins offshore, but very little is known about their stratigraphy or depositional environment. Up to 5 km of post-Eocene strata form a seaward-thickening wedge west of Banks Island, and up to 3 km of sediment (probably including equivalents of the Eureka Sound Group) occur in Lancaster Sound (Hea et al., 1980). G. Norris (pers. comm., 1987) and Miall proposed some tentative correlations of the Arctic continental margin sediment wedge with the Beaufort-Mackenzie Basin, based on the well section at Meighen Island.

By Early Miocene time, sedimentation on the northwest margin of the Arctic Islands had backfilled on to what is now the Arctic Coastal Plain, and formed a veneer of sediment of mid-Miocene to early Pliocene age, named the Beaufort Formation by Tozer (1956). This unit rests disconformably on the Eureka Sound Group in the western Arctic, but may be conformable with the Oligocene to Lower Miocene succession offshore. Similar deposits on eastern Axel Heiberg Island, including strata previously considered as Miocene in age and assigned to the Beaufort Formation (Balkwill and Bustin, 1975; Bustin, 1982), have been reassigned to the Buchanan Lake Formation by Ricketts and McIntyre (1986). The latter is synonymous with the Boulder Hills Formation of this report, and is mid- to Late Eocene, and possibly earliest Oligocene, in age.

Sedimentation has undoubtedly continued from Pliocene to the present in offshore regions, but virtually nothing is known about this part of the stratigraphic succession.

Eureka Sound Group

Brief descriptions are given below of the nine constituent formations of the Eureka Sound Group, based on lithostratigraphic characteristics in their type areas. Other occurrences of the formations are noted. Most of these are physically separated in other basins, and are of similar lithology and depositional environment, but may be of different age. Additional details were provided by Miall (1986).

Table 15.2. Lithofacies assemblages in the Eureka Sound Group (Miall, 1981, 1984a).

Assemblage	Description	Interpretation
A	Mainly very fine grained sandstone, marine fish and invertebrates	Estuarine to shallow marine
B	Mudstone, siltstone, marine microplankton	Prodeltaic
C	Fine grained glauconitic sandstone	Marine shoreline
D	Mudstone, siltstone, thin sandstones, thin coals, coarsening-upward cycles	Distal delta front
E	Mudstone, siltstone, fine to very fine sandstone, coal rare to abundant, coarsening upward cycles	Proximal delta front
F	Very fine to very coarse sandstone, mudstone, siltstone, rare conglomerate, crossbedding common, including lateral accretion sets, fining upward cycles	High sinuosity fluvial
G	Fine to coarse crossbedded sandstone, minor conglomerate, siltstone, mudstone, coal, fining-upward cycles locally present	Low sinuosity fluvial (Platte and Donjek-type)
H	Fine to medium sandstone predominant, plane lamination common	Flood deposits of ephemeral streams (Bijou Creek-type braided)
J	Conglomerate, breccia, minor sandstone, siltstone	Proximal alluvial fan (Scott-type braided)
K	Thin bedded fine clastics, well-preserved plants	Distal alluvial plain and lacustrine

Summarized from Miall (1981, 1984a)

Table 15.3. Stratigraphic subdivision of the Eureka Sound Group.

*Formation	Type area	Other references for type area	Lithofacies assemblage(s)	Other occurrences	Local map unit	References
BOULDER HILLS	Mokka Fiord	Miall (1984a)	J	Lake Hazen Basin	Conglomerate mbr.	Miall (1979b)
MARGARET	Strand Fiord		Mainly E, Minor F Interbeds	Strathcona Fiord Banks Basin	Mbr. IV Cyclic mbr.	West et al. (1981) Miall (1979a)
MOKKA FIORD	Mokka Fiord	Miall (1984a)	F, G, H	Remus Creek Judge Daly Basin Lake Hazen Basin Eclipse Trough Somerset Island	— Mbr. I Ss-mudstone mbr. Upper ss mbr. —	Miall (1984a) Miall (1982) Miall (1979b) Miall et al. (1980) Dineley and Rust (1968)
CAPE LAWRENCE	Cape Back	Mayr and de Vries (1982) Miall (1982, 1984a)	J	Cape Lawrence	—	Mayr and de Vries (1982)
CAPE BACK	Judge Daly Basin (mbrs. 2, 3)	Miall (1982)	K		—	
MOUNT MOORE	Strathcona Fiord (mbr. III)	West et al. (1981)	A	Cañon Fiord	—	Miall (1986)
MOUNT LAWSON	Strathcona Fiord (mbr. II)	West et al. (1981)	D, B	Banks Basin Eclipse Trough Vesle Fiord	Shale mbr. Lower mudstone mbr.	Miall (1979a) Miall et al. (1980) Miall (1986)
VESLE FIORD	Vesle Fiord	—	E	—		
MOUNT BELL	Strathcona Fiord	West et al. (1981)	G	Cañon Fiord	—	Miall (1986)

*Formations are listed in approximate chronostratigraphic order, but no more than four units are present in any one area.
Formal descriptions are given by Miall (1986).

Table 15.4. Chronostratigraphy and correlation, Upper Cretaceous to Oligocene(?) strata of the Innuitian Orogen.

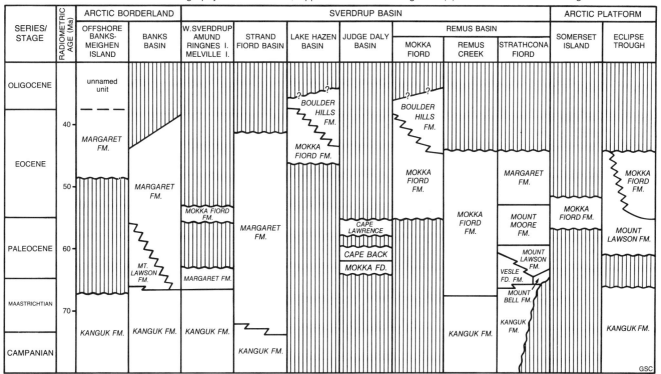

Mount Bell Formation

This unit comprises the basal beds of the Eureka Sound Group throughout the southern and eastern part of Remus Basin. South of Strathcona Fiord, the formation rests conformably on the Kanguk Formation, but in the Wolf Valley and Cañon Fiord areas there is a disconformable contact with older Mesozoic and Paleozoic units. The type section is located south of Strathcona Fiord. There, a well exposed section 194 m thick is present, but the base of the formation is not exposed. In the extreme northeast corner of Remus Basin, near the south shore of Cañon Fiord, the formation has a variable thickness, ranging from 10 to 40 m. Flanking a large inlier of Ordovician carbonates south of central Cañon Fiord, the Mount Bell Formation is absent, the Mount Lawson Formation resting directly on Paleozoic basement. In the central and eastern part of Remus Basin, the Mount Bell Formation therefore appears to comprise a basal unit of variable thickness and facies. It also varies in age.

In western Remus Basin (Remus Creek, Mokka Fiord) the formation cannot be distinguished from the Mokka Fiord Formation, and is not separately mapped. The Mount Bell Formation is not recognized elsewhere in the Arctic Islands.

The formation consists mainly of fine- to medium-grained quartzose sandstone with abundant planar crossbedding. Locally the crossbeds are deformed and overturned. Trough crossbeds and ripple marks are locally present. Interbedded coarse grained sandstone, siltstone and mudstone comprise a minor part of the formation.

Few palynological recoveries have been made from the Mount Bell Formation.

The formation probably is fluvial in origin. The variable thickness of the formation suggests that it is filling erosional relief around the margins of Remus Basin.

Vesle Fiord Formation

The formation is named for a variegated sandstone-mudstone-coal succession 650 m thick outcropping north of Vesle Fiord (Fig. 15.2). The formation forms a wedge, thinning and passing laterally into the Kanguk Formation toward the south, and into the Mokka Fiord Formation in the vicinity of Vesle Fiord Thrust.

The formation consists of superimposed cyclic, sandstone-dominated sequences of facies assemblage E. It is Paleocene in age, according to palynostratigraphic determinations. The unit is very similar lithologically to the Margaret Formation.

Mount Lawson Formation

This formation is named after a prominent topographic feature on Fosheim Peninsula. The formation corresponds to member II of West et al. (1981) and has a widespread distribution in Remus Basin between Strathcona Fiord and Cañon Fiord (Fig. 15.2, 15.3). The unit corresponds to the Strand Bay Formation of Ricketts (1986). The type section is located on the north shore of Strathcona Fiord. There the base is not exposed, but only a few metres of the lower part of the section, above the Mount Bell Formation, are thought to be covered. The exposed section is 576 m thick, and the contact with the unit above is drawn at the first massive, sideritic sandstone of the Mount Moore Formation. A complete but poorly exposed section 227 m thick is exposed above the type section of the Mount Bell Formation. Contacts with units above and below are not well exposed but are assumed to be conformable. A well exposed section 314 m thick is present near Mount Moore. The formation rests directly on Paleozoic basement, and the top is not exposed. At Vesle Fiord, 980 m of beds are tentatively assigned to the Mount Lawson Formation, but the upper part of this incomplete section is not well exposed and may include part of the overlying Mount Moore Formation. The contact with the underlying Vesle Fiord Formation at Vesle Fiord is gradational and is drawn at the top of a sandstone unit above which mudstone becomes the dominant lithology.

Comparable beds in other parts of the Arctic Islands are also assigned to the Mount Lawson Formation. The Shale Member of the Eureka Sound Formation in Banks Basin (map unit KTe[1] of Miall, 1979a) rests conformably on the Kanguk Formation and ranges up to 240 m in thickness (Miall, 1979a). The lower mudstone member (map unit Te[2]) in Eclipse Trough is also comparable in lithology and depositional environment. It reaches at least 500 m in thickness (Miall et al., 1980).

The formation is locally diachronous. In Eclipse Trough, near the "Twosnout Creek" section of Miall et al. (1980), a lateral facies change into deltaic and fluvial rocks of the Remus Formation can be seen in hillside exposures.

The dominant lithology throughout is mudstone (Fig. 15.4A). This is variably silty and carbonaceous. Thin coals are common, as are thin silty to fine grained sandstone beds containing ripples and rare large-scale crossbedding. Plant debris is common throughout. The well exposed section near Mount Moore shows several coarsening-upward cycles up to 55 m thick. One of these culminates in a pebble conglomerate, 4 cm thick, above which the section fines up through crossbedded sandstone, silty sandstone, and coal, over a thickness of 16 m. Bioturbation and horizontal feeding trails are common throughout this section. In Eclipse Trough, the upper part of the formation is characterized by coarsening-upward cycles that become thicker and coarsen upward to the contact with the overlying Mokka Fiord Formation. A similar succession and contact occurs throughout Banks Basin.

In Remus Basin, the formation is composed mainly of lithofacies assemblage D, which is interpreted as the deposit of a low-energy distal delta front and associated interdistributary bays (Fig. 15.4A). Rare pelecypods are probably fresh water in origin. Coarsening-upward cycles record the progradation of distributary mouth sand bodies. Coals record the development of swamps in bay areas. Fluvial delta plain conditions were rarely established.

In Strand Fiord Basin, Ricketts (1984, 1986) mapped a shale-dominated unit 287 m thick, which he designated as the type section of his Strand Bay Formation (informal map unit 3 in Ricketts, 1984; regarded here as closely comparable to the Mount Lawson Formation). The same beds form part of the Margaret Formation of Miall (1986). The relationship of map unit 3 of Ricketts (1984) to the Margaret Formation in Strand Fiord Basin is shown by Miall (1986, Fig. 3).

In Banks Basin and Eclipse Trough, the beds have been assigned mainly to lithofacies assemblage B, a slightly different assemblage, characterized by a scarcity of sandstone and coal beds. Marine microplankton (in Banks

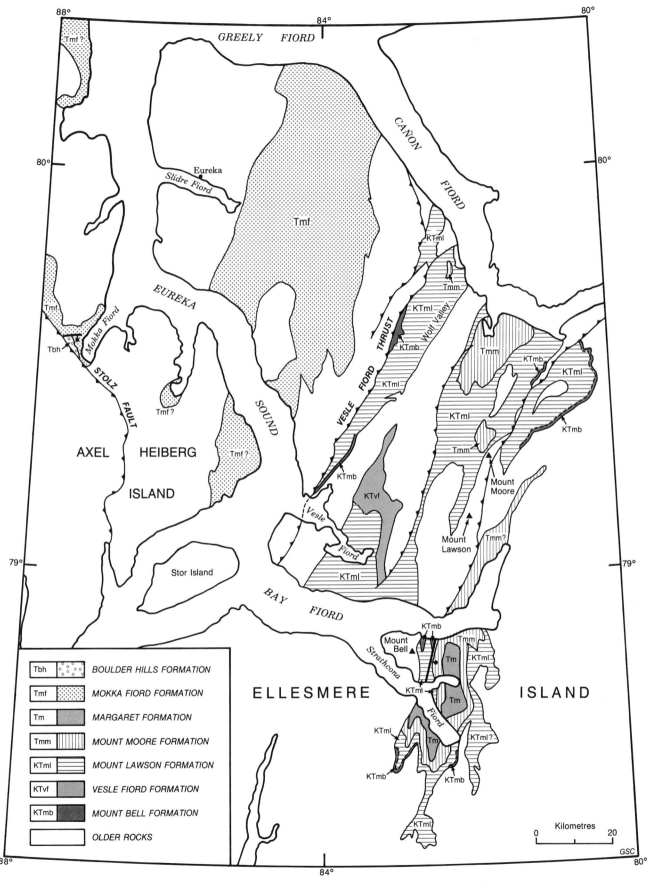

Figure 15.2. Sketch geological map of the Eureka Sound Group in central Ellesmere Island.

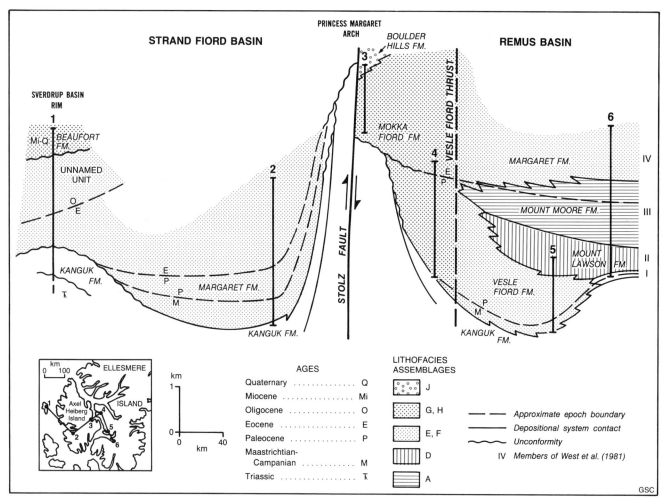

Figure 15.3. Reconstructed stratigraphic cross-section, Eureka Sound Group of Strand Fiord and Remus basins. For code of lithofacies assemblages see Table 15.2.

Basin) and scattered glauconite (in Eclipse Trough) suggest a more open marine environment. In these two basins, the Mount Lawson Formation forms part of major regressive sequences, whereas in Remus Basin it occurs in the middle of a transgressive sequence that reaches its maximum extent with the development of the Mount Moore Formation. In Eclipse Trough, there is an upward transition through assemblages D and E at the top of the formation, to assemblage F comprising the base of the Mokka Fiord Formation.

Mount Moore Formation

This unit corresponds to member III of West et al. (1981). It has a widespread distribution between Strathcona Fiord and Cañon Fiord (Fig. 15.2, 15.3). The type section, at Strathcona Fiord, is 1060 m thick, but the formation reaches at least 1400 m in thickness in exposures a few kilometres to the north. "Member 1" of the Eureka Sound Formation at Stenkul Fiord in southern Ellesmere Island, which is 90 m thick, was correlated with the Mount Moore Formation by Riediger and Bustin (1987).

The formation is characterized by fine- to very fine-grained, poorly consolidated, pale weathering quartzose sandstone. Commonly, beds are siderite-cemented and form prominent ridges, between which hillside exposures are poor. Rare ripple marks and rare pelecypods are present. Feeding trails, coaly lenses, and petrified wood fragments are also characteristic. Ricketts (1986) noted the presence of hummocky cross-stratification in the sandstone beds (he included these beds in his Iceberg Bay Formation). The sandstones are interbedded with subordinate siltstones and silty mudstones. A thin pebble conglomerate is used to define the base of the formation in the type area. West et al. (1981) recorded the presence of a marine fauna of scaphopods, marine to brackish water pelecypods, sharks and marine bony fishes.

Most macrofossils and palynomorphs recovered from this unit have a wide age range. The combined faunal and floral evidence, and regional stratigraphic relationships, suggest a mid-Paleocene to Early Eocene age for the formation.

The Mount Moore Formation constitutes the only occurrence of lithofacies assemblage A in the Eureka Sound Group. This is interpreted as the product of a generally low-energy marine or estuarine environment. The predominance of sand indicates an abundant sediment supply, but the absence of channels, scour surfaces or large-scale

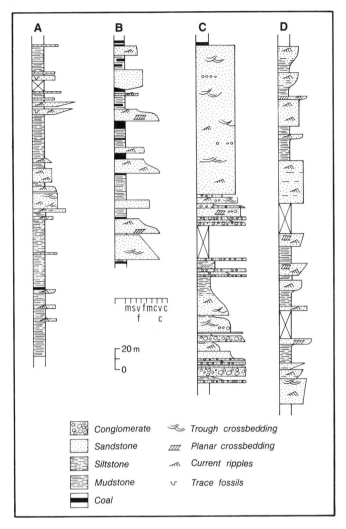

Figure 15.4. Selected stratigraphic sections showing characteristic lithofacies associations within the Eureka Sound Group: A. Mount Lawson Formation, Mount Moore; B. Mokka Fiord Formation, Remus Creek; C. Mokka Fiord Formation, Mokka Fiord (part of the type section); D. Margaret Formation, Strand Fiord (part of the type section).

crossbedding suggests a low-energy environment. Hummocky cross-stratification indicates the occasional passage of high-energy storm waves.

Margaret Formation

The entire Upper Cretaceous – Cenozoic section of the Strand Fiord – Expedition Fiord area of western Axel Heiberg Island is assigned to the Margaret Formation. The unit is named after the Princess Margaret Range, a mountain range extending the length of Axel Heiberg Island. The type section of the formation is on the south shore of Strand Fiord, and is 3200 m thick. There is a gradational contact with the the underlying Kanguk Formation, but the upper contact is not preserved. Ricketts (1984, 1986) showed that these beds could be subdivided into at least three mappable units, and assigned separate formation names to each. Alternatively these units could be designated as members of the Margaret Formation. The Margaret Formation also occurs at Strathcona Fiord, where it constitutes member IV of West et al. (1981). Up to 1160 m of beds overlie the Mount Moore Formation above a thin conglomerate unit which provides a local marker horizon. At Stenkul Fiord in southern Ellesmere Island, Riediger and Bustin (1987) correlated 360 m of beds with the Margaret Formation (their Members 2, 3 and 4). On Banks Island, the Cyclic Member of Miall (1979a) is assigned to the Margaret Formation. There is a gradational lower contact with the Mount Lawson Formation, and here the Margaret Formation reaches a preserved maximum thickness of 1100 m.

The formation consists of interbedded sandstone, siltstone, mudstone, and coal arranged into a series of coarsening-upward cycles up to 160 m thick (Fig. 15.4D). The Margaret Formation is distinguished from the Mokka Fiord by its distinctive cyclic character, and from the Mount Lawson by the greater abundance of sandstone and coal. Sandstone units in the Margaret Formation are abundantly crossbedded. Large exposures reveal broad, shallow channels up to 100 m wide and 3 m deep. Coals are of variable abundance. They are sparse in the type section of the formation, but are common in Banks Basin and abundant in the Strathcona Fiord area. There the formation has 26 coal seams, including one 45 m thick. A washout channel 37 m deep has been mapped at one locality. Ricketts (1984) also mapped a coal-bearing upper member of the formation on the north shore of Strand Fiord.

Rare marine fossils have been recovered from the Margaret Formation in Banks Basin, including foraminifera and radiolaria, and the trace fossil *Ophiomorpha*. Elsewhere the formation has yielded only nonmarine palynomorphs and, at Strathcona Fiord, a remarkable vertebrate fauna. West et al. (1977, 1981) recorded the presence of tortoises, lizards, crocodiles and lemurs in the Margaret Formation there, a fauna that has not been observed elsewhere in the Eureka Sound Group. They indicate an Early to Middle Eocene age for the formation in this area.

At the type section of the formation, the age ranges from Campanian or Maastrichtian to Late Eocene. In Banks Basin the formation forms the upper part of the Eureka Sound Group and is Paleocene to Eocene in age (Miall, 1979a).

Most of the Margaret Formation consists of lithofacies assemblage E, which is interpreted as having formed in proximal delta front to lower delta plain environments. Coarsening-upward cycles represent prograding distributary mouth sands and crevasse splays. Fluvial delta plain conditions were locally established, forming fluvial channel fills and point bars of assemblage F. These have not yet been separately mapped.

Mokka Fiord Formation

This name is applied to distinctive cyclic successions of sandstone and pebbly sandstone, with local thin conglomerate beds, that occur in many parts of the Arctic Islands. The formation has a variable thickness and variable stratigraphic relationships, but is distinguished by its lithofacies characteristics and interpreted depositional environment. The type section is located at Mokka Fiord, where an incomplete section 1310 m thick is exposed (Fig. 15.4C). The base is not exposed, and the formation grades up into the conglomerate of the Boulder Hills Formation. At Remus Creek the formation is at least 2550 m thick. In Judge Daly Basin, the Mokka Fiord Formation corresponds to map unit Te^1 of Miall (1982) and is 650 m thick. The base is not exposed, and there is a gradational upper contact with the Cape Back Formation. In Lake Hazen Basin, the sandstone-mudstone member of Miall (1979b) is assigned to the Mokka Fiord Formation. Complete sections through the formation are not available, but structural considerations lead to an estimated thickness of 450 m. The basal contact is disconformable to unconformable on Paleozoic and Mesozoic rocks, whereas the upper contact is gradational with the Boulder Hills Formation, as at Mokka Fiord.

In Eclipse Trough, the Mokka Fiord Formation corresponds to map unit Te^3 of Miall et al. (1980). There are gradational contacts with the Mount Lawson Formation below, and with the unnamed upper mudstone member above. The Mokka Fiord Formation there is at least 1370 m thick. Exposures of the Eureka Sound Group north of Stanwell Fletcher Lake, Somerset Island, were named the Idlorak Formation by Dineley and Rust (1968), and are reassigned to the Mokka Fiord Formation here. About 300 m of beds are preserved, resting unconformably on Precambrian metamorphic rocks. In southeast Banks Island, the formation rests directly on the Kanguk Formation.

The characteristic lithofacies assemblage of the Mokka Fiord Formation consists of interbedded sandstone, siltstone and mudstone. Thin coal seams are locally abundant, and pebbly sandstones and thin conglomerate beds may also be present (Fig. 15.4B, 15.4C). Relative abundance of these lithofacies varies considerably, and the rocks have been subdivided into four assemblages: F, G_1, G_2, and H, which may be interbedded with each other within the formation. The sedimentology of these rocks has been described in detail elsewhere (Miall, 1984a), and is summarized here.

The section at Remus Creek is a good example of assemblage F. There, the lithofacies are arranged into a series of fining-upward cycles with sharp, erosional bases. Channels may be present, cutting down several metres into underlying coal or fine grained facies. Low angle crossbedding, representing lateral accretion of point bars, is present in other exposures of this assemblage in Lake Hazen Basin and Eclipse Trough. The overall interpretation of lithofacies assemblage F is that it represents the deposits of high-sinuosity sandy to muddy rivers.

Assemblage G_1 is characterized by crossbedded sandstone, with other lithofacies rare to absent. Few cyclic sequences are present. Typical occurrences include scattered outcrops near Remus Creek and much of the section near Stanwell Fletcher Lake, Somerset Island. The deposits probably were formed by the migration of dunes and linguoid or transverse sand waves in shallow, low-sinuosity rivers (cf. Platte model of Miall, 1977, 1978).

Assemblage G_2 consists of cyclic sequences of sandstone, pebbly sandstone and conglomerate with minor silty mudstone. Cycles fine upward and range in thickness from 2.5 to 37.0 m. Typical occurrences include much of the Mokka Fiord Formation at Mokka Fiord, and scattered outcrops in Lake Hazen Basin. The cycles are of several origins. Thinner cycles probably represent channel and bar aggradation (cf. Donjek model of Miall, 1977, 1978), thicker cycles may represent aggradation and progressive abandonment of alluvial fan lobes, or they could be tectonically controlled.

Much of the Mokka Fiord Formation in Eclipse Trough consists of laminated, fine- to coarse-grained sandstone. Crossbedding is rare and no cyclicity is detectable. Other lithologies are virtually absent. This is typical of lithofacies assemblage H. Other examples of this assemblage occur near the top of the formation at Mokka Fiord and Lake Hazen. The deposits are thought to have been deposited by flash floods in ephemeral streams (cf. Bijou Creek model of Miall, 1977, 1978).

The Mokka Fiord Formation represents a particular suite of depositional environments. It formed whenever local paleogeographic and tectonic conditions were suitable, and thus had a variable age range (Table 15.4).

Boulder Hills Formation

The formation is named for a distinctive conglomerate succession 450 m thick that forms the uppermost part of the Eureka Sound Group at Boulder Hills, near Lake Hazen. A similar section at least 200 m thick occurs west of Mokka Fiord. A few kilometres to the north, at Geodetic Hills, Bustin (1982) mapped a conglomerate section up to 1 km thick which he assigned to the Beaufort Formation, but Ricketts and McIntyre (1986) showed from palynological evidence that these beds are Middle Eocene in age, and they are here reassigned to the Boulder Hills Formation (Ricketts, 1986, erected a new unit, the Buchanan Lake Formation, for these conglomerates).

The predominant lithology is cobble and boulder conglomerate, clasts locally reaching 1.4 m in diameter. Minor interbedded lenses of crossbedded and plane bedded sandstone are present. The sediments form lithofacies assemblage J, and are interpreted as the deposits of gravelly braided distributaries within alluvial fans (cf. Scott model of Miall, 1977, 1978).

At Geodetic Hills, the beds are Middle Eocene in age (Ricketts and McIntyre, 1986). As noted by Miall (1986), palynological evidence reported by Miall (1979b) and Sepulveda and Norris (1982) shows that elsewhere the beds may range up to Late Eocene or earliest Oligocene in age.

Cape Back Formation

This formation has been mapped only in the Cape Back area of Judge Daly Basin. It constitutes map units Te^2 and Te^3 of the Carl Ritter Bay outlier (Miall, 1982). These are 630 and 460 m thick, respectively, giving a combined total of 1090 m. Map unit Te^2 is designated as a lower member of the Cape Back Formation. It consists of thinly interbedded, fine- to very fine-grained sandstone, siltstone and mudstone. Ripples are common, and many of the fine grained beds contain well preserved deciduous leaves. The member rests with a gradational contact on the underlying Mokka Fiord Formation. It grades up into map unit Te^3, which consists mainly of dark grey siltstone, thin argillaceous units, and beds of sideritic mudstone. Well preserved plant material is abundant and gastropods are present. This upper member of the Cape Back Formation has an unconformable contact with the overlying Cape Lawrence Formation.

Miall (1982) assigned a Paleocene age to the Cape Back Formation, mainly on the basis of macroflora identified by

L.V. Hills. The sediments of this formation comprise lithofacies assemblage K (see Table 15.2). The lower member is interpreted as the deposits of a distal alluvial plain, possibly containing shallow lakes. The section at Cape Back may have been deposited on a lake margin, where sheet floods inundated subaerial sand flats, and shallow subaqueous environments were subjected to wind-driven currents alternating with quiet water conditions. The finer grained deposits of the upper member probably indicate water deepening and the more widespread development of shallow lacustrine conditions.

Cape Lawrence Formation

This unit is exposed in three isolated areas of northeast Ellesmere Island, where it was mapped as "Tertiary conglomerate" by Mayr and de Vries (1982). The formation is 630+ m thick in Judge Daly Basin, where it rests with angular unconformity on the Cape Back Formation, and is 1 km thick at Cape Lawrence.

The formation consists predominantly of boulder conglomerate and is similar lithologically to the Boulder Hills Formation. Plant specimens were dated by G.E. Rouse (reported in Mayr and deVries, 1982) as Paleocene in age.

Undifferentiated Eureka Sound Group of Northen Ellesmere Island

Outcrops of the Eureka Sound Group in several small areas of northern Ellesmere Island (Fig. 15.5) have not yet been assigned to formations (and will not be included in the discussion of the regional depositional history below) because of uncertainties about their age and setting. Available information can be summarized as follows.

South of western Phillips Inlet (loc. 1, Fig. 15.5) Upper Cretaceous basalt (Chapter 14) is overlain by an assemblage of sandstone, conglomerate, and coal that has yielded plant fossils of undifferentiated late Cretaceous – Paleogene age (K.G. Osadetz, pers. comm., 1988).

A fault block of the Eureka Sound Formation on the south side of Emma Fiord, east of Fire Bay (loc. 2), comprises a lower unit of mudrock and sandstone and an upper unit of conglomerate (R. Thorsteinsson, pers. comm., 1987). Spores of Maastrichtian age were extracted from samples of the lower unit.

An outcrop area of Eureka Sound strata on the west side of Yelverton Bay (loc. 3) is bounded by a glacier on the west and is in fault contact with Upper Proterozoic or Cambrian strata on the south and east (cf. Trettin and Frisch, 1987). The exposed succession (Wilson, 1976) is divisible into a lower unit of sandstone and siltstone, about 500 m thick, and an upper unit of sandstone and pebble to boulder conglomerate, about 200 m thick. The lower unit has yielded florules of early Tertiary and undivided Maastrichtian – early Tertiary ages, and was interpreted to be of nearshore marine origin. The upper unit probably was deposited by a northward-flowing braided river.

A gently- to steeply-dipping conglomerate east of the mouth of Disraeli Fiord (loc. 6), that lies unconformably on the Ordovician M'Clintock Formation, is tentatively assigned to the Eureka Sound Group. It consists largely of rounded carbonate clasts of pebble to boulder grade.

Unnamed Oligocene units

Seismic data obtained off the southwest coast of Banks Island (Hea et al., 1980) indicate the presence of at least 5 km of sediment overlying the Eureka Sound Group unconformably, and seaward-prograding clinoform reflections are locally identifiable. This seismic unit has been penetrated by two wells, Isachsen J-37 on northern Ellef Ringnes Island, and Crocker I-53 on Meighen Island. The beds are unconsolidated and well cuttings show extensive cavings, so that lithological and age determinations

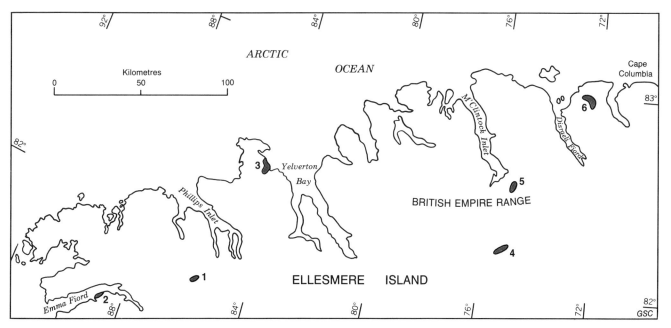

Figure 15.5. Occurrences of the Eureka Sound Group in the western and central part of northernmost Ellesmere Island. Numbers refer to localities discussed in the text.

are extremely difficult.

In the Crocker I-53 well the section consists mainly of interbedded sandstone, siltstone and mudstone with minor coal. Geophysical logs have the blocky to bell-shaped character typical of fluvial cycles. The upper and lower contacts of this unit are drawn tentatively at 1900 ft (580 m) and 6000 ft (1830 m), respectively, based on log character and palynological data (G. Norris, pers. comm., 1987). The unit is therefore 1250 m thick at this locality.

Distinctive palynomorphs are few, and are mixed with numerous taxa recycled from older rocks. Miall and G. Norris (pers. comm., 1987) assigned an Early to Middle Oligocene age to the succession. Regional considerations suggest that this unit extends throughout the length of the Arctic borderland, and probably continues into the Beaufort-Mackenzie Basin, where it correlates with the Kugmallit sequence (Dietrich et al., 1985).

South of upper M'Clintock Inlet (Fig. 15.5, loc. 4), a partly covered Tertiary unit, more than 170 m thick, has been thrust over upper Paleozoic strata. It consists mainly of a breccia that is largely composed of carbonate clasts (U. Mayr, pers. comm., 1988). Spores from an interbedded wood-bearing siltstone (Christie, 1964, p. 55) were re-examined by D.J. McIntyre (pers. comm., 1985), who stated that they are undiagnostic but suggestive of an Oligocene or possibly Miocene age. The same breccia appears to be exposed southeast of M'Clintock Inlet (loc. 5) where it lies on undisturbed upper Paleozoic strata. Here the unit has a minimum thickness of 133 m and its top is not preserved (U. Mayr, pers. comm., 1988).

Strata in Haughton impact crater, Devon Island

An impact crater, about 16 km in diameter, occurs in Ordovician-Silurian carbonate rocks of west-central Devon Island (Frisch and Thorsteinsson, 1978; Thorsteinsson and Mayr, 1987).Within the crater, a polymict impact breccia up to 90 m thick is overlain by about 28 m of lake sediments. The top of the succession is not preserved. The breccia consists mainly of lower Paleozoic sedimentary rocks, predominantly carbonates, but also includes fragments of Precambrian metamorphic rocks. The lake sediments are composed mainly of thinly stratified silt and minor sand of detrital dolomite. They have yielded a Miocene vertebrate fauna as well as ostracodes, pollen and a megaflora. A fission track analysis on apatite from shocked gneiss indicates that the impact occurred 22.4 ± 1.4 Ma ago, i. e. in the Early Miocene (Omar et al., 1987).

Beaufort Formation

The type area of this formation is located on Prince Patrick Island, where it consists of at least 75 m of coarse grained, crossbedded quartzose sand with abundant gravel beds. Large, unaltered logs and tree stumps are common (Tozer and Thorsteinsson, 1964). The formation extends throughout the Arctic Coastal Plain, from southwest Banks Island to Meighen Island. On Banks Island the unit forms a thin veneer a few metres thick over the central part of the island, thickening to 100 m at Ballast Brook in the northwest corner of the island (Hills, 1969), and to more than 260 m on the west coast, where it has been penetrated by several exploration wells (Miall, 1979a). Hills (1969) recognized two members at Ballast Brook, a lower member of medium- to coarse-grained, crossbedded sand, clay, silty clay, and peat, and an upper member of pebble to cobble gravel and medium- to coarse-grained sand with numerous lenses of uncompressed wood.

In the Crocker I-53 well on Meighen Island, the base of the Beaufort Formation is tentatively placed at a depth of 1900 feet (580 m). Geophysical logs and samples indicate that the formation consists mainly of gravel and coarse grained sand. A marine tongue exposed on the island has yielded diagnostic foraminifera, including *Cibicides grossa* of Early or earliest Late Pliocene age (D.H. McNeil, pers. comm., 1988). Palynomorphs and macroscopic plants from Banks and Meighen islands have a possible age range approximately from mid-Miocene to Early Pliocene (Matthews, 1987).

A clastic succession exposed at Geodetic Hills on eastern Axel Heiberg Island was formerly assigned to the Beaufort Formation by Bustin (1982), but has now been shown to be Eocene in age (Ricketts and McIntyre, 1986), and is part of the Eureka Sound Group, as noted above. The same revision may also be necessary for a presumed outcrop of Beaufort Formation at Mackinson Inlet (Riediger et al., 1984).

The Beaufort Formation is probably mainly fluvial in origin, derived from rivers flowing northwest across the Arctic Coastal Plain into the Arctic Ocean. A marine tongue is exposed on Meighen Island. Offshore, the beds probaby contain a deltaic to marine transition.

DEPOSITIONAL HISTORY
Tectonic framework

The Kanguk Formation, underlying the Eureka Sound Group, is a lithologically rather uniform unit that extends throughout the Arctic Islands. The Eureka Sound Group, by contrast, is highly variable in lithology and thickness. Facies and paleocurrent data indicate that it formed in a series of isolated or semi-isolated basins separated by intrabasin upwarps. Several of the latter underwent significant uplift and erosion during deposition of the Eureka Sound Group. Stratigraphic evidence reveals that in some cases several kilometres of Mesozoic sediment were removed, and this material presumably was the source for most or all of the Eureka Sound sediments in the adjacent basins.

The marked contrast in structural setting and depositional style that exists between the Eureka Sound Group and older sediments reflects the commencement of the Eurekan Orogeny (Chapters 17, 18), which was the response to a complex series of plate adjustments brought about by the opening of Baffin Bay – Labrador Sea and consequent relative movements between Canada and Greenland. Balkwill (1978) divided the orogeny into three phases and Miall (1984b) revised the timing of the phases to accord with additional stratigraphic evidence and improved geophysical documentation of the timing of seafloor spreading events. Phase 1 of the orogeny lasted from late Campanian or early Maastrichtian time until the Late Paleocene. Sverdrup Basin was fragmented into a series of smaller basins and intervening upwarps. Toward the end of this phase some of the upwarps appear to have become inactive and were covered, unconformably, by basal sediments of the Eureka Sound Group. Phase 1 culminated with an episode of thrust faulting in eastern Ellesmere Island.

Phase 2 of the orogeny records a change in regional stress patterns as a result of a change in plate trajectories in the Arctic – North Atlantic region (Srivastava, 1985; Srivastava and Tapscott, 1986). However, facies and paleocurrent patterns in the Eureka Sound Group did not show any marked change in style until near the end of phase 2. At this time, during the mid-Eocene to earliest Oligocene, phase 2 culminated with a major compressive episode, and Eureka Sound Group sedimentation was terminated.

Seafloor spreading ceased in Baffin Bay in the Early Oligocene, and phase 3 of the Eurekan Orogeny is now interpreted as post-orogenic isostatic uplift in response to erosion and crustal thickening (Chapters 17, 18). From Early Oligocene or possibly Late Eocene time to Early or mid-Miocene time, the Arctic Islands were uplifted and underwent erosion. The main depocentres shifted to the continental margins.

The basins and upwarps, active during deposition of the Eureka Sound Group, are shown in the accompanying paleogeographic maps (Fig. 15.6-15.11). Sverdrup Basin Rim was named by Meneley et al. (1975) and was intermittently active during much of Sverdrup Basin sedimentation. It did not become an important sediment source until uplifted during phase 1 of the Eurekan Orogeny. Storkerson Uplift, in Banks Basin (Miall, 1975) is physically separated from Sverdrup Basin Rim, but has a similar trend and occupies the same structural position, parallel to the continental margin. It was also active for a short time during the early stages of Eureka Sound sedimentation. Both uplifts are partly fault-bounded.

Princess Margaret Arch (Gould and DeMille, 1964) and Cornwall Arch (Balkwill, 1973) are subparallel arches that trend north-northwest to south-southeast. Both were uplifted about 4.5 km during the Eurekan Orogeny. The Cornwall Arch was uplifted mainly during phase 1 of the orogeny (Balkwill, 1978), but the timing of uplift of Princess Margaret Arch is less well constrained. On Cornwall Island, the Eureka Sound Group ("unnamed Paleocene-Eocene sandstones" of Balkwill, 1983) rests unconformably on rocks as old as Triassic, indicating that up to at least 3.6 km of erosion took place during the latest Cretaceous and the Paleocene (Balkwill, 1983). Triassic rocks are extensively exposed in the core of Princess Margaret Arch. However, the Eureka Sound Group does not cut down lower than the Upper Jurassic (at Flat Sound, eastern Axel Heiberg Island), so that the deeper levels of erosion exposed along the axis of the uplift may have developed following the main

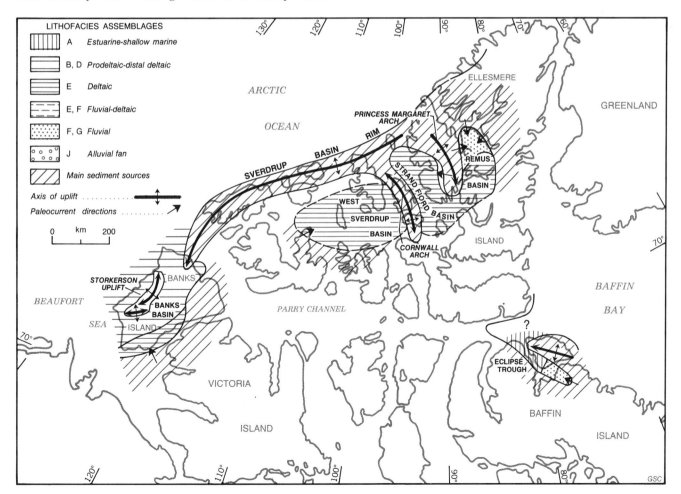

Figure 15.6. Paleogeography, mid- to late Maastrichtian. See text regarding Princess Margaret Arch, which may not have become active until the Eocene (Ricketts, 1987).

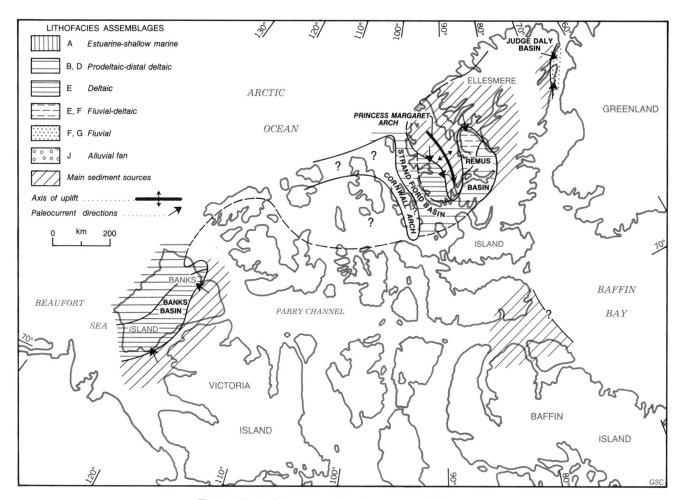

Figure 15.7. Paleogeography, Early to mid-Paleocene.

compressive phase of the orogeny (post Late Eocene). Ricketts (1987) suggested that the onlap of the Eureka Sound Group onto Jurassic strata on eastern Axel Heiberg Island is related to diapiric uplift of Late Paleozoic evaporites and overlying sediments, not to uplift during the Eurekan Orogeny, and he pointed out that that there is little evidence of derivation of Eureka Sound detritus from the rocks of the uplift (e.g. little mafic igneous material from the dykes and sills is present). He suggested that Princess Margaret Arch may not have been uplifted during phase 1, but was formed primarily during the closing compressive movements of phase 2. Farther work will be needed to resolve this point.

Eastern and northern Ellesmere Island were (mountainous?) source areas throughout Eureka Sound sedimentation. Cretaceous units thin dramatically toward the edge of Sverdrup Basin (e.g. at Strathcona Fiord; Balkwill and Bustin, 1975), and throughout eastern Remus Basin the Eureka Sound Group onlaps Mesozoic and Paleozoic rocks. The small Cenozoic basins at Judge Daly Promontory and Lake Hazen were short-lived and contain entirely nonmarine sedimentary fills.

The Franklinian Shelf and Arctic Platform areas also remained mildly positive during much of Eureka Sound sedimentation. Fault-bounded patches of Eureka Sound Group occur in many areas, including Bathurst, Cornwallis, Devon and Somerset islands, and the Bache Peninsula area of Ellesmere Island, but most of these probably represent erosional remnants of local fluvial veneers that have been preserved by later faulting, presumably of Miocene or Pliocene age. The largest outcrop area of Eureka Sound strata within the Arctic Platform is at Stanwell Fletcher Lake, Somerset Island (Dineley and Rust, 1968). The lack of marginal facies changes within this graben structure suggests that the bounding faults are postdepositional.

Within the present continental area of the Arctic Islands, and between the uplifts and landmasses described above, the Eureka Sound Group accumulated in seven major basins. The thickest succession occurs in Remus Basin (named by Bustin, 1977), where stratigraphic reconstructions suggest that more than 4 km of sediment may have accumulated, although no more than 2.5 km are now preserved in any part of the basin. At the centre of the basin, the Eureka Sound Group rests conformably on the Kanguk Formation, but it onlaps older rocks to the west, east and north. The basin is truncated to the north, along Cañon and Greely fiords, possibly by major faults in these seaways.

The Eureka Sound Group sediments of western Axel Heiberg Island were formerly assigned to the Meighen Basin (Miall, 1981), but recently-acquired data indicate that the stratigraphy there and at Meighen Island are quite different. Accordingly this basin is renamed the Strand

Fiord Basin, after the location of the thickest preserved section (3.2 km). The Meighen Island section (Crocker I-53 well) is part of the continental margin sediment wedge, and the name Meighen Basin is abandoned.

West Sverdrup Basin (Miall, 1981) is represented by remnants of Eureka Sound Group preserved at Amund Ringnes Island (300 m; Balkwill, 1983), southern Ellef Ringnes Island (60 m; Stott, 1969), Lougheed Island (60 m; Balkwill et al., 1982), and Sabine Peninsula, Melville Island (33 m; Rahmani and Hopkins, 1977). It is unlikely that this was ever a major depocentre.

Banks Basin (Miall, 1975) was initiated in Jurassic time as a partly fault-bounded trough on the edge of the craton, bordered and partly closed to the west by Storkerson Uplift (Miall, 1975, 1979a). Eclipse Trough (Jackson and Davidson, 1975; Jackson et al., 1975) is a smaller but similar pericratonic feature. It is separated from Baffin Bay by Byam Martin High. The stratigraphic and structural evolution of this basin is probably similar to that of Lancaster Sound, directly to the north (Miall et al., 1980).

Lastly, Judge Daly and Lake Hazen basins were small, short-lived intermontane basins that received sediment during the Paleocene and Eocene, respectively. In both, the Eureka Sound Group is entirely nonmarine (Miall, 1979b, 1982).

Geophysical data do not indicate thick Tertiary strata underlying any of the major inter-island seaways (H.R. Balkwill, pers. comm., 1980), with the exception of eastern Lancaster Sound (Hea et al., 1980; Rice and Shade, 1982).

Paleogeographic evolution
Campanian-Maastrichtian

In Banks, West Sverdrup, Strand Fiord and Remus basins and in Eclipse Trough, the marine mudstones and siltstones of the Kanguk Formation grade up into the deltaic sandstone-mudstone sequences of the Eureka Sound Group. In most localities, this transition is dated as late Campanian or early Maastrichtian, although the deltaic transition took place somewhat later in Banks Island (Maastrichtian to Late Paleocene), and the lowermost beds of the Eureka Sound Group there (Mount Lawson Formation) are similar in lithology and origin to the Kanguk Formation.

The appearance of abundant sandy detritus in the section records the emergence of local sediment sources, and therefore provides a minimum age for the initial differentiation of the Sverdrup Basin into smaller basins and arches by phase 1 of the Eurekan Orogeny. Initially, many local sediment sources may have been present. For example, Kerr (1974) recorded typical Eureka Sound sediments containing marine dinoflagellates and nonmarine flora of Maastrichtian age in southeastern Bathurst Island, and similar marine to nonmarine transition beds may have been widely deposited throughout the Arctic, just as the underlying Kanguk Formation formerly extended much farther than its present confines within Sverdrup Basin (e.g. Dixon et al., 1973). However, by mid- to late Maastrichtian, the sea had probably withdrawn from most of the southern Arctic Islands, and the pattern of basins and arches had become well established (Fig. 15.6). As noted above, the evidence for the existence of Princess Margaret Arch at this time is equivocal. Strand Fiord Basin and Remus Basin may have formed one continuous, southward- to westward-prograding fluvial-deltaic complex (Ricketts, 1987). A major fluvial wedge prograded into Eclipse Trough, and, elsewhere, deltaic conditions were established. Deltaic progradation probably also occurred along the Arctic continental margin beyond Sverdrup Basin Rim, where the Eureka Sound Group rests on progressively older rocks toward the northwest, across the tilted edge of Sverdrup Basin (Meneley et al., 1975). The edge of this sediment wedge may be exposed at Yelverton Bay in northern Ellesmere Island (Wilson, 1976) but elsewhere is deeply buried beneath Oligocene to Pliocene strata.

Based on present evidence, it seems unlikely that a direct marine connection between the Arctic Ocean and Baffin Bay existed after the beginning of the Eurekan Orogeny. In southern Ellesmere and Devon islands, the most likely site for such a seaway, nonmarine strata of Paleocene and/or Eocene age rest unconformably on Paleozoic rocks.

Early to mid-Paleocene

Cratonic sources (and possibly Princess Margaret Arch) continued to supply abundant sandy detritus for deltaic progradation into Strand Fiord Basin and Banks Basin. Judge Daly Basin was initiated, with a flood of pebbly sand detritus from the southwest. Elsewhere, the initial regressive pulse of the Eureka Sound Group appears to have slackened. There is no evidence of sediments younger than Maastrichtian in West Sverdrup Basin, and both Cornwall Arch and Sverdrup Basin Rim may have become tectonically inactive and subject to peneplanation.

In Remus Basin, the transgressive (retrograding) distal deltaic Mount Lawson Formation (Strand Bay Formation of Ricketts, 1986) overlies coarser deltaic sediments of the Vesle Fiord Formation (Fig. 15.3). Regional correlations of the Mount Lawson/Strand Bay with similar beds elsewhere in the Arctic Islands were pointed out by Miall (1986) and Ricketts (1986), although these probably do not indicate a widespread marine transgression at this time, because the beds appear to be of slightly different age in different areas. For example Ricketts (1986) suggested that equivalents of the Mount Lawson/Strand Bay beds, which rest on an unconformity in Eclipse Trough and (according to him) in Strand Fiord Basin, indicate a regional fall, followed by a rise in sea level. However, the unconformity in Eclipse Trough was termed the Bylot Unconformity by McWhae (1981), and is dated as mid-Paleocene, and the succeeding beds are mid-Paleocene to Eocene in age, younger than the Lower to mid-Paleocene age of the Strand Bay Formation in its type area.

Late Paleocene

Sedimentation commenced anew in Eclipse Trough (overlying Bylot Unconformity). The first sediments are quartzose barrier sands of lithofacies assemblage C (lower sandstone member of Miall et al., 1980, and the only well documented occurrence of this assemblage in the Eureka Sound Group). This unit onlaps the Kanguk Formation disconformably and is followed by open marine to prodeltaic muds of the Mount Lawson Formation. In Banks Basin, deltaic progradation probably continued without interruption.

Elsewhere in the Arctic, the latter part of the Paleocene was characterized by peneplanation, and by retrogression or transgression (Fig. 15.8). This is particularly apparent in Remus Basin which became occupied, except for its northwest corner, by a marine embayment. Here, the fine sands and silts of the Mount Moore Formation accumulated. At Stenkul Fiord in southern Ellesmere Island, this unit rests directly on Paleozoic rocks, recording a transgressive enlargement of the southern end of the Remus Basin (Riediger and Bustin, 1987). Westward fluvio-deltaic progradation of the Margaret Formation began there in the mid- to Late Paleocene, considerably earlier than in the Strathcona Fiord area of Remus Basin (Riediger and Bustin, 1987). The Cornwall Arch was overlain by fluvial sediments (the "unnamed Paleocene-Eocene sandstones" of Balkwill, 1983), indicating peneplanation of this area.

The reason for this regional reduction in relief and renewed transgression may be a change in response to the stresses of the Eurekan Orogeny (Miall, 1984b, 1985). Plate tectonic reconstructions by Srivastava (1985) and Srivastava and Tapscott (1986) imply that Sverdrup Basin underwent east-west compression during phase 1 of the Eurekan Orogeny. At first this took the form of broad warping into arches and basins, as described above. Princess Margaret Arch is an example of tectonic "inversion" of a major Mesozoic depocentre into a structural culmination. Cornwall Arch is cored by basement rocks and may be bounded by large reverse faults (Balkwill, 1983, p. 75). In Late Paleocene time, however, it seems likely that the main locus of Eurekan strain shifted to eastern Ellesmere Island. There several patches of conglomerate occur, the Cape Lawrence Formation, that locally reach 1 km in thickness and contain Paleocene plants (Mayr and deVries, 1982). These rocks flank, and are cut by, major thrust faults, including Parrish Glacier, Rawlings Bay and Cape Back thrusts. The conglomerates are interpreted as fanglomerates derived from rising uplifts associated with the thrust faulting. Fault movement may have been contemporaneous with sedimentation and certainly postdated it in part, as conglomerates commonly occupy the footwall of the thrusts. Localization of strain along these structures may have been accompanied by cessation of compression to the west, permitting subsidence and transgression in Remus Basin, Cornwall Arch, and elsewhere.

Latest Paleocene to Middle Eocene

Remus Basin became the site of renewed deltaic progradation (Fig. 15.9). The location of sediment sources and direction of progradation are largely unknown, as reliable paleocurrent data could not be obtained from these rocks. At Strathcona Fiord, the Mount Moore Formation passes up into the

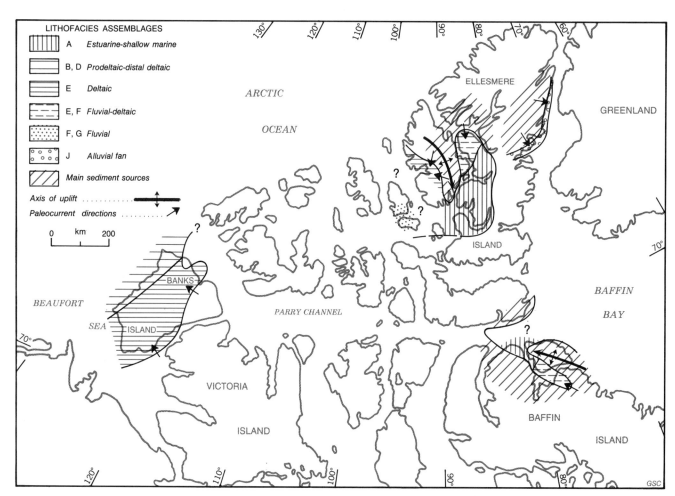

Figure 15.8. Paleogeography, Late Paleocene.

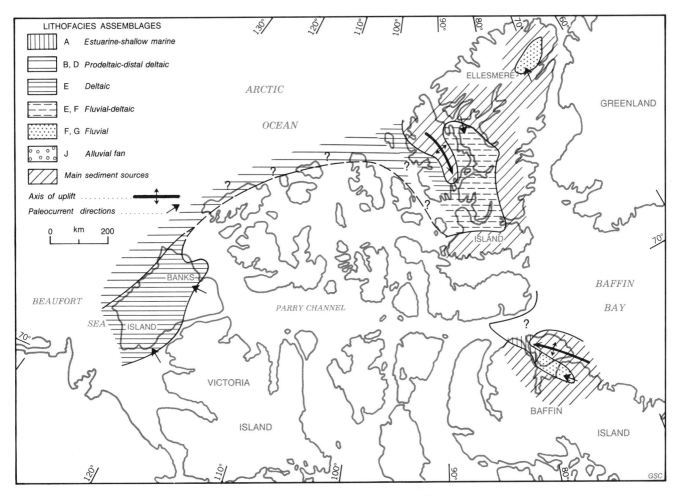

Figure 15.9. Paleogeography, Early Eocene.

vertebrate-bearing Margaret Formation, and to the south, in the Makinson Inlet area, the Margaret Formation rests directly on Paleozoic basement. At Stenkul Fiord, transgressive marine deltaic strata of Member 3 of Riediger and Bustin (1987) are interbedded with nonmarine fluvio-deltaic deposits of Members 2 and 4, suggesting local transgressive-regressive fluctuations, possibly of tectonic origin. In northwest Remus Basin, at Mokka Fiord and Flat Sound, Eocene fluvio-deltaic sediments, here assigned to the Mokka Fiord Formation, rest unconformably on Cretaceous and Upper Jurassic strata (Balkwill et al., 1975). Remus Basin was therefore considerably expanded during this phase.

Deltaic progradation continued westward into Strand Fiord Basin, and onlapped Sverdrup Basin Rim in the vicinity of Meighen Island. To the north and west, a transition into marine facies probably occurred, and similar marine rocks probably underlie the entire continental margin as far as Banks Island (G. Norris, pers. comm., 1987). West Sverdrup Basin may have become uplifted and inactive at this time.

In Eclipse Trough, a broad fluvial plain became established throughout the southeast part of the basin. The deltaic facies is capped by fining-upward cycles of assemblage F, and then by thick, largely plane bedded or massive sandstone of assemblage H, suggesting the development of an ephemeral river system subject to flash floods.

The oldest exposed sediments in Lake Hazen Basin are Eocene in age. They consist of interbedded lenses of assemblages F and G, and were derived by rivers flowing northward across a source terrain consisting of uplifted lower Paleozoic rocks of Hazen Fold Belt.

Many fault-bounded outliers of nonmarine, coal-bearing sandstone and siltstone of Paleocene and/or Eocene age rest on Paleozoic rocks in the Arctic Platform, notably in Somerset and Devon islands (Dineley and Rust, 1968; Hopkins, 1971). The age of some of these outliers is unknown, but they are probably mainly Eocene. The expansion of Remus Basin suggests a regional increase in relief which may have been widespread in the eastern Arctic in response to changed stress patterns of phase 2 of the Eurekan Orogeny (Srivastava, 1985; Srivastava and Tapscott, 1986). At this time, Ellesmere Island may have been undergoing sinistral shear across a broad region corresponding to the present Eurekan Fold Belt (Miall, 1983, 1984b, 1985), although there is no stratigraphic evidence for contemporaneous strike-slip faulting, such as distinctive facies changes or dispersal patterns (cf. Crowell and Link, 1982; Steel and Gloppen, 1980).

Middle Eocene to Early Oligocene(?)

In eastern Axel Heiberg Island and at Lake Hazen, the Eureka Sound Group is capped conformably and gradationally by conglomerate of the Boulder Hills Formation. This formation represents deposition on alluvial fans, and is interpreted as a syntectonic deposit, reflecting uplift of two major axes, Princess Margaret Arch and Grantland Uplift (Fig. 15.10). Uplift was localized along Stolz Fault and Lake Hazen Fault Zone, respectively, and movement on these faults probably was partly contemporaneous with sedimentation.

These conglomerate wedges, of Middle Eocene to possibly Early Oligocene age (W.S. Hopkins Jr., in Miall, 1979b; Sepulveda and Norris, 1982; Ricketts and McIntyre, 1986), are the youngest preserved rocks of the Eureka Sound Group. After their deposition, the entire Arctic Islands area probably underwent uplift, following the climactic phase 2 of the Eurekan Orogeny.

The main locus of sedimentation shifted to the offshore, where thick fluvial sediments continued to prograde toward the Arctic Ocean (Fig. 15.10). No seismic or well data are available to map this wedge between Meighen Island (Crocker I-53 well) and southwest Banks Island (seismic data), but presumably the sediments pass into a marine facies toward the north. Palynostratigraphic data from Meighen Island suggest that the "unnamed Oligocene unit" spans the Early and Middle Oligocene (G. Norris, pers. comm., 1987). A similar stratigraphic wedge probably occurs in eastern Lancaster Sound (McWhae, 1981; Rice and Shade, 1982), but no stratigraphic or facies data are yet available for this area.

Palynostratigraphic data from Meighen Island indicate the presence of a disconformity corresponding to the Late Oligocene and Early Miocene. This disconformity probably extends throughout the Arctic borderland, as a major unconformity occurs at the same stratigraphic position in the southern Beaufort Sea (Willumsen and Côté, 1982). There, the structure is attributed to an episode of uplift, tilting and faulting. The same break also occurs in Baffin Bay, where it has been termed the Beaufort Unconformity (McWhae, 1981).

The "Beaufort Delta" prograded across this unconformity during the Miocene to Pliocene/Pleistocene interval in the Beaufort-Mackenzie Basin, and this depositional system corresponds partly with the Beaufort Formation of the Arctic Islands. The latter unit, of Pliocene and possibly Late Miocene age, represents the development of a broad alluvial plain draining across the Arctic Coastal Plain toward the northwest (Fig. 15.11). Post-tectonic isostatic uplift (Chapter 18) provided the initial impetus for this depositional episode. Sedimentation backfilled onto the Arctic borderland, forming the present Arctic Coastal Plain.

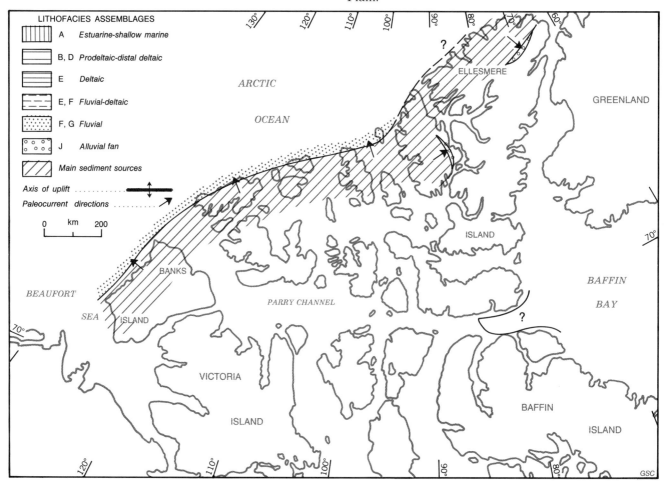

Figure 15.10. Paleogeography, Middle Eocene to Early Oligocene(?).

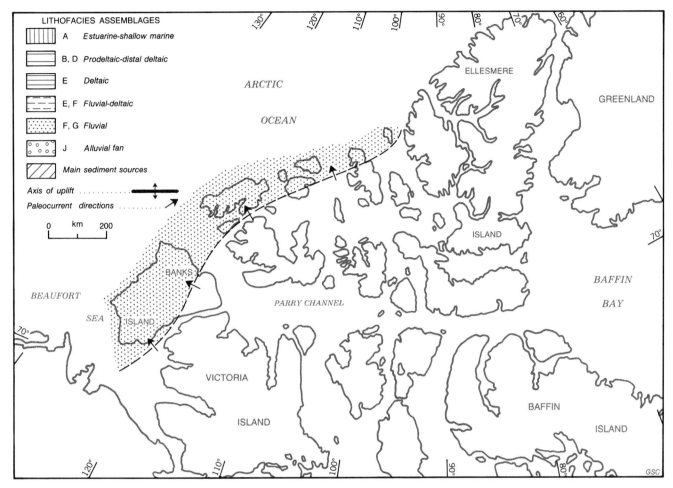

Figure 15.11. Paleogeography, Early Pliocene.

Paleoclimate and paleolatitude

Rouse (1977) interpreted the available palynological evidence for the Paleocene to Eocene of the Canadian Arctic:

Early to Middle Paleocene: cool, wet conditions are indicated by a preponderance of triporates assignable mainly to the Betulaceae and Juglandaceae, both bladdered and taxodiaceous conifer pollen, ericaceous tetrads, and *Lycopodium* and polypodiaceous fern spores.

Middle to Late Paleocene: a marked warming is indicated by an increase in tricolporates and the appearance of the characteristic Gulf Coast genus *Pistillipollenites*.

Early to Middle Eocene: subtropical conditions are suggested by the presence of *Pistillipollenites* (cf. *Tournefortia*, a tropical Central American form), *Rhoiipites latus* (cf. *Triumfetta*, a tropical genus of Tiliaceae) and *Momipites coryloides*.

Throughout the Maastrichtian to Eocene, humid conditions are indicated by the presence of locally abundant coal seams and large fossil logs and tree stumps. Pronounced growth rings in the wood indicate a markedly seasonal climate, which is inconsistent with a paleolatitude lower than about 25° (J. White, in Donn, 1982).

Independent evidence for paleoclimate comes from the rich Early to Middle Eocene vertebrate and invertebrate faunas described by Dawson et al. (1976) and West et al. (1977, 1981) from the Strathcona Fiord area, Ellesmere Island. These occur in association with unusually thick and abundant coals (Miall, 1984a) and indicate "an equable climate with little likelihood of more than brief periods of sub zero (°C) winter temperatures" (West et al., 1981). Modern representatives of many of the fossil groups "live only in tropical and subtropical areas". A mean sea water paleotemperature of 15°C is indicated by oxygen isotope data from a scaphopod collected at this locality (Donn, 1982).

The Strathcona Fiord area is located at 78°45'N, well within the region of continuous winter night and summer day delimited by the Arctic Circle. Several workers have tackled the problem of reconciling a temperate to tropical Paleogene climate with present day climate and latitude. McKenna (1980) evaluated available paleopole determinations, and concluded that the paleolatitude of the Arctic Islands was not significantly different in the Eocene from the latitude at the present day. However, Irving and Sweeney (1982), using a more complete catalogue of paleomagnetic data (Irving and Irving, 1982), showed that the Arctic drifted northward by about 10° between the Eocene and Oligocene. The precise timing of this shift cannot be determined because of statistical imprecision, but if it occurred relatively rapidly after the early Middle Eocene (the age of the youngest vertebrates), the Strathcona

Fiord area may have been located at a latitude of less than 70° during most of the Paleogene. The Arctic lacked a permanent ice cap at this time (Clark, 1971, 1974), and global climates were therefore probably more equable than at present. The evidence for a temperate to tropical climate in Ellesmere Island therefore does not seem so unreasonable.

Miall (1984a) pointed out that local microclimates may also have been strongly influenced by the rugged but variable topography generated during the Eurekan Orogeny. The abundance of coal and fossil vertebrates at Strathcona Fiord is unusual for the Eureka Sound Group, and may indicate the development of local rain forest conditions on the windward flanks of the uplifted craton and Paleozoic fold belt immediately to the east. A coal-bearing member at the top of the Margaret Formation at Strand Fiord, west of Princess Margaret Arch (Ricketts, 1984; Ricketts and Embry, 1984) may be similarly explained.

These considerations seem to explain the existing climatic evidence satifactorily, and there is no support for such far-reaching hypotheses as changes in the obliquity of the earth's spin axis or in solar radiation, as discussed by McKenna (1980).

Neogene climatic cooling is indicated by the flora of the Pliocene (and possibly Late Miocene) Beaufort Formation on Meighen Island. The fossil trees are more diverse taxonomically than modern boreal forest, but are similar to present forest near the treeline with respect to their stunted growth. The associated plant fossils (eg. *Saxifrage oppositifolia*, *Dryas integrifolia*, and *Oxyria digyna*) and insects are characteristic of modern tundra (Matthews, 1987).

REFERENCES

Balkwill, H.R.
1974: Structure and tectonics of Cornwall Arch, Amund Ringnes and Cornwall Islands, Arctic Archipelago; in Proceedings of the Symposium on the Geology of the Canadian Arctic, J.D. Aitken and D.J. Glass (ed.); Geological Association of Canada and Canadian Society of Petroleum Geologists, p. 39-62.
1978: Evolution of Sverdrup Basin, Arctic Canada; The American Association of Petroleum Geologists Bulletin, v. 62, p. 1004-1028.
1983: Geology of Amund Ringnes, Cornwall and Haig Thomas islands, District of Franklin; Geological Survey of Canada, Memoir 390, 76 p.

Balkwill H.R. and Bustin, R.M.
1975: Stratigraphic and structural studies, central Ellesmere Island and eastern Axel Heiberg Island, District of Franklin; in Report of Activities, Part A, Geological Survey of Canada, Paper 75-1A, p. 513-517.

Balkwill, H.R., Bustin, R.M., and Hopkins, W.S., Jr.
1975: Eureka Sound Formation at Flat Sound, Axel Heiberg Island, and chronology of the Eurekan Orogeny; in Report of Activities, Part B, Geological Survey of Canada, Paper 75-1B, p. 205-207.

Balkwill, H.R., Hopkins, W.S., Jr., and Wall, J.H.
1982: Geology of Lougheed Island and nearby small islands, District of Franklin (Parts of 69C, 79D); Geological Survey of Canada, Memoir 395, 22 p.

Bustin, R.M.
1977: The Eureka Sound and Beaufort formations, Axel Heiberg and west central Ellesmere islands, District of Franklin; unpublished M.Sc. thesis, University of Calgary, Calgary, Alberta, 208 p.
1982: Beaufort Formation, eastern Axel Heiberg Island, Canadian Arctic Archipelago; Bulletin of Canadian Petroleum Geology, v. 30, p. 140-149.

Christie, R.L.
1964: Geological reconnaissance of northeastern Ellesmere Island, District of Franklin; Geological Survey of Canada, Memoir 331, 79 p.

Clark, D.L.
1971: Arctic Ocean ice cover and its late Cenozoic history; Geological Society of America Bulletin, v. 82, p. 3313-3324.
1974: Late Mesozoic and early Cenozoic sediment cores from the Arctic Ocean; Geology, v. 2, p. 41-44.

Crowell, J.C. and Link, M.H. (ed.)
1982: Geologic history of Ridge Basin, southern California; Pacific Section, Society of Economic Paleontologists and Mineralogists, Los Angeles, California, 304 p.

Dawson, M.R., West, R.M., Langston, W., Jr., and Hutchison, J.H.
1976: Paleogene terrestrial vertebrates: northernmost occurrence, Ellesmere Island, Canada; Science, v. 192, p. 781-782.

Dietrich, J.R., Dixon, J., and McNeil, D.H.
1985: Sequence analysis and nomenclature of Upper Cretaceous to Holocene strata in the Beaufort-Mackenzie Basin; in Current Research, Part A, Geological Survey of Canada, Paper 85-1A, p. 613-628.

Dineley, D.L. and Rust, B.R.
1968: Sedimentology and paleontological features of the Tertiary-Cretaceous rocks of Somerset Island, Arctic Canada; Canadian Journal of Earth Sciences, v. 5, p. 791-799.

Dixon, J., Hopkins, W.S., Jr., and Dixon, O.A.
1973: Upper Cretaceous marine strata on Somerset Island; Canadian Journal of Earth Sciences, v. 10, p. 1337-1339.

Doerenkamp, A., Jardine, S., and Moreau, P.
1976: Cretaceous and Tertiary palynomorph assemblages from Banks Island and adjacent areas (N.W.T.); Bulletin of Canadian Petroleum Geology, v. 24, p. 372-417.

Donn, W.L.
1982: The enigma of high-latitude paleoclimate; Palaeogeography, Palaeoclimatology, Palaeoecology, v. 40, p. 199-212.

Felix, C.J. and Burbridge, P.P.
1973: A Maastrichtian age microflora from Arctic Canada; Geosciene and Man, v. 7, p. 1-30.

Flynn, J.J., MacFadden, B.J., and McKenna, M.C.
1984: Land-mammal ages, faunal heterochrony, and temporal resolution in Cenozoic terrestrial sequences; Journal of Geology, v. 92, p. 687-705.

Frisch, T. and Thorsteinsson, R.
1978: Haughton Astrobleme: a mid-Cenozoic impact crater, Devon Island, Canadian Arctic Archipelago; Arctic, v. 31, p. 108-124.

Gould, D.R. and DeMille, G.
1964: Piercement structures in the Arctic Islands; Bulletin of Canadian Petroleum Geology, v. 12, p. 719-753.

Hea, J.P., Arcuri, J., Campbell, F.R., Fuglem, M.O., Fraser, I., O'Bertos, J.J., and Smith, D.R.
1980: Post Ellesmerian basins of Arctic Canada: their depocentres, rates of sedimentation and petroleum potential; in Facts and Principles of World Petroleum Occurrence, A.D. Miall (ed.); Canadian Society of Petroleum Geologists, Memoir 6, p. 447-488.

Hickey, L.J., West, R.M., Dawson, R.M., and Choi, D.K.
1983: Arctic terrestrial biota: paleomagnetic evidence of age disparity with mid-northern latitudes during the Late Cretaceous and Early Tertiary; Science, v. 221, p. 1153-1156.

Hills, L.V.
1969: Beaufort Formation, northwest Banks Island, District of Franklin; in Report of Activities, Part A, Geological Survey of Canada, Paper 69-1A, p. 204-207.

Hills, L.V. and Bustin, R.M.
1976: *Picea banksii* Hills and Ogilvie from Axel Heiberg Island, District of Franklin, in Report of Activities, Part B, Geological Survey of Canada, Paper 76-1B, p. 61-63.

Hills, L.V. and Fyles, J.G.
1973: The Beaufort Formation, Canadian Arctic Islands, abstract; in Symposium on the Geology of the Canadian Arctic; Canadian Society of Petroleum Geologists and Geological Association of Canada, Saskatoon, May 1973, Program and Abstracts, p. 11.

Hills, L.V. and Ogilvie, R.T.
1970: *Picea banksii* n. sp., Beaufort Formation (Tertiary), northwestern Banks Island, Arctic Canada; Canadian Journal of Botany, v. 48, p. 457-464.

Hopkins, W.S., Jr.
1971: Cretaceous and/or Tertiary rocks, northern Somerset Island, District of Franklin; in Report of Activities, Part B, Geological Survey of Canada, Paper 71-1B, p. 102-104.

Irving, E. and Irving, G.A.
1982: Apparent polar wander paths Carboniferous through Cenozoic and the assembly of Gondwana; Geophysical Surveys, v. 5, p. 141-188.

Irving, E. and Sweeney, J.F.
1982: Origin of Arctic Basin; Transactions of the Royal Society, Canada, Series IV, v. XX, p. l409-416.

Jackson, G.D. and Davidson, A.
1975: Bylot Island map area, District of Franklin; Geological Survey of Canada, Paper 74-29, 12 p.

Jackson, G.D., Davidson, A., and Morgan, W.C.
1975: Geology of the Pond Inlet map-area, Baffin Island, District of Franklin; Geological Survey of Canada, Paper 74-25, 33 p.

Kent, D.V., McKenna, M.C., Opdyke, N.D., Flynn, J.J., and MacFadden, B.J.
1984: Arctic biostratigraphic heterochroneity; Science, v. 224, p. 173-174.

Kerr, J.W.
1974: Geology of Bathurst Island Group and Byam Martin Island, Arctic Canada; Geological Survey of Canada, Memoir 378, 152 p.

Matthews, J.V., Jr.
1987: Plant macrofossils from the Neogene Beaufort Formation on Banks and Meighen islands, District of Franklin; in Current Research, Part A, Geological Survey of Canada, Paper 87-1A, p. 73-87.

Mayr, U. and de Vries, C.D.S.
1982: Reconnaissance of Tertiary structures along Nares Strait, Ellesmere Island, Canadian Arctic Archipelago; in Nares Strait and the Drift of Greenland: a Conflict in Plate Tectonics, P.R. Dawes and J.W. Kerr (ed.); Meddelelser om Grønland, Geoscience 8, p. 167-175.

McIntyre, D.J.
1984: Paleocene palynological assemblages from the Eureka Sound Formation, Somerset Island, N.W.T., Canada; American Association of Stratigraphic Palynologists, 17th Annual Meeting, abstract.

McKenna, M.C.
1980: Eocene paleolatitudes, climate and mammals of Ellesmere Island; Palaeogeography, Palaeoclimatology, Palaeoecology, v. 30, p. 349-362.

McWhae, J.R.H.
1981: Structure and spreading history of the northwestern Atlantic region from the Scotian Shelf to Baffin Bay; in Geology of the North Atlantic Borderlands, J.W. Kerr and A.J. Fergusson (ed.); Canadian Society of Petroleum Geologists, Memoir 7, p. 299-332.

Meneley, R.A., Henao, D., and Merritt, R.K.
1975: The northwest margin of the Sverdrup Basin; in Canada's Continental Margins and Offshore Petroleum Exploration, C.J. Yorath, E.R. Parker and D.J. Glass (ed.); Canadian Society of Petroleum Geologists, Memoir 4, p. 531-544.

Miall, A.D.
1975: Post-Paleozoic geology of Banks, Prince Patrick and Eglinton Islands, Arctic Canada; in Canada's Continental Margins and Offshore Petroleum Exploration, C.J. Yorath, E.R. Parker and D.J. Glass (ed.); Canadian Society of Petroleum Geologists, Memoir 4, p. 557-588.
1977: A review of the braided river depositional environments; Earth Science Reviews, v. 13, p. 1-62.
1978: Lithofacies types and vertical profile models in braided river deposits: a summary; in Fluvial Sedimentology, A.D. Miall (ed.); Canadian Society of Petroleum Geologists Memoir, 5, p. 597-604.
1979a: Mesozoic and Tertiary geology of Banks Island, Arctic Canada: the history of an unstable craton margin; Geological Survey of Canada, Memoir 387, 235 p.
1979b: Tertiary fluvial sediments in the Lake Hazen intermontane basin, Arctic Canada; Geological Survey of Canada, Paper 79-9, 25 p.
1981: Late Cretaceous and Paleogene sedimentation and tectonics in the Canadian Arctic Islands; in Sedimentation and Tectonics in Alluvial Basins, A.D. Miall (ed.); Geological Association of Canada, Special Paper 23, p. 221-272.
1982: Tertiary sedimentation and tectonics in the Judge Daly Basin, northeast Ellesmere Island, Arctic Canada; Geological Survey of Canada, Paper 80-30, 17 p.
1983: The Nares Strait problem: a re-evaluation of the geological evidence in terms of a diffuse, oblique-slip plate boundary between Greenland and the Canadian Arctic Islands; Tectonophysics, v. 100, p. 227-239.
1984a: Variations in fluvial style in the Lower Cenozoic synorogenic sediments of the Canadian Arctic Islands; Sedimentary Geology, v. 38, p. 499-523.
1984b: Sedimentation and tectonics of a diffuse plate boundary: the Canadian Arctic Islands from 80 Ma B.P. to the present; Tectonophysics, v. 107, p. 261-277.
1985: Stratigraphic and structural predictions from a plate-tectonic model of an oblique-slip orogen: The Eureka Sound Formation (Campanian-Oligocene), northeast Canadian Arctic Islands; in Strike-Slip Deformation, Basin Formation, and Sedimentation, K.T. Biddle and N. Christie-Blick (ed.); Society of Economic Paleontologists and Mineralogists, Special Publication 37, p. 361-374.
1986: The Eureka Sound Group (Upper Cretaceous – Oligocene), Canadian Arctic Islands; Bulletin of Canadian Petroleum Geology, v. 34, p. 240-270.
1988: The Eureka Sound Group: alternative interpretations of the stratigraphy and paleogeographic evolution; in Current Research, Part D, Geological Survey of Canada, Paper 88-1D, p. 143-147.

Miall, A.D., Balkwill, H.R., and Hopkins, W.S., Jr.
1980: The Cretaceous-Tertiary sediments of Eclipse Trough, Bylot Island area, Arctic Canada, and their regional setting; Geological Survey of Canada, Paper 79-23, 20 p.

Norris, G.
1986: Systematic and stratigraphic palynology of Eocene to Pliocene strata in Imperial Nuktak C-22 well, Mackenzie Delta region, District of Mackenzie; Geological Survey of Canada, Bulletin 340, 89 p.

Norris, G. and Miall, A.D.
1984: Arctic terrestrial biota: paleomagnetic evidence of age disparity with mid-northern latitudes during the Late Cretaceous and Early Tertiary: discussion; Science, v. 224, p. 174-175.

Omar, G., Johnson, K.R., Hickey, L.J., Robertson, P.B., Dawson, M.R., and Barnosky, C.W.
1987: Fission-track dating of Haughton Astrobleme and included biota, Devon Island, Canada; Science, v. 237, p. 1603-1605.

Rahmani, R.A. and Hopkins, W.S., Jr.
1977: Geological and palynological interpretation of Eureka Sound Formation on Sabine Peninsula, northern Melville Island, District of Franklin; in Report of Activities, Part B, Geological Survey of Canada, Paper 77-1B, p. 185-189.

Rice, P.D. and Shade, B.D.
1982: Reflection seismic interpretation and seafloor spreading history of Baffin Bay; in Arctic Geology and Geophysics, A.F. Embry and H.R. Balkwill (ed.); Canadian Society of Petroleum Geologists, Memoir 8, p. 245-265.

Ricketts, B.D.
1984: Strand Fiord and Middle Fiord map sheets. Parts of NTS 59G/1, 8; 59H/3, 4, 5, 6. Scale 1:50 000; Geological Survey of Canada, Open File 1147.
1986: New formations in the Eureka Sound Group, Canadian Arctic Islands; in Current Research, Part B, Geological Survey of Canada, Paper 86-1B, p. 363-374.
1987: Princess Margaret Arch: re-evaluation of an element of the Eurekan Orogeny, Axel Heiberg Island, Arctic Archipelago; Canadian Journal of Earth Sciences, v. 24, p. 2499-2505.
1988: The Eureka Sound Group: alternative interpretations of the stratigraphy and paleogeographic evolution — Reply; in Current Research, Part D, Geological Survey of Canada, Paper 88-1D, p. 149-152.

Ricketts, B.D. and Embry, A.F.
1984: Summary of geology and resource potential of coal deposits in the Canadian Arctic Archipelago; Bulletin of Canadian Petroleum Geology, v. 32, p. 359-371.

Ricketts, B.D. and McIntyre, D.J.
1986: The Eureka Sound Group of eastern Axel Heiberg Island: new data on the Eurekan Orogeny; in Current Research, Part B, Geological Survey of Canada, Paper 86-1B, p. 405-410.

Riediger, C.L. and Bustin, R.M.
1987: The Eureka Sound Formation, southern Ellesmere Island; Bulletin of Canadian Petroleum Geology, v. 35, p. 123-142.

Riediger, C.L., Bustin, R.M., and Rouse, G.E.
1984: New evidence for the chronology of the Eurekan Orogeny from south-central Ellesmere Island; Canadian Journal of Earth Sciences, v. 21, p. 1286-1295.

Rouse, G.E.
1977: Paleogene palynomorph ranges in western and northern Canada; in Cenozoic Palynology, American Association of Stratigraphic Palynologists, Contributions Series No. 5A, Contributions of Stratigraphic Palynology, v. 1, p. 48-65.

Sepulveda, E.G. and Norris, G.
1982: A comparison of Paleogene fungal spores from northern Canada and Patagonia, Argentina; Ameghiniana, v. 19, p. 319-334.

Srivastava, S.P.
1985: Evolution of the Eurasian Basin and its implications to the motion of Greenland along Nares Strait; Tectonophysics, v. 114, p. 29-53.

Srivastava, S.P. and Tapscott, C.R.
1986: Plate kinematics in the North Atlantic; in Geology of the Western Atlantic; Geological Society of America, v. M, p. 379-404.

Steel, R.J. and Gloppen, T.G.
1980: Late Caledonian (Devonian) basin formation, western Norway: signs of strike-slip tectonics during infilling; in Sedimentation in Oblique-Slip Mobile Zones, P.F. Ballance and H.G. Reading, (ed.); International Association of Sedimentologists, Special Publication 4, p. 79-104.

Stott, D.F.
1969: Ellef Ringnes Island, Canadian Arctic Archipelago; Geological Survey of Canada, Paper 68-16, 38 p.

Tauxe, L. and Clark, D.R.
1987: New paleomagnetic results from the Eureka Sound Group: implications for the age of early Tertiary biota; Geological Society of America Bulletin, v. 99, p. 739-747.

Thorsteinsson, R. and Mayr, U.
1987: The sedimentary rocks of Devon Island, Canadian Arctic Archipelago; Geological Survey of Canada, Memoir 411, 182 p.

Thorsteinsson, R. and Tozer, E.T.
1957: Geological investigations in Ellesmere and Axel Heiberg Islands, 1956; Arctic, v. 10, p. 2-31.

Tozer, E.T.
1956: Geological reconnaissance, Prince Patrick, Eglinton and western Melville Islands, Arctic Archipelago, Northwest Territories; Geological Survey of Canada, Paper 55-5, 32 p.
1963: Mesozoic and Tertiary stratigraphy; in Geology of the North-Central Part of the Arctic Archipelago, Northwest Territories (Operation Franklin), Y.O. Fortier et al.; Geological Survey of Canada, Memoir 320, p. 74-95.

Tozer, E.T. and Thorsteinsson, R.
1964: Western Queen Elizabeth Islands; Geological Survey of Canada, Memoir 332, 242 p.

Trettin, H.P. and Frisch, T.
1987: Bedrock geology, Yelverton Inlet map area, northern Ellesmere Island; Geological Survey of Canada, Open File 1651, 98 p.

Troelsen, J.C.
1950: Contributions to the geology of north-west Greenland, Ellesmere Island and Axel Heiberg Island; Meddelelser om Grønland, v. 149, no. 7, 85 p.

Vinson, T.E.
1981: A paleomagnetic study of the Upper Cretaceous to Early Eocene Eureka Sound Formation, Strathcona Fiord, Ellesmere Island, Canada; unpublished M.Sc. thesis, University of Wisconsin-Milwaukee, Wisconsin, 125p.

West, R.M., Dawson, M.R., Hickey, L.J., and Miall, A.D.
1981: Upper Cretaceous and Paleogene sedimentary rocks, eastern Canadian Arctic and related North Atlantic areas; in Geology of the North Atlantic Borderlands, J.W. Kerr and A.J. Fergusson (ed.); Canadian Society of Petroleum Geologists, Memoir 7, p. 279-298.

West, R.M., Dawson, M.R., and Hutchison, J.H.
1977: Fossils from the Paleogene Eureka Sound Formation, N.W.T., Canada: occurrence, climate and paleogeographic implications; in Paleontology and Plate Tectonics, R.M. West (ed.); Milwaukee Public Museum, Special Publications in Biology and Geology, no. 2, p. 77-93.

Willumsen, P.S. and Côté, R.P.
1982: Tertiary sedimentation in the southern Beaufort Sea, Canada; in Arctic Geology and Geophysics, A.F. Embry and H.R. Balkwill (ed.); Canadian Society of Petroleum Geologists, Memoir 8, p. 43-53.

Wilson, D.G.
1976: Eureka Sound and Beaufort formations, Yelverton Bay, Ellesmere Island, District of Franklin; in Report of Activities, Part A, Geological Survey of Canada, Paper 76-1A, p. 453-456.

Author's address

A. D. Miall
Department of Geology
University of Toronto
Toronto, Ontario
M5S 1A1

Chapter 16
LATE CRETACEOUS – EARLY TERTIARY DEFORMATION, NORTH GREENLAND

Introduction

Upper Cretaceous dyke swarms

Harder Fjord fault zone

Kap Washington Group

Kap Cannon thrust zone

Wandel Sea structures

References

Chapter 16

LATE CRETACEOUS – EARLY TERTIARY DEFORMATION, NORTH GREENLAND

N.J. Soper and A.K. Higgins

INTRODUCTION

Northern Greenland underwent within-plate deformations in Tertiary time in response to a rather complex and incompletely understood displacement history associated with the opening of Labrador Sea – Baffin Bay, the Arctic Ocean basins, and the North Atlantic. Because post-Ellesmerian cover sequences are not widely preserved in the region, it is frequently difficult to differentiate these "Eurekan" structures from earlier deformations, or to establish their precise age. Nonetheless, three sets of structures and two geotectonically important magmatic events of Cretaceous-Paleogene age have been recognized (Fig. 16.1, Table 16.1). Since these have been subject to speculative and often conflicting geotectonic interpretations, in this account we concentrate on the available factual information, following the review by Soper et al. (1982) and references therein, plus new information derived from 1984 fieldwork. We allude briefly to geotectonic implications, referring the reader to the Nares Strait symposium volume (Dawes and Kerr, 1982) for fuller discussion.

UPPER CRETACEOUS DYKE SWARMS

A dense swarm of approximately north-south trending (coast-normal) dolerite dykes occurs along the north coast of Greenland between about 38°W and 48°W (Nansen Land, westernmost Johannes V. Jensen Land and the intervening islands). Scattered examples are also present throughout Johannes V. Jensen Land. The swarm is most dense at sea level, approaching 50 per cent of the rock in places, with individual dykes up to 25 m in width and occasionally much larger. Dyke density diminishes both upward and to the south, the swarm being confined to the region north of the Harder Fjord fault zone (see below). The dykes cut lower Paleozoic metasediments deformed in the Ellesmerian Orogeny, and post-Ellesmerian Permian and Cretaceous strata, but not the uppermost Cretaceous Kap Washington Group volcanics (see below). Their age is therefore Late Cretaceous. Compositionally they are alkalic, with exceptionally high TiO_2 (usually >4%) and "within-plate" immobile element characteristics. Crystalline gneiss xenoliths reported from one dyke (J.D. Friderichsen, pers. comm., 1985) confirm that they were erupted through continental crust (and incidentally provide evidence of the ensialic nature of the exposed portion of the deep water basin in North Greenland; cf. Chapter 11).

The dyke swarm indicates east-west extension of continental lithosphere. On some pre-drift reconstructions (e.g. Soper et al., 1982, Fig. 7) it is aligned with the initial position of the Nansen spreading axis. The outermost ridge flank anomaly identified in the Eurasian Basin is 24 (52 Ma), but spreading may have begun as early as anomaly 27 (61 Ma, Paleocene) (Vogt et al., 1979). On other reconstructions (e.g. Srivastava, 1985) the dyke swarm lines up better with the Makarov Basin whose age is controversial but which may have developed by Late Cretaceous spreading (Taylor et al., 1981; Kovacs, 1982). The swarm would then represent a direct onshore continuation of an oceanic spreading axis. The alternative possibilities are discussed more fully by Brown et al. (1987).

South of the Harder Fjord fault zone, the north-south swarm is replaced by a less dense suite of west-northwest trending dolerite dykes, and farther south, in Peary Land, a northwest-trending set lies parallel to the Wandel Sea structures described below.

HARDER FJORD FAULT ZONE (HFFZ)

This major east-west fault zone has been traced 250 km from the Wandel Sea coast westward to Nansen Land (Fig. 16.1). Its eastern portion takes the form of a narrow graben (partly occupied by Frederick E. Hyde Fjord) in which representatives of the post-Ellesmerian cover sequences are preserved. These include nonmarine Upper Permian (Wagner et al., 1982), marine Upper Cretaceous (Håkansson et al., 1981; Birkelund and Håkansson, 1983), and nonmarine lower Tertiary sediments (Croxton et al., 1980). The central section of the HFFZ in southern Johannes V. Jensen Land has thick dolerite dykes emplaced along it, some of which are brecciated; greenstones are present, one example of which shows pillow structures and is therefore extrusive; and spatially associated with the fault zone are a number of volcanic necks, which contain basement gneiss xenoliths. Only on the islands west of Harder Fjord is the fault zone adequately exposed. In southern Hazen Land, the northern branch is an extensional fault, inclined steeply to the south, with a vertical displacement of 1.5-2 km. In Nansen Land, offset of Ellesmerian fold trains and stratigraphic boundaries indicates a dextral strike-slip displacement of approximately 20 km. The braided nature of the fault zone, where discernable, indicates a pattern of dextral riedel shears (Fig. 16.1). The fault zone becomes indistinct in western Nansen Land.

Soper, N.J. and Higgins, A.K.
1991: Late Cretaceous – Early Tertiary deformation, North Greenland; Chapter 16 in Geology of the Innuitian Orogen and Arctic Platform of Canada and Greenland, H.P. Trettin (ed.); Geological Survey of Canada, Geology of Canada, no. 3; (also Geological Society of America, The Geology of North America, v. E).

Table 16.1. Cretaceous-Tertiary events and their significance.

EVENT	AGE	DISPLACEMENT, TECTONIC ENVIRONMENT	GEOTECTONIC SIGNIFICANCE
E-W Harder Fjord fault zone	Mid? Late Tertiary-Recent (brecciates dykes; Recent solfataras & earthquakes)	1-2 km uplift to north	Epeirogenic
Wandel Sea faults (NW-SE)	Post-Paleocene	NE-SW extension	Oblique extension
Wandel Sea folds (E-W trending domes)	Late Cretaceous-Paleocene	NW-SE dextral strike-slip	Dextral transpression in Greenland-Barents sector until anomaly 13
Kap Cannon thrust zone	Paleocene-Eocene (post KWG, pre- 32-35 Ma)	N-S compression, >15 km northward overthrusting	Within-plate accommodation of anticlockwise displacement of Greenland
Kap Washington Group volcanics	End-Cretaceous (pollen; Rb-Sr age)	Extension, alkaline magmatism in continental rift	Spreading commenced on Nansen axis at anomaly 24-27 (52-62 Ma, Paleocene)
E-W Harder Fjord fault zone	Late Cretaceous (dykes along fault; Cretaceous pull apart basins?)	c. 20 km dextral strike-slip, accommodating E-W extension to north	Pre-spreading extensional rifting in Eurasia basin
N-S dyke swarm	Late Cretaceous (cuts Cretaceous sediments but not KWG)	E-W extension in continental rift	

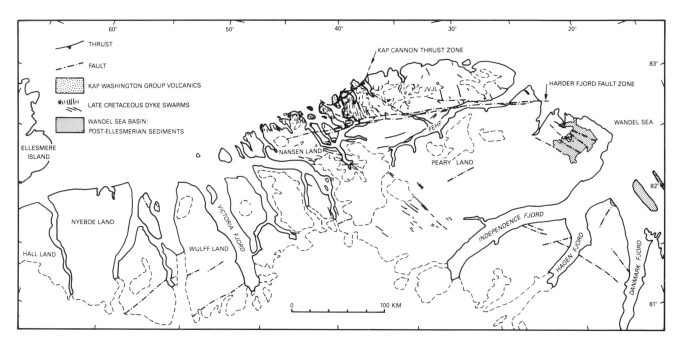

Figure 16.1. Cretaceous-Tertiary magmatic and tectonic features in North Greenland. JVJL = Johannes V. Jensen Land; FEHF = Frederick E. Hyde Fjord.

A conservative interpretation of the displacement history requires two phases — a Cretaceous dextral transtensile movement of a few tens of kilometres associated with dyke emplacement and volcanic activity, and a post-early Tertiary up-to-the-north movement of 1.2 km. We speculate that the Cretaceous displacement accommodated crustal extension north of the fault during emplacement of the north-south dyke swarm. The fault would thus represent a type of continental transform associated with crustal stretching rather than rigid plate displacement. The Tertiary movements presumably relate to a period of stress relaxation after the Kap Cannon thrusting episode and are epeirogenic in nature. Recent sulphurous fumarole and earthquake activity along the fault indicate continuing movement.

There has been much speculation about the earlier history of the HFFZ. Surlyk and Hurst (1984) have proposed that the precursor to this fault zone controlled Early Cambrian sedimentation at the southern margin of the deep water basin. The position of this hypothetical basin margin fault is unclear, since the facies changes in question (Paradisfjeld Group — Portfjeld Formation, and Polkorridoren Group — Buen Formation) occur well to the south of the surface trace of the HFFZ, and also considerable north-south truncation of the trough-platform transition region took place during the Ellesmerian Orogeny. Further, the other Tertiary fault interpreted as a reactivated Paleozoic basin margin structure, the Nyeboe Land fault zone (Dawes, 1982), proves to be non-existent (Soper and Higgins, 1985). However, in support of the idea, it must be said that the linearity and continuity of the HFFZ, and particularly the control it has exerted on magmatism, do seem to require the reactivation of a pre-existing deep crustal fracture. An interpretation which embodies all these considerations is illustrated in a cross-section of the North Greenland fold belt (Fig. 3A, in pocket). A shallow detachment is inferred to exist beneath the HFFZ, associated with thin skinned Ellesmerian deformation of the deep water basin. As shown by Soper and Higgins (1987), the Early Cambrian platform margin, marked by extensional basin margin faults, is probably located in an autochthonous position beneath the footwall of the detachment. Strike-slip reactivation of the faults in Cretaceous-Tertiary time may then have produced the HFFZ in the hanging wall rocks, well to the north of the post-deformation position of the Lower Cambrian basin-shelf transition.

Surlyk and Hurst (1984) have also associated the westerly thrusting in Amundsen Land (Chapter 11) with sinistral strike slip on the HFFZ during the Ellesmerian Orogeny. Håkansson and Pedersen (1982) envisage hundreds of kilometres of sinistral displacement, shifting part of the Amundsen Land structures to northern Ellesmere Island, and making the HFFZ in effect a terrane boundary. These suggestions can be dismissed: there is no evidence that the HFFZ, as now seen, was in existence during the Ellesmerian folding; there is no evidence for sinistral movements at any time; the continuity of lower Paleozoic stratigraphic units across the fault (Surlyk et al., 1980; Higgins et al., 1981; Higgins and Soper, 1983) precludes large post-Ellesmerian displacements; and there is no compelling similarity between the geology of Amundsen Land and that of northern Ellesmere Island.

KAP WASHINGTON GROUP

A thick pile of post-Paleozoic volcanic rocks outcrops on the north coast of Greenland between Kap Cannon and Lockwood Ø (Dawes and Soper, 1970, 1973; Brown and Parsons, 1981; Brown et al., 1987; Fig. 16.2). Over 3 km of extrusive rocks are exposed in southeasterly inclined thrust wedges, together with Permo-Carboniferous and Tertiary sediments beneath the volcanics. A variety of pyroclastic rocks are present, together with basic and felsic flows and minor intrusives. The suite is peralkaline and markedly bimodal in character, with basalts, riebeckite trachytes, and comendites prominent.

The volcanic rocks overlie, apparently conformably, shales and sandstones which have yielded Early Cretaceous and Late Cretaceous leaf fragments. Shales interbedded with the volcanics contain Lake Cretaceous (Campanian-Maastrichtian) spores and pollen (Batten et al., 1981; Batten, 1982). The age of the volcanic episode is thus latest Cretaceous, perhaps extending across the Cretaceous-Tertiary boundary, in close agreement with a Rb-Sr whole-rock determination of 64 ± 3 Ma (Larsen, 1982). The volcanicity followed immediately upon emplacement of the north-south dyke swarm described above, and may be assigned to a similar geotectonic environment — extensional rifting of continental crust. This extensional regime in North Greenland terminated at about the Cretaceous-Tertiary boundary, to be followed by compressional thrusting as described below.

KAP CANNON THRUST ZONE (KCTZ)

A section across the thrust system, which cuts the volcanic pile, the underlying Permo-Carboniferous sediments, and previously deformed metasediments of the fold belt, is shown in Figure 3A (in pocket). The main thrust dips at about 30°SE at Kap Cannon and steepens as traced westward to Lockwood Ø. Metasediments in the hanging wall are mylonitized and Cretaceous dykes are sheared and converted to greenstones. Two thrusts are exposed to the north and others may be present offshore. The volcanic rocks are rarely folded, but the tuffs are locally cleaved and the flows sheared and brecciated.

The cross-section indicates a northward displacement of about 15 km on the thrusts and this may be regarded as a minimum. Little can be inferred about the deep trajectory of the thrust zone. There are no grounds to suppose that it is thin skinned, or forms a listric system linked to extensional faults as depicted by Håkansson and Pedersen (1982, Fig. 19). It may continue to depth at approximately 30°, the regional dip of the fold belt metasediments.

The timing of this thrusting episode is bracketed by the latest Cretaceous age of the volcanic rocks and K-Ar whole-rock dates of 32 and 35 Ma (Early Oligocene) on cleaved felsic flows (Dawes and Soper, 1971), which may reflect erosional unloading after the thrusting. It is thus of Paleocene-Eocene age, and corresponds broadly with Eurekan events in adjacent parts of the Canadian Arctic Islands. Dating is insufficiently precise to permit a finer correlation, either with Eurekan phase 1 — the change from continuous subsidence in the Sverdrup Basin to a "basin

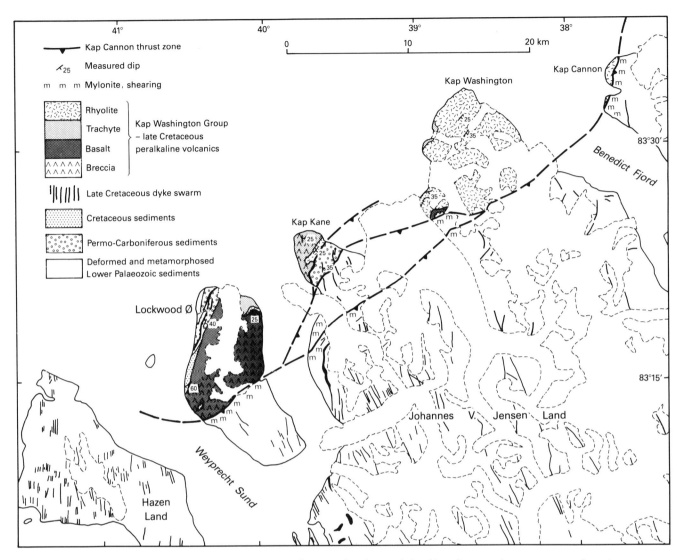

Figure 16.2. Outcrops of Kap Washington Group volcanics and the Kap Cannon thrust zone, northwest Johannes V. Jensen Land. Dips in the volcanics and sediments are to the southwest.

and arch" regime (Balkwill, 1978) — as suggested by Soper et al. (1982), or with phase 2, the main Eurekan deformation, as suggested by Miall (1984). Suffice it to say that the KCTZ compressional phase represents a within-plate response to the anticlockwise rotation of Greenland due to spreading in the Labrador Sea.

WANDEL SEA STRUCTURES

Southeast-trending block faults associated with east-west trending domal folds affect the Wandel Sea Basin cover sequences of eastern Peary Land and Kronprins Christian Land (Haller, 1970; Dawes, 1976; Håkansson, 1979). Håkansson et al. (1981) recognized a Late Cretaceous – Paleocene folding episode, followed by post-Paleocene faulting and thermal events. The faults are parallel to the Wandel Sea continental margin and Spitzbergen fracture zone.

Håkansson and Pedersen (1982) attempted to combine the Kap Cannon thrusting, compression across the HFFZ, and dextral strike slip on the southeasterly-trending faults into a grand "Wandel Hav Strike-Slip Mobile Belt". However, the timing of these displacements is too poorly constrained to be able to say that they are synchronous; substantial strike slip has yet to be demonstrated on the Wandel Sea faults; and the compressional interpretation of the HFFZ stems from the misidentification of a low-angled extensional fault at Depotbugt as a thrust (Håkansson and Pedersen, 1982, Fig. 10). It is more realistic to associate the KCTZ with Eurekan structures of similar trend, and the Wandel Sea structures with the interaction of the Greenland and Barents Shelf continental margins. Reconstruction of the relative displacements by magnetic anomaly matching (Talwani and Eldholm, 1977; Phillips et al., 1978) indicates oblique compression until about the time of anomaly 13 (Oligocene), followed by extension. The Wandel Sea folding and faulting may thus represent weak counterparts of the West Spitzbergen Orogeny (Lowell, 1972; Harland and Horsfield, 1974) and formation of the Forlandsundet graben (Harland, 1969), respectively.

REFERENCES

Balkwill, H.R.
1978: Evolution of Sverdrup Basin, Arctic Canada; The American Association of Petroleum Geologists Bulletin, v. 62, p. 1004-1028.

Batten, D.J.
1982: Palynology of shales associated with the Kap Washington Group volcanics, central North Greenland; Grønlands Geologiske Undersøgelse, Report no. 108, p. 15-23.

Batten, D.J., Brown, P.E., Dawes, P.R., Higgins, A.K., Koch, B.E., Parsons, I., and Soper, N.J.
1981: Peralkaline volcanicity on the Eurasia Basin margin; Nature, v. 294, p. 150-152.

Birkelund, T. and Håkansson, E.
1983: The Cretaceous of North Greenland — a stratigraphic and biogeographical analysis; Zitteliana, v. 10, p. 7-25.

Brown, P.E. and Parsons, I.
1981: The Kap Washington volcanics; Grønlands Geologiske Undersøgelse, Report no. 106, p. 65-68.

Brown, P.E., Parsons, I., and Becker, S.M.
1987: Peralkaline volcanicity in the Arctic basin — the Kap Washington volcanics, petrology and palaeotectonics; Journal of the Geological Society, London, v. 144, p. 707-715.

Croxton, C.A., Dawes, P.R., Soper, N.J., and Thomsen, E.
1980: An occurrence of Tertiary shales from the Harder Fjord Fault, North Greenland fold belt, Peary Land; Grønlands Geologiske Undersøgelse, Report no. 101, p. 61-64.

Dawes, P.R.
1976: Precambrian to Tertiary of northern Greenland; in Geology of Greenland, A. Escher and W.S. Watt, (ed.); The Geological Survey of Greenland, Copenhagen, p. 248-303.
1982: The Nyboe Land fault zone: a major dislocation on the Greenland coast along Nares Strait; in Nares Strait and the Drift of Greenland: a Conflict in Plate Tectonics, P.R. Dawes and J.W. Kerr (ed.); Meddelelser om Grønland, Geoscience 8, p. 177-192.

Dawes, P.R. and Kerr, J.W. (ed.)
1982: Nares Strait and the Drift of Greenland: a Conflict in Plate Tectonics; Meddelelser om Grønland, Geoscience 8, 392 p.

Dawes, P.R. and Soper, N.J.
1970: Geological investigations in northern Peary Land; Grønlands Geologiske Undersøgelse, Report no. 28, p. 9-15.
1971: Significance of K/Ar age determinations from northern Peary Land; Grønlands Geologiske Undersøgelse, Report no. 35, p. 60-62.
1973: Pre-Quaternary history of Greenland; in Arctic Geology, M.G. Pitcher (ed.); American Association of Petroleum Geologists, Memoir 19, p. 117-134.

Håkansson, E.
1979: Carboniferous to Tertiary development of the Wandel Sea Basin, eastern North Greenland; Grønlands Geologiske Undersøgelse, Report no. 88, p. 73-83.

Håkansson, E., Heinberg, C., and Stemmerik, L.
1981: The Wandel Sea Basin from Holm Land to Lockwood Ø, eastern North Greenland; Grønlands Geologiske Undersøgelse, Report no. 106, p. 47-63.

Håkansson, E. and Pedersen, S.A.S.
1982: Late Paleozoic to Tertiary tectonic evolution of the continental margin in North Greenland; in Arctic Geology and Geophysics, A.F. Embry and H.R. Balkwill (ed.); Canadian Society of Petroleum Geologists, Memoir 8, p. 331-348.

Haller, J.
1970: Tectonic map of East Greenland (1:500 000). An account of tectonism, plutonism and volcanism in East Greenland; Meddelelser om Grønland, v. 171, no. 5, 286 p.

Harland, W.B.
1969: Contribution of Spitsbergen to understanding of tectonic evolution of North Atlantic region; in North Atlantic — Geology and Continental Drift, a Symposium, M. Kay (ed.); American Association of Petroleum Geologists, Memoir 12, p. 817-851.

Harland, W.B. and Horsfield, W.T.
1974: West Spitsbergen orogenic belts: data for orogenic studies; in Mesozoic-Cenozoic Orogenic Belts, Data for Orogenic Studies, A.M. Spencer (ed.); Geological Society of London, Special Publication no. 4, p. 747-755.

Higgins, A.K., Friderichsen, J.D., and Soper, N.J.
1981: The North Greenland fold belt between central Johannes V. Jensen Land and eastern Nansen Land; Grønlands Geologiske Undersøgelse, Report no. 106, p. 35-45.

Higgins, A.K. and Soper, N.J.
1983: The Lake Hazen fault zone: a transpressional upthrust?; in Current Research, Part B, Geological Survey of Canada, Paper 83-1B, p. 215-221.

Kovacs, L.C.
1982: Motion along Nares Strait recorded in the Lincoln Sea: aeromagnetic evidence; in Nares Strait and the Drift of Greenland: a Conflict in Plate Tectonics, P.R. Dawes and J.W. Kerr (ed.); Meddelelser om Grønland, Geoscience 8, p. 275-290.

Larsen, O.
1982: The age of the Kap Washington Group volcanics, North Greenland; Bulletin of the Geological Society of Denmark, v. 31, p. 49-55.

Lowell, J.D.
1972: Spitsbergen Tertiary orogenic belt and the Spitsbergen fracture zone; Geological Society of America Bulletin, v. 83, p. 3091-3102.

Miall, A.D.
1984: Sedimentation and tectonics of a diffuse plate boundary: the Canadian Arctic Islands from 80 Ma B.P. to the present; Tectonophysics, v. 107, p. 261-277.

Phillips, J.D., Feden, R., Fleming, H.S., and Tapscott, C.
1978: Aeromagnetic studies of the Greenland-Norwegian Sea and Arctic Ocean; unpublished manuscript, 64 p.

Soper, N.J., Dawes, P.R., and Higgins, A.K.
1982: Cretaceous magmatic and tectonic events in North Greenland and the history of adjacent ocean basins; in Nares Strait and the Drift of Greenland: a Conflict in Plate Tectonics, P.R. Dawes and J.W. Kerr (ed.); Meddelelser om Grønland, Geoscience 8, p. 205-220.

Soper, N.J. and Higgins, A.K.
1985: Thin-skinned structures at the trough-platform transition in North Greenland; Grønlands Geologiske Undersøgelse, Report no. 126, p. 87-94.
1987: A shallow detachment beneath the North Greenland fold belt: implications for sedimentation and tectonics; Geological Magazine, v. 124, p. 441-450.

Srivastava, S.P.
1985: Evolution of the Eurasian Basin and its implications to the motion of Greenland along Nares Strait; Tectonophysics, v. 114, p. 29-53.

Surlyk, F. and Hurst, J.M.
1984: The evolution of the early Paleozoic deep-water basin of North Greenland; Geological Society of America Bulletin, v. 95, p. 131-154.

Surlyk, F., Hurst, J.M., and Bjerreskov, M.
1980: First age-diagnostic fossils from the central part of the North Greenland foldbelt; Nature, v. 286, p. 800-803.

Talwani, M. and Eldholm, O.
1977: Evolution of the Norwegian-Greenland Sea; Geological Society of America Bulletin, v. 88, p. 969-999.

Vogt, P.R., Taylor, P.T., Kovacs, L.C., and Johnson, G.L.
1979: Detailed aeromagnetic investigation of the Arctic Basin; Journal of Geophysical Research, v. 84, p. 1071-1089.

Wagner, R.H.. Soper, N.J., and Higgins, A.K.
1982: A Late Permian flora of Pechora affinity in North Greenland; Grønlands Geologiske Undersøgelse, Report no. 108, p. 5-13.

Authors' addresses

N.J. Soper
Department of Earth Sciences
University of Leeds
Leeds LS2 9JT
U.K.

A.K. Higgins
Geological Survey of Greenland
Øster Voldgade 10
DK-1350, Copenhagen K
Denmark

CHAPTER 16

Printed in Canada

Chapter 17
LATE CRETACEOUS – EARLY TERTIARY DEFORMATION, ARCTIC ISLANDS

Introduction

Eurekan Orogeny

Faulting in the southeastern Arctic Platform and in the Boothia Uplift

Tectonic interpretation of Eurekan Deformation

References

Chapter 17

LATE CRETACEOUS – EARLY TERTIARY DEFORMATION, ARCTIC ISLANDS

A.V. Okulitch and H.P. Trettin

INTRODUCTION

Four major phases of deformation with partly overlapping age ranges affected different parts of the Innuitian Orogen and Arctic Platform in late Mesozoic – Tertiary time:

(1) Middle Jurassic (Aalenian) to Early Cretaceous (Barremian): mild uplift and extensional faulting (i.e. rifting) along the northwestern margin of the Arctic Islands;

(2) Early Cretaceous (Neocomian) to earliest Oligocene: normal faulting in the southeastern part of the Arctic Platform and in the Boothia Uplift, accompanied by local transcurrent faulting;

(3) Paleogene (and possibly latest Cretaceous): compressional deformation in the northeastern part of the Arctic Islands (Eurekan Orogeny);

(4) Middle and Late Tertiary (Oligocene-Pliocene): widespread differential uplift, most pronounced in the eastern part of the Eurekan Orogen and around Baffin Bay, accompanied by some normal faulting.

Event 1, known mainly from the stratigraphic record, is discussed in Chapter 14 by Embry, who relates it to rifting in the Amerasian Basin (Fig. 14.60). Structural treatment of this region, including Late Cretaceous and Tertiary events, will have to await systematic analysis of available geophysical information. However, normal faulting, such as on Banks Island (Miall, 1979a), recurred in the Tertiary.

Event 4, inferred from limited geological and geomorphic information, is discussed in Chapter 18.

This chapter is concerned with the Eurekan Orogeny (event 3) and the partly coeval but longer ranging extensional event 2. Kerr's (1967) hypothesis that the two are related in origin — having been caused by counterclockwise rotation of Greenland — now is widely accepted, and therefore his term, Eurekan Deformation (Kerr, 1977), is here retained for the combined compressional and extensional events. (For geographic names in this chapter, see Fig. 1, in pocket).

EUREKAN OROGENY
Introduction

The term, Eurekan Orogeny, was introduced by Thorsteinsson and Tozer (1970) for major compressive deformation that followed the deposition of the Eureka Sound Formation (now Group), then regarded as Paleocene to Eocene in age, and preceded the deposition of the Beaufort Formation, then regarded as Pliocene.

Two different lines of investigation have led to revisions in the age range of the deformation.

(1) Stratigraphic studies have shown that significant deformation occurred during the deposition of the Eureka Sound Group, now known to range in age from latest Cretaceous (late Campanian) to Late Eocene or possibly Oligocene (Chapter 15). The most cogent evidence consists of unconformities, such as a sub-Middle Eocene unconformity that truncates the Cornwall Arch, an important Eurekan structure (Balkwill, 1974), and of syntectonic conglomerates of Middle Eocene and possibly Paleocene and Oligocene ages (Miall, 1986 and Chapter 15, this volume; Ricketts and McIntyre, 1986). In addition, Late Cretaceous and early Paleogene sandstones of the Eureka Sound Group that are older than the conglomerates may indicate incipient Eurekan movements.

Stratigraphic evidence also led to a temporary revision of the upper age limit of the orogeny when syntectonic conglomerates on eastern Axel Heiberg Island, previously regarded as part of the Eureka Sound Formation (Group), were reassigned to the Beaufort Formation on the basis of spruce cones that suggested a Miocene age (Balkwill and Bustin, 1975, 1980; Bustin, 1982). However, more recent palynological and lithological studies indicate a Middle and possibly Late Eocene age (Ricketts and McIntyre, 1986) and thus restore these deposits to the Eureka Sound Group. Consequently, stratigraphic-structural studies indicate only than that the Eurekan Orogeny terminated some time during the wide hiatus between the Eureka Sound Group and the Beaufort Formation (Late Eocene or Oligocene to Miocene or Pliocene).

(2) Plate-tectonic considerations (discussed at the end of this chapter) also have a bearing on the age range of the compressive deformation. As mentioned, Kerr (1967) proposed that the Eurekan Orogeny was caused by counterclockwise rotation of Greenland, which he attributed to rifting in the Labrador Sea – Baffin Bay region (the Eurekan Rifting Episode of Kerr, 1977). Subsequent geological and geophysical studies, especially analyses of magnetic seafloor anomalies, have shown that both rifting and seafloor spreading have occurred, the latter extending from late Early Cretaceous to earliest Oligocene. This would imply that the Eurekan compression terminated in the earliest Oligocene, if age and tectonic significance of the anomalies have been interpreted correctly. On the other hand, the onset of the inferred spreading and rotation is much older than any known compressive deformation in the Arctic Islands and this poses a problem.

Okulitch, A.V. and Trettin, H.P.
1991: Late Cretaceous – Early Tertiary deformation, Arctic Islands; Chapter 17 in Geology of the Innuitian Orogen and Arctic Platform of Canada and Greenland, H.P. Trettin (ed.); Geological Survey of Canada, Geology of Canada, no. 3; (also Geological Society of America, The Geology of North America, v. E).

Several authors have attempted to establish the sequence of events during the Eurekan Orogeny on the basis of stratigraphic-sedimentological and/or plate tectonic information (e.g. Balkwill, 1978; Miall, 1984, 1986 and Chapter 15; Srivastava, 1985). However, such attempts must necessarily remain speculative until the ages of unconformities and syntectonic conglomerates in the Eureka Sound Group have been established more firmly. Moreover, structures lying within the Eurekan Orogen, apart from those which evidently have formed during the Ellesmerian Orogeny or Mesozoic diapirism, can rarely be dated with assurance because few affect the Eureka Sound Group itself; most involve only Mesozoic or Paleozoic beds. Major structural elements of proven or assumed Late Cretaceous to Tertiary age are therefore discussed by structural province, not in an evolutionary order.

Setting of orogen, trends, and major domains

The Eurekan Orogen (Fig. 17.1) is developed mainly in rocks of the early Paleozoic deep water basin and Pearya Terrane and in unconformably overlying strata of the Sverdrup Basin, but also includes a part of the lower Paleozoic Shelf Province in southeastern Ellesmere Island.

Eurekan structural trends are highly variable but parallel with the highly variable pre-Carboniferous trends in given areas (cf. Fig. 12A.1 and 17.1). Four first-order domains, distinguished mainly on the basis of trend, are briefly characterized below with respect to trend, structural style, and present topographic relief. The latter probably is indicative of Eurekan crustal shortening, although in some areas post-Eurekan thermal events may have been influential (Chapter 18). Also described is a "stable block" that forms a fifth domain within the orogen. The most important structural elements are further discussed in subsequent sections of this chapter.

The **Central Ellesmere Domain** is characterized by extensive thrust faults with sinuous southwesterly to southerly strike and northwesterly to westerly dip that have folds and normal faults associated with them. Most important is a belt of thrust faults that parallels an uplifted salient of the Canadian Shield to the southeast (the Inglefield Uplift of Chapter 12D). The faults dip beneath the Victoria and Albert Mountains, and imbrication on them has likely produced the crustal roots of this range, which is more than 2 km in height. The style of faulting and folding has been influenced by Ordovician evaporites between Cañon and Baumann fiords, and by a Carboniferous evaporite (Otto Fiord Formation) on east-central Axel Heiberg Island where diapirs occur. Strike-slip motion has been inferred for some faults on Judge Daly Promontory adjacent to Nares Strait (sinistral) and on Raanes Peninsula, west-central Ellesmere Island (dextral).

The **Hazen Plateau Stable Block** separates eastern parts of the Central Ellesmere and Northern Ellesmere domains. On the Hazen Plateau, horizontal strata of the Eureka Sound Formation overlie steeply-dipping beds of the Hazen Fold Belt with a high-angle unconformity (Chapter 12E). This region appears to have been stabilized to a degree by tight folding during the Ellesmerian Orogeny, but may be underlain at depth by subhorizontal Eurekan thrusts. Topographic relief is about 1 km.

The **Northern Ellesmere Domain** is characterized by arcuate, southwesterly- to westerly-striking thrust faults and associated folds. Most thrust faults (e.g. Lake Hazen Fault Zone) dip northwest or north, but some (e.g. Mitchell Point Fault) dip southeast or south. Upper Paleozoic evaporites (Otto Fiord and Mount Bailey formations) have influenced the style of deformation in southeastern parts of the domain. Dextral strike slip is inferred for the Feilden Fault Zone, and a component of dextral slip may also be present in the Lake Hazen Fault Zone. Normal faults are common, especially in the north coast region. Topographic relief is highest (up to 2.6 km) in the southeastern part, which is underlain by the Lake Hazen Fault Zone.

The **Sverdrup Islands Domain** has northwesterly- to northerly-trending structures that are aligned with the Boothia Uplift to the south and the Northern Heiberg Fold Belt to the northeast. The most important structures are the Cornwall Arch of Cornwall and Amund Ringnes islands, and the Princess Margaret Arch of Axel Heiberg Island, the southern part of which is bounded on the east by the west-dipping Stolz Thrust. The Carboniferous Otto Fiord evaporite, which underlies much of this domain, has given rise to numerous diapirs and forms an important décollement. The fact that topographic relief is high on central Axel Heiberg Island (Princess Margaret Range) and moderate to predominantly low in the western part of the domain suggests that crustal shortening diminishes in a southwesterly direction; this inference is in accord with structural observations.

The **Prince Patrick Domain** has very gentle, southeasterly-trending folds, a northeast-dipping thrust fault (Harrison et al., 1988), and low topographic relief.

Arches

Cornwall Arch

This major anticline, named for Cornwall Island (Balkwill, 1983), plunges northward through Cornwall and Amund Ringnes islands and is at least 200 km long and 70 km wide. It has a structural relief of about 4 km across its crest on Cornwall Island and of about 5 km along the plunge from central Cornwall Island to the northwestern tip of Amund Ringnes Island. The arch is asymmetric in cross-section: strata on the western flank on both Cornwall and Amund Ringnes islands dip westward at 4-7°, strata on the eastern flank on Cornwall Island dip eastward at 10-16°. Beds on the eastern flank on Amund Ringnes Island are disrupted by faults and minor folds.

On the western flank of the arch on southwestern Amund Ringnes Island, strata of the Eureka Sound Group of late Campanian and/or early Maastrichtian age lie conformably on Upper Cretaceous shale of the Kanguk Formation, whereas, on the crest of the arch, nonmarine sandstone of undifferentiated late Paleocene–middle Eocene age lies unconformably on Middle Triassic to Lower Cretaceous sediments. These relationships indicate that at least 4.5 km of Mesozoic section was eroded from Cornwall Island and at least 2.7 km from south-central Amund Ringnes Island in latest Cretaceous and/or early Tertiary time.

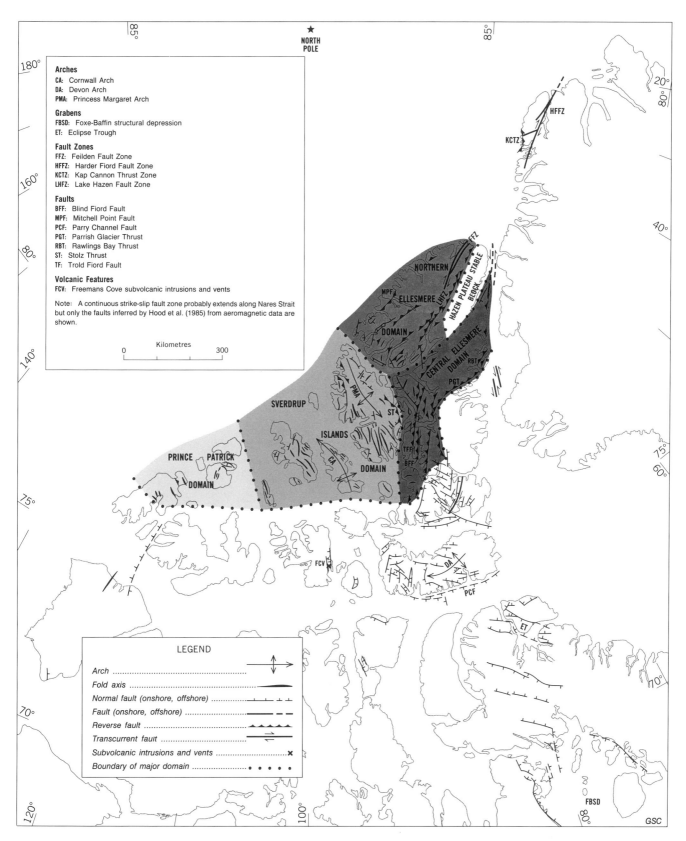

Figure 17.1. Elements of the Eurekan Orogen and related extensional structures.

Surface exposures and available subsurface information do not reveal the deep structure of the arch. Balkwill (1974; 1983, Fig. 4, 26) speculated that the surface structures formed in response to tilting of a basement block bounded on the east by normal faults. More recently, Balkwill (1983, p. 75, Addendum) has interpreted the arch as a response to regional compression, and proposed that it is "underlain by large reverse faults that may penetrate to a mobilized zone deep in the crust". These reverse faults were visualized as having originated as extensional listric faults during early stages in the development of the Sverdrup Basin.

Princess Margaret Arch

This structural and topographic high, named by Gould and de Mille (1964) after the Princess Margaret Range, can be traced for about 275 km along the centre of Axel Heiberg Island (Thorsteinsson, 1974) and may be part of an extensive elevated region extending to the northwest. The age of units exposed along the axis of the arch increases from Late Triassic in southern and central parts, to Early Cambrian or older in the northwest, reflecting a southeastward plunge combined with a northwestward decrease in altitude of the northern half where the pre-Carboniferous Northern Heiberg Fold Belt is exposed (Chapters 12B, 12E). North of about 80°30'N latitude, the arch is a broad, faulted anticlinorium, the axis of which coincides with a structural high within the pre-Carboniferous basement complex. To the south, the structure is partly obscured by ice fields but consists of numerous elongate, doubly-plunging folds. In this region the arch is flanked on the northeast by the arcuate, southwest-dipping Stolz Thrust Fault that locally places Carboniferous units upon Tertiary strata. Along most of its length (except for the northernmost part), the structures apparent on the surface probably are bounded at depth by a décollement within Carboniferous evaporite of the Otto Fiord Formation (Ricketts, 1987), but the arch itself may be controlled by structures beneath the décollement, similar to those proposed for the Cornwall Arch. It is possible that the Princess Margaret Arch dates back to the earliest Tertiary, as suggested by Miall (1986), but the oldest sediments unquestionably derived from the arch are Middle and(?) Upper Eocene boulder conglomerates in the footwall of the Stolz Thrust (Ricketts and McIntyre, 1986).

Reverse faults and folds

The style of reverse faulting and folding has been affected significantly by Paleozoic evaporites, except where these are thin. It is necessary therefore to describe separately areas with or without significant evaporite influence.

Structural style not influenced by evaporites

Southeastern parts of Central Ellesmere Domain

This region is characterized by extensive thrust faults with a sinuous trend that parallel the boundary of an uplifted salient (Inglefield Uplift) of the Canadian Shield on the southeast. Eurekan thrust faults have been recognized at Cape Back, Rawlings Bay and Copes Bay (Mayr and de Vries, 1982). The well exposed Parrish Glacier Thrust (Fig. 17.1) dips north and northwest 30-50° and has a stratigraphic displacement of 4.5 km. Displacement could be as much as several tens of kilometres, if significant portions of the fault dip at very shallow angles, but neither the true displacement nor the geometry of the fault are known. The thrust plate consists of slightly undulating Lower Cambrian to Silurian strata that dip northwest at 20-40°. In front of the main thrust are several imbricate thrust sheets with incomplete stratigraphic sequences, and a klippe with an almost horizontal thrust plane. The footwall of the Parrish Glacier Thrust consists of Paleogene (probably Paleocene or Eocene) conglomerate in the east and Silurian carbonate farther west. The structural configuration of the Rawlings Bay Thrust is similar to that of the Parrish Glacier Thrust.

Northern Ellesmere Domain

The **Lake Hazen** zone of imbricate, southeast-directed, en echelon thrust faults extends from the coast of the Lincoln Sea to Hare Fiord for about 375 km. Its trend is southwesterly in the east half and becomes westerly in the west half. The main faults place member B of the Lower Cambrian Grant Land Formation on strata ranging in age from Cambrian (Hazen Formation) to Cretaceous. Faults within the Grant Land Formation to the northwest of the Lake Hazen Fault Zone are difficult to delineate in the absence of distinct marker horizons. Southeast of the main thrust, northeast of Lake Hazen, a thrust sheet composed of upper Paleozoic to Upper Cretaceous strata has overridden the Paleogene Eureka Sound Group. Footwall strata have been deformed into an asymmetric syncline whose northwestern limb is locally overturned to the southeast (Fig. 17.2). Syntectonic sediments in this area date the initiation of reverse faulting as Eocene or possibly Oligocene (Chapter 15).

The Lake Hazen Fault Zone is best exposed in the vicinity of upper Tanquary Fiord (Fig. 17.3). The following summary for this area is based on a detailed investigation by Higgins and Soper (1983; see also Osadetz, 1982). Faults at the surface dip mainly 30-62°; steeper dips are present where they are confined to member B of the Grant Land Formation. Apparent vertical displacement is about 3.5 km at section a, and 2.5 km at section b (Fig. 17.4), inferred from the elevation of the unconformity between the Grant Land Formation and strata of the Sverdrup Basin. That only moderate horizontal displacement has occurred is suggested by the facts that individual faults are not connected, and that some die out laterally into folds (cf. sections d, e and f

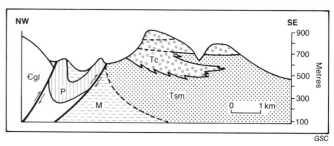

Figure 17.2. Diagrammatic cross-section of Lake Hazen Fault Zone northwest of Lake Hazen at Boulder Hills (Miall, 1979b, Fig. 3). €gl = Grant Land Formation; P = Permian; M = Mesozoic; Tsm = Eureka Sound Formation (now Group), sandstone-mudstone member; Tc = conglomerate member.

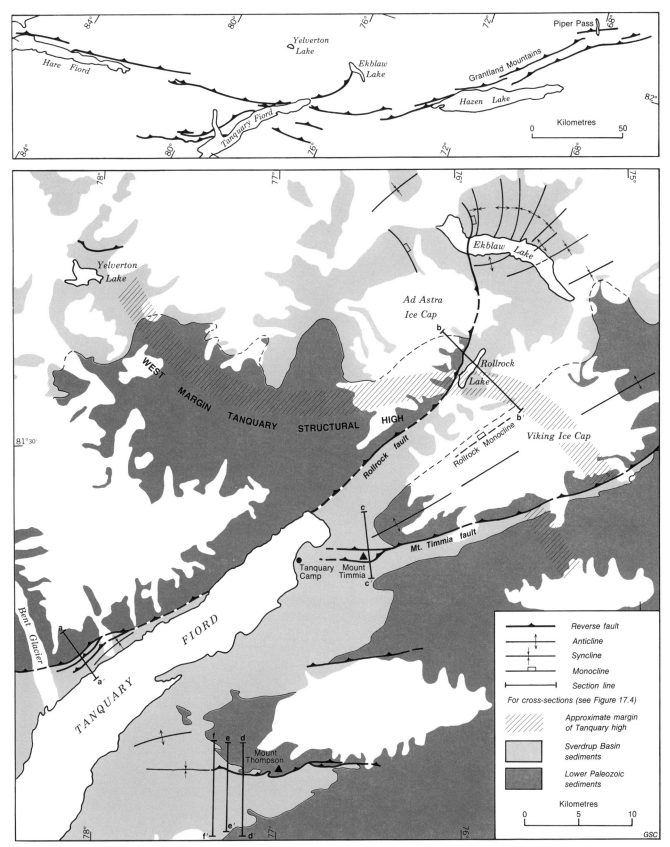

Figure 17.3. Inset at top: en echelen elements of Lake Hazen Fault Zone, Hare Fiord to Piper Pass; below: generalized geology at the head of Tanquary Fiord and index for Figure 17.4. (Higgins and Soper, 1983, Fig. 1.)

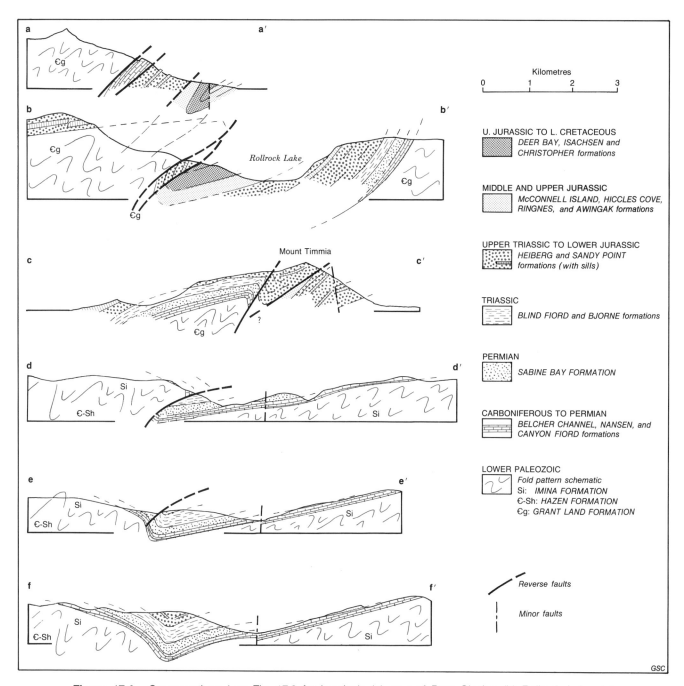

Figure 17.4. Cross-sections (see Fig. 17.3 for location): (a) east of Bent Glacier; (b) Rollrock Lake; (c) Mount Timmia; (d), (e), and (f) Mount Thompson (Higgins and Soper, 1983, Fig. 2).

from the Mount Thompson area and the northern end of the Rollrock Fault, Fig. 17.4). Bed-length comparisons across Rollrock Lake and Mount Timmia (sections b and c) indicate shortening of 7-8 km along a cross-strike section 34 km long (21-25%).

Important for the understanding of the style and evolution of this fault zone is the common occurrence of overturned upper Paleozoic or Mesozoic strata in the hanging wall of the thrusts, because the base of these units probably outlines the sub-Sverdrup Basin unconformity (sections a, b in Fig. 17.4). Higgins and Soper "envisage that southeast facing drape monoclines developed in the cover sequence above the upthrusts and that their steep limbs were subsequently disrupted and locally inverted by continued displacement on the reverse faults, together with footwall imbrication. The essentially monoclinal nature of the Lake Hazen fault zone seems to persist along the southern frontal scarp of the Grantland Mountains, at least as far as Piper Pass, with a thin veneer of steeply inclined cover sediments resting unconformably on the lower Paleozoic rocks and passing southeast into the Lake Hazen Syncline" (Fig. 17.5). Different stages in this evolution are portrayed by cross-sections f-b of Figure 17.4 (in that order).

A small dextral component of movement on the fault zone is suggested by subsidiary, northerly-trending compressive features in the Rollrock Lake area (Fig. 17.3). The western margin of the Tanquary Structural High (marked by the absence of Carboniferous to Middle Triassic strata at the base of the Sverdrup Basin succession; Chapter 14) trends across the fault zone and limits to 12 km the possible dextral displacement.

The configuration of the faults at depth remains uncertain in the absence of seismic or drilling data. Osadetz (1982) inferred a gently-dipping décollement at a depth of about 3 km beneath the sub-Sverdrup Basin unconformity, but the observations of Higgins and Soper permit continuation of relatively steep faults to greater depth (Fig. 17.6).

The **Mitchell Point Fault** is exposed for about 100 km in the vicinity of Yelverton Bay and Phillips Inlet. It bounds the Mitchell Point gneiss belt on its northwestern side and, like the Porter Bay Fault, is inferred to have had sinistral motion in the Late Silurian (Chapter 12B). An extensional phase in the early Late Cretaceous is apparent from the emplacement of the Wootton intrusion and Hansen Point volcanics at that time (Trettin and Parrish, 1987). Subsequent compressive motion on the Mitchell Point Fault is evident from the fact that the gneiss belt has been thrust over the Hansen Point volcanics southwest and northeast of Yelverton Bay. The southeast dip of this fault conforms with the attitude of the gneissic layering (Fig. 9.3, 9.4), which probably formed during early Paleozoic or Proterozoic events.

Figure 17.5. Part of the Lake Hazen Fault Zone northeast of Lake Hazen near southern end of Piper Pass. The fault plane dips 60°N. Grant Land Formation in the hanging wall is faulted against Permian sandstones. (Higgins and Soper, 1983, Fig. 4.)

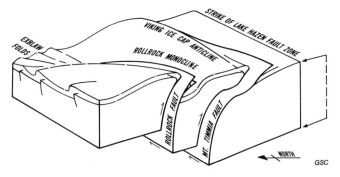

Figure 17.6. Schematic block diagram for the Rollrock and Mount Timmia faults, associated folds, and inferred displacements, according to a transpressional upthrust interpretation (Higgins and Soper, 1983, Fig. 5).

Structural style influenced by evaporites

The distribution of major Paleozoic evaporite units is shown in Figures 8B.31, 8B.32, 13.1, and 13.6. The Ordovician Baumann Fiord Formation and Carboniferous Otto Fiord Formation are the principal units affecting the Eurekan structural style, but the Ordovician Bay Fiord and Lower Permian Mount Bailey formations likely had some influence also. The Otto Fiord Formation has influenced the structural style by (largely pre-tectonic) diapirism and by constituting a décollement, the other units influenced mainly or exclusively by the second mechanism.

Western parts of Central Ellesmere Domain

The thrust and fold belt of west-central Ellesmere Island between Cañon and Baumann fiords contains structures of both Ellesmerian and Eurekan age. Both orogenies produced east-directed, north-northeast trending, compressive features, and these commonly cannot be distinguished except where strata of the Sverdrup Basin succession unconformably overlie structures, or where the Eureka Sound Group is affected by them (Chapter 12F). Where such distinctions have been made, it appears that Eurekan structures are primarily thrust faults and associated folds, whereas Ellesmerian structures are predominantly folds, and that most of the shortening in the belt was accomplished during the Eurekan Orogeny. This generalization is based on limited data and may not be entirely correct.

The style of deformation (Okulitch, 1982) is illustrated on two cross-sections between Troll Bay and Vendom Fiord (Fig. 3B [in pocket], section A-B) and east of the north end of Vendom Fiord (Fig. 3B, section C-D). Major Eurekan structures are gently-dipping thrust faults that sole in a detachment near crystalline basement, with some detachment also probable within the evaporitic Baumann Fiord Formation. Primary control of structural style by this evaporite unit, however, is in the development of disharmonic folds in massive carbonate units adjacent to it (section A, west half). The age of most of such structures remains uncertain. Estimates of horizontal shortening, based on preliminary studies of available data, range from 25 to 50% within the illustrated sections, suggesting about 20 to 40 km of movement. Displacement on individual faults is in the order of 5 km.

The south end of the thrust and fold belt lies covered in Baumann Fiord. To the south of the fiord, only east-west trending folds of Ellesmerian age and block faults of post-Permian age are present. Immediately north of the fiord, Eurekan folds are more open and thrust faults dip more steeply than still farther north. In the region between Vendom Fiord and Makinson Inlet, compressive structures are rare and of limited displacement.

Northern Ellesmere Domain

In the Greely Fiord – Hare Fiord area are several northwest- and southeast-verging thrust sheets that carry Otto Fiord evaporite at their base. The largest structure is the Blue Mountain Thrust which places Otto Fiord evaporite northwestward on the Triassic Blind Fiord Formation. The thrust appears to diminish eastward in keeping with the disappearance of evaporitic strata. Farther east, on the east side of lower Tanquary Fiord, thrust faults are evident where the Permian Mount Bailey evaporite outcrops; elsewhere, in the absence of these evaporitic strata, normal faults and open folds prevail (Thorsteinsson, 1974, Map 1306A).

Axel Heiberg Island (Sverdrup Islands Domain)

Eurekan structures on Axel Heiberg Island were superimposed on halokinetic features that formed by upward movement of the Otto Fiord evaporite throughout the Mesozoic (Chapters 13, 14). On eastern Axel Heiberg Island, a major anticline is composed of a chain of coalesced domes, at least one of which is cored by an evaporite diapir (van Berkel et al., 1983). The sense of bedding-parallel shear is opposite to that characteristic of buckle folds and indicates upward movement of the core of the fold relative to its flanks. Van Berkel et al. (1983) proposed that: "... all large anticlines on east-central Axel Heiberg Island mantle first-order salt walls (Trusheim, 1960) and that the coalescent domes are hoods of second-order diapirs".

In the Strand Fiord region of western Axel Heiberg Island (Thorsteinsson, 1974, maps 1292A, 1301A), narrow anticlines, cored by evaporite diapirs, separate much wider, crudely oval synclines, a configuration named "wall-and-basin" structure by van Berkel et al. (1984). Most wall segments strike in two, nearly perpendicular directions, suggesting rise of diapirs into two sets of fractures; however, the structures could also be explained by two episodes of buckle folding above a décollement in the Otto Fiord evaporite (op. cit.).

The Stolz Fault Zone, an arcuate zone of reverse faults at least 140 km long, contains diapirs as well as complexly deformed anticlines of evaporites and limestone (Thorsteinsson, 1974, map 1302A; van Berkel et al., 1984). It is uncertain whether the diapirs intruded the thrust faults, whether the latter developed along a chain of diapirs localized by normal faulting, or if diapirs grew during compression and were affected by continuing thrusting. Numerous offsets on the main fault either represent tear faults or lateral ramps (van Berkel et al., 1984). The arcuate shape of the Stolz Thrust conforms with a general change in trend in this region of transition from the Sverdrup Islands Domain, with its southeasterly to southerly trends, to the Central Ellesmere Domain with its southwesterly trends.

On northwestern Axel Heiberg Island, Eurekan deformation consisted of post-Late Cretaceous, early and late phases of normal faulting and an intervening phase of reverse faulting (Fischer, 1985). Otto Fiord evaporite forms the detachment zone of reverse faults and has also intruded the fault planes (Fig. 3E, in pocket). Competent limestone of the Nansen Formation, brecciated in the lower 10-100 m, forms the basal unit of the thrust sheets. The thrust faults cut upsection and downsection and can be described as a series of flats at different levels that are linked by oblique and lateral ramps. In one area, development of a steep ramp was caused by the "buttressing" effect of a horst elevated during the first phase of normal faulting. Strata moved northeast on three major reverse faults (Bukken Fiord, Bjarnason Island, and Bunde Fiord faults), but southwest on a fourth (Grisebach Creek Fault). About 25 km of shortening in the southern part of the area have been inferred (Fischer, 1985 and pers. comm., 1986).

Transcurrent faults

A sinistral transcurrent fault in Nares Strait has been postulated to account for the opening of Baffin Bay, and subsidiary sinistral faults are recognized in east-central Ellesmere Island. Dextral transcurrent faults have been recognized in northeasternmost and west-central Ellesmere Island.

Inferred sinistral motions
Nares Strait

Geological information that limits possible displacements along the hypothetical Wegener Fault in Nares Strait has been compiled by Dawes and Kerr (1982b). Consideration of nine regional stratigraphic and structural features and of twenty geological or geophysical markers suggested to them that sinistral motion along Nares Strait was in the range 0-25 km. Few of these markers, however, provide conclusive evidence. Markers in the northern part of the area, for example, cross the strait at such acute angles that a small change in trend would result in great changes in apparent offset. In the southern part, the reliability of such markers as marbles in the crystalline basement, isopachs of the Thule Group, and the axis of the Bache Peninsula Arch has been questioned on geological grounds (Miall, 1983). A Lower Silurian facies change from shelf carbonates to basinal mudrock (marker 10 of Dawes and Kerr, 1982b, Fig. 4) could have had 0-50 km of offset, depending on the interpretation of its trend in Nares Strait and the adjacent part of Ellesmere Island. The same data also show that little or no vertical displacement, of either normal or reverse sense, has occurred.

More precise information has come from aeromagnetic investigations. A widely spaced, high-level survey revealed an anomaly that appeared to cross the southern part of the strait without displacement (Riddihough et al., 1973; marker 18 of Dawes and Kerr, 1982b). A closely spaced, low-level survey of the same feature, however, showed that strike-slip faults with about 25 km sinistral displacement are present in the centre of the strait, in an area 60 km east of Bache Peninsula. The magnetic anomalies are attributed to late Middle Proterozoic dykes trending about N75°E.

The strait is aseismic at present, an anomalous condition in the Arctic Archipelago, where many Phanerozoic structures exhibit mild seismicity (Chapter 5C).

East-central Ellesmere Island (Central Ellesmere Domain)

The Judge Daly Fault Zone in the northeasternmost part of Judge Daly Promontory is about 40 km long and its main branch is marked by a linear valley (Mayr and de Vries, 1982). Sinistral strike-slip faulting in this area has been inferred primarily from angular relationships between the main fault zone and conjugate "synthetic" and "antithetic" strike-slip faults with small displacements. Offset on the main fault may be about 19 km, based on matching of overturned panels of Lower Cambrian formations on either side of the fault, but has not been established with certainty. Normal faulting that affected Paleocene(?) strata of the Eureka Sound Group on the northwest side of the main fault has been attributed by Mayr and de Vries (1982) to a divergent strike-slip regime. The northern part of the fault is characterized by a magnetic anomaly that continues along Robeson Channel to the Lincoln Sea (Hood et al., 1985). The anomaly appears to have been caused by southeasterly-dipping dykes, the tops of which are not far below the surface.

Conjugate faults suggesting sinistral strike slip also were recognized by Mayr and de Vries (1982) in lower Paleozoic strata near Carl Ritter Bay and on Darling Peninsula to the southwest on the east coast of central Ellesmere Island.

Inferred dextral motions
Northeasternmost Ellesmere Island (Northern Ellesmere Domain)

The Feilden Fault Zone, named for Feilden Peninsula in northeasternmost Ellesmere Island, is composed of the extensive Porter Bay and Guide Hill faults and associated minor faults (Fig. 17.7).

The Porter Bay Fault is exposed for 120 km between the coast of Lincoln Sea and a major ice cap, and can be traced for another 40 km in sparse nunataks. It does not reappear on the other side of the ice cap, about 130 km farther southwest. The sinuous trend of the fault parallels structural trends in the Lower Cambrian Grant Land Formation, which lies on its southeast side. The attitude of the fault plane cannot be determined precisely because of insufficient exposure, but appears to be very steep. The history of this fault is prolonged and complex and details of its motion remain uncertain. The fault marks the boundary between the Hazen and Clements Markham fold belts and is interpreted to have been a sinistral strike-slip fault in the Late Silurian (Chapter 12B). Structures in the Grant Land Formation adjacent to the fault suggest northwestward-directed thrust faulting, probably in the Late Silurian to Early Carboniferous. The present structural and topographic configuration indicates that the Grant Land Formation on the southeast side of the fault has been elevated with respect to the Lands Lokk Formation and overlying upper Paleozoic units in post-Permian time. The indirect evidence for strike-slip motion is discussed below.

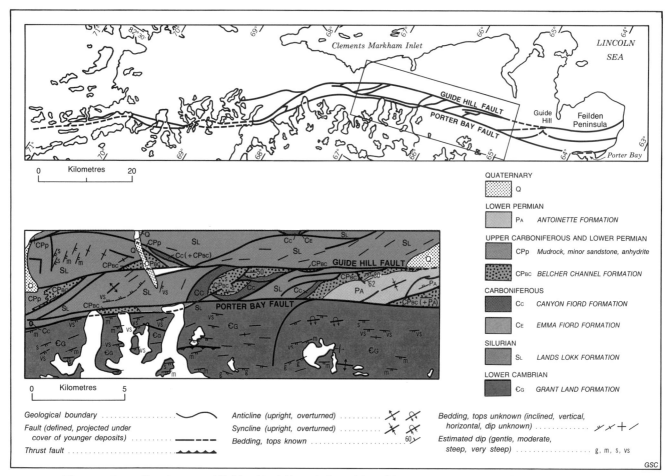

Figure 17.7. Sketch maps of Feilden Fault Zone (adapted from Mayr et al., 1982 and U. Mayr, pers. comm., 1986).

The Guide Hill Fault, which can be traced for 125 km, parallels the Porter Bay Fault but cuts across Late Silurian to Devonian structural trends in the Silurian Lands Lokk Formation along its eastern segment.

The associated minor faults generally have more northerly trends than the Porter Bay and Guide Hill faults, but conform with structural trends in the Lands Lokk Formation.

The array of these major and minor faults suggests dextral transcurrent motion in post-Early Permian, presumably Cretaceous or Paleogene time (cf. Ogawa et al., 1985; Woodcock and Fischer, 1986). The oblique minor faults may represent "thrust shears", the development of which was favoured by pre-existing trends in the Lands Lokk Formation, or they may have developed in the complex fashion illustrated by Ogawa et al. (1985, Fig. 11).

Structures north and south of the Guide Hill Fault cannot be matched. This implies either offset in excess of tens of kilometres, or development of different structures during or after the postulated transcurrent motion on opposite sides of the fault. On the other hand, the fact that the same formations and facies appear north and south of the Guide Hill fault shows that the postulated strike slip could not have been in the order of hundreds of kilometres. No estimates of offset are possible for the Porter Bay Fault.

West-central Ellesmere Island (Central Ellesmere Domain)

On Raanes Peninsula, two southeastward-directed thrust faults named Blind Fiord Fault and Trold Fiord Fault also have components of dextral slip of 8 km and 16 km, respectively. This offset was inferred originally by N. Haimila (pers. comm., 1985) from offset on mafic dykes with distinct petrographic characteristics and was subsequently confirmed by upper Paleozoic facies relationships (Beauchamp, 1987).

Normal faults

Innumerable normal faults of diverse orientation and related grabens and horsts are present throughout the orogen on Axel Heiberg and Ellesmere islands (cf. Thorsteinsson, 1974). Some of these faults, in particular those parallel to compressive structures of the orogen, have been reinterpreted as reverse and thrust faults (Okulitch, 1982; Fischer, 1985). In many cases, only the relative stratigraphic throw of such faults is known, as the faults are not exposed. The kinematic relationships between reverse and normal faults are largely unknown. Some normal faults may have formed in response to stresses within overthrust sheets; others may be reactivated faults that originated during the

Ellesmerian Orogeny or formation of the Sverdrup Basin. Still others may be the product of distal effects of rifting in the Baffin Bay region. It also is probable that compression in the Eurekan Orogen was followed by general extension when the regional stress field changed. In the only detailed study of normal faults made in the Eurekan Orogen, Fischer (1985) concluded that normal faults formed after the early Late Cretaceous and both preceded and followed reverse faulting.

Dykes

A southeasterly-trending dyke swarm in northwestern Ellesmere Island (mapped from air photographs) cuts across southwesterly-trending folds and thrust panels composed of upper Paleozoic and Mesozoic strata (cf. Thorsteinsson, 1974, geological maps of Greeley Fiord West and Otto Fiord areas). The angle of intersection is generally high and close to 90° at some localities north of central Hare Fiord and southeast of lower Hare Fiord. The relationships portrayed on the map imply that the dyke swarm is coeval with or younger than the folds and thrusts, an inference that should be followed up by structural analyses and isotopic age determinations.

FAULTING IN THE SOUTHEASTERN ARCTIC PLATFORM AND IN THE BOOTHIA UPLIFT
Introduction

Normal faults of late Mesozoic – Tertiary aspect are common in the region south of the Eurekan Orogen but their specific ages and modes of origin are difficult to determine. An exception — at least as far as mode of origin is concerned — are fault systems in the vicinity of Baffin Bay, Jones Sound, Lancaster Sound, and Foxe Basin that appear to be related to the development of the Labrador Sea and Baffin Bay, i.e. the Labrador Basin of Balkwill (1987). The age control for these faults also is very limited, but well logs and geophysical data indicate that the combined rift and drift stages of the Labrador Basin extended from Early Cretaceous to early Oligocene; it is reasonable to assume that most faults in the southeastern Arctic Platform originated during this broad interval.

The western extent of the rifting that is related to the development of the Labrador Basin is problematic. Isotopically-dated alkaline magmatism on southeastern Bathurst Island indicates that rifting occurred in the Boothia Uplift in early Eocene time and therefore this region is included in the following discussion. On the other hand, Late Cretaceous – Paleogene faults farther west, for example on Banks Island (Miall, 1979a), are excluded from this account of the Eurekan Deformation because they belong to a different and poorly known topic — the evolution of the Canada Basin continental margin.

In addition to the ubiquitous normal faults, transcurrent faults are recognized in one area (southeastern Ellesmere Island).

Normal faults

The following summary of normal faults is restricted to structures exposed on land and to offshore structures known from seismic investigations. Linear coasts are common in this region, but it is difficult to determine whether they are related to major faults (as fault line scarps) or to minor faults and fractures. Structural interpretations of these coasts emphasizing the first alternative are included in syntheses by Kerr (1980, 1981).

Southeastern Ellesmere Island

Normal faults of several trends form part of the complex array of structures on southeastern Ellesmere Island (Mayr and Okulitch, 1984). The predominant fault trends are northwest and east-west, parallel to major extensional structures offshore in Baffin Bay and Jones Sound. Northwest-southeast trending normal faults are concentrated in the region between South Cape and Makinson Inlet on the south and east, and Baumann and Vendom fiords on the northwest (Fig. 3B [in pocket], sections A-B and C-D). Displacement on most faults is small, in the order of hundreds of metres, but stratigraphic throw is as much as 2 km in a few grabens. Near Baumann and Vendom fiords, some grabens contain deformed strata of the Eureka Sound Group (Fig. 3B, section C-D).

East-west trending normal faults occur primarily along the south coast of Ellesmere Island, near Jones Sound. The nature of structures within the sound is unknown, but preliminary geophysical investigations (MacLean et al., 1984) suggest the presence of a few faults and of Paleozoic and younger strata, which appear to have been downthrown relative to Precambrian basement exposed on adjacent coastlines. The structure may be similar to that in Lancaster Sound but with less displacement.

Devon Island

Devon Island is broken by numerous, steeply-dipping normal faults that vary in length from a few hundred metres to about 100 km (Thorsteinsson and Mayr, 1987; Fig. 17.8). The stratigraphic displacement ranges up to 600 m but is generally in the order of a few metres to about 100 m. Westerly- to northwesterly-trending faults in the eastern and central parts of the island step down toward Jones Sound in the north and toward Lancaster Sound in the south. Combined with gentle flexuring in these directions, and with a regional dip to the west, this pattern of faulting has produced a westerly-plunging arch (the Devon Arch) in the eastern and central part of the island. Northeasterly-trending faults occur in the central and western parts of the island and increase in density to the west. Aeromagnetic anomalies, and exposures on eastern Devon Island (Fig. 17.9) and in the Boothia Uplift suggest that the two sets of faults reflect basement provinces with different structural trends. The youngest preserved strata affected by the faulting are lower Maastrichtian beds of the Eureka Sound Formation.

Lancaster Sound

In eastern Lancaster Sound, two half-grabens containing Mesozoic to Cenozoic sediments (Lancaster Sound Basin and West Basin) are separated by a north-trending structural high with a faulted western margin (Daae, 1983). The Parry Channel Fault, which bounds the half-grabens on the north, had a vertical displacement of at least 7.6 km. Subsidiary horsts and grabens are present within the Lancaster Sound

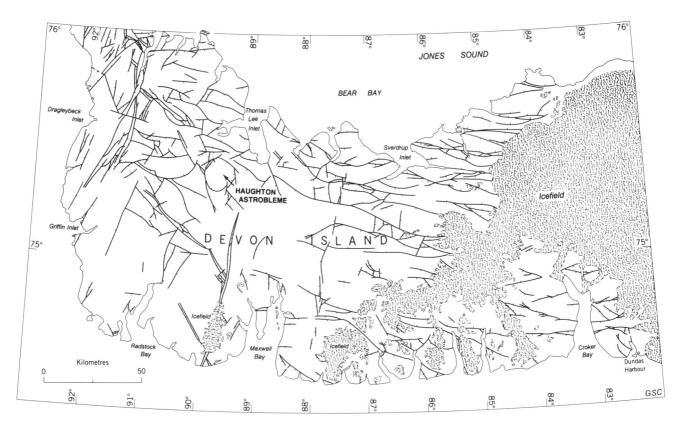

Figure 17.8. Fault systems on Devon Island (Thorsteinsson and Mayr, 1987, Fig. 82).

Basin. Crystalline basement and lower Paleozoic carbonate strata likely underlie the basins, but the age range of the overlying succession with lesser seismic velocities is uncertain. Seismic velocities suggest that Jurassic to Oligocene sediments are present (Daae, 1983) whereas the tectonic evolution of the region and comparisons with the Eclipse Trough of Baffin and Bylot islands (Miall et al., 1980) suggest that the base of the overlying succession is no older than late Early Cretaceous (Kerr, 1980).

The time of movement is not well defined because of uncertainty about the age and composition of the sediments in Lancaster Sound. It is possible that the bounding faults were active during rifting and/or spreading in Baffin Bay, although evidence for growth faulting has not yet been reported. If so, the faults were reactivated during the middle Tertiary (late Oligocene and Miocene) when the crystalline terranes around Baffin Bay experienced thermal(?) uplift (Chapter 18).

Eclipse Trough

The Eclipse Trough of northeastern Baffin and southwestern Bylot islands is an elongate, northwesterly-trending synclinal downwarp with linear, partly-faulted boundaries. It contains more than 3 km of clastic sediments of late Early Cretaceous to early Eocene age that appear to have been deposited within a predecessor of the present structural depression (Miall et al., 1980 and Chapter 15). The deformation of these strata probably occurred in latest Oligocene and Miocene time, concomitantly with the uplift of crystalline terranes surrounding Baffin Bay (Chapter 18).

Foxe-Baffin structural depression and faults on northwestern Baffin Island

Normal faults bound the lower Paleozoic strata in Foxe Basin and adjacent parts of Baffin Island and Melville Peninsula (Trettin, 1975). The Foxe-Baffin structural depression, a large but relatively shallow graben, extends 810 km in a northwesterly direction from the head of Frobisher Bay, which opens into the Labrador Sea, to northeastern Melville Peninsula, and is up to 295 km wide. At its centre, the structurally deepest part contains Silurian strata exposed on Prince Charles Island, Rowley Island, and possibly the Spicer Islands. Structural relief between Rowley Island and eastern Melville Peninsula is in the order of 750 m. On Melville Peninsula, the western boundary of the graben, is a zone of normal faults, 225 km long, that has an overall southerly to southeasterly trend, although individual faults show considerable variation in strike (Fig. 17.10).

Related half-grabens on northwestern Baffin Island, east of Admiralty Inlet, that preserve lower Paleozoic strata, are bounded by northwest-trending normal faults. These faults do not extend beyond Admiralty Inlet, as correlative strata on Brodeur Peninsula are undisturbed.

Direct evidence indicates only that normal faulting in this region was post-Early Silurian. Proterozoic normal faults bounding the Steensby and Devon highs (Jackson and Iannelli, 1978) may have been reactivated during Siluro-Devonian tectonism (Chapter 12C) but more likely were affected by Late Cretaceous to Paleogene extension during the opening of the Labrador Sea and Baffin Bay.

Figure 17.9. Westerly-striking faults on north-central Devon Island, extending from the Precambrian crystalline basement (Agn) into the lower Paleozoic cover (ЄBP-RP = Bear Point and Rabbit Point formations; ЄOCF = Cass Fjord Formation; oblique aerial photograph T345L-38, looking west; Thorsteinsson and Mayr, 1987, Fig. 83).

Figure 17.10. Western margin of Foxe-Baffin structural depression on Melville Peninsula. Upper Ordovician limestone (Ols) and Lower-Middle Ordovician Ship Point Formation (Os) in fault contact with Lower Proterozoic (Aphebian) or older crystalline basement (pЄ) (Trettin, 1975, Plate 2).

New faults may have formed at this time also. The Foxe-Baffin structural depression is considered analogous to the Lancaster Sound grabens.

Boothia Uplift

Several grabens containing Tertiary strata are preserved in the region of the Boothia Uplift and Cornwallis Fold Belt. The largest, at Stanwell Fletcher Lake, west-central Somerset Island, trends northerly, parallel to the regional structural grain of the underlying Precambrian Shield. Similar structures may lie within Peel Sound immediately to the west. On northern Somerset Island, Tertiary rocks lie within a northwest-trending graben. No direct link to regional extension or earlier structures is evident, except that this graben is parallel to possible extensions of Proterozoic horsts and grabens noted above on Baffin Island and may be related to faults bounding the basin of deposition of Proterozoic strata to the southwest at Aston Bay.

On Cornwallis and western Devon islands, several north-northwest and north-northeast trending grabens contain Tertiary strata and older rocks. These grabens are likely related to pre-existing structures of the Boothia Uplift and Cornwallis Fold Belt. Other similar features may lie offshore in Wellington Channel between Cornwallis and Devon islands and in Crozier Strait, west of Cornwallis Island (Kerr and Ruffman, 1979).

Transcurrent faults, inferred dextral motion

Gently arcuate, northeast to north-northeast trending lineaments, along which basement and lower Paleozoic cover have been offset, extend from Jones Sound to under the ice cap of southeastern Ellesmere Island. North-south folds, east-west normal faults, and low-angle detachment faults spatially associated with these lineaments suggest that dextral strike slip with associated minor transpression and transtension may have occurred along them. Definitive resolution of true slip and amount of displacement remain uncertain, but given the overall consistence of platform stratigraphy in the area, offsets are not likely to have exceeded a few kilometres.

Magmatism

On southeastern Bathurst Island, dykes, sills, small plugs, and agglomeratic vents intrude Lower and Middle Devonian units and a downfaulted block of Upper Cretaceous strata assigned to the Eureka Sound Formation (Kerr, 1974; Mitchell and Platt, 1984). These rocks evidently represent feeder systems to now-eroded lavas, remnants of which are preserved as clasts in the vents. The suite is bimodal and alkaline in composition, consisting mainly of nephelinite or larnite-normative nephelinite and basanite, with subordinate proportions of olivine melilite nephelinite, phonolite, and tholeiitic and alkali basalt. It has the petrological characteristics of intraplate magmatism associated with continental rifting and doming (Mitchell and Platt, 1984). A Rb-Sr isochron age of 47.1 ± 4 Ma indicates emplacement in the late Early Eocene (late Ypresian) according to the time scale of Snelling (1987).

TECTONIC INTERPRETATION OF EUREKAN DEFORMATION

The Eurekan Deformation is the only event in the entire history of the Innuitian region amenable to a quantitative plate-tectonic analysis based on seafloor magnetic anomalies. These anomalies suggest that the Eurekan Deformation was caused by counterclockwise rotation (Kerr, 1967) and northwestward translation of Greenland relative to North America, in response to seafloor spreading in the Labrador Sea and adjacent seaways. This concept is widely accepted (e.g. Soper et al., 1982; Peirce, 1982; Miall, 1984) because the two processes are spatially and temporally related; no other plausible mechanisms are presently known. Nonetheless, some quantitative discrepancies between the plate-tectonic reconstructions and structural observations on land have yet to be resolved.

Another aspect must be emphasized here. Throughout the Innuitian Orogen, Eurekan structures are parallel in trend with Ellesmerian and older structures, demonstrating that the latter had a profound influence on the Eurekan Deformation. This implies that caution must be exercised when interpreting Eurekan trends as indicators of the contemporaneous stress system.

Geophysical evidence concerning the origin of the Labrador Basin

In recent years, most authors have interpreted the crust of the Labrador Sea – Baffin Bay region as a truly oceanic crust that originated by seafloor spreading, but some have regarded it as attenuated continental crust. This question is important in the present context because the first alternative would imply a considerably larger angle of rotation of Greenland than the second.

That the Labrador Sea has oceanic crust is apparent not only from well developed, paired magnetic anomalies but also from evidence for an extinct mid-oceanic ridge with a fault-bounded median valley; the ridge is recognizable in a seismic cross-section and marked by a negative gravity anomaly (Hinz et al., 1979; Srivastava et al., 1981).

Davis Strait lacks magnetic anomalies correlatable with seafloor anomalies (Srivastava et al., 1982). In the central region of the strait, mantle (seismic velocity: 8.5 km/s) is overlain by an 18 km thick crust (6.2 km/s), in turn overlain by a 2 km thick layer with velocities of 4.2 to 4.8 km/s. This region differs from both typical oceanic and continental crust. Some authors (Keen et al., 1974; Hyndman, 1975) have interpreted it as an extinct hot spot, comparable to the Iceland-Faroe Ridge; others (Srivastava et al., 1982; Menzies, 1982) as a mixture of oceanic and continental rocks. A fault-bounded, northeast-trending structural high, for example, appears to be a fragment of the Baffin Shelf.

In Baffin Bay, sediments (seismic velocities: 2.0-4.2 km/s) are underlain by a crust (5.0-7.0 km/s) that presumably represents the undifferentiated oceanic crust layers 2 and 3 (Keen et al., 1974; Jackson et al., 1979; Srivastava et al., 1981). Layer 2, if present, either is too thin to be recognized (less than 2 km) or is masked by high-velocity sediments. The main crustal layer (6.5-7.0 km/s) varies in thickness from 4 to 7 km. A deeply buried median valley, analogous to that in the Labrador Sea, but smaller, has been inferred from a gravity anomaly. It is flanked by weak magnetic anomalies, reminiscent of anomalies 24-13 (Jackson et al., 1979). Overall, the crust of Baffin Bay has more oceanic than continental characteristics (Keen and Peirce, 1982) but differs from normal oceanic crust by its thinness, the absence of well defined magnetic anomalies, and the possible absence of layer 2.

Plate-tectonic reconstructions based on magnetic anomalies

Integrated, computer-based studies of paired magnetic anomalies in the Labrador Sea, North Atlantic, and eastern Arctic Ocean indicate that seafloor spreading extended from more southerly parts of the North Atlantic into the Labrador Sea in Late Cretaceous, probably Campanian time (anomaly 34), a conclusion supported by sedimentological and tectonic interpretation of the subsurface record of the Labrador Shelf (Balkwill, 1987). Greenland was attached to Eurasia at that time. Anomalies 34-25 (Campanian to late Paleocene) are about parallel with the present coast of the Labrador Sea, but a profound reorganization of spreading patterns occurred near the Paleocene-Eocene boundary. A triple junction developed south of Greenland, Greenland separated from Eurasia, and anomalies in the Labrador Sea attained a more northwesterly orientation that forms an acute angle with the present coast line. The spreading in the Labrador Sea seems to have terminated in the Early Oligocene with anomaly 13.

If seafloor spreading is accepted for the Atlantic region and the Labrador Sea, geometric constraints require that this process also has occurred in Davis Strait and Baffin Bay, and plausible spreading models for this region have been developed (Srivastava et al., 1981; Menzies, 1982). Mere rifting in Baffin Bay would require an unusual degree of crustal attenuation (Keen and Peirce, 1982). However, the continuity of cratonic features and of platformal strata across Nares Strait demonstrates that Greenland did not become completely detached from North America, and did not appear to form a discrete plate.

In conventional reconstructions based on magnetic anomalies, plates are treated as rigid bodies — a geometric, rather than a geological requirement. Most of these reconstructions have treated Greenland and North America as separate plates and placed the boundary between them in Nares Strait. These assumptions have led to the conclusion that Greenland was separated from Ellesmere Island by a wide gap in the Cretaceous, which gradually closed in early Tertiary time. Such reconstructions, however, are unacceptable from a geological point of view, because they confine the deformation to Nares Strait and leave unexplained the deformation on land. If the Eurekan Orogeny is interpreted as the result of the motions of Greenland relative to Arctic North America, then the Eurekan Orogen may be regarded as the non-rigid buffer zone between two more or less rigid plates, but is perhaps better viewed as an intracratonic orogen within one (North America-Greenland) plate. Paleogeographic reconstructions based on such a model must restore (stretch) the geographic outlines within the Eurekan Orogen for Cretaceous and earlier times, and thus close the artificial gap between Greenland and Ellesmere Island. The problem, to what extent the cratonic plate was rigid, is addressed below.

The rigid-plate reconstruction of Srivastava (1985, Fig. 17.11) showed calculated trajectories for two points in northwestern Greenland. These trajectories are valid, to some extent, as indicators of the motions that caused the Eurekan Orogeny, if disassociated from the paleogeographic base, and from the assumption of plate rigidity (cf. Srivastava and Falconer, 1982). The theoretical plate motions and inferred structural effects can be summarized as follows (cf. Srivastava, 1985, Table 3).

Anomaly 34-25 (early Campanian to Late Paleocene): Movements of Greenland (relative to North America but not necessarily to Ellesmere Island) during this interval are constrained only by sparse data from the southern Labrador Sea and cannot be quantified with confidence. The spreading in the Labrador Sea implies a counterclockwise rotation of Greenland that resulted in compression in the Arctic Islands, perhaps with a sinistral strike-slip component within the orogen and/or Nares Strait.

Anomaly 25-21 (Late Paleocene to mid-Eocene): From anomaly 25 onward, inferences about plate motions are constrained not only in the Labrador Sea but also in the Greenland-Norwegian Sea and thus are relatively accurate. Oblique motion of Greenland during this interval can be resolved geometrically into about 125 km of sinistral strike slip displacement parallel with Nares Strait, and 90 km of compression normal to the strait. Most of the inferred strike slip occurred in anomaly 25-24 time (latest Paleocene – earliest Eocene).

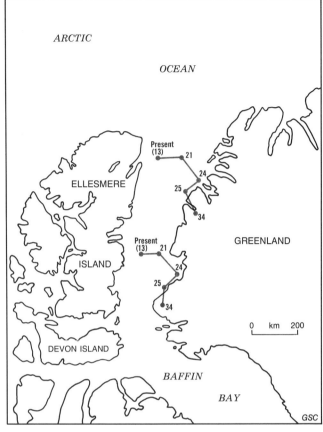

Figure 17.11. Inferred motions of Greenland relative to Ellesmere Island (from Srivastava, 1985, Fig. 9 and pers. comm., 1986). Numbers refer to magnetic anomalies.

Anomaly 21-13 (mid-Eocene to Early Oligocene): About 100 km of inferred motion perpendicular to Nares Strait resulted in compression.

The reconstructions based on magnetic anomalies are broadly compatible with previous geological conclusions, but add to them, or contradict them in the following respects:

(1) The anomalies demonstrate that the plate motions extended from Late Cretaceous to early Oligocene time. Any movements after anomaly 13 time therefore are probably not compressional in nature but are isostatic adjustments to crustal thickening.

(2) Geological evidence from eastern Axel Heiberg Island (Ricketts and McIntyre, 1986) suggests that Late Cretaceous to mid-Eocene compressive deformation was minor compared with Middle Eocene and later compression, whereas the magnetic evidence indicates greater amounts of shortening during pre-Middle Eocene time. If the magnetic evidence is reliable, then severe compressive deformation must have occurred in other parts of the Eurekan Orogen during pre-Middle Eocene time. Solution of this problem, as pointed out above, requires better age determinations for syntectonic conglomerates of the Eureka Sound Formation.

(3) No reliable field evidence concerning the age of possible strike slip in Nares Strait is available. The present trajectories suggest that it occurred mainly in latest Paleocene – early Eocene (anomaly 25-24) time.

(4) According to the calculated trajectories, total convergence between Greenland and Ellesmere Island is in the order of 200 km. This amount is similar to the upper limit of estimates of crustal shortening within the Eurekan Orogen.

(5) The estimate of sinistral strike slip derived from the trajectories (125 km during anomaly 25-21 time plus an uncertain amount during anomaly 34-25 time) exceeds previous field estimates of possible strike slip in Nares Strait (see above) and poses a major problem that has been debated extensively (cf. Dawes and Kerr, 1982a). Attempts to resolve this discrepancy have attempted either to raise the estimates of sinistral translation within the Eurekan Orogen (including Nares Strait); or to reduce the theoretically required strike slip by questioning the assumption of the rigidity of the two parts of the cratonic plate.

Changing estimates of sinistral translation has been attempted in three ways. The first is a revision of estimates of possible strike slip along Nares Strait, but, as pointed out above, the upper limit for such estimates is about 50 km. The second is the hypothesis that the strike slip is distributed through a broad sinistral shear zone in the Precambrian basement (Hugon, 1983). This hypothesis is partly contradicted by evidence for dextral shear in the crystalline basement of southeastern Ellesmere Island (Okulitch and Mayr, 1984). A third possible solution is that sinistral strike slip is not confined to the crystalline basement but extends into the cover and is distributed throughout the Eurekan Orogen, especially the Central Ellesmere Fold Belt (Miall, 1984). Apart from areas immediately adjacent to Nares Strait (Judge Daly Promontory, Carl Ritter Bay and Dobbin Peninsula) where perhaps 20-30 km of slip may have occurred (Mayr and de Vries, 1982), no proof for such motion has been found as yet.

Alternatively, the estimate of strike slip can be reduced if it is accepted that cratonic North America and possibly also cratonic Greenland were not entirely rigid (cf. Srivastava and Falconer, 1982). The trajectories calculated by Srivastava (1985) describe the movement of Greenland relative to North America, but can only be compared to Ellesmere Island if it was stationary relative to the North American craton. However, an extensive array of normal faults on southern Ellesmere Island, Jones Sound, Devon Island, Lancaster Sound, Baffin Island and Hudson Strait, most of demonstrable or probable Cretaceous-Tertiary age, suggest that rifting was distributed across this broad zone (Rice and Shade, 1982). This extension may be reasonably assumed to have resulted from northward movement of the northeastern edge of North America. At least 10 km of extension can be documented on southern Ellesmere Island, Jones Sound and Lancaster Sound alone. Net displacement of the part of Ellesmere Island bordering Nares Strait could be in the order of 30-50 km (3-5% extension). These tentative estimates, together with those for sinistral slip in the Nares Strait/east-central Ellesmere Island region (40-80 km), approximately balance the lower estimates derived from geophysical data (125-190 km).

The theoretically required strike slip (and compressional shortening) would also be smaller if part of the seafloor spreading in the Labrador Sea and Baffin Bay was accommodated by strike slip in Greenland (Beh, 1975). This solution, however, has generally not been accepted because no significant cross-faults of Late Cretaceous – Tertiary age have been described on the east side of Greenland.

Significance of dextral motions

Dextral faulting on southeastern Ellesmere Island has been interpreted (Okulitch and Mayr, 1984) as being related to rifting in Baffin Bay (Fig. 17.12). Rifting may have produced, in continental crust, a fault pattern analogous to spreading centres and transform faults in oceanic crust. In this area, northwest-trending grabens and normal faults (rifts) are intersected by northeast-trending dextral faults (transforms). The sense of displacement implies that the axis of Baffin Bay rifting emerges on Ellesmere Island northeast of the study area, perhaps near Makinson Inlet. Sinistral displacement, predicted by this model to lie northeast of the rift axis, is found in the Nares Strait region (see above).

Far more structural information is required before the inferred dextral motions in the eastern parts of the Eurekan Orogen can be interpreted with confidence. One possible explanation, arising from the angular relationship between Ellesmerian and older faults and the stress created by the rotation of Greenland (cf. Fig. 19.9), is discussed briefly in the concluding section of this chapter.

Paleomagnetic evidence for rotation

Determinations on the Lower Permian Esayoo Formation at two localities in northwestern Ellesmere Island indicated a counterclockwise rotation of $36 \pm 8°$ relative to stable North America (Wynne et al., 1983). A counterclockwise rotation of $33 \pm 29°$ was obtained for the combined data from the Isachsen and Strand Fiord formations at several localities on central and northern Axel Heiberg Island (Wynne et al., in press). In both cases the inferred paleolatitude coincided with the expected paleolatitude within confidence limits. The results from Axel Heiberg Island coincide, within the very wide confidence limits stated, with the counterclockwise rotation of Greenland inferred from magnetic striping (about 11°), but indicate a much larger angle of rotation if taken at face value. The cause of this apparent discrepancy is not clear.

Summary of plate-tectonic model and structural implications

Throughout the time under consideration, the Arctic Islands and Greenland essentially formed a single plate composed of two elements with different mechanical properties: the relatively rigid Shield in the south and southeast, and those relatively flexible parts of the Innuitian Orogen that lie on oceanic crust or on thinned and disrupted parts of the Shield.

After initial rifting, seafloor spreading occurred in the Labrador Basin from early Late Cretaceous to earliest Oligocene time, which caused rotation and northeastward translation of the Precambrian Shield of Greenland relative to North America. The stress produced by early phases of rotation must have been absorbed in Greenland or in the North Atlantic region. Later phases caused compressive deformation mainly in those parts of the Innuitian Orogen developed outboard of the integer Shield.

The most important structural implications of the rotation model can be summarized as follows.

(1) If the rock mass affected by the rotational compression had been isotropic, a fan-shaped array of folds

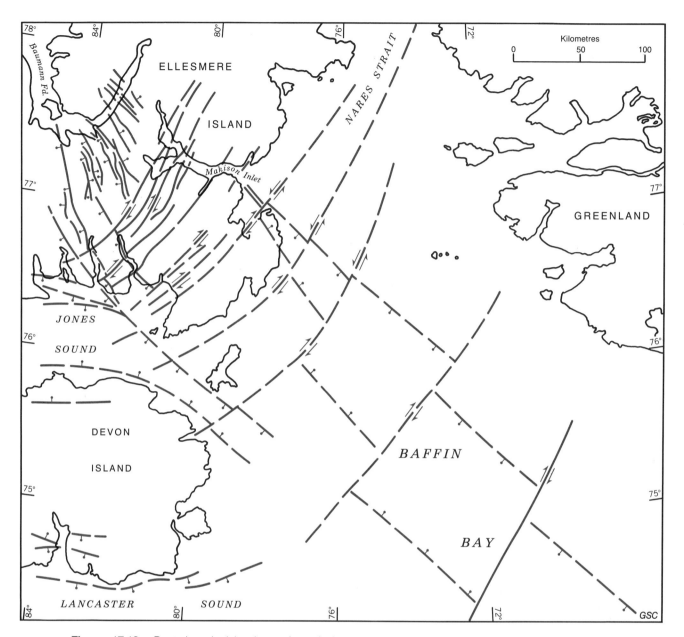

Figure 17.12. Dextral and sinistral transform faults associated with the Baffin Bay spreading centre. Structures on southeastern Ellesmere Island have been observed; all offshore structures are inferred or hypothetical. Structures in Baffin Bay are schematic, following Srivastava and Tapscott (1986, Plate 10).

or reverse faults, radiating from the pivot should have originated: that such a pattern is developed only very imperfectly is attributed to the anisotropy created by earlier deformations. The obliquity between rotational stress and pre-existing structures should result in dextral strike slip faulting in Ellesmere Island (cf. Fig. 17.11). Far more fieldwork is required to establish whether or not such strike slip is common, but it has been confirmed for a few areas.

(2) The rotational model implies that the absolute amount of shortening in a concentric (circular) direction increases proportionally with distance from the pivot whereas the percentage of shortening remains constant. This model, however, does not demand that the percentage of shortening is uniform in a concentric sense around the 110-120° sector affected. In fact, it seems to decrease in a counterclockwise direction and most of it is probably concentrated in the northeastern 65° or so of the sector. Assuming a rotational compression of about 11°, 90% of which is accommodated within a sector of 65°, a value of 15% of shortening (in a circular sense) is obtained.

(3) The present model implies that the shortening extends vertically throughout the lithosphere of the orogen although thin skinned deformation has occurred along its southeastern margin. To date we have no deep crustal

sections showing how this was accomplished precisely, but the following mechanisms are known, or can be inferred, to have operated at different depths:

(a) Movement on thin skinned thrust faults (with associated folds) above décollements at various stratigraphic levels (cf. Ricketts, 1987). Well known décollements are associated with Carboniferous and Ordovician evaporites (Otto Fiord, Bay Fiord, and Baumann Fiord formations); others probably are present within incompetent lower Paleozoic (e.g. Hazen Formation) and Proterozoic units.

(b) Reverse motion on listric normal faults that developed during Carboniferous to Cretaceous extensional phases (Balkwill, 1983, p. 75). A good example is the Mitchell Point Fault, which was a site of extension during the early Late Cretaceous and of reverse faulting in latest Cretaceous or Paleogene time. The Lake Hazen Fault Zone, which is close to the southeastern margin of the Sverdrup Basin, may also have originated as a belt of normal faults. Such faults can extend to depths of 6 to 10 km before levelling out.

(c) Ductile flow, accompanied by mylonization and metamorphism, at deep lithospheric levels that are not exposed and have not been penetrated by wells.

(4) Relationships with the Arctic Ocean Basin are unknown but two different effects can be evisaged: expulsion of parts of the Sverdrup Islands Domain into this region and/or dextral transform faulting along the continental margin.

REFERENCES

Balkwill, H.R.
1974: Structure and tectonics of Cornwall Arch, Amund Ringnes and Cornwall Islands, Arctic Archipelago; in Proceedings of the Symposium on the Geology of the Canadian Arctic, J.D. Aitken and D.J. Glass (ed.); Geological Association of Canada and Canadian Society of Petroleum Geologists, p. 39-62.
1978: Evolution of Sverdrup Basin, Arctic Canada; The American Association of Petroleum Geologists Bulletin; v. 62, p. 1004-1028.
1983: Geology of Amund Ringnes, Cornwall, and Haig-Thomas islands, District of Franklin; Geological Survey of Canada, Memoir 390, 76 p.
1987: Labrador Basin: structural and stratigraphic style; in Sedimentary Basins and Basin-Forming Mechanisms, C. Beaumont and A.J. Tankard (ed.); Canadian Society of Petroleum Geologists, Memoir 12, p. 17-43.

Balkwill, H.R. and Bustin, R.M.
1975: Stratigraphic and structural studies, central Ellesmere Island and eastern Axel Heiberg Island, District of Franklin; in Report of Activities, Part A, Geological Survey of Canada, Paper 75-1A, p. 513-517.
1980: Late Phanerozoic structures, Canadian Arctic Archipelago; Palaeogeography, Palaeoclimatology, Palaeoecology, v. 30, p. 219-227.

Beauchamp, B.
1987: Stratigraphy and facies analysis of the Upper Carboniferous to Lower Permian Canyon Fiord, Belcher Channel and Nansen formations, southwestern Ellesmere Island; unpublished Ph.D. thesis, Department of Geology and Geophysics, University of Calgary, Calgary, Alberta, 370 p.

Beh, R.L.
1975: Evolution and geology of western Baffin Bay and Davis Strait, Canada; in Canada's Continental Margins and Offshore Petroleum Exploration, C.J. Yorath, E.R. Parker, and D.J. Glass (ed.); Canadian Society of Petroleum Geologists, Memoir 4, p. 453-476.

Berkel, J.T. van, Hugon, H., Schwerdtner, W.M., and Bouchez, J.L.
1983: Study of anticlines, faults and diapirs in the Central Eureka Sound Fold Belt, Canadian Arctic Islands: preliminary results; Bulletin of Canadian Petroleum Geology, v. 31, p. 109-116.

Berkel, J.T. van, Schwerdtner, W.M., and Torrance, J.G.
1984: Wall-and-basin structure: an intriguing tectonic prototype in the central Sverdrup Basin, Canadian Arctic Archipelago; Bulletin of Canadian Petroleum Geology, v. 32, p. 343-358.

Bustin, R.M.
1982: Beaufort Formation, eastern Axel Heiberg Island, Canadian Arctic Archipelago; Bulletin of Canadian Petroleum Geology, v. 30, p. 140-149.

Daae, H.D.
1983: The geological history and evaluation of the Lancaster Sound, N.W.T. with specific reference to the Dundas structure; Joint Consolidex Magnorth Oakwood Lancaster Sound Venture, Resource Management Plan Support Document 1 (unpublished), 25 p.

Dawes, P.R. and Kerr, J.W. (ed.)
1982a: Nares Strait and the Drift of Greenland: a Conflict in Plate Tectonics; Meddelelser om Grønland, Geoscience 8, 392 p.
1982b: The case against major displacement along Nares Strait; in Nares Strait and the Drift of Greenland: a Conflict in Plate Tectonics, P.R. Dawes and J.W. Kerr (ed.); Meddelelser om Grønland, Geoscience 8, p. 369-386.

Fischer, B.F.G.
1985: The geology of the area surrounding Bunde and Bukken fiords, northwestern Axel Heiberg Island, Canadian Arctic Archipelago; unpublished M.Sc. thesis, Department of Geology and Geophysics, University of Calgary, Calgary, Alberta, 114 p.

Gould, D.B. and de Mille, G.
1964: Piercement structures in the Arctic Islands; Bulletin of Canadian Petroleum Geology, v. 12, p. 719-753.

Harrison, J.C., Embry, A.F., and Poulton, T.P.
1988: Field observations on the structural and depositional history of Prince Patrick Island and adjacent areas, Canadian Arctic Islands; in Current Research, Part D, Geological Survey of Canada, Paper 88-1D, p. 41-49.

Higgins, A.K. and Soper, N.J.
1983: The Lake Hazen fault zone: a transpressional upthrust?; in Current Research, Part B, Geological Survey of Canada, Paper 83-1B, p. 215-221.

Hinz, K., Schlüter, H-U., Grant, A.C., Srivastava, S.P., Umpleby, D., and Woodside, J.
1979: Geophysical transects of the Labrador Sea: Labrador to southwest Greenland; Tectonophysics, v. 59, p. 151-183.

Hood, P.J., Bower, M.E., Hardwick, C.D., and Teskey, D.J.
1985: Direct geophysical evidence for displacement along Nares Strait (Canada-Greenland) from low-level aeromagnetic data: a progress report; in Current Research, Part A, Geological Survey of Canada, Paper 85-1A, p. 517-522.

Hugon, H.
1983: Ellesmere-Greenland Fold Belt: structural evidence for left-lateral shearing; Tectonophysics, v. 100, p. 215-225.

Hyndman, R.D.
1975: Marginal basins of the Labrador Sea and the Davis Strait hot spot; Canadian Journal of Earth Sciences, v. 12, p. 1041-1045.

Jackson, G.D., Davidson, A., and Morgan, W.C.
1975: Geology of the Pond Inlet map-area, Baffin Island, District of Franklin; Geological Survey of Canada, Paper 74-25, 33 p.

Jackson, G.D. and Ianelli, T.R.
1981: Rift-related cyclic sedimentation in the Neohelikian Borden Basin, northern Baffin Island; in Proterozoic Basins of Canada, F.H.A. Campbell (ed.); Geological Survey of Canada, Paper 81-10, p. 269-302.

Jackson, H.R., Keen, C.E., Falconer, R.K.H., and Appleton, K.P.
1979: New geophysical evidence for sea-floor spreading in central Baffin Bay; Canadian Journal of Earth Sciences, v. 16, p. 2122-2135.

Keen, C.E., Keen, M.J., Ross, D.I., and Lack, M.
1974: Baffin Bay: small ocean basin formed by sea-floor spreading; The American Association of Petroleum Geologists Bulletin, v. 58, p. 1089-1108.

Keen, C.E. and Peirce, J.W.
1982: The geophysical implications of minimal Tertiary motion along Nares Strait; in Nares Strait and the Drift of Greenland: a Conflict in Plate Tectonics, P.R. Dawes and J.W. Kerr (ed.); Meddelelser om Grønland, Geoscience 8, p. 327-337.

Kerr, J.W.
1967: Nares submarine rift valley and the relative rotation of north Greenland; Bulletin of Canadian Petroleum Geology, v. 15, p. 483-520.

1974: Geology of the Bathurst Island Group and Byam Martin Island, Arctic Canada (Operation Bathurst Island); Geological Survey of Canada, Memoir 378, 152 p.
1977: Cornwallis Fold Belt and the mechanism of basement uplift; Canadian Journal of Earth Sciences, v. 14, p. 1374-1401.
1980: Structural framework of Lancaster Aulacogen, Arctic Canada; Geological Survey of Canada, Bulletin 319, 24 p.
1981: Evolution of the Canadian Arctic Islands: a transition between the Atlantic and Arctic oceans; in The Ocean Basins and Margins, v. 5, The Arctic Ocean, A.E.M. Nairn, M. Churkin, Jr., and F.G. Stehli (ed.); Plenum Press, New York and London, p. 105-199.

Kerr, J. W. and Ruffman, A.
1979: The Crozier Strait fault zone, Arctic Archipelago, Northwest Territories, Canada; Bulletin of Canadian Petroleum Geology, v. 27, p. 39-52.

MacLean, B., Woodside, J.M., and Girouard, P.
1984: Geological and geophysical investigations in Jones Sound, District of Franklin; in Current Research, Part A, Geological Survey of Canada, Paper 84-1A, p. 359-365.

Mayr, U. and de Vries, C.D.S.
1982: Reconnaissance of Tertiary structures along Nares Strait, Ellesmere Island, Canadian Arctic Archipelago; in Nares Strait and the Drift of Greenland: a Conflict in Plate Tectonics, P.R. Dawes and J.W. Kerr (ed.); Meddelelser om Grønland, Geoscience 8, p. 167-175.

Mayr, U. and Okulitch, A.V.
1984: Geological maps (1:125 000) of North Kent Island and southern Ellesmere Island; Geological Survey of Canada, Open File 1036.

Mayr, U., Trettin, H.P., and Embry, A.F.
1982: Preliminary geological map and notes, Clements Markham Inlet and Robeson Channel map-areas, District of Franklin (NTS 120E, F, G); Geological Survey of Canada, Open File 833, 40 p.

Menzies, A.W.
1982: Crustal history and basin development of Baffin Bay; in Nares Strait and the Drift of Greenland: a Conflict in Plate Tectonics, P.R. Dawes and J.W. Kerr (ed.); Meddelelser om Grønland, Geoscience 8, p. 295-312.

Miall, A.D.
1979a: Mesozoic and Tertiary geology of Banks Island, Arctic Canada: the history of an unstable cratonic margin; Geological Survey of Canada, Memoir 387, 235 p.
1979b: Tertiary fluvial sediments in the Lake Hazen intermontane basin; Geological Survey of Canada, Paper 79-9, 25 p.
1983: The Nares Strait problem: a re-evaluation of the geological evidence in terms of a diffuse oblique-slip plate boundary between Greenland and the Canadian Arctic Islands; Tectonophysics, v. 100, p. 227-239.
1984: Sedimentation and tectonics of a diffuse plate boundary: the Canadian Arctic Islands from 80 Ma B.P. to the present; Tectonophysics, v. 107, p. 261-277.
1986: The Eureka Sound Group (Upper Cretaceous-Oligocene), Canadian Arctic Islands; Bulletin of Canadian Petroleum Geology, v. 34, p. 240-270.

Miall, A.D., Balkwill, H.R., and Hopkins, W.S., Jr.
1980: Cretaceous and Tertiary sediments of Eclipse Trough, Bylot Island area, Arctic Canada, and their regional setting; Geological Survey of Canada, Paper 79-23, 20 p.

Mitchell, R.H. and Platt, R.G.
1984: The Freemans Cove volcanic suite: field relations, petrochemistry, and tectonic setting of nephelite – basanite volcanism associated with rifting in the Canadian Arctic Archipelago; Canadian Journal of Earth Sciences, v. 21, p. 428-436.

Ogawa, Y., Horiuchi, K., Taniguchi, H., and Naka, J.
1985: Collision of the Izu Arc with Honshu and the effects of oblique subduction in the Miura–Boso Peninsulas; Tectonophysics, v. 119, p. 349-379.

Okulitch, A.V.
1982: Preliminary structure sections, southern Ellesmere Island, District of Franklin; in Current Research, Part A, Geological Survey of Canada, Paper 82-1A, p. 55-62.

Okulitch, A.V. and Mayr, U.
1984: Compressive and extensional tectonics near dextral continental transform faults within the Arctic Platform, southeastern Ellesmere Island, Canada (abstract); in The Geological Society of America, 97th meeting, Abstracts with Programs, v. 16, p. 613.

Osadetz, K.G.
1982: Eurekan structures of the Ekblaw Lake area, Ellesmere Island, Canada; in Arctic Geology and Geophysics, H.R. Balkwill and A.F. Embry (ed.); Canadian Society of Petroleum Geologists, Memoir 8, p. 219-232.

Peirce, J.W.
1982: The evolution of the Nares Strait lineament and its relation to the Eurekan orogeny; in Nares Strait and the Drift of Greenland: a Conflict in Plate Tectonics, P.R. Dawes and J.W. Kerr (ed.); Meddelelser om Grønland, Geoscience 8, p. 237-251.

Rice, P.D. and Shade, B.D.
1982: Reflection seismic interpretation and seafloor spreading history of Baffin Bay; in Arctic Geology and Geophysics, A.F. Embry and H.R. Balkwill (ed.); Canadian Society of Petroleum Geologists, Memoir 8, p. 245-267.

Ricketts, D.B.
1987: Preliminary structural cross-sections across Fosheim Peninsula and Axel Heiberg Island, Arctic Archipelago; in Current Research, Part A, Geological Survey of Canada, Paper 87-1A, p. 369-374.

Ricketts, B.D. and McIntyre, D.J.
1986: The Eureka Sound Group of eastern Axel Heiberg Island; new data on the Eurekan Orogeny; in Current Research, Part B, Geological Survey of Canada, Paper 86-1B, p. 405-410.

Riddihough, R.P., Haines, G.V., and Hannaford, W.
1973: Regional magnetic anomalies of the Canadian Arctic; Canadian Journal of Earth Sciences, v. 10, p. 157-163.

Snelling, N.J.
1987: Measurement of geological time and the geological time scale; Modern Geology, v. 11, p. 365-374.

Srivastava, S.P.
1985: Evolution of the Eurasian Basin and its implications to the motion of Greenland along Nares Strait; Tectonophysics, v. 114, p. 29-53.

Srivastava, S.P. and Falconer, R.H.
1982: Nares Strait: a conflict between plate tectonic predictions and geological interpretation; in Nares Strait and the Drift of Greenland: a Conflict in Plate Tectonics, P.R. Dawes and J.W. Kerr (ed.); Meddelelser om Grønland, Geoscience 8, p. 339-352.

Srivastava, S.P., Falconer, R.K.H., and MacLean, B.
1981: Labrador Sea, Davis Strait, Baffin Bay: geology and geophysics — a review; in Geology of the North Atlantic Borderlands, J.W. Kerr and A.J. Ferguson (ed.); Canadian Society of Petroleum Geologists, Memoir 7, p. 333-398.

Srivastava, S.P., MacLean, B., Macnab, R.F., and Jackson, H.R.
1982: Davis Strait: structure and evolution as obtained from a systematic geophysical survey; in Arctic Geology and Geophysics, A.F. Embry and H.R. Balkwill (ed.); Canadian Society of Petroleum Geologists, Memoir 8, p. 267-278.

Srivastava, S.P. and Tapscott, C.R.
1986: Plate kinematics of the North Atlantic; in The Western North Atlantic Region, P.R. Vogt and B.E. Tucholke (ed.); Geological Society of America, The Geology of North America, v. M, p. 379-404.

Soper, N.J., Dawes, P.R., and Higgins, A.K.
1982: Cretaceous magmatic and tectonic events in North Greenland and the history of adjacent ocean basins; in Nares Strait and the Drift of Greenland: a Conflict in Plate Tectonics, P.R. Dawes and J.W. Kerr (ed.); Meddelelser om Grønland, Geoscience 8, p. 205-220.

Thorsteinsson, R.
1974: Carboniferous and Permian stratigraphy of Axel Heiberg Island and western Ellesmere Island, Canadian Arctic Archipelago; Geological Survey of Canada, Bulletin 224, 115 p.

Thorsteinsson, R. and Mayr, U.
1987: The sedimentary rocks of Devon Island, Canadian Arctic Archipelago; Geological Survey of Canada, Memoir 411, 182 p.

Thorsteinsson, R. and Tozer, E.T.
1970: Geology of the Arctic Archipelago; in Geology and Economic Minerals of Canada, R.J.W. Douglas (ed.); Geological Survey of Canada, Economic Geology Report no.1, p. 547-590.

Trettin, H.P.
1975: Investigations of lower Paleozoic geology, Foxe Basin, northeastern Melville Peninsula, and parts of northwestern and central Baffin Island; Geological Survey of Canada, Bulletin 251, 177 p.
1978: Devonian stratigraphy, west-central Ellesmere Island; Geological Survey of Canada, Bulletin 302, 119 p.

Trettin, H.P. and Parrish, R.
1987: Late Cretaceous bimodal magmatism, northern Ellesmere Island: isotopic age and origin; Canadian Journal of Earth Sciences, v. 24, p. 257-265.

Trusheim, F.
1960: Mechanism of salt migration in northern Germany; Bulletin of the American Association of Petroleum Geologists, v. 44, p. 1519-1540.

Woodcock, N.H. and Fischer, M.
1986: Strike-slip duplexes; Journal of Structural Geology, v. 8, p. 725-735.

Wynne, P.J., Irving, E., and Osadetz, K.
1983: Paleomagnetism of the Esayoo Formation (Permian) of northern Ellesmere Island: possible clue to the solution of the Nares Strait problem; Tectonophysics, v. 100, p. 241-256.
in press: Paleomagnetism of Cretaceous volcanic rocks of the Sverdrup Basin — magnetostratigraphy, paleolatitudes and rotations; Canadian Journal of Earth Sciences.

Authors' addresses

A.V. Okulitch
Geological Survey of Canada
3303-33rd Street N.W.
Calgary, Alberta
T2L 2A7

H.P. Trettin
Geological Survey of Canada
3303-33rd Street N.W.
Calgary, Alberta
T2L 2A7

CHAPTER 17

Printed in Canada

Chapter 18
MIDDLE AND LATE TERTIARY TECTONIC AND PHYSIOGRAPHIC DEVELOPMENTS

Introduction

Uplift

Erosion

References

Chapter 18

MIDDLE AND LATE TERTIARY TECTONIC AND PHYSIOGRAPHIC DEVELOPMENTS

H.P. Trettin

INTRODUCTION

The post-Eurekan pre-Pleistocene history of the Arctic Islands is poorly known because the stratigraphic record of this interval is limited mainly to the northwestern margin of the Queen Elizabeth Islands (Chapter 15). The lack or scarcity of preserved sediments in other areas indicates uplifts, which, in interaction with erosion and eustatic sea level changes, gave rise to the present landscape (cf. Thorsteinsson and Tozer, 1970).

UPLIFT

Middle or Late Tertiary positive movements were most pronounced within the Eurekan Orogen and a belt surrounding the Labrador Basin, and of lesser magnitude in the area south of the Eurekan Orogen. The uplift of the Eurekan Orogen represents isostatic adjustments to post-Eurekan erosion; the causes of the uplifts in the other areas are uncertain although a thermal model appears to be applicable to the mountains around Labrador Basin. The positive movements have been amplified to a minor extent by an overall drop in sea level since early Tertiary time (Vail and Hardenbol, 1979). The Holocene postglacial rebound will not be considered in this context because it merely tends to restore conditions that existed before the Pleistocene.

Eurekan Orogen

Crustal thickening during the Eurekan Orogeny resulted in high topographic relief that produced the syntectonic clastic sediments of the Eureka Sound Group (Chapter 15). Post-Eurekan erosion must have caused an isostatic disequilibrium that was compensated by upward movement. The presence of high mountain ranges and matching negative Bouguer anomalies shows that the crustal roots of the orogen have not yet been eliminated by these processes. It is generally assumed that isostatic uplift is intermittent, long periods of erosion alternating with brief periods of relatively rapid uplift and comcomitant erosion (e.g. Chorley et al., 1984, Fig. 3.30), a pattern that has been inferred from elevated relict erosion surfaces.

Little information is available about the time-variation of isostatic uplift in the Eurekan Orogen. It is generally agreed, however, that the present mountains existed prior to the onset of extensive alpine glaciation in the Pleistocene or latest Tertiary. At least some of the coarse clastic sediments in the Beaufort Formation were probably derived from the Princess Margaret Range and Grantland Mountains, suggesting that these ranges stood high during Late Miocene(?) – Early Pliocene time.

The earlier record is even more uncertain. High rates of sedimentation, suggesting rapid erosion, prevailed during the Eurekan Orogeny in intermontane basins (Chapter 15). It is possible that this erosion resulted in the development of a system of pediments in mid-Tertiary time (cf. Thorsteinsson and Tozer, 1970, p. 588) but this has not yet been documented.

The present physiography shows that the isostatic uplift must have been most pronounced in the Grantland Mountains, the Victoria and Albert Mountains, and the Princess Margaret Range. It is not likely that the rejuvenation of the first two ranges was accomplished by renewed motion on the moderately inclined thrust faults that produced them (Chapter 17), because the principal stress had changed from horizontal to vertical. Movement on steeply-dipping faults, reverse or normal, is ruled out because none of the known faults account for the present relief. Thus the only remaining mechanism for them is doming without significant faulting. This conclusion gains some support from the Foothills of the Rocky Mountains in Alberta where no steeply-dipping faults are recognized that are related to the post-Laramide uplift of the Front Ranges.

The attitude at depth, and movement history of the Stolz Thrust, which borders the Princess Margaret Range on the east, are uncertain (Chapter 17) and reactivation of this fault in mid-Tertiary time cannot be excluded.

Environs of Labrador Basin

Labrador Basin is surrounded by mountains, locally more than 2 km high, that are carved in Archean – Lower Proterozoic crystalline rocks (cf. Chapter 3, Baffin Region, Thule Upland, and Thule-Nûgssuaq Region). The transition zone between these mountains and a relatively gently inclined cratonic monocline that underlies the continental shelf and slope (Balkwill, 1987), is characterized by uplands, plateaus, and lowlands in which successions of Middle and Late Proterozoic, early Paleozoic, and Cretaceous-Paleogene age are preserved locally. High structural relief, largely due to faulting, exists between the highlands and this inner belt.

From southeastern Ellesmere Island to Bylot Island, the central mountains are flanked on the outside by an arcuate, northwesterly- to southwesterly-dipping

---

H.P. Trettin
1991: Middle and Late Tertiary tectonic and physiographic developments; Chapter 18 in Geology of the Innuitian Orogen and Arctic Platform of Canada and Greenland, H.P. Trettin (ed.); Geological Survey of Canada, Geology of Canada, no. 3; (also Geological Society of America, The Geology of North America, v. E).

monocline of Phanerozoic strata that lie unconformably on Precambrian basement. The Phanerozoic succession is early Paleozoic in age on Ellesmere and Devon islands, and Late Cretaceous to probably Early Eocene on Bylot Island, where it comprises the Hassel and Kanguk formations and the Eureka Sound Group (Jackson and Davidson, 1975; Jackson et al., 1975; Miall et al., 1980). The latter three units were deposited in shallow marine and nonmarine environments, generally close to sea level. Their surface now rises from sea level in the southwest to an altitude of about 760 m in the northeast near the contact with the Precambrian rocks, and their southwesterly dip increases from 2 to 40° in the same direction. Thus tilting, caused by rise of the central mountains, occurred in post-Early Eocene time. The stratigraphic record of the shelf adjacent to Labrador and southern Baffin Island suggests that the high relief was established in Late Oligocene time and rejuvenated in Late Miocene time (Balkwill and McMillan, in press).

These uplifts did not involve Lancaster Sound and Jones Sound. If the normal faults bounding these depressions had originated during the Cretaceous-Paleogene rifting event, they must have been reactivated in Late Oligocene and Late Miocene time.

The timing of the uplift around Labrador Basin is somewhat anomalous, as in most coastal areas thermal subsidence prevails after the onset of spreading, but there are exceptions to this rule. Spreading in the North Atlantic, for example, commenced in the Eocene but in East Greenland about 2 km of plateau uplift occurred in Oligocene-Miocene time, after extensive mafic and alkalic magmatism that culminated in the Eocene (Gleadow and Brooks, 1979). Brooks and Nielsen (1982) stated that the uplift is probably not thermal in origin because of its "long term stability" but due to "some addition of light material at depth, possibly by either metasomatism or by underplating" with gabbro. Tertiary magmatism appears to have been far less intensive in the mountains around the Labrador Basin so that the latter explanation may not be applicable there. More recently, Fleitout at al. (1986) have demonstrated by computer simulation that passive rifting can induce small-scale convection that leads to a delayed shoulder uplift. In their model the uplift occurs 20 Ma after crustal thinning and persists for more than 80 Ma.

Region south of Eurekan Orogen

Moderate relief exists in all those parts of the archipelago lying south of the Eurekan Orogen and west of the Baffin region. The fact that it is no older than Oligocene is apparent on Banks Island where Eocene strata of the Eureka Sound Group that were deposited close to sea level now occur at altitudes up to 480 m (Chapter 15, and Miall, 1979, Map 1454A). It appears that this uplift is not confined to the Arctic Islands but extends far into the continental interior.

EROSION
Erosional regimes

Rates of erosion are dependent mainly on relief, climate (especially humidity), vegetation, and rock type. High relief, causing fast erosion, has existed around Baffin Bay and within the Eurekan Orogen through most of the Cenozoic Era with the possible exception of parts of the Oligocene and Miocene epochs. The climate (Chapter 15) cooled in the Late Tertiary but still permitted forest growth at low altitudes in Early Pliocene time. The Arctic Ocean, which was open until the end of Tertiary time, and Baffin Bay provided moisture that precipitated on the surrounding mountains but intermontane basins such as the present Hazen Plateau and west-central Ellesmere Island may have been relatively arid. Resistance to weathering varied widely with composition and metamorphic state of the rocks. Overall, conditions probably were comparable to present regimes in the northernmost parts of the Cordillera. Unfortunately, estimates of denudation rates for this kind of setting vary so widely that this information is insufficient to establish the probable age of the present mountain ranges: some permit Oligocene and even older ages, whereas others imply Miocene or Pliocene maximum ages (cf. Selby, 1982, Fig. 11.2 and 11.3; Chorley et al., 1984, Tables 3.13 and 3.14).

Development of drainage, and origin of seaways and fiords

Fortier and Morley (1956) proposed that the Tertiary drainage pattern can be reconstructed from the present fiords and seaways (Fig. 18.1). Two major systems, separated by an extensive divide, were inferred: a northerly and northwesterly system draining into the Arctic Ocean, and an easterly and southeasterly system draining into Baffin Bay. During the Pleistocene, alpine glaciers advanced into lower lying regions along the river valleys, deepening and straightening them in the process. In the early Holocene, when the glaciers receded and sea level rose, the glacial troughs were invaded by the sea, a process that has been reversed to some extent by the rebound of the crust.

This hypothesis has been confirmed, to some degree, by bathymetric mapping and coring (Pelletier, 1966). Glacial modification of the original channels is apparent from U-shaped transverse profiles with depths locally in excess of 900 m, longitudinal profiles that rise seaward, hanging tributaries, drowned cirques and other features (Pelletier, 1966).

Objections to the erosional hypothesis have arisen from studies in glaciology and Quaternary geology. First, it has become known that glaciers in the high Arctic do not deepen the valleys along which they move: the flow occurs above a layer of ice, frozen to the ground, that protects it. Second, there is sufficient evidence that the Late Wisconsinan glaciers also were "cold-based", i.e. non-erosive. Moreover, these glaciers were not as extensive as once assumed; at least some of the present fiords and seaways were occupied by the sea (England, 1987; cf. Chapter 19).

However, high-level erratics of older age were deposited by thicker ice sheets that must have filled at least some fiords and seaways at an earlier time. This, for example, has been inferred for Clements Markham Inlet in northeasternmost Ellesmere Island (Bednarski, 1986; for location see Fig. 1, in pocket) and for Nares Strait, which was overridden by Greenland ice during the earliest advance known in northern Ellesmere Island, perhaps 0.5-1 Ma ago (England, 1987). In contrast to the Holocene and Late Wisconsinan glaciers, the early glaciers must have been "warm-based" if they indeed have caused the erosion attributed to them. This assumption is questioned by England (1987) who sees no evidence on land for the kind of erosion and deposition associated with temperate glaciers.

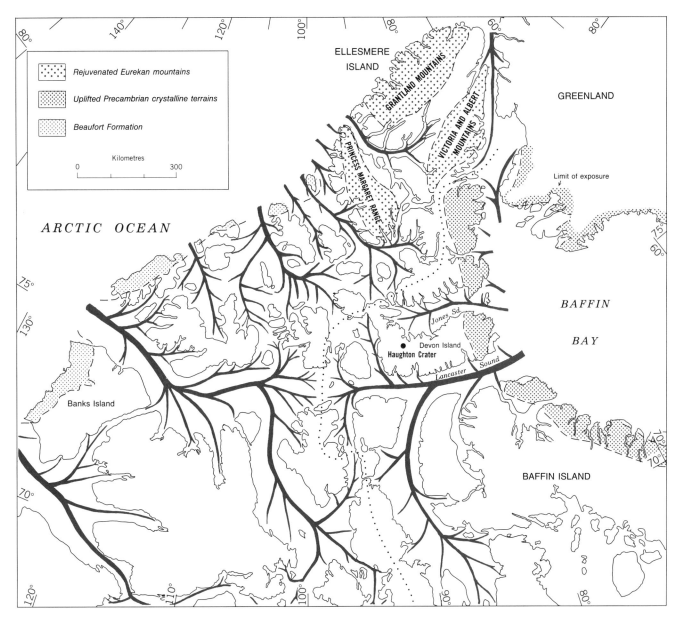

Figure 18.1. Middle-Late Tertiary physiography and deposits; drainage adapted from Fortier and Morley (1956) and Pelletier (1966).

To provide an alternative explanation, he proposed that the seaways and fiords formed by (graben) faulting — after the early, extensive advances, but before the Late Wisconsinan. Unfortunately, this hypothesis conflicts with structural evidence or considerations that can be summarized as follows:

(1) Only a few of the seaways with known structure are grabens (e.g. Lancaster Sound, Jones Sound, Foxe Basin; Chapter 17), many others are not (e.g. fiords in northern Greenland [see Chapter 3, Kane Basin – Independence Fiord region]; Disraeli, Archer, Trold and Bay fiords in Ellesmere Island; Glacier and Strand fiords in Axel Heiberg Island; Young, May and Erskine inlets in Bathurst Island; cf. Fig. 3).

(2) Most of the relatively young normal faulting occurred in Cretaceous to Miocene time. The Lower Pliocene (and possibly Upper Miocene) Beaufort Formation has minor faults and lineaments, and lineaments also occur in Quaternary deposits but there is no evidence for a major extensional event in mid-Pleistocene time.

(3) Graben faulting usually is accompanied by rapid clastic sedimentation, which keeps the grabens filled with nonmarine sediments. Submergence below sea level may occur later — perhaps as a result of thermal subsidence (as in the Red Sea), or of glacial erosion of the poorly consolidated graben fill, followed by a sea level rise (as in the Arctic Islands).

(4) There is ample evidence for lithological and/or structural control of the postulated erosive processes. In some instances the fiords coincide with linear belts of relatively weakly resistant rock types such as lower Paleozoic – Devonian mudrocks (e.g. Archer and Trold fiords in Ellesmere Island; cf. Fig. 3) or poorly consolidated Tertiary sediments (e.g. Glacier Fiord in Axel Heiberg Island and Lancaster Sound). In other instances there is indirect evidence for major faults or fault zones, i.e. fractured rocks, beneath them (e.g. Nares Strait, Cañon Fiord).

In conclusion, no viable alternative is known to the hypothesis that the fiords and seaways represent Tertiary river valleys that were deepened and straightened by pre-Late Wisconsinan glaciers. It is established that these glaciers extended across at least some of the present fiords and channels. Evidence for erosion in the subsea environment has been presented by Pelletier (1966); evidence for erosion on land may have been eliminated by prolonged weathering in subsequent periglacial settings (cf. Barsch, 1981) — generally, the Early and Middle Pleistocene record is poorly known (Chapter 19).

Comparison of Fortier and Morley's drainage reconstruction (Fig. 18.1) with a map of Eurekan structural elements (Fig. 17.1) suggests that the drainage was established during late phases of the Eurekan Orogeny and remained relatively stable.

REFERENCES

Balkwill, H.R.
1987: Labrador Basin: structural and stratigraphic style; in Sedimentary Basins and Basin-Forming Processes, C. Beaumont and A.J. Tankard (ed.); Canadian Society of Petroleum Geologists, Memoir 12, p. 17-43.

Balkwill, H.R. and McMillan, N.J.C.
in press: Mesozoic-Cenozoic stratigraphy of the Labrador Shelf; in Geology of the Continental Margin of Eastern Canada, M.J. Keen and G.L. Williams (ed.); Geological Survey of Canada, Geology of Canada, no. 2 (also Geological Society of America, The Geology of North America, v. I-1).

Barsch, D.
1981: Zur Geomorphologie des Expeditionsgebietes Oobloyah Bay, N-Ellesmere Island, N.W.T., Kanada; in Ergebnisse der Heidelberg-Ellesmere Island-Expedition (Results of the Heidelberg Ellesmere Island Expedition), D. Barsch and L. King (ed.); Department of Geography, University of Heidelberg, Federal Republic Germany, p. 109-122.

Bednarski, J.
1986: Late Quaternary glacial and sea-level events, Clements Markham Inlet, northern Ellesmere Island; Canadian Journal of Earth Sciences, v. 23, p. 1343-1355.

Brooks, C.K. and Nielsen, T.F.D.
1982: The E. Greenland continental margin: a transition between oceanic and continental magmatism; Geological Society of London Journal, v. 139, p. 265-275.

Chorley, R.J., Schumm, S.A., and Sugden, D.E.
1984: Geomorphology; Methuen, London and New York, 605 p.

England, J.
1987: Glaciation and the evolution of the Canadian high arctic landscape; Geology, v. 15, p. 419-424.

Fleitout, L., Froidevaux, C., and Yuen, D.
1986: Active lithospheric thinning; Tectonophysics, v. 132, p. 271-278.

Fortier, Y.O. and Morley, L.W.
1956: Geological unity of the Arctic Islands; Royal Society of Canada Transactions, v. 50, p. 3-12.

Gleadow, A.J.W. and Brooks, C.K.
1979: Fission track dating, thermal histories and tectonics of igneous intrusions in East Greenland; Contributions to Mineralogy and Petrology; v. 71, p. 45-60.

Jackson, G.D. and Davidson, A.
1975: Bylot Island map-area, District of Franklin; Geological Survey of Canada, Paper 74-29, 12 p.

Jackson, G.D., Davidson, A., and Morgan, W.C.
1975: Geology of the Pond Inlet map-area, Baffin Island, District of Franklin; Geological Survey of Canada, Paper 74-25, 33 p.

Miall, A.D.
1979: Mesozoic and Tertiary geology of Banks Island, Arctic Canada: the history of an unstable cratonic margin; Geological Survey of Canada, Memoir 387, 235 p.

Miall, A.D., Balkwill, H.R., and Hopkins, W.S., Jr.
1980: Cretaceous and Tertiary sediments of Eclipse Trough, Bylot Island area, Arctic Canada, and their regional setting; Geological Survey of Canada, Paper 79-23, 20 p.

Pelletier, B.R.
1966: Development of submarine physiography in the Canadian Arctic and its relation to crustal movements; in Continental Drift, G.D. Garland (ed.); The Royal Society of Canada, Special Publication no. 9, p. 77-101.

Selby, M.J.
1982: Hillslope Materials and Processes; Oxford University Press, Oxford, England, 264 p.

Thorsteinsson, R. and Tozer, E.T.
1970: Geology of the Arctic Archipelago; in Geology and Economic Minerals of Canada, R.J.W. Douglas (ed.); Geological Survey of Canada, Economic Geology Report no. 1, p. 547-590.

Vail, P.R. and Hardenbol, J.
1979: Sea-level changes during the Tertiary; Oceanus, v. 22, p. 71-79.

Author's address

H.P. Trettin
Geological Survey of Canada
3303-33rd Street N.W.
Calgary, Alberta
T2L 2A7

Printed in Canada

Chapter 19
THE QUATERNARY RECORD

Introduction

Quaternary deposits

Quaternary stratigraphy and chronology

References

Chapter 19

THE QUATERNARY RECORD

D.A. Hodgson

INTRODUCTION

Quaternary geology studies in most of the Arctic Archipelago inevitably have focused on the age and extent of Pleistocene glaciations. In the islands south of Parry Channel (Fig. 19.1), repeated advances of temperate glaciers spreading from continental dispersal centres have eroded and deposited a wide range of landforms. There, till is the most widespread surficial material; marine deposits, raised with a crust rebounding from the last glaciation, are also conspicuous. In the Queen Elizabeth Islands to the north, the mountainous eastern rim (Fig. 19.2) remains half glacierized and has attracted the interest of the glacial geologist as well as glaciologist; drift landforms are scarcer than to the south. In the western islands, drift is sparse and glacial studies are retarded in comparison with the rest of the archipelago. Suitably, this area has drawn some studies of Holocene and older nonglacial deposits and processes, including the thermal history of near-surface sediments.

The Quaternary of the archipelago is discussed at greater length in Fulton (1989), by Hodgson (1989) for the Queen Elizabeth Islands, by Funder (1989) for north Greenland, and, in three separate articles by Andrews (1989), Dyke and Dredge (1989), and Vincent (1989), respectively, for the islands south of Parry Channel.

The Quaternary climate of the Arctic Islands is poorly understood — in particular, the polar weather patterns that resulted when continental ice domes lay to the south. The present climate of most of the northern islands is dominated year-round by the anticyclonic air of the central Arctic Ocean, and thus differs significantly from the more continental climate of the southern islands and mainland and from the shores of northern Baffin Bay, which receive some maritime air. In the mountainous northeastern islands, the cool northwesterly summer air flow is blocked, and isotherms (as well as floral zones; Edlund, 1987) trend north, in contrast to the predominantly latitudinal trends that are prevalent in the southwestern and southern islands. Mean July temperatures for northern islands range from 10 to 18°C (all stations are coastal) and annual mean daily temperatures from -16 to -19°C. Precipitation is low enough to make this an arid zone, but low evaporation rates permit sporadic wetlands.

The low annual mean temperatures imply that permafrost is universally present under land. In the Arctic Islands, permafrost thickness, determined in 37 drill holes at 30 sites (Taylor et al., 1982), varies from 143 m on western Sabine Peninsula, Melville Island to 726 m on Cameron Island (Fig. 19.3). Thicknesses of this magnitude indicate mean annual surface temperatures below freezing for most of the Late Quaternary. The variations in permafrost thickness between measuring sites are due to a variety of factors such as heat flow (Chapter 5G), geological setting and rock conductivity, and Quaternary history including shore line regression (Taylor et al., 1983).

The northernmost limits of a number of major vascular plant families and species (mainly dwarf woody plants and sedges) presently cross the Queen Elizabeth Islands (Fig. 19.4), following the isotherm patterns described above. Surface materials are a significant secondary influence on plant distribution — especially strong alkalinity associated with weathered carbonate rocks, strong acidity of some shale, and salinity of fine grained raised marine deposits (Edlund, 1983). The Arctic Basin shores have supported tundra vegetation in the Pliocene or even Miocene (Funder et al., 1985; Matthews, 1987), though little is known from the northern archipelago from then until the Late Pleistocene (last interglacial?) when low-Arctic species were present (Blake, 1974). In mid-Arctic Banks Island, several Quaternary intervals, warmer than the present, have been recorded (see below).

QUATERNARY DEPOSITS
Glacial deposits

Glacial deposits are sparse on many islands north of Parry Channel (Fig. 19.5). The most continuous and thickest drift occurs within 50 km of margins of present ice caps on Ellesmere Island, and where the Wisconsinan Laurentide Ice Sheet impinged on southern Melville Island. Locally thick drift of undetermined age and origin has been identified in channels between central islands by shipborne geophysical surveys (McLean et al., in press). Scattered Precambrian Shield erratics at all elevations indicate invasion by continental ice of the western and central islands; a few degraded eskers, kames, and glacially deformed sediments (Fyles, 1965; Hodgson, 1982) remain from this event. Erratics on eastern islands record a long history of local ice sources, except where Greenland ice and erratics overran the northeast coast of Ellesmere Island (Christie, 1967). Subglacial landforms were rarely developed by ice flows originating in the Queen Elizabeth Islands. This absence of interaction between glacier and bed indicates basal temperatures below the pressure melting point, except where strong compressive flow and scouring occurred in deep valleys on Ellesmere and Axel Heiberg islands. A similar limitation of scouring is found on north Greenland.

To the south of Parry Channel, most islands are veneered or blanketed by till from continental ice sheets (Fig 19.5). Till composition is broadly related to underlying bedrock

---

Hodgson, D.A.
1991: The Quaternary record; Chapter 19 in Geology of the Innuitian Orogen and Arctic Platform of Canada and Greenland, H.P. Trettin (ed.); Geological Survey of Canada, Geology of Canada, no. 3; (also Geological Society of America, The Geology of North America, v. E).

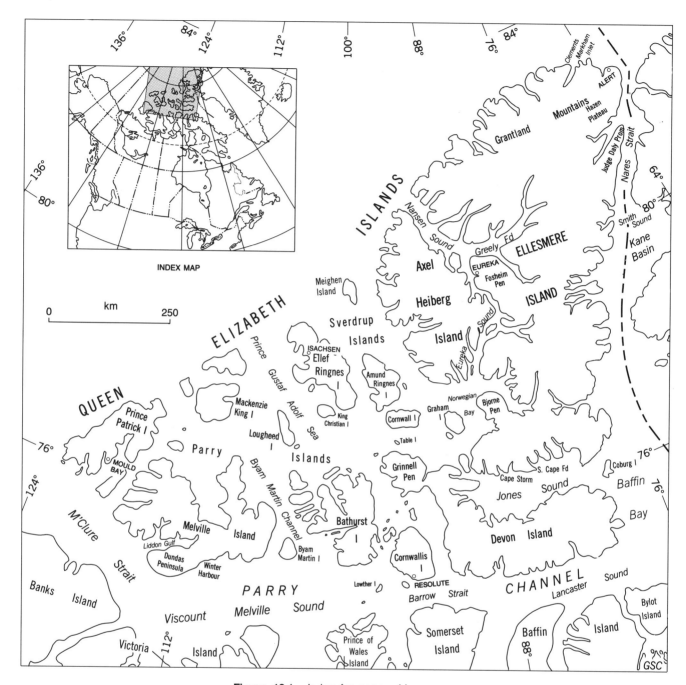

Figure 19.1. Index for geographic names.

lithology (e.g. silty calcareous till on carbonates), but fast-flowing ice streams within ice sheets did disperse foreign material widely (Dyke, 1984). Streamlined bedforms (including spectacular drumlin fields), subglacial meltwater landforms (including eskers), massive end moraines and outwash trains terminating in thick bodies of glaciomarine sediments are widely present. Whereas Laurentide ice overlapped some parts of the southern islands, others remained outside the temperate Laurentide Ice Sheet (Fig. 19.5). Northern Somerset Island (Dyke, 1983) and northwest Baffin Island lay under independent ice caps, in part with protective frozen beds, Bylot Island retained its alpine glaciers and landforms, while most of Banks Island was unglaciated, exposing relatively featureless (degraded?) till from earlier glaciations.

Nonglacial deposits and processes

Much of the surficial material of the Queen Elizabeth Islands is weathered rock; a variety of processes (chiefly physical) contributing to disaggregation are reviewed by French (1976) and Washburn (1980), though this topic

THE QUATERNARY RECORD

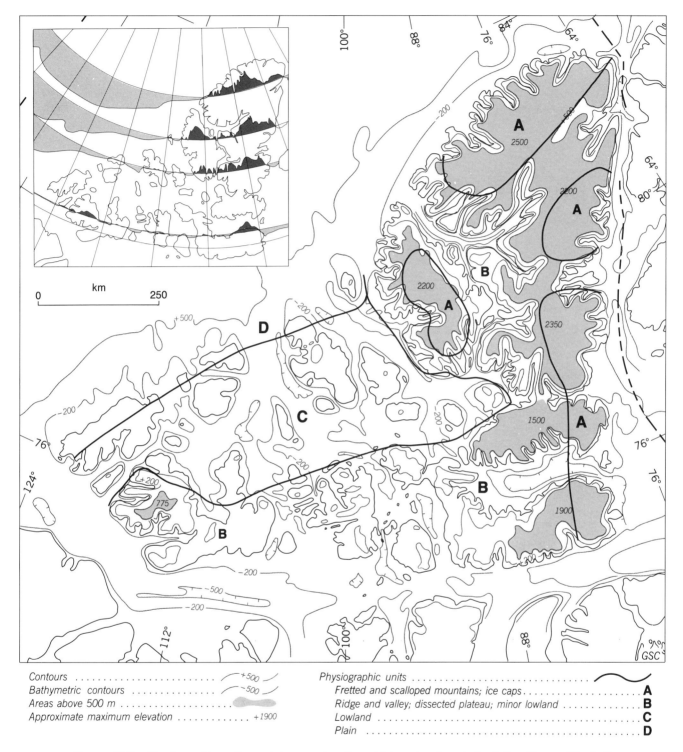

Figure 19.2. Physiographic units and topography (terrestrial and submarine) of the Queen Elizabeth Islands (after GEBCO 5-17, Canadian Hydrographic Service, 1979). Inset shows four profiles (from National Atlas of Canada, 1973).

remains poorly understood. The weak rocks common in the Sverdrup Basin of the western and central Queen Elizabeth Islands disaggregate to sand, silt, and more rarely clay, whereas more resistant rocks of the Arctic Platform and Franklinian mobile belt break into pebble- to boulder-size gravel, and silt or sand.

Permafrost, which is manifested conspicuously by ubiquitous ice-cored frost-fissure systems, permits

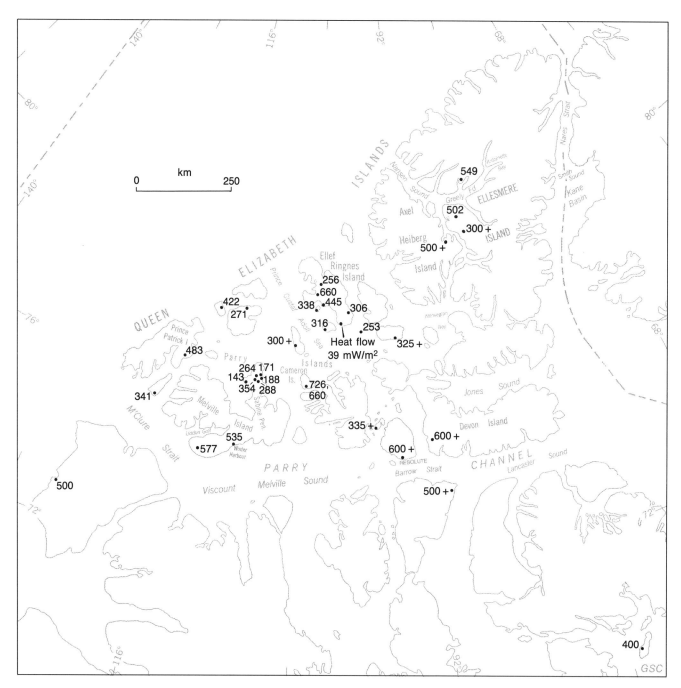

Figure 19.3. Permafrost thickness (in metres) from drill holes in the Queen Elizabeth Islands (from Taylor et al., 1982, Fig. 2-5). Also shown is the site of a heat flow determination (discussed in Chapter 5G).

segregation of the ice that represents more than 10% of uppermost surficial deposits (including weathered rock).

Although permafrost and associated periglacial processes and landforms particularly characterize the central and northern archipelago (French, 1976; Washburn, 1980), it is fluvial processes, concentrated by snowmelt runoff and the thin seasonal pervious layer, that presently are most effective in shaping the landscape (Carter et al., 1987). Most of the archipelago is subject to this nival regime, which is extreme in the western and central Queen Elizabeth Islands where there are a few ponds or lakes to regulate flow. In the glacierized eastern islands, some of the largest rivers receive a seasonally more regular flow under the glacial regime.

Raised marine deposits are widespread, as expected in an archipelago recovering from glacially induced lithospheric flexing. Typically, offlap flights of gravelly and sandy beach ridges overlie silty or sandy nearshore sediments in the eastern and southern islands, where a wide range of clast sizes has been available for reworking during the seasonal

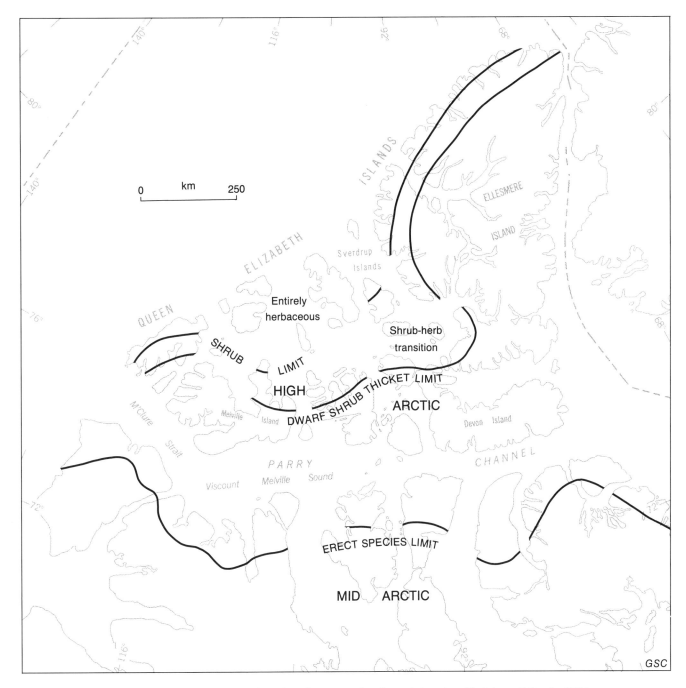

Figure 19.4. Regional distribution of major vegetation boundaries (modified from Edlund, 1987).

breakup of the sea ice throughout the Holocene. Marine deposits are dominantly fine grained in the northwestern islands, where a relatively permanent sea ice cover has been maintained and coarse materials are less common. Modern shorelines reflecting these characteristics are described by Taylor and McCann (1983). Less widespread but nevertheless thick (to 100 m) deposits of massive to rhythmically bedded glaciomarine silt occur in the central and eastern islands.

Marine bivalve molluscs rapidly colonized nearshore and deltaic deposits of nutritionally favourable postglacial seas in the central and eastern islands. Distribution of macrofauna in the west is sporadic. The most common species are *Hiatella arctica*, *Mya truncata*, *Astarte* sp., *Macoma* sp., and *Portlandia arctica*. The Late Quaternary glacial chronology is based largely on ^{14}C dating of molluscs from marine sediments in the rare localities where they are intercalated with drift. The scattered raised marine deposits, which have provided "infinite" ^{14}C ages (i.e. ages above the detection limit of this method), can rarely be related to a direction in change of sea level (transgression or regression),

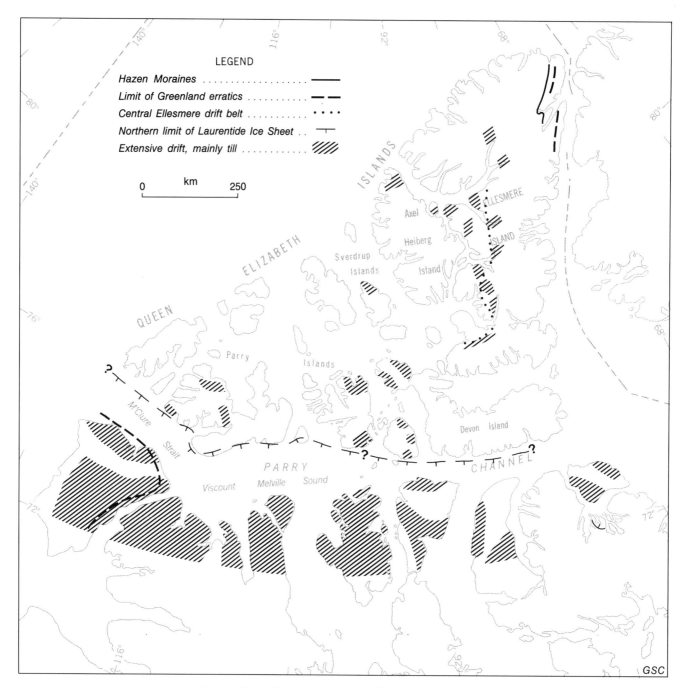

Figure 19.5. Generalized distribution of glacial deposits.

and can be related to former glacier margins only in northeastern Ellesmere Island.

The lack of recognition of marine deposits coeval with the Late Wisconsin Glaciation once led to the assumption that the latest Pleistocene/Holocene marine record started after retreat of an omnipresent ice cover. Subsequently it was emphasized that direct evidence of glaciation also was necessary (Andrews and Miller, 1976). More recently, England (1983) suggested that throughout the last glaciation, certainly in northern Ellesmere Island, and perhaps over much of the Queen Elizabeth Islands, marine deposition occurred in topographically low areas that in many places were separated from ice capped uplands by nonmarine periglacial belts (see below). However, a proglacial/marine boundary has been recognized in this region only at a few localities peripheral to existing ice caps. In the southern archipelago, a well defined marine limit, synchronous with the retreat of the Laurentide Ice Sheet, is exposed at many localities.

Glacial ice

In the Arctic Archipelago, glaciers (which cover about 25% of the Queen Elizabeth Islands) are restricted to the mountainous eastern part, except for several small, thin, stagnant ice caps on Melville and Meighen islands (Fig. 19.1), and stagnant glacierets in gullies. Ice caps are up to 800 m thick and thickest around Baffin Bay, the most open of the surrounding seas (Koerner, 1977). Because of their relatively small accumulation areas, glaciers in the Queen Elizabeth Islands move relatively slowly, are little crevassed, and calve few icebergs where they reach tidewater. In contrast to the Arctic Archipelago, the huge ice cap of Greenland, known as the Inland Ice, is rather dynamic. The presence of an unglaciated foreland on the northwest side of the Inland Ice, which is up to 100 km wide, is due to the aridity of this belt.

QUATERNARY STRATIGRAPHY AND CHRONOLOGY

Chronostratigraphic terminology and methods

In the past, the stratigraphic terminology for the Quaternary was based entirely on glacial advances and retreats, which are highly diachronous events, but in recent years chronostratigraphic terms have found increasing acceptance. In this chapter a terminology recommended by Fulton (1984), Fulton and Prest (1987), and St-Onge (1987) is used, which is related to oxygen isotope stages of deep sea sediments and magnetostratigraphic time units. The ages assigned to the older boundaries are tentative because of dating problems (see below).

In the present time scale (Fig. 19.6) the beginning of the Quaternary is placed at the end of the Olduvai Normal-Polarity Subchron. The Quaternary is divided into the Pleistocene and Holocene epochs, the latter beginning at 10 ka. The Pleistocene is split into three subdivisions, with the Early-Middle boundary, at 790 ka, corresponding to the boundary between the Brunhes Normal-Polarity and Matayuma Reversed-Polarity chrons. The Middle-Late Pleistocene boundary, at 130 ka, corresponds to the approximate beginning of the last major warm interval, dated from oxygen isotope curves for deep sea cores. The Late Pleistocene is further subdivided into the Sangamonian and Wisconsinan stages. The boundary between these two stages is set at 80 ka, the approximate beginning of the last main phase of glaciation, also dated from oxygen isotope curves. The Wisconsinan itself is split into three substages by a minor warm interlude at approximately 65-28 ka. The terms Sangamon Interglaciation and Wisconsin Glaciation refer to time-transgressive events that occurred mainly during the corresponding stages but are not confined to them.

In the absence of volcanic material in the Arctic Islands, the only reliable method for absolute dating is the ^{14}C method, which has an upper limit of about 45 ka for wood and of 20 ka for shells according to Vincent and Prest (1987; higher values are cited by other authors). Beyond these limits, ages are presently determined chiefly from the rate of amino acid racemization of wood and shells. This method requires a rather detailed knowledge of post-depositional temperature changes that generally is difficult to obtain.

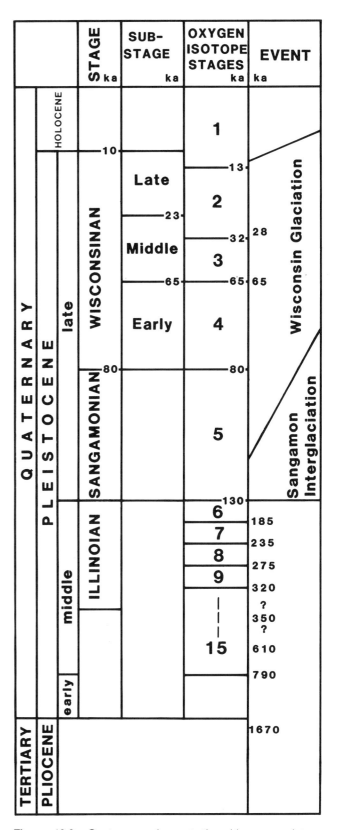

Figure 19.6. Quaternary chronostratigraphic nomenclature (from Fulton and Prest, 1987, Fig. 2).

Paleomagnetic determinations are useful in combination with absolute age determinations.

Onset of glaciation

In Pliocene time, the Arctic Ocean was still open (at least until 1.7 Ma; Funder et al., 1985) and forest tundra grew along its margin (Funder et al., 1985) but glaciers, that locally extended to the sea, probably existed in the uplands of Alaska, Ellesmere Island, Greenland, and Baffin Island. The indirect evidence for such glaciers consists of ice-rafted debris in sediments of the Atlantic (Berggren, 1972) and Arctic (Herman and Hopkins, 1980) oceans, and in raised marine beds of the Upper Pliocene Kap København Formation of northernmost Greenland (Funder et al., 1985).

Early and Middle Pleistocene

The scattered, degraded drift of the western islands and North Greenland was deposited during at least one invasion by ice from the south. In northern Greenland, this till is treated as a product of a single event, informally named the Bliss Bugt glaciation, which is believed to be correlative with an advance of Greenland ice onto northeast Ellesmere Island (Kelly and Bennike, 1985). That advance is recorded by sparse Greenland erratics up to 800 m above sea level (asl) (Christie, 1967). The upper limit of this till dips southward on the west shore of Nares Strait, indicating that the glacier issued from Greenland (England et al., 1981). Deeply weathered erratic-free uplands project locally above this zone. During the retreat of this Greenland ice, an ice-contact marine terrace was deposited at 285 m asl that has been dated at 0.5-1 Ma on the basis of amino acid ratios in shells (Retelle, 1986; England, 1987). The subsequent most extensive eastward advance of Ellesmere Island ice left a till that crosscuts the Greenland erratics. Bracketing marine sediments from seas as much as 175 m asl date beyond the range of the ^{14}C method. The fact that these two older shorelines are higher than the Late Wisconsinan shoreline (at 120 m asl) demonstrates that the related glaciations were more extensive, i.e. resulted in a deeper isostatic depression and correspondingly higher rebound. The western limit of local ice advances on Ellesmere Island (which rise to 250 m) and Axel Heiberg Island (2200 m), and the relationship of this ice to the continental ice that covered the western Queen Elizabeth Islands are unknown.

Most of the area south of Parry Channel was overrun more than once by glaciers from continental dispersal centres. Events predating the last glaciation have been described from three regions:

(1) Banks Island, where one of the most complete records of the Quaternary in North America is preserved in nonglacial, mainly terrestrial sediments, in drift from three glaciations (Fig. 19.7a), and in proglacial lacustrine and marine sediments (Vincent, 1983). The oldest possibly Quaternary deposit recognized in section is the Worth Point Formation, a nonglacial forest-tundra deposit that overlies the fluvial deposits of the Neogene Beaufort Formation (Fig. 19.7b). The overlying Duck Hawk Bluffs Formation includes till from the oldest and most extensive recognized glaciation — the Banks Glaciation — as well as coeval glaciolacustrine sediments and pre- and postglacial marine sediments. Ice from a continental source covered most of the island at this time.

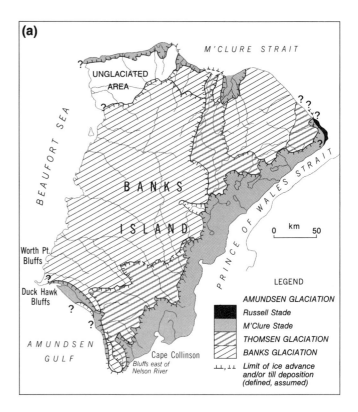

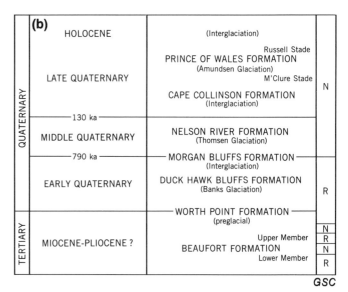

Figure 19.7. Quaternary events on Banks Island (modified from Vincent, 1989). (a) Limit of glaciations and distribution of till sheets. (b) Age of formations and events. Right column indicates magnetization (N = normal, R = reverse).

Vincent (1984) correlated a belt of prominent kames and associated till on southern Melville Island — the Dundas Till of Hodgson et al. (1984) — with the Banks Glaciation. If this correlation is correct, then the degraded drift of the western Queen Elizabeth Islands, which clearly is older than the Dundas Till, represents an earlier and more extensive glacial advance that has not yet been recognized on Banks Island.

In the succeeding interglacial, the Morgan Bluffs Formation was deposited in a climate slightly warmer than at present, when sea level stood at 20-30 m asl. The Early-Middle Pleistocene boundary (at 790 ka) occurs within this nonglacial unit acccording to paleomagnetic determinations.

In the second (Thomsen) glaciation, till of the Nelson River Formation was deposited on the eastern half of the island. This event was followed by an interglacial, again warmer than at present, the deposits of which are rich in organic material (Cape Collinson Formation). Radiocarbon dates greater than 49 and 61 ka (i.e. beyond detection limits) suggest that this interglacial is of Sangamonian age (Fig. 19.6, 19.7b).

(2) In the central part of the area south of Parry Channel, only one Pleistocene stratigraphic succession has been described — at Pasley River on western Boothia Peninsula (Dyke and Mathews, 1987). A fluvial gravel between upper and lower glaciogenic assemblages is probably Sangamonian in age and contains plant and insect fossils indicating a climate at least as warm as the present.

(3) On Bylot Island (Fig. 19.1), scattered erratics record an invasion of much of the island, to an elevation of at least 1100 m, by northward flowing continental ice (Klassen, 1985). This event was followed by an expansion of local glaciers. Sediments found in a section near Pond Inlet, believed to be from this glaciation, are capped by interglacial organic deposits from a climate warmer than present (Sangamonian Stage?).

Late Pleistocene glaciation
Queen Elizabeth Islands and Greenland

Analysis of cores from ice caps in the eastern archipelago suggests regeneration of the ice at about 100 ka, toward the end of an ice-free interglacial (Fig. 19.8; Koerner, 1989). However, no terrestrial evidence has been found of extensive glaciation in the Early Wisconsinan (80-65 ka) though perhaps an advance of upland ice on the northeast coast of Ellesmere Island dates from this time (England et al., 1981).

A number of terrestrial organic deposits containing fauna and flora typical of middle or low arctic environments have yielded ^{14}C dates between 20 and 45 ka (Blake, 1982; Hodgson, 1985); some or all of them may have true ages greater than 50 ka (Blake, 1974). It has not been determined whether these deposits are interglacial, from an interstadial ice-free embayment, or from a nonglacial environment that persisted throughout the Late Pleistocene. Some deposits are related to sea levels as low as at present, others to levels higher than Holocene maxima (Washburn and Stuiver, 1985). Shells at elevations as high as 600 m asl were probably transported by ice.

The record remains equally sparse into the Late Wisconsinan (23-10 ka) in most of the northern archipelago, allowing the development of several controversial hypotheses. Very broadly, these concepts developed in three stages.

(1) In the first half of the 20th century, the dearth of glacial landforms encouraged the hypothesis of a limited last glaciation, or even no expansion beyond the present margins in places (see review in Washburn, 1947).

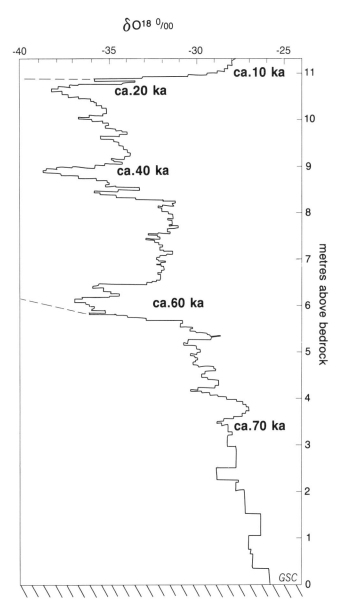

Figure 19.8. Oxygen isotope profile from the base of the Agassiz ice cap, indicating temperature changes (cooler to the left; from Koerner, 1989).

(2) In the middle of the century the theory of glacial isostasy was applied to the results of reconnaissance mapping and ^{14}C dating of raised shorelines, and a widespread Late Pleistocene ice cover was invoked. This culminated in Blake's (1970) hypothesis of the "Innuitian Ice Sheet", portrayed by Prest (1969) as a Late Wisconsinan ice dome over the Queen Elizabeth Islands (Fig. 19.9).

(3) Subsequently, it was realized that lithospheric flexuring extends beyond the margins of an ice sheet; that it may be magnified by interaction between adjacent independent ice caps; and hence, that widespread postglacial uplift and shoreline regression can result from a much less extensive ice cover. This theoretical possibility was confirmed for northeastern Ellesmere Island by England (1974, 1976)

Figure 19.9. Different concepts of the extent of the last glaciation. The maximum corresponds to the concept of a contiguous ice sheet (adapted from Prest, 1969, 1970) and the minimum to a noncontiguous glacier complex (adapted from Prest, 1984).

who inferred a Late Wisconsinan glacial limit at the Hazen Moraines (Fig. 19.5). Consequently, he replaced the concept of a single "Innuitian Ice Sheet" by that of a noncontiguous "Franklin Ice Complex" (composed of a number of separate sheets), and thus re-established the "minimalist" concept of the first half of this century (cf. Paterson, 1977). Nevertheless, a ridge of maximum emergence, which trends from west-central Ellesmere Island southwestward to Bathurst Island (Blake, 1970, Fig. 12; Andrews, 1970, Fig. 7-8 and 7-9) and crosses the hypothetical centre of the "Innuitian Ice Sheet",

has yet to be satisfactorily explained in terms of the "minimalist" model.

Evidence of the last glaciation is best preserved on Ellesmere Island, where the most detailed studies have been made in the northeast. England (1983) suggested that throughout Late Wisconsinan time an ice-free corridor persisted between Ellesmere Island and Greenland, which was partly occupied by a "full glacial" sea with a shoreline higher than at present — a sea that presumably also surrounded unglaciated terrain beyond the Franklin Ice

Complex. Its shoreline is tilted upward to the southeast, climbing from 120 m asl at the Hazen Moraines to 150 m in northeastern Greenland (Fig. 19.10). The tilt resulted from the rebound that followed depression by the Greenland ice load (England, 1982, 1983, 1985). On Ellesmere Island north of the Grantland Mountains, beyond the influence of Greenland ice, deglaciation started 2 ka earlier (10 ka ago) than at the Hazen Moraines to the southeast (England, 1978, 1983, 1985; Bednarski, 1984, 1986).

In central Ellesmere Island, west of the present large ice caps, a 500 km long belt of prominent drift and ice-marginal landforms was deposited by a fluctuating ice margin between 9 and 7 ka ago (Fig. 19.5; Hodgson, 1973, 1985). Hodgson inferred that the last glacial limit lay an unknown distance west of this belt and that much of central and southern Ellesmere Island, together with Grinnell Peninsula of Devon Island, were glaciated at that time.

Elsewhere in the Queen Elizabeth Islands, glacial features of definitely Late Wisconsinan age have not been reported except for Axel Heiberg Island (Boesch, 1963), and southern Melville Island, which was overlapped by the Laurentide Ice Sheet (see below).

On northern Greenland, an extensive glaciation that occurred between the Bliss Bugt glaciation and Holocene marine transgression was speculatively assigned to the Late Wisconsinan Substage and informally named the Independence Fjord glaciation by Funder (1989). This ice sheet was thin and topographically controlled and, in the northwest, possibly produced an ice shelf stretching to Ellesmere Island (Funder and Larsen, 1982). A dynamically independent ice cap covered mountains of Peary Land at this time. Protagonists of the extensive glaciation hypothesis also believe that Greenland ice filled Nares Strait (Fig. 19.11), coalescing with ice from Ellesmere Island and flowing south toward Baffin Bay (Blake, 1977). This conflicts with England's concept of the "full glacial sea", which he sees supported by field evidence from Hall Land, northwest Greenland (England, 1985, 1987), while Bennike and others (1987) disagree.

South of Parry Channel: Laurentide Ice Sheet

The contiguous ice domes that occupied much of northern North America during the last glaciation formed the Laurentide Ice Sheet (Fulton and Prest, 1987). The Wisconsin Glaciation includes the regional Amundsen Glaciation of the northwest, when ice flowed from a continental source area to Banks Island, and the Foxe Glaciation of the northeast, when a dynamically distinct body of ice existed over Baffin Island and Foxe Basin.

There is general agreement that ice margins retreated during the Middle Wisconsinan Substage; but whether the Laurentide Ice Sheet then largely disintegrated, or whether a further glacial stade developed, is highly contentious (Dredge and Thorleifson, 1987; Dyke and Prest, 1987).

The extent of the Late Wisconsinan ice sheet, even though it falls into the range of ^{14}C dating, is no less controversial (Dyke and Prest, 1987, Fig. 1). Currently, an ice sheet margin well behind that of the Early Wisconsinan is proposed for the northeast; in the northwest it either was attenuated (Vincent, 1984) or expanded (Dyke, 1987). It does seem clear, however, that the northern limit of Laurentide ice, whether thick and grounded or floating as an ice shelf, lay along Parry Channel, abutting or even overlapping the southern Queen Elizabeth Islands.

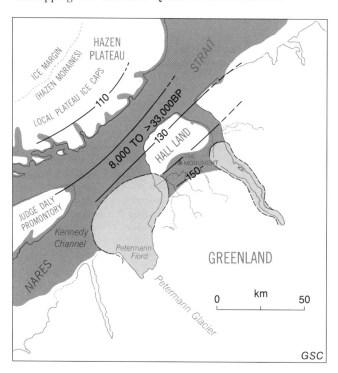

Figure 19.10. Inferred extent of the full glacial sea (darkest tone) in relation to the last ice limit in northwestern Greenland and adjacent parts of Ellesmere Island. Isobases show tilt of the upper limit of the full glacial sea in 10 m intervals (modified from England, 1985, Fig. 10).

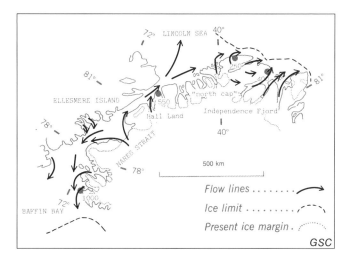

Figure 19.11. Past and present glaciation in North Greenland. Conceptual model of Independence Fjord glaciation (Late Wisconsinan), proposed location of ice margins, major flow lines, and ice surface elevation (m above present sea level). The model is based entirely on land evidence. (After Funder, 1989.)

The regional events of the Wisconsinan Stage are as follows. In the extreme northeast, at the beginning of the Foxe Glaciation (locally the informal "Eclipse glaciation"), a major ice stream filled Lancaster Sound and overlapped Bylot Island to 600 m on the north coast (Klassen, 1985). This event was closely followed by the last major expansion of local ice caps. A less extensive advance, probably in Late Wisconsinan time, overlapped northern Bylot Island only to 15 m above present sea level. The horizontality of this glacier surface indicates that it was part of an ice shelf in Lancaster Sound, likely issuing from Prince Regent Inlet (Fig. 19.12; Dyke and Prest, 1987).

For the northwest, Vincent (1983) suggested that the Amundsen Glaciation included two distinct stades recorded by the undated Prince of Wales Formation (Fig. 19.7). The first expansion of ice, in the M'Clure Stade (Early Wisconsinan?), overlapped most coasts of Banks Island, plus southern Melville Island, while leaving nunataks in northwest Victoria Island. The last advance, in the Late Wisconsinan Russell Stade, possibly impinged on Banks Island only in the extreme northeast. Dyke's reinterpretation of these data places both stades of the Amundsen Glaciation in Middle or Late Wisconsinan time and suggests correlation with events in Bylot Island.

Hodgson et al. (1984) suggested that partially grounded Laurentide ice in Parry Channel deposited a till found along the southern coast of Melville Island. A re-advance of floating ice, the Viscount Melville Sound Ice Shelf, deposited the featureless Winter Harbour Till on all shores of Viscount Melville Sound around 10 ka (Hodgson and Vincent, 1984). This was a minor reversal in the overall retreat of this sector of the Laurentide Ice Sheet, resulting from instability of glaciers calving into inter-island marine channels. By 8.5 ka, the ice sheet margin had retreated to the mainland, except in the east, where ice remained in the Foxe Basin until 7 ka (Dyke and Prest, 1987).

Latest Pleistocene and Holocene

The latest Pleistocene and early Holocene are characterized by climatic warming that resulted in an overall retreat of the ice (with local reversals) and concomitant crustal rebound and marine regression. These events were diachronous — earlier in the western than in the eastern Arctic.

Marine regression was underway by 12 ka in the western islands, compared with only 6 or 7 ka at the Hazen Moraines. Emergence exceeds 160 m in the central Queen Elizabeth Islands and falls to several tens of metres or less at both ends of Parry Channel, where a (eustatic?) transgression may be in progress now. South of central Parry Channel, strandlines of 250 m elevation have been recorded. Similar emergence curves, apparently based on the same function, have been drawn for many sites, but some significant regional differences have been detected (Fig. 19.13). The flat tops of curves 1 and 2 possibly trace the pattern of emergence for areas beyond the limit of glaciation (England, 1983). Otherwise these curves remain smooth even where a high density of data has been obtained at a specific site (Blake, 1975).

Climate changes during the Holocene have been determined from changes in terrestrial and marine fauna and flora, from fluctuations in margins of glaciers, and from physical properties of ice cores through stable summit

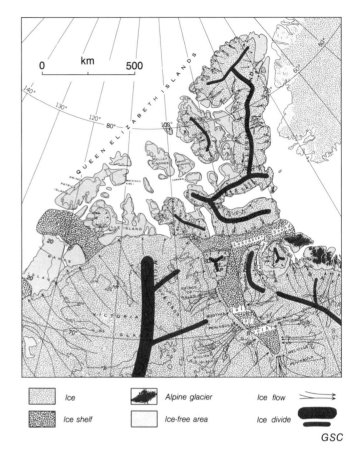

Figure 19.12. Paleogeography 18 ka ago (modified from Dyke and Prest, 1987).

divides of ice caps. The glacier record from the northeastern part of the archipelago (Koerner, 1989) shows a warming at about 10 ka (Fig. 19.6) peaking between 9 and 7 ka on the Agassiz Ice Cap and at 5 ka on the Devon Ice Cap where it coincided with particularly ice-free seas. A gradual cooling around 4 ka was followed by a generally cold period, including one minimum in the Neoglacial or Little Ice Age between A.D. 1570 and 1850. Many outlet glaciers and ice cap margins presently lie at their most advanced positions since the early Holocene retreat (England, 1978; Blake, 1981); others, including ice shelves off northern Ellesmere Island, have retreated in this century.

Figure 19.13. Holocene and latest Pleistocene emergence curves. (1) Northeastern and (2) northern Ellesmere Island (modified from England, 1983, Fig. 7); (3) central Ellesmere Island, combined with western Ellesmere Island at Eureka Sound (adapted from Hodgson, 1985, Fig. 8); (4) Cape Storm and (5) South Cape Fiord (after Blake, 1975, Fig. 22 and 25); (6) Cornwallis Island area (modified from Washburn and Stuiver, 1985, Fig. 2); (7) Melville Island (after Henoch, 1964, Fig. 6); (8) northeast Devon Island, Truelove Inlet (after Barr, 1971, Fig. 5); (9) eastern and (10) southern Melville Island (after McLaren and Barnett, 1978, Fig. 3 and 4); (11) southern Melville Island (adapted from Hodgson et al., 1984).

THE QUATERNARY RECORD

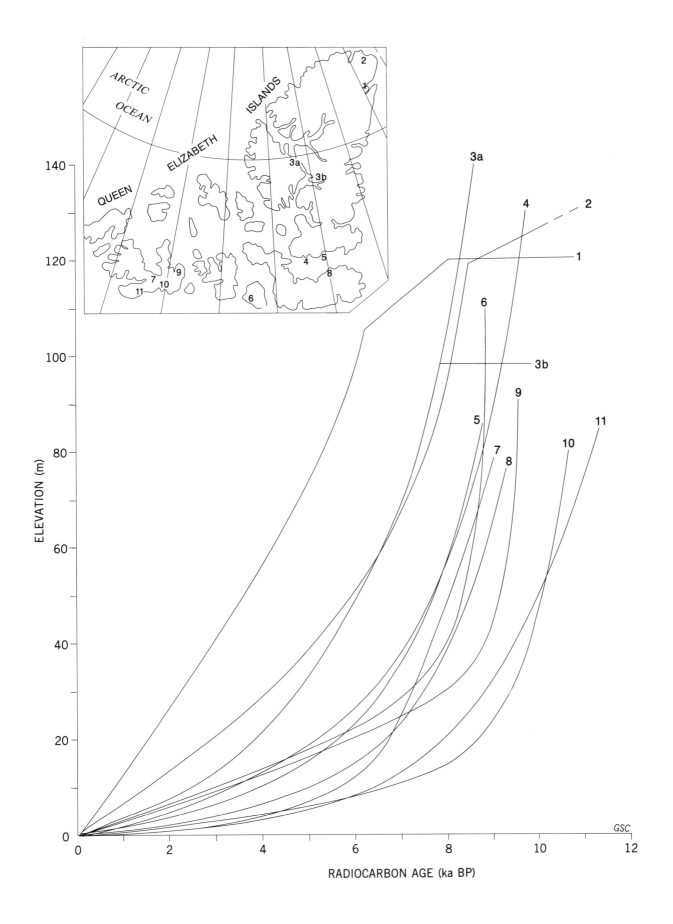

The terrestrial and marine record of the Holocene warming generally coincides with the glacier record: e.g. increased peat accumulation in the middle Holocene, lichen kill on high plateaus during the Neoglacial (Blake, 1974; Brassard and Blake, 1978; Koerner, 1980; Edlund, 1985). A general warming, freeing the Arctic Ocean from some of its ice, may explain the arrival of driftwood (from the western Arctic and Siberia), beginning after 9 ka and culminating between 6 and 4 ka (Blake, 1972; Stewart and England, 1983). The influx of driftwood in the last several thousand years may coincide with the intermittent climatic ameliorations which permitted expansion of paleo-eskimo cultures (see Chapter 2).

REFERENCES

Andrews, J.T.
1970:   A geomorphological study of post-glacial uplift with particular reference to Arctic Canada; Institute of British Geographers, Special Publication no. 2.
1989:   Quaternary geology of the northeastern Canadian Shield; in Chapter 3 of Quaternary Geology of Canada and Greenland, R.J. Fulton (ed.); Geological Survey of Canada, Geology of Canada, no. 1 (also Geological Society of America, the Geology of North America, v. K-1).

Andrews, J.T. and Miller G.H.
1976:   Quaternary glacial chronology of the eastern Canadian Arctic: a review and a contribution on amino acid dating of Quaternary molluscs from the Clyde cliffs; in Quaternary Stratigraphy of North America, W.C. Mahaney (ed.); Dowden, Hutchinson and Ross, Inc., Stroudsburg, Pennsylvania, p. 1-32.

Bednarski, J.
1984:   Glacier fluctuations and sea level history of Clements Markham Inlet, northern Ellesmere Island; unpublished Ph.D. thesis, The University of Alberta, Edmonton, Alberta, 232 p.
1986:   Late Quaternary glacial and sea-level events, Clements Markham Inlet, northern Ellesmere Island; Canadian Journal of Earth Sciences, v. 23, p. 1343-1355.

Barr, W.
1971:   Postglacial isostatic movement in northeastern Devon Island: a reappraisal; Arctic, v. 24, p. 249-268.

Bennike, O., Dawes, P.R., Funder, S., Kelly, M., and Weidick, A.
1987:   The late Quaternary history of Hall Land, northwest Greenland: discussion; Canadian Journal of Earth Sciences, v. 24, p. 370-374.

Berggren, W.A.
1972:   Late Pliocene-Pleistocene glaciation; in Initial Reports of the Deep Sea Drilling Project, v. 12, T.A. Davies (ed.); U.S. Government Printing Office, p. 953-963.

Blake, W., Jr.
1970:   Studies of glacial history in Arctic Canada. I. Pumice, radiocarbon dates, and differential postglacial uplift in the eastern Queen Elizabeth Islands; Canadian Journal of Earth Sciences, v. 7, p. 634-664.
1972:   Climatic implications of radiocarbon-dated driftwood in the Queen Elizabeth Islands, Arctic Canada; in Climatic Changes in Arctic Areas During the Last Ten-Thousand Years, Y. Vasari, H. Hyvärinen, and S. Hicks (ed.); Acta Universitatis Ouluensis, Ser. A, Scientiae Rerum Naturalium, no. 3, Geologica no. 1, p. 77-104.
1974:   Studies of glacial history in Arctic Canada. II. Interglacial peat deposits in Bathurst Island; Canadian Journal of Earth Sciences, v. 11, p. 1025-1042.
1975:   Radiocarbon age determinations and postglacial emergence at Cape Storm, southern Ellesmere Island, Arctic Canada; Geografiska Annaler, v. 57A, p. 1-71.
1977:   Glacial sculpture along the east-central coast of Ellesmere Island, Arctic Archipelago; in Report of Activities, Part C, Geological Survey of Canada, Paper 77-1C, p. 107-115.
1981:   Lake sediment coring along Smith Sound, Ellesmere Island and Greenland; in Current Research, Part A, Geological Survey of Canada, Paper 81-1A, p. 191-200.
1982:   Terrestrial interstadial deposits, Ellesmere Island, N.W.T., Canada (abstract); in Character and Timing of Rapid Enviromental and Climatic Changes; American Quaternary Association, National Conference (7th Biennial) Abstract; Seattle, Washington, p. 73.

Boesch, H.
1963:   Notes on the geomorphological history; in Axel Heiberg Research Reports, Preliminary Report 1961-62, F. Müller et al.; McGill University, Montreal, p. 163-167.

Brassard, G.R. and Blake, W., Jr.
1978:   An extensive subfossil deposit of the arctic moss *Aplodon wormskioldii*; Canadian Journal of Botany, v. 56, p. 1852-1859.

Canadian Hydrographic Service
1979:   General Bathymetric Chart of the Oceans; Chart 5-17, 5th edition; Canada Department of Fisheries and Oceans, scale 1: 6 000 000.

Carter, L.D., Heginbottom, J.A., and Woo, M.
1987:   Arctic Lowlands; in Geomorphic Systems of North America, W.L. Graf (ed.); Geological Society of America, Decade of North American Geology, Centennial Special Volume 2, p. 583-628.

Christie, R.L.
1967:   Reconnaissance of the surficial geology of northeastern Ellesmere Island, Arctic Archipelago; Geological Survey of Canada, Bulletin 138, 50 p.

Dredge, L.A. and Thorleifson, L.H.
1987:   The Middle Wisconsinan history of the Laurentide Ice Sheet; Géographie physique et Quaternaire, v. 41, p. 215-235.

Dyke, A.S.
1983:   Quaternary geology of Somerset Island, District of Franklin; Geological Survey of Canada, Memoir 404, 32 p.
1984:   Quaternary geology of Boothia Peninsula and northern District of Keewatin, central Canadian Arctic; Geological Survey of Canada, Memoir 407, 25, 26, p.
1987:   A reinterpretation of glacial marine limits around the northwestern Laurentide Ice Sheet; Canadian Journal of Earth Sciences, v. 24, p. 591-601.

Dyke, A.S. and Dredge, L.A.
1989:   Quarternary geology of the northwestern Shield; in Chapter 3 of Quaternary Geology of Canada and Greenland, R.J. Fulton (ed.); Geological Survey of Canada, Geology of Canada, no. 1 (also Geological Society of America, The Geology of North America, v. K-1).

Dyke, A.S. and Matthews, J.V., Jr.
1987:   Stratigraphy of Quaternary sediments along Pasley River, Boothia Peninsula, central Canadian Arctic; Géographie physique et Quaternaire, v. 41, p. 23-344.

Dyke, A.S. and Prest, V.K.
1987:   Late Wisconsinan and Holocene record of the Laurentide Ice Sheet; Géographie physique et Quaternaire, v. 41, p. 237-263.

Edlund, S.A.
1983:   Bioclimatic zonation in a High Arctic region: central Queen Elizabeth Islands; in Current Research, Part A, Geological Survey of Canada, Paper 83-1A, p. 381-390.
1985:   Lichen-free zones as neoglacial indicators on western Melville Island, District of Franklin; in Current Research, Part A, Geological Survey of Canada, Paper 85-1A, p. 709-712.
1987:   Plants: living weather stations; Geos, v. 16, p. 9-13.

England, J.
1974:   The glacial geology of the Archer Fiord – Lady Franklin Bay area, northeastern Ellesmere Island, N.W.T., Canada; unpublished Ph.D. thesis, Department of Geography, University of Colorado, Boulder, Colorado, 234 p.
1976:   Late Quaternary glaciation of the eastern Queen Elizabeth Islands, Northwest Territories, Canada: alternative models; Quaternary Research, v. 6, p. 185-202.
1978:   The glacial geology of northeastern Ellesmere Island, Northwest Territories, Canada; Canadian Journal of Earth Sciences, v. 15, p. 603-617.
1982:   Postglacial emergence along northern Nares Strait; in Nares Strait and the Drift of Greenland, a Conflict in Plate Tectonics, P.R. Dawes and J.W. Kerr (ed.); Meddelelser om Grønland, Geoscience 8, p. 65-75.
1983:   Isostatic adjustments in a full glacial sea; Canadian Journal of Earth Sciences, v. 20, p. 895-917.
1985:   The late Quaternary history of Hall Land, northwest Greenland; Canadian Journal of Earth Sciences; v. 22, p. 1394-1408.
1987:   The late Quaternary history of Hall Land, northwest Greenland: reply; Canadian Journal of Earth Sciences, v. 24, p. 374-380.

England, J., Bradley, R.S., and Stuckenrath, R.
1981:   Multiple glaciations and marine transgressions, western Kennedy Channel, Northwest Territories, Canada; Boreas, v. 10, p. 71-89.

French, H.M.
1976:   The Periglacial Environment; Longmans, London, 309 p.

Fulton, R.J.
1984: Summary: Quaternary stratigraphy of Canada; in Quaternary Stratigraphy of Canada — A Canadian Contribution to IGC Project 24, R.J. Fulton (ed.); Geological Survey of Canada, Paper 84-10, p. 1-5.

Fulton R.J. (ed.)
1989: Quaternary geology of Canada and Greenland; Geological Survey of Canada, Geology of Canada, no. 1 (also Geological Society of America, The Geology of North America, v. K-1.)

Fulton, R.J. and Prest, V.K.
1987: The Laurentide Ice Sheet and its significance; Géographie physique et Quaternaire, v. 41, p. 181-186.

Funder, S.
1989: Quaternary geology of North Greenland; in Chapter 13 of Quaternary Geology of Canada and Greenland, R.J. Fulton (ed.); Geological Survey of Canada, Geology of Canada, no. 1 (also Geological Society of America, The Geology of North America, v. K-1).

Funder, S. and Larsen, O.
1982: Implications of volcanic erratics in Quaternary deposits of North Greenland; Geological Society of Denmark, Bulletin, v. 31, p. 57-61.

Funder, S., Abrahamsen, H., Bennike, O., and Feyling-Hanssen, R.W.
1985: Forested arctic: evidence from North Greenland; Geology, v. 13, p. 542-546.

Fyles, J.G.
1965: Surficial geology, western Queen Elizabeth Islands, in Report of Activities: Field, 1964, Geological Survey of Canada, Paper 65-1, p. 3-5.

Henoch, W.E.S.
1964: Postglacial marine submergence and emergence of Melville Island, N.W.T.; Geographical Bulletin, no. 22, p. 105-126.

Herman, Y. and Hopkins, D.M.
1980: Arctic Ocean climate in Late Cenozoic time; Science, v. 209, p. 557-562.

Hodgson, D.A.
1973: Landscape, and late-glacial history, head of Vendom Fiord, Ellesmere Island; in Report of Activities, Part B, Geological Survey of Canada, Paper 73-1B, p. 129-136.

1982: Surficial materials and geomorphic processes, western Sverdrup and adjacent islands; Geological Survey of Canada, Paper 81-9, 37 p.

1985: The last glaciation of west-central Ellesmere Island, Arctic Archipelago, Canada; Canadian Journal of Earth Sciences, v. 22, p. 347-368.

Hodgson, D.A. (co-ordinator)
1989: Quaternary geology of the Queen Elizabeth Islands; in Chapter 6 of Quaternary geology of Canada and Greenland, R.J. Fulton (ed.); Geological Survey of Canada, Geology of Canada, no. 1 (also Geological Society of America, The Geology of North America, v. K-1).

Hodgson, D.A. and Vincent, J-S.
1984: A 10,000 yr. B.P. extensive ice shelf over Viscount Melville Sound, Arctic Canada; Quaternary Research, v. 22, p. 18-30.

Hodgson, D.A., Vincent, J-S., and Fyles, J.G.
1984: Quaternary geology of central Melville Island, Northwest Territories; Geological Survey of Canada, Paper 83-16, 25 p.

Kelly, M. and Bennike, O.
1985: Quaternary geology of parts of central and western North Greenland; Grønlands Geologiske Undersøgelse, Report no. 126, p. 111-116.

Klassen, R.A.
1985: An outline of glacial history of Bylot Island, District of Franklin, N.W.T.; in Quaternary Environments: Eastern Canadian Arctic, Baffin Bay and Western Greenland, J.T. Andrews (ed.); Allen and Unwin, Boston, U.S.A., p. 309-327.

Koerner, R.M.
1977: Ice thickness measurements and their implications with respect to past and present ice volumes in the Canadian High Arctic ice caps; Canadian Journal of Earth Sciences, v. 14, p. 2697-2705.

1980: The problem of lichen-free zones in Arctic Canada; Arctic and Alpine Research, v. 12, p. 87-94.

1989: Queen Elizabeth Islands glaciers; in Chapter 6 of Quaternary Geology of Canada and Greenland, R.J. Fulton (ed.); Geological Survey of Canada, Geology of Canada, no. 1 (also Geological Society of America, The Geology of North America, v. K-1).

McLaren, P. and Barnett, D.M.
1978: Holocene emergence of the south and east coasts of Melville Island, Queen Elizabeth Islands, Northwest Territories, Canada; Arctic, v. 31, p. 415-427.

McLean, B., Sonnichsen, G., Vilks, G., Powell, C., Moran, K., Jennings, A., Hodgson, D., and Deonarine, B.
in press: Marine geological and geotechnical investigations in Wellington, Byam Martin, Austin and adjacent channels, Canadian Arctic Archipelago; Geological Survey of Canada, Paper.

Matthews, J.V., Jr.
1987: Plant macrofossils from the Neogene Beaufort Formation on Banks and Meighen islands, District of Franklin; in Current Research, Part A, Geological Survey of Canada, Paper 87-1A, p. 73-87.

National Atlas of Canada
1973: Relief profiles; National Atlas of Canada, Physical Geography Section, Map 3-4, 4th edition; Surveys and Mapping Branch, Canada Department of Energy, Mines and Resources, scale 1:15 000 000.

Paterson, W.S.B.
1977: Extent of Late-Wisconsin glaciation in northwest Greenland and northern Ellesmere Island: a review of the glaciological and geological evidence; Quaternary Research, v. 8, p. 180-190.

Prest, V.K.
1969: Retreat of Wisconsin and recent ice in North America; Geological Survey of Canada, Map 1257A, scale 1:5 000 000.

1970: Quaternary Geology; in Geology and Economic Minerals of Canada, R.J.W. Douglas (ed.); Geological Survey of Canada, Economic Geology Report no. 1, p. 675-764.

1984: Late Wisconsinan glacier complex; Geological Survey of Canada, Map 1584A, scale 1:7 500 000.

Retelle, M.J.
1986: Glacial geology and Quaternary marine stratigraphy of the Robeson Channel area, northeastern Ellesmere Island, Northwest Territories; Canadian Journal of Earth Sciences, v. 23, p. 1001-1012.

Stewart, T.G. and England, J.
1983: Holocene sea-ice variations and paleoenvironmental change, northernmost Ellesmere Island, N.W.T., Canada; Arctic and Alpine Research, v. 15, p. 1-17.

St-Onge, D.A.
1987: An introduction to the continental record of the Laurentide Ice Sheet; Géographie physique et Quaternaire, v. 41, p. 187-188.

Taylor, A.E., Burgess, M., Judge, A.S., and Allen, V.S.
1982: Canadian geothermal data collection — northern wells 1981; Energy, Mines and Resources Canada, Earth Physics Branch, Geothermal Service of Canada, Geothermal Series no. 13, 153 p.

Taylor, A. Judge, A., and Desrochers, D.
1983: Shoreline regression: its effect on permafrost and the geothermal regime, Canadian Arctic Archipelago; in Permafrost: Fourth International Conference, Proceedings; National Academy Press, Washington, D.C., p. 1239-1244.

Taylor, R.B. and McCann, S.B.
1983: Coastal depositional landforms in northern Canada; in Shorelines and Isostasy, D.E. Smith and A.G. Dawson (ed.); Institute of British Geographers, Special Publication no. 16, Academic Press, p. 53-75.

Vincent, J-S.
1983: La géologie quarternaire et la géomorphologie de l'île Banks, Arctique Canadien; Commission géologique du Canada, Mémoire 405, 118 p.

1984: Quaternary stratigraphy of the western Canadian Arctic Archipelago; in Quaternary Stratigraphy of Canada — A Canadian Contribution to IGC Project 24, R.J. Fulton (ed.); Geological Survey of Canada, Paper 84-10, p. 87-100.

1989: Quaternary geology of the northern Canadian Plains; in Chapter 2 of Quaternary Geology of Canada and Greenland, R.J. Fulton (ed.); Geological Survey of Canada, Geology of Canada, no. 1 (also Geological Society of America, The Geology of North America, v. K-1).

Vincent, J-S. and Prest, V.K.
1987: The Early Wisconsinan history of the Laurentide Ice Sheet; Géographie physique et Quaternaire, v. 41, p. 199-213.

Washburn, A.L.
1947: Reconnaissance geology of portions of Victoria Island and adjacent regions, Arctic Canada; Geological Society of America, Memoir 22, 142 p.

1980: Geocryology: a Survey of Periglacial Processes and Environments; John Wiley and Sons, New York, 406 p.

Washburn, A.L. and Stuiver, M.
1985: Radiocarbon dates from Cornwallis Island area, Arctic Canada — an interim report; Canadian Journal of Earth Sciences, v. 22, p. 630-636.

CHAPTER 19

Author's address

D.A. Hodgson
Geological Survey of Canada
601 Booth Street
Ottawa, Ontario
K1A 0E8

Printed in Canada

Chapter 20
RESOURCES

A. Petroleum resources, Arctic Islands — *A.F. Embry, T.G. Powell, and U. Mayr*

B. Petroleum resources, North Greenland — *F. G. Christiansen, S. Piasecki, and L. Stemmerik*

C. Coal resources, Arctic Islands — *R.M. Bustin and A.D. Miall*

D. Economic mineral resources, Arctic Islands — *Walter A. Gibbins*

E. Economic mineral resources, North Greenland — *A. Steenfelt*

References

Chapter 20

RESOURCES

A. PETROLEUM RESOURCES, ARCTIC ISLANDS

A.F. Embry, T.G. Powell, and U. Mayr

INTRODUCTION

The early geological studies of the Canadian Arctic Archipelago by the Geological Survey of Canada indicated that the area had good petroleum potential (Fortier et al., 1954; Douglas et al., 1963). The first well, Dome Winter Harbour No. 1, was completed in 1962, and it tested a thick lower Paleozoic section on an Ellesmerian anticline on southern Melville Island. The well was dry and abandoned although good gas shows were encountered in Middle Devonian sandstones. Two more wells were drilled in the early 1960s in the Cornwallis – eastern Bathurst Island area, but they too were dry.

In 1967, Panarctic Oils Limited, a partnership of the Government of Canada and private industry, was formed, and shortly afterward the giant Prudhoe Bay oil field was found in Arctic Alaska. These two events resulted in a rapid expansion of petroleum exploration in the Canadian Arctic. The first Panarctic well, Sandy Point L-46, was drilled in the summer of 1969 and Panarctic's third well that year, Drake N-67, discovered a giant gas field on Sabine Peninsula, Melville Island. More gas discoveries quickly followed in the early 1970s — King Christian in 1970, Kristoffer Bay in 1971, Thor and Hecla in 1972 and Wallis in 1973 — and exploration reached its zenith in 1973 when 23 wells were completed and over 60 000 m of bedrock were drilled. (For geographic names and well locations, see Fig. 1, in pocket).

Since 1973, drilling activity has gradually declined, despite a number of significant discoveries of oil and gas, much higher petroleum prices, and government incentives. Only one well was drilled in 1987, the lowest number since Panarctic began drilling. As of April 1988, 176 wells have been drilled in the Arctic Islands, and 18 hydrocarbon fields have been discovered (Fig. 20A.1).

Previous work

Published information on the petroleum geology of the Arctic Islands is not plentiful. A number of authors including Drummond (1973), Meneley (1977), Stuart Smith and Wennekers (1977), Rayer (1981), Hamilton and Varney (1982) and Nassichuk (1983) have described the general geology of the area and commented on petroleum discoveries and potential. Waylett (1979) described the six gas fields which had been discovered by 1978, Henao-Londoño (1977a) discussed the reservoir strata in several fields, and Crain (1977) outlined some of the logging problems experienced with these reservoirs. Embry (1982) described the hydrocarbon discoveries within Upper Triassic – Lower Jurassic sandstones and commented on the potential for further discoveries in these strata.

Estimates of oil and gas reserves in the discovered fields have been published by Waylett (1979), Rayer (1981), Hamilton and Varney (1982), Panarctic (1983), and Procter et al. (1984). Estimates of discovered and ultimate reserves have been made by Drummond (1973), McCrossan and Porter (1973), Nassichuk (1983) and Procter et al. (1984). The Geological Survey of Canada presently estimates the total of discoveries and untapped recoverable reserves at 686×10^6 m^3 (4.3 B bbl) of oil and 2257×10^9 m^3 (79.7 TCF) of gas (Procter et al., 1984).

Early geochemical work (to 1974) in the Arctic Islands consisted chiefly of reconnaissance geochemical studies by analysis of cuttings for light hydrocarbon gases and organic carbon, and led to the recognition of regional maturation trends (Snowdon and Roy, 1975). Subsequently, Baker et al. (1975) suggested that the Triassic-Jurassic section in the Sverdrup Basin has good potential for gas, but that certain intervals, particularly the Awingak-Savik (now Jameson Bay Formation), have a consistent potential for liquid generation. Henao-Londoño (1977b) published a preliminary geochemical evaluation of the Arctic Islands, in which the source richness was assessed in terms of organic carbon content and petrographic type, and maturation levels were assessed using vitrinite reflectance data. Powell (1978) reviewed the results of previous workers and carried out a systematic evaluation of the maturation and source rock potential of each of the major stratigraphic units. Recently, Brooks et al. (in press) correlated recovered oils with organic-rich Triassic shale units of the Schei Point Group on the basis of bio-markers.

Present work

This summary of the petroleum geology of the Arctic Islands is based mainly on stratigraphic, structural, geochemical and geophysical studies carried out by the Geological Survey of Canada over the past thirty years, as well as on published

---

Embry, A.F., Powell, T.G., and Mayr, U.
1991: Petroleum resources, Arctic Islands; *in* Chapter 20 of Geology of the Innuitian Orogen and Arctic Platform of Canada and Greenland, H.P. Trettin (ed.); Geological Survey of Canada, Geology of Canada, no. 3; (also Geological Society of America, The Geology of North America, v. E).

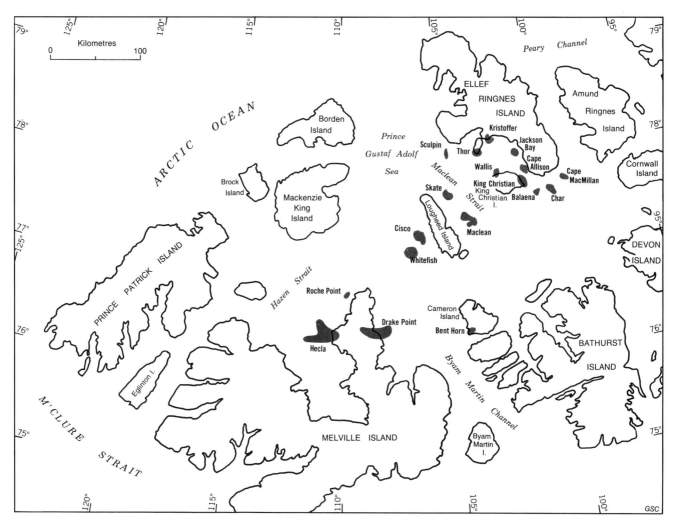

Figure 20A.1. Hydrocarbon fields, Canadian Arctic Archipelago.

information quoted above. U. Mayr contributed the section on the lower Paleozoic petroleum geology and A.F. Embry is responsible for the sections on the upper Paleozoic, Mesozoic and Tertiary. T.G. Powell has summarized his work on organic geochemistry, which is described in more detail in the previously quoted paper.

LOWER PALEOZOIC
Discoveries

Only one field has been discovered in lower Paleozoic strata, despite the fact that over 50 wells have been drilled on lower Paleozoic prospects. That discovery is the Bent Horn oil field on southwestern Cameron Island (Fig. 20A.1). The oil occurs in bioclastic limestones of the upper portion of the Blue Fiord Formation (Mayr, 1980). Blue Fiord limestones usually are not porous in the Cameron Island area, but at Bent Horn porosity is present in the form of vugs and caverns created by vadose processes (Hamilton and Varney, 1982) and fractures related to folding and faulting. However, even in the vicinity of the field, porosity development is very sporadic.

The field occurs at the intersection of a carbonate shelf margin, which trends northeast, and an Ellesmerian fold, which trends east-west (Fig. 20A.2a). At least two northward-directed thrust faults cut the upper Blue Fiord Formation, creating over 400 m of structural relief (Fig. 20A.2b). Large, down-to-the-south, upper Paleozoic normal faults occur within, and south of, the field (Fig. 20A.2b). Oil occurs both in thrust slices and in the footwall block of the Blue Fiord Formation.

The recovered oil is 43° API gravity, and tests have indicated production capabilities up to 5300 barrels a day. However, due to the very sporadic porosity and the small areal extent of the field, recoverable reserves are low and are presently estimated to be 5.5×10^6 m^3 (35.5 MM bbl) (D. Smith, pers. comm., 1983).

Organic geochemistry

Geochemical studies of the lower Paleozoic strata of the Franklinian Shelf and the Arctic Platform have consisted of gas analyses and organic carbon measurements on well cuttings, supplemented by extract analyses and elemental

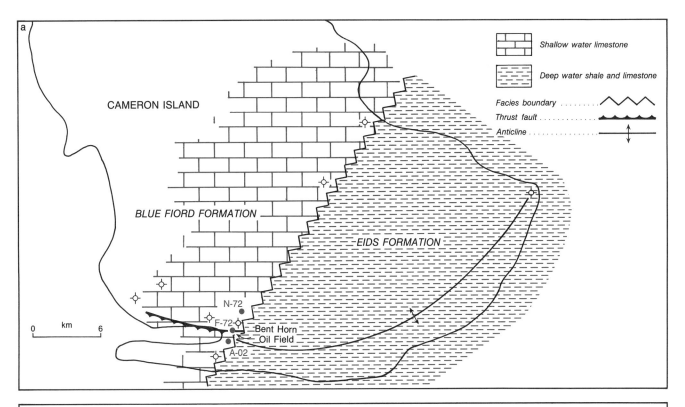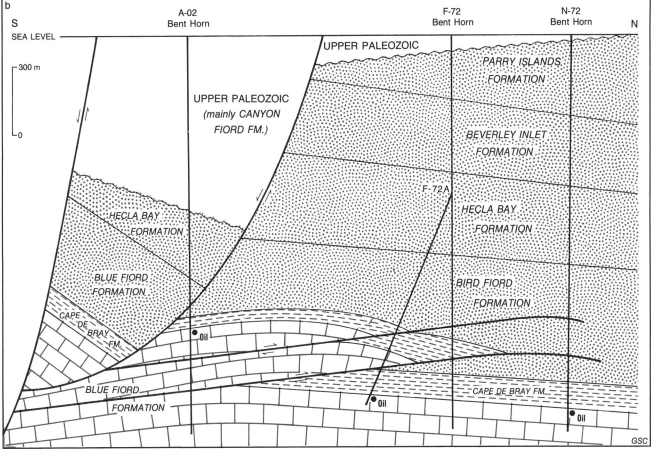

Figure 20A.2. Features of the Bent Horn oil field. (a) Localization at the intersection of a carbonate margin and an Ellesmerian structure. (b) Structural cross-section.

analyses of isolated kerogens (Powell, 1978). The sparsity of wells, structural complexities, and large thicknesses of organically barren carbonate all make it difficult to establish regional trends of maturation and source rock potential, and so only a broad description will be attempted here.

Two potential source rock intervals have been recognized in the lower Paleozoic, namely the fine grained sediments (Weatherall, Bird Fiord, Blackley and Cape de Bray formations) of the Middle-Upper Devonian clastic wedge, and black, graptolitic shales of the Cape Phillips and Kitson formations (Upper Ordovician to Lower Devonian). The former are relatively depleted in organic carbon (ca. 0.5%), but extract data indicate that the organic matter is oil-prone over a wide area from western Melville Island to Cameron Island. Parts of the Bird Fiord Formation on Cameron Island are stained with migrated hydrocarbons, which are thought to have originated from within the formation. The shales of the Kitson and Cape Phillips formations are characterized by high organic carbon contents (3-5%) and high gas yields (>50 000 ppm) where overmature. Where the shales fall within the mature zone, the hydrocarbon yields show good to excellent source potential.

Geochemical studies have shown that all the boreholes of the Franklinian Shelf succession commence in either the mature or overmature zones from the standpoint of hydrocarbon generation. In the western part of Melville Island, the transition from the mature to overmature zone occurs in the lower part of the Weatherall Formation. On Cameron Island the transition lies within or below the Blue Fiord Formation. Throughout the remainder of the area the transition takes place within the Cape Phillips or equivalent formations. On Banks Island the Paleozoic rocks are mostly overmature with only Middle-Upper Devonian clastics on eastern Banks Island lying within the mature zone.

Since the sediments of the Devonian clastic wedge are either in the mature or overmature zones at surface, it is reasonable to assume that maximum burial occurred in latest Devonian time, prior to uplift and erosion during the Ellesmerian Orogeny (Famennian to Viséan). In the Parry Islands Fold Belt, hydrocarbon generation probably took place during this time. Circumstantial evidence suggests that the oil in the Bent Horn field is derived from sources within the Bird Fiord Formation, although no conclusive oil-source correlation studies have been carried out.

Prospects

Meneley (1977) and Rayer (1981) have outlined the four types of plays in lower Paleozoic strata which have been tested and which still have potential. These plays are: (1) Lower Ordovician carbonates in sub-Bay Fiord salt anticlines; (2) Upper Ordovician carbonates (Thumb Mountain Formation) in Ellesmerian anticlines and thrust plates; (3) Silurian to Middle Devonian shelf margins and isolated carbonate buildups; and (4) Middle Devonian sandstones in Ellesmerian anticlines.

The sub-salt play has been tested at a few localities on Melville Island, but the target reservoir, the Eleanor River Formation, was not porous. Untested closures exist (e.g. Rayer, 1981), but the great depth of burial of the potential reservoirs, which usually implies negligible porosity and very high thermal maturity, makes this a gas play and puts a very high risk on it.

The Thumb Mountain Formation has been tested on a number of Ellesmerian structures, but the formation was either tight or yielded water. These failures have diminished the prospectiveness of this play, but the presence of porosity, oil staining, and gas shows, combined with the close proximity to source rocks (Cape Phillips Formation), indicate that undrilled closures still have some potential to contain petroleum (probably gas).

The Silurian – Middle Devonian carbonate margin has been traced through the Arctic Islands (Chapter 8) and porous, oil-stained, bioclastic limestones and dolomites are common along the margin. Organic buildups, up to 13 km long, which also have porosity and bitumen, are found at or near the margin at various localities. The margin and a few isolated carbonate buildups have been drilled at several places from southern Ellesmere Island to central Banks Island, but the carbonates have either been water-bearing or plugged with bitumen. The geology of the shelf margin has not been explored sufficiently in the subsurface, and despite the lack of success, this play is still viable and has the best potential for hydrocarbon discoveries in lower Paleozoic strata.

Middle Devonian sandstones of the Bird Fiord and Weatherall formations are present in the subsurface over Ellesmerian anticlines in the western Arctic, and gas shows were encountered in these strata in the Winter Harbour well. Porosity is usually less than 10% (Embry and Klovan, 1976), and this, combined with the lack of success, dampens this play. However, oil staining has been noted within the sandstones both in surface and subsurface sections, and the proximity to source rocks (interbedded and underlying shales) and a favourable maturation level over much of the area all indicate some potential for this play.

Overall, the petroleum potential of the lower Paleozoic strata is rated as fair, with isolated carbonate buildups having the best chance of harbouring large fields. The ultimate recoverable reserves have been estimated to be 0.25×10^9 m^3 (1.5×10^9 bbl) of oil and 0.46×10^{12} m^3 (16 TCF) of gas (Nassichuk, 1983).

UPPER PALEOZOIC
Discoveries and shows

Twenty-nine wells have penetrated upper Paleozoic strata in the Sverdrup Basin, but no hydrocarbon discoveries have been made in these strata. However, good shows of oil and gas have been encountered, indicating that the strata have potential to contain significant reserves. Oil was recovered from a sandstone in the Canyon Fiord Formation in a well drilled on the southeastern basin margin, and from a carbonate unit in the Belcher Channel Formation on Sabine Peninsula, Melville Island. Bitumen-impregnated sandstones occur in outcropping Canyon Fiord Formation on southern Sabine Peninsula. Gas shows have been encountered in carbonates of the Belcher Channel Formation and turbiditic sandstones of the van Hauen Formation on Sabine Peninsula.

Organic geochemistry

Studies of source rock quality and thermal maturity are of a broad reconnaissance nature only because few wells have penetrated the upper Paleozoic. Initial results suggest that the strata are dominated by gas-prone organic matter (Powell, 1978). The presence of oil shows, and the occurrence in outcrop of dark brown to black, marine shales in the Hare Fiord and van Hauen formations suggest that oil-prone source rocks are also present in the succession. Recently, oil shales with organic carbon contents up to 50% have been recognized in the Emma Fiord Formation on Grinnell Peninsula (Goodarzi et al., 1987). The strata are of lacustrine origin and have a very limited extent in outcrop (Davies and Nassichuk, 1988). Similar deposits may be present in the subsurface (Utting et al., in press).

Upper Paleozoic strata are mainly within the mature zone along the margins of the Sverdrup Basin and are overmature over the central portion of the basin (Powell, 1978).

Prospects

Upper Paleozoic strata are prospective along the margins of the Sverdrup Basin where potential reservoirs exist and oil window maturation levels are present. Structural traps in this area include shallow water carbonates and clastics in anticlines or faulted uplifts. Stratigraphic traps potentially occur within isolated carbonate buildups, along shelf margins, and within deep water sandstones and carbonates. Combination traps, involving normal faults and shallow water clastics and carbonates, are also conceivable along the basin margin.

All the necessary ingredients for petroleum occurrence appear to be present in the upper Paleozoic succession along the margins of the basin. Much more data on the stratigraphy and sedimentology of the upper Paleozoic succession, as well as on organic geochemistry and seismic stratigraphy of the strata, are required before a reasonable assessment of the petroleum potential can be made.

MESOZOIC
Discoveries

Almost all of the hydrocarbons discovered so far in the Arctic Islands occur within the Mesozoic succession. Seventeen fields have been found (Fig. 20A.1) and reservoirs consist of shallow marine to fluvial sandstones within the Roche Point, Pat Bay, Heiberg, Skybattle, Maclean Strait, King Christian, Awingak and Isachsen formations (Table 20A.1). All of the fields occur on anticlines in the western part of the Sverdrup Basin and most of the structures are salt-cored.

The fields are conveniently divided into three groups on the basis of geographic location and structural style. Ten fields have been discovered in the Ellef Ringnes Island area, and they all occur on the crests of high-amplitude, salt-cored anticlines. The structures were growing by Triassic time (Embry, 1982), and the last surge of growth coincided with the Eurekan Orogeny in mid-Tertiary. Five fields, King Christian, Wallis, Sculpin, Jackson Bay, and Kristoffer Bay, contain only dry gas; three, Thor, Char and Cape MacMillan, contain mainly gas with minor oil; one, Cape Allison, contains significant oil and gas; and one, Balaena, contains only highly biodegraded oil.

The main reservoirs in the Ellef Ringnes area lie within sandstones of the Heiberg Formation and its lateral equivalent, the King Christian Formation (Embry, 1982).

Table 20A.1. Hydrocarbon fields in Sverdrup Basin.

	FIELD	HYDROCARBON-BEARING UNITS	HYDROCARBONS
1	Drake Point	Maclean Strait Formation	Gas
2	Hecla	Maclean Strait and Awingak formations	Gas
3	Roche Point	Pat Bay and Roche Point formations	Gas
4	Whitefish	King Christian, Awingak and Isachsen formations	Gas
5	Cisco	King Christian and Awingak formations	Oil and gas
6	Maclean	Skybattle Formation	Oil and gas
7	Skate	King Christian Formation	Oil and gas
8	Sculpin	King Christian Formation	Gas
9	Thor	King Christian Formation	Gas
10	Kristoffer	King Christian Formation	Gas
11	Jackson Bay	King Christian Formation	Gas
12	Wallis	King Christian Formation	Gas
13	King Christian	King Christian Formation	Gas
14	Cape Allison	King Christian Formation	Oil and gas
15	Balaena	Isachsen Formation	Oil
16	Char	Heiberg and Awingak formations	Gas
17	Cape MacMillan	Heiberg and Awingak formations	Oil and gas

Minor reserves occur in the Awingak and Isachsen formations (Table 20A.1). The traps are less than 10% filled, and the main reasons for this are: (1) considerable structural growth followed by hydrocarbon migration; (2) relatively small hydrocarbon drainage areas, restricted mainly to adjacent synclines; and (3) hydrocarbon leakage through fractures on the crest of the structures.

Four offshore fields have been discovered near Lougheed Island, and both oil and gas are present in significant quantities in three of these fields (Skate, Maclean, Cisco) with the fourth (Whitefish) containing only gas. The anticlinal traps in this area have much lower relief than those in the Ellef Ringnes area (Fig. 20A.3) but are filled much closer to spill point. The main reservoir strata are sandstones in the King Christian and Awingak formations with lesser reserves occurring in the Pat Bay, Skybattle, Maclean Strait and Isachsen formations.

The third area, Melville Island, is dominated by two giant natural gas fields, Drake Point and Hecla, which occur on Sabine Peninsula. The anticlines on which the fields are located are very broad, low-amplitude, en echelon folds which are cut by northeast-trending normal faults. The genesis of these structures is not clear and they may have formed in response to minor strike-slip faulting during the Eurekan Orogeny (Hamilton and Varney, 1982). It is not known if salt was a factor in the development of these folds.

Both fields are filled to spill point, and the main reservoir strata are beach and nearshore sandstones of the Maclean Strait Formation (Table 20A.1). These strata subcrop along the southern edge of the Hecla field and thus the trap is in part stratigraphic. Other minor occurrences of petroleum in these fields include gas in the Awingak Formation at Hecla, and gas and minor oil in the Roche Point Formation at both Hecla and Drake Point.

A third field in the Melville area, Roche Point, lies west of northern Sabine Peninsula on a small, salt-cored uplift. Natural gas occurs in the Pat Bay and Roche Point formations (Table 20A.1).

Gas and oil shows are found within Mesozoic sandstones at many localities within the western Sverdrup Basin, with the best known being the Marie Bay tar sands in the Bjorne Formation of western Melville Island (Trettin and Hills, 1966). No fields have been discovered in the eastern Sverdrup

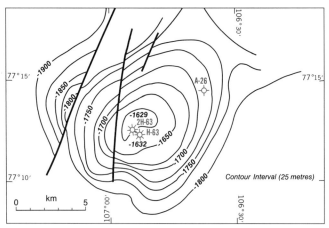

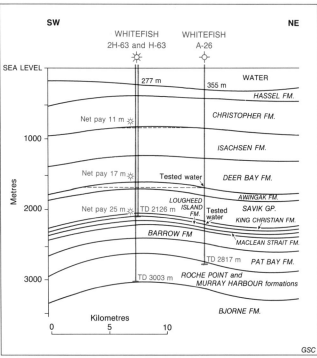

Figure 20A.3. Structure of Whitefish gas field. (Diagrams supplied by D. Smith, COGLA). (a) Structural contours on top of Awingak Formation (contours in metres). (b) Structural cross-section.

Basin, and the best oil and gas shows are in the Romulus C-42 well on Fosheim Peninsula. In this well, oil was tested in the uppermost sandstone of the Awingak Formation, and gas is present in the Heiberg and Bjorne formations.

Organic geochemistry

Despite there being an abundance of organic matter in the Mesozoic section, with three exceptions the organic matter is gas-prone. The three exceptions are the lower Jameson Bay Formation (Lower Jurassic), the lower part of the Hoyle Bay Formation (Upper Triassic), and the Murray Harbour Formation (Middle Triassic). The Jameson Bay Formation is immature or only marginally mature west of Lougheed Island, where it contains Type II kerogen, but in the mature zone to the east it contains largely gas-prone Type III organic matter. The bituminous shales in the Hoyle Bay and Murray Harbour formations contain abundant (up to 10% organic carbon) Type II kerogen and have good to excellent hydrocarbon yields in the mature zone. These formations are in the mature zone over much of the western Sverdrup Basin (Fig. 20A.4). Burial history curves for the Schei Point Group source rocks show that in the axis of the basin, the main phases of hydrocarbon generation took place in Late Cretaceous, whereas farther west, hydrocarbon generation occurred later, in early Tertiary. Irrespective of the stratigraphic or geographic position, the geochemical evidence suggests that the oils and condensates belong to the same genetic family, and that they have been derived from the shales of the Murray Harbour and Hoyle Bay formations (Brooks et al., in press). This indicates that considerable vertical and lateral migration has occurred. However, the composition of gasoline-range hydrocarbons indicates considerable variation in the mature level at which they were generated. Methane in associated and non-associated gases shows a wide range of isotopic composition ($\delta^{13}C_1$ -35.0 to -54.1 per mil relative to the PDB standard), again indicating considerable variation in the maturation levels of the source. Hydrocarbons in the Isachsen and Awingak reservoirs have commonly experienced varying degrees of biodegradation, even in the centre of the basin, and locally, gas occurs in the same reservoir as biodegraded oil (e.g. Isachsen Formation, Whitefish field). This latter phenomenon suggests that more than one phase of hydrocarbon generation and accumulation has occurred.

The distribution of oil and gas fields east of, and including, the Cisco field, is entirely consistent with the observed maturation patterns in the underlying Triassic source rocks. South and west of Cisco, the major accumulations are of gas and gas-condensate, suggesting a somewhat higher maturation level for their source than is indicated from regional maturation considerations. These variations are reflected in the isotopic composition of methane in the gases. The $\delta^{13}C_1$ values for methane in the Cisco, Maclean, Skate, Char and Balaena discoveries fall in the range -48.0 to -54.0 mil, and are consistent with having been generated with oil in the mature zone. The heavy isotope values of methane ($\delta^{13}C_1$ -35.6 to -37.0 per mil) in gases from the vicinity of Ellef Ringnes Island are consistent with hydrocarbon generation from the overmature Triassic shales. The situation is more complex in the Whitefish field. The isotopic data on methanes from the Awingak and Heiberg reservoirs ($\delta^{13}C_1 = -38.2$ per mil) suggest an overmature source. However, the gasoline-range composition of the Whitefish condensates indicates a similar maturation level to the nearby Cisco oil. The isotopic value for the Cisco methane is at a somewhat lower maturation level than the associated liquids. Consideration of the distribution of carbon isotopic compositions between different components of the natural gases, using the principles laid down by James (1983), suggests there has been a mixing in both the Whitefish and Jackson Bay fields of late supermature gas with earlier generated mature gas and associated liquids.

This situation can be readily understood in terms of the tectonic history of the area. Extensive faulting, salt tectonism and igneous activity has occurred in the western Sverdrup Basin (Balkwill and Fox, 1982). There are two areas of salt tectonism, one to the west and one to the east of Lougheed

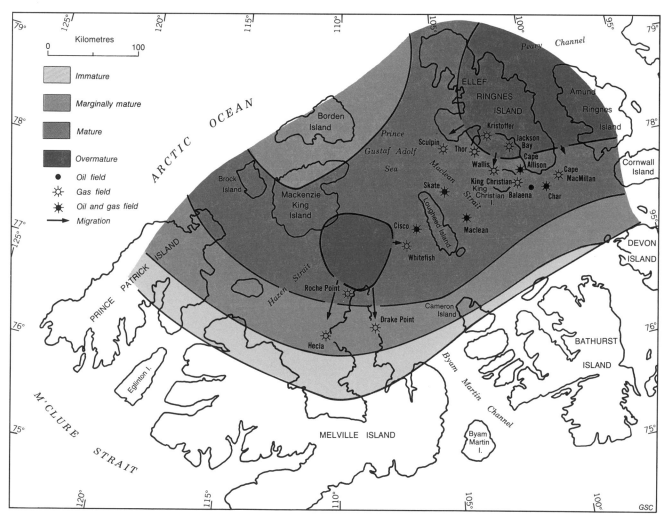

Figure 20A.4. Thermal maturation zones of Middle-Upper Triassic shales.

Island. The oil fields occur between, or on, the margins of these two areas of salt tectonism but are not obviously associated with them. Their hydrocarbons have clearly been generated by normal burial processes. The Whitefish field lies within the western zone of salt diapirism. It is consistent with the observed isotope ratios and the composition of the liquid hydrocarbons that a reservoir of pre-existing mature hydrocarbons has been invaded by gas from an overmature source. This overmature source could be deep-seated or occur in shallower sediments affected by salt diapirism and igneous intrusion (Balkwill and Fox, 1982). Such a scenario would also explain the occurrence of dry gas with intermediate isotopic values ($\delta^{13}C_1$ -47.2 to -42.5 per mil) in the Drake Point and Hecla fields. Gas generated by relatively rapid heating would not show the normal isotopic fractionation effects that are encountered during burial maturation. Gas generated by such a mechanism and migrating to the basin margin would be trapped in the Drake and Hecla fields at the edge of the basin. Data from wells within the western area of salt tectonism show enhanced maturation when compared with wells outside the area of salt tectonism.

Prospects

The main prospects for the Mesozoic succession are salt-cored anticlines that developed early. Structures in the central and northeastern Sverdrup Basin are considered to have low potential because of very high thermal maturation and poor reservoir characteristics of the Heiberg Formation, the main potential reservoir. More favourable conditions in terms of thermal maturity and reservoir development occur along the southeastern and eastern margins of the basin. Nine wells were drilled on anticlines in the vicinity of Fosheim Peninsula, Ellesmere Island, which is on the eastern margin of the basin. Good oil and gas shows were encountered in two wells, Romulus C-42 and Taleman J-34, but no significant hydrocarbon accumulations were found. These failures severely downgrade anticlinal prospects in this area.

Several anticlinal closures occur in the Norwegian Bay area (southeastern Sverdrup Basin) and no diagnostic tests exist in the area. These structures must be considered prospective, although possible lack of seal due to the thinness of shale units and the occurrence of fractures over the high-amplitude structures puts a high risk on these plays.

Conditions for petroleum occurrence in structures of the western Sverdrup Basin are very good, as demonstrated by the seventeen fields discovered so far in this area. However, some large anticlines did not contain petroleum (e.g. Sutherland, Elve) and the most plausible reason for those failures is the lack of seal due to fracturing over the structures. Most undrilled structural closures in this area are considerably smaller than those already drilled, although a few big anomalies still exist (e.g. Cape Norem). All structures are certainly prospective, owing to favourable maturity, reservoir and source relationships, but, due to the small size, most are presently uneconomic, especially if they are offshore. The "big structure" play has almost run its course and future structural prospects will have to be either on much smaller or much higher risk anticlines.

Another prospect type for Mesozoic strata is "sandstone pinch-out" along the basin margins or the flanks of salt structures. Embry (Chapter 14) has demonstrated that many unconformities occur along the flanks of the basin, and numerous sandstone units are overstepped by shales. An example of such pinch-outs are shown in Figure 20A.5. Because of the favourable thermal maturity and reservoir characteristics along the basin margin, the early nature of such traps, and the large potential hydrocarbon drainage areas, this prospect type is considered to have excellent hydrocarbon potential with the possibility of large fields being good. Detailed seismic interpretation in conjunction with facies analysis and stratigraphic modelling will be needed to explore such prospects.

Fault traps also have potential in the western Sverdrup Basin, especially near shale-out edges of sandstone units where faults are likely to act as seals. Mesozoic strata may also be petroleum-bearing in Lancaster Sound, where depth of burial may be as much as 5000 m. A large structure, the Dundas structure, has been identified in southern Lancaster Sound (Hea et al., 1980), and Mesozoic sandstones may occur on the crest and/or flank of this uplift.

Mesozoic strata on Banks Island and in the adjacent offshore are judged to have very low potential owing to low thermal maturity levels (Powell, 1978).

TERTIARY

Tertiary strata occur in Sverdrup Basin, Lancaster Sound, and on the continental shelf northwest of the Arctic Islands. Very few wells penetrate Tertiary strata and no wells have been drilled with the Tertiary as a target.

Tertiary deposits in the Sverdrup Basin, preserved mainly in synclinoria in the eastern part of the basin, are up to 3000 m thick. The petroleum potential of these strata is poor due to immaturity, very thin sections on anticlines, and probable freshwater flushing of potential reservoirs.

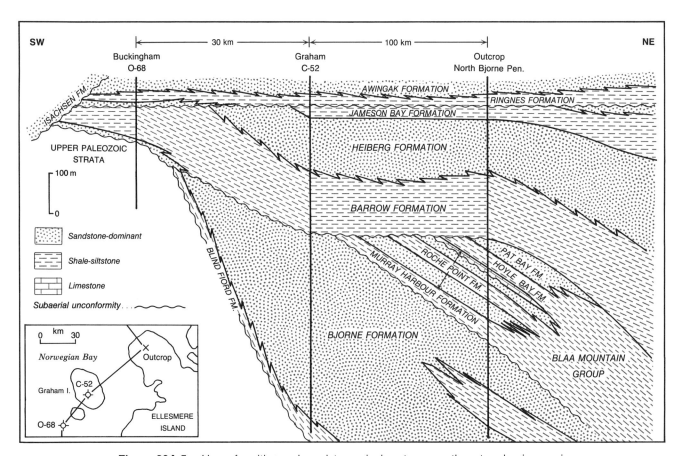

Figure 20A.5. Unconformities and sandstone pinch-outs on southeastern basin margin.

Upper Mesozoic and Tertiary deposits in Lancaster Sound are up to 5000 m thick (Hea et al., 1980) and on this basis Tertiary strata are considered to have some potential in this area. Much more data on the structure, potential seals, and source rocks of the succession are needed before a realistic evaluation can be made.

A thick succession of Tertiary strata probably exists on parts of the continental shelf northwest of the Arctic Islands. Seismic data are available for offshore Banks Island. There, the deposits are only about 2000 m thick and have a gentle, homoclinal dip toward the Arctic Ocean. Thicker, more structurally complex sections are expected northeastward, toward the area affected by the Eurekan Orogeny. Up to 11 km of Tertiary and Cretaceous sediments are interpreted to be present on the outer shelf north of Axel Heiberg Island (Asudeh et al., in press). However, the staggering logistical problems of drilling a well on this shelf, which is continually covered by the shifting Polar Ice Pack, make any petroleum assessment of this area an academic exercise for the foreseeable future.

B. PETROLEUM RESOURCES, NORTH GREENLAND

F.G. Christiansen, S. Piasecki, and L. Stemmerik

INTRODUCTION

Two of the world's most remote and inaccessible onshore and partly offshore sedimentary basins constitute the main part of northern Greenland: the lower Paleozoic Franklinian Basin and the upper Paleozoic–Tertiary Wandel Sea Basin (Fig. 20B.1a). The region forms an interesting geological link between petroliferous basins in the Canadian Arctic Islands and in the Barents Sea and northern North Sea. Technical challenges and economic risks, however, defer far into the next century any production of petroleum from such areas.

Commercial petroleum exploration has been restricted to a single period from 1968 to 1973 when a group of companies, the Greenarctic Consortium, headed by Edmonton-based Pondary Exploration Company, held a non-exclusive five-year concession. Some results of this work were included in previous papers (Dawes, 1976; Henderson, 1976; Stuart Smith and Wennekers, 1977; Nassichuk, 1983). In 1978-1980 and 1984-1985, the Geological Survey of Greenland carried out systematic regional mapping and general studies covering most geological aspects, especially of the Franklinian Basin (see numerous papers in Grønlands Geologiske Undersøgelse Report, numbers 88, 99, 106, 126, 133, published in 1979, 1980, 1981, 1985, 1987 respectively). Of great importance for the assessment of the petroleum potential is the well understood tectonic-sedimentological evolution of the Franklinian Basin (Surlyk and Hurst, 1984; Higgins et al., Chapter 7). The Wandel Sea Basin has been studied in much less detail, but a litho- and biostratigraphic frame is established (Håkansson and Stemmerik, 1984; Stemmerik and Håkansson, in press).

Recent petroleum-related studies of the region have concentrated on distribution of potential source rocks and evaluation of the thermal maturity pattern. Promising analytical results of scattered samples from the lower Paleozoic of eastern Peary Land (Rolle, 1981; Rolle and Wrang, 1981) led to the initiation of a source rock project of the lower Paleozoic strata in central and western North Greenland. Systematic sampling of all organic-rich units of this region in 1984 (Christiansen and Rolle, 1985; Christiansen et al., 1985) was followed by detailed studies and shallow core drilling of the most interesting units in 1985 (Christiansen et al., 1986), and a comprehensive analytical program is still in progress.

The knowledge of the source rock potential of the Wandel Sea Basin is based on analytical work of a limited number of samples employing Rock Eval pyrolysis, vitrinite reflectance, palynological studies, and gas chromatography (Bjorøy et al., 1982) combined with unpublished Grønlands Geologiske Undersøgelse (GGU) data.

Due to lack of wells and seismic work the potential prospects of both basins are highly speculative.

FRANKLINIAN BASIN
Source rocks

Organic-rich sediments of potential hydrocarbon source rock quality constitute thick intervals in the Cambrian shelf sequence and in the Ordovician-Silurian deep water sequence (Fig. 20B.2a). Ordovician shales occur only in the metamorphic fold belt and are not likely to have an appropriate maturity for petroleum generation anywhere in the region.

Lime mudstone and siliciclastic shale in the Lower to Middle Cambrian Henson Gletscher Formation form a 25-40 m thick sequence of potential source rock for oil, which outcrops over a length of approximately 100 km (Fig. 20B.1b). The TOC (Total Organic Carbon) values typically range from 2 to 5%, and immature or early mature samples often have a generative potential in excess of 10 mg HC/g rock (Christansen et al., 1987). However, most of the potential drainage area is mature to postmature at the surface (Fig. 20B.1c).

The Silurian shales can be traced for more than 700 km along the east-west strike of the basin and form a subcrop area of at least 10 000 km^2 (Fig. 20B.1b). Only a limited part of the shales, the upper Lower Silurian Thors Fjord Member and the age-equivalent part of the Lafayette Bugt Formation, is of source rock quality (Fig. 20B.2a). This 50-100 m thick sequence is dominated by oil-prone organic matter and typically contains 2 to 6% TOC. The immature or early

---

Christiansen, F.G., Piasecki, S., and Stemmerik, L
1991: Petroleum resources, North Greenland; in Chapter 20 of Geology of the Innuitian Orogen and Arctic Platform of Canada and Greenland, H.P. Trettin (ed.); Geological Survey of Canada, Geology of Canada, no. 3; (also Geological Society of America, The Geology of North America, v. E).

CHAPTER 20

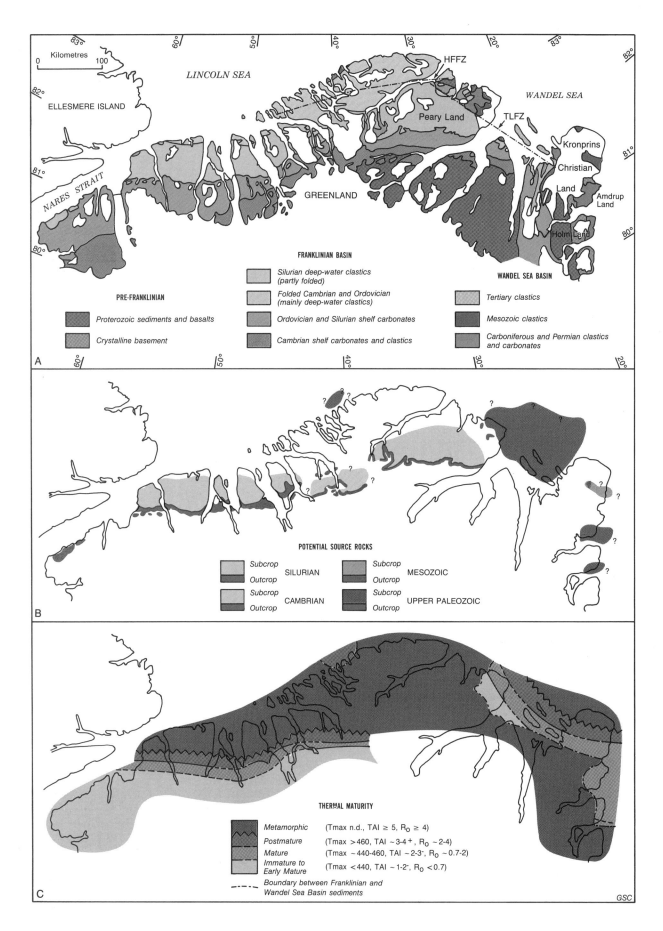

mature parts have generative potentials of 5 to 30 mg HC/g rock (Christiansen and Nøhr-Hansen, in press) but most of the Silurian shales are thermally postmature (Fig. 20B.1c).

The thermal maturity pattern is generally well defined in the exposed lower Paleozoic rocks except in areas that are lean in organic-rich units. There is a drastic change in maturity from south to north (Fig. 20B.1c). The southern part is immature or early mature and changes to postmature within a narrow zone, which is only 10 to 20 km across. All of the northern, folded part of the region has undergone low-grade metamorphism and has virtually no hydrocarbon potential (Fig. 20B.1c). This thermal alteration probably took place during the Ellesmerian Orogeny in Late Devonian – Early Carboniferous time, and the postmature lower Paleozoic sequence is overlain or truncated by immature upper Paleozoic and Mesozoic sediments toward the east (Fig. 20B.1c). Little is known about the thermal maturity of the lower Paleozoic strata in the eastern part of the region. Unpublished scattered values of the conodont alteration index (Aldridge et al., 1983) indicate that most of the lower Paleozoic sediments in Kronprins Christians Land are postmature or of low metamorphic grade, probably due to Caledonian deformation.

Reservoirs

Evidence of hydrocarbon generation is widespread in the region. Bitumen and asphalt occurrences as well as oil-stained carbonates and sandstones have been reported at many locations, both in conjunction with Cambrian and Silurian source rocks and far south of known source rocks (Fig. 20B.2a) (Christiansen and Rolle, 1985; Christiansen et al., 1986). The structural simplicity of the southern half of the basin, with few degrees northerly dip, favours hydrocarbon migration toward the south, away from the once generating and now mature to postmature source rocks. Permeable carrier beds are closely related to both Cambrian and Silurian source rocks.

The most obvious Silurian play with shelf margin limestones and reefs as reservoirs, also known in the Arctic Islands (Chapter 20A), is judged to be little prospective. Most reefs in the early mature or mature zone are too deeply eroded and commonly exposed, whereas the deeper buried ones toward the north are strongly thermally altered.

The most promising Cambrian reservoirs include sandstones in the Henson Gletscher and Sæterdal formations, which are closely associated with the source rocks and sandstones of the underlying Buen Formation (Fig. 20B.2a). Traps might occur along major northeast-striking faults. The main problem in these areas is not thermal maturity but efficiency of migration conduits and the deep erosional level. Subtle stratigraphic traps may occur in Upper Cambrian and Ordovician carbonates, but little is known about the porosity and permeability of these rocks. Weathered basement or Proterozoic sandstones cannot be excluded as potential reservoir rocks.

WANDEL SEA BASIN
Source rocks

The available source rock analyses from the Wandel Sea Basin are rather limited and were performed on samples collected for other purposes. Some of the most interesting units have only been studied in postmature areas, others have not been analyzed at all.

A number of fine grained, organic-rich intervals occur throughout the upper Paleozoic and Mesozoic sequence, which consists of continental or shallow marine deposits dominated by terrestrial gas-prone organic material (Fig. 20B.2b). Deep water shales with oil-prone organic material may occur in adjacent offshore areas.

The shales of the Lower Carboniferous Sortebakker Formation and the Upper Carboniferous part of the Mallemuk Mountain Group have not been geochemically analyzed. Their regional occurrence in the basin is restricted, but they may extend eastward offshore (Fig. 20B.1b).

Shales of the Upper Permian Midnatfjeldet Formation, the Upper Jurassic and Lower Cretaceous parts of the Ladegårdsåen Formation, and the Lower-Upper Cretaceous sequence are widely distributed in the basin (Fig. 20B.1b). With the exception of a few samples, the analytical data suggest only gas potential. The Upper Permian shale might be buried at depths of 1000-2000 m in the central part of the region where they may act as a source for hydrocarbons.

The regional maturity pattern of the Wandel Sea Basin shows that most of the sedimentary sequence has not subsided to sufficient depth to reach the mature zone with respect to oil generation. All units along the Trolle Land fault zone, which forms the basin margin toward the southwest, are thermally immature or early mature.

In Peary Land, along the western margin of the Wandel Sea Basin, a vertical maturity gradient is observed between early mature upper Paleozoic and immature Mesozoic sediments. A Tertiary thermal event has superimposed a conspicuous semicircular maturity pattern, with low-grade metamorphic sediments in the centre, at northern Kronprins Christian Land and on progressively less mature rocks toward the west and south (Fig. 20B.1c). The cause of this thermal event has not been identified, but unpublished aeromagnetic data reveal no evidence of major subsurface intrusions and the heating was probably associated with Tertiary rifting of the Arctic Ocean.

The oldest sediments of the Wandel Sea Basin, which occur in Holm Land, display an interesting jump in maturity where postmature Lower Carboniferous sediments are unconformably overlain by mature Upper Carboniferous deposits. The outcrops of Permian to Cretaceous sediments along the Harder Fjord fault zone are postmature, possibly as a result of repeated activation of the fault zone. A similar occurrence of lower Tertiary sediments is immature. Farther westward at Kap Washington and Lockwood Ø, thrusted upper Paleozoic and Mesozoic sediments are postmature (Fig. 20B.1c).

Figure 20B.1. Maps summarizing the petroleum geology of North Greenland. (a) Simplified geological map (based on Higgins et al., Chapter 7, and various GGU maps). HFFZ = Harder Fjord fault zone; TLFZ = Trolle Land fault zone. (b) Distribution of outcrop (exaggerated) and subcrop of potential source rocks. (c) Simplified thermal maturity pattern (based on Christiansen et al., 1987; Christiansen and Nøhr-Hansen, in press; and various unpublished GGU data).

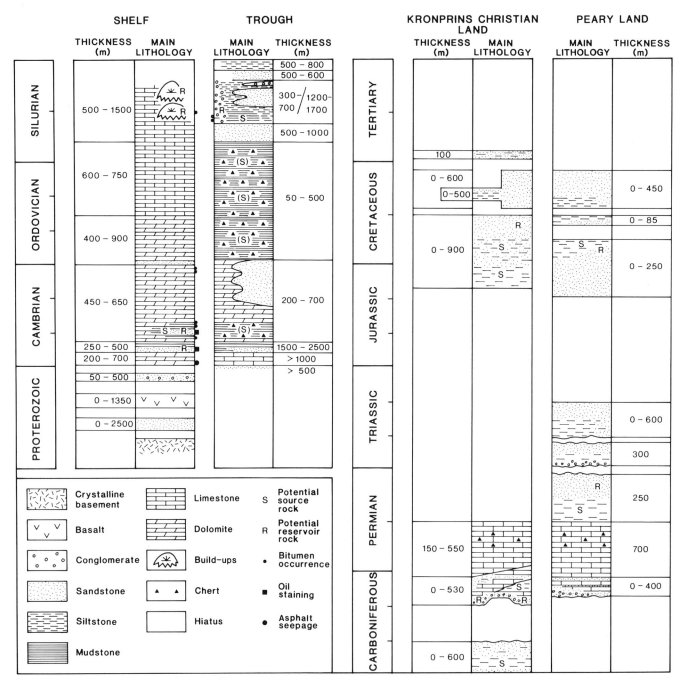

Figure 20B.2. Simplified stratigraphy of North Greenland indicating main lithology, thickness, potential reservoir and source rocks, and evidence of hydrocarbon generation. (a) The Franklinian Basin (based on Higgins et al., Chapter 7). (b) The Wandel Sea Basin (based on Håkansson and Stemmerik, 1984; Stemmerik and Håkansson, unpub. ms.).

Reservoirs

No significant evidence of hydrocarbon generation, migration, or trapping has so far been reported from the Wandel Sea Basin. The maturity pattern indicates that prospects are restricted to offshore areas east of Holm Land and Amdrup Land and east of Peary Land. The complex tectonic and depositional history of the basin during Mesozoic times (Håkansson and Pedersen, 1982; Håkansson and

Stemmerik, 1984) makes estimates of cumulative thickness and the presence or absence of stratigraphic intervals outside the exposed areas highly speculative, and also delimits the size of both drainage areas and traps.

Potential reservoir rocks in conjunction with potential source intervals are found in the Upper Carboniferous part of the Mallemuk Mountain Group, the Midnatfjeld Formation, and in the Upper Jurassic to Lower Cretaceous Ladegårdsåen Formation and time-equivalent units (Fig. 20B.2b). Most promising appears to be the upper Paleozoic sequence, which includes three possible types of traps: (1) Upper Carboniferous fault-controlled wedges of conglomerate and sandstone, overlain by tight limestones, that pinch out against unconformities; (2) Upper Carboniferous mounds and local carbonate platforms embodied in shale; and (3) Upper Permian sandstones unconformably overlain by Triassic shales.

The northern part of the basin is subdivided by northwestward-trending transcurrent faults into a series of fault blocks (Håkansson and Pedersen, 1982). The strata are folded in gentle anticlines in eastern Peary Land and in open domes in northern Kronprins Christian Land (Håkansson, 1979; Håkansson et al., 1981). The southern part of the area, Holm Land and Amdrup Land, is dominated by extensional faulting, which may have produced structural traps in addition to the already mentioned stratigraphic traps.

ACKNOWLEDGMENTS

Fieldwork and analytical work in central and western Greenland was supported by the Danish Ministry of Energy through project no. EFP 83-2251-305. E. Thomas is thanked for supplying unpublished vitrinite data of the Wandel Sea Basin. H.C. Larsen and F. Surlyk commented on an earlier draft of this paper, and M. Larsen and B.S. Hansen assisted technically.

C. COAL RESOURCES, ARCTIC ISLANDS

R.M. Bustin and A.D. Miall

INTRODUCTION

Coal deposits in the Arctic Islands were first described in the late 19th century, but had been known by the Inuit much earlier. Coal on northern Ellesmere Island was discovered and mined by the Nares (1875-1876) and Greeley (1881-83) expeditions and coal deposits on southeastern Banks Island and near Pond Inlet, Baffin Island, have been used by the native population. Only in the last decade, however, has an effort been made to assess the distribution and resources of coal in the Arctic Islands. In this region, unlike southern Canada, coal seams are commonly the most resistant rock unit within a poorly consolidated section, which, coupled with the paucity of vegetation, has facilitated their recognition and evaluation. Thus, although a comprehensive evaluation of the coal resources has only been completed for the most promising areas, much is now known about the occurrence and distribution of coal in this region.

STRATIGRAPHIC AND GEOGRAPHIC DISTRIBUTION

Coal deposits in the Arctic Islands range in age from Late Devonian to Late Eocene or Oligocene, and occur in strata of: (1) the Devonian clastic wedge, (2) the Sverdrup Basin, (3) Maastrichtian-Paleogene basins that succeeded the Sverdrup Basin, and (4) the Arctic Continental Terrace Wedge, major resources being restricted to the third category.

Bustin, R.M. and Miall, A.D.
1991: Coal resources, Arctic Islands; in Chapter 20 of Geology of the Innuitian Orogen and Arctic Platform of Canada and Greenland, H.P. Trettin (ed.); Geological Survey of Canada, Geology of Canada, no. 3; (also Geological Society of America, The Geology of North America, v. E).

The following discussion is restricted to coal occurrences; the stratigraphy and paleogeographic and tectonic setting of the host formations have been discussed in previous chapters. The results presented here are based on earlier studies and summarized by Bustin (1981), Mackay (1947), Caine (1973), observations and analyses included in geological studies mainly by officers of the Geological Survey of Canada, and previously unpublished field investigations and coal analyses by the authors.

Devonian clastic wedge

Carbonaceous shales and thin coal seams have been described from the Fram, Nordstrand Point, Beverly Inlet, Parry Islands, and Griper Bay formations on Ellesmere, Bathurst, Melville and Prince Patrick islands (Tozer and Thorsteinsson, 1964; Embry and Klovan, 1976; Chapter 10). Coal, however, comprises an exceedingly minor portion of these units and most seams are highly argillaceous and generally less than 20 cm thick. Coals from the Griper Bay Subgroup on Prince Patrick and Melville islands are high-volatile bituminous in rank (Tozer and Thorsteinsson, 1964).

Sverdrup Basin

Nonmarine and transitional marine strata in the Sverdrup Basin (Chapters 13, 14) commonly include carbonaceous laminae, and locally thin seams of bituminous coal. No major resources are known, but the following occurrences are noteworthy.

(1) On northern Axel Heiberg Island, the Lower Carboniferous Emma Fiord Formation includes three seams of coal of unknown rank that in the average are about 1 m thick (Thorsteinsson, 1974).

(2) The Fosheim Member of the Upper Triassic – Lower Jurassic Heiberg Formation contains carbonaceous shale

and thin coal seams (less than 0.5 m) at most localities. Analyses on samples from Axel Heiberg Island indicate a medium- to low-volatile bituminous rank (Fortier et al., 1963, Table IX). In the western Sverdrup Basin, the King Christian Island Formation of the Heiberg Group includes thin coals up to 1 m thick.

(3) The Middle Jurassic Hiccles Cove Formation on Prince Patrick Island is composed in part of carbonaceous shale. A 1 m thick seam of subbituminous coal is exposed on the south side on Intrepid Inlet (Tozer and Thorsteinsson, 1964).

(4) The Upper Jurassic – Lower Cretaceous Awingak Formation has carbonaceous laminae and thin coal seams (less than 0.5 m thick) at many localities.

(5) The Lower Cretaceous Isachsen Formation contains carbonaceous laminae and seams of lignite and subbituminous coal up to 0.5 m thick on Cornwall, Amund Ringnes and Ellef Ringnes islands. On Prince Patrick, Eglington, Melville, McKenzie King, and Banks islands, seams of subbituminous coal up to 1.5 m thick have been described (Tozer and Thorsteinsson, 1964; Bustin, unpublished data). The thickest coal-bearing interval is at Buchanan Lake, Axel Heiberg Island, where five seams, up to 2 m thick, of bituminous coal and semianthracite occur (Fortier et al., 1963, Table IX).

(6) In the Eclipse Trough of northern Baffin Island, strata assigned to the Upper Cretaceous Hassel Formation (Miall et al., 1980) include up to five seams of high-volatile bituminous coal that range in thickness up to 2 m. These coals were mined by the residents of Pond Inlet until 1959. Thin seams of subbituminous to high-volatile bituminous coal also occur on Ellesmere Island.

Maastrichtian-Paleogene basins

Significant resources of coal occur in the Eureka Sound Group of Banks Island (Banks Basin), western Axel Heiberg Island (Strand Fiord Basin), eastern Axel Heiberg and central and southern Ellesmere islands (Remus Basin), northern Ellesmere Island (Hazen and Judge Daly basins), and Bache Peninsula (Chapter 15).

Banks Basin. Coal seams are present at the base of the Mount Lawson Formation in the subsurface, but coal in outcrop is restricted to the Margaret Formation (terms of Miall, 1986; these formations correspond to the Shale Member and Cyclic Member, respectively, of the Eureka Sound Formation in Miall, 1979). Coal seams up to 30 cm thick occur throughout the unit; seams 0.5 m thick at several localities; and a 3.3 m thick seam at one locality. The rank of these deposits ranges from lignite to subbituminous B.

West Sverdrup Basin. Coal seams up to 1.5 m thick are exposed on Melville Island; thin coal seams and carbonaceous laminae (partings) are found throughout the basin. In this basin also, coal rank ranges from lignite to subbituminous B.

Strand Fiord Basin. On western Axel Heiberg Island, in the area of Strand and Expedition fiords, the lower part of the Eureka Sound Group contains only thin laminae of coal. The upper part includes at least 10 seams ranging in thickness from 1 to 9 m. The rank of these coals ranges from volatile bituminous A to high-volatile bituminous C.

On southern Axel Heiberg Island, at Glacier Fiord, only the basal 300 m of the Eureka Sound Group is preserved. It includes several seams of subbituminous coal that are up to 1.5 m thick.

Remus Basin contains the most important coal deposits in the Arctic Islands. Occurrences in five different areas can be summarized as follows.

(1) The central part of the basin, between the Sawtooth Mountains and Eureka Sound, contains over 90 seams of lignite and subbituminous and high-volatile bituminous coal in a 3300 m thick succession. Most seams are less than 1 m thick, but several seams in the upper part of the section are up to 15 m thick.

(2) Few seams occur east of the Sawtooth Mountains and north of Vesle Fiord. Several seams of high-volatile bituminous coal, up to 3.5 m thick, oucrop in Wolf Valley, southeast of central Cañon Fiord; farther east, no seams thicker than about 0.5 m have been observed.

(3) Coal is found throughout the Eureka Sound Formation on eastern Axel Heiberg Island. Forty seams of subbituminous coal, 18 of which are greater than 1 m thick, are present at Mokka Fiord. The thickest measured seam is 6 m thick. Ten seams of lignite, up to 2 m thick, outcrop at Flat Sound. One to three seams, up to 4 m thick, also are exposed at May Point, on Schei Peninsula, and near Gibbs Fiord Diapir. Elsewhere exposures are poor and the total number and thickness of coal seams are unknown.

(4) In the southern part of the Remus Basin, between Vesle and Stenkul fiords, the Eureka Sound Formation includes major resources of coal. In the area of Vesle and Strathcona fiords the Group is up to 2500 m thick and contains substantial thicknesses of coal in the upper part of the section, and locally also in the lower part. Several coal seams adjacent to Vesle Fiord range in thickness from 1 to 4 m. Between Bay and Strathcona fiords, at least 15 seams of lignite are present that range in thickness from 1 to 25 m, with the thicker seams occurring in the upper part of the section (Fig. 20C.1). Several seams of lignite to subbituminous coal, ranging in thickness from 6 to 25 m, occur in the Stenkul Fiord area, southeast of Bay Fiord.

Lake Hazen Basin. Rare thin seams of high-volatile bituminous coal occur throughout the succession, and up to 5 seams, 0.5 to 2.5 m thick, are present locally.

Judge Daly Basin. A 9 m thick seam of high-volatile bituminous coal at Watercourse Bay, adjacent to Fort Conger (Fig. 20C.2) is the only significant coal occurrence in this area.

Eclipse Trough. On southern Bylot Island, thin seams of subbituminous coal outcrop in strata assigned to the Eureka Sound Group but no seams thicker than 0.5 m have been described.

Other localities. Erosional remnants of the Eureka Sound Group exist at several localities, mostly in fault-bounded basins superimposed on lower Paleozoic units. On Somerset Island, thin seams of subbituminous coal are known from strata assigned to the Idlorak Formation (Dineley and Rust, 1968) or Mokka Fiord Formation (Miall, 1986). At Bache Peninsula, eastern Ellesmere Island, three seams of subbituminous coal, 1 to 4 m thick, are exposed. On Devon Island, several thin (less than 20 cm thick) seams of subbituminous coal occur in the Eureka Sound Group in the vicinity of Viks Fiord, and on southern Bathurst Island 1 to

Figure 20C.1. Coal-bearing upper part of a section of the Eureka Sound Group north of Strathcona Fiord, Ellesmere Island.

2 m of coal have been reported (Kerr, 1974). At Intrepid Bay, Cornwallis Island, 12 seams of subbituminous coal ranging from a few centimetres to 1.5 m, have been described from a 600 m thick succession (Thorsteinsson, 1974). Although the strata were initially thought to be Pennsylvanian in age, they are now regarded as Tertiary and equivalent to the Eureka Sound Group (Caine, 1973).

Arctic Continental Terrace Wedge

In the Crocker I-53 well on Meighen Island, unnamed Oligocene and older sediments include numerous seams ranging in thickness from 0.001 to 10 m. Coal rank increases progressively through the sequence from lignite near the surface to high-volatile bituminous coal at a depth of 3200 m.

The Miocene(?)-Pliocene Beaufort Formation consists in part of "peaty" material resembling compressed wood, but has little true lignite.

RESOURCES

In reporting coal resources, two factors must be considered (Bielenstein et al., 1979): (1) the level of assurance that the resources exist; and (2) the feasibility of their exploitation. Four categories are recognized with regard to the first factor: measured, indicated, inferred and speculative. The coal resources reported here are classified as "inferred" because the estimates are based on outcrop information and preliminary mapping together with a broad knowledge of the geology.

The second factor, feasibility of exploitation, is dependent on a combination of coal quality, thickness, depth, and location (Bielenstein et al., 1979). In the Arctic Islands, "resources of future interest" are considered to include:

(1) all seams greater than 1 m thick, to a depth of 200 m, in the deformed areas of Axel Heiberg Island and part of Ellesmere Island;

(2) all seams greater than 1 m thick, to a depth of 400 m, in all other areas.

The results are listed in Table 20C.1 with the exception of central and southeastern Banks Island, where major "speculative" resources may exist, and some minor areas with seams thicker than 1 m.

With the exception of the Isachsen Formation on Banks and Axel Heiberg islands and the Hassel Formation on northern Baffin Island, all major resources occur in the Eureka Sound Formation. About 80% are lignite or subbituminous coal and the remaining 20% are high-volatile bituminous coal, i.e. all are thermal coal, and no significant deposits of metallurgical (medium- or low-volatile bituminous) coal are known. The total coal resources calculated for this area (51 000 megatonnes) make up about 20% of the total inferred coal resources of Canada.

The quality of coal has only been determined in a few areas. Analyses indicate that, with few exceptions, the sulphur content is low (less than 1.0%, averaging about 0.5%). The ash content is highly variable but it is noteworthy that seams of clean coal (less than 10% ash) occur in many deposits. Measured calorific values are similarly variable, ranging from 23 000 to 32 500 kJ/kg. All major deposits consist exclusively of humic coal, comprising mainly the macerals vitrinite (70-90%) with minor inertinite, semi-inertinite and liptinite.

Figure 20C.2. Seam of high-volatile bituminous coal in the Eureka Sound Group at Watercourse Valley near Fort Conger, north of Archer Fiord, Ellesmere Island.

Table 20C.1. Coal resources of the Arctic Islands.

LOCATION	FORMATION	RANK			TOTAL (in millions of tonnes)
		Lignite	Subbituminous	High-volatile bituminous	
BANKS ISLAND	Eureka Sound		(>) 5000*		5000
	Isachsen		2		2
AXEL HEIBERG ISLAND					
Strand Fiord area	Eureka Sound			230	230
Glacier Fiord area	Eureka Sound			2	2
Eastern Axel Heiberg	Eureka Sound	5000	4000	300	9300
	Isachsen			5	5
ELLESMERE ISLAND					
West Fosheim Peninsula	Eureka Sound	10 000	7000	4000	21 000
East Fosheim Peninsula	Eureka Sound	2500	500	200	3200
Southern Ellesmere	Eureka Sound	8000	3500		11 500
Lake Hazen	Eureka Sound			600	600
Judge Daly Promontory	Eureka Sound			10	10
Bache Peninsula	Eureka Sound		14		14
BAFFIN ISLAND	Eureka Sound		5		5
Eclipse Trough	Hassel			200	200
*does not include speculative resources from Southern Banks Island		25 500	20 021	5547	51 068

GSC

D. ECONOMIC MINERAL RESOURCES, ARCTIC ISLANDS

Walter A. Gibbins

INTRODUCTION

Mineral exploration in the Arctic Islands has been impeded by the remoteness of the region and the severity of its climate. Nevertheless, large, high-grade, Mississippi Valley-type (MVT), lead-zinc deposits have been discovered during the past three decades. Underground mining has been underway since 1982 at Polaris and since 1976 at Nanisivik. Both deposits are on tidewater and are accessible during the short Arctic shipping season.

This subchapter is concerned mainly with the Polaris mine on Little Cornwallis Island and to a lesser extent with the Nanisivik mine at Strathcona Sound, northern Baffin Island (Fig. 20D.1). These are the world's most northerly metal mines, and account for almost all of the non-government economic activity, and all of the profitable mineral production in the region. More than 12 million tonnes of ore, grading more than 15% lead-zinc, had been mined from these deposits by March, 1988.

Non-hydrocarbon mineral exploration in the Arctic is summarized regularly in Northwest Territories (NWT) Mineral Industry reports produced by the NWT Geology Division of the Department of Indian Affairs and Northern Development (e.g. Laporte, 1974; Gibbins et al., 1977).

CORNWALLIS LEAD-ZINC DISTRICT

General geology. The Cornwallis Lead-Zinc District (Kerr, 1977b) is a 125 by 275 km area that corresponds to the northern Boothia Uplift (northern Cornwallis Fold Belt of Kerr, 1977a) and comprises Cornwallis Island, eastern Bathurst Island, most of Grinnell Peninsula of northwestern Devon Island, as well as several smaller islands, including Little Cornwallis and Truro (Fig. 20D.2). Geological maps and reports have been published by Thorsteinsson (1958, 1980, 1984, 1988), Thorsteinsson and Kerr (1968), Kerr (1974; 1977a, b), Morrow and Kerr (1977, 1986), and Thorsteinsson and Mayr (1987).

In Late Cambrian to latest Middle Ordovician time, the area formed part of the Franklinian Shelf (or miogeocline), which received mainly carbonate sediments (Cass Fjord, Cape Clay, Eleanor River, Thumb Mountain, and Irene Bay formations) but also subtidal evaporites of late Early and early Middle Ordovician age (Baumann Fiord and Bay Fiord formations)(Chapter 8B, Fig. 8B.30-8B.33). A different facies configuration existed from Late Ordovician to earliest Devonian time (Fig. 8B.34, 8B.35): downwarping of outer parts of the shelf resulted in the deposition of slope and basin deposits (graptolitic shales and carbonates of Cape Phillips Formation) in northern and western parts of the area, while shelf carbonate deposition (Allen Bay Formation and Read Bay Group) continued in southern and eastern parts (Fig. 20D.2).

The Cornwallis Disturbance affected northern parts of the Boothia Uplift in Early Devonian time (Kerr, 1977a; Okulitch et al., 1986; Chapter 12C). Faulting of the crystalline basement along northerly trends caused faulting and folding of the lower Paleozoic cover. After an interval of uplift and erosion, the northern Boothia Uplift was unconformably overlapped by strata of the upper Lower Devonian Disappointment Bay Formation. The syntectonic and post-tectonic deposits of this unit are overlain by a thick carbonate and clastic succession of latest Early Devonian to Late Devonian age (Blue Fiord Formation, Okse Bay Group, etc.). The Ellesmerian Orogeny of latest Devonian – Early Carboniferous age caused renewed faulting and folding. The post-Devonian stratigraphic record is restricted to local outcrops of Carboniferous, Cretaceous and Tertiary strata, commonly preserved in grabens of Tertiary age.

Controls on mineralization. A few empirical controls or guides to mineralization account for most of 15 known MVT showings in the region (Fig. 20D.2). Most of the showings, including the three main deposits, are in the upper bioclastic part of the Ordovician **Thumb Mountain Formation** (Kerr, 1977b). The country rock is invariably **brecciated dolostone**; veins or veinlets with sparry dolomite referred to as "pseudobreccia" are common. Most showings lie **within the bounds of the Cape Phillips shale basin** (Fig. 20D.2) and are structurally higher, but stratigraphically lower, than the shale. In most cases where stratigraphic relationships can be determined, the showings are in areas where the Thumb Mountain Formation is **unconformably overlain by the Disappointment Bay Formation**. Solution and collapse features related to this surface may have provided conduits for the transportation and deposition of ore fluids (Kerr, 1977b).

A fifth control may be northerly-trending, **basement-controlled faults or fractures and anticlines** related to the development of the Boothia Uplift. Kerr (1977b) noted that the Thumb Mountain Formation was exposed during the Cornwallis Disturbance only where erosion cut deeply, primarily into anticlines, like the Crozier Strait Anticline.

The Little Cornwallis Subdistrict. Despite the impressive size of the Cornwallis Lead-Zinc District, all of the economic resources, all of the drill-indicated resources, and the best potential for additional resources are limited to Little Cornwallis and Truro islands, which together make up less than 5% of the area. Two deposits, Truro and Polaris (4 and 5 in Fig. 20D.2), lie on opposite limbs of the Crozier Strait Anticline/Crozier Strait Graben (Kerr, 1977b, p. 1411). The third, Eclipse (6 in Fig. 20D.2), is in the western limb of a second anticline that underlies eastern Little Cornwallis Island (Kerr, 1977b).

The Polaris Deposit is an MVT deposit of exceptional size and grade. Preproduction ore reserves amounted to 23 million tonnes of 4.3% lead and 14.1% zinc. The orebody is a north-trending lens with a thick eastern section, called the Keel Zone, and a thinner, conformable wing extending updip to the west, known as the Panhandle Zone (Fig. 20D.3).

Gibbins, Walter A.
1991: Economic mineral resources, Arctic Islands; *in* Chapter 20 of Geology of the Innuitian Orogen and Arctic Platform of Canada and Greenland, H.P. Trettin (ed.); Geological Survey of Canada, Geology of Canada, no. 3; (*also* Geological Society of America, The Geology of North America, v. E).

The top of the orebody is 80 to 180 m below the surface and has a maximum stratigraphic thickness of 120 m, a maximum width of 400 m, and a length of 1000 m.

Host rock stratigraphy and lithology are particularly important at the Polaris mine, as they control and influence styles of mineralization. The upper Thumb Mountain Formation consists of nodular, dolomitized limestone, containing variable amounts of argillaceous and bioclastic material in the matrix between nodules. In mineralized areas, sulphides and sparry dolomite preferentially replace the muddy matrix, forming a breccia containing subrounded nodules of unaltered dolostone.

The lower Thumb Mountain Formation contains extensive, large lenses of dark brown, thinly laminated argillaceous and petroliferous dolostone, rich in fragments of the colonial coral *Tetradium cellulosum* (Fig. 20D.3). These lenses are preferentially replaced *in situ* by sulphides, most commonly without signs of brecciation. The remaining lower Thumb Mountain Formation is a beige, well bedded, sucrosic dolostone. Mineralized portions are characterized by ramifying veinlets of sphalerite, galena, and sparry dolomite, or less commonly by breccia zones.

The ore consists of galena and sphalerite, with gangue of sparry dolomite, marcasite, and ice with minor calcite and traces of bitumen. Typical ore textures are illustrated in Figure 20D.4. The sphalerite is mainly colloform (encrusting, massive, botryoidal or pisolitic) and to a lesser extent crystalline; the galena occurs as disseminated or vug-lining cubes, as skeletal or hopper-shaped crystals, or in radiating or dendritic intergrowths with sphalerite.

Ore textures, mineralogy, fluid inclusion temperatures (ranging from 50 to 105° [Jowett, 1975]), widespread evidence of dolomitization, karst development, and regional tectonics found at the Polaris deposit and in the district as a whole, are remarkably similar to those at Pine Point and typical of MVT deposits in general. However, Pine Point and Polaris-Truro do not have anomalous J-type lead isotope ratios characteristic of most MVT deposits (Heal, 1976).

Heal (1976) studied sulphur and lead isotopes of ore samples from Polaris and Truro. He showed that the sulphur is in isotopic disequilibrium and suggested that it was produced by bacterial reduction of Ordovician (Baumann Fiord and/or Bay Fiord formations) evaporitic sulphate and transported to depositional sites as reduced sulphide in a brine. A lead model age of 525 Ma suggests that the source of the metals was Precambrian rock, but the conformable nature of the Polaris-Truro lead may signify a deep-seated (upper mantle) source (Heal, 1976, p. 135). He concluded that precipitation of the sulphide minerals occurred upon the meeting of the two brines.

Kerr (1977b) proposed a similar genetic model in which a metal-bearing brine and a sulphide-bearing brine migrated to caverns and pores, which served as traps, and, upon meeting, precipitated sulphide minerals, probably during the Devonian. However, Kerr suggested that one solution travelled mainly laterally from a shale source (e.g. Cape Phillips Formation), while the other came from depth.

Anderson and Macqueen (1982, p. 114) pointed out that "Mixing models which involve adding H_2S to metal bearing brines at the site of ore deposition are much more attractive hydrodynamically than the mixing of two transported solutions". Anderson (1983) concluded that it is unlikely that ore-forming quantities of metal and H_2S could be transported in the same aqueous solution in the MVT situation. Therefore, sulphide must be supplied at the depositional site by the reduction of sulphate already in the brine, or, if bacterial reduction is involved, by diffusion of fluid flow from a cooler source area some distance from the ore deposition site. In this case, the precipitation of sulphides releases acid that would inevitably dissolve the carbonate host rocks, causing brecciation or aiding in replacement (Anderson, 1983). Such a process may be applicable, at least in part, at Polaris, where widespread brecciation is found and replacement is indicated by *in situ* replacement of *Tetradium* beds by sulphide, and by the continuity of a chert-nodule marker unit across the deposit (Fig. 20D.3). Anderson also stressed the importance of an impermeable formation such as shale overlying the ore zone (Irene Bay or Cape Phillips Formation at Polaris?) to contain the H_2S and retard oxidation of organic matter.

Callahan (1978, 1979) questioned the timing of Kerr's sequence of events and proposed that the sulphides may have been deposited prior to the Cornwallis Disturbance in cavities related to a postulated unconformity between the Thumb Mountain and Irene Bay formations. So far, no evidence has been recorded for a significant unconformity at that stratigraphic level. A limonite-stained hardground at the top of the Thumb Mountain Formation on Devon Island probably indicates a hiatus in sedimentation, but it must have been shorter than the age span of one conodont zone, and no karst features are apparent (Thorsteinsson and Mayr, 1987, p. 58).

The **Eclipse** deposit contains more than 1 million tonnes averaging 13% zinc in three zones of brecciated and faulted dolostone (Laporte, 1974, p. 166). Sphalerite, galena, and pyrite with smithsonite and iron oxide fill space between breccia fragments.

Figure 20D.1. Index of mineral deposits or occurrences discussed.
1: Cornwallis Lead-Zinc District.

2: Lead-zinc-silver deposits of the Milne Inlet Trough – Society Cliffs outcrop belt, Borden Basin.

2a: Hawker Creek showing.

3: Natkusiak Formation
(Upper Proterozoic basalt with native Cu).

4: Area with kimberlite pipes.

5: Central Ellesmere Fold Belt.
5a: Northeastern Judge Daly Promontory (Pb-Zn in Upper Cambrian – Lower Ordovician shelf margin carbonates).

6: North of upper Hare Fiord (Cu in Cambrian – Lower Silurian deep-water mudrock etc. of Hazen Fold Belt).

7: Southwest of Yelverton Inlet (Cu in Proterozoic or lower Paleozoic metabasalt of Clements Markham Fold Belt).

8: Southwest of Disraeli Fiord (Cu in Upper Ordovician shelf carbonates of Pearya).

RESOURCES

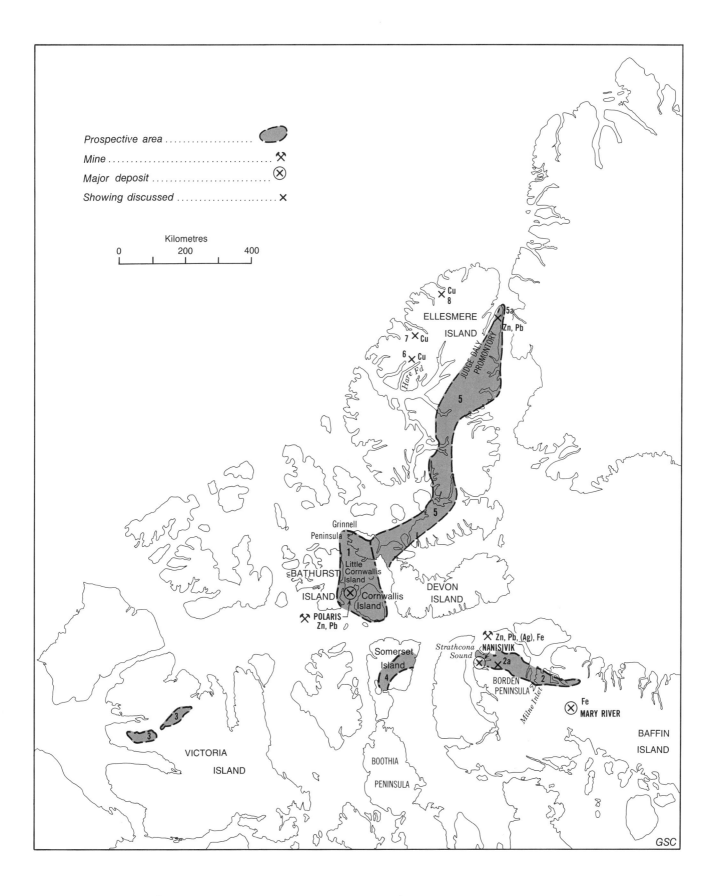

Truro Island (Fig. 20D.2, loc. 4) hosts a spectacular showing of high-grade galena, 1-2 m wide, that outcrops along the northeast shore for 10 m. Gravity surveys have outlined a large anomaly, and limited drilling gave some good but short intersections.

Other mineral showings in the region include an MVT zinc showing in Lower Ordovician sandy dolostone near the northeast coast of Aston Bay, Somerset Island (Gibbins et al., 1977, p. 42), and copper-lead showings in limestone of the Allen Bay Formation on southeastern Grinnell Peninsula (G and H in Fig. 20D.2) (Gibbins, 1984, p. 106-108).

PROTEROZOIC LEAD-ZINC DEPOSITS OF BORDEN BASIN, NORTHWESTERN BAFFIN ISLAND

The Borden Basin of northern Baffin Island originated by rifting in late Middle Proterozoic time (Jackson and Ianelli, 1981; Chapter 6). The Nanisivik lead-zinc-silver deposit and several smaller sulphide deposits occur in stromatolitic dolostones of the Society Cliffs Formation, which outcrops along a 150 km belt, extending from Admiralty Inlet to Tay Sound in a structure known as the Milne Inlet Trough (Fig. 20D.1, area 2). The deposits cluster at the northwest end of the belt, which has been uptilted and hence was most susceptible to karst processes (Geldsetzer, 1973a, 1973b). An uneconomic sulphide deposit at Hawker Creek, 40 km east-southeast of Nanisivik, contains a small amount of zinc sulphide similar to Nanisivik ore (Olson 1977, 1984) while other showings are very small and low grade.

The Nanisivik deposit at Strathcona Sound (Fig. 20D.5) has some characteristics of an MVT deposit but is unusual in several respects. It has a high proportion of iron sulphide (pyrite and marcasite), a high silver content in sphalerite, a Precambrian age for host rock and ore, and high fluid inclusion temperatures. Past production plus current proven or probable ore reserves (January, 1988) are 10 million tonnes of ore containing more than 10% zinc, 1% lead, 40g/t silver, and minor cadmium.

The deposit is shaped like a flattened tube. In cross-section, the "main" or "upper lens" is 60 to 120 m wide and 2 to 20 m thick. Commonly a second, more irregular, "lower lens" of sulphide is present below the main lens and connected to it by a vertical to subvertical keel. In plan view, it looks like a gently meandering river channel over its 3 km length (Fig. 20D.5). Typical ore consists of euhedral to subhedral grains of sphalerite and galena with anhedral pyrite, sparry dolomite, and ice as gangue. Sulphide ore is normally well bedded, but very coarse grained sulphide breccias are locally developed. Both pyrite and sphalerite stalactites are present but very rare. Massive argentiferous pyrite is extensive in the main orebody and its immediate vicinity (Fig. 20D.5).

Several genetic models for the Nanisivik deposit have been developed. All recognize the importance of karst development, the high temperature and salinity of ore fluids inferred from fluid inclusions (Arne et al., 1987), and of the underlying shales (Arctic Bay Formation) as the probable source of the metals. Each model, however, envisages different processes. Clayton and Thorpe (1982) proposed that horizontal caves were formed by meteoric water at or above the water table, and were later filled by sulphides at greater depth, under Victor Bay Formation cover.

Olson (1977, 1984) inferred four separate karst events, the second of which produced the ore-bearing caves. Hydrocarbons and metal-bearing brines, released during subsequent thermal metamorphism and dewatering of underlying shales, migrated to large integrated cave systems, where carbon oxidation, sulphate reduction, and metal deposition took place.

Ford (1981) recognized notches, fins, and other corrosion features at Nanisivik, which he attributed to hydrodynamic processes. He estimated that more than 95% of the main ore lens was created by a fast-flowing ore fluid that dissolved dolomite as it precipitated sulphides. Secondary cavities grew upwards, towards a piezometric surface (water table) that may have been rising or fixed.

Curtis (1984) agreed that much of the main ore lens was created by ore fluids dissolving dolostone. However, he interpreted the flat top of the main lens as a gas-liquid interface rather than a meteoric water table and questioned the primary, hydrodynamic origin of fins and corrosion notches.

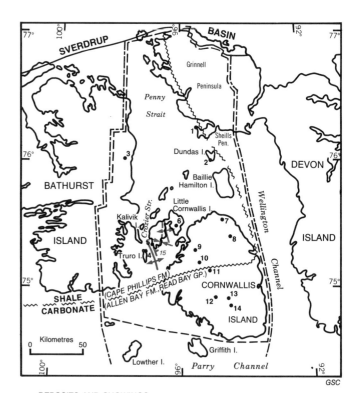

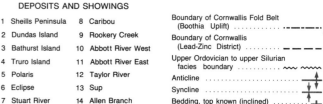

Figure 20D.2. Index map of the Cornwallis Lead-Zinc District (modified from Kerr, 1977b).

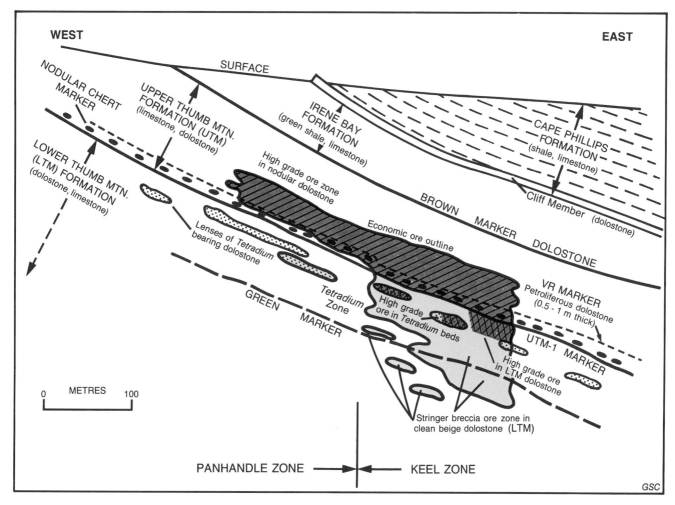

Figure 20D.3. West-east section through the Polaris ore body.

INDICATIONS OF MINERAL RESOURCES IN OTHER AREAS

The crystalline basement rocks of eastern Ellesmere Island (Chapter 6) were explored by geologists of Petro-Canada Ltd. in 1982 and 1983. They identified numerous sulphide-facies iron-formation showings, several malachite occurrences and skarn-related showings, and one barite showing (Gibbins, 1985). South of the region discussed in this subchapter, large high-grade iron deposits occur in metamorphosed Lower Proterozoic rocks in the Mary River area, Baffin Island (Fig. 20D.1) (Gross, 1966; Jackson, 1966).

Noteworthy in the Upper Proterozoic strata underlying the Arctic Platform are copper occurrences in the Natkusiak Formation of Victoria Island (Jefferson et al., 1985; Fig. 20D.1, area 3).

Within the Arctic Platform, several kimberlite pipes and dykes cut Silurian strata of Somerset Island (Fig. 20D.1, area 4; Mitchell, 1976) and a few micro-diamonds have been recovered. Kimberlite indicator minerals also have been identified in heavy mineral samples from southwestern Ellesmere Island (Gibbins, 1985).

The lower Paleozoic rocks of the Central Ellesmere Fold Belt and adjacent Arctic Platform (Fig. 20D.1, area 5) have attracted several combined prospecting and geochemistry projects aimed at the discovery of lead-zinc in carbonates and/or shales. Favorable geological criteria are: (1) slow sedimentation rates, euxinic conditions, and possible synsedimentary faulting in the "Cape Phillips shale basin"; (2) shelf-basin hinge lines; (3) carbonate-shale facies changes; and (4) reefal, brecciated and dolomitized carbonates (Sangster, 1981; cf. Chapter 8, this volume). The most spectacular discovery to date is the main showing found by Great Plains Development Co. Ltd. in Upper Cambrian or Lower Ordovician strata of the Copes Bay Formation on northeastern Judge Daly Promontory, a few kilometres southeast of the facies boundary with deep water sediments of the Hazen Formation (Fig. 20D.1, loc. 5a). There, galena and sphalerite occur in lenses and disseminated crystals over a strike length of 300 m (Gibbins et al., 1977, p. 63-65).

In 1982, Petro-Canada geologists discovered 130 minor showings in west-central Ellesmere Island. They included 87 formational and fracture-related hydrozincite showings in mudrocks of the Cape Phillips and Devon Island formations; 16 MVT showings of galena, sphalerite, and

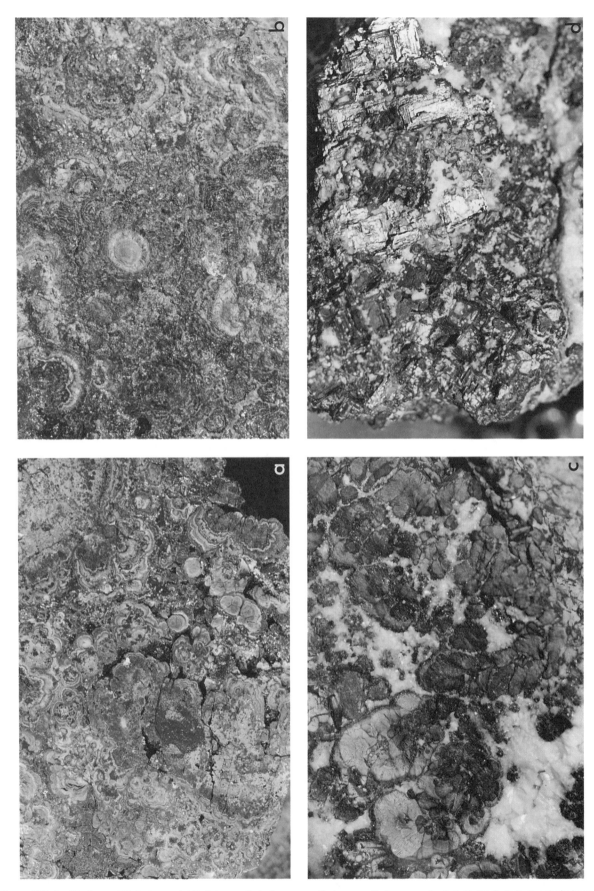

Figure 20D.4. Textures of Polaris ore. (a) Colloform sphalerite; marcasite fragment in lower centre (x 1.1). (b) Colloform sphalerite; radial intergrowth of galena in light sphalerite ring near centre (x 1.4). (c) Pale brown sphalerite spheroids in sparry dolomite; note thin colloform sphalerite encrustation and galena intergrowths (x 1.3). (d) Galena crystals and sparry dolomite; skeletal to hopper-shaped crystals (x 1.4).

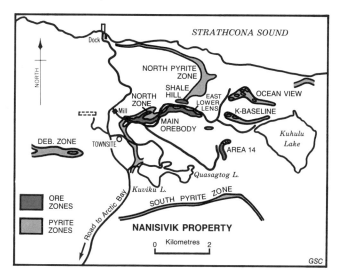

Figure 20D.5. Nanisivik sulphide zones, Strathcona Sound, Baffin Island.

barite in Allen Bay dolomite; 14 vein or fault-related showings of galena, sphalerite, and barite; and nine showings of nodular or bedded barite in the Devon Island Formation (Gibbins, 1985).

In northern Ellesmere Island, malachite and chalcopyrite occur in mudrock and chert of the Hazen Formation (Lower Cambrian to Lower Silurian; Chapter 8C, Fig. 8C.7, 8C.8) north of upper Hare Fiord (Geological Survey of Canada, Economic Geology Division, 1980 and H.P. Trettin, pers. comm., 1986; Fig. 20D.1, loc. 6). Malachite is common in mafic rocks of the Yelverton assemblage (Upper Proterozoic or lower Paleozoic; Fig. 8C.3) of the western Clements Markham Fold Belt (Fig. 20D.1, loc.7). Tennantite ($Cu_{12}As_4S_{13}$) was found in dolomitized strata of the Zebra Cliffs Formation of the Pearya Terrane (Upper Ordovician) close to a major fault interpreted to have originated as a Middle Ordovician suture (Trettin, 1981; Chapter 9; Fig. 20D.1, loc. 8). The only mineral resources known in the Sverdrup Basin are salt and sulphur associated with diapirs of the Otto Fiord Formation (Chapter 13).

ACKNOWLEDGMENTS

I would like to acknowledge the help, co-operation and hospitality of the Polar Continental Shelf Project (Department of Energy, Mines and Resources) and numerous geologists who have contributed to my knowledge of Arctic geology and have made my work enjoyable. W.A. Padgham read and made valuable suggestions to improve this report.

E. ECONOMIC MINERAL RESOURCES, NORTH GREENLAND

A. Steenfelt

Mineral exploration has been very limited in North Greenland, and economic mineral occurrences are not known. An evaluation of the mineral potential must therefore be based upon reconnaissance geochemical surveys and observed indications of mineralization.

Geochemical surveys using drainage samples collected at very low density have been carried out over most of North Greenland (Steenfelt, 1980, 1985; Jakobsen and Stendal, 1987). An area of 22 400 km² in central and western North Greenland (Fig. 20E.1) was investigated by stream sediment sampling at a density of one sample per 30 km², and the <0.1 mm grain size fraction was analyzed by X-ray fluorescence for 27 elements. Outside this area the stream sediment sites were very irregularly distributed, and on average with a smaller density; these samples were analyzed for fewer elements.

Preliminary presentations and evaluations of the surveys are given by Ghisler and Stendal (1980) and Steenfelt

Steenfelt, A.
1991: Economic mineral resources, North Greenland; in Chapter 20 of Geology of the Innuitian Orogen and Arctic Platform of Canada and Greenland, H.P. Trettin (ed.); Geological Survey of Canada, Geology of Canada, no. 3; (also Geological Society of America, The Geology of North America, v. E).

(1980, 1985, 1987). The results for central and western North Greenland indicate a potential for Zn and Ba deposits as illustrated by Figure 20E.1. On a regional scale, two zones are marked by high values of BaO and/or Zn. The zones are parallel to the axis of the basin, and a comparison with the evolutionary stages of the Franklinian Basin in North Greenland (Fig. 7.4) shows that the northernmost anomalous zone (mainly Zn anomalies) follows the position of the shelf margin at stages 3 and 4 (Early Cambrian to Early Ordovician), whereas the southern zone, dominated by high Ba values, follows the shelf margin and reef zone at stage 7 (mid- to Late Silurian).

Follow-up work has shown that there is syngenetic as well as epigenetic Ba enrichment in both anomalous zones, and epigenetic Zn mineralization in the northern zone. At Navarana Fjord, (N in Fig. 20E.1), samples from certain horizons, particularly of cherty mudstone, of the Ordovician starved slope sequence (stage 5 S_2), yielded Ba contents in the order of 0.4-6%. Barium is disseminated in the chert as barite and as barium silicate (cymrite?) (U.H. Jakobsen, pers. comm. 1987).

The high BaO values along the Silurian shelf margin are mostly ascribed to syngenetic Ba enrichment in the black mudstone deposited during stage 7T (early Late Llandovery). However, some stream sediment samples from Nyboe Land contain up to 2% barite, which is taken to

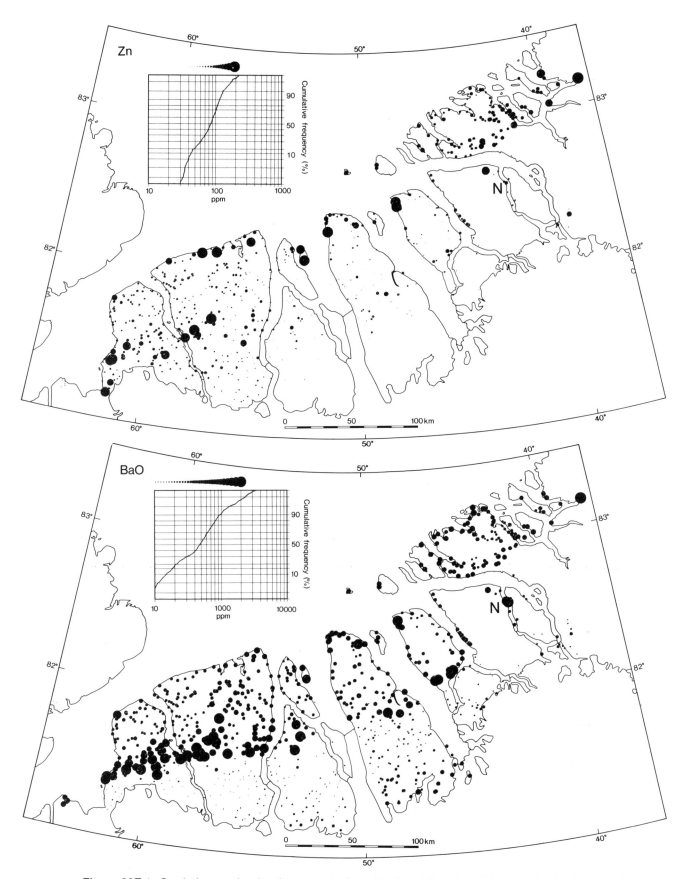

Figure 20E.1. Symbol maps showing the concentration of BaO and Zn in the <0.1 mm grain size fraction of stream sediment samples. N marks the location of breccia-type sphalerite mineralization. The increase in symbol size with element concentration is greater for Zn than for BaO. This is chosen to emphasize the background variation from BaO and the anomalies for Zn.

indicate the existence of vein-type barite mineralization. The anomalous sites have not been studied in the field.

Epigenetic sphalerite occurs in the Navarana Fjord Anticline (Fig. 7.7). The sphalerite and associated subordinate barite constitute 60-70% of the matrix of a 1 m wide breccia zone situated centrally in a 5-7 m wide vertical calcite vein. The vein is emplaced in a fault fracture intersecting dolomites of the Portfjeld Formation. The age of the mineralization is not known, but is believed to be contemporaneous with the faulting (Ellesmerian?) (Jakobsen and Steenfelt, 1985).

Geochemical surveys in eastern North Greenland have indicated a potential for Cu mineralization in the Upper Precambrian sedimentary and volcanic rocks south of the Franklinian Basin (Ghisler and Stendal, 1980). Minor chalcopyrite occurs in sandstone members, and native copper is found in vesicles in the basaltic sills.

REFERENCES

Aldridge, R.J., Armstrong, H.A., and Smith, M.P.
1983: Conodont colour alteration as a guide to thermal history of the Lower Paleozoic strata of North and East Greenland; Geological Survey of Greenland, unpublished report, 12 p.

Anderson, G.M.
1983: Some geochemical aspects of sulfide precipitation in carbonate rocks; in International Conference on Mississippi Valley-Type Lead-Zinc Deposits, Proceedings Volume, G. Kisvarsanyi, S.K. Grant, W.P. Pratt, and J.W. Koenig (ed.); University of Missouri-Rolla, Rolla, Missouri, p. 61-76.

Anderson, G.M. and Macqueen, R.W.
1982: Ore deposit models — 6. Mississippi Valley-type lead-zinc deposits; Geoscience Canada, v. 9, p. 108-117.

Arne, D.C., Kissin, S.A., and Curtis, L.W.
1987: A fluid inclusion study of sphalerite and dolomite from the Nanisivik lead-zinc deposit, Baffin Island, Northwest Territories — a discussion; Economic Geology, v. 82, p. 1968-1971.

Asudeh, I., Forsyth, D., Stephenson, R., Embry, A., Jackson, R., and White, D.
in press: Crustal structure of the Canadian Polar Margin: Part I; Canadian Journal of Earth Sciences.

Baker, D.A., Illich, H.A., Martin, S.J., and Landin, R.R.
1975: Hydrocarbon source potential of sediments in the Sverdrup Basin; in Canada's Continental Margins, C. Yorath, E.R. Parker and D.J. Glass (ed.); Canadian Society of Petroleum Geologists, Memoir 4, p. 545-556.

Balkwill, H.R. and Fox, F.G.
1982: Incipient rift zone, western Sverdrup Basin, Arctic Canada; in Arctic Geology and Geophysics, A.F. Embry and H.R. Balkwill (ed.); Canadian Society of Petroleum Geologists, Memoir 8, p. 171-188.

Bielenstein, H.U., Chrismas, L.P., Latour, B.A., and Tibbets, T.E.
1979: Coal resources and reserves of Canada; Energy, Mines and Resources Canada, Report ER 79-9, 37 p.

Bjorøy, M., Mills, N., Mørk, A., and Vigran, J.P.
1982: Organic geochemical analyses on samples from North-East Greenland; Continental Shelf Institute, Trondheim, Norway, unpublished report, 283 p.

Brooks, P.W., Embry, A.F., Goodarzi, F., and Stewart, R.
in press: Geochemical studies of Sverdrup Basin (Arctic Islands): Part I, Organic geochemistry and biological marker geochemistry of the Schei Point Group (Triassic) and recovered oils; Bulletin of Canadian Petroleum Geology.

Bustin, R.M.
1981: Tertiary coal resources, eastern Arctic Archipelago; Arctic, v. 33, p. 38-49.

Caine, T.W.
1973: Coal in the Arctic Archipelago; Canada, Department of Indian Affairs and Northern Development, unpublished report, 27 p.

Callahan, W.H.
1978: Cornwallis Lead-Zinc District; Mississippi Valley-type deposits controlled by stratigraphy and tectonics: Discussion; Canadian Journal of Earth Sciences, v. 15, p. 459-460.

1979: Cornwallis Lead-Zinc District; Mississippi Valley-type deposits controlled by stratigraphy and structure: Discussion II; Canadian Journal of Earth Sciences, v. 16, p. 614-615.

Christiansen, F.G., Nykjær, O., and Nøhr-Hansen, H.
1986: Source rock investigations and shallow core drilling in central and western North Greenland — project "Nordolie"; Grønlands Geologiske Undersøgelse, Report no. 130, p. 17-23.

Christiansen, F.G. and Nøhr-Hansen, H.
in press: The Silurian shales of North Greenland: a mature to postmature hydrocarbon source rock sequence; Grønlands Geologiske Undersøgelse, Report.

Christiansen, F.G., Nøhr-Hansen, H., and Nykjær, O.
1987: The Cambrian Henson Gletscher Formation: a mature to postmature source rock sequence from North Greenland; Grønlands Geologiske Undersøgelse, Report no. 133, p. 141-157.

Christiansen, F.G., Nøhr-Hansen, H., Rolle, F., and Wrang, P.
1985: Preliminary analysis of the hydrocarbon source rock potential of central and western North Greenland; Grønlands Geologiske Undersøgelse, Report no. 126, p. 117-128.

Christiansen, F.G. and Rolle, F.
1985: Project "Nordolie": hydrocarbon source rock investigations in central North Greenland; Grønlands Geologiske Undersøgelse, Report no. 125, p. 17-21.

Clayton, R.H. and Thorpe, L.
1982: Geology of the Nanisivik zinc-lead deposit; in Precambrian Sulphide Deposits, H.S. Robinson Memorial Volume, R.W. Hutchinson, C.D. Spence, and J.M. Franklin (ed.); Geological Association of Canada, Special Paper 25, p. 738-758.

Crain, E.R.
1977: Log interpretation in the High Arctic; in Transactions of the CWLS Sixth Formation Evaluation Symposium, Canadian Well Logging Society, p. S1-S32.

Curtis, L.W.
1984: Nanisivik Report; unpublished report prepared for Nanisivik Mines Ltd., 4 volumes.

Davies, G.R. and Nassichuk, W.W.
1988: An Early Carboniferous (Viséan) lacustrine oil shale in Canadian Arctic Archipelago; The American Association of Petroleum Geologists Bulletin, v. 72, p. 8-20.

Dawes, P.R.
1976: Precambrian to Tertiary of northern Greenland; in Geology of Greenland, A. Escher and W.S. Watt (ed.); Geological Survey of Greenland, Copenhagen, p. 248-303.

Dineley, D.L. and Rust, B.R.
1968: Sedimentary and paleontological features of the Tertiary-Cretaceous rocks of Somerset Island; Canadian Journal of Earth Sciences, v. 5, p. 791-799.

Douglas, R.J.W., Norris, D.K., Thorsteinsson, R., and Tozer, E.T.
1963: Geology and petroleum potentialities of northern Canada; World Petroleum Congress Proceedings, Frankfurt, Section 1, Paper 7, p. 519-571.

Drummond, K.J.
1973: Canadian Arctic Islands; in Future Petroleum Provinces in Canada, R.G. McCrossan (ed.); Canadian Society of Petroleum Geologists, Memoir 1, p. 443-472.

Economic Geology Division, Geological Survey of Canada
1980: Non-hydrocarbon mineral resource potential of parts of northern Canada; Geological Survey of Canada, Open File 716, 377 p.

Embry, A.F.
1982: The Upper Triassic – Lower Jurassic Heiberg Deltaic Complex of the Sverdrup Basin; in Arctic Geology and Geophysics, A.F. Embry and H.R. Balkwill (ed.); Canadian Society of Petroleum Geologists, Memoir 8, p. 189-217.

Embry, A.F. and Klovan, J.E.
1976: The Middle-Upper Devonian Clastic Wedge of the Franklinian Geosyncline; Bulletin of Canadian Petroleum Geology, v. 24, p. 485-639.

Ford, D.C.
1981: Report upon karstic features of the sulphide deposits at Nanisivik, Baffin Island; Strathcona Mineral Services Ltd., Toronto, Ontario, unpublished report, 28 p.

Fortier, Y.O., Blackadar, R.G., Glenister, B.F., Greiner, H.R., McLaren, D.J., McMillan, N.J., Norris, A.W., Roots, E.F., Souther, J.G., Thorsteinsson, R., and Tozer, E.T.
1963: Geology of the north-central part of the Arctic Archipelago, Northwest Territories (Operation Franklin); Geological Survey of Canada, Memoir 320, 671 p.

Fortier, Y.O, McNair, A.H., and Thorsteinsson, R.
1954: Geology and petroleum possibilites in Canadian Arctic Islands; Bulletin of the American Association of Petroleum Geologists, v. 38, p. 2075-2109.

Geldsetzer, H.
1973a: Syngenetic dolomitization and sulfide mineralization; in Oresin Sediments, G.G. Amstutz and A.J. Bernard (ed.); Springer-Verlag, p. 115-127.
1973b: The tectono-sedimentary development of an algal-dominated Helikian succession on northern Baffin Island, N.W.T.; in Proceedings of the Symposium on Geology of the Canadian Arctic; Geological Association of Canada and Canadian Society of Petroleum Geologists, p. 99-126.

Ghisler, M. and Stendal, H.
1980: Geochemical and ore microscopic investigations on drainage sands from the Peary Land region; Grønlands Geologiske Undersøgelse, Report no. 99, p. 121-128.

Gibbins, Walter A.
1984: Arctic Islands region; in Mineral Industry Report 1980-81, Northwest Territories; Canada, Department of Indian Affairs and Northern Development, EGS 1984-5, p. 83-130.
1985: Arctic Islands region; in Mineral Industry Report 1982-83, Northwest Territories; Canada, Department of Indian and Northern Affairs, EGS 1985-4, p. 95-155.

Gibbins, Walter A., Seaton, J.B., Laporte, J.P., Murphy, J.D., Hurdle, E.J., and Padgham, W.A.
1977: Mineral Industry Report 1974, Northwest Territories; Canada, Department of Indian and Northern Affairs, EGS 1977-5, 267 p.

Goodarzi, F., Davies, G.R., Nassichuk, W.W., and Snowdon, L.R.
1987: Organic petrology and rock evaluation analysis of Lower Carboniferous Emma Fiord Formation in the Sverdrup Basin, Canadian Arctic Archipelago; Marine and Petroleum Geology, v. 4, p. 132-145.

Gross, G.A.
1966: The origin of high grade iron deposits on Baffin Island, Northwest Territories; Canadian Mining Journal, v. 87, p. 111-114.

Hamilton, J.K. and Varney, G.R.
1982: Cores from the Canadian Arctic Archipelago: Geology and hydrocarbon occurrences; in Canada's Giant Hydrocarbon Reservoirs, W.G. Cutler (ed.); Canadian Society of Petroleum Geologists, Core Conference, p. 5-9.

Hea, J.P., Arcuri, J., Campbell, G.R., Fraser, I., Fuglem, M.O., O'Bertos, J.J., Smith, D.R., and Zayar, M.
1980: Post-Ellesmerian basins of Arctic Canada: their depocentres, rates of sedimentation and petroleum potential; in Facts and Principles of World Oil Occurrence, A.D. Miall (ed.); Canadian Society of Petroleum Geologists, Memoir 6, p. 447-488.

Heal, G.E.N.
1976: The Wrigley-Lou and Polaris-Truro lead zinc deposits, N.W.T.; unpublished M.Sc. thesis, Department of Geology, University of Alberta, Edmonton, Alberta, 172 p.

Henao-Londoño, D.
1977a: Correlation of producing formations in the Sverdrup Basin; Bulletin of Canadian Petroleum Geology, v. 25, p. 969-980.
1977b: A preliminary geochemical evaluation of the Arctic Islands; Bulletin of Canadian Petroleum Geology, v. 25, p. 1059-1084.

Henderson, G.
1976: Petroleum Geology; in Geology of Greenland, A. Escher and W.S. Watt (ed.); Geological Survey of Greenland, Copenhagen, p. 489-505.

Håkansson, E.
1979: Carboniferous to Tertiary development of the Wandel Sea Basin, eastern North Greenland; Grønlands Geologiske Undersøgelse, Report no. 88, p. 73-83.

Håkansson, E., Heinberg, C., and Stemmerik, L.
1981: The Wandel Sea Basin from Holm Land to Lockwood Ø, eastern North Greenland; Grønlands Geologiske Undersøgelse, Report no. 106, p. 47-63.

Håkansson, E. and Pedersen, S.A.S.
1982: Late Paleozoic to Tertiary tectonic evolution of the continental margin in North Greenland; in Arctic Geology and Geophysics, A.F. Embry and H.R. Balkwill (ed.); Canadian Society of Petroleum Geologists, Memoir 8, p. 331-348.

Håkansson, E. and Stemmerik, L.
1984: Wandel Sea Basin — the North Greenland equivalent to Svalbard and the Barents Shelf; in Petroleum Geology of North European Margin, A.M. Spencer et al. (ed.); Norwegian Petroleum Society, p. 97-107.

Jackson, G.D.
1966: Geology and mineral possibilities of the Mary River region, northern Baffin Island; Canadian Mining Journal, v. 87, no. 6, p. 57-61.

Jackson, G.D. and Ianelli, T.R.
1981: Rift-related cyclic sedimentation in the Neohelikian Borden Basin, northern Baffin Island; in Proterozoic Basins of Canada, F.H.A. Campbell (ed.); Geological Survey of Canada, Paper 81-10, p. 269-302.

Jakobsen, U.H. and Steenfelt, A.
1985: Zinc mineralization at Navarana Fjord, central North Greenland; Grønlands Geologiske Undersøgelse, Report no. 126, p. 105-109.

Jakobsen, U.H. and Stendal, H.
1987: Geochemical exploration in central and western North Greenland; Grønlands Geologiske Undersøgelse, Report no. 133, p. 113-121.

James, A.T.
1983: Correlation of natural gases by use of carbon isotopic distribution between organic components; The American Association of Petroleum Geologists Bulletin, v. 67, p. 1176-1191.

Jefferson, C.W., Nelson, W.E., Kirkham, R.Y., Reedman, J.H., and Scoates, R.F.J.
1985: Geology and copper occurrences of the Natkusiak basalts, Victoria Island; in Current Research, Part A, Geological Survey of Canada, Paper 85-1A, p. 203-214.

Jowett, E.C.
1975: Nature of the ore-forming fluids of the Polaris lead-zinc deposit, Little Cornwallis Island, N.W.T, from fluid inclusion studies; Canadian Institute of Mining and Metallurgy Bulletin, v. 68, no. 754, p. 124-129.

Kerr, J.W.
1974: Geology of the Bathurst Island group and Byam Martin Island, Arctic Canada (Operation Bathurst Island); Geological Survey of Canada, Memoir 378, 152 p.
1977a: Cornwallis Fold Belt and the mechanism of basement uplift; Canadian Journal of Earth Sciences, v. 14, p. 1374-1401.
1977b: Cornwallis Lead-Zinc District; Mississippi Valley-type deposits controlled by stratigraphy and tectonics; Canadian Journal of Earth Sciences, v. 14, p. 1402-1426.

Laporte, P.J.
1974: Mineral Industry Report, 1969 and 1970, Northwest Territories east of 104° West longitude; Canada, Department of Indian and Northern Affairs, v. 2, 191 p.

Mayr, U.
1980: Stratigraphy and correlation of Lower Paleozoic formations, subsurface of Bathurst Island and adjacent smaller islands, Canadian Arctic Archipelago; Geological Survey of Canada, Bulletin 306, 52 p.

MacKay, B.R.
1947: Coal reserves of Canada; Report of the Royal Commission on Coal, 1946, Chapter 1 and Appendix A (reprint), 113 p.

McCrossan, R.G. and Porter, J.W.
1973: The geology and petroleum potential of the Canadian sedimentary basins — a synthesis; in Future Petroleum Provinces of Canada, R.G. McCrossan (ed.); Canadian Society of Petroleum Geologists, Memoir 1, p. 589-720.

Meneley, R.
1977: Exploration Prospects in the Canadian Arctic Islands; Panarctic Oils Ltd., Calgary, Alberta, 16 p.

Miall, A.D.
1979: Mesozoic and Tertiary geology of Banks Island, Arctic Canada: the history of an unstable craton margin; Geological Survey of Canada, Memoir 387, 235 p.
1986: The Eureka Sound Group (Upper Cretaceous–Oligocene), Canadian Arctic Islands; Bulletin of Canadian Petroleum Geology, v. 34, p. 240-270.

Miall, A.D., Balkwill, H.R., and Hopkins, W.S., Jr.
1980: Cretaceous and Tertiary sediments of Eclipse Trough, Bylot Island area, Arctic Canada, and their regional setting; Geological Survey of Canada, Paper 79-23, 20 p.

Mitchell, R.H.
1976: Kimberlites of Somerset Island, District of Franklin; in Report of Activities, Part A; Geological Survey of Canada, Paper 76-1A, p. 501-502.

Morrow, D.W. and Kerr, J.W.
1977: Stratigraphy and sedimentology of lower Paleozoic formations near Prince Alfred Bay, Devon Island; Geological Survey of Canada, Bulletin 254, 122 p.
1986: Geology of Grinnell Peninsula and the Prince Alfred Bay area, Devon Island, District of Franklin, N.W.T.; Geological Survey of Canada, Open File 1325, 45 p.

Nassichuk, W.W.
1983: Petroleum potential in Arctic North America and Greenland; Cold Regions Science and Technology, v. 7, p. 51-81.

Okulitch, A.V., Packard, J.J., and Zolnai, A.I.
1986: Evolution of the Boothia Uplift, arctic Canada; Canadian Journal of Earth Sciences, v. 23, p. 350-358.

Olson, R.A.
1977: Geology and genesis of zinc-lead deposits within a late Precambrian dolomite, northern Baffin Island, N.W.T.; unpublished Ph.D. thesis, University of British Columbia, Vancouver, B.C., 371 p.
1984: Genesis of paleokarst and strata-bound zinc-lead sulfide deposits in a Proterozoic dolostone, northern Baffin Island, Canada; Economic Geology, v. 79, p. 1056-1103.

Panarctic Oils Ltd.
1983: Sixteenth Annual Report; Panarctic Oils Ltd., Calgary, Alberta, 28 p.

Powell, T.G.
1978: An assessment of the hydrocarbon source rock potential of the Canadian Arctic Islands; Geological Survey of Canada, Paper 78-12, 82 p.

Procter, R.M., Taylor, G.C., and Wade, J.A.
1984: Oil and natural gas resources of Canada; Geological Survey of Canada, Paper 83-31, 57 p.

Rayer, F.G.
1981: Exploration prospects and future petroleum potential of the Canadian Arctic Islands; Journal of Petroleum Geology, v. 3, p. 367-412.

Rolle, F.
1981: Hydrocarbon source rock sampling in Peary Land 1980; Grønlands Geologiske Undersøgelse, Report no. 106, p. 99-103.

Rolle, F. and Wrang, P.
1981: En foreløbig oliegeologisk vurdering af Peary Land området i Nordgrønland [in Danish]; Geological Survey of Greenland, unpublished report, 30 p.

Sangster, D.F.
1981: Three potential sites for the occurrence of stratiform, shale-hosted lead-zinc deposits in the Canadian Arctic; in Current Research, Part A, Geological Survey of Canada, Paper 81-1A, p. 1-8.

Snowdon, L.R. and Roy, J.K.
1975: Regional organic metamorphism in the Mesozoic strata of the Sverdrup Basin; Bulletin of Canadian Petroleum Geology, v. 23, p. 131-148.

Steenfelt, A.
1980: The geochemistry of stream silt, North Greenland; Grønlands Geologiske Undersøgelse, Report no. 99, p. 129-135.
1985: Reconnaissance scale geochemical survey in central and western North Greenland. Preliminary results concerning zinc and barium; Grønlands Geologiske Undersøgelse, Report no. 126, p. 95-104.
1987: Geochemical trends in central and western North Greenland; Grønlands Geologiske Undersøgelse, Report no. 133, p. 123-132.

Stemmerik, L. and Håkansson, E.
in press: Stratigraphy and depositional history of the Upper Palaeozoic and Triassic sediments of the Wandel Sea Basin, eastern North Greenland; Grønlands Geologiske Undersøgelse, Bulletin.

Stuart Smith, J.H. and Wennekers, J.H.N.
1977: Geology and hydrocarbon discoveries of Canadian Arctic Islands; The American Association of Petroleum Geologists Bulletin, v. 61, p. 1-27.

Surlyk, F. and Hurst, J.M.
1984: The evolution of the early Paleozoic deep-water basin of North Greenland; Geological Society of America Bulletin, v. 95, p. 131-154.

Tozer, E.T. and Thorsteinsson, R.
1964: Western Queen Elizabeth Islands, Arctic Archipelago; Geological Survey of Canada, Memoir 332, 242 p.

Thorsteinsson, R.
1958: Cornwallis and Little Cornwallis Islands, District of Franklin, Northwest Territories; Geological Survey of Canada, Memoir 294, 134 p.
1974: Carboniferous and Permian stratigraphy of Axel Heiberg Island and western Ellesmere Island, Canadian Arctic Archipelago; Geological Survey of Canada, Bulletin 224, 115 p.
1980: Part I. Contributions to stratigraphy (with contributions by T.T. Uyeno); in Stratigraphy and conodonts of Upper Silurian and Lower Devonian rocks in the environs of the Boothia Uplift, Canadian Arctic Archipelago; Geological Survey of Canada, Bulletin 292, p. 1-38.
1984: A sulphide deposit containing galena in the lower Devonian Disappointment Bay Formation on Baillie Hamilton Island, Canadian Arctic Archipelago; in Current Research, Part B, Geological Survey of Canada, Paper 84-1B, p. 269-274.
1988: Geology of Cornwallis Island and neighbouring smaller islands, Canadian Arctic Archipelago, District of Franklin, Northwest Territories; Geological Survey of Canada, Map 1626A (scale 1: 250 000)

Thorsteinsson, R. and Kerr, J.W.
1968: Cornwallis Island and adjacent smaller islands, Canadian Arctic Archipelago; Geological Survey of Canada, Paper 67-64, 16 p.

Thorsteinsson, R. and Mayr, U.
1987: The sedimentary rocks of Devon Island, Canadian Arctic Archipelago; Geological Survey of Canada, Memoir 411, 182 p.

Trettin, H.P.
1981: A tennantite deposit in the M'Clintock Inlet area, northern Ellesmere Island, District of Franklin; in Current Research, Part A, Geological Survey of Canada, Paper 81-1A, p. 103-106.

Trettin, H.P. and Hills, L.V.
1966: Lower Triassic tar sands of northwestern Melville Island, Arctic Archipelago; Geological Survey of Canada, Paper 66-34, 122 p.

Utting, J., Goodarzi, F., Dougherty, B.J., and Henderson, C.
1989: Thermal maturity of Carboniferous and Permian rocks in the Sverdrup Basin, Canadian Arctic Archipelago; Geological Survey of Canada, Paper 87-19.

Waylett, D.C.
1979: Natural gas in the Arctic Islands: discovered reserves and future potential; Journal of Petroleum Geology, v. 1, no. 3, p. 21-34.

Authors' addresses

R.M. Bustin
Department of Geology
University of British Columbia
Vancouver, B.C.
V6T 2B4

F.G. Christiansen
Geological Survey of Greenland
Øster Voldgade 10
DK-1350 Copenhagen K
Denmark

A.F. Embry
Geological Survey of Canada
3303-33rd Street N.W.
Calgary, Alberta
T2L 2A7

Walter A. Gibbins
NWT Geology Division
Department of Indian Affairs and
Northern Development
Yellowknife, N.W.T.
X1A 2R3

U. Mayr
Geological Survey of Canada
3303-33rd Street N.W.
Calgary, Alberta
T2L 2A7

CHAPTER 20

A.D. Miall
Department of Geological Sciences
University of Toronto
Toronto, Ontario
M5S 1A1

S. Piasecki
Geological Survey of Greenland
Øster Voldgade 10
DK-1350 Copenhagen K
Denmark

T.G. Powell
Bureau of Mineral Resources
Geology and Geophysics
P.O. Box 378
Canberra ACT, 2601
Australia

Agnete Steenfelt
Geological Survey of Greenland
Øster Voldgade 10
DK-1350 Copenhagen K
Denmark

L. Stemmerik
Geological Survey of Greenland
Øster Voldgade 10
DK-1350 Copenhagen K
Denmark

Printed in Canada

Chapter 21
SUMMARY AND REMAINING PROBLEMS

Introduction

Outline of geological history

Resources

References

Chapter 21

SUMMARY AND REMAINING PROBLEMS

H.P. Trettin

INTRODUCTION

This concluding chapter provides a brief, interpretative overview of the geological history of Innuitian Orogen and Arctic Platform with emphasis on unsolved major problems, and a summary of their resources. References quoted in previous chapters will not be repeated; for geographic names, see Figure 1 (in pocket).

OUTLINE OF GEOLOGICAL HISTORY
Early Proterozoic orogenesis

The northern part of the Canadian Shield (including Greenland) consists of a number of relatively small continental plates of Archean age that were welded together into a single supercontinent in a series of Early Proterozoic orogenies. The outlines of the Archean cratons and intervening Early Proterozoic orogenic belts have not yet been established for the High Arctic, but an orogen 1.9-1.95 Ga old is recognized in southeastern Ellesmere Island and adjacent parts of Greenland (Chapter 6).

Middle-Late Proterozoic basin development

An extensive late Middle Proterozoic rifting event is indicated by mafic volcanic and intrusive rocks and associated coarse clastic sediments exposed on Melville Peninsula (Fury and Hecla Basin), northern Baffin Island and Bylot Island (Borden Basin), southeastern Ellesmere Island and adjacent parts of West Greenland (Thule Basin), and northeastern Greenland (Chapter 6). Radiometric ages suggest correlation with the extensive Mackenzie dyke swarm of the eastern Cordillera, dated at about 1.2 Ga. The rift deposits are succeeded by shallow marine clastic and carbonate sediments that also are represented on Victoria Island (Shaler Group). Age and correlation of the sedimentary units are problematic. Lithological relationships suggest that the sedimentary units in the Borden and Thule basins are correlative, but the former have yielded late Middle Proterozoic paleomagnetic ages, whereas the latter have yielded Late Proterozoic (Riphean and Vendian) acritarch ages. The Shaler Group has been correlated with the Mackenzie Mountains Supergroup dated at 880-770 Ma on the basis of paleomagnetism.

Combined, these exposures represent the oldest succession that shows the overall southwesterly to westerly Innuitian trend, but their tectonic significance — whether they represent a passive margin related to a proto-Arctic Ocean or an intra-cratonic basin — remains uncertain.

Initiation of Franklinian mobile belt

The early history of the Franklinian mobile belt is poorly known. Analogy with the Cordilleran and Caledonian mobile belts suggests that it was initiated by a rifting event in the Late Proterozoic, probably represented by the Franklin diabase dyke swarm, tentatively dated at 750 Ma, and basic volcanics and intrusions on Victoria Island (Natkusiak Formation) that have given K-Ar ages of 635-640 Ma (Chapter 6). Poorly dated mafic volcanics and associated sediments of possible Late Proterozoic or earliest Cambrian age occur also in northern Axel Heiberg and northwestern Ellesmere islands (Chapter 8C). Apart from these predominantly volcanic units, and glaciogenic sediments in northeastern Greenland, the Late Proterozoic record of the Franklinian mobile belt is concealed.

Early Cambrian to Early Devonian basin development

The Lower Cambrian to Lower Devonian strata of the Franklinian mobile belt were deposited in two major provinces: a southeastern and southern unstable shelf contiguous with the relatively stable Arctic Platform, and a northwestern deep water basin (Chapters 4, 7, 8). The shelf margin retreated cratonward during the Early Cambrian, and again in latest Ordovician to Early Devonian time.

Franklinian Shelf and Arctic Platform

The record of the Franklinian Shelf (Chapters 7, 8B) dates back to the earlier part of the Early Cambrian, but the pre-*Olenellus* Zone stratigraphic record is poorly dated and the boundary with the Upper Proterozoic rocks is concealed. The Cambrian to lowermost Ordovician succession is characterized by clastic and carbonate sediments with a thick clastic unit in the upper part of the Lower Cambrian Series; the Lower to lower Middle Ordovician succession by carbonates and evaporites; and the upper Middle Ordovician to Upper Silurian succession by carbonates with clastic sediments in areas affected by Late Silurian – Early Devonian tectonism.

The sedimentary record indicates a series of transgressions and regressions that probably reflect both vertical oscillations of the continent caused by deep-seated processes and eustatic sea level changes. The transgressions increased in extent from Cambrian to Late Ordovician time,

---

Trettin, H.P.
1991: Summary and remaining problems; Chapter 21 in Geology of the Innuitian Orogen and Arctic Platform of Canada and Greenland, H.P. Trettin (ed.); Geological Survey of Canada, Geology of Canada, no. 3; (also Geological Society of America, The Geology of North America, v. E).

attaining an all-time maximum in the late Middle and Late Ordovician. The regressions are marked by disconformities that commonly die out on the outer shelf or upper slope. Most problematic is an extensive(?) disconformity near the Ordovician-Silurian boundary, inferred primarily from biostratigraphic evidence; it has been attributed to a sea level fall caused by glaciation in the Sahara.

Shelf architecture was characterized by alternations of two basic modes, the rimmed shelf and the distally steepened ramp (Fig. 8C.27). The rimmed shelf is best documented for the middle Tremadoc and late Arenig – Llanvirn intervals when evaporites were deposited in an extensive, subtidal intra-shelf basin (Fig. 8B.31, 8B.32).

The thickness of the deposits increases markedly from the craton toward the shelf edge, where it probably is greater than 9 km. The causes of the subsidence are uncertain as quantitative studies have not yet been made. Extension and lithospheric cooling, as well as rapid loading of the basin floor with turbidites in Early Cambrian and Early Silurian to Middle Devonian time, will have to be considered in such analyses.

During the second (Late Ordovician to Early Devonian) phase of shelf retreat, isolated carbonate buildups, surrounded and overlain by deep water sediments, developed in some areas near the previous shelf edge. The large Bent Horn – Towson Point buildup of Cameron and northeastern Melville islands (Fig. 8B.34-8B.36) persisted from Late Ordovician to early Eifelian time, whereas others were terminated earlier by clastic sedimentation or drowning.

Deep Water Basin

The deep water basin is divisible into a southeastern sedimentary subprovince and a northwestern sedimentary and volcanic subprovince. The sedimentary subprovince is comparable in setting and stratigraphy to the Richardson Trough of the northern Cordillera (Cecile and Norford, in press) and may have been connected with it, but the junction is covered by the Beaufort Sea.

The sedimentary subprovince, which is exposed from northwestern Melville Island to northeastern Greenland, is linked with the shelf province by interlocking facies changes. Four major units (Fig. 8C.1, 8C.2, Table 8C.1) are recognized: (1) resedimented carbonates of Early Cambrian age; (2) craton-derived turbidites of Early Cambrian age; (3) a highly diachronous and lithologically variable succession of resedimented carbonates, mudrock, and chert with an overall age range from latest Early Cambrian to early Middle Devonian (early Eifelian); (4) highly diachronous turbidites (flysch) with an overall age range from Late Ordovician or Early Silurian to Eifelian; and (5) a mudrock unit of Early-Middle Devonian age deposited in transitional slope to outer shelf environments. Units 1 and 2 are exposed only in northern Ellesmere Island (Hazen Fold Belt) and northeastern Greenland, in the "axial" region of the basin. In the marginal region, units 3 and 4 overlie shelf carbonates of late Middle Ordovician and younger ages. The diachronism of units 3 and 4 is due to southwestward progradation of the flysch, which was derived mainly from the northern Caledonides and sources in northern Ellesmere Island or north of it; and to southeasterly and southerly retreat of the shelf margin.

The sedimentary and volcanic subprovince, confined to northern Ellesmere and Axel Heiberg islands, contains arc-type volcanics and volcanic-derived sediments in addition to strata broadly comparable to units 2, 3 and 4 of the sedimentary subprovince, but with some significant differences. This subprovince has been divided into the Clements Markham and Northern Heiberg fold belts, which differ in stratigraphy and structural trend (Fig. 4.3 and 12A.1). The original location of these two belts is problematic. The Silurian flysch of the Clements Markham Fold Belt is more comparable to flysch in North Greenland than to strata in the Hazen Fold Belt, suggesting that it was involved in the sinistral motion inferred for Pearya. The Silurian volcanics in northwestern Ellesmere and northern Axel Heiberg islands indicate subduction regimes that preceded major deformation in the Late Silurian – Early Devonian (see below).

The Middle Proterozoic to Late Silurian record of Pearya

The composite Pearya Terrane of northern Ellesmere Island consists of four major successions with an overall age range from Middle Proterozoic to Late Silurian (Chapter 9). Succession I consists of sedimentary and volcanic(?) rocks, deformed, metamorphosed to amphibolite grade, and intruded by granitic rocks between 1.0 and 1.1 Ga.

Succession II is composed mainly of shelf carbonates, phyllite, schist, and quartzite with lesser proportions of mafic and siliceous volcanics, diamictite, and chert. Its stratigraphy is poorly known owing to complex structure and lack of fossils, but the diamictites provide an internal marker and a link with uppermost Proterozoic glaciogenic deposits in other regions. The succession probably lies unconformably on the Middle Proterozoic crystalline basement and ranges in age from Late Proterozoic to earliest Ordovician. It appears to have been deposited in rift-related and miogeoclinal settings.

Succession III comprises arc-type and ocean floor volcanics, carbonate, mudrock, and chert of assumed early Middle Ordovician age, and is associated with fault slices of late Arenig ultramafic and mafic plutonic rocks, tentatively interpreted as dismembered ophiolites.

The faulted contact between successions II and III is overlapped by succession IV of late Middle Ordovician to Late Silurian (late Ludlow or Pridoli) age. The angular unconformity at the base of this succession represents the mid-Ordovician (Llandeilo – early Caradoc) M'Clintock Orogeny, which was accompanied by metamorphism up to amphibolite grade and emplacement of a variety of granitic plutons. The event is interpreted as a collision that placed arc-type and oceanic assemblages (succession III and ultramafic-mafic complexes) on continental crust (successions I and II).

Succession IV consists of more than 7 km of clastic, carbonate, and volcanic rocks. The sedimentary strata are mainly of shallow marine origin but also include deep water and nonmarine facies; the volcanic rocks, which are confined to the Middle and early Late Ordovician (Caradoc and early Ashgill), are mainly of arc-type.

Pearya is related to the Caledonides by the Grenville age of its metamorphic-plutonic basement; the presence of ultramafic complexes with Arenig U-Pb ages; and the presence of an early Middle Ordovician collisional orogen, comparable in age and character to the Taconian. By contrast, the Franklinian mobile belt proper has a crystalline basement of Archean – Early Proterozoic age and was not deformed in the Ordovician. The original position of Pearya within the northern Caledonides remains to be clarified by further studies of successions II and III and systematic regional comparisons.

Late Silurian – Early Carboniferous deformational pulses and associated metamorphism, plutonism, and deposition
Deformation

The Late Silurian to Early Carboniferous deformational history can be summarized in terms of five major events, or groups of events (Chapters 11, 12).

(1) The Pearya Terrane appears to have been transported by sinistral strike slip as three or more slices. Intense deformation, including horizontal flexuring, occurred when the motion was arrested and the terrane was accreted to the Clements Markham Fold Belt. The latter must have been affected by this deformation and probably participated in the sinistral motion to some extent; if so, northwestern parts of the Hazen Fold Belt may also have been deformed at that time. These events are tentatively dated as Late Silurian – Early Devonian on the basis of rather indirect evidence.

(2) A well established Late Silurian – Early Devonian deformation in northern Axel Heiberg Island was characterized by southwest-northeast compression, possibly caused by southwestward motion of northern Ellesmere Island (Clements Markham Fold Belt and(?) northwestern Hazen Fold Belt). Alternatively, it may be due to the same stress field that caused the contemporaneous deformation of the Boothia Uplift (see below).

(3) Late Silurian – Early Devonian movements of the Boothia Uplift, previously attributed to mantle-controlled basement uplift, are now interpreted as thrust faulting. If so they are due to an east-west compressional regime that affected areas with northerly basement trends. This interpretation has also been applied to Early Devonian movements of the Inglefield Uplift.

(4) A Givetian(?)-Frasnian uplift of the northwestern part of the Hazen Fold Belt, inferred from syntectonic clastic sediments in the Clements Markham Fold Belt, may be due to renewed sinistral motion at the faulted boundary between the two belts. Such motion may also explain a second compressional deformation in northern Axel Heiberg Island that occurred between mid-Early Devonian (Pragian) and late Early Carboniferous (Viséan) times.

(5) A final event of Late Devonian – Early Carboniferous age, known as the Ellesmerian Orogeny, affected the entire mobile belt, from North Greenland to Prince Patrick Island (Chapters 11, 12), creating a highly sinuous assemblage of fold belts that is more than 375 km wide. Deformation was most intense in previously undeformed areas. Thin skinned deformation is inferred for the shelf province and adjacent parts of the deep water basin (North Greenland, Central Ellesmere, and Parry Islands fold belts; southeastern and central parts of Hazen Fold Belt), and basement involvement for northern Ellesmere Island (Pearya, Clements Markham Fold Belt, northwestern Hazen Fold Belt[?]) and southeastern Ellesmere Island (Jones Sound Fold Belt).

This event probably was caused by convergence of North America with another plate in the present offshore area that has not yet been identified with assurance.

Regional metamorphism and granitic plutonism

Regional metamorphism of greenschist and amphibolite grade is restricted to relatively small areas in North Greenland and Ellesmere Island (Chapters 11B, 12B). It clearly is related to the Late Silurian – Early Carboniferous orogenic events but has not yet been dated precisely.

Granitic plutons occur in northwestern Ellesmere and northern Axel Heiberg islands (Chapter 12B). Zircon analysis indicates that the largest of these bodies was derived from the Middle Proterozoic basement of Pearya and emplaced in about late Early Devonian time (390 ± 10 Ma). It is interpreted as a post-tectonic intrusion related to the Late Silurian – Early Devonian accretion of Pearya.

Syntectonic basin development and deposition

The Late Silurian – Early Devonian phases of deformation produced coarse clastic sediments of alluvial to shallow marine origin (molasse) as well as sediment gravity flows (flysch) in adjacent basinal settings (Chapter 8). About 3-4 km of molasse are preserved on northern Axel Heiberg Island, more than 2 km on the northwestern flank of the Inglefield Uplift in central Ellesmere Island, and 600-750 m on the flanks of the Boothia Uplift.

By early Middle Devonian time, a foreland basin had been created in the northeastern part of the mobile belt that was fringed by uplands on the northwest, northeast and southeast. Out of this region, an enormous clastic wedge prograded to the southwest in Middle and Late Devonian time that filled the remnants of the deep water basin to shelf level and transformed the shelf into an alluvial plain (Chapter 10). The preserved maximum thickness of the wedge is about 5 km, the inferred original thickness (including eroded Early Carboniferous strata) in the order of 10 km.

A minor nonmarine basin, separated from the main foreland basin by an uplift in the northwestern part of the Hazen Fold Belt, is represented by perhaps 1 km of Givetian(?) and Frasnian strata in the Clements Markham Fold Belt.

Carboniferous to Early Cretaceous development of the Sverdrup Basin

The Sverdrup Basin (Chapters 13, 14) originated during a rifting event, inferred from listric faults and related coarse clastic sediments, in late Early Carboniferous (late Viséan) to Early Permian time. Rift volcanism, in part alkaline, was restricted to parts of northwestern Ellesmere and northern Axel Heiberg islands. The extensional origin of the basin is also apparent from the fact that the crust is thinner in the axial region than at the margins (37-41 km versus 48 km; Chapter 5B). The cause of the rifting is uncertain, but on northwestern Melville Island there is evidence for dextral transcurrent motion in Late Carboniferous time. Permian to Early Cretaceous subsidence is attributed to lithospheric cooling (Chapter 14).

The basin is about 1300 km long, up to 400 km wide (after compressive deformation), and bounded on the northwest by a horst, (Sverdrup Rim), that subsided relatively slowly. Up to 3 km of upper Paleozoic strata and up to 9 km of Mesozoic strata were deposited in the axial region. Following the initial rifting, and, concomitantly with later stages of rifting, concentric facies patterns evolved. A peripheral clastic belt enveloped a carbonate belt, which, in turn, surrounded an axial region in which subsidence exceeded deposition so that deep water sediments (mudrock, evaporite, chert) prevailed. Permian marine sedimentation was terminated by a eustatic fall in sea level that resulted in an extensive disconformity on the shelves. The effects of this regression in the axial region of the basin are controversial because of a lack of diagnostic fossils from the top of the Permian succession in this area. The absence of fossils of latest Permian age, here as in most other parts of the world, and the physical stratigraphy established by Thorsteinsson suggest a disconformity (Chapter 13); the seemingly gradational character of the contact and the presence of turbidites in the lowest Triassic strata suggest continuous sedimentation in a deep water setting (Chapter 14). In order to resolve this problem, intensive efforts are being made to obtain conodonts from the critical strata.

Mesozoic sedimentation differed from late Paleozoic sedimentation by lack of evaporites and carbonates, except for some impure limestones of Middle Triassic age. This change reflects a drift from equatorial to intermediate latitudes — an inference based on paleomagnetic data.

The Mesozoic clastic sediments were derived mainly from southerly and southeasterly sources, which included Paleozoic sedimentary rocks at the margin of the basin, the Canadian Shield, and uplands on the Atlantic margin; only a small proportion of the sediments was derived from northwesterly metamorphic and sedimentary sources. The succession consists of alternating sandy and argillaceous units that record numerous transgressions and regressions, commonly terminated by disconformities at the basin margin. The intermittent rise of Carboniferous evaporite diapirs, from Early Triassic time onward, produced structural highs within the basin that are overlain by relatively thin successions with local disconformities.

Cretaceous-Cenozoic development of the present continental margin – effects in the Arctic Islands and North Greenland

The present architecture of the Arctic Ocean Basin (Fig. 20.1) developed in Cretaceous-Cenozoic time and has affected the North American continental margin in different ways.

Banks Island to Meighen Island

The history of the southwestern sector of the Arctic Islands continental margin has been profoundly affected by the development of the Amerasian Basin, which includes Canada Basin, the Alpha and Medeleyev ridges, and Makarov Basin (Fig. 21.1). Seismic data indicate that Canada Basin is floored by oceanic crust, but in the absence of drill cores and well defined magnetic anomalies, inferences about age and origin of this region, based largely on the geology of the surrounding lands, are controversial. Suffice it to state that the oldest mobilistic hypothesis (Carey, 1958), in its present version (e.g. Jackson et al., 1986), envisages counterclockwise rotation of an "Arctic Alaska Plate" (comprising northern Alaska and parts of northeastern Siberia with adjacents shelves) about a pivot west of the Mackenzie Delta. Although this hypothesis is widely accepted, its implications for the pre-Tertiary history of the Arctic Islands, Alaska, and Siberia have not yet been adequately assessed. A rival hypothesis (e.g. Smith, 1987; Hubbard, 1987) proposes that Arctic Alaska attained its present position by sinistral transcurrent motion along the present margin of the Arctic Islands. However, this reconstruction is unacceptable from a stratigraphic point of view. The Carboniferous succession of Arctic Alaska, for example, is closely related to that of the northeastern Cordillera (Richards et al., in press) but different from that of the Arctic Islands.

Regardless of the mode of formation of Canada Basin, stratigraphic observations in the western Sverdrup Basin indicate Middle Jurassic to Early Cretaceous uplifts near the present continental margin, followed by Late Cretaceous subsidence (Chapter 14). The uplifts can be attributed to rifting and the subsidence to lithospheric cooling that coincided with seafloor spreading in Canada Basin. The newly formed passive margin is overlain with unconformity by a landward thinning and younging prism of clastic sediments — the Arctic Continental Terrace Wedge (Chapter 15). From these relationships, Embry (Chapter 14, Fig. 14.60) infers an early rift phase from early Middle Jurassic (Bajocian) to earliest Cretaceous (Berriasian); a main rift phase from Early Cretaceous (Berriasian) to early Late Cretaceous (Cenomanian); and a seafloor spreading phase during the Late Cretaceous (late Cenomanian to Maastrichtian[?]). On the other hand, if the absence of distinct magnetic anomalies indicates seafloor spreading during the extended interval of normal polarity before anomaly 33 (Sweeney, 1985), then this process occurred in late Early Cretaceous (Aptian) to early Late Cretaceous (Santonian) time.

Axel Heiberg and northwestern Ellesmere islands

This area, which lies adjacent to the Alpha Ridge, lacks the Arctic Continental Terrace Wedge, but contains three suites of Cretaceous igneous rocks: upper Lower Cretaceous tholeiitic volcanics; a major gabbro pluton with a granitic component of early Late Cretaceous (92 ± 1 Ma) age; and Late Cretaceous, bimodal, partly alkaline volcanics. The volcanic and intrusive activity in Axel Heiberg and Ellesmere islands may be related to the development of the Alpha Ridge, which has the characteristics of an oceanic plateau and has recently been interpreted as the track of a hot spot (Jackson et al., 1986).

Northeastern Ellesmere Island

Volcanics and intrusions are present in northeastern Ellesmere Island but are less abundant than in the northwestern part of the island. This area lies adjacent to the Lomonosov Ridge, generally interpreted as a continental fragment split off the Barents Shelf by Tertiary seafloor spreading (Wilson, 1963; Forsyth and Mair, 1984). Most

SUMMARY AND REMAINING PROBLEMS

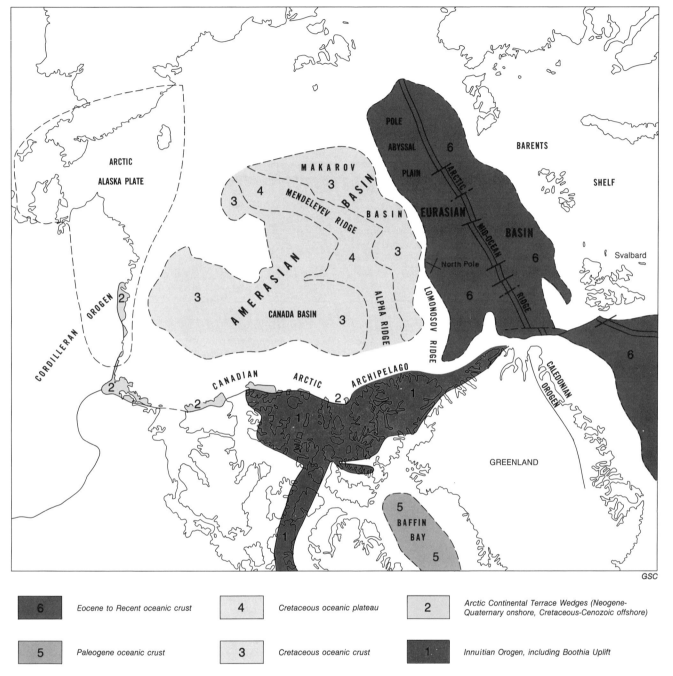

Figure 21.1. Cretaceous – Tertiary elements of Arctic Ocean region; geology and topography from Churkin and Johnson (1983), names from Perry and Fleming (1986).

plate tectonic analyses conclude or assume that the Lomonosov Ridge remained attached to Ellesmere Island throughout the Cretaceous and Cenozoic.

North Greenland

North Greenland lies adjacent to the Pole Abyssal Plain, produced by spreading on the Arctic mid-Ocean Ridge. Emplacement of a dense swarm of north-trending alkalic dykes in the Late Cretaceous was followed by extrusion of about 3 km of peralkaline volcanics in latest Cretaceous (Campanian-Maastrichtian) and earliest(?) Tertiary (Paleocene) time.

Cretaceous-Paleogene Eurekan deformation

Compressive deformation, known as the Eurekan Orogeny, occurred in the northeastern Queen Elizabeth Islands and in North Greenland in Paleogene (probably Paleocene or Eocene to earliest Oligocene) time (Chapters 16, 17). It was

551

characterized by thrust faulting and associated folding and minor strike-slip faulting along diverse trends that are parallel with Paleozoic and older trends in any given area.

Poorly dated normal faulting in the environs of Baffin Bay, Jones Sound, Lancaster Sound, and Foxe Basin (included in the Eurekan Deformation together with the Eurekan Orogeny) appears to be related to the development of the Labrador Basin (Labrador Sea, Davis Strait, and Baffin Bay). There, Early Cretaceous rifting was followed by Late Cretaceous to earliest Oligocene spreading, inferred from magnetic anomalies and the stratigraphic record of the adjacent shelves.

Tectonic analysis of the magnetic anomalies in the Labrador Sea suggests counterclockwise rotation and sinistral displacement of Greenland relative to the Arctic Islands. Treatment of both regions as separate rigid plates inevitably suggested that Greenland and Ellesmere Island were initially separated by an ocean that was subducted during the rotation. The fact that there is no geological evidence for this scenario led to the alternative hypothesis that spreading, rotation and deformation all occurred within a single, nonrigid plate. This model has several important implications.

(1) The most intensive deformation occurred in the least rigid part of the plate, i.e. that region where the Precambrian crystalline basement is thin or absent.

(2) The compressional deformation was not thin skinned although detachment surfaces occur at shallow levels, for example within a Carboniferous evaporite. Reverse motion on listric normal faults, formed during the initiation of the Sverdrup Basin, probably played a major role.

(3) Deviation of trends from the expected fan pattern must be due to the influence of the inherited fabric. Obliquity of the rotational stress with respect to the latter may explain the dextral strike-slip faulting inferred for parts of Ellesmere Island.

(4) Absolute shortening, in a circular sense, should be proportional to distance from the pivotal area.

Middle Tertiary to Recent tectonic and physiographic developments

The middle and late Tertiary sedimentary record of the region is mostly restricted to Miocene(?) and Pliocene deposits on the Arctic Islands and Wandel Sea coastal plains (Chapter 15); the scarcity of preserved sediments in other parts reflects widespread uplift and erosion (Chapter 18) that was most pronounced in parts of the Eurekan Orogen and around Baffin Bay.

Differential uplift in the northeastern Queen Elizabeth Islands and North Greenland can be interpreted as isostatic response to Eurekan crustal thickening combined with post-Eurekan erosion. The fact that mountains and negative Bouguer anomalies still are present indicates that the crustal roots of the orogen have not yet been eliminated by the interplay of uplift and erosion. Analogy with other mountain belts suggests that the isostatic uplift was intermittent and rapid, but little information is available on related relict erosion surfaces. These cycles of uplift and erosion presumably were of large amplitude (in the order of kilometres) and long duration (millions or tens of millions of years). By comparison, superimposed Quaternary oscillations of the land surface, caused by elastic response to glacial loading and unloading, were of low amplitude (hundreds of metres) and short duration (<1 Ma).

The mid-Tertiary rise of Precambrian crystalline terranes around Baffin Bay is tentatively interpreted as a delayed shoulder uplift, caused by thermal perturbations in the mantle after the rifting. These positive movements must have been accompanied by renewed normal faulting in Jones and Lancaster sounds, which remained submerged.

The present fiords and seaways probably originated as a Tertiary drainage system that followed faults, fractures, and weakly resistant rock types exposed in folds and grabens. The system was deepened and straightened by extensive, presumably warm-based, ice sheets during the earlier part of the Pleistocene. Later glacial episodes were restricted to present land areas and cold-based, i.e. incapable of deep erosion.

RESOURCES

The Phanerozoic strata of the Arctic Islands contain significant proportions of Canada's resources of petroleum, coal, and economic minerals. North Greenland has potential mainly for economic minerals.

Petroleum. The petroleum potential of the Arctic Islands is known from 176 exploration wells (drilled from 1962 to 1987), and extensive seismic and geochemical investigations (Chapter 20A). To date, 18 hydrocarbon fields have been discovered. One small oil field is contained in Devonian (Emsian-Eifelian) strata of Cameron Island, at the top of the extensive, long-lived Bent Horn – Towson Point carbonate buildup. The oil is localized by vuggy porosity, developed by vadose processes and fractures. The remaining 17 fields occur in the western part of the Sverdrup Basin, in sandstones of Late Triassic to Early Cretaceous, predominantly Jurassic, age. All are in anticlines, most of which are salt-cored. The petroleum was generated from Middle and Upper Triassic shales in Late Cretaceous – early Tertiary time.

Apart from these discoveries, there are various prospects in the Cambrian to Devonian succession of the Arctic Platform and Franklinian Shelf, and in the Carboniferous to Cretaceous succession of the Sverdrup Basin. Tertiary prospects are limited to Lancaster Sound and the Arctic Islands Shelf. In 1984, the total of discovered and untapped, recoverable resources in the Arctic Islands was estimated at 686 $10^6 m^3$ (4.3 B bbl) of oil and 2257 x $10^9 m^3$ (79.7 TCF) of gas.

Stratigraphic-structural surface investigations and geochemical studies indicate a rather limited petroleum potential for North Greenland (Chapter 20B).

Coal. The most important coal resources are in the Maastrichtian-Paleogene Eureka Sound Group of eastern Axel Heiberg and west-central Ellesmere islands, with lesser occurrences in Cretaceous and Tertiary strata on Banks and Baffin islands (Chapter 20C). Surface reconnaissance indicates about 25 x 10^9 tonnes of lignite, 20 x 10^9 tonnes of subbituminous coal, and 5 x 10^9 tonnes of high-volatile bituminous coal.

Economic minerals. Zinc and lead deposits have been been mined since 1976 at Nanisivik, northern Baffin Island, and since 1982 at Polaris, Little Cornwallis Island — the most important mineral deposit in the entire region discussed in this volume (Chapter 20D).

The Polaris mine, like other deposits of the Cornwallis Lead-Zinc District, is of Mississippi Valley-type and hosted by an Ordovician shelf carbonate unit that was subjected to karsting during Early Devonian movements of the Boothia Uplift. The pre-production ore reserves amounted to 23 million tonnes of 4.3% lead (galena) and 14.1% zinc (sphalerite).

The Nanisivik deposit occurs in karst formed in a dolostone unit of the Proterozoic Borden Basin (Chapter 6). Past production and current proven and probable ore reserves add up to 10 million tonnes of ore containining more than 10% zinc (sphalerite), 1.5% lead (galena), and 40 g/t silver (in sphalerite).

Other discoveries in the Arctic Islands include: additional zinc-lead (silver) showings in the northern Boothia Uplift, especially on Little Cornwallis Island and adjacent Truro Island; additional lead-zinc showings in the Borden Basin; lead-zinc showings in lower Paleozoic carbonates of the Central Ellesmere Fold Belt; native copper in Upper Proterozoic basalt on Victoria Island; and copper showings in the Franklinian deep water basin and Pearya Terrane.

Geochemical reconnaissance work has shown that the lower Paleozoic sediments of northern Greenland have a potential for barium and zinc deposits, and this is supported by some mineral occurrences (Chapter 20E). Minor showings of copper have been discovered in Upper Proterozoic volcanic and sedimentary strata of eastern North Greenland.

Author's address

H.P. Trettin
Geological Survey of Canada
3303-33rd Street N.W.
Calgary, Alberta
T2L 2A7

REFERENCES

Carey, S.W.
1958: A tectonic approach to continental drift; in Continental Drift. A symposium, S.W. Carey (convener); Geology Department, University of Tasmania, Hobart, Australia, p. 177-355.

Cecile, M.P. and Norford, B.S.
in press: Ordovician and Silurian; in Sedimentary Cover of the North American Craton in Canada, D.F. Stott and J.D. Aitken (ed.); Geological Survey of Canada, Geology of Canada, no.5 (also Geological Society of America, The Geology of North America, v. D-1).

Forsyth, D.A. and Mair, J.A.
1984: Crustal structure of the Lomonosov Ridge and the Fram and Makarov basins near the North Pole; Journal of Geophysical Research, v. 89, p. 473-481.

Hubbard, R.J., Edrich, S.P., and Rattey, R.P.
1987: Geologic evolution and hydrocarbon habitat of the "Arctic Alaska Microplate"; Marine and Petroleum Geology, v. 4, p. 2-34.

Jackson, H.R., Forsyth, D.A., and Johnson, G.L.
1986: Oceanic affinities of the Alpha Ridge, Arctic Ocean; Marine Geology, v. 73, p. 237-261.

Richards, B.C., Bamber, E.W., Higgins, A.C., and Utting, J.
in press: Carboniferous; in Sedimentary Cover of the Craton in Canada, D.F. Stott and J.D. Aitken (ed.); Geological Survey of Canada, Geology of Canada, no. 5 (also Geological Society of America, The Geology of North America, v. D-1).

Smith, D.G.
1987: Late Paleozoic to Cenozoic reconstructions of the Arctic; in Alaskan North Slope Geology, I. Tailleur and P. Weimer (ed.); The Pacific Section, Society of Economic Paleontologists and Mineralogists, and The Alaska Geological Society, v. 2, p. 785-795.

Sweeney, J.F.
1985: Comments about the age of the Canada Basin; in Geophysics of the Polar Regions, E.S. Husebye, G.L. Johnson, and Y. Kristoffersen (ed.); Tectonophysics, v. 114, p. 1-10.

Wilson, J.T.
1963: Hypothesis of Earth's behaviour; Nature, v. 198, p. 925-929.

Printed in Canada

INDEX

Adams, P.J.
  Historical geological account 20
Age determination
  Boothia Peninsula 103
  Cape Columbia Belt, Pearya 243
  Cape Richards Plutonic Complex,
  Pearya 244,249
  Cape Woods Pluton, Pearya 244,298,301
  Coppermine River basalt, Victoria Island 106
  Deuchars Glacier Belt, Pearya 244
  Devon Island 106
  Granodiorite, Marvin Peninsula,
  Ellesmere Island 426
  Grant Land Formation, Axel Heiberg Island .. 301
  Independence Fiord Group
  -northeast Greenland 59,106
  Kap Washington Group, North Greenland 463
  M'Clintock West body, Pearya 247
  Markhan Fiord pluton, Pearya 249
  Melville Peninsula 103
  Milne Fiord assemblage, Pearya 244
  Minto Arch -granodiorite, Victoria Island 103
  Mount Rawlinson Formation,
  Ellesmere Island 225
  Natkusiak Formation 107
  Ophiolites -Newfoundland and Scotland 257
  Zig-Zag Dal Basalt Formation, Greenland 106
Alaskan type ultramafics
  Thores Suite, Pearya 247
Allaaart, J.H.
  Historical geological account 21
Amerasian Basin 377,378,428
Amund Ringes Island
  Eureka Sound Group -Kanguk Formation 470
Amundsen Basin
  Rae Group 106
Amundsen Land
  Structure -Harder Fjord Fault Zone 463
Amundsen, R.
  Explorer -history 14
Anderson, R.M.
  Historical geological account 16
Arctic Archipelago
  Quaternary deposits, stratigraphy
  and chronology 505-506
Arctic Archipelago and northern Greenland
  Geological provinces -map 60,63
Arctic Canada
  Distribution and zonation of
  Paleogene palynomorphs -table 438
Arctic Coastal Plain 440
Arctic Continental Terrace Wedge 61,65
Arctic Islands
  Age -Cambrian to Early Devonian
  Basin development 165-238

Age -Correlation -Beaufort-
  Mackenzie Basin 440
Age -Late Cretaceous -Early
  Tertiary deformation -Arctic Islands 469-487
Age -Late Cretaceous and Tertiary basin
  development and sedimentation 437-456
Age -Middle and Late Tertiary tectonic
  developments 493-496
Age -Middle-Upper Devonian
  clastic wedge 263-274
  Beverley Inlet Formation 264
  Bird Fiord Formation 264,266
  Blackley Formation 205,264
  Blue Fiord Formation 205,264,266
  Cape de Bray Formation 205,264,266
  Fram Formation 267
  Hecla Bay Formation 264,266,267
  Hell Gate Formation 269
  Nordstrand Point Formation 269
  Okse Bay Formation 269
  Parry Islands Formation 264,270,272,273
Quaternary deposits
  -Arctic Archipelago 499-512
  Resources -coal 529-532
  Strathcona Fiord Formation 264,266
  Stratigraphic framework -clastic wedge 263
  Tectonics -summary of external factors
  -clastic wedge 273,274
  Weatherall Formation 264,266
Arctic Islands -Sverdrup Basin
  Age -Mesozoic history of Arctic Islands ... 370-430
  Age -Mesozoic stratigraphy 423-429
Arctic Platform
  Axel Heiberg -Ellesmere Region 46
  Banks -Victoria Region 44
  Coastal plains and continental shelves 50
  Foxe Plain (outlier) 44
  Geophysical characteristics -gravity 69-84
  Inliers -Shaler Upland 35
  Kane Basin -Independence Fjord Region ... 38,40
  Lancaster Region 40,41
  Magmatism -alkaline 483
  Minto Arch 35,44,59,84,106
  Plate tectonic reconstructions
  -magnetics 483-487
  Polaris Highlands 38
  Shaler Upland 35
  Structure -normal faults 479,480,482
  Structure -transcurrent faults
  -Ellesmere Island 483
  Tectonic framework 60,61
Arctic Islands
  Franklinian Shelf and Arctic Platform 165
Asudeh, I.
  Crustal structure -seismic and gravity
  -Arctic Platform 71-78

Austin, H.
  Historical geological account 10
Axel Heiberg -Ellesmere Region
  Axel Heiberg Island 46
  Eureka Upland 46
  Grantland Mountains 46
  Structure 46,315
  Victoria and Albert Mountains 46
Axel Heiberg Island
  Arctic Islands -resources -coal -distribution ... 529
  Awingak Formation 406,408
  Blind Fiord Formation 378
  Buchanan Lake Formation 440
  Bukken Fiord beds 231
  Deer Bay Formation 406,408
  Eetookashoo Bay beds 231,300,315
  Emma Fiord Formation 315
  Eureka Sound Group -lithofacies
    (Late Cretaceous -Tertiary) 440-448
  Eureka Sound Group
    -Margaret Formation 442,445,452
  Granitic plutonism 301
  Mackenzie King Formation 406
  Metamorphism -regional 300
  Northern Heiberg Fold Belt 222,295,298,300
  Paleogeographic evolution -Middle Eocene to
    Early Oligocene (?) 454
  Palynostratigraphy - lithostratigraphic
    framework
    -Eureka Sound Group 439
  Princess Margaret Arch 472
  Ringnes Formation 406
  Stallworthy Formation 232,300,301,315
  Svartevaeg Formation 231,232,315
  Sverdrup Basin -Carboniferous
    and Permian 345-366
  Tectonic interpretation 300
  Van Hauen Formation 378
Axel Heiberg Islands
  Deformation -Middle Devonian to
    early Carboniferous 300,309-317
Bache Peninsula 170,172
Baffin Island
  Arctic Islands -resources -coal
    -distribution 530
  Baillarge Formation 180,182
  Bay Fiord Formation180
  Cape Crauford Formation 183
  Deposits -minerals -lead-zinc
    -Borden Basin -Nanisivik deposit 536
  Gallery Formation 175
  Resources -minerals -lead-zinc
    -Borden Basin 536
  Ship Point Formation 177,179
  Structure -Foxe-Baffin depression 480
Baffin Region
  Baffin Coastal Lowland, Borden Upland,
    Eastern Coastal Uplift, Penny Surface 32,34
Baffin, W.
  Explorer -history 7
Banks -Victoria Region 44

Banks Basin
  Arctic Islands -resources -coal -distribution ...530
  Awingak Formation 406,408
  Christopher Formation 422
  Deer Bay Formation 406,408
  Development, evolution 442,451
  Eureka Sound Formation
    -Shale Member 442
  Eureka Sound Group -lithofacies
    (Late Cretaceous -Tertiary) 439-448
  Hassel Formation 422,426
  Kanguk Formation 422
  Mesozoic history of Arctic Islands 371-430
  Ringnes Formation 406
Banks Island
  Eureka Sound Group 447,448
  Late Devonian -Early Carboniferous
    deformation 321-333
  Weatherall Formation
    -Mercy Bay Member 270
Basin
  Amerasian basin 376,377,428
  Banks Basin -Late Cretaceous and
    Tertiary development 442,451
  Banks Basin -paleogeographic evolution 451
  Canada Basin 336
  Lake Hazen Basin -Eureka Sound Group
    -Mokka Fiord Formation 442,445,446
  Remus Basin -Late Cretaceous and
    Tertiary development 442,451,452
  Strand Fiord Basin -Late Cretaceous and
    Tertiary development 442,451
  West Sverdrup Basin -Late Cretaceous and
    Tertiary development 451
Basin Development
  Cambrian to Early Devonian
    -Arctic Islands 165-238
  Cambrian to Silurian Basin
    -North Greenland 111-164
Bathurst Island
  Arctic Islands -resources
    -coal -distribution 529
  Arctic Islands -resources -minerals
    -Cornwallis lead-zinc 533
  Bathurst Island Formation 187
  Bay Fiord Formation 180
  Beverley Inlet Formation 270
  Canrobert Hills Fold Belt 321,322,332
  Disappointment Bay Formation 185
  Lower Devonian -siliciclastic sediments 222
  Parry Islands Fold Belt 321-333
  Stewart Bay Formation 187
Beck, G.
  Explorer -history 9
Bell, R.
  Historical geological account 11
Bentham, R.
  Historical geological account 17
Bessels, E.
  Historical geological account 10

Biostratigraphy
  *see also Franklinian Basin -biostratigraphic
  -time rock slices 179
Blackadar, R.G.
  Historical geological account 20,21
Boothia Peninsula
  Turner Cliffs Formation 175
Boothia Uplift 35,62,64,70,166,182,300,302,337
  Analogy -Wyoming Province 304
  Aston Formation 302
  Barlow Inlet Formation 303
  Caledonian Orogeny 306
  Cape Storm Formation 302,303
  Eastern flank -syntectonic sediments 186
  Geophysics -crustal structure -seismicity 79
  Geophysics -isostasy 70
  Hunting Formation 302
  Late Silurian -Early Devonian
  deformation 302-307
  Parry Upland 50,62,64,70
  Sophia Lake Formation 185
  Structure 482
  Structure -deformation -early history 302
  Tectonic models 304
  Tectonics 35,62,64,70,470,482,533
  Tectonostratigraphic elements map 303
  Upper Silurian -lower Devonian 186
  Western flank -syntectonic sediments 187
Boothia Uplift -north of Barrow Strait
  Franklinian Shelf -carbonates 187
Boothia Uplift -south of Barrow Strait
  Franklinian Shelf -carbonates 186
Borden Basin
  Paleomagnetics 106,302
Borup, G.
  Historical geological account 11
Bowers, A.F.
  Historical geophysical account 21
Brainard, D.L.
  Historical geological account 11,12
Brent, T.A.
  Late Devonian -Early Carboniferous deformation
  -Prince Patrick Island,Ba 335-336
Bronlund, J.
  Explorer -history 14
Bustin, R.M.
  Resources -coal Arctic Islands 529-532
Caledonian Mobile Belt
  Pearya -comparison Franklinian and
  Caledonian Mobile Belts 256,257,258
Cambrian -Lower Cambrian
  Paralleldal Formation 132
Cambrian -Upper Cambrian
  Londal Formation 133
  Perssuaq Gletscher Formation 133,136
Cambrian to Early Devonian
  Arctic Islands -correlation
  biostratigraphy 165-238
Cambrian to Middle Devonian
  Arctic Islands -sedimentary subprovince
  -depositional phases BI-BV 233

Cameron Island
  Resources -petroleum -Bent Horn Oil field 518
  Sverdrup Basin -Carboniferous
  and Permian 345-366
Canadian Arctic
  Paleoclimate and paleolatitude 455,456
Canadian Arctic Archipelago
  Distribution and isopach map 370
  Extent 16,29
  Seismicity 79-83
Canadian Shield
  Extent 59
Canadian Shield -northernmost part
  Lithology 103,106
  Metamorphism -granulite facies 103,106
  Metamorphism -plutonic basement 103-105
  Precambrian successions 103-110
  Proterozoic sedimentary and
  volcanic successions 106
  Proterozoic volcanic/sedimentary
  succession -map 105
  Victoria Island -Proterozoic successions 106
Canrobert Hills Fold Belt
  Blackley Formation 332
  Canrobert Formation 332
  Cape de Bray Formation 332
  Ibbett Bay Formation 332
  Parry Upland 50,64,321-336
  Structure 321-336
  Tectonics 332
  Weatherall 332
Central Ellesmere Fold Belt
  Structure 312,318,320
Challenger Basin
  Yelverton Thrust 298,299
Christiansen, F.
  Historical geological account 11
Christiansen, F.G.
  Resources -petroleum,
  North Greenland 525-529
Christie, R.L.
  Geomorphic regions 29-56
  Historical geological account 20,21
  Innuitian Region -geographic and
  geological exploration 7-25
Clements Markham Fold Belt
  *see also Ellesmere Island, Pearya
  *see also structure .. 64,205,222,224,225,229,241,
  .... 250,254,283,298, 299, 300, 309, 312
  Accretion -Pearya 337
  Structure 64,205,222,224,225
Coastal plains
  Arctic Islands -Coastal Plain 50
  Davis Shelf 53
  Lincoln -Morris Jesup Shelf 50
  Melville Bugt -Disko Shelf 53
  Wandel Sea Coastal Plain 50
Coles, R.L.
  Aeromagnetic field 84-88
Cook, F.A.
  Explorer -history 11

Cornwallis Island
  Arctic Islands -resources -minerals
  -Cornwallis lead-zinc 533
  Bay Fiord Formation 180
  Cape Phillips Formation 303
  Cornwall Arch........................306
  Disappointment Bay Formation 185
  Eleanor River Formation 180
  Prince Alfred Formation 187
  Resources -minerals -lead-zinc 533,534
  Snowblind Bay Formation 187
  Sophia Lake Formation 187
Correlation chart
  Arctic Islands, Greenland,
  Ellesmere Island ....Fig. 4 (pocket)
Cowie, J.W.
  Historical geological account 20
Crozier, F.R.M.
  Explorer -history 9
Crust
  Mohorovicic discontinuity 71
Davies, G.R.
  Carboniferous and Permian history
  Sverdrup Basin 345-366
Davies, W.E.
  Historical geological account 20
Davis, J.
  Explorer -history 7
Davison, W.L.
  Historical geological account 20
Dawes, P.R.
  Geomorphic regions 29-56
  Innuitian Region -geographic and
  geological exploration 7-25
De Rance, C.E.
  Historical geological account 12
Dease, P.
  Explorer -history 9
Deformation
  Arctic Islands -Late Cretaceous
  -Early Tertiary deformation 469-487
  Axel Heiberg Island -Middle Devonian to
  Early Carboniferous 300,309-317
  Bathurst Island -Late Devonian
  -Early Carboniferous deformation 321-333
  Boothia Uplift -Late Silurian
  -Early Devonian deformation 302-307
  Devonian -early Carboniferous
  North Greenland 283-290
  Ellesmere Island -age of accretion
  of Pearya 295
  Eurekan deformation -tectonic
  interpretation 483
  Hazen Fold Belt -Devonian
  -Early Carbonferous deformation 283-290
  North Greenland -Devonian
  -Early Carboniferous deformation 283-290
  North Greenland -Late Cretaceous
  -Early Tertiary deformation 461-464
  Prince Patrick Island, Banks Island
  -deformation 334-336
  Summary outline of Eurekan
  deformation 551,552
Devon Island
  Arctic Islands -resources -minerals
  -Cornwallis lead-zinc 533
  Bay Fiord Formation 180
  Biostratigraphy -time rock slices
  -Lower-Middle Cambrian P2b 171
  Blanley Bay Formation 177,179
  Disappointment Bay Formation 185
  Eleanor River Formation 180
  Haughton impact crater -strata 448
  Structure 479
  Sverdrup Basin -Carboniferous
  and Permian 345-366
Drever, H.
  Historical geological account 17
Eclipse Trough
  Arctic Islands -resources -coal -distribution ...530
  Eureka Sound Group -lithofacies
  (Late Cretaceous -Tertiary) 440-448
  Eureka Sound Group -Mokka
  Fiord Formation 442,445,446,453
  Late Cretaceous and
  Tertiary development 442
  Mudstone -comparable to Mount
  Lawson Formation 442
  Paleogeographic evolution
  -Late Paleocene 451,452,453
  Remus Formation 442
  structure 480
Egede, H.
  Explorer -history 8
Eglinton Island
  Kanguk Formation -Eglinton Member ...422
Ekblaw, W.E.
  Historical geological account 15
Ellesmere Group
  *see also Franklinian Basin 169,170,209
Ellesmere Island
  *see also Clements Markham Fold Belt
  Arctic Islands -resources
  -coal -distribution 529
  Baumann Fiord Formation 130,179,318,476
  Bay Fiord Formation 180,318
  Beaufort Formation 440
  Biostratigraphy -time rock slices
  -Lower-Middle Cambrian P2b 172
  Blue Fiord Formation 185,264
  Boulder Hills Formation 440
  Buildups -isolated 189
  Caledonian Bay Conglomerate Member ...219
  Cape Phillips Formation 216,219
  Cape Wood Formation 173
  Central Ellesmere Fold Belt 318,320
  Christian Elv Formation 177
  Clements Markham
  Fold Belt 64,205,222,224,225,230,283,285
  Copes Bay Formation 177,178
  Danish River Formation 152,220,225,226
  Deformation -age of accretion of Pearya 295

Dyke swarm 479	Ellitsgaard-Rasmussen, K.
Disappointment Bay Formation 186	Historical geological account 19
Eids Formation 219,318	Embry, A.F.
Eleanor River Formation 131,180	Mesozoic history of Arctic Islands 371-430
Ella Bay Formation 120,205	Middle-Upper Devonian clastic wedge,
Ellesmere Group...................... 312,315	Arctic Islands 263-274
Eureka Sound Group 439-448,450,451,452	Resources -petroleum -Arctic Islands 517-525
Fairman Point Formation 174	Erichsen, M.
Fram Formation 267	Explorer -history 14
Grant Land Formation	Eureka Sound Group
-Member A,B,C 205,209,224,231,477,478	*see also Ellesmere Island 439-448
Hansen Point Formation 426,428	Ellesmere Island -geological -map 443
Hazen Fold Belt 309,312,315	Eurekan Orogeny
Hazen Formation 152,224,285,289,295,309	Domain -Axel Heiberg Island (Sverdrup
Hell Gate Formation 269	Islands) -structure -evaporites 476
Imina Formation 226,254,300	Domain -Central Ellesmere Domain
Jones Sound Fold Belt 320	-structure 470,472,476
Kennedy Channel Formation 115	Domain -Hazen Plateau Stable Block 470
Kitson Formation 216	Domain -Northern Ellesmere
Lands Lokk Formation 226,297,477,478	Domain -structure 470,472-475,476
Late Pleistocene glaciation	Domain -Prince Patrick Domain
-Hazen Moraines 508,510,512	-structure 470
Late Silurian -Early Devonian	Domain -Sverdrup Island Domain
deformation, metamorphism, plutonism 295	-structure 470
Lower Cambrian carbonates B1	Eurekan deformation -tectonic
-Nesmith beds 205	interpretation 483,484
Lower Cambrian siliclastic sediments	Middle and Late Tertiary tectonic
-Grant Land Formation 205	developments -uplift 493
Lower Cambrian to Lower Silurian	Setting of orogen, trends and major
sediments -Hazen Formation 209,210,212	domains -Arctic Islands 470
Middle Devonian to early Carboniferous	Structure -normal faults -Axel Heiberg
deformations 309-317	Island and Ellesmere Island 478,479
Mount Rawlinson Formation 225	Structure -transcurrent faults 477,478
Nordstrand Point Formation 269	Expeditions
Nygaard Bay Formation 130	Bicentenary Jubilee Expedition 16
Okse Bay Formation 269,298,309,312	British Joint Services Expedition 21
Otto Fiord Formation 476,477	Canadian Arctic 16
Palynostratigraphy - lithostratigraphic	Crocker Land 15,16
classification 438,439	Danish Geodetic Institute 19
Parrish Glacier Formation 173,174	Danish Peary Land Expedition 19
Quaternary deposits -glacial deposits 499,500	Danish Thule 17
Rabbit Point Formation 170	Eastern Arctic Patrol 17
Read Bay Formation 182	Ellesmere Land Expedition 17
Resources -minerals -sulphides, barite, iron,	Fifth Thule 17
kimberlites 537	First Thule 15
Scoresby Bay Formation 173	MacGregor Arctic Expedition 19
Stratigraphic section 167,168,169,171	Operation Admiralty 21
Sverdrup Basin -Carboniferous	Operation Amadjuak 21
and Permian 345-366	Operation Bathurst 21
Tellevak limestone, Blue Mountains 354	Operation Bylot.........................21
Upper Ordovician to Lower Devonian	Operation Ellef Ringes Island 21
-carbonates, mudrock, chert 216	Operation Eureka 21
Upper Ordovician to Lower Devonian	Operation Franklin......................20
-siliclastic sediments 216	Operation Grant Land 21
Ellesmere Island -Axel Heiberg Island	Operation Grinnel 21
Structural subdivision -map 297	Operation Peel Sound 21
Ellesmere Island -Franklinian Shelf	Operation Prince of Wales 21
Allen Bay Formation 181,182,220	Oxford University Ellesmere
Ellesmerian Orogeny	Land Expedition 17
North Greenland Fold Belt 283,288	Polar Continental Shelf Project 21,69
	Putnam Baffin Island Expedition 17

Second Thule 16
Third Thule 16
Feilden, H.W.
   Historical geological account 12
Forsyth, D.A.
   Canadian Arctic Archipelago -seismicity ... 79-83
   Crustal structure -seismic and gravity
   -Arctic Platform 71-78
Fortier, Y.O.
   Historical geological account 19
Fox, F.G.
   Late Devonian -Early Carboniferous
   deformation Bathurst Island, Melville ... 321-333
Foxe Basin
   Turner Cliffs Formation 175
Foxe Plain (outlier)
   Foxe Basin 44
   Great Plain of the Koukdjuak 44
Frankl, E.
   Historical geological account 20
Franklin, J.
   Explorer -history 9
Franklinian Basin
   Biostratigraphy 179
   Bronlund Fjord Group
   -Buen Formation 127,131,132,133
   Extent 111
   Harder Fjord Formation 145
   Henson Gletscher Formation 134
   Humboldt Formation 123
   Kap Lucie Marie Formation 153
   Kap Mjolner Formation 145
   Lafayette Bugt Formation 153
   Lauge Koch Land Formation
   -Profifjeldet Member 153,158
   Megabreccia unit 120
   Nordpasset Formation 145
   Peary Land Group 115,139,152,289
   Polkorridoren
   Group 120,122,123,125,126,127,137
   Ryder Gletscher Group 127,128,130,131,132
   Saeterdal Formation 134
   Skagen Group
   -Buen Formation 112,120,123,125,285,286
   Skagen Group
   -Paradisfjeld Group 112,115,120,122,123,125
   Skagen Group
   -Portfjeld Formation .112,115,119,120,122,125,127
   Structure 112,123
   Tavsens Iskappe Group . 128,131,132,133,134,174
   Volvedal Group 114,130,136,137,143
   Wulff Land Formation -Thors Fjord Member ..153
Franklinian Basin -evolution
   Stage 1: Late Proterozoic -Early
   Cambrian -basin initiation 115
   Stage 2: Early Cambrian stable platform
   -incipient trough 119
   Stage 3: Early Cambrian siliciclastic
   shelf -turbidit trough 123
   Stage 4: Late Early Cambrian -Middle
   Ordovician platform -starved basin 127

Stage 4: Regressive unstable shelf 131,132,133
Stage 4: Trough 137
Stage 5: Middle Ordovician and Early
Silurian stable platform 139
Stage 5: Middle Ordovician and Early
Silurian starved slope 143
Stage 5: Middle Ordovician and Early
Silurian trough 143
Stage 6: Early Silurian longitudinal trough ...149
Stage 6: Early Silurian ramp 145
Stage 7: final drowning of the platform 153
Stage 7: Late Llandovery -Early
Ludlow reef belt 153,156
Stage 7: trough expansion 157
Franklinian Deep Water Basin
   Age relationships of major rock units
   and carbonates -table 206
   Blackley Formation 205,264
   Cape de Bray Formation 205,264
   Hazen Trough 283
   Sedimentary subprovince 205
Franklinian Mobile Belt
   Danish River Formation 297
   Geophysics -crustal structure -geomagnetic
   anomalies 72
   Geosynclinal concepts 61
   Paleozoic sequences and depositional
   provinces 62
   Parry Islands Fold Belt 321-333
   Pearya -comparison Franklinian and
   Caledonian Mobile Belts 256,257,258,298
   Structure 1,62,256-258,297
Franklinian Shelf
   Barlow Inlet Formation 182,184
   Baumann Fiord Formation 174
   Biostratigraphic standards 165,166,169
   Biostratigraphy -time rock slices
   -Lower Cambrian 166
   Biostratigraphy -time rock slices
   -Lower-Middle Cambrian 166,169,171
   Biostratigraphy -time rock slices
   -Mid-Lower Ordovician P3 173,174
   Biostratigraphy -time rock slices
   -Middle and Upper Ordovician P5 181
   Biostratigraphy -time rock slices
   -Silurian and Lower Devonian P6 182
   Biostratigraphy -time rock slices
   -Sloss Vail 166
   Biostratigraphy -time rock slices
   -Upper Middle Cambrian 175
   Boothia Uplift -south of Barrow Strait
   - sediments 186
   Buildups -isolated 188,189
   Bulleys Lump Formation 174
   Cape Crauford Formation 183
   Cape Storm Formation 182,183
   Correlation -Lower-Middle Cambrian
   carbonate unit P2b 171,172,173
   Devon Island Formation 185
   Douro Formation 182,184
   Ella Bay Formation 205,209

Ellesmere Group 169,170,209
   Goose Fiord Formation .185
   Irene Bay Formation 181
   Main shelf - Upper Ordovician -Lower
   Middle Devonian -isolated buildups 188
   Prince Alfred Formation 185
   Sutherland River Formation 185
   Thumb Mountain Formation 181
   Vendom Fiord Formation 185
Frebold, H.
   Historical geological account 17
Freuchen Land
   Borglum River Formation 139
   Morris Bugt Group 139
   Sjaelland Fjelde Formation 131
   Tureso Formation 139
   Wandel Valley Formation 131,145
   Washington Land Group 139
   Ymers Gletscher Formation 139
Freuchen, P.
   Explorer -history 15
Frisch, T.
   Precambrian successions
   -Canadian Shield 103-110
Fristrup, B.
   Historical geological account 19
Frobisher, M.
   Explorer -history 7
Geochemistry
   Organic geochemistry
   -petroleum resources 518,520,521,522
Geomorphic regions
   Canadian Arctic Archipelago -map 30
Geomorphology
   Glacial deposit 506-508
   Glacial events -Latest Pleistocene
   and Holocene 510,512
   Glaciation 506,507,509
   Laurentide Ice Sheet 500,509,510
   Quaternary deposits
   -Arctic Archipelago 499-512
   Wisconsin Stage -Amundsen Glaciation 510
   Wisconsin Stage -Foxe (Eclipse)
Geophysics -crustal structure
   Aeromagnetic field 84-88
   Arctic Archipelago 71-94
   Arctic Continental Terrace Wedge 73
   Canadian Arctic Margin Basin 73
   Depth to mantle 72
   Franklinian Mobile Belt .72
   Geomagnetic anomalies 89-94
   North Magnetic Pole 96
   Paleomagnetics -Borden Basin 106
   Parry Islands Fold Belt .74
   Seismic refraction surveys 71,72,74
   Seismicity 79-83
Geophysics -gravity anomalies
   Arctic Platform 69-84
   Bouguer and free-air offshore 69
   Character of gravity field 70

Geophysics -isostasy
   Sverdrup Basin, Gustaf-Lougheed Arch,
   Boothia Uplift 70
Gibbins, W.A.
   Arctic Islands -resources -minerals 533-539
Giesecke, K.L.
   Explorer -history 8
Graben
   Eglinton Graben 331
Greeley, A.W.
   Explorer -history .10,11,12
Greenland
   *see North Greenland
Gregory, A.F.
   Historical geophysical account 21
Gustaf-Lougheed Arch
   Geophysics -isostasy 70
Hall, C.F.
   Explorer -history 10
Hansen, G.
   Historical paleontology .16
Harder Fiord Fault Zone
   Structure 123,145,283,286,461
Harley River Group
   *see also Pearya (northernmost
   Ellesmere Island) 251,252,253
Harrison, J.C.
   Late Devonian -Early Carboniferous
   deformation -Prince Patrick Island, B 334-336
   Late Devonian -Early Carboniferous
   deformation Bathurst Island, Melville . . . 321-333
Hasegawa, H.S.
   Seismicity -Arctic Platform 79-83
Hattersley-Smith, G.
   Historical geological account 20
Haughton, S.
   Historical geological account 10
Haycock, M.H.
   Historical geological account 17
Hayes, I.I.
   Explorer -history 10
Hazen Fold Belt
   *see also Ellesmere Island 309,312,315
   *see also structure 62,93,205
   Devonian -early Carboniferous
   deformation 283-290
   Geophysics -crustal structure
   -geomagnetic anomalies 93
   Structure 62,205,283,286,295
   Tectonics 93
Hazen Land
   Structure -Harder Fjord Fault Zone 461
Hearne, S.
   Explorer -history 8
Heat flow
   Sverdrup Basin 96
Heiberg Group
   *see also Sverdrup Basin 388,391,397
Helena Island
   Sverdrup Basin -Carboniferous
   and Permian 345-366

561

Henriksen, N.
   Historical geological account 21
Heywood, W.W.
   Historical geological account 20
Higgins, A.K.
   Cambrian to Silurian Basin sedimentation,
   North Greenland 111-164
   Devonian -early Carboniferous deformation/
   metamorphism -North Greenland 283-290
   Late Cretaceous -Early Tertiary deformation
   -North Greenland 461-464
Hobson, G.D.
   Historical geophysical account 21
Hodgson, D.A.
   Quaternary deposits -Arctic Archipelago . 499-512
Hoeg Hagen, N.P.
   Explorer -history 14
Hood, P.J.
   Historical geophysical account 21
Hot spot
   Labrador Basin 483
Hudson Platform
   Correlation -Arctic Islands 183
Hudson's Bay Company
   Explorer -history 8
Hudson, H.
   Explorer -history 7
Ineson, J.R.
   Cambrian to Silurian Basin sedimentation,
   North Greenland 111-164
Inglefield Land
   Cape Ingersoll Formation 130
   Cape Kent Formation 130
   Cape Leiper Formation 130
   Cape Wood Formation 130
   Dallas Bugt Formation . 123,125,130,166,169,182
   Disappointment Bay Formation 309
   Franklinian Basin development 111,124
   Permin Land Formation 130
   Rensselaer Bay Formation 308
   Thule Group 308
   Vendom Fiord Formation 186,308,309
   Wulff River Formation 130
Inglefield Uplift
   Lower Devonian red beds 187,188,308,309
   Silurian and lower Devonian
   -upper carbonate units 186
   Structure 186-188,308,309,337
   Tectonics 470,472
   Tectonics -early Devonian movements 308
Inland Ice
   Extent 29,32
Innuitian Orogen
   Summary and remaining problem -history
   Innuitian Orogen/Arctic Platform 547-553
Innuitian Region
   Aircraft age of geological study -history 18,19
   Early mariners and the Franklin
   search 1576-1859 7-10
   Geological exploration by ship and
   dog sledge 1903-1947 13-18

Mineral deposits -brief overview 22
Paleo-Eskimo habitation and Norsemen
2000 B.C.-A.D. 1500 7
Innuitian Tectonic Province
   Aeromagnetic field 84-88
   Cretaceous-Paleogene deposition 65
   Franklinian Mobile Belt 61
   Mountain structures 61
   Paleozoic sequences and
   depositional provinces 62
   Pearya Terrane 62
   Tertiary deformation 62
Iversen, I.
   Historical geological account 15
Jackson, G.D.
   Historical geological account 21
Jepsen, H.F.
   Historical geological account 21
Jessop, A.M.
   Heat flow 96
Johannes V. Jensen Land
   Dyke swarms -Upper Cretaceous 461
   Orthotectonic zone (Hazen Trough) 283,289
Jones Sound Fold Belt
   *see also structure 40
   Parry Upland 50,64,321-333
Judge Daly Basin
   Arctic Islands -resources -coal -distribution ...530
   Eureka Sound Group
   -Cape Back Formation 446,447
   Eureka Sound Group -lithofacies
   (Late Cretaceous -Tertiary) 447
Judge Daly Promontory
   Cass Fiord Formation 175
   Scoresby Bay Formation 173
   Ninnis Glacier beds 177
Kane Basin 38,40,41
Kane Basin -Independence Fjord Region
   Bache Peninsula -southern sub-region 38
   Freuchen Land 38,40
   Hall Land 38,40
   J.C. Christensen Land -eastern sub-region 40
   Kronsprins Christian Land
   -southern sub-region 39
   Nares Land 38,40
   Nyeboe Land -southern sub-region 38
   Peary Land 38,40
   Polaris Plain -northern sub-region 40
   Structure 40
   Vlademar Glikstadt Land
   -eastern sub-region 40
   Warming Land -southern sub-region 38
   Washington Land -southern sub-region 38
   Wulff Land -southern sub-region 38
Kane, E.K.
   Explorer -history 10
Kerr, J.W.
   Historical geological account 21
Kindle, E.M.
   Historical geological account 17

King Christian Basin
  Arctic Islands -resources -coal -distribution . . . 530
Kingigstorssuaq runic stone 7
Knuth, E.
  Historical archeologist 16,19
Koch, L.
  Explorer -history 16,17
Konig, C.
  Historical geological account 10
Krinsley, D.B.
  Historical geological account 20
Kronprins Christian Land
  Franklinian Basin development 111
  Mound Formation 156
  Sjaelland Fjelde Formation 131
  Wandel Valley Formation 131
Kurtz, R.D.
  Conductive anomalies -Arctic Platform 89-94
Labrador Basin
  Eurekan deformation -tectonic interpretation
    -geophysics 483
  Middle and Late Tertiary tectonic
    developments -extent, uplift 493,494
Lake Hazen Basin
  Arctic Islands -resources -coal -distribution . . . 530
  Development, evolution 442,446
  Eureka Sound Group
    -Boulder Hills Formation 446
  Eureka Sound Group -lithofacies (Late
    Cretaceous -Tertiary) 440-448
  Paleogeographic evolution 453,454
Lake Hazen Fault Zone
  Structure 46,454,470,472
Lake Hazen Zone
  Eurekan Orogeny 472-475
Lancaster Region 35,40,41
Lancaster Sound Basin
  Structure 479,480
Larsen, H.
  Explorer -history 17
Lithology
  M'Clintock plutonic bodies, Pearya 246,247
  Sverdrup Basin 347-360
Lockwood, J.B.
  Historical geological account 11
Long, G.D.F.
  Cambrian to Early Devonian Basin
    development -Arctic Islands 165-238
Low, A.P.
  Historical geological account 14
  Lowther and Young Islands
  Disappointment Bay Formation 185
M'Culloch, J.
  Historical paleontology 10
MacMillan, D.B.
  Historical geological account 11,15,16
Manning, T.H.
  Historical geological account 20
Mathiassen, T.
  Historical geological account 17

Mayr, U.
  Cambrian to Early Devonian Basin
    development -Arctic Islands 165-238
  Resources -petroleum -Arctic Islands 517-525
McMillan, J.G.
  Historical geological account 14
Meek, J.B.
  Historical paleontology 12
Meighen Island
  Arctic Islands -resources -coal -distribution . . . 531
  Eureka Sound Group -Beaufort Formation 448
Melville Island
  Arctic Islands -resources -coal -distribution . . . 529
  Bay Fiord Formation 180
  Beverley Inlet Formation 270
  Blue Hills Fault Belt 331
  Canyon Fiord Formation 345
  Devon Island Formation 216
  Eleanor River Formation 180
  Late Devonian -Early Carboniferous
    deformation 322-333
  Lower Ordovician to Lower
    Devonian sediments 216
  Sverdrup Basin -Carboniferous
    and Permian 345-366
  Trold Fiord Formation 345
Melville Upland
  Melville Arch (Horst) 35,59,61
  Melville Peninsula 35
Melvillian Disturbance
  Canyon Fiord Formation,
    Melville Island 345,362
  Sverdrup Basin . . . . . . , ... 345,362
Mesozoic history Arctic Islands
  Transgressive-regressive cycles 373
Metamorphism
  Devonian -early Carboniferous
    North Greenland 283-290
  North Greenland Fold Belt 288,289,290
Miall, A.D.
  Arctic Islands -resources -coal 529-531
  Late Cretaceous and Tertiary Basin
    development -Arctic Islands 437-456
Mikkselsen, E.
  Explorer -history 15
Minto Inlier
  *see also Arctic Platform -Minto Arch 31,59
Morley, L.W.
  Historical geophysical account 21
Muller, F.
  Historical geological account 20
Munck, E.
  Explorer -history 16
Nansen Land
  Orthotectonic zone (Hazen Trough) 283
  Paradisfjeld Group 286
  Polkorridoren Group -Frigg Fjord
    Formation 283,285,286,288,289
Nares Strait
  Eurekan Orogeny -Structure
    -transcurrent faults 477

Nares, G.S.
  Explorer -history 10
Nassichuk, W.W.
  Carboniferous and Permian history
  Sverdrup Basin 345-366
Newitt, L.R.
  North Magnetic Pole 96
Niblett, E.R.
  Conductive anomalies -Arctic Platform 89-94
Nichols, D.A.
  Historical paleontology 17
North Greenland
  *see also Franklinian Basin
  *see also Kane Basin -Independence
  Fjord Region 38-40
  Amundsen Land Group 224
  Caledonian Orogeny 159
  Cambrian to Silurian
  Basin development 111-164,169
  Campanuladal Formation 106
  Cape Clay Formation 177
  Cape Webster Formation 131
  Chester Bjerg Formation 159
  Christian Elv Formation 177
  Correlation -Arctic Islands -Lower-
  Middle Cambrian P2a 169
  Devonian -early Carboniferous deformation
  and metamorphism 283-290
  Dyke swarms -Upper Cretaceous 461
  Fins So Formation 106
  Franklinian Basin -evolution 116-117
  Franklinian Basin -geological map 112
  Franklinian Basin -stratigraphy map 113
  Johansen Land Fromation 130
  Kap Holbaek Formation 123,124
  Kap Washington Group 463
  Late Cretaceous -Early
  Tertiary deformation 461-464
  Lauge Koch Formation 228
  Merqujoq Formation 115,152,228
  Metamorphic zones -map 290
  Mound Formation 156
  Nordkronen Formation 228,297
  Nyeboe Land Formation 159
  Odins Fiord Formation 182
  Paradisfjeld Group -Portfjeld Formation 463
  Permin Land Formation 174
  Polkorridoren Group -Buen Formation . . . 209,463
  Portfjeld Formation 286
  Quaternary stratigraphy -Early and
  Middle Pleistocene 506
  Resources -minerals -lead-zinc-barite-
  copper mineralization 539,541
  Resources -petroleum 525-529
  Ryder Gletscher Group 174
  Samuelson Hoj Formation 182
  Steensby Gletscher Formation 131
  Sydgletscher Formation 152
  Tectonic zones and stratigraphy
  -map 284-285,462
  Tureso Formation 182
  Volvedal Group 224
  Wandel Valley Formation 131
  Wulff Land Formation -Thors Fjord Member . . 157
  Ymers Gletscher Formation 182
North Greenland Foreland Plateau
  Kane Basin -Independence Fjord Region 38
North Greenland Region
  Herluf Trolle upland -eastern
  upland sub-region 46
  Lincoln Sea 46
  Roosevelt Fjelde Formation
  -northern sub-region 46
  Wandel Sea -eastern upland sub-region 46
Northern Heiberg Fold Belt
  *see also Structure 222
  Aurland Fiord beds 231
  Jasper Lake assemblages 231
  Structure 222,295,298,300,315,470,472
Nyeboe Land
  Nordkronen Formation 124,158
O'Neill, J.J.
  Historical geological account 17
O'Reilly, B.
  Explorer -history 8
Okulitch, A.V.
  Central Ellesmere Fold Belt 319-320
  Devonian movements Inglefield Uplift 309
  Late Cretaceous -Early Tertiary deformation
  -Arctic Islands 469-487
  Late Devonian -Early Carboniferous deformation
  Bathurst Island, Melville 321-333
  Late Silurian -Early Devonian deformation
  -Boothia Uplift 302-307
Organic mounds and reefs
  *see also Sverdrup Basin 363-366
Orogeny
  *see also Ellesmerian, Eurekan, Innuitian
  Caledonian 111,159,306
  Ellesmerian 111,112,283,304,322,323,
  336,345,461,463
  Eurekan 300,315,347,374,439,440,448,449,451
  Grenville-Sveconorwegian Orogen 256
  Innuitian 29,38,44,61,440,483
  M'Clintock 241,249
  Taconian 257
  West Spitzbergen 464
Overton, A.
  Crustal structure -seismic and gravity
  -Arctic Platform 71-78
  Historical geophysical account 21
Packard, J.J.
  Cambrian to Early Devonian Basin
  development -Arctic Islands 165-238
  Late Silurian -Early Devonian deformation
  -Boothia Uplift 302-307
Paleozoic tectonic elements
  North America -map 59
Palynostratigraphy
  Campanian-Maastrichtian -Banks Island 437
  Campanian-Maastrichtian
  -Ellef Ringnes Island 437

Campanian-Maastrichtian -Ellesmere Island .437
Late Cretaceous and Tertiary Basin
   development -Arctic Islands 437-456
Miocene-Pliocene palynomorphs
   -Arctic Islands 439
Oligocene -palynomorphs -Arctic Islands 438
Paleocene-Eocene -palynomorph distribution
   -Arctic Islands -table 438
Parry Islands Fold Belt
   *see also Franklinian Mobile Belt 74,321-333
   Baumann Fiord Formation 325
   Bay Fiord Formation -structure 322,325,331
   Bird Fiord Formation .322
   Cape de Bray Formation -structure 331
   Cape Phillips Formation 322
   Eids Formation 322
   Eleanor River Formation 322,325
   Geophysics -crustal structure 74
   Griper Bay Subgroup .322
   Hecla Bay Formation 322
   Kitson Formation 322
   Structure 322-333
   Thumb Mountain Formation 322
   Weatherall Formation 322
Parry Upland
   Boothia Uplift 50,62,64,70
   Canrobert Hills Fold Belt 50,64,216,321-333
   Jones Sound Fold Belt 50,64,321-333
Parry, W.E.
   Explorer -history 8
Peacock, J.D.
   Historical geological account 21
Peary Land
   Aleqatsiaq Fjord Formation 140,141
   Borglum River Formation 140,181
   Cape Calhoun Formation 140
   Citronens Fjord Member 139
   Gonioceras Bay Formation 140
   Kap Jackson Formation 140
   Koch Vaeg Formation 128
   Mound Formation 156
   Paradisfield Group 112,205
   Peary Land Group 146,147,148,149
   Petermann Halvo Formation 142
   Sjaelland Fjelde Formation 131
   Troedsson Cliff Formation 140
   Tureso Formation 141
   Wandel Valley Formation 131,181
   Ymers Gletscher Formation 142
Peary Land Group
   *see also Franklinian Basin 115
Peary, R.E.
   Historical geological account 10,11
Pearya (northernmost Ellesmere Island)
   Correlation -Caledonian Mobile Belt 241-259
   Correlation -Franklinian
   Mobile Belt .241-259,297
   Granitoid basement -succession I 241,243,244
   Metamorphism 249
   Metasediments and metavolcanics 244,245
   Proterozoic to Late Silurian record
   of Pearya 241-259
   Sedimentary and volcanic rocks 250,251
   Structure -belts * see also structure and
   belts listed by name 241,243,249
   Structures -description and interpretation 298
   Tectonic interpretation 255,256,257
   Thores Suite 246,251
   Upper Middle Proterozoic
   -granitoid basement .241
Pearya Terrane
   Exotic interpretation 62,295,298,426
Peel, J.S.
   Cambrian to Silurian Basin sedimentation,
   North Greenland 111-164
Penny, W.
   Historical geological account 10
Permian -Triassic boundary
   Sverdrup Basin 362
Piasecki, S.
   Resources -petroleum, North Greenland . .525-529
Plate tectonics
   Arctic Platform -interpretations 483-487
Powell, T.G.
   Resources -petroleum -Arctic Islands 517-525
Precambrian basement rocks
   *see also Canadian Shield 31
   Baffin Region 32
   Thule -Nugssuaq Region 31,32
Prince of Wales Island
   Drake Bay Formation 187
   Peel Sound Formation 304
   Stratigraphic section 168,170
Prince Patrick Island
   Arctic Islands -resources -coal -distribution . . .529
   Canrobert Hills Fold Belt 334,336
   Cape de Bray Formation 334
   Eureka Sound Group -Beaufort Formation 448
   Late Devonian -Early Carboniferous
   deformation 334-336
   Weatherall Formation 270
Quaternary deposits
   Glacial deposit -Ellesmere Island 499,500
Queen Elizabeth Islands
   Extent 29
   Late Pleistocene glaciation 507
   Quaternary deposits -nonglacial deposits
   and processes 500-504
Rasmussen, K.
   Explorer -history 15
Remus Basin
   Arctic Islands -resources -coal -distribution . . .530
   Late Cretaceous and Tertiary
   development 442,450
   Paleogeographic evolution -Latest Paleocene
   to Middle Eocene 452,453
Rens Fiord Uplift
   Axel Heiberg Island .315
Resources
   Arctic Islands -minerals -sulphides, barite,
   iron, kimberlites 537

565

Arctic Islands -petroleum 517-525
Arctic Islands -resources -coal 529-532
Arctic Islands -resources -minerals 533-539
Deposits -minerals -lead-zinc 533,534,535
Franklinian Basin
-petroleum resources 525,527
Mississippi Valley Type deposits
-Cornwallis lead-zinc 533,534,535, 536
North Greenland -petroleum 525-529
Wandel Sea Basin -petroleum resources . . 527-529
Richardson, J.
Explorer -history 9
Riley, G.C.
Historical geological account 20
Roots, E.F.
Historical geological account 21
Ross, J.
Explorer -history 8
Russel islands
Drake Bay Formation 187
Ryder Gletscher Group
*see also Franklinian Basin 127,128,131,132
Salter, W.
Historical geological account 10
Sander, G.W.
Historical geophysical account 21
Scoresby, W.Jr.
Explorer -history 8
Sedimentation
Cambrian to Silurian Basin
-North Greenland 111-164
Shei, P.
Historical geological account 12,13
Simpson, T.
Explorer -history 9
Skagen Group
*see also Franklinian Basin .112,120,123,125,285
Smith, G.P.
Devonian movements Inglefield Uplift 308
Sobczak, L.W.
Arctic Archipelago -geophysical
characteristics -gravity 69-84
Somerset Island
Bay Fiord Formation 180
Cape Crauford Formation 183
Eureka Sound Formation 304
Hunting Formation 106
Peel Sound Formation 187,304
Shaler Group 106
Ship Point Formation 179
Somerset Island Formation 186
Turner Cliffs Formation 175
Sonderholm, M.
Cambrian to Silurian Basin sedimentation, North
Greenland 111-164
Soper, N.J.
Devonian -early Carboniferous deformation/
metamorphism -North Greenland 283-290
Historical geological account 21
Late Cretaceous -Early Tertiary deformation
-North Greenland 461-464

Steenfelt, A.
North Greenland -resources -minerals . . . 539-541
Stefansson, V.
Historical geological account 16
Stemmerik, L.
Resources -petroleum, North Greenland . . 525-529
Stott, D.F.
Historical geological account 21
Strand Fiord Basin
Arctic Islands -resources -coal -distribution . . . 530
Late Cretaceous and Tertiary
development 442,450,451
Structure
Amundsen Land -divergent zone 288
Arctic Platform 60,61,479
Arctic Platform -normal faults -Devon Island . . 479
Audhild Bay Belt, Pearya 244,298,301
Beaufort Unconformity 454
Bjarnason Island Fault 477
Blind Fiord Fault 478
Boothia Uplift . . .35,62,64,70,166,182,300,302,338
Bromley Island Belt, Pearya 245,249,255,298
Bukken Fiord Fault 477
Canrobert Hills Fold Belt 321-333
Cap Back Fault 452
Cape Alfred Ernest Belt 241-259,298,301
Cape Columbia Belt, Pearya 241,243,298
Central Ellesmere Domain -structure
-Grant Land Formation 472
Central Ellesmere Fold Belt 312,318,320
Clements Markham Fold Belt 64,205,222,224,
. . . . 225,229,241,250,254,283,
. . . . 285,298,299,300,309,312,337
Cornwall Arch 472
Cornwallis Fold Belt 185,302,305,322,482
Deuchars Glacier Belt, Pearya 243,298
Disraeli Glacier Belt, Pearya 244
Empire Belt -Pearya 241,244,249,250,298
Eurekan fault systems 112
Eurekan Fold Belt 453
Eurekan Orogeny -Arctic Island 470
Feilden Fault Zone -Guide Hill Fault .470,477,478
Feilden Fault Zone -Porter Bay Fault .470,477,478
Franklinian Mobile Belt 61,62,256-258,297
Freuchen Land 286
Halokinetic features -evaporites 476
Harder Fjord Fault Zone 123,145,283,286,461
Hazen Fold Belt . 62,205,283,286,295,309,312,338
Inglefield Uplift 186-188,308,309,337
J.P. Koch Fjord 286
Jones Sound Fold Belt 320
Kap Bridgman fault 50
Kap Cannon Thrust Zone 290,463
Kulutingwak Belt 299
Lake Hazen Fault Zone 46,454,470,472
M'Clintock Glacier Fault 298,299
M'Clintock plutonic bodies, Pearya . . .246,247,255
Mitchell Point Belt -Pearya . .241,243,244,298,299
Mitchell Point Fault 470,475
Mount Disraeli Belt -Pearya 241,244,245,250
Mount Rawlinson Fault 298,299

Nares Land 286
Navarana Fjord lineament 283,288
North Greenland -fold and thrust zone ... 285,286
North Greenland Fold Belt 283,288
Northern Heiberg
  Fold Belt 222,295,298,300,315,337,470,472
Nyeboe Land fault zone 286,463
Parrish Glacier Fault 452,472
Parry Channel Fault 479
Parry Islands Fold Belt 321-333
Petersen Bay Fault 298,300
Porter Bay Fault...................299,300,309
Purchase Bay Homocline 332,334
Rawlings Bay Fault 452,472
Rollrock Fault 474
Salt movements -Otto
  Fiord Formation 347,374,428,430,476
Schei Syncline...................40,318,320
Stephenson O 286
Stolz Thrust Fault 454,470,472,476
Sverdrup Basin 50,61,64,65,300,324,345
Trold Fiord Fault 478
Wulff Land 286
Yelverton Thrust 298,299
Structure -lithosphere stretching model
  North Greenland Fold Belt 283,288
Suess, E.
  Historical geological account 17
Surlyk, F.
  Cambrian to Silurian Basin sedimentation,
  North Greenland 111-164
Sutherland, P.C.
  Historical geological account 10
Sverdrup Basin
  Antoinette Formation......................350
  Arctic Islands -resources -coal -distribution ...529
  Artinskian-Wordian lithofacies 355-359
  Assistance Formation 355,361
  Audhild Formation 348
  Barrow Formation 388,397
  Bastion Ridge Formation 422,426
  Belcher Channel Formation 350,352
  Bjorne Formation 379,380,382
  Blind Fiord Formation 379
  Borup Fiord Formation 348,349,352
  Canyon Fiord Formation 349,352
  Carboniferous and Permian history 345-367
  Christopher Formation 422
  Correlations -Carboniferous and Permian
  -Greenland 362,363
  Deer Bay Formation
  -Glacier Fiord Member 418,421
  Degerbols Formation 357,360
  Emma Fiord Formation
  -Visean lithofacies 345,348,349
  Esayoo Formation 348
  Eureka Sound Group -Expedition Formation ..439
  Hare Fiord Formation 352,353,354
  Hassel Formation 422,423,426
  Heiberg Formation 388,397,402
  Heiberg Group 388

Hiccles Cove Formation, Prince Patrick
  -Emerald Island 403
Hoyle Bay Formation 385,388
Isachsen Formation 408,418,421,426
Jameson Bay Formation 398
Kanguk Formation 422,423
Late Cretaceous and Tertiary development
  -paleogeographic evolution 451
McConnell Island Formation,
  Axel Heiberg Island 403
Mesozoic -stratigraphy -Early Triassic ... 377,379
Mesozoic -stratigraphy -Middle Triassic 382
Mesozoic -stratigraphy -Permian
  -Triassic boundary 377
Mesozoic history of Arctic Islands 371-433
Mount Bayley Formation 351,354
Murray Harbour Formation 382,385
Namurian to Artinskian lithofacies 350
Nansen Formation 347
Organic mounds and reefs 363,365,366
Otto Fiord Formation 346,352,354,374
Pat Bay Formation 385,388
Roche Point Formation 382,385,388
Sabine Bay Formation 355,361
Salt tectonics 346,428-430,476
Strand Fiord Formation 423
Stratigraphy and lithofacies 348-363
Structure 50,61,64,65,300,324,345
Subsurface facies 359
Tanquary High 376,379,380,382,385
Tanquery Formation 351
Tectonics 345
Trold Fiord Formation 357,361
Van Hauen Formation 356,361
Sverdrup Rim
  Geophysics -isostasy 70
Sverdrup, O.
  Explorer -history 12
Tavsens Iskappe Group
  *see also Franklinian Basin 133,134
Tectonics
  Alpha Ridge 428
  Axel Heiberg Island 315
  Bache Peninsula Arch 308
  Boothia Uplift 35,62,64,70,79,470,482,533
  Byam Martin High 451
  Bylot Unconformity 451
  Clastic wedge -Arctic Islands 273,274
  Cornwall Arch 449,451,452,470
  Cornwallis Disturbance 182,303,533
  Devon High 47,302,480
  Duke of York Topographic High 38,44,59,61
  Eetookashoo Bay Thrust 315
  Eglinton Graben 377,421
  Grantland Uplift 454
  Greenstone Lake Thrust 315
  Gustaf-Lougheed Arch 70,79,88
  Hazen Fold Belt 93
  Inglefield Uplift..................... 470,472
  Innuitian Orogeny 29,38,44,61,440,483
  Jaeger Lake Thrust 315

Kap Cannon Thrust Zone 463
Lake Hazen Syncline 475
Melvillian Disturbance 332,336,345,362
Ninnis Glacier Syncline 312,320
Pearya -comparison Franklinian and
Caledonian Mobile Belts 256,257,258
Princess Margaret
Arch 88,449,451,452,454,470,472
Rens Fiord Uplift 315
Salt tectonics
-Sverdrup Basin 346,377,428,430,476
Steensby High 302,480
Storkerson Uplift 449,451
Svartevaeg Thrust 315
Sverdrup Basin 345
Sverdrup Rim (Ridge) 50,61,64,65,72,
. 79,88,345,354,377,451,453
Tanquary High 376,379,380,382,385,475
Wellington Topographic High 35,59,61,106
Yelverton Thrust 298,299
Tectonics -lineaments
Navarana Fjord lineament 114
Teichert, K.
Historical geological account 17
Terrane -exotic
Pearya (northernmost
Ellesmere Island) 62,295,298,428
Thores Suite
Pearya (northernmost
Ellesmere Island) 246,247,251
Thorsteinsson, R.
Historical geological account 20,21
Thorvaldsson, E.
Explorer -history 7
Thrust zone *see structure
Thule -Nugssuaq Region
Inglefield Land 32
Nunataks 32
Paleo-erosion surfaces 32
Thule Upland 32
West Greenland Volcanic Province 32
Tozer, E.T.
Historical geological account 21
Trettin, H.P.
Cambrian to Early Devonian Basin
development -Arctic Islands 165-238
Late Cretaceous -Early Tertiary
deformation -Arctic Islands 469-487
Middle and Late Tertiary tectonic
developments 493-496
Middle Devonian to early Carboniferous
deformations, Ellesmere Island 309-317
Precambrian successions
-Canadian Shield 103-110
Proterozoic to Late Silurian
record of Pearya 241-259
Silurian -early Carboniferous deformational
phases Arctic Islands 295-341
Summary and remaining problems
-history Innuitian Orogen/
Arctic Platform 547-553

Tectonic framework -Arctic Platform 57-66
Troelsen, J.C.
Historical geological account 17,19
Valdemar Gluckstadt Land
Mound Formation 156
Victoria Island
Goulbourn Group 321
Natkusiak Formation 59,106
Shaler Group 321
Volcanics
Canadian Shield 105,106
Ellesmere Island 224,225
Pearya (northernmost
Ellesmere Island) 250-254
Sverdrup Basin 347
West Greenland Volcanic Province 32
Volvedal Group
*see also Franklinian Basin 137,143
Wandel Sea Basin
Structure 61,65,461
Warming Land -Nares Land
Cape Webster Formation 180
Steensby Gletscher Formation 180
Warming Land Formation 130,180
Washburn, A.L.
Historical geological account 17
Washington Land
Adams Bjerg Formation 142
Aleqatsiak Fiord Formation 182
Canyon Elv Formation 131
Cape Calhoun Formation 181
Cape Clay Formation 130
Cape Schuert Formation 142
Cass Fjord Formation 128,130,175
Christian Elv Formation 130
Franklinian Basin development 111
Gonioceras Bay Formation 181
Kap Godfred Hansen Formation 142
Kastrup Elv Formation 130,173
Morris Bugt Group 139
Nunatami Formation 131
Pentamerus Formation 142
Poulsen Cliff Formation 179
Telt Bugt Formation 130,173
Troedsson Cliff Formation 181
Washington Land Group 139
Weeks, L.J.
Historical geological account 17
West Sverdrup Basin
Late Cretaceous and Tertiary development . . .451
Wetmiller, R.J.
Seismicity -Arctic Platform 79-83
Weymouth, G.
Explorer -history 7
Wordie, J.M.
Historical geological account 17
Wulff Land 124
Zolnai, A.I.
Late Silurian -Early Devonian deformation
-Boothia Uplift 302-307

PHOTO INDEX

Fig. no.	Photo number	Page	Fig. no.	Photo number	Page
7.5	GSC 1991-318	118	8B.22	ISPG-2489-17	185
7.6	GSC 1991-319	118	8B.23	ISPG-2489-16	186
7.7	GSC 1991-320	119	8B.24	ISPG-925-3	188
7.8a	GSC 1991-321	121	8B.26	ISPG-2489-18	190
7.8b	GSC 1991-322	121	8C.9	ISPG-978-39	214
7.9	GSC 1991-323	122	8C.12a	ISPG-2444-329	218
7.10	GSC 1991-324	123	8C.12b	ISPG-2444-344	218
7.11	GSC 1991-325	124	8C.12c	ISPG-2533-12	218
7.12	GSC 1991-326	125	8C.12d	ISPG-2444-30	218
7.13	GSC 1991-327	126	8C.12e	ISPG-2425-77	218
7.14	GSC 1991-328	126	8C.12f	ISPG-2533-11	218
7.15	GSC 1991-329	129	8C.16	ISPG-003-9	222
7.16	GSC 1991-330	130	8C.18	ISPG-1878-1	224
7.17	GSC 1991-331	130	8C.19	ISPG-648-21	224
7.19	GSC 1991-332	133	8C.25	ISPG	228
7.21a	GSC 1991-333	135	9.3	ISPG-648-5	243
7.21b	GSC 1991-334	135	9.4	ISPG-1505-9	243
7.22	GSC 1991-335	135	9.6	ISPG-1025-2	245
7.23	GSC 1991-336	136	9.7	ISPG-978-7	246
7.24	GSC 1991-337	137	9.11	ISPG-978-18	250
7.25	GSC 1991-338	138	9.12	ISPG-978-2	251
7.26	GSC 1991-339	139	10.4	ISPG-2615-6	266
7.27	GSC 1991-340	140	10.5	ISPG-2615-15	266
7.28	GSC 1991-341	141	10.6	ISPG-2709-2	266
7.29	GSC 1991-342	142	10.11	ISPG-2615-17	269
7.30	GSC 1991-343	143	10.12	ISPG-2709-1	269
7.31	GSC 1991-344	144	10.16	ISPG-2615-20	272
7.32a	GSC 1991-345	144	11.3a	GSC 1991-365	287
7.32b	GSC 1991-346	145	11.3b	GSC 1991-366	287
7.33	GSC 1991-347	146	12B.4	ISPG-2761-A	299
7.34	GSC 1991-348	147	12B.5	ISPG-1360-39	300
7.35	GSC 1991-349	148	12E.2	ISPG-1505-7	310
7.36	GSC 1991-350	149	12G.1a	ISPG-2887-2	322
7.37	GSC 1991-351	150	12G.1b	ISPG-2887-6	322
7.38	GSC 1991-352	150	12G.17	GSC 1991-367	331
7.39a	GSC 1991-353	151	12G.18	ISPG-2887-50	332
7.39b	GSC 1991-354	151	13.7	ISPG-2393-5	352
7.40	GSC 1991-355	152	13.8	ISPG-2393-1	353
7.41	GSC 1991-356	153	13.9	ISPG-279-47	354
7.42a	GSC 1991-357	154	13.11	ISPG-2393-4	356
7.43a	GSC 1991-358	154	13.12	GSC 1991-368	358
7.42b	GSC 1991-359	155	13.15	GSC 1991-369	361
7.43b	GSC 1991-360	155	14.9	ISPG-2615-1	380
7.44	GSC 1991-361	156	14.11	ISPG-2615-2	381
7.45	GSC 1991-362	157	14.15	ISPG-2615-3	383
7.46	GSC 1991-363	158	14.19	ISPG-2615-4	388
8B.6	ISPG-2658-3	172	14.22	ISPG-2709-5	390
8B.7	ISPG-1563-7	172	14.23	ISPG-2615-5	390
8B.8	GSC 1991-364	172	14.24	ISPG-2615-6	391
8B.9	ISPG-2658-1	172	14.25	ISPG-2615-7	391
8B.10	ISPG-2489-26	173	14.37	ISPG-2709-3	403
8B.11	ISPG-2489-207	174	14.40	ISPG-2615-8	405
8B.13	ISPG-2489-2	177	14.41	ISPG-2615-18	406
8B.14	ISPG-2489-183	177	14.48	ISPG-2615-10	413
8B.15	ISPG-2489-208	179	14.53	ISPG-2615-11	417
8B.16	ISPG-2489-184	179	14.57	ISPG-2615-22	420
8B.17	ISPG-2489-9	180	14.58	ISPG-2615-19	420
8B.18	ISPG-347-2	180	14.62	ISPG-2615-13	424
8B.20	ISPG-2489-190	184	14.66	ISPG-2615-14	427
8B.21	ISPG-2489-195	184